FUNDAMENTALS OF AEROACOUSTICS WITH APPLICATIONS TO AEROPROPULSION SYSTEMS

FUNDAMENTALS OF AEROACOUSTICS WITH APPLICATIONS TO AEROPROPULSION SYSTEMS

Elsevier and Shanghai Jiao Tong University Press Aerospace Series

XIAOFENG SUN
Fluid and Acoustic Engineering Laboratory,
School of Energy and Power Engineering,
Beihang University, Beijing, China

XIAOYU WANG
Fluid and Acoustic Engineering Laboratory,
Research Institute of Aero-Engine,
Beihang University, Beijing, China

ELSEVIER

ACADEMIC PRESS
An imprint of Elsevier

Academic Press is an imprint of Elsevier
125 London Wall, London EC2Y 5AS, United Kingdom
525 B Street, Suite 1650, San Diego, CA 92101, United States
50 Hampshire Street, 5th Floor, Cambridge, MA 02139, United States
The Boulevard, Langford Lane, Kidlington, Oxford OX5 1GB, United Kingdom

Library of Congress Cataloging-in-Publication Data
A catalog record for this book is available from the Library of Congress

British Library Cataloguing-in-Publication Data
A catalogue record for this book is available from the British Library

ISBN 978-0-12-408069-0

For information on all Academic Press publications
visit our website at https://www.elsevier.com/books-and-journals

Publisher: Matthew Deans
Acquisitions Editor: Glyn Jones
Editorial Project Manager: Lindsay Lawrence
Production Project Manager: Poulouse Joseph
Cover Designer: Christian J Bilbow

Typeset by SPi Global, India

Contents

Preface

Aeroacoustics had its beginnings in the early 1950s thanks to the pioneering work of Sir James Lighthill about aerodynamic sound. Since then, this discipline has concentrated on understanding noise generation by either turbulent fluid motion or aerodynamic forces interacting with surfaces. In particular, emphasis has also been placed on aeronautical applications for noise reduction of both aircraft and aeropropulsion systems. At present, there are already several books available that introduce basic knowledge and progress in its various aspects. It is noted, however, that the most prominent noise sources of modern aircrafts originate from the aeropropulsion system in many situations; meanwhile, the thermoacoustic oscillations occurring in the combustion chamber of the aeroengine has received great attention on its prediction and control, which is equally related to aeroacoustics. Therefore, it is necessary to have a book dedicated to aeroengine acoustic problems.

As is well known, there are various scales, such as different vortex structures, entropy, and pressure fluctuations in moving flow media, which make it extremely difficult to accurately calculate and analyze all flow and acoustic field details under the condition of multiscales. There are thus two ways to get work done. One method is Lighthill's acoustic analogy theory, which is still widely applied in various aerodynamic noise calculations, especially for rotating source prediction. The other method is numerical techniques like computational aeroacoustics (CAA), which suppresses both dissipation and dispersion on the basis of the differential schemes. There is no doubt that CAA has shown great potential for the study of jet noise and other flow-induced acoustic problems. Still, given the limits of current computational capabilities, acoustic computation for a problem of practical interest is still out of reach by directly solving Navier-Stokes equations, particularly for a numerical simulation of rotating or moving sound source problems in association with aeroengine noise generation.

With reference to the context, the aim of this book is to develop a unified framework to handle the acoustic problem of aeropropulsion systems for the design phase of an aeroengine; meanwhile, it is also intended for scientists, engineers, and graduates who are interested in modeling and doing it right. Therefore, the core parts of this book are still based on the acoustic analogy theory or the unified solution approach of flow and acoustic fields under the linearizing assumption. More importantly, the theoretical work described herein is substantially a compilation of our research publications, except the basic knowledge of aeroacoustic equations. In addition, it is worth noting that understanding the mechanism of the aeroengine noise generation and what methods can be used to predict or suppress the resulting noise have been the core problem of aeroacoustic research. This book will focus on presenting our research results and other important progress in these regards.

The outline of the book is as follows: Chapter 1 consists of a colloquial introduction to basic equations of aeroacoustics with an emphasis on the application of Green's function method. Advanced propeller noise generation and prediction are given in Chapter 2, accompanied by the solution of Ffowcs Williams-Hawkings equation using both frequency and time domain methods. This is followed in Chapter 3 by a brief introduction of various aerodynamic noise sources in the aeroengine. This chapter also involves certain algorithm details for both rotor/stator interaction and shock noise predictions. Chapter 4 is composed of a detailed description of unsteady aerodynamics of compressor cascade, which shows how to calculate aerodynamic blade loading under linearizing assumption. A description of vortex sound theory is provided in Chapter 5, which introduces discussion and analysis on how to use the discrete vortex method to study the energy exchange between vortex and acoustic waves. Then in Chapter 6, the Transfer Element Method (TEM) and its applications are introduced in detail. The TEM can be applied to study various interacting problems like the interaction between a sound source and acoustic treatment in flow ducts. This is also one of the core parts of our research work presented in this book. Chapter 7 deals with the thermoacoustic problem or combustion instabilities in an aeropropulsion system. This chapter is particularly devoted to the recent progress, both algorithmical and theoretical, in applying the three-dimensional thermoacoustic model, including acoustic treatment, to a more complex combustion system. Finally, Chapter 8 touches upon the eduction of wall acoustic impedance, which is considered one of the most important aeroacoustic tests. The straightforward impedance eduction method and its latest developments are expounded along with the relevant experimental and computational results.

The writing of this book tremendously benefits from the existing theories and various research results openly published in this discipline. We are particularly grateful to these authors for their valuable contributions together with various citations marked in this book. Professor Vigor Yang introduced us to the field of thermoacoustics, and some joint research with his help has been reflected in Chapter 7. His precious contribution is very much appreciated. Moreover, the core contents of this book originate from the work of our faculty and graduates in different periods of time at the Fluid and Acoustic Engineering Laboratory, BUAA. Accordingly, we would like to take this opportunity to thank these people who partly contributed and read the manuscript and suggested definite improvements, including Professor Xiaodong Jing, Dr. Zhiliang Hong, Mr. Guangyu Zhang, Mr. Lingfeng Chen, Dr. An Liang, Dr. Weiguang Zhang, Ms. Yu Sun, Mr. Zhuo Wang, Dr. Lin Du, Dr. Lei Li, Dr. Xiwen Dai, Mr. Wei Dai, Ms. Lei Qin, and Mr. Shuo Tian.

Furthermore, we are indebted to the continuous support of our research by the National Natural Science Foundation of China (NSFC), and by the 973 projects of the Ministry of Science and Technology, China.

Finally, we wish to express our gratitude to Dr. Fangzhen Qian of Shanghai Jiao Tong University Press for her indispensable enthusiasm, patience, and care throughout this project.

Xiaofeng Sun
Xiaoyu Wang
Beihang University
Beijing, China
February 2020

CHAPTER 1

Basic equations of aeroacoustics

Aeroacoustics is a discipline that investigates noise generation by either turbulent fluid motion or aerodynamic forces interacting with surfaces, which first originated from the publication of Sir James Lighthill in the early 1950s [1], when aerodynamic noise associated with an aeroengine was beginning to receive great attention due to the application of new jet aircraft [2]. Relevant knowledge has been developed since then, which can be applied in aeronautical engineering to study how to quiet aircraft, including airframe and aeroengine noise. Understanding the mechanism of aeroengine noise generation and what methods can be used to predict or suppress the resulting noise have been core problems of aeroacoustic research. This chapter will focus on introducing the fundamental knowledge of aeroacoustics and relevant progress in this regard.

As is well known, there are various scales such as different vortex structures, entropy, and pressure fluctuations in moving flow media, which make it extremely difficult to accurately calculate and analyze all the flow details under the condition of multiscales. Lighthill [1] found a smart way to overcome the difficulties of theoretical analysis and numerical simulation due to the multiscale effect based on the assumption of "acoustic analogy". By proper manipulation of the Navier-Stokes equations, a wave equation was derived with the stress tensor as its source term. The resulting wave equation can then be integrated with the help of Green's Function, or it can be solved numerically, whereas the source term can be independently obtained either by experimental measurement or numerical approach. Thus, this equation can represent the sound propagation from an independent flow sound source in an ambient condition. With the success of the acoustic analogy, many improvements and developments were made on the derivation of the wave equation for the situation of flow-body interaction. In particular, Curle [3] extended Lighthill's equation to deal with sound generation by a static body in flow, whereas the most general equation for sound generation by a moving body was

Fundamentals of Aeroacoustics with Applications to Aeropropulsion Systems
https://doi.org/10.1016/B978-0-12-408069-0.00001-4

obtained by Ffowcs Williams and Hawkings [4] via application of generalized function.

There is no doubt that computing techniques will play a more and more important role in the course of aeroacoustic development. One of the main difficulties in computational techniques related to aeroacoustics is still the treatment of a multiscale problem in flow. An acoustic wave has a high propagation velocity relative to the flow structures and, at the same time, its amplitude is nearly 10 orders of magnitude smaller than the average quantity in flow or atmospheric environment. Also, it was found that the numerical schemes for an aeroacoustic problem must be capable of suppressing both dissipation and dispersion [5, 6]. This also means that it is difficult to apply an ordinary computational fluid dynamics (CFD) solver to solve an aeroacoustic problem. With the appearance of various new schemes such as Dispersion Relation Preserving schemes [6] and compact schemes [5] aimed at less dissipative and dispersive solutions, computational aeroacoustics (CAA) have shown great potential for the study of jet noise and other flow-induced acoustic problems [7, 8]. Still, given the limits of current computational capability, acoustic computation for a problem of practical interest is still out of reach, especially for the numerical simulation of a rotating or moving sound source problem in association with aeroengine noise generation [9, 10].

In scientific terms, no complete theory of noise generation by aerodynamic flows has been established up to now. In contrast, most practical aeroacoustic analysis still relies upon "acoustic analogy" methodology. More importantly, the main methods widely used in the prediction of aeroengine noise are also based on acoustic analogy, especially for rotating sources.

Therefore, an emphasis in this book will be placed on the mechanism of aerodynamic sound generation and the relevant analytical method for practical application.

It is noted that many concepts and techniques used in aeroacoustics have been taken directly from the acoustics in moving media. Section 1.1 is devoted to a brief introduction to those aspects of these subjects necessary for understanding the theory of aerodynamic sound. Section 1.2 is used to develop certain mathematical tools with an emphasis on the application of Green's function, which is required in succeeding Section 1.3 on the derivation of aeroacoustic equations.

Table 1.1 Common notations and special terminologies.

ρ	Density
$\mathbf{v}$	Velocity
p	Pressure
S	Entropy
$p_0,\ \vec{v}_0,\ \rho_0$	Mean flow quantities
$p', \vec{v}',\ \rho'$	Fluctuations
q	An external volume flow source within the fluid
c_0	Speed of sound
f	An externally applied volume force
U	Mean flow velocity
E	Acoustic energy density
$\vec{I}$	Acoustic energy flux
τ	Time associated with emission of sound wave
t	Time associated with the arrival of sound wave at observation point
G	Green's function, i.e., fundamental solution of wave function
G_0	Free-space Green's function
R	Distance between the sound source and the observation point
ψ	Velocity potential
M	Mach number, $M = U/c_0$
T_{ij}	Lighthill stress tensor

1.1 Sound sources in moving media

1.1.1 Basic equations of sound propagation

Considering that the effects of viscosity and heat conduction can be ignored, fluid motion can be determined by solving Euler's equation. Besides, we assume that the following inequalities exist:

$$|p'|/p_0 << 1,\ \left|\vec{v}'\right|/\vec{v}_0 << 1,\ |\rho'|/\rho_0 << 1, \tag{1.1}$$

where p_0, $\vec{v}_0$, and ρ_0 represent average pressure, velocity, and density, respectively, whereas p', $\vec{v}'$, and ρ' denote the relevant fluctuations. For convenience, the common notations and special terminologies that used in this chapter are listed in Table 1.1.

Euler's equations are described as:

$$\begin{cases} \left(\dfrac{\partial \rho}{\partial t} + \mathbf{v}\cdot\nabla\rho\right) + \rho\cdot\nabla\mathbf{v} = \rho\cdot q \\ \rho\left(\dfrac{\partial \mathbf{v}}{\partial t} + \mathbf{v}\cdot\nabla\mathbf{v}\right) = -\nabla p + f \\ \dfrac{\partial S}{\partial t} + \mathbf{v}\cdot\nabla S = 0 \end{cases}, \tag{1.2}$$

where ρ, $\mathbf{v}$, p, and S stand for fluid density, velocity, pressure, and entropy, respectively, whereas q and f denote an external volume source and an externally applied volume force, respectively. There are four unknowns in Eq. (1.2), and an additional equation is required to close the equation. Assume that:

$$\rho = \rho(p, \ S),$$

thus,

$$d\rho = \left(\frac{\partial \rho}{\partial p}\right)_S \cdot dp + \left(\frac{\partial \rho}{\partial S}\right)_p \cdot dS. \tag{1.3}$$

Considering $c_0^2 = \left(\frac{\partial p}{\partial \rho}\right)_S$, and for an isotropic motion, $\frac{DS}{Dt} = 0$, therefore it follows from Eq. (1.3) that:

$$\left(\frac{\partial \rho}{\partial t} + \mathbf{v} \cdot \nabla \rho\right) = \frac{1}{c_0^2} \cdot \left(\frac{\partial p}{\partial t} + \mathbf{v} \cdot \nabla p\right). \tag{1.4}$$

Applying the following relations:

$$p = p_0 + p', \ \mathrm{v} = \mathrm{v}_0 + \mathrm{v}', \ \rho = \rho' + \rho_0, \ S = s' + S_0,$$

and the steady Euler's equation:

$$\begin{cases} \nabla \cdot \rho_0 \mathbf{v}_0 = 0 \\ \rho_0 \mathbf{v}_0 \cdot \nabla \mathbf{v}_0 = -\nabla p_0 \\ \mathbf{v}_0 \cdot \nabla S_0 = 0 \\ c_0^2 \mathbf{v}_0 \cdot \nabla \rho_0 = \mathbf{v}_0 \cdot \nabla p_0 \end{cases}, \tag{1.5}$$

the basic equations of sound propagation in moving media can be derived as:

$$\begin{cases} \dfrac{\partial \rho'}{\partial t} + \nabla \cdot (\rho' \mathbf{v}_0 + \mathbf{v}' \rho_0) = \rho_0 \cdot q \\ \rho_0 \left(\dfrac{\partial \mathbf{v}'}{\partial t} + \mathbf{v}_0 \cdot \nabla \mathbf{v}' + \mathbf{v}' \cdot \nabla \mathbf{v}_0\right) + \rho' \mathbf{v}_0 \cdot \nabla \mathbf{v}_0 = -\nabla p' + f \\ \dfrac{\partial s'}{\partial t} + \mathbf{v}_0 \cdot \nabla s' + \mathbf{v}' \cdot \nabla S_0 = 0 \\ c_0^2 \left(\dfrac{\partial \rho'}{\partial t} + \mathbf{v}' \cdot \nabla \rho_0 + \mathbf{v}_0 \cdot \nabla \rho'\right) = \dfrac{\partial p'}{\partial t} + \mathbf{v}' \cdot \nabla p_0 + \mathbf{v}_0 \cdot \nabla p' \end{cases} \tag{1.6}$$

The results are actually consistent with linearized aerodynamic equations. Under different simplified conditions, the equations can be used to

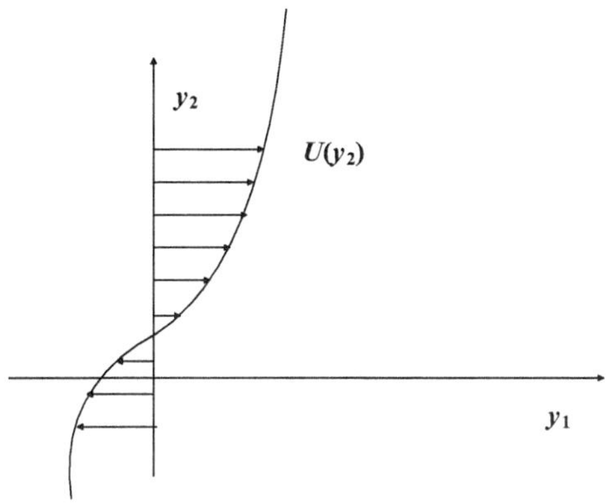

Fig. 1.1 Schematic of parallel shear, mean flow.

study the sound propagation in ducts. For example, if a parallel shear flow is considered as shown in Fig. 1.1, along with the assumption that

$$\mathbf{v}_0 = \mathbf{i} \cdot U[\mathrm{r}(y_2, y_3)], \quad \rho_0 = \text{constant}, \quad p_0 = \text{constant}, \quad S_0 = \text{constant}. \tag{1.7}$$

Substituting Eq. (1.7) into Eq. (1.6) yields:

$$\begin{cases} \dfrac{1}{c_0^2}\dfrac{D_0 p'}{Dt} + \rho_0 \nabla \cdot \mathbf{v}' = \rho_0 \cdot q \\ \rho_0 \left(\dfrac{D_0 \mathbf{v}'}{Dt} + i\dfrac{dU}{dr} v_r \cdot g \right) = -\nabla p' + f, \\ \dfrac{D_0 s'}{Dt} = 0 \end{cases} \tag{1.8}$$

where

$$\frac{D_0}{Dt} = \frac{\partial}{\partial t} + U\frac{\partial}{\partial y_1}, \quad v_r = v' \cdot \frac{\nabla r}{|\nabla r|}, \quad g(y_2, y_3) = |\nabla r|, \quad r = \sqrt{y_2^2 + y_3^2}.$$

Applying $\dfrac{D_0}{Dt}$ to the first equation of Eq. (1.8), and imposing ∇ to the second equation in Eq. (1.8) yields:

$$\nabla^2 p' - \frac{1}{c_0^2}\frac{D_0^2 p'}{Dt^2} + 2\rho_0 \frac{dU}{dr} g \frac{\partial v_r}{\partial y_1} = \nabla \cdot f - \rho_0 \frac{D_0 q}{Dt}. \tag{1.9}$$

Because:

$$\nabla \cdot \frac{D_0 v'}{Dt} = \nabla \cdot \left(\frac{\partial v'}{\partial t} + U\frac{\partial v'}{\partial y_1} \right) = \frac{D_0}{Dt}(\nabla \cdot v') + \frac{dU}{dr} g \frac{\partial v_r}{\partial y_1},$$

$$\nabla\cdot\left(i\frac{dU}{dr}v_r\cdot g\right)=\frac{\partial}{\partial y_1}\left(\frac{dU}{dr}v_r\cdot g\right)=\left(\frac{dU}{dr}g\frac{\partial v_r}{\partial y_1}\right)+v_r\cdot g\left(\frac{\partial}{\partial y_1}\frac{dU}{dr}\right)$$

$$=\frac{dU}{dr}g\frac{\partial v_r}{\partial y_1},$$

and for uniform flow, $\mathbf{v}_0=U=$ constant, Eq. (1.9) can be written as:

$$\nabla^2 p'-\frac{1}{c_0^2}\frac{D_0^2 p'}{Dt^2}=\nabla\cdot f-\rho_0\frac{D_0 q}{Dt}=-\gamma. \tag{1.10}$$

This is the basic equation used to study sound propagation in ducts with uniform flow [11]. When $U=0$, the resulting equation becomes:

$$\nabla^2 p'-\frac{1}{c_0^2}\frac{\partial^2 p'}{\partial t^2}=\nabla\cdot f-\rho_0\frac{\partial q}{\partial t}. \tag{1.11}$$

This is the basic equation for classical acoustics.

1.1.2 Energy relations in moving media

For classical acoustics, the energy flux is defined by:

$$E=\frac{1}{2}\rho_0\left|\vec{u}\right|^2+\frac{1}{2}\frac{c_0^2}{\rho_0}\rho'^2=\frac{1}{2}\rho_0\left|\vec{u}\right|^2+\frac{1}{2}\frac{p'^2}{\rho_0 c_0^2}, \tag{1.12}$$

where the first term is the kinetic energy per unit volume carried by the wave, and the second term denotes the potential energy per unit volume associated with the acoustic filed.

Considering that:

$$\vec{I}=p'\,\vec{u}=\rho' c_0^2\,\vec{u}\,, \tag{1.13}$$

a conservation relation can be given in the form of:

$$\frac{\partial}{\partial t}\int_V E dy+\int_S \vec{I}\cdot\vec{n}\,dS=0, \tag{1.14}$$

or

$$\frac{\partial E}{\partial t}+\nabla\cdot\vec{I}=0.$$

Eq. (1.14) means that, no matter what form of definition of sound energy flux is given, it must satisfy the conservation law. However, as far as moving media are concerned, it is still very difficult to give a reasonable definition for a general case [12, 13]. But for an isentropic irrotational flow, sound energy flux and sound intensity can be given by:

$$E = \frac{p'\rho'}{2\rho_0} + \left(\frac{\rho_0|\vec{u}|^2}{2} + \rho'\,\vec{u}\cdot\vec{\mathrm{v}}_0\right), \tag{1.15}$$

$$\vec{I} = \left(\frac{p'}{\rho_0} + \vec{u}\cdot\vec{\mathrm{v}}_0\right)\left(\rho_0\,\vec{u} + \rho'\vec{\mathrm{v}}_0\right). \tag{1.16}$$

It can be verified that Eqs. (1.16) and (1.15) satisfy the conservation law in Eq. (1.14). When $\vec{\mathrm{v}}_0 = 0$, the relevant equations reduce to Eqs. (1.12) and (1.13) for classical acoustics.

1.1.3 Sound field of moving sound sources

To aim at a more general case, emphasis is placed on sound generation by monopole and dipole sources with arbitrary motion. The source path and observer are described in Fig. 1.2.

(1) Monopole source with arbitrary motion

Giving:

$$\frac{1}{c_0^2}\frac{\partial^2 p'}{\partial \tau^2} - \nabla^2 p' = \frac{\partial}{\partial \tau}\{\rho_0 q_0(\tau)\delta[\mathbf{y} - \mathbf{y}_0(\tau)]\},$$

where $\mathbf{V}$ is the instantaneous velocity of a moving source and $\mathbf{y}_0 = \int \mathbf{V} d\tau$. Introducing ψ and $p' = \frac{\partial \psi}{\partial \tau}$ yields:

$$\frac{1}{c_0^2}\frac{\partial^2 \psi}{\partial \tau^2} - \nabla^2 \psi = \rho_0 q_0(\tau)\delta[\mathbf{y} - \mathbf{y}_0(\tau)]. \tag{1.17}$$

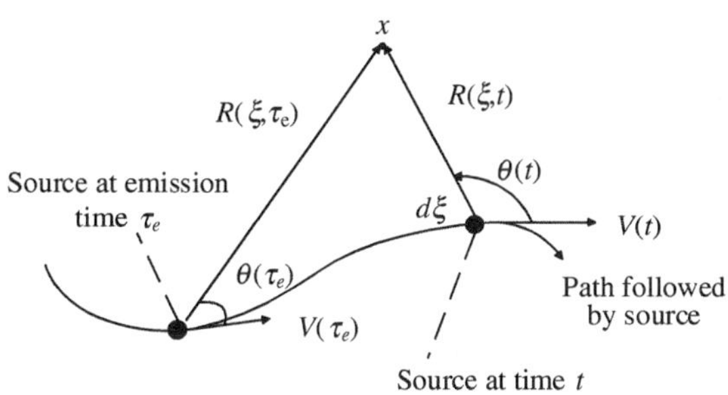

Fig. 1.2 Geometrical relation of a moving point source.

In terms of generalized Green's function theory, the solution of Eq. (1.17) for free space can be written as:

$$\psi\left(t, \vec{x}\right) = \int_{-T}^{T}\int_{v(\tau)} G_0(\mathbf{y}, \tau/\mathbf{x}, t)\cdot\gamma(\mathbf{y}, \tau)d\mathbf{y}d\tau, \tag{1.18}$$

where $\gamma(\mathbf{y}, \tau)$ is the source term, whereas $G_0(\mathbf{y}, \tau/\mathbf{x}, t)$ denotes Green's function for free space, which can be expressed as:

$$G_0 = \frac{1}{4\pi R}\delta\left(\tau - t + \frac{R}{c_0}\right). \tag{1.19}$$

Hence the solution of Eq. (1.17) is:

$$\psi\left(t, \vec{x}\right) = \frac{\rho_0 q_0(\tau^*)}{4\pi R^*(1 - M_r{}^*)}, \tag{1.20}$$

where

$$R = |\mathbf{x} - \mathbf{y}_0(\tau)|, \quad M_r = \frac{R_i}{R}\cdot\frac{\mathbf{V}}{c_0}, \quad R_i = x_i - y_{0i},$$

and τ^* is the root of the retarded time equation $g = \tau - t + |\mathbf{x} - \mathbf{y}_0(\tau)|/c_0$ (noting that τ^* can have several different choices, and superscript "*" denotes the value of the corresponding quantity at time τ^*), whereas the derivative of g is:

$$\frac{dg}{d\tau} = 1 - \frac{1}{c_0}\frac{[\mathbf{x} - \mathbf{y}_0(\tau)]}{|\mathbf{x} - \mathbf{y}_0(\tau)|}\cdot\frac{d\mathbf{y}_0}{d\tau} = 1 - \frac{\mathbf{R}}{R\cdot c_0}\cdot\mathbf{V} = 1 - \frac{R_i}{R\cdot c_0}\cdot V_i = 1 - M_r \tag{1.21}$$

Consequently, the solution of sound pressure is:

$$p' = \frac{\partial\psi}{\partial t} = \frac{\rho_0 q_0'(\tau^*)}{4\pi R^*(1 - M_r{}^*)}\frac{\partial\tau^*}{\partial t} - \frac{\rho_0 q_0(\tau^*)}{4\pi R^{*2}(1 - M_r{}^*)^2}\frac{\partial}{\partial t}[R^*(1 - M_r{}^*)]. \tag{1.22}$$

In consideration of

$$\frac{\partial\tau^*}{\partial t} = \frac{\partial}{\partial t}\left(t - \frac{R^*}{c_0}\right) = 1 - \frac{1}{c_0}\frac{\partial R^*}{\partial t} = 1 - \frac{1}{c_0}\frac{\partial R^*}{\partial\tau^*}\frac{\partial\tau^*}{\partial t} = 1 - \frac{1}{c_0}\frac{R_i{}^*}{R^*}(-V_i)\frac{\partial\tau^*}{\partial t}$$

$$= 1 + M_r{}^*\frac{\partial\tau^*}{\partial t}$$

i.e.,

$$\frac{\partial \tau^*}{\partial t} = \frac{1}{1 - M_r^*}, \tag{1.23}$$

and

$$\frac{\partial}{\partial t}[R^*(1 - M_r^*)] = (1 - M_r^*)\frac{\partial R^*}{\partial \tau^*}\frac{\partial \tau^*}{\partial t} - R^*\frac{\partial M_r^*}{\partial \tau^*}\frac{\partial \tau^*}{\partial t},$$

where

$$\frac{\partial R^*}{\partial \tau^*} = \frac{-R_i^* \cdot V}{R^*} = -c_0 \cdot M_r^*$$

and

$$\begin{aligned}\frac{\partial M_r^*}{\partial \tau^*} &= \frac{\partial}{\partial \tau^*}\left[\frac{R_i^*}{R^*} \cdot M_i^*\right] = \frac{R_i^*}{R^*}\frac{\partial M_i^*}{\partial \tau^*} + M_i^*\frac{1}{R^{*2}}\left[R^*\frac{\partial}{\partial \tau^*}R_i^* - R_i^*\frac{\partial}{\partial \tau^*}R^*\right] \\ &= \frac{R_i^*}{R^*}\frac{\partial M_i^*}{\partial \tau^*} - \frac{c_0}{R^*}\left[M^{*2} - M_r^{*2}\right]\end{aligned}$$

we obtain

$$\frac{\partial}{\partial t}[R^*(1 - M_r^*)] = -\frac{R_i^*}{1 - M_r^*}\frac{\partial M_i^*}{\partial \tau^*} + c_0\frac{M^{*2} - M_r^*}{1 - M_r^*} \tag{1.24}$$

Substituting Eqs. (1.23) and (1.24) into Eq. (1.22) yields:

$$\begin{aligned}p'(\mathbf{x}, t) = &\frac{\rho_0 q_0'(\tau^*)}{4\pi R^*(1 - M_r^*)^2} + \frac{\rho_0 q_0(\tau^*)}{4\pi R^*(1 - M_r^*)^3}\frac{R_i^*}{R^*}\frac{\partial M_i^*}{\partial \tau^*} \\ &- \frac{\rho_0 q_0(\tau^*) c_0}{4\pi R^{*2}(1 - M_r^*)^3}\left(M^{*2} - M_r^*\right)\end{aligned} \tag{1.25}$$

Eq. (1.25) denotes the sound field generated by a monopole source with arbitrary motion. For a far-field solution, the third term including $1/R^{*2}$ can be ignored. Compared with the first term and second term, the solution is thus expressed as:

$$p'(\mathbf{x}, t) = \frac{\rho_0}{4\pi R^*(1 - M_r^*)^2}\left[\left.\frac{\partial q_0}{\partial \tau}\right|_{\tau = \tau^*} + \frac{q_0(\tau^*)}{(1 - M_r^*)}\frac{R_i^*}{R^*}\left.\frac{\partial M_i}{\partial \tau}\right|_{\tau = \tau^*}\right], \tag{1.26}$$

and the near-field solution becomes:

$$p'(\mathbf{x}, t) = \frac{\rho_0 c_0 q_0(\tau^*)}{4\pi R^{*2}(1 - M_r^*)^3}\left(M_r^* - M^{*2}\right). \tag{1.27}$$

It is shown that, if the intensity of sound source does not depend on time and the acceleration is zero, the sound pressure is also zero in the far field.

When the monopole source has a rectilinear motion described by $\mathbf{y}_0(\tau) = V_0 \cdot \tau \cdot \vec{i}$, where $\vec{i}$ stands for the unit coordinate vector which is the moving direction of the point source, whereas $M = M_0 = V_0/c_0$, $M_r = \frac{R_i}{R} \cdot \frac{V_0}{c_0} = M_0 \cdot \cos\theta$, and θ is the angle between the observer and moving direction, the relevant M_r is the projection of the moving Mach number in the direction of the observer. Under the condition of knowing the point source path, the root of the retarded time equation can be solved with the help of $g = \tau - t + \left|\mathbf{x} - V_0 \cdot \tau \cdot \vec{\mathbf{i}}\right| / c_0 = 0$, and the distance between the observer and source can also be given. In fact, the retarded time equation gives:

$$[c_0(t-\tau)]^2 = (x_1 - V_0 \cdot \tau)^2 + x_2^2 + x_3^2. \tag{1.28}$$

Taking $(t - \tau)$ as an unknown, Eq. (1.28) becomes:

$$\left(1 - M_0^2\right)[c_0(t-\tau)]^2 - 2(x_1 - V_0 t)M_0[c_0(t-\tau)] - \left|\vec{x} - V_0 t \cdot \vec{\mathbf{i}}\right|^2 = 0 \tag{1.29}$$

For an observer at time t, the solution of Eq. (1.29) may give two different source radiation time$\tau^{\pm}$; thus, the distance between the observer and source can be expressed as:

$$\begin{aligned} R^{\pm} &= c\left(t - \tau^{\pm}\right) \\ &= \frac{M_0(x_1 - V_0 t) \pm \sqrt{[M_0(x_1 - V_0 t)]^2 + \left(1 - M_0^2\right)\left|\vec{x} - V_0 t \cdot \vec{i}\right|^2}}{1 - M_0^2}. \end{aligned} \tag{1.30}$$

Obviously, only the positive real root of Eq. (1.30) has an exact physical meaning. On the other hand, for subsonic motion, i.e., $V_0 < c_0$, Eq. (1.30) has only one real root, and R^+ is a physical solution. This can also be seen with the help of the geometrical relation described in Fig. 1.3. Therefore, when the point source is rectilinearly moved in subsonic speed, the expression for the sound field reduces to:

$$p'(\mathbf{x}, t) = \frac{\rho_0 q_0'(\tau)}{4\pi R(1 - M_0 \cos\theta)^2} + \frac{\rho_0 q_0(\tau) V_0}{4\pi R^2(1 - M_0 \cos\theta)^3}(\cos\theta - M_0). \tag{1.31}$$

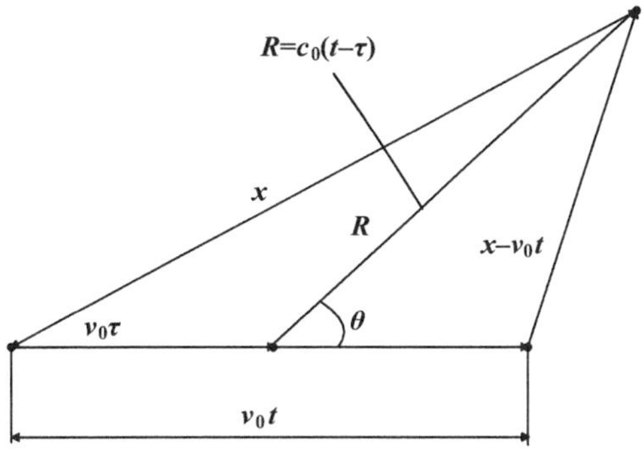

Fig. 1.3 Rectilinear motion of point source.

Besides, when the moving speed is zero, i.e., $V_0 = 0$, the previous equation further becomes:

$$p'(\mathbf{x}, t) = \frac{\rho_0 q_0'(\tau)}{4\pi R}. \tag{1.32}$$

Eq. (1.32) is the expression of the sound field generated by monopole for classic acoustics.

When the point source moves in supersonic speed, i.e., $V_0 > c_0$, Eq. (1.30) may have two positive real roots. This means that the observer can receive two signals at time t, which radiate from two different positions. But the condition for two positions of R is:

$$[M_0(x_1 - V_0 t)]^2 + \left(1 - M_0^2\right)\left|\vec{x} - V_0 t \cdot \vec{i}\right|^2 > 0, \tag{1.33}$$

or

$$\frac{\sqrt{M_0^2 - 1}}{M_0} < \frac{|x_1 - V_0 t|}{\left|\vec{x} - V_0 t \cdot \vec{i}\right|} \tag{1.34}$$

It is seen that the left side of Eq. (1.34) denotes the cosine relation of the Mach angle, i.e., $\alpha = \cos^{-1}\left(\sqrt{M_0^2 - 1}/M_0\right)$. As shown in Fig. 1.4, the right side represents the cosine relation of the angle between the moving direction and the connecting line from the observer to the vertex of the Mach cone, i.e., $\delta = \cos^{-1}\left[(x_1 - V_0 t)/\left|\vec{x} - V_0 t \cdot \vec{i}\right|\right]$. As shown in Fig. 1.5, Eq. (1.34) can be expressed as $\delta < \alpha$. This indicates that the condition that R has two positive real roots is that the observer must be located inside the Mach cone.

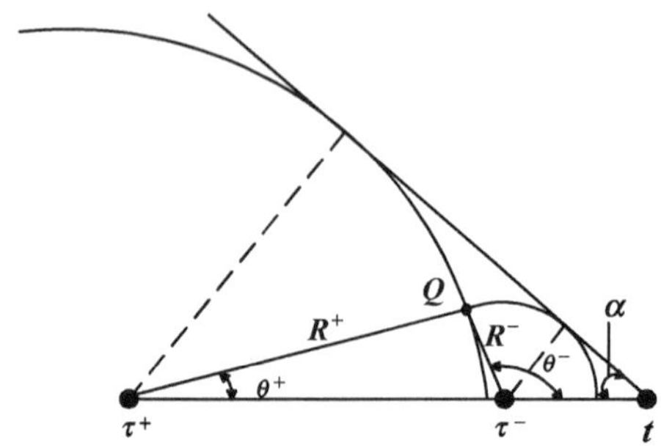

Fig. 1.4 Geometrical relation for an observer receiving two signals.

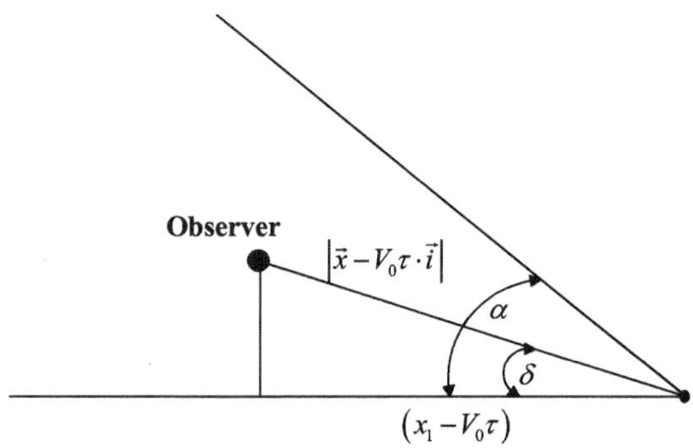

Fig. 1.5 Observer within the Mach cone will hear the sound coming from two different points.

If the condition is satisfied, the observer can hear two different signals at time t from different positions as shown in Fig. 1.5.

When the point source is moving at supersonic speed, the acoustic field can described as:

$$\begin{aligned} p'(\mathbf{x},t) = & \frac{-\rho_0 q_0'(t-R^+/c_0)}{4\pi R^+(M_0\cos\theta^+-1)^2}+\frac{\rho_0 q_0'(t-R^-/c_0)}{4\pi R^-(M_0\cos\theta^--1)^2} \\ & -\frac{\rho_0 V_0 q_0(t-R^+/c_0)(M_0-\cos\theta^+)}{4\pi(R^+)^2(M_0\cos\theta^+-1)^3} \\ & +\frac{\rho_0 V_0 q_0(t-R^-/c_0)(M_0-\cos\theta^-)}{4\pi(R^-)^2(M_0\cos\theta^--1)^3} \end{aligned} \quad (1.35)$$

It is seen that, compared with the subsonic case, the sound pressure in Eq. (1.35) has some additional terms. In addition, if the strength of sound source $q_0(\tau)$ is constant, the sound pressure in far field will be approximately zero. However, for some practical situations related to accelerating motion, such as the rotating motion of a propeller, the sound generation will be inevitable, even though the sound source $q_0(\tau)$ is not dependent on time. This means that the acceleration of a sound source is one of the key factors to determine the sound field.

(2) Dipole source with arbitrary motion
A dipole point source in free space can be described by:

$$\frac{1}{c_0^2}\frac{\partial^2 p'}{\partial \tau^2} - \nabla^2 p' = -\frac{\partial}{\partial x_i}\{f_i(\tau)\delta[\mathbf{y}-\mathbf{y}_0(\tau)]\}. \tag{1.36}$$

Introducing variable $\mathbf{A}$ and assuming $p' = -\frac{\partial A_i}{\partial x_i}$, Eq. (1.36) is changed to:

$$\nabla^2 A_i - \frac{1}{c_0^2}\frac{\partial^2 A_i}{\partial \tau^2} = f_i(\tau)\delta[\mathbf{y}-\mathbf{y}_0(\tau)]. \tag{1.37}$$

Its solution is similar to that of the monopole source mentioned earlier. According to generalized Green's function formula, the solution of Eq. (1.37) can be expressed as:

$$A_i = \left.\frac{f_i(\tau)}{4\pi R(1-M_r)}\right|_{\tau=\tau^*} \cdot s \tag{1.38}$$

For simplicity, in the following content we drop the superscript "*," so it should be noted that the corresponding values at τ^* are assigned to those relevant variables. One obtains:

$$\begin{aligned} p'(\mathbf{x}, t) &= -\frac{\partial}{\partial x_i}\left[\frac{f_i(\tau)}{4\pi R(1-M_r)}\right] \\ &= -\frac{1}{4\pi R(1-M_r)}\frac{\partial f_i(\tau)}{\partial \tau}\frac{\partial \tau}{\partial x_i} + \frac{f_i(\tau)}{4\pi[R(1-M_r)]^2}\frac{\partial[R(1-M_r)]}{\partial x_i}, \end{aligned} \tag{1.39}$$

where $\frac{\partial \tau}{\partial x_i}$ and $\frac{\partial[R(1-M_r)]}{\partial x_i}$ are all derivatives of the compound function. In terms of the relevant derivative rule, the following relations can be obtained:

$$\frac{\partial \tau}{\partial x_i} = \frac{\partial}{\partial x_i}\left[t - \frac{R(\mathbf{x}, \tau)}{c_0}\right] = -\frac{1}{c_0}\left\{\frac{\partial}{\partial x_i}[R(\mathbf{x}, \tau)]_{\tau=const} + \frac{\partial}{\partial \tau}[R(\mathbf{x}, \tau)]_{\mathbf{x}=const}\frac{\partial \tau}{\partial x_i}\right\},$$

and

$$\frac{\partial \tau}{\partial x_i} = -\frac{R_i}{c_0 R(1-M_r)}. \tag{1.40}$$

Similarly,

$$\frac{\partial[R(1-M_r)]}{\partial x_i}=\frac{\partial}{\partial x_i}[R(1-M_r)]_{\tau=const}+\frac{\partial}{\partial \tau}[R(1-M_r)]_{\mathbf{x}=const}\frac{\partial \tau}{\partial x_i}. \tag{1.41}$$

The first term of Eq. (1.41) can be written as:

$$\begin{aligned}\frac{\partial}{\partial x_i}[R(1-M_r)]_{\tau=const}&=(1-M_r)\frac{\partial R}{\partial x_i}-R\frac{\partial M_r}{\partial x_i}=(1-M_r)\frac{R_i}{R}-RM_j\frac{\partial}{\partial x_i}\left[\frac{R_j}{R}\right]\\&=(1-M_r)\frac{R_i}{R}-M_j\left[\delta_{ij}-\frac{R_i}{R}\cdot\frac{R_j}{R}\right]=\frac{R_i}{R}-M_i.\end{aligned}$$

The second term of Eq. (1.41) deals with:

$$\begin{aligned}\frac{\partial}{\partial \tau}[R(1-M_r)]_{\mathbf{x}=const}&=(1-M_r)\frac{\partial R}{\partial \tau}-R\frac{\partial M_r}{\partial \tau}\\&=-(1-M_r)\frac{R_i}{R}V_i-R\left[\frac{R_i}{R}\frac{\partial M_i}{\partial \tau}+M_i\frac{\partial}{\partial \tau}\frac{R_i}{R}\right]\\&=-(1-M_r)M_rc_0-R_i\frac{\partial M_i}{\partial \tau}-RM_ic_0\frac{-RM_i+R_i\frac{R_j}{R}M_j}{R^2}\\&=-(1-M_r)M_rc_0+c_0M_i^2-c_0M_r^2-R_i\frac{\partial M_i}{\partial \tau}=c_0\left(M_i^2-M_r\right)-R_i\frac{\partial M_i}{\partial \tau}.\end{aligned}$$

Substituting these relations and Eq. (1.40) into Eq. (1.41) leads to:

$$\begin{aligned}\frac{\partial[R(1-M_r)]}{\partial x_i}&=\frac{R_i}{R}-M_i-\frac{R_ic_0\left(M_i^2-M_r\right)}{c_0R(1-M_r)}+\frac{R_iR_j}{c_0R(1-M_r)}\frac{\partial M_j}{\partial \tau}\\&=\frac{R_i\left(1-M_i^2\right)}{R(1-M_r)}-M_i+\frac{R_iR_j}{c_0R(1-M_r)}\frac{\partial M_j}{\partial \tau}.\end{aligned} \tag{1.42}$$

In addition, substituting Eqs. (1.40) and (1.42) into Eq. (1.39) results in:

$$\begin{aligned}p'(\mathbf{x},t)=&\frac{1}{4\pi c_0R(1-M_r)^2}\left[\frac{R_i}{R}\frac{\partial f_i}{\partial \tau}+\frac{f_i(\tau)}{(1-M_r)}\frac{R_i}{R}\frac{R_j}{R}\frac{\partial M_j}{\partial \tau}\right]\\&+\frac{1}{4\pi R^2(1-M_r)^2}\left[\frac{f_i}{(1-M_r)}\frac{R_i}{R}\left(1-M_i^2\right)-f_iM_i\right].\end{aligned} \tag{1.43}$$

In view of different influences for the terms related to $1/R$ and $1/R^2$, the solution of a far sound field is:

$$p'(\mathbf{x},t)=\frac{1}{4\pi c_0 R(1-M_r)^2}\left[\frac{R_i}{R}\frac{\partial f_i}{\partial\tau}+\frac{f_i(\tau)}{(1-M_r)}\frac{R_i}{R}\frac{R_j}{R}\frac{\partial M_j}{\partial\tau}\right], \tag{1.44}$$

whereas the solution of the near sound field is:

$$p'(\mathbf{x},t)=\frac{1}{4\pi R^2(1-M_r)^2}\left[\frac{f_i}{(1-M_r)}\frac{R_i}{R}\left(1-M_i^2\right)-f_iM_i\right]. \tag{1.45}$$

From Eq. (1.44), if the required solution of a far sound field is zero, the relevant conditions are that the dipole source must be steady, i.e., $\partial f_i/\partial\tau=0$, and at the same time there is no accelerating motion for the source, i.e., $\partial M_j/\partial\tau=0$.

For uniform rectilinear motion of a dipole source, this means that $\mathbf{y}_0(\tau)=V_0\tau$, $M_1=M_0=V_0/c_0$, and $M_r=\frac{R_1}{R}\cdot\frac{V_0}{c_0}=M_0\cdot\cos\theta$, where θ is the included angle between the moving direction and observer point, whereas M_r is the projection of the Mach number of the moving direction with respect to the direction of the observer point. Eq. (1.43) thus reduces to:

$$p'(\mathbf{x},t)=\frac{f_1'(\tau)\cos\theta}{4\pi c_0R(1-M_0\cdot\cos\theta)^2}+\frac{f_1(\tau)(\cos\theta-M_0)}{4\pi R^2(1-M_0\cdot\cos\theta)^3}. \tag{1.46}$$

This is the solution of sound pressure for a point dipole source with uniform rectilinear motion when the moving direction of the dipole is consistent with the pole of the sound source. When $V_0=0$, Eq. (1.46) becomes:

$$p'(\mathbf{x},t)=\frac{f_1'(\tau)\cos\theta}{4\pi c_0R}+\frac{f_1(\tau)\cos\theta}{4\pi R^2}. \tag{1.47}$$

This is the expression of sound pressure for a longitudinal dipole.

As for a transversal dipole that is perpendicular to the moving direction, considering $R_2/R=\sin\theta$ and $f_iM_i=0$, Eq. (1.43) reduces to:

$$p'(\mathbf{x},t)=\frac{f_2'(\tau)\sin\theta}{4\pi c_0R(1-M_0\cdot\cos\theta)^2}+\frac{f_2(\tau)\sin\theta\left(1-M_0^2\right)}{4\pi R^2(1-M_0\cdot\cos\theta)^3}. \tag{1.48}$$

When the moving speed of the source is zero, this formula will become:

$$p'(\mathbf{x},t)=\frac{f_2'(\tau)\sin\theta}{4\pi c_0R}+\frac{f_2(\tau)\sin\theta}{4\pi R^2}. \tag{1.49}$$

This is also a classical expression for a point dipole source.

1.1.4 Frequency features of moving sound source—Doppler effect

Consider a point monopole with uniform rectilinear motion. Assume that the strength of the source is given by

$$q(\tau)=q_0 \sin \omega_0 \tau$$

where ω_0 is the radian frequency of the source. Therefore, the phase at the observer point is:

$$\varphi=\omega_0\left(t-\frac{R^{\pm}}{c_0}\right). \tag{1.50}$$

For a point source with subsonic speed, the receiving frequency at the observer point is:

$$\omega=\frac{d\varphi}{dt}=\omega_0-\frac{\omega_0}{c_0}\frac{dR}{dt}=\frac{\omega_0}{1-M_0\cos\theta}. \tag{1.51}$$

From this formula, it can be seen that, if the observer lies in front of the sound source, the frequency will increase due to positive cosθ, and in contrast, if the observer is behind the sound source, the frequency will decrease due to negative cosθ. This is called a typical Doppler effect. Also, it is found that the range of the receiving frequency is:

$$\frac{\omega_0}{1+M_0}<\omega<\frac{\omega_0}{1-M_0}. \tag{1.52}$$

For a point source with supersonic speed, the phase at the observer point has different values, i.e.,

$$\varphi^{+}=\omega_0\left(t-\frac{R^{+}}{c_0}\right), \varphi^{-}=\omega_0\left(t-\frac{R^{-}}{c_0}\right) \tag{1.53}$$

The corresponding frequency is, respectively,

$$\omega^{+}=\frac{\omega_0}{1-M_0\cos\theta^{+}}, \omega^{-}=\frac{\omega_0}{1-M_0\cos\theta^{-}}, \tag{1.54}$$

where

$$\frac{1}{c_0}\frac{dR^{+}}{dt}=\frac{1}{M_0^2-1}\left[M_0^2+\frac{(M_0-\cos\theta^{+})M_0}{M_0\cos\theta^{+}-1}\right]=\frac{M_0\cos\theta^{+}}{M_0\cos\theta^{+}-1},$$

and

$$\frac{1}{c_0}\frac{dR^-}{dt}=\frac{1}{M_0^2-1}\left[M_0^2+\frac{(M_0-\cos\theta^-)M_0}{M_0\cos\theta^- -1}\right]=\frac{M_0\cos\theta^-}{M_0\cos\theta^- -1}.$$

It can be seen that, for the supersonic case, the range of receiving frequency at the observer point corresponds to two different intervals, i.e.,

$$-\infty<\omega<\frac{\omega_0}{1-M_0},\ \frac{\omega_0}{M_0+1}<\omega<+\infty.$$

The results also indicate that the observer will receive different frequencies simultaneously if the sound source moves at supersonic speed. The interaction between different frequencies will cause dramatic pressure fluctuations.

It can be verified that [14], no matter how the sound source moves, supersonically or subsonically, the sound radiation power in far field is:

$$E=\frac{E_s}{\left(1-M_0^2\right)^2}, \tag{1.55}$$

where E_s represents the sound power for a static source; if the source is harmonic with the frequency f_0, its expression is:

$$E_s=\frac{\pi\left(q_0 f_0\right)^2}{2\rho_0 c_0}. \tag{1.56}$$

1.2 Generalized Green's formula

Consider a solution of wave equation with retarded time in form of Eq. (1.10). The corresponding Green's function is given by:

$$\nabla^2 G-\frac{1}{c_0^2}\frac{D_0^2 G}{D\tau^2}=-\delta(t-\tau)\delta\left(\vec{x}-\vec{y}\right), \tag{1.57}$$

which satisfies the condition:

$$G=\frac{D_0}{D\tau}G=0\quad(t<\tau). \tag{1.58}$$

Let $v(\tau)$ denote an arbitrary region of space bounded by $S(\tau)$; $\mathbf{V}_S$ stands for the velocity of the arbitrary point on the boundary surface, $\mathbf{A}$ is an arbitrary vector in defined on$v(\tau)$, and Ψ, Φ are any two functions defined on $v(\tau)$. Then the divergence theorem states that:

$$\int_{S(\tau)}\mathbf{A}\cdot\vec{n}\ dS(\mathbf{y})=\int_{v(\tau)}\nabla\cdot\mathbf{A}\ d\mathbf{y}. \tag{1.59}$$

The three-dimensional Leibniz's rule shows that:

$$\frac{d}{d\tau}\int_{v(\tau)} \Psi \, d\mathbf{y} = \int_{v(\tau)} \frac{\partial \Psi}{\partial \tau} \, d\mathbf{y} + \int_{S(\tau)} \mathbf{V}_S \cdot \vec{n} \, \Psi \, dS(\mathbf{y}). \tag{1.60}$$

In addition, the Green's theorem states that:

$$\int_{S(\tau)} \left(\Phi \frac{\partial \Psi}{\partial n} - \Psi \frac{\partial \Phi}{\partial n} \right) dS(\mathbf{y}) = \int_{v(\tau)} \left(\Phi \nabla^2 \Psi - \Psi \nabla^2 \Phi \right) \, d\mathbf{y}, \tag{1.61}$$

where $\frac{\partial \Psi}{\partial n} = \vec{n} \cdot \nabla \Psi$. By applying these relations, Eq. (1.57) can be rewritten as:

$$\begin{aligned}
\int_{-T}^{T}\int_{S(\tau)} \left(G\frac{\partial p'}{\partial n} - p'\frac{\partial G}{\partial n} \right) dS(\mathbf{y}) d\tau &= \int_{-T}^{T}\int_{v(\tau)} \left(G\nabla^2 p' - p'\nabla^2 G \right) d\mathbf{y} d\tau \\
&= \frac{1}{c_0^2}\int_{-T}^{T}\int_{v(\tau)} \left(G\frac{D_0^2 p'}{D\tau^2} - p'\frac{D_0^2 G}{D\tau^2} \right) d\mathbf{y} d\tau \\
&\quad - \int_{-T}^{T}\int_{v(\tau)} \left(G\gamma - p'\delta(t-\tau)\delta(\vec{x} - \vec{y}) \right) d\mathbf{y} d\tau.
\end{aligned} \tag{1.62}$$

However, because:

$$\begin{aligned}
G\frac{D_0^2 p'}{D\tau^2} - p'\frac{D_0^2 G}{D\tau^2} &= \frac{D_0}{D\tau}\left(G\frac{D_0 p'}{D\tau} - p'\frac{D_0 G}{D\tau} \right) \\
&= \frac{\partial}{\partial \tau}\left(G\frac{D_0 p'}{D\tau} - p'\frac{D_0 G}{D\tau} \right) + U\frac{\partial}{\partial y_1}\left(G\frac{D_0 p'}{D\tau} - p'\frac{D_0 G}{D\tau} \right),
\end{aligned}$$

it can be shown that:

$$\begin{aligned}
&\int_{v(\tau)} \left(G\frac{D_0^2 p'}{D\tau^2} - p'\frac{D_0^2 G}{D\tau^2} \right) d\mathbf{y} \\
&= \int_{v(\tau)} \frac{\partial}{\partial \tau}\left(G\frac{D_0 p'}{D\tau} - p'\frac{D_0 G}{D\tau} \right) d\mathbf{y} + \int_{v(\tau)} U\frac{\partial}{\partial y_1}\left(G\frac{D_0 p'}{D\tau} - p'\frac{D_0 G}{D\tau} \right) d\mathbf{y} \\
&= \frac{d}{d\tau}\int_{v(\tau)} \left(G\frac{D_0 p'}{D\tau} - p'\frac{D_0 G}{D\tau} \right) \, d\mathbf{y} + \int_{S(\tau)} (iU - \mathbf{V}_S) \cdot \vec{n} \, \left(G\frac{D_0 p'}{D\tau} - p'\frac{D_0 G}{D\tau} \right) \, dS(\mathbf{y}).
\end{aligned}$$

Integrating the result with respect to time τ yields:

$$\int_{-T}^{T}\int_{v(\tau)}\left(G\frac{D_0^2p'}{D\tau^2}-p'\frac{D_0^2G}{D\tau^2}\right)d\mathbf{y}d\tau$$
$$=\int_{v(\tau)}\left(G\frac{D_0p'}{D\tau}-p'\frac{D_0G}{D\tau}\right)d\mathbf{y}\Bigg|_{\tau=-T}^{\tau=T} \tag{1.63}$$
$$-\int_{-T}^{T}\int_{S(\tau)}V_n'\left(G\frac{D_0p'}{D\tau}-p'\frac{D_0G}{D\tau}\right)dS(\mathbf{y})d\tau.$$

where $V_n' = (\mathbf{V}_S - iU)\cdot\vec{n}$ denotes the projection of the relative velocity on the normal direction of $v(\tau)$. Considering the causality condition, the first integrated term vanishes at the upper limit ($\tau = T$), whereas at the lower limit ($\tau = -T$), it stands for the effect of initial conditions in the remote past. This factor should be possible to ignore. Then we have:

$$\int_{-T}^{T}\int_{v(\tau)}\left(G\frac{D_0^2p'}{D\tau^2}-p'\frac{D_0^2G}{D\tau^2}\right)d\mathbf{y}d\tau=-\int_{-T}^{T}\int_{S(\tau)}V_n'\left(G\frac{D_0p'}{D\tau}-p'\frac{D_0G}{D\tau}\right)dS(\mathbf{y})d\tau. \tag{1.64}$$

Consider the boundary condition and domains shown in Fig. 1.6 and the relation given by:

$$\int_{-T}^{T}\int_{v(\tau)}p'\delta(t-\tau)\delta(\vec{\mathbf{x}}-\vec{\mathbf{y}})d\mathbf{y}d\tau=\begin{cases}p'(t,\vec{\mathbf{x}}), & \text{for } \vec{\mathbf{x}} \text{ is in } v(\tau)\\ 0, & \text{for } \vec{\mathbf{x}} \text{ is out of } v(\tau)\end{cases}.$$

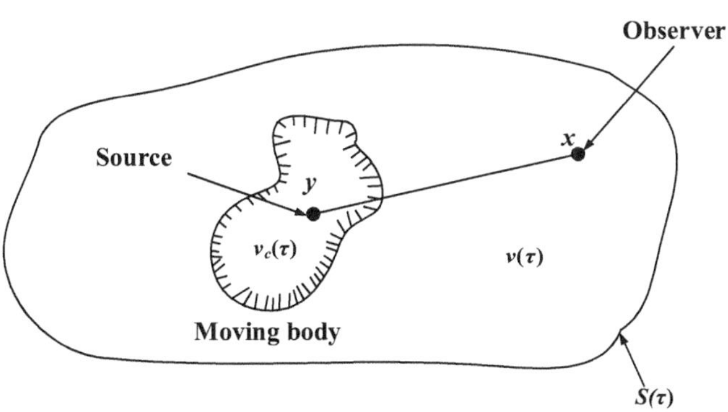

Fig. 1.6 Sound radiation from bounded source region.

Substituting Eq. (1.64) into Eq. (1.62) yields:

$$p'(t, \vec{x}) = \int_{-T}^{T}\int_{v(\tau)} G(\mathbf{y}, \tau/\mathbf{x}, t)\cdot \gamma(\mathbf{y}, \tau) d\mathbf{y} d\tau$$
$$+ \int_{-T}^{T} d\tau \int_{S(\tau)} \left[G\left(\frac{\partial}{\partial n} + \frac{V_n'}{c_0^2}\frac{D_0}{D\tau}\right) p'(\mathbf{y}, \tau) \right. \qquad \text{for } \vec{\mathbf{x}} \text{ is in } v(\tau). \quad (1.65)$$
$$\left. - p'(\mathbf{y}, \tau)\left(\frac{\partial}{\partial n} + \frac{V_n'}{c_0^2}\frac{D_0}{D\tau}\right) G(\mathbf{y}, \tau/\mathbf{x}, t) \right] dS(\mathbf{y}),$$

This is called the generalized Green's formula, which plays a very important role in the derivation of aeroacoustic equations.

1.3 Lighthill equation

1.3.1 Derivation of basic equations

In the first section, emphasis was placed on how a specific source (dipole or monopole) radiates acoustic waves in moving media. As for the practical problems, there are more concerns to describe the sound generation by turbulent motion itself and its interaction with a static or moving body. In this section, we begin with the Navier-Stokes equation and then discuss how to introduce Lighthill's acoustic analogy theory.

Consider the following continuity and momentum equations,

$$\frac{\partial \rho}{\partial t} + \frac{\partial \rho u_i}{\partial y_i} = 0, \tag{1.66}$$

$$\frac{\partial \rho u_i}{\partial t} + \frac{\partial \rho u_i u_j}{\partial y_j} = -\frac{\partial p}{\partial y_i} + \frac{\partial e_{ij}}{\partial y_j}, \tag{1.67}$$

where e_{ij} denotes viscous stress tensor. Taking the derivative of Eq. (1.66) with respect to time t, and the divergence of Eq. (1.67), it shows that:

$$\frac{\partial^2 \rho}{\partial t^2} - \frac{\partial^2 p}{\partial y_i^2} = \frac{\partial^2 \rho u_i u_j}{\partial y_i \partial y_j} - \frac{\partial^2 e_{ij}}{\partial y_i \partial y_j}. \tag{1.68}$$

Subtracting $c_0^2\nabla^2\rho$ on both sides of Eq. (1.68) yields:

$$\frac{\partial^2 \rho}{\partial t^2} - c_0^2\nabla^2\rho = \frac{\partial^2 T_{ij}}{\partial y_i \partial y_j}, \tag{1.69}$$

where

$$T_{ij} = \rho u_i u_j - e_{ij} + \delta_{ij}\left(p - c_0^2 \rho\right),$$

and

$$\delta_{ij} = \begin{cases} 0 & i = j \\ 1 & i \neq j \end{cases}.$$

After linearizing the equation, the wave equation with fluctuating density as a variable can be given in the form of:

$$\frac{\partial^2 \rho'}{\partial t^2} - c_0^2 \nabla^2 \rho' = \frac{\partial^2 T'_{ij}}{\partial y_i \partial y_j}, \tag{1.70}$$

where $\rho' = \rho - \rho_0$, whereas $T'_{ij} = \rho u_i u_j - e_{ij} + \delta_{ij}[(p - p_0) - c_0^2(\rho - \rho_0)]$ is called Lighthill's turbulence stress tensor.

Eq. (1.70) is Lighthill's equation on the basis of the concept of acoustic analogy. However, it is difficult to find its analytic solution because the two sides of Eq. (1.70) all contain unknown variables. The only possibility is to directly solve Eq. (1.70) using a numerical approach. However, this equation actually deals with various flow scales; it was found that the numerical solution with such a multiscale problem was also very difficult in a long period of time. However, in terms of Lighthill' theory, the right term of Eq. (1.70), i.e., the source term, can be independently obtained either by numerical simulation with a single-flow scale or by an experimental approach. With a known source term, the sound pressure can be given by solving the conventional wave equation. In fact, Lighthill's acoustic analogy offers a very successful application for various practical problems.

1.3.2 Effect of solid boundary on sound generation

Note that the solution of the wave equation similar to Eq. (1.70) sometimes needs to consider the effect of a solid boundary on the sound field. In this section, emphasis is placed on how to include the effect of a solid boundary using Green's function formula.

Consider a moving body with volume $v(\tau)$ bounded by surface $S(\tau)$, and let $\mathbf{V}_S$ denote the velocity of an arbitrary surface point of the body. Suppose the media is static, i.e., $U = 0$, then introducing generalized Green's function formula yields:

$$\rho'(t, \vec{x}) = \frac{1}{c_0^2}\int_{-T}^{T}\int_{v(\tau)} G\cdot\frac{\partial^2 T'_{ij}}{\partial y_i \partial y_j}d\mathbf{y}d\tau + \int_{-T}^{T}\int_{S(\tau)}\left[G\left(\frac{\partial}{\partial n}+\frac{V'_n}{c_0^2}\frac{\partial}{\partial \tau}\right)\rho' - \rho'\left(\frac{\partial}{\partial n}+\frac{V'_n}{c_0^2}\frac{\partial}{\partial \tau}\right)G\right]dS(\mathbf{y})d\tau. \tag{1.71}$$

It is known that:

$$\frac{\partial}{\partial y_i}G\frac{\partial T'_{ij}}{\partial y_j} - \frac{\partial}{\partial y_j}T'_{ij}\frac{\partial G}{\partial y_i} = G\frac{\partial^2 T'_{ij}}{\partial y_i \partial y_j} - T'_{ij}\frac{\partial^2 G}{\partial y_i \partial y_j},$$

and using $n_i\frac{\partial}{\partial y_i}$ to replace $\frac{\partial}{\partial n}$, the application of the divergence theorem results in:

$$\int_{v(\tau)}\frac{\partial}{\partial y_i}G\frac{\partial T'_{ij}}{\partial y_j}d\mathbf{y} = \int_{S(\tau)} G\frac{\partial T'_{ij}}{\partial y_j}\cdot\vec{n}\ dS(\mathbf{y}),$$

$$\int_{v(\tau)}\frac{\partial}{\partial y_j}T'_{ij}\frac{\partial G}{\partial y_i}d\mathbf{y} = \int_{S(\tau)} T'_{ij}\frac{\partial G}{\partial y_i}\cdot\vec{n}\ dS(\mathbf{y}).$$

Hence,

$$\begin{aligned}\rho'(t, \vec{x}) = &\frac{1}{c_0^2}\int_{-T}^{T}\int_{v(\tau)} T'_{ij}\cdot\frac{\partial^2 G}{\partial y_i \partial y_j}d\mathbf{y}d\tau \\ &+\frac{1}{c_0^2}\int_{-T}^{T}\int_{S(\tau)} Gn_i\left[\frac{\partial}{\partial y_j}\left(c_0^2\delta_{ij}\rho' + T'_{ij}\right) + V_i^s\frac{\partial \rho'}{\partial \tau}\right]dS(\mathbf{y})d\tau \\ &-\frac{1}{c_0^2}\int_{-T}^{T}\int_{S(\tau)} n_j\left[\left(T'_{ij}+c_0^2\delta_{ij}\rho'\right)\frac{\partial G}{\partial y_i} + V_j^s\rho'\frac{\partial G}{\partial \tau}\right]dS(\mathbf{y})d\tau,\end{aligned} \tag{1.72}$$

where $n_i\frac{\partial}{\partial y_j}\left(\delta_{ij}\rho'\right) = n_i\frac{\partial \rho'}{\partial y_i}$. Replacing the second and third terms on the right side of Eq. (1.72) using Lighthill's stress tensor, the corresponding solution becomes:

$$\begin{aligned}\rho'(t, \vec{x}) = &\frac{1}{c_0^2}\int_{-T}^{T}\int_{v(\tau)} T'_{ij}\cdot\frac{\partial^2 G}{\partial y_i \partial y_j}d\mathbf{y}d\tau + \frac{1}{c_0^2}\int_{-T}^{T}\int_{S(\tau)} f_i\frac{\partial G}{\partial y_i}dS(\mathbf{y})d\tau \\ &-\frac{1}{c_0^2}\int_{-T}^{T}\int_{S(\tau)} n_ih_i dS(\mathbf{y})d\tau,\end{aligned} \tag{1.73}$$

where

$$f_i = n_j e_{ij} - n_i(p - p_0), \tag{1.74}$$

which denotes force per unit area acting on the body. Besides,

$$\begin{aligned} n_i h_i = & -G n_i \left\{ \frac{\partial}{\partial y_j} \left[\rho u_i u_j - e_{ij} + \delta_{ij}(p - p_0) \right] + V_i^s \frac{\partial \rho'}{\partial \tau} \right\} \\ & + n_j \left[V_j^s \rho' \frac{\partial G}{\partial \tau} + (\rho u_i u_j) \frac{\partial G}{\partial y_i} \right]. \end{aligned} \tag{1.75}$$

In terms of Eq. (1.67), it is known that:

$$\frac{\partial \rho u_i u_j}{\partial y_j} - \frac{\partial e_{ij}}{\partial y_j} + \frac{\partial p}{\partial y_i} = -\frac{\partial \rho u_i}{\partial t}.$$

Introducing this relation into Eq. (1.75) yields:

$$h_i = G\left[\frac{\partial \rho u_i}{\partial \tau} - V_i^s \frac{\partial \rho'}{\partial \tau}\right] + (\rho u_i u_j)\frac{\partial G}{\partial y_j} + \rho' V_i^s \frac{\partial G}{\partial \tau}. \tag{1.76}$$

Assume the body is impenetrable, so this means $n_i u_i = n_i V_i^s$. Naturally,

$$n_i h_i = n_i \left[G \frac{\partial \rho u_i}{\partial \tau} - G u_i \frac{\partial \rho'}{\partial \tau} + (\rho u_i u_j) \frac{\partial G}{\partial y_j} \right] + n_i u_i \rho' \frac{\partial G}{\partial \tau}.$$

Notice that $\rho' = \rho - \rho_0$, hence:

$$n_i h_i = n_i \left[\frac{\partial \rho u_i G}{\partial \tau} - G u_i \frac{\partial \rho}{\partial \tau} + (\rho u_i u_j) \frac{\partial G}{\partial y_j} \right] - n_i \rho_0 u_i \frac{\partial G}{\partial \tau}.$$

Substituting Eq. (1.66) into the previous equation yields:

$$n_i h_i = n_i \left[\frac{\partial \rho u_i G}{\partial \tau} + u_i \frac{\partial \rho u_j G}{\partial y_j} \right] - n_i \rho_0 u_i \frac{\partial G}{\partial \tau}. \tag{1.77}$$

The application of Leibniz's rule results in:

$$\frac{d}{d\tau}\int_{v(\tau)} \frac{\partial \rho u_j G}{\partial y_j} d\mathbf{y} = \int_{v(\tau)} \frac{\partial}{\partial \tau}\frac{\partial \rho u_j G}{\partial y_j} d\mathbf{y} + \int_{S(\tau)} n_i u_i \frac{\partial \rho u_j G}{\partial y_j} dS(\mathbf{y}), \tag{1.78}$$

and by applying the divergence theorem, we have:

$$\frac{d}{d\tau}\int_{v(\tau)} \frac{\partial \rho u_j G}{\partial y_j} d\mathbf{y} = \int_{S(\tau)} n_i \left(\frac{\partial \rho u_i G}{\partial \tau} + u_i \frac{\partial \rho u_j G}{\partial y_j} \right) dS(\mathbf{y}). \tag{1.79}$$

Hence,

$$\rho'(t,\vec{x}) = \frac{1}{c_0^2}\int_{-T}^{T}\int_{v(\tau)} T'_{ij}\frac{\partial^2 G}{\partial y_i \partial y_j}\mathbf{dy}d\tau + \frac{1}{c_0^2}\int_{-T}^{T}\int_{S(\tau)} f_i\frac{\partial G}{\partial y_i}dS(\mathbf{y})d\tau \\ + \frac{1}{c_0^2}\int_{-T}^{T}\int_{S(\tau)} \rho_0 V_n \frac{\partial G}{\partial \tau}dS(\mathbf{y})d\tau, \tag{1.80}$$

where the following relation has been used,

$$\int_{-T}^{T}\frac{d}{d\tau}\int_{v(\tau)}\frac{\partial \rho u_j G}{\partial y_j}\mathbf{dy} = \int_{v(\tau)}\frac{\partial \rho u_j G}{\partial y_j}\mathbf{dy}\Bigg|_{\tau=-T}^{\tau=T} = 0,$$

whose physical explanation has been given during the derivation of generalized Green's function formula.

Eq. (1.80) is a basic expression including the effect of a solid boundary on sound generation. Among which, the first term denotes the sound source due to the moving volume belonging to a quadruple source, whereas the second term represents the source due to the force belonging to dipole source, and the last term stands for the source due to the displacement of the moving body belonging to a monopole source. This equation can be applied to any aeroacoustic problem with the effect of a solid boundary.

1.4 Ffowcs Williams-Hawkings equation

With a general expression similar to Eq. (1.80), it is convenient to consider a solution in free space for a moving body denoted by volume $v_c(\tau)$ and bounded by $S(\tau)$. In fact, the Green's function for free space is:

$$G = \frac{\delta\left(\tau - t + \dfrac{R}{c_0}\right)}{4\pi R}, \tag{1.81}$$

where $R = |\mathbf{x} - \mathbf{y}|$ denotes the distance from the source point $\mathbf{y}(\tau)$ to an arbitrary observer point $\mathbf{x}(t)$, τstands for emission time, and t represents receiving time. It is noted that:

$$\frac{\partial G}{\partial x_i} = -\frac{\partial G}{\partial y_i}.$$

Introducing this relation into Eq. (1.80) yields:

$$\begin{aligned}
\rho'\left(t, \vec{x}\right) = & \frac{1}{c_0^2}\frac{\partial^2}{\partial x_i \partial x_j}\int_{-T}^{T}\int_{v(\tau)}\frac{\delta\left(\tau - t + \frac{R}{c_0}\right)}{4\pi R}T'_{ij}d\mathbf{y}d\tau \\
& - \frac{1}{c_0^2}\frac{\partial}{\partial x_i}\int_{-T}^{T}\int_{S(\tau)} f_i \frac{\delta\left(\tau - t + \frac{R}{c_0}\right)}{4\pi R}dS(\mathbf{y})d\tau \\
& + \frac{1}{c_0^2}\int_{-T}^{T}\int_{S(\tau)} \rho_0 V_n \frac{\partial}{\partial \tau}\left[\frac{\delta\left(\tau - t + \frac{R}{c_0}\right)}{4\pi R}\right]dS(\mathbf{y})d\tau.
\end{aligned} \tag{1.82}$$

Especially, for the third term, the partial derivative with respect to time needs to be moved out of the integration. Substituting $\Psi = V_i \cdot \frac{\partial G}{\partial x_i}$ into Eq. (1.60) results in:

$$\begin{aligned}
0 = & \int_{-T}^{T}\frac{d}{d\tau}\int_{v_c(\tau)} V_i \cdot \frac{\partial G}{\partial x_i}\, d\mathbf{y}d\tau = \int_{-T}^{T}\int_{v_c(\tau)}\frac{\partial}{\partial \tau}V_i \cdot \frac{\partial G}{\partial x_i}\, d\mathbf{y}d\tau \\
& - \int_{-T}^{T}\int_{S(\tau)} V_j n_j V_i \cdot \frac{\partial G}{\partial x_i}\, dS(\mathbf{y})d\tau,
\end{aligned} \tag{1.83}$$

where the second term takes a negative sign because n_j is the outward normal vector of $v(\tau)$ and the outward normal vectors of $v_c(\tau)$ and $v(\tau)$ oppose each other. Again, in terms of the divergence theorem, the third term of Eq. (1.82) can be expressed as:

$$\begin{aligned}
\int_{-T}^{T}\int_{S(\tau)} \rho_0 V_n \frac{\partial G}{\partial \tau} dS(\mathbf{y})d\tau = & \int_{-T}^{T}\int_{S(\tau)} \rho_0 n_i V_i \frac{\partial G}{\partial \tau} dS(\mathbf{y})d\tau \\
= & - \int_{-T}^{T}\int_{v_c(\tau)} \rho_0 \frac{\partial}{\partial y_i}\left(V_i \frac{\partial G}{\partial \tau}\right)d\mathbf{y}d\tau.
\end{aligned} \tag{1.84}$$

Moreover,

$$\begin{aligned}
& \int_{-T}^{T}\int_{S(\tau)} \rho_0 V_n \frac{\partial G}{\partial \tau} dS(\mathbf{y})d\tau \\
& = - \int_{-T}^{T}\int_{v_c(\tau)} \rho_0 \left[\frac{\partial}{\partial y_j}V_j\left(\frac{\partial G}{\partial \tau} + V_i \cdot \frac{\partial G}{\partial x_i}\right) + \frac{\partial}{\partial \tau}V_i \cdot \frac{\partial G}{\partial x_i}\right]d\mathbf{y}d\tau.
\end{aligned} \tag{1.85}$$

For a rigid body, the projection of the velocity on the body is zero, i.e., $\frac{\partial V_i}{\partial y_i}=0$, hence,

$$\frac{\partial}{\partial y_j}V_j\left(\frac{\partial G}{\partial \tau}+V_i\cdot\frac{\partial G}{\partial x_i}\right)+\frac{\partial}{\partial \tau}V_i\cdot\frac{\partial G}{\partial x_i}$$
$$=V_j\frac{\partial^2 G}{\partial y_j\partial \tau}+V_j\frac{\partial V_i\partial G}{\partial y_j\,\partial x_i}+V_jV_i\frac{\partial^2 G}{\partial y_j\partial x_i}+\frac{\partial V_i\partial G}{\partial \tau\,\partial x_i}+V_i\frac{\partial^2 G}{\partial \tau\partial x_i}.$$

Considering $\frac{\partial G}{\partial x_i}=-\frac{\partial G}{\partial y_i}$, it is shown that:

$$\frac{\partial}{\partial y_j}V_j\left(\frac{\partial G}{\partial \tau}+V_i\cdot\frac{\partial G}{\partial x_i}\right)+\frac{\partial}{\partial \tau}V_i\cdot\frac{\partial G}{\partial x_i}=\frac{\partial G}{\partial x_i}\frac{DV_i}{D\tau}-V_jV_i\frac{\partial^2 G}{\partial x_j\partial x_i},\tag{1.86}$$

where the total derivative $\frac{DV_i}{D\tau}=a_i$denotes the accelerated velocity of the moving body. Notice that:

$$a_i\frac{\partial G}{\partial x_i}-V_jV_i\frac{\partial^2 G}{\partial x_j\partial x_i}=\frac{\partial}{\partial x_i}(Ga_i)-\frac{\partial^2}{\partial x_j\partial x_i}\{GV_jV_i\},$$

so, introducing this expression into Eq. (1.82) yields:

$$\begin{aligned}\rho'(t,\vec{x})=&\frac{1}{c_0^2}\frac{\partial^2}{\partial x_i\partial x_j}\int_{v(t_0)}\int_{-T}^{T}\frac{\delta\left(\tau-t+\frac{R}{c_0}\right)}{4\pi R}T'_{ij}d\tau d\mathbf{y}\\&-\frac{1}{c_0^2}\frac{\partial}{\partial x_i}\int_{S(t_0)}\int_{-T}^{T}\frac{\delta\left(\tau-t+\frac{R}{c_0}\right)}{4\pi R}f_i d\tau dS(\mathbf{y})\\&-\frac{1}{c_0^2}\frac{\partial}{\partial x_i}\int_{v_c(t_0)}\int_{-T}^{T}\frac{\delta\left(\tau-t+\frac{R}{c_0}\right)}{4\pi R}\rho_0 a_i d\tau d\mathbf{y}\\&+\frac{1}{c_0^2}\frac{\partial^2}{\partial x_i\partial x_j}\int_{v_c(t_0)}\int_{-T}^{T}\frac{\delta\left(\tau-t+\frac{R}{c_0}\right)}{4\pi R}\rho_0 V_iV_j d\tau d\mathbf{y}.\end{aligned}\tag{1.87}$$

Notice that the integration of any function $f(\tau)$ with δ function gives:

$$\int_{-T}^{T}f(\tau)\delta[g(\tau)]d\tau=\sum_i\frac{f(\tau_i)}{\left|\frac{dg(\tau_i)}{d\tau}\right|},\tag{1.88}$$

where τ_i is the ith root determined by the retarded time equation $g(\tau_i) = 0$, whereas:

$$g(\tau) = \tau - t + \frac{R}{c_0} = \tau - t + \frac{|\mathbf{x} - \mathbf{y}|}{c_0}. \tag{1.89}$$

Its derivative is:

$$\frac{dg}{d\tau} = 1 - \frac{1}{c_0}\frac{\mathbf{R}}{R}\frac{d\mathbf{y}}{d\tau} = 1 - \frac{\mathbf{R}}{R}\mathbf{M}, \tag{1.90}$$

where $\mathbf{R} = \mathbf{x} - \mathbf{y}$, $\mathbf{M} = \frac{\mathbf{V}}{c_0}$. Integrating Eq. (1.87) with respect to time gives:

$$\begin{aligned}
\rho'(t, \vec{x}) = & \frac{1}{4\pi c_0^2}\frac{\partial^2}{\partial x_i \partial x_j}\int_{v(t_0)}\left[\frac{T'_{ij}}{R|1 - M_r|}\right]_{\tau=\tau_e} d\boldsymbol{\xi} \\
& - \frac{1}{4\pi c_0^2}\frac{\partial}{\partial x_i}\int_{S(t_0)}\left[\frac{f_i}{R|1 - M_r|}\right]_{\tau=\tau_e} dS(\boldsymbol{\xi}) \\
& - \frac{1}{4\pi c_0^2}\frac{\partial}{\partial x_i}\int_{v_c(t_0)}\left[\frac{\rho_0 a_i}{R|1 - M_r|}\right]_{\tau=\tau_e} d\boldsymbol{\xi} \\
& + \frac{1}{4\pi c_0^2}\frac{\partial^2}{\partial x_i \partial x_j}\int_{v_c(t_0)}\left[\frac{\rho_0 V_i V_j}{R|1 - M_r|}\right]_{\tau=\tau_e} d\boldsymbol{\xi},
\end{aligned} \tag{1.91}$$

where $M_r = \frac{\mathbf{R}}{R}\mathbf{M}$ is the projection of the Mach number of a moving body in the direction of the observer, whereas the emission time is determined by:

$$g(\tau_e) \equiv \tau_e - t + \frac{|\mathbf{x} - \mathbf{y}|}{c_0} = 0.$$

This equation was presented by Ffowcs Williams and Hawkings in 1969, which is now named as Ffowcs Williams-Hawkings equation or shortened as an FW-H equation. In Eq. (1.91), $v(\tau)$ is the volume without the body, whereas $v_c(\tau)$ is the volume of the moving body. It is also seen that the first term is a quadruple source due to the moving effect, the second term is a dipole source due to the force on the body, and the third and fourth terms represent the dipole and quadruple sources due to the displacement and acceleration of the body, respectively. It is shown that the last two terms are equivalent to a monopole source. In this case, the FW-H equation can also be expressed as:

$$\begin{aligned}
\rho'(t, \vec{x}) = & \frac{1}{4\pi c_0^2}\frac{\partial^2}{\partial x_i \partial x_j}\int_{v(t_0)}\left[\frac{T'_{ij}}{R|1 - M_r|}\right]_{\tau=\tau_e} d\boldsymbol{\xi} \\
& - \frac{1}{4\pi c_0^2}\frac{\partial}{\partial x_i}\int_{S(t_0)}\left[\frac{f_i}{R|1 - M_r|}\right]_{\tau=\tau_e} dS(\boldsymbol{\xi}) \\
& + \frac{\partial}{\partial t}\int_{S(t_0)}\left[\frac{\rho_0 V'_n}{R|1 - M_r|}\right]_{\tau=\tau_e} dS(\boldsymbol{\xi}),
\end{aligned}$$

where V_n' denotes the dot product of the outward normal vector of $v_c(\tau)$and the moving velocity of the body, i.e., $V_n' = -V_n$.

When the body is static, i.e., $\mathbf{a} = \mathbf{M} = \mathbf{V} = \mathbf{0}$, then Eq. (1.91) becomes:

$$\begin{aligned}\rho'\left(t, \vec{x}\right) = & \frac{1}{4\pi c_0^2}\frac{\partial^2}{\partial x_i \partial x_j}\int_{v(t_0)}\frac{T_{ij}'}{R}\left(\mathbf{y}, t-\frac{R}{c_0}\right)d\mathbf{y} \\ & -\frac{1}{4\pi c_0^2}\frac{\partial}{\partial x_i}\int_{S(t_0)}\frac{f_i}{R}\left(\mathbf{y}, t-\frac{R}{c_0}\right)dS(\mathbf{y}),\end{aligned} \tag{1.92}$$

which was presented by Curle [3]. Now it is named Curle's equation.

For some specific cases, its far-field solution is interesting. Consider any time-correlation function $A(\tau)$, similar to the integration function in Eq. (1.91), by introducing

$$\breve{\mathbf{A}} = \left[\frac{A(\tau)}{R|C^\dagger|}\right]_{\tau=\tau_e},$$

where $C^\dagger = 1 - M_r$ stands for the Doppler factor. Notice that:

$$\left(\frac{\partial g}{\partial x_i}\right)_{\tau_e} + \left(\frac{\partial g}{\partial \tau_e}\right)_{\mathbf{x}}\left(\frac{\partial \tau_e}{\partial x_i}\right) = 0. \tag{1.93}$$

It is shown that:

$$\frac{\partial \tau_e}{\partial x_i} = -\left[\frac{R_i}{c_0 C^\dagger R}\right]_{\tau=\tau_e}.$$

Then for any function $f(\mathbf{x}, \tau_e)$, we obtain:

$$\frac{\partial f}{\partial x_i} = \left(\frac{\partial f}{\partial x_i}\right)_{\tau_e} - \left(\frac{\partial f}{\partial \tau_e}\right)_{\mathbf{x}}\left[\frac{R_i}{c_0 C^\dagger R}\right]_{\tau=\tau_e},$$

whereas for a specific function R, i.e., the distance from the observer to the source point, it is shown that:

$$\frac{\partial R}{\partial x_i} = \frac{R_i}{R} - \frac{R_i}{c_0 C^\dagger R}\frac{\partial R}{\partial y_j}\frac{\partial y_j}{\partial \tau_e} = \frac{R_i}{RC^\dagger}.$$

In this case, for any function similar to $\breve{\mathbf{A}}$, when $R \to \infty$, it is easy to verify that $(\partial\breve{\mathbf{A}}/\partial x_i)_{\tau_e} = O(R^{-2})$. Therefore, the first and second order partial derivatives with respect to x_i for function $\breve{\mathbf{A}}$ are:

$$\frac{\partial \breve{\mathbf{A}}}{\partial x_i} = -\left[\frac{R_i}{c_0 C^\dagger R^2}\frac{\partial}{\partial \tau}\frac{A}{|C^\dagger|}\right]_{\tau=\tau_e} + O(R^{-2}),$$

and

$$\frac{\partial \breve{\mathbf{A}}}{\partial x_i \partial x_j} = \left[\frac{R_i R_j}{c_0^2 C^\dagger R^3} \frac{\partial}{\partial \tau} \frac{1}{C^\dagger} \frac{\partial}{\partial \tau} \frac{A}{|C^\dagger|}\right]_{\tau=\tau_e} + O\left(R^{-2}\right)$$

Then, the far-field solution for FW-H can be written as:

$$\begin{aligned}
\rho'\left(t, \vec{x}\right) \approx & \frac{1}{4\pi c_0^4} \int_{v(t_0)} \left[\frac{R_i R_j}{C^\dagger R^3} \frac{\partial}{\partial \tau} \frac{1}{C^\dagger} \frac{\partial}{\partial \tau} \frac{T'_{ij}}{|C^\dagger|}\right]_{\tau=\tau_e} d\boldsymbol{\xi} \\
& + \frac{1}{4\pi c_0^3} \int_{S(t_0)} \left[\frac{R_i}{C^\dagger R^2} \frac{\partial}{\partial \tau} \frac{f_i}{|C^\dagger|}\right]_{\tau=\tau_e} dS(\boldsymbol{\xi}) \\
& + \frac{1}{4\pi c_0^3} \int_{v_c(t_0)} \left[\frac{R_i}{C^\dagger R^2} \frac{\partial}{\partial \tau} \frac{\rho_0 a_i}{|C^\dagger|}\right]_{\tau=\tau_e} d\boldsymbol{\xi} \\
& + \frac{1}{4\pi c_0^4} \int_{v_c(t_0)} \left[\frac{R_i R_j}{C^\dagger R^3} \frac{\partial}{\partial \tau} \frac{1}{C^\dagger} \frac{\partial}{\partial \tau} \frac{\rho_0 V_i V_j}{|C^\dagger|}\right]_{\tau=\tau_e} d\boldsymbol{\xi}.
\end{aligned} \tag{1.94}$$

The corresponding far-field solution for Curle's equation is given by:

$$\rho'\left(t, \vec{x}\right) \approx \frac{x_i x_j}{4\pi c_0^4 x^3} \int_{v(t_0)} \frac{\partial^2 T'_{ij}}{\partial t^2} d\mathbf{y} + \frac{x_i}{4\pi c_0^3 x^2} \int_{S(t_0)} \frac{\partial f_i}{\partial t} dS(\mathbf{y}) \tag{1.95}$$

1.5 Generalized Lighthill's equation

Lighthill's equation exists under the assumption of static media with respect to the observer. However, in some cases, it is necessary to consider the situation with both the moving body and the flow media. For example, for any reasonable noise prediction for a fan/compressor, it is inevitable to include the effect of duct flow on the sound generation. For convenience, a typical treatment is to assume the media has a uniform flow. Now, emphasis is placed on how to extend Lighthill's equation to be suitable for uniform moving media.

Assume that the media has a uniform velocity U in the parallel direction of y_1. Now, introducing a moving coordinate, $y'_i = y_i - \delta_{1i} U\tau$, and considering that the derivation of Lighthill's equation is based on conservation laws, it is shown that, in the moving coordinate system, Lighthill's equation is still:

$$\frac{\partial^2 \rho'}{\partial \tau^2} - c_0^2 \frac{\partial^2 \rho'}{\partial y'_i \partial y'_j} = \frac{\partial^2 T'_{ij}}{\partial y'_i \partial y'_j}, \tag{1.96}$$

where the Lighthill's stress tensor $T_{ij}' = \rho u_i' u_j' - e_{ij} + \delta_{ij}[(p - p_0) - c_0^2 \rho']$ is defined by the relative velocity

$$u_i' = u_i - \delta_{1i} U$$

In practical problems, it is more convenient to use a ground-fixed coordinate. It is shown that Eq. (1.96) can be written by:

$$\frac{D_0^2 \rho'}{D\tau^2} - c_0^2 \frac{\partial^2 \rho'}{\partial y_i \partial y_j} = \frac{\partial^2 T_{ij}'}{\partial y_i \partial y_j}, \tag{1.97}$$

where

$$\frac{D_0}{D\tau} = \frac{\partial}{\partial \tau} + U \frac{\partial}{\partial y_1}.$$

Eq. (1.97) is named a generalized Lighthill equation, which is an extension of Lighthill's equation in moving media.

To derive the solution of generalized Lighthill's equation, consider sound generation by a moving body with velocity $\mathbf{V}_s$ and volume $v_c(\tau)$ bounded by $S(\tau)$. Now the right term of Eq. (1.97) is assumed to be known, and in terms of generalized Green's function formula, it is shown that:

$$\rho'(t, \vec{x}) = \frac{1}{c_0^2} \int_{-T}^{T} \int_{v(\tau)} G \cdot \frac{\partial^2 T_{ij}'}{\partial y_i \partial y_j} d\mathbf{y} d\tau + \int_{-T}^{T} \int_{S(\tau)} \left[G \left(\frac{\partial}{\partial n} + \frac{V_n'}{c_0^2} \frac{D_0}{D\tau} \right) \rho' - \rho' \left(\frac{\partial}{\partial n} + \frac{V_n'}{c_0^2} \frac{D_0}{D\tau} \right) G \right] dS(\mathbf{y}) d\tau,$$

where Green's function satisfies Eq. (1.57) and the causality condition of Eq. (1.58), besides, $V_n' = \mathbf{V}' \cdot \mathbf{n}$, $\mathbf{V}' = \mathbf{V}_s - \mathbf{i}U$, and $\frac{\partial}{\partial \mathbf{n}} = n_i \frac{\partial}{\partial y_i}$. Applying the same approach within Section 1.1.3 yields:

$$\rho'(t, \vec{x}) = \frac{1}{c_0^2} \int_{-T}^{T} \int_{v(\tau)} T_{ij}' \cdot \frac{\partial^2 G}{\partial y_i \partial y_j} d\mathbf{y} d\tau + \frac{1}{c_0^2} \int_{-T}^{T} \int_{S(\tau)} f_i \frac{\partial G}{\partial y_i} dS(\mathbf{y}) d\tau - \frac{1}{c_0^2} \int_{-T}^{T} \int_{S(\tau)} n_i h_i' dS(\mathbf{y}) d\tau, \tag{1.98}$$

where

$$h_i' = -G \left\{ \frac{\partial}{\partial y_j} \left[\rho u_i' u_j' - e_{ij} + \delta_{ij}(p - p_0) \right] + V_i' \frac{D_0 \rho'}{D\tau} \right\} + V_i' \rho' \frac{D_0 G}{D\tau} + \left(\rho u_i' u_j' \right) \frac{\partial G}{\partial y_j}. \tag{1.99}$$

Considering the different forms of the continuity and momentum equations, i.e.,

$$\begin{cases} \dfrac{D_0\rho'}{D\tau} = \dfrac{D_0\rho}{D\tau} = -\dfrac{\partial \rho u_j'}{\partial y_j} \\ \dfrac{D_0\rho u_i'}{D\tau} = -\dfrac{\partial}{\partial y_j}\left[\rho u_i' u_j' - e_{ij} + \delta_{ij}(p - p_0)\right] \end{cases}, \tag{1.100}$$

and using the impenetrable condition $n_i V_i' = n_i u_i'$, we have:

$$\begin{aligned} n_i h_i' &= n_i\left[G\frac{D_0\rho u_i'}{D\tau} + \rho u_i' \frac{D_0 G}{D\tau}\right] \\ &\quad + n_i u_i'\left[G\frac{\partial \rho u_j'}{\partial y_j} + \left(\rho u_j'\right)\frac{\partial G}{\partial y_j}\right] - n_i u_i' \rho_0 \frac{D_0 G}{D\tau} \\ &= n_i\left(\frac{D_0 G\rho u_i'}{D\tau} + u_i'\frac{\partial G\rho u_j'}{\partial y_j}\right) - n_i u_i' \rho_0 \frac{D_0 G}{D\tau}. \end{aligned} \tag{1.101}$$

Combining the divergence theorem and Leibniz rule yields:

$$\frac{d}{d\tau}\int_{v(\tau)} \Psi \, d\mathbf{y} = \int_{v(\tau)} \frac{D_0\Psi}{D\tau}\, d\mathbf{y} + \int_{S(\tau)} V_n'\Psi \, dS(\mathbf{y}).$$

Hence,

$$\begin{aligned} \frac{d}{d\tau}\int_{v(\tau)} \frac{\partial G\rho u_j'}{\partial y_j}\, d\mathbf{y} &= \int_{v(\tau)} \frac{D_0}{D\tau}\frac{\partial G\rho u_j'}{\partial y_j}\, d\mathbf{y} + \int_{S(\tau)} V_n' \frac{\partial G\rho u_j'}{\partial y_j}\, dS(\mathbf{y}) \\ &= \int_{S(\tau)} n_i\left(\frac{D_0 G\rho u_i'}{D\tau} + u_i'\frac{\partial G\rho u_j'}{\partial y_j}\right) dS(\mathbf{y}). \end{aligned}$$

This expression is zero after the integration with respect to time τ. Therefore, it is shown that only the last term of Eq. (1.101) remains after some mathematical treatment, and Eq. (1.98) becomes:

$$\begin{aligned} \rho'\left(t, \vec{x}\right) &= \frac{1}{c_0^2}\int_{-T}^{T}\int_{v(\tau)} T_{ij}' \cdot \frac{\partial^2 G}{\partial y_i \partial y_j} d\mathbf{y} d\tau + \frac{1}{c_0^2}\int_{-T}^{T}\int_{S(\tau)} f_i \frac{\partial G}{\partial y_i} dS(\mathbf{y}) d\tau \\ &\quad + \frac{1}{c_0^2}\int_{-T}^{T}\int_{S(\tau)} \rho_0 V_n' \frac{D_0 G}{D\tau} dS(\mathbf{y}) d\tau, \end{aligned} \tag{1.102}$$

where V_n' is the projection of the relative velocity in the normal direction of the body. Eq. (1.102) is a general expression for a moving body with flow media. The first term denotes the sound caused by a volume source belonging to a quadruple source, the second terms stands for the sound due to the force on the body belonging to a dipole source, and the last term represents the sound due to the displacement of the volume of a moving body belonging to a monopole source.

References

[1] J. Lighthill, On sound generated aerodynamically I: general theory, Proc. Roy. Soc. A 211 (1952) 564–587.

[2] J. Lighthill, Early development of an "acoustic analogy" approach to aeroacoustic theory, AIAA J. 20 (1982) 449–450.

[3] N. Curle, The influence of solid boundaries upon aerodynamic sound, Proc. Roy. Soc. A 231 (1955) 505–514.

[4] J.E. Ffowcs Williams, D.L. Hawkings, Sound generation by turbulence and surfaces in arbitrary motion, Philos. Trans. R. Soc. Lond. Ser. A 264 (1969) 321–342.

[5] S.K. Lele, Compact finite difference schemes with spectral-like resolution, J. Comput. Phys. 103 (1992) 16–42.

[6] C.K.W. Tam, J.C. Webb, Dispersion-relation-preserving finite difference schemes for computational acoustics, J. Comput. Phys. 107 (1993) 262–281.

[7] T. Colonius, S.K. Lele, Computational aeroacoustics: progress on nonlinear problems of sound generation, Prog. Aerosp. Sci. 40 (2004) 345–416.

[8] C.K.W. Tam, Recent advances in computational aeroacoustics, Fluid Dyn. Res. 38 (2006) 591–615.

[9] E. Envia, A.G. Wilson, D.L. Huff, Fan noise: A challenge to CAA, Int. J. Comput. Fluid Dyn. 18 (6) (2004) 471–480.

[10] N. Peake, A. Parry, Modern challenges facing turbomachinery aeroacoustics, Annu. Rev. Fluid Mech. 44 (2012) 227–248.

[11] M. Goldstein, Aeroacoustics, McGraw-Hill International Book Company, 1976.

[12] C. Morfey, Acoustic energy in non-uniform flows, J. Sound Vib. 14 (1971) 159–170.

[13] M. Myers, Transport of energy by disturbances in arbitrary steady flows, J. Fluid Mech. 226 (1991) 383–400.

[14] M. Lowson, The sound field for singularities in motion, Proc. Roy. Soc. A 286 (1965) 559–572.

CHAPTER 2

Propeller noise: Prediction and control

2.1 Noise sources of propeller

2.1.1 An overview, the developing history of propeller noise prediction

Propeller noise can be classified into two categories according to the characteristics of the sound pressure spectrums: rotational noise (i.e., pure tones) and broadband noise. The interaction between the rotor blades and the periodic oncoming flow gives rise to the pure tones; whereas broadband noise results from the interaction between the blades and a randomly fluctuating part of the surrounding flow field. It is the pure tones, compared with broadband noise, that play the predominant role in the total sound field generated by airplane propellers. Therefore, most of the studies on numerical prediction of propeller noise focus on the pure tones [1].

2.1.1.1 Noise generation due to steady blade force and blade thickness effect

Both the blade forces and the volume displacement effect contribute to the typical rotational noise, which are the most dominant sound sources of propeller noise. Thus, most of the efforts in the early stage are devoted to the study of these two kinds of sound sources. Lynan [2] presented the first attempt to build a theoretical model for propeller noise radiation. It is believed that the rotation of propeller blades will cause periodic fluctuation in the flow passing the rotor, which can cause sound perturbation to the far field. Consequently, we are able to use a continuous ring of stationary sound sources and a rotating point source to model the propeller blades' effect. Under some fundamental simplification assumptions, Bryan [3] obtained the solution of the sound field due to a circumferentially moving point source, which attracted great attention for solving the retarded time equation at that time.

Fundamentals of Aeroacoustics with Applications to Aeropropulsion Systems
https://doi.org/10.1016/B978-0-12-408069-0.00002-6

Both of these papers did not predict the absolute amplitude of a radiated sound field. This instead was accomplished by Gutin [4] in 1948. He pointed out that, based on the linearized acoustic analysis, the steady aerodynamic force exerted on each propeller blade can be represented by a dipole distribution on the blade surface. To generate the same pressure distribution caused by the rotating blades, it is assumed that the stationary disk surface is covered with the sound sources radiating with specific phase and strength. Then, by conducting the Fourier analysis for the solution, one can obtain the amplitudes of the noise due to the torsion and the thrust, respectively. It is worth mentioning that, although the solution was obtained under many simplifications, the basic ideas revealed in this work still have a great effect on the subsequent researches.

With the development of computers in the 1950s, more complicated numerical computations could be realized. Hence, many simplifications are not necessary anymore.

Hubbard [5] further developed Gutin's theory to calculate the fluctuating amplitude and sound field at the points that are one chord-length away from the blade tip. Also, he conducted some parameter studies and compared the theoretical calculation with the experimental results. Moreover, Garrick and Watkins [6] generalized Gutin's theory to consider the effect of forward flight. In their approach, the sound field was calculated by an integration over the propeller disk surface, this method was further simplified as the integration over an effective ring, thus the formulae for the near field as well as the far field could be obtained. The analysis for a specific propeller indicates the maximum noise point would move forward as the forward flight speed increases.

Furthermore, Watkins and Deming [7] extended the work of Garrick and Watkins to investigate the influence of spanwise and chordwise load distribution on the harmonic distribution as well as the total sound pressure level. According to the calculation and analysis, with the radial distribution of thrust and torque prescribed, the noise distributions showed negligible difference compared with the results obtained through an effective radius simplification. Therefore, it is appropriate to concentrate propeller force on an effective radius to take into the effect of real load distribution when concerning the noise variation trend. Inevitably, this simplification will bring about certain errors in the prediction of total sound pressure level. On the other hand, it was found that the chordwise load distribution had a more significant effect on the sound field. For instance, the propeller might cause great noise,especially the high-order harmonics,even operating at zero resultant force.

Kemp and Aronldi [8] extended the work of Garrick and Watkins to consider the correction of varying loads along the blade chord but still had not investigated the noise generated by the thickness effect. In 1954 Shashady and Kemp [9] checked the validity of adopting an effective radius in sound field calculation. Further, Shashady [10] computed the sound pressure level for the typical propeller under specific conditions. The results indicated, with the forward flight speed increasing, the maximum noise position would move forward, and the sound intensity of each harmonic would increase at the same time.

In 1936, Ernsthausen [11] first carried on research about propeller noise due to the blade thickness effect. On the assumption of symmetrical sections at zero blade angle, Deming [12] proposed the first theory of thickness noise. In fact, this theory was derived from Gutin's propeller loading noise theory, more specifically, which replaced the blade profile with a point source. The strengths of these sound sources reflect the normal velocity component distribution on the blade surface, thus making it possible to determine the velocity potential at an arbitrary point. Nevertheless, by using this model, only the sound pressure level of the fundamental frequency and a few low-order harmonics can be determined. Then, in a subsequent work by Ernsthausen [13, 14], some limitations of Deming's theory were successfully overcome, and it emphasized the remarkable influence of higher-order harmonics on the total sound pressure level when the relative Mach number at the blade tip became close to unity. Later, Diprose [15] further extended the theory for the propeller thickness noise by considering the forward flight effect. His computation indicates, for high relative Mach number at the blade tip, the thickness noise, especially for higher-order harmonic components, may dominate over the sound generated by thrust and torque. Through theoretical analysis, Arnoldi [16, 17] thoroughly studied the difference between thickness noise and loading noise, respectively. Moreover, he proposed the corresponding numerical method for noise generation by various blade profiles. Some of those computation results had been compared with the relevant flight records. These results indicate that some tendencies observed during flight (i.e., the sound field moves forward and noise increases) might result from the blade thickness noise rather than the thrust or torque noise. However, this method is only used to calculate the noise radiation at the first few harmonics and also is of limited accuracy.

All the previously mentioned results are of relatively complicated forms as they all concern the propeller's location at the moment when the sound is received rather than associated with the time at which the sound wave was emitted. Morfey [18] presented another method to consider the forward flight effect, which is to associate the sound field with the time at which the sound was emitted, and it had a remarkable effect on the various time-domain methods developed later.

In the early 1960s, a solution for the sound field due to a singularity moving along a helicoidal line was derived by Vooren and Zandbergen [19], which had been used for the calculation of the propeller loading noise and thickness noise generated during flight. Although the initial results obtained are for a moving singularity, this method could be extended to consider the distribution of point sources on the blade surfaces by applying the superposition principle.

Arnoldi [16] proposed a correction method for the nonlinear thickness effect. He suggested deriving an equivalent thickness distribution on the blade section's plane by considering the real surface pressure distribution (taking the average of the values on the upper and lower surfaces). Nevertheless, the near-field pressure measurement results of the propeller during the flight reveals, if the relative Mach number is not greater than unity, the discrepancy between the calculation results obtained from the simple linear model based on blade section area and the experimental measurement results is within 3 dB. Therefore, in view of the complexity of the propeller's nonlinear theory, at least in the range of subsonic speed, linear theory is an appropriate choice with acceptable accuracy.

2.1.1.2 Noise generation due to unsteady blade force

The researches concerning propeller noise generation due to the unsteady blade force started relatively late. But as early as the 1920s, it was noticed that the propeller noise generated under the ground stationary condition is different from that produced under flight conditions. Through the experimental measurements, it was found that the noise during the flight, compared with the noise in the ground tests, is greatly reduced [20, 21].

With the appearance of advanced instruments in the 1960s, deeper analysis for propeller noise spectrums became much more convenient.

With the assistance of these instruments, one can distinguish sound waves at the blade-passing frequency and its harmonics. The results presented in the reference actually indicates, with the inlet disturbance increasing, the difference between the measurement results and the prediction, only considers the steady loading noise and the thickness noise, also become greater. This also illustrates the importance of research on noise generation due to unsteady loads.

Barry and Magliozzi [22] achieved satisfactory agreements between the measurement and prediction by taking the unsteady loading noise into consideration. For helicopter rotor noise, Lowson and Ollerhead [23] carried out a similar analysis. The results of prediction, compared with the measured values, was greatly improved by including unsteady blade forces as sound sources.

Some researchers have studied the generating mechanism of the unsteady loading noise. In 1974, Hanson published some experimental results [24]. In his experiments, the inlet turbulent flow field was measured under stationary conditions, and the results manifested the existence of turbulent vortices with small circumferential and large axial scales. When fans or propellers are operating at zero flight speed, this kind of naturally generating turbulence would turn out to become the high-intensity turbulent vortices under the compression by the compressive flow, which would result in fluctuating forces on the blade surface. Furthermore, when the vortices in this kind of shape pass through the propellers, each of them would be "cut" several times, thus generating considerable blade loads. And this will also generate the prominent random noise in a narrow band with several peaks appearing at multiples of the blade-passing frequency on the spectrum.

Comparing the ground static tests with the high-altitude flow condition, there is a significant difference in the shapes as well as the scales of the vortices in the incoming turbulence. Hence the fluctuating blade forces can be different for the same propeller, which leads to a change on the propeller noise spectrum. Based on the previous analysis, Magliozzi [25] gave a series of innovative experimental results. The magnitude of the propeller noise generated by unsteady loads could be affected by many factors, for instance, the ground vortex, the fuselage, the strut, and the incoming flow condition. However, Pegg et al. [26] performed a comparison of the computation and measurement results which reveal that, for the prediction of the propeller noise emitted under flight conditions, a computation whose results are close to the experimental measurements can be achieved,by treating the unsteady loads as compact sound sources while viewing the volume displacement and the steady loads as incompact sound sources. But there was still difficulty in the propeller noise prediction for the static condition.

2.1.2 Advanced propeller noise (Propfan noise)

The major differences between advanced propellers and traditional propellers are the following aspects. First, advanced propellers have more blades with longer blade chord, which calls for applying acoustic models able to consider the incompact effect, that is to say, the sound sources have a certain axial and spanwise distribution rather than concentrating at a single point so they can interfere with each other. Second, advanced propellers move with the Mach number $Ma_H = 0.7 \sim 0.8$ at high altitude, which leads to high relative speed on the blade section surfaces (supersonic relative speed in the blade tip region in most cases), which may enhance the major nonlinear effect (i.e., increase the quadrupole noise). Moreover, advanced propellers are equipped with back-sweep blades, which are seldom used in traditional propellers, and aim to improve aerodynamic performance and reduce noise, but at the same time increasing the complexity of theoretical computations. Whether it is for traditional propellers or advanced propellers, the calculation of far-field noise serves to check the period of take-off and landing to judge if the aircraft can satisfy the airworthiness requirement. Notably, the results of some tests [27] indicate, even for the case where the relative speed exceeds unity in part of the blade tip region, advanced propellers would not produce a "sonic boom," so we can still treat them as sound sources to deal with the problem of acoustic propagation in the atmosphere.

The near-field noise mainly involves the problems of cabin noise. We hope advanced propellers can provide a kind of high-quality cabin environment comparable with that provided by current turbofan engines. In view of the high-intensity noise generated by advanced propellers, it is necessary to combine the control of the sound sources with effective acoustic treatment of the fuselage.

Undoubtedly, the most economical and efficient method of noise reduction is to reduce the strength of the sound sources. So, for advanced propellers, a reasonable design for propeller blade thickness and pressure distribution is indispensable. To meet this requirement, establishing a strict acoustic model becomes very crucial. However, studies before the 1970s could not meet these demands. Twenty years later, new calculation methods of the open rotor's pure tones began to spring up. Farassat [28] made an excellent review of the most commonly used methods. Although these methods were developed after the 1970s, they can be applied not only for advanced propellers but also for normal propellers. For example, Hawkings and Lowson [29] proposed a method suitable for subsonic and supersonic rotors but limited to the case of static operation or when the observer is in the far field. Obviously, this method cannot be used to predict cabin noise. Subsequently, Jou [30] extended Hawkings and Lowson's method to be applicable for forward flight problems. Some intriguing as well as essential theoretical results concerning high-speed propeller noise have been obtained based on this method. However, there are still many problems to solve before it can be directly applied to numerical computations.

Furthermore, there are some other methods, such as those developed by Woan and Gregorek [31] and Succi [32], able to consider the forward flight effect, but all of them are limited to the case of subsonic relative speed at the blade tip. A method proposed by Nystrom and Farassat [33], which is a time-domain solution for the Ffowcs Williams-Hawkings (FW-H) equation, has been widely employed in calculating the open rotor noise and completely fits the advanced propeller problem. In this method, the geometry of each blade has been strictly specified in the coordinate system fixed to the rotor, and then each blade is divided into strips. Subsequently, two different acoustic formulations are utilized for two different cases to solve the problem: if the local relative speed is supersonic, choose collapsing sphere formulation; and for the subsonic relative speed condition, apply the direct solution. Nystrom and Farassat's technique can be used to evaluate the load noise both in the near field and the far field. So far, this method has been applied to the aircraft noise prediction program [34].

Another kind of method widely used for turbofan noise prediction is Hanson's frequency-domain method [35–39]. The basic ideas to deduce this method are organized as follows. First, take the Fourier transform of the Goldstein equation, then make an integration in the given helical coordinates. The results obtained not only involve the load noise and the noise generated due to the thickness effect but also include the quadrupole term in the FW-H equation. Note that the quadrupole term is ignored for subsonic propellers because it is much smaller than the terms concerning the load noise and the volume displacement effect. Nevertheless, Hanson [38] pointed out the noise caused by the quadrupoles is of vital importance for the slight back-sweep or back-sweep blades under the transonic condition.

It turned out that the accuracy of the computation relative to the measurement results would be improved taking into account the quadrupole term.

The essence of propeller noise study is to figure out the interaction mechanism between the propeller flow field and the sound field. To simplify the problem, the existing numerical methods of propeller noise prediction solve the flow field and the sound field separately in most cases. However, aerodynamics and aeroacoustics are indeed two sides of the same problem, which are tightly bound to each other. There are already many meaningful efforts that have been devoted to seeking a unified solution for both the flow field and the sound field [37, 40–44]. By using the Eluer equations for moving medium, Bober [44] made a prediction of the propeller flow field. However, the application of these methods to propeller flow field and noise prediction was still in the discovery phase. It cost considerable time and computer capacity to conduct numerical computations based on the Euler equations. Therefore, to overcome these deficiencies Hanson established a unified theory for both the aerodynamic prediction and acoustic prediction, which considered the blade interference effect, the thickness effect, and the three-dimensional effect. This theory has been used for the design and evaluation of propellers. In addition, by applying the time-domain solution, Farassat [41] and Long [42] have developed similar aerodynamic and acoustic integrated analytical methods as well.

Contra-rotating propellers achieve higher efficiency compared with traditional propellers, but meanwhile they are accompanied with a more complex noise generation problem. Theoretically the aeroacoustic theories developed for propellers with a single rotational direction should be valid for the contra-rotating problems as well. However, there are still some additional source terms that should be considered. The most essential effect is the aerodynamic interaction between the potential flow and the viscous wake. So far, there has been no completely numerical theoretical method to solve the noise problem of a contra-rotating propeller, but many researchers have been attracted to conduct relevant theoretical and experimental studies in this field [45–47]. Also, recent numerical simulations for counter-rotating propeller have been conducted by getting the unsteady loading through the non-linear harmonic method, which shows a kind of new trend for such noise prediction.

In the following sections, we will introduce the widely applied frequency domain and time-domain methods of propeller noise and discuss in detail how to realize the numerical solution for the time-domain method.

2.2 Propeller noise prediction in frequency domain

2.2.1 The basic equations

In the previous chapter, we developed the formulation of the sound propagation problem with the effect of solid boundaries included. The sound generation in free space by an object with arbitrary motion is formulated by:

$$\begin{aligned}\rho'\left(t,\vec{x}\right) = & \frac{1}{c_0^2}\int_{-T}^{T}\int_{v(\tau)} T'_{ij}\frac{\partial^2 G}{\partial y_i \partial y_j}d\mathbf{y}d\tau - \frac{1}{c_0^2}\int_{-T}^{T}\int_{S(\tau)} f_i\frac{\partial G}{\partial y_i}dS(\mathbf{y})d\tau \\ & -\frac{1}{c_0^2}\int_{-T}^{T}\int_{S(\tau)} \rho_0 V_n\frac{\partial G}{\partial \tau}dS(\mathbf{y})d\tau,\end{aligned} \tag{2.1}$$

where the minus signs of the last two terms on the right-hand side arise due to the convention that both V_n and f_i are taken to be positive when they are in the direction of the outward normal to the solid boundary $S(\tau)$, i.e., from the solid boundary to the external flow field.

Here, even in the presence of the moving object with the solid surface, we can still choose G as the free-space Green's function:

$$G = \frac{\delta(\tau - t + R/c_0)}{4\pi R},$$

where

$$R = \left|\vec{x} - \vec{y}\right|,$$

with $\vec{x}$ and $\vec{y}$ denoting the coordinates of observation point and source point, respectively.

One of the fundamental problems in propeller noise prediction is how to describe the motion trace of each source point on the blade surface and relate it to the location of the observation point. To address this problem, we first introduce three coordinate systems as shown in Fig. 2.1:

x, r, φ—the stationary cylindrical coordinates, i.e., the cylindrical coordinates for the observation point, where the x-axis points in the direction of forward flight.

x_1, r, φ—the rotor-fixed cylindrical coordinates with $x = x_1 + V \cdot t$ where V is the forward flight speed.

γ, r, ξ—the blade-fixed helical coordinates, relative to which the location of the source point is denoted by γ_0, r_0, ξ_0.

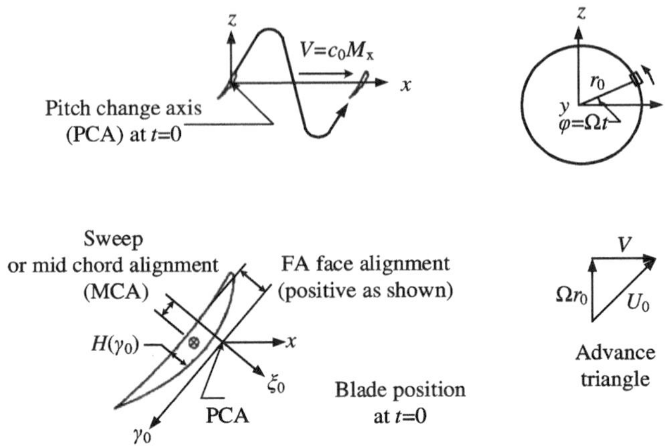

Fig. 2.1 Sketch of the stationary coordinates and the blade-fixed coordinates.

As the first step in establishing the relation between the source point coordinates γ_0, r_0, ξ_0 and the observation point x, r, φ, we consider the coordinate transform depicted by Fig. 2.2, which can be formulated by:

$$y_1{}' = \xi_0 \cos\theta - \gamma_0 \sin\theta = \frac{\Omega r_0 \xi_0}{U_0} - \frac{V\gamma_0}{U_0},$$

$$y_2{}' = -\gamma_0 \cos\theta - \xi_0 \sin\theta = -\frac{\Omega r_0 \gamma_0}{U_0} - \frac{V\xi_0}{U_0},$$

where $y_1{}'$, $y_2{}'$, r_0 denote the location of the source point relative to the rotor-fixed coordinates x_1, r, φ and $\varphi' = y_2{}'/r_0$.

Furthermore, the location of the source point relative to the rotor-fixed rectangular coordinates y_1, y_2, y_3 can be expressed in terms of its helical coordinates γ_0, r_0, ξ_0 in the blade-fixed frame of reference by:

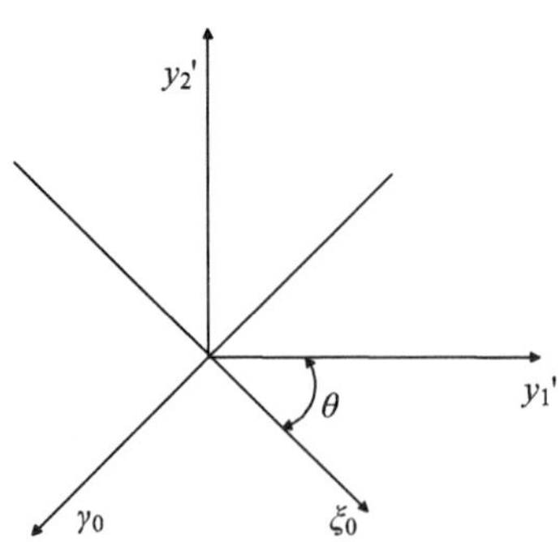

Fig. 2.2 Sketch of the conversion of coordinates.

$$\begin{cases} y_1 = y_1' = \dfrac{\Omega r_0 \xi_0}{U_0} - \dfrac{V\gamma_0}{U_0} \\ y_2 = r_0 \cos\varphi = r_0 \cos\left(\dfrac{y_2'}{r_0}\right) = r_0 \cos\left(\dfrac{\Omega r_0}{U_0} + \dfrac{V\xi_0}{r_0 U_0}\right) \\ y_3 = r_0 \sin\varphi = r_0 \sin\left(\dfrac{y_2'}{r_0}\right) = -r_0 \sin\left(\dfrac{\Omega r_0}{U_0} + \dfrac{V\xi_0}{r_0 U_0}\right) \end{cases}.$$

Relative to the stationary rectangular coordinates, the location of the observation point is denoted by $(x,\ r\cdot\cos\varphi,\ r\cdot\sin\varphi)$. The distance between the observation point and the source point can then be given by:

$$\begin{aligned} R &= |\mathbf{x}-\mathbf{y}| \\ &= \sqrt{(x_1-y_1)^2+(r\cos\varphi-y_2)^2+(r\sin\varphi-y_3)^2} \\ &= \sqrt{\left(x_1-\frac{\Omega r_0\xi_0}{U_0}+\frac{V\gamma_0}{U_0}\right)^2+r^2+r_0{}^2-2r_0 r\cos\left(\varphi+\frac{\Omega\gamma_0}{U_0}+\frac{V\xi_0}{r_0U_0}\right)}. \end{aligned} \tag{2.2}$$

For simplicity, the integral over the upper and lower blade surfaces is transformed into the integral over the middle surface in-between. In the case of thin airfoils, the normal component of the blade motion velocity correspondingly becomes:

$$V_n = U_0\frac{h'}{\sqrt{1+h'^2}} \approx U_0 h'.$$

where $h' = \tan\alpha = \partial h/\partial\gamma_0$ denotes the slope of the middle surface in the blade motion direction (see Fig. 2.3).

Substituting the relations

$$d\mathbf{y} = d\gamma_0 d\xi_0 dr_0,\ dS(\mathbf{y}) = \frac{1}{\sqrt{1+h'^2}} d\gamma_0 dr_0 \approx d\gamma_0 dr_0.$$

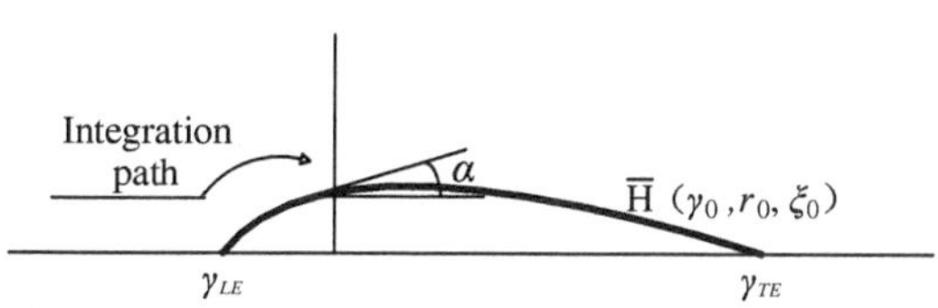

Fig. 2.3 The integration path.

into Eq. (2.1) yields:

$$
\begin{aligned}
c_0^2\rho'(\vec{x},t) = &-\int_{-\infty}^{\infty}\int_{r_0}\int_{\gamma_0}\left[\rho_0 U_0 h'\frac{\partial G}{\partial \tau}+f_i\frac{\partial G}{\partial y_i}\right]d\gamma_0 dr_0 d\tau \\
&+\int_{-\infty}^{\infty}\int_{r_0}\int_{\xi_0}\int_{\gamma_0} T'_{ij}\frac{\partial^2 G}{\partial y_i \partial y_j}d\gamma_0 d\xi_0 dr_0 d\tau.
\end{aligned} \tag{2.3}
$$

Then we attempt to transform the surface integral in the previous equation into the volume integral. With FA being the circumferential blade spacing of the propeller, as shown in Fig. 2.3, we have:

$$h'(\mathbf{y},\tau) = \int h'(\gamma_0, r_0, \xi_0, \tau)\delta(\xi_0 + FA)d\xi_0.$$

In a similar fashion, the aerodynamic force acting on rotor blade $f_i(\gamma_0, r_0)$ $e^{-i\omega_0\tau}$ can be written in the form of:

$$f_i(\mathbf{y},\tau) = \int f_i(\gamma_0, r_0, \xi_0)e^{-i\omega_0\tau}\delta(\xi_0 + FA)d\xi_0,$$

where ω_0 is the circular frequency of the blade force and $\omega_0 = 0$ corresponds to the steady blade force.

Substituting these two relations into Eq. (2.3), one can obtain the sound generation formula in the form of the following volume integral:

$$
\begin{aligned}
c_0^2\rho'(\vec{x},t) = &\int_{-\infty}^{\infty}\int_{r_0}\int_{\xi_0}\int_{\gamma_0}\left\{-\left[\rho_0 U_0 h'\frac{\partial G}{\partial \tau}+D\frac{\partial G}{\partial \gamma_0}+\Delta P\frac{\partial G}{\partial \xi_0}\right.\right. \\
&\left.\left.+F_r\frac{\partial G}{\partial r_0}\right]\delta(\xi_0+FA)+T'_{ij}\frac{\partial^2 G}{\partial y_i \partial y_j}\right\}e^{-i\omega_0\tau}d\gamma_0 d\xi_0 dr_0 d\tau,
\end{aligned} \tag{2.4}
$$

where D and ΔP, respectively, denote the drag and the lift acting on the blade surface per unit area and F_r is the radial component of the blade force per unit area.

Moreover, as the sound source is moving in the opposite direction of γ with velocity U_0, for a source point located at (γ_0, r_0, ξ_0) at $\tau = 0$, its coordinates become $(\gamma_0 + U_0\tau, r_0, \xi_0)$ after time interval τ. Hence, the simultaneous γ coordinate of the source point should be $\gamma_0 + U_0\tau$. For instance, we have $f_i(\mathbf{y},\tau) = f_i(\gamma_0 + U_0\tau, r_0, \xi_0)$.

Furthermore, the integration by parts is applied to simplify the integrand involving Green's function in Eq. (2.4):

$$\int_{-\infty}^{\infty} h'\frac{\partial G}{\partial \tau}d\tau = h'G\Big|_{-\infty}^{\infty} - \int_{-\infty}^{\infty} G\frac{\partial h'}{\partial \tau}d\tau = \int_{-\infty}^{\infty} GU_0\frac{\partial h'}{\partial \gamma_0}d\tau,$$

$$\int_{-\infty}^{\infty} D\frac{\partial G}{\partial \gamma_0}d\gamma_0 = DG\Big|_{-\infty}^{\infty} - \int_{-\infty}^{\infty} G\frac{\partial D}{\partial \gamma_0}d\gamma_0 = -\int_{-\infty}^{\infty} G\frac{\partial D}{\partial \gamma_0}d\gamma_0,$$

$$\int_{-\infty}^{\infty} \Delta P\frac{\partial G}{\partial \xi_0}\delta(\xi_0 + FA)d\xi_0 = -\int_{-\infty}^{\infty} \Delta P\cdot G\cdot\delta'(\xi_0 + FA)d\xi_0,$$

$$\int_{0}^{\infty} F_r\frac{\partial G}{\partial r_0}dr_0 = F_rG\Big|_{0}^{\infty} - \int_{0}^{\infty} G\frac{\partial F_r}{\partial r_0}dr_0 = -\int_{0}^{\infty} G\frac{\partial F_r}{\partial r_0}dr_0,$$

where we have used the relations $\frac{Dh'}{D\tau} = 0$, $h'G|_{\tau=-\infty}^{\tau=\infty} = 0$, $DG|_{\gamma_0=-\infty}^{\gamma_0=\infty} = 0$, $\Delta P\cdot G\cdot\delta(\xi_0 + FA)|_{\xi_0=-\infty}^{\xi_0=\infty} = 0$, $\int_{-\infty}^{\infty} G\frac{\partial \Delta P}{\partial \xi_0}\delta(\xi_0 + FA)d\xi_0 = 0$, and $F_rG|_{r_0=0}^{r_0=\infty} = 0$.

With the previous formulae, Eq. (2.4) is further reduced to:

$$c_0^2\rho'(\vec{x}, t) = \iiiint g(\gamma_0 + U_0\tau, \xi_0, r_0)e^{-i\omega_0\tau}\frac{\delta\left(t - \tau - R/_{c_0}\right)}{4\pi R}d\gamma_0 d\xi_0 dr_0 d\tau, \tag{2.5}$$

where we have introduced the source function $g(\gamma_0 + U_0\tau, \xi_0, r_0)$ in the form of:

$$\begin{aligned} g(\gamma_0, \xi_0, r_0, \tau) = &\left[\rho_0 U_0^2\frac{\partial^2 h(\gamma_0, r_0)}{\partial \gamma_0{}^2} + \frac{\partial D(\gamma_0, r_0)}{\partial \gamma_0} + \frac{\partial F_r(\gamma_0, r_0)}{\partial r_0}\right]\delta(\xi_0 + FA) \\ &+ \Delta P(\gamma_0, r_0)\delta'(\xi_0 + FA) + \frac{\partial^2 T_{ij}(\gamma_0, \xi_0, r_0)}{\partial \gamma_i \partial \gamma_j}. \end{aligned} \tag{2.6}$$

Because τ and γ_0, r_0, ξ_0 are independent of each other in the sound source coordinates, for $\tau = t - R/_{c_0}$, conducting the integration over τ for Eq. (2.5) leads to:

$$c_0^2\rho'(x, t) = \iiiint g\left[\gamma_0 + U_0\left(t - R/_{c_0}\right), \xi_0, r_0\right]\frac{e^{-i\omega_0\left(t - R/_{c_0}\right)}}{4\pi R}d\gamma_0 d\xi_0 dr_0. \tag{2.7}$$

Then, applying the Fourier transform and the inverse Fourier transform

$$\psi\left(\frac{\omega'}{U_0}, \xi_0, r_0\right) = \int_{-\infty}^{\infty} g(\gamma_0, \xi_0, r_0) e^{i\frac{\omega'}{U_0}\gamma'_0} d\gamma'_0,$$

$$g(\gamma'_0, \xi_0, r_0) = \frac{1}{2\pi U_0} \int_{-\infty}^{\infty} \psi\left(\frac{\omega'}{U_0}, \xi_0, r_0\right) e^{-i\frac{\omega'}{U_0}\gamma_0} d\omega',$$

to Eq. (2.7):

$$c_0{}^2\rho'(\vec{x}, t) = \iiiint \frac{1}{8\pi^2 U_0 R} \int \psi\left(\frac{\omega'}{U_0}, \xi_0, r_0\right) e^{\left\{i\left[\frac{\omega' + \omega_0}{c_0}R - (\omega' + \omega_0)t - \frac{\omega'}{U_0}\gamma_0\right]\right\}} d\omega' d\gamma_0 d\xi_0 dr_0. \quad (2.8)$$

Considering an observer moving with the forward flight speed, i.e., $x = x_1 + V \cdot t$, the distance between the source point and the observation point becomes:

$$R = \left[\left(x_1 + V \cdot t - \frac{\Omega r_0 \xi_0}{U_0} + \frac{V\gamma_0}{U_0}\right)^2 + r^2 + r_0{}^2 - 2r_0 r \cos\left(\varphi + \frac{\Omega\gamma_0}{U_0} + \frac{V\xi_0}{r_0 U_0}\right)\right]^{\frac{1}{2}}.$$

To evaluate the integral, here we introduce two important formulae as follows:

(1) Weyric's formula

$$\frac{\exp\left(\pm ik\sqrt{X^2 + R_0^2}\right)}{\sqrt{X^2 + R_0{}^2}} = \pm\frac{1}{2} \int_{-\infty}^{\infty} e^{\pm i\tau X} H_0{}^{(1)}\left(R_0\sqrt{k^2 - \tau^2}\right) d\tau, \quad (2.9)$$

(2) Bessel function supplementary theorem

$$H_0{}^{(1)}\left(K\sqrt{r^2 + r_0{}^2 - 2r_0 r\cos(\varphi)}\right) = \sum_{-\infty}^{\infty} J_n(Kr_<) H_n{}^{(1)}(Kr_>) e^{in\varphi}, \quad (2.10)$$

where H and J are the Hankel function and the Bessel function, respectively. $r_<$ represents the smaller value taken from r and r_0, whereas $r_>$ represents the larger value. With

$$X^2 = \left(x_1 + Vt - \frac{\Omega r_0 \xi_0}{U_0} + \frac{V\gamma_0}{U_0}\right)^2,$$

$$R_0^2 = r^2 + r_0{}^2 - 2r_0 r\cos\left(\varphi + \frac{\Omega\gamma_0}{U_0} + \frac{V\xi_0}{r_0 U_0}\right).$$

one can obtain the following expansion:

$$\frac{\exp\left(i\left(\frac{\omega' + \omega_0}{c_0}\right)\sqrt{X^2 + R_0^2}\right)}{\sqrt{X^2 + R_0{}^2}} = \frac{1}{2}\int_{-\infty}^{\infty} e^{i\tau\left(x_1 + Vt - \frac{\Omega r_0\xi_0}{U_0} + \frac{V\gamma_0}{U_0}\right)}$$

$$\sum_{-\infty}^{\infty} J_n\left(r_<\sqrt{\left(\frac{\omega' + \omega_0}{c_0}\right)^2 - \tau^2}\right) H_n{}^{(1)}\left(r_>\sqrt{\left(\frac{\omega' + \omega_0}{c_0}\right)^2 - \tau^2}\right)$$

$$e^{in\left(\varphi + \frac{\Omega\gamma_0}{U_0} + \frac{V\xi_0}{r_0 U_0}\right)} d\tau.$$

Substituting this expansion and $\tau = \widehat{\omega}/V$ into Eq. (2.8):

$$c_0{}^2\rho' = \iiiint \frac{1}{16\pi^2 U_0 V}\int \psi\left(\frac{\omega'}{U_0}, \xi_0, r_0\right)$$

$$\int_{-\infty}^{\infty}\sum_{-\infty}^{\infty} J_n\left(r_<\sqrt{\left(\frac{\omega' + \omega_0}{c_0}\right)^2 - \left(\frac{\widehat{\omega}}{V}\right)^2}\right)$$

$$H_n{}^{(1)}\left(r_>\sqrt{\left(\frac{\omega' + \omega_0}{c_0}\right)^2 - \left(\frac{\widehat{\omega}}{V}\right)^2}\right)$$

$$e^{i\frac{\widehat{\omega}}{V}\left(x_1 + Vt - \frac{\Omega r_0\xi_0}{U_0}\right)} e^{in\left(\varphi + \frac{V\xi_0}{r_0 U_0}\right)} e^{\{i[-(\omega' + \omega_0)t]\}} e^{i\frac{\gamma_0}{U_0}\left[\widehat{\omega} + n\Omega - \omega'\right]} d\widehat{\omega}\, d\omega'\, d\gamma_0\, d\xi_0\, dr_0. \tag{2.11}$$

for which it is convenient to first integrate it with respect to γ_0. Noting that $\delta(\alpha) = \frac{1}{2\pi}\int_{-\infty}^{\infty} e^{i\omega\alpha} d\omega$, we arrive at the relation:

$$\int e^{i\left(n\Omega - \omega' + \widehat{\omega}\right)\frac{\gamma_0}{U_0}} d\gamma_0 = 2\pi U_0 \delta\left(n\Omega - \omega' + \widehat{\omega}\right). \tag{2.12}$$

Substituting this relation into Eq. (2.11) and integrating it with respect to ω' yields that, for $\omega' = n\Omega + \widehat{\omega}$,

$$c_0{}^2\rho' = \frac{i}{8\pi V}\iint \sum_{-\infty}^{\infty} e^{\{i[n(\varphi-\Omega t)-\omega_0 t]\}} e^{i\frac{\widehat{\omega}}{V}x_1} J_n(\mu r_<) H_n{}^{(1)}(\mu r_>) \int_{-\infty}^{\infty} \psi\left(\frac{n\Omega+\widehat{\omega}}{U_0}, \xi_0, r_0\right) e^{i\left(n\frac{V}{r_0}-\frac{\widehat{\omega}\Omega r_0}{V}\right)\frac{\xi_0}{U_0}} d\widehat{\omega}\, d\xi_0\, dr_0, \tag{2.13}$$

where $\mu = \sqrt{\left(\frac{n\Omega+\widehat{\omega}+\omega_0}{c_0}\right)^2 - \left(\frac{\widehat{\omega}}{V}\right)^2}$.

Letting $\overline{\omega} = n\Omega + \widehat{\omega}$, it follows:

$$c_0{}^2\rho'(x_1, r, \varphi, t) = \frac{i}{8\pi V}\iint \sum_{-\infty}^{\infty} e^{\{i[n(\varphi-\Omega t)-\omega_0 t]\}} e^{i\frac{\overline{\omega}-n\Omega}{V}x_1} J_n(\mu r_<) H_n{}^{(1)}(\mu r_>) \int_{-\infty}^{\infty} \psi\left(\frac{\overline{\omega}}{U_0}, \xi_0, r_0\right) e^{-i(\overline{\omega}-n\Omega)\frac{\Omega r_0 \xi_0}{V U_0}} e^{in\frac{V\xi_0}{r_0 U_0}} d\xi_0\, d\overline{\omega}\, dr_0, \tag{2.14}$$

and

$$\mu = \sqrt{\left(\frac{\overline{\omega}+\omega_0}{c_0}\right)^2 - \left(\frac{\overline{\omega}-n\Omega}{V}\right)^2}. \tag{2.15}$$

2.2.2 Aerodynamic performance prediction

It should be noted that Eq. (2.14) can also be used to predict the aerodynamic performance or characteristics of the propeller. By converting the observation point coordinates (x_1, r, φ) into the blade-fixed coordinates, with $\varphi = \varphi_1 + \Omega t$ and

$$\begin{cases} x_1 = \dfrac{\Omega r}{U_0}\xi - \dfrac{V}{U_0}\gamma \\ \varphi_1 = -\dfrac{V}{U_0 r}\xi - \dfrac{\Omega}{U_0}\gamma \end{cases}.$$

one can reformulate Eq. (2.14) as:

$$p(\gamma, \xi, r, t) = \frac{i\exp(-i\omega_0 t)}{8\pi V}\iiint g(\gamma_0, \xi_0, r_0) \sum_{n=-\infty}^{\infty} \int J_n(\mu r_>) H_n{}^{(1)}(\mu r_>) e^{-i\left[\frac{(\omega'-\omega_0)(\gamma-\gamma_0)}{U_0}\right]} e^{i\left\{[\Omega r^2(\omega'-n\Omega)-nV^2]\frac{\xi}{U_0 V r}\right\}} e^{-i\left\{[\Omega r_0{}^2(\omega'-\omega_0-n\Omega)-nV^2]\frac{\xi_0}{U_0 V r_0}\right\}} d\omega'\, d\gamma_0\, d\xi_0\, dr_0. \tag{2.16}$$

This formula has been applied to the prediction of the steady and unsteady aerodynamic loading distribution on a propeller blade as well as the evaluation of propeller aerodynamic performance.

2.2.3 The near-field solution of propeller noise

For steady blade force, $\omega_0 = 0$, and consequently Eq. (2.14) reduces to:

$$\begin{aligned} c_0^2\rho'(x_1, r, \varphi, t) = & \frac{i}{8\pi v}\sum_{n=-\infty}^{\infty} \exp\left[in(\varphi - \Omega t)\right] \\ & \cdot \int_0^{-\infty}\int_{-\infty}^{\infty} \exp\left[i(\overline{\omega} - n\Omega)x_1/v\right] J_n(\mu r_0) H_n^{(1)}(\mu r) \\ & \cdot \int_{-\infty}^{\infty} \psi\left(\frac{\overline{\omega}}{U_0}, \xi_0, r_0\right) \exp\left(-i(\overline{\omega} - n\Omega)\frac{\Omega r_0 \xi_0}{v U_0}\right) \\ & \cdot \exp\left[in(v\xi_0/r_0 U_0)\right] d\xi_0 d\overline{\omega} dr_0. \end{aligned} \tag{2.17}$$

where

$$\mu = \sqrt{\left(\frac{\overline{\omega}}{c_0}\right)^2 - \left(\frac{\overline{\omega} - n\Omega}{v}\right)^2}. \tag{2.18}$$

Note that Eq. (2.17) is derived for a single blade. For the sound generation by a B-bladed propeller, it can be proven [48] that one can obtain the solution from Eq. (2.17) by changing n into m and then multiplying the resulting expression by B, i.e.,

$$\begin{aligned} c_0^2\rho'(x_1, r, \varphi, t) = & \frac{iB}{8\pi v}\sum_{n=-\infty}^{\infty} \exp\left[imB(\varphi - \Omega t)\right] \\ & \cdot \int_0^{-\infty}\int_{-\infty}^{\infty} \exp\left[i(\overline{\omega} - mB\Omega)x_1/v\right] J_n(\mu r_0) H_n^{(1)}(\mu r) \\ & \cdot \int_{-\infty}^{\infty} \psi\left(\frac{\overline{\omega}}{U_0}, \xi_0, r_0\right) \exp\left(-i(\overline{\omega} - mB\Omega)\frac{\Omega r_0 \xi_0}{v U_0}\right) \\ & \cdot \exp\left[imB(v\xi_0/r_0 U_0)\right] d\xi_0 d\overline{\omega} dr_0. \end{aligned} \tag{2.19}$$

For the convenience of integration, we let $\overline{\omega} = mB\Omega k$. Then the respective wavenumbers in the γ_0 and ξ_0 directions as well as their product become:

$$\frac{\overline{\omega}}{U_0}\gamma_0 = \frac{mB\Omega k}{U_0}\frac{\gamma_0}{b} = \frac{2mBB_D Ma_T k}{Ma_r}\frac{\gamma_0}{b} = k_x X,$$

where

$$k_x = \frac{2mBB_D Ma_T k}{Ma_r}, \quad X = \frac{\gamma_0}{b},$$

and

$$B_D = \frac{b}{2r_T}, \quad Ma_T = r_T\Omega/c_0, \quad Ma_r = U_0/c_0,$$

$$\begin{aligned}
&\left[-(\overline{\omega} - mB\Omega)\frac{\Omega r_0}{vU_0} + mB\frac{v}{r_0 U_0}\right]\xi_0 \\
&= \left[-\frac{2mBB_D Ma_1^2 z(k-1)}{Ma_x Ma_r} + \frac{2mBB_D Ma_1^2}{zMa_x Ma_r^2}\right]\frac{\xi_0}{b} \\
&= \frac{2mBB_D}{zMa_x Ma_r}\left(Ma_r^2 - kz^2 Ma_T^2\right) \\
&= k_y Y,
\end{aligned}$$

with

$$k_y = \frac{2mBB_D}{zMa_x Ma_r}\left(Ma_r^2 - kz^2 Ma_T^2\right),$$

$$z = \frac{r_0}{r_T}, \quad Y = \frac{\xi_0}{b}.$$

Furthermore, let:

$$\varphi_x = \frac{2mBMa_T z(k-1)}{Ma_x}\frac{x_1}{D},$$

where D is the diameter of the propeller, i.e., $D = 2r_T$, and we have:

$$p = c_0^2\, p' = \sum_{n=-\infty}^{\infty} p_{mB} \exp\left[imB(\varphi - \Omega t)\right],$$

with

$$p_{mB} = \frac{imB^2 Ma_T}{8\pi Ma_x}\int_0^{\infty}\int_{-\infty}^{\infty} e^{i\varphi_x} J_n H_n^{(1)} \int_{-\infty}^{\infty} \psi \exp\left[ik_y\xi_0/b\right] d\xi_0\, dk\, dz. \tag{2.20}$$

First, we evaluate the integral in Eq. (2.20) with respect to ξ_0, which leads to:

$$I_s = \int_{-\infty}^{\infty} \psi \exp\left[ik_y \xi_0 / b\right] d\xi_0. \tag{2.21}$$

Taking the Fourier transform with respect to ψ yields:

$$I_s = \iint g(\gamma_0, \xi_0, r_0) \exp\left[ik_x \gamma_0 / b\right] \exp\left[ik_y \xi_0 / b\right] d\gamma_0 d\xi_0. \tag{2.22}$$

Then, by substituting Eq. (2.6) into the previous equation and further ignoring the term concerning the radial force, F_r and that concerns the radial derivative of T_{ij}, Eq. (2.22) becomes:

$$\begin{aligned}
I_s = {} & \rho_0 U_0^2 \int \frac{\partial^2}{\partial \gamma^2} h(\gamma_0, r_0) \exp\left(ik_x \gamma_0 / b\right) d\gamma_0 \int \delta(\xi_0 + FA) \exp\left(ik_y \xi_0 / b\right) d\xi_0 \\
& + \int \frac{\partial}{\partial \gamma_2} D(\gamma_0, r_0) \exp\left(ik_x \gamma_0 / b\right) d\gamma_0 \int \delta(\xi_0 + FA) \exp\left(ik_y \xi_0 / b\right) d\xi_0 \\
& + \int \Delta P(\gamma_0, r_0) \exp\left(ik_x \gamma_0 / b\right) d\gamma_0 \int \delta(\xi_0 + FA) \exp\left(ik_y \xi_0 / b\right) d\xi_0 \\
& + \iint \frac{\partial^2}{\partial \gamma_i \partial \gamma_i} T_{ij}(\gamma_0, \xi_0, r_0) \exp\left(ik_x \gamma_0 / b\right) \exp\left(ik_y \xi_0 / b\right) d\xi_0.
\end{aligned} \tag{2.23}$$

Eliminating the δ function in the integrals in Eq. (2.23) by using the integration by parts, we arrive at:

$$\begin{aligned}
I_s = {} & \rho_0 U_0^2 \exp\left[i\varphi_0\right] \left(\frac{-k_x^2}{b^2}\right) \int h(\gamma_0, r_0) \exp\left(ik_x \gamma_0 / b\right) d\gamma_0 \\
& - \exp\left[i\varphi_0\right] \left(\frac{-k_x}{b^2}\right) \int D(\gamma_0, r_0) \exp\left(ik_x \gamma_0 / b\right) d\gamma_0 \\
& - \exp\left[i\varphi_0\right] \left(\frac{-k_y}{b^2}\right) \int \Delta P(\gamma_0, r_0) \exp\left(ik_x \gamma_0 / b\right) d\gamma_0 \\
& - \frac{k_x^2}{b^2} \int \int_{-T}^{T} T_{11}(\gamma_0, \xi_0, r_0) \exp\left(ik_x \gamma_0 / b\right) \exp\left(ik_y \xi_0 / b\right) d\gamma_0 d\xi_0 \\
& - 2 \frac{k_x k_y}{b^2} \iint T_{12}(\gamma_0, \xi_0, r_0) \exp\left(ik_x \gamma_0 / b\right) \exp\left(ik_y \xi_0 / b\right) d\gamma_0 d\xi_0 \\
& - \frac{k_y^2}{b^2} \int \int_{-T}^{T} T_{22}(\gamma_0, \xi_0, r_0) \exp\left(ik_x \gamma_0 / b\right) \exp\left(ik_y \xi_0 / b\right) d\gamma_0 d\xi_0,
\end{aligned} \tag{2.24}$$

where

$$\varphi_0 = -k_y \frac{FA}{b},$$

represents the phase variation due to the circumferential inclination of propeller blades. And the blade thickness distribution can be expressed in terms of the regular shape function:

$$h(\gamma_0, r_0) = bt_b H\left(X - \frac{MCA}{b}\right) \tag{2.25}$$

where t_b denotes the ratio of the maximum blade thickness to the blade chord length and MCA is the sweepback spacing of the propeller (see Figs. 2.1 and 2.4).

In a similar fashion, the drag D and the lift ΔP can also be represented by the regular shape functions in the following form of:

$$D(\gamma_0, r_0) = \frac{\rho_0 U_0^2}{2} C_D f_D\left(X - \frac{MCA}{b}\right), \tag{2.26}$$

$$\Delta P(\gamma_0, r_0) = \frac{\rho_0 U_0^2}{2} C_L f_L\left(X - \frac{MCA}{b}\right), \tag{2.27}$$

where C_D and C_L are the drag and lift coefficients for the propeller blades, and both f_D and f_L satisfy that their respective chordwise integrals equal unity. Correspondingly, the quadrupole source T_{ij} can be expressed as:

$$T_{ij}(\gamma_0, \xi_0, r_0) = T_{ij}\left(X - \frac{MCA}{b}, Y + \frac{FA}{b}, z\right). \tag{2.28}$$

Substituting Eqs. (2.25) to (2.28) into Eq. (2.24) gives:

$$\begin{aligned} I_s = -\rho_0 U_0^2 \exp\left[i(\varphi_0 + \varphi_s)\right] \Big[& k_x^2 t_b \psi_v(k_x) + ik_x \frac{C_D}{2} \psi_D(k_x) \\ & + ik_y \frac{C_L}{2} \psi_L(k_x) + \psi_Q(k_x, k_y) \Big]. \end{aligned} \tag{2.29}$$

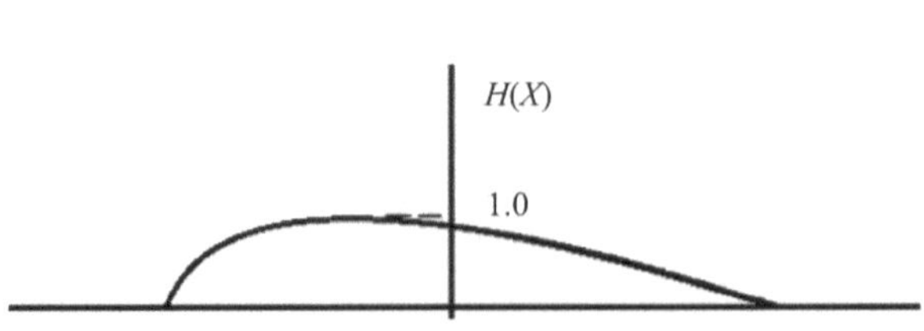

Fig. 2.4 Normalized thickness distributions [39].

where

$$\varphi_s = k_x \frac{MCA}{b}$$

represents the phase variation due to the sweepback.

The monopole and dipole terms are in the form of:

$$\begin{bmatrix} \psi_V(k_x) \\ \psi_D(k_x) \\ \psi_L(k_x) \end{bmatrix} = \int_{-1/2}^{1/2} \begin{bmatrix} H(x) \\ f_D(x) \\ f_L(x) \end{bmatrix} \exp\left[ik_x X\right] dX, \tag{2.30}$$

And correspondingly, the quadrupole term is:

$$\psi_Q = k_{xx}^2 \psi_{11} + 2k_x k_y \psi_{12} + k_y^2 \psi_{22}, \tag{2.31}$$

with

$$\psi_{ij}\left(k_x, k_y\right) = \iint \frac{T_{ij}}{\rho_0 U_0^2} \exp\left[ik_z X\right] \exp\left[ik_y Y\right] dXdY. \tag{2.32}$$

Finally, by putting Eq. (2.29) into Eq. (2.19), one can obtain the sound radiation formula as:

$$\begin{aligned} \begin{bmatrix} P_{vm} \\ P_{Dm} \\ P_{Lm} \\ P_{Qm} \end{bmatrix} &= \frac{-i\rho_0 c_0^2 mB^2 Ma_T}{8\pi Ma_x} \int_{x_{root}}^{x_{tip}} Ma_r^2 \int_{-\infty}^{\infty} \exp\left[i(\varphi_0 + \varphi_s + \varphi_x)\right] \\ &\cdot J_{mB}\left[\frac{mBzMa_T}{Ma_x} \sqrt{Ma_x^2 k^2 - (k-1)^2}\right] \\ &\cdot H_{mB}^{(1)}\left[\frac{2mBMa_T}{Ma_s} \frac{y}{D} \sqrt{Ma_x^2 k^2 - (k-1)^2}\right] \\ &\cdot \begin{bmatrix} k_x^2 t_b \psi_v(k_x) \\ ik_z(C_D/2)\psi_D(k_x) \\ ik_y(C_L/2)\psi_L(k_x) \\ \psi_Q\left(k_x, k_y\right) \end{bmatrix} d_k d_z, \end{aligned} \tag{2.33}$$

where

$$P_{mB} = P_{Vm} + P_{Dm} + P_{Lm} + P_{Qm}.$$

On the other hand, the phase variation due to the circumferential and backward inclination of the propeller can also be formulated by:

$$\varphi_0 = \frac{2mBkz^2 Ma_T^2 - Ma_r^2}{z \quad Ma_x Ma_y} \frac{FA}{D}, \tag{2.34}$$

$$\varphi_s = \frac{2mBMa_T k}{Ma_r} \frac{MCA}{D}. \tag{2.35}$$

With the effect of the quadrupole sources ignored, Eq. (2.33) has been used to evaluate the near-field propeller noise [39].

2.2.4 The far-field solution of propeller noise

When it comes to far-field propeller noise prediction, there are two typical approaches. The first approach derived form Eq. (2.33). It first finds the asymptotic expansion of the Hankel function, $H_{mB}^{(1)}$, and then evaluate the resulting integral to predict the radiation sound field. The second approach directly solves Eq. (2.8) by deriving the approximate expression of the distance between the source point and the observation point. To make the solution more intuitive in physics, here we will derive the far-field solution of propeller noise based on the second approach.

For Eq. (2.2), with the small terms of the second order ignored, the distance between the source point and the observation point can be approximated by:

$$R = r_s - \frac{\Omega r_0}{U_0}\xi_0 \cos\theta + \frac{v}{U_0}\gamma_0 \cos\theta - r_0 \sin\theta \cos\left(\varphi + \frac{v\xi_0}{U_0 r_0} + \frac{\Omega\gamma_0}{U_0}\right), \tag{2.36}$$

where $r_s = \sqrt{r^2 + x^2}$ is the distance between the observation point and the origin of the stationary coordinates, and we have $\cos\theta = \frac{x}{r_s}$ and $\sin\theta = \frac{r}{r_s}$ with θ being the angle between r_s and the flight direction. Also,

$$\frac{1}{R} \approx \frac{1}{r_s}.$$

Thus for the sound field resulting from the steady aerodynamic loads ($\omega_0 = 0$), Eq. (2.8) reduces to:

$$p(x,t) = \frac{1}{8\pi^2 r_s}\iiint \frac{1}{U_0}\int \psi\left(\frac{\omega}{U_0'}, r_0, \xi_0\right) \cdot \exp\left[-i\frac{\omega}{U_0}\left(\gamma_0 + U_0 t - \frac{U_0 R}{c_0}\right)\right] d\omega d\xi_0 dr_0. \tag{2.37}$$

Take the Fourier transform of the previous equation, i.e.,

$$p(x,t)=\int P(x,\omega)e^{-i\omega_0\tau}d\omega, \tag{2.38}$$

where

$$\begin{aligned}P(x,\omega)=&\frac{1}{8\pi^2 r_s}\iiint\frac{1}{U_0}\int\psi\left(\frac{\omega}{U_0},r_0,\xi_0\right)\\&\cdot\exp\left[-i\frac{\omega}{U_0}\left(\gamma_0-\frac{U_0R}{c_0}\right)\right]d\omega d\xi_0 d\gamma_0 dr_0.\end{aligned} \tag{2.39}$$

Substituting Eq. (2.36) into Eq. (2.39):

$$P(x,\omega)=\frac{\exp\left[i\frac{\omega r_s}{c_0}\right]}{4\pi r_s}\iint\psi\left(\frac{\omega}{U_0},r_0,\xi_0\right)\cdot\exp\left[-i\frac{\omega\Omega r_0}{c_0 U_0}\xi_0\cos\theta\right]Id\xi_0 dr_0, \tag{2.40}$$

where

$$\begin{aligned}I=&\frac{1}{2\pi U_0}\int_{-\infty}^{\infty}\exp\left[-i\frac{\omega}{U_0}(1-Ma_x\cos\theta)\gamma_0\right]\\&\cdot\exp\left[-i\frac{\omega r_0}{c_0}\sin\theta\cos\left(\frac{\Omega\gamma_0}{U_0}+\frac{\nu\xi_0}{U_0 r_0}+\varphi\right)\right]d\gamma_0.\end{aligned} \tag{2.41}$$

In view of:

$$\exp[iz\cos x]=\sum_{n=-\infty}^{\infty}J_n(z)\exp[in(x+\pi/2)],$$

we have

$$\begin{aligned}&\exp\left[-i\frac{\omega r_0}{c_0}\sin\theta\cos\left(\frac{\Omega\gamma_0}{U_0}+\frac{\nu\xi_0}{U_0 r_0}+\varphi\right)\right]\\&=\exp\left[-i\frac{\omega r_0}{c_0}\sin\theta\cos\left(\frac{\Omega\gamma_0}{U_0}+\frac{\nu\xi_0}{U_0 r_0}+\varphi-\pi\right)\right]\\&=\sum_n J_n\left(\frac{\omega r_0}{c_0}\sin\theta\right)\exp\left[in\left(\frac{\Omega\gamma_0}{U_0}+\frac{\nu\xi_0}{U_0 r_0}+\varphi-\pi/2\right)\right],\end{aligned}$$

and thus

$$\begin{aligned}I_n =& \sum_{n=-\infty}^{\infty} J_n\left(\frac{\omega r_0}{c_0}\sin\theta\right)\exp\left[in\left(\frac{\Omega\gamma_0}{U_0}+\frac{\nu\xi_0}{U_0 r_0}-\pi/2\right)\right]\\
&\cdot\frac{1}{2\pi U_0}\int\exp\left\{i\left[\frac{n\Omega}{U_0}-\frac{\omega}{U_0}(1-Ma_x\cos\theta)\right]\gamma_0\right\}d\gamma_0\\
=&\frac{1}{1-Ma_x\cos\theta}\sum_{n=-\infty}^{\infty} J_n\left(\frac{\omega r_0}{c_0}\sin\theta\right)\exp\left[in\left(\frac{\Omega\gamma_0}{U_0}+\frac{\nu\xi_0}{U_0 r_0}-\pi/2\right)\right]\\
&\cdot\delta\left(\omega-\frac{n\Omega}{1-Ma_x\cos\theta}\right).\end{aligned} \tag{2.42}$$

Then substituting, Eq. (2.42) into Eq. (2.38):

$$P(x,t)=\sum_{-\infty}^{\infty}P_n(x)\exp[-in\Omega_0 t], \tag{2.43}$$

where

$$\begin{aligned}P_n(x)=&\frac{\exp\left[in\left(\frac{\Omega r_0}{c_0}+\frac{\nu\xi_0}{U_0 r_0}-\pi/2\right)\right]}{4\pi r_s(1-Ma_x\cos\theta)}\int_0^{\infty}J_n\left(\frac{n\Omega_D r_0}{c_0}\sin\theta\right)\\
&\cdot\int_{-\infty}^{\infty}\psi\left(\frac{\omega}{U_0},r_0,\xi_0\right)\cdot\exp\left[-i\frac{n}{U_0}\left(\frac{\Omega\Omega_D r_0}{c_0}\cos\theta-\frac{\nu}{r_0}\right)\xi_0\right]d\xi_0 dr_0,\end{aligned} \tag{2.44}$$

$$\Omega_D=\Omega/(1-Ma_x\cos\theta). \tag{2.45}$$

For B blades,

$$p(x,t)=\sum_{n=-\infty}^{\infty}p_{mB}(x)\exp[-imB\Omega_D t]. \tag{2.46}$$

Finally, through the derivation similar to Eqs. (2.21) to (2.35), one can obtain the formula for the far-field noise prediction as:

$$\begin{bmatrix}P_{Vm}\\P_{Dm}\\P_{Lm}\\p_{11m}\\p_{12m}\\p_{22m}\end{bmatrix}=-\frac{\rho_0 c_0^2 B\sin\theta\exp\left[imB\left(\frac{\Omega c_0}{c_0}+\varphi-\pi/2\right)\right]}{8\pi\frac{r_s}{D}(1-Ma_x\cos\theta)} \tag{2.47}$$

$$\cdot \int Ma_r^2 e^{i(\varphi_0+\varphi_s)} J_{mB}\left(\frac{mBzMa_T\sin\theta}{1-Ma_x\cos\theta}\right)\begin{bmatrix} k_x^2 t_b \psi_v(k_x) \\ ik_x(C_D/2)\psi_D(k_x) \\ -ik_y(C_L/2)\psi_L(k_x) \\ k_x^2\psi_{11}(k_x,k_y) \\ 2k_xk_y\psi_{12}(k_x,k_y) \\ k_y^2\psi_{22}(k_x,k_y) \end{bmatrix}.$$

where

$$P_{mB} = P_{Vm} + P_{Dm} + P_{Lm} + p_{11m} + p_{12m} + p_{22m},$$

and

$$k_x = \frac{2mBB_D Ma_T}{Ma_r(1-Ma_x\cos\theta)}, \quad (2.48)$$

$$k_y = \frac{2mBB_D}{zMa_r}\left(\frac{Ma_r^2\cos\theta - Ma_x}{1-Ma_x\cos\theta}\right), \quad (2.49)$$

$$\varphi_0 = \frac{2mB}{zMa_r}\left(\frac{Ma_r^2\cos\theta - Ma_x}{1-Ma_x\cos\theta}\right)\frac{FA}{D}, \quad (2.50)$$

$$\varphi_s = \frac{2mBMa_T}{Ma_r(1-Ma_x\cos\theta)}\frac{MCA}{D}, \quad (2.51)$$

It can be found from Eqs. (2.47) to (2.51) that the same formulae can be obtained by substituting the relation $k = 1/(1 - Ma_x \cos\theta)$ into the expressions for the near-field sound pressure.

Fig. 2.5 shows the near-field noise prediction by the frequency-domain method in contrast with the experimental data [39]. Fig. 2.6 gives the comparison of the noise prediction results respectively obtained based on the near-field and the far-field theories, it shows that under the same physical condition the discrepancies of the prediction results mainly arise within the low-order harmonics.

2.3 Propeller noise prediction in time domain

The time-domain methods of propeller noise prediction are established on the basis of the FW-H equation. One of the FW-H equation's formulas was presented in the last chapter. Indeed, for the propeller noise prediction methods currently, there are multiple forms of the FW-H equation involved, which causes great difficulty in numerical computation. And it is noteworthy that

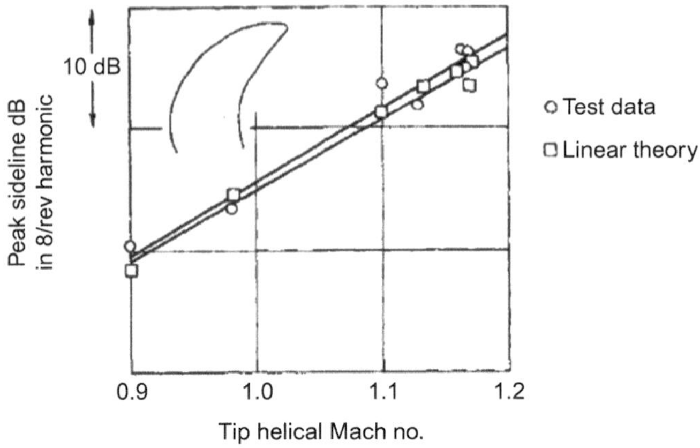

Fig. 2.5 Comparison of the near-field sound prediction by the frequency-domain method and the measured data [39].

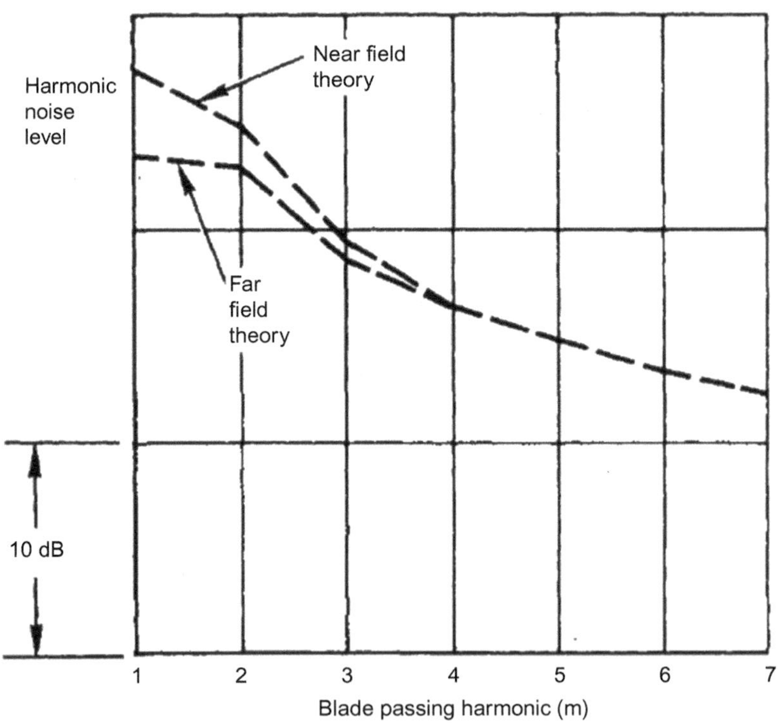

Fig. 2.6 Comparison of the noise prediction results by the near-field and the far-field theories in frequency domain [39].

they were all obtained by introducing the generalized function. Therefore, in this section we will first investigate how to use the generalized functions to consider the effect of solid boundaries and deduce the free-space wave equation, and then various formulas of the FW-H equation will be given and their computation methods in the time domain will be discussed.

2.3.1 The basic equations

With the existence of solid boundaries, the influence of the boundaries should be considered by imposing the boundary conditions. Combined with these boundary conditions, the problem would turn out to be the boundary value problem of a partial equation. Generally, the exact Green's function for this kind of boundary problem is hard to obtain. Hence, it is not easy to solve this problem directly. To use the free-space Green's function to solve this boundary problem, we should first extend the function's definitional domain to entire space by introducing generalized functions. First of all, the Heaviside function is defined as follows:

$$H(f)=\begin{cases}1 & (f(x,t)>0, \quad \text{for } x \text{ is inside the body surface})\\ 0 & (f(x,t)<0, \quad \text{for } x \text{ is outside the body surface})\end{cases}.$$

The governing equations of fluid outside the body surfaces are:

$$\frac{\partial}{\partial t}(\rho-\rho_0)+\frac{\partial}{\partial x_i}(\rho u_i)=0, \tag{2.52}$$

$$\frac{\partial}{\partial t}(\rho u_i)+\frac{\partial}{\partial x_i}\left(p_{ij}+\rho u_i u_j\right)=0, \tag{2.53}$$

p_{ij} denotes the deforming tensor, i.e., the interaction force between the fluid. With the Heaviside function, the continuity equation can be rewritten as:

$$\begin{aligned}&\frac{\partial}{\partial t}[\rho' H(f)]+\frac{\partial}{\partial x_i}[\rho u_i H(f)]\\&=H(f)\left[\frac{\partial\rho'}{\partial t}+\frac{\partial}{\partial x_i}(\rho u_i)\right]+\rho'\left(\frac{\partial H}{\partial t}+u_i\frac{\partial H}{\partial x_i}\right)+\rho_0 u_i\frac{\partial H}{\partial x_i}.\end{aligned} \tag{2.54}$$

By substituting Eq. (2.52) into this equation, the continuity equation can be further simplified as:

$$\frac{\partial}{\partial t}[\rho' H(f)]+\frac{\partial}{\partial x_i}[\rho u_i H(f)]=\rho'\left[\frac{\partial H}{\partial t}+u_i\frac{\partial H}{\partial x_i}\right]+\rho_0 u_i\frac{\partial H}{\partial x_i}. \tag{2.55}$$

According to the relation of the body surfaces as $Df/Dt = 0$, we can obtain:

$$\begin{aligned}\frac{\partial H}{\partial t}+u_i\frac{\partial H}{\partial x_i}&=\frac{\partial H}{\partial f}\left(\frac{\partial f}{\partial t}+u_i\frac{\partial f}{\partial x_i}\right)\\&=\delta(f)\left(-\frac{\partial f}{\partial x_i}V_i+u_i\frac{\partial f}{\partial x_i}\right)=\delta(f)(u_i-V_i)\frac{\partial f}{\partial x_i}=0,\end{aligned}\tag{2.56}$$

which means, on the body surface, we have $u_i = V_i$. Substituting Eq. (2.56) into Eq. (2.55) leads to:

$$\frac{\partial}{\partial t}[\rho' H(f)]+\frac{\partial}{\partial x_i}[\rho u_i H(f)]=\rho_0 u_i\delta(f)\frac{\partial f}{\partial x_i}.\tag{2.57}$$

Similarly, we can formulate the moment equation as:

$$\frac{\partial}{\partial t}(\rho u_i H(f))+\frac{\partial}{\partial x_i}\left[\left(p_{ij}+\rho u_i u_j\right)H(f)\right]=p_{ij}\frac{\partial f}{\partial x_i}\delta(f).\tag{2.58}$$

Take the partial derivative of Eq. (2.57) with respect to t, while taking the partial derivative of Eq. (2.58) with respect to x_i' and then compare the results. Furthermore, subtracting $c_0^2\frac{\partial^2}{\partial x_i^2}[\rho' H(f)]$ from both hands of the resultant equation and rearranging the result:

$$\begin{aligned}&\frac{\partial^2}{\partial t^2}[\rho' H(f)]-c_0^2\frac{\partial^2}{\partial x_i{}^2}[\rho' H(f)]\\&=\frac{\partial}{\partial t}[\rho_0 V_n|\nabla f|\delta(f)]-\frac{\partial}{\partial x_i}\left[p_{ij}\frac{\partial f}{\partial x_j}\delta(f)\right]+\frac{\partial^2}{\partial x_i x_j}\left[T_{ij}\cdot H(f)\right],\end{aligned}\tag{2.59}$$

where $T_{ij} = p_{ij} + \rho u_i u_j - c_0^2\rho'\delta_{ij}$ is Lighthill's turbulence stress tensor.

For an adiabatic isentropic process, we have $p - p_0 = c_0^2(\rho - \rho_0)'$ and substituting this into the previous equation:

$$\begin{aligned}&\left[\frac{1}{c_0^2}\frac{\partial^2}{\partial t^2}-\frac{\partial^2}{\partial x_i{}^2}\right][(p-p_0)H(f)]\\&=\frac{\partial}{\partial t}[\rho_0 V_n|\nabla f|\delta(f)]-\frac{\partial}{\partial x_i}\left[p_{ij}\frac{\partial f}{\partial x_j}\delta(f)\right]+\frac{\partial^2}{\partial x_i x_j}\left[T_{ij}\cdot H(f)\right].\end{aligned}\tag{2.60}$$

In view of the relation $n_i = \frac{1}{|\nabla f|}\frac{\partial f}{\partial x_i}$, the equation can be further simplified to:

$$\square^2 p' = \frac{\partial}{\partial t}[\rho_0 V_n|\nabla f|\delta(f)]-\frac{\partial}{\partial x_i}[l_i|\nabla f|\delta(f)]+\frac{\partial^2 T_{ij}}{\partial x_i x_j},\tag{2.61}$$

where $\Box^2 = \frac{1}{c_0^2}\frac{\partial^2}{\partial t^2} - \nabla^2$ is the D'Alembert operator and $l_i = p_{ij}\frac{\partial f}{\partial x_j} = p_{ij}n_j|\nabla f|$. When the viscous effect is negligible, the body surface equation can be represented by $l_i = pn_i; f(\vec{x}, t) = 0$ with $\mathbf{n} = (n_1, n_2, n_3)$ being the outer unit normal vector and $V_n = -\frac{\partial f}{\partial t}/|\nabla f|$ being the outer normal velocity of the body surface, respectively.

2.3.2 The solution of the free-space generalized wave equation

For the inhomogeneous wave equation:

$$\Box^2\Phi = Q(\vec{x}, t)|\nabla f|\delta(f), \tag{2.62}$$

the solution obtained based on the Green's function is:

$$4\pi\cdot\Phi(\vec{x}, t) = \int\frac{Q(\vec{y}, \tau)}{R}|\nabla f|\delta(f)\delta(g)d\vec{y}\,d\tau. \tag{2.63}$$

The limit of integration with respect to τ is $(-\infty, t)$, whereas the domain of integration with respect to the variable $\mathbf{y}$ is entire space. Note that both $\vec{x}$ and t appearing in Eq. (2.63) are variable parameters. Therefore, the solution to the FW-H equation is formulated as:

$$\begin{aligned}4\pi p'(\vec{x}, t) &= \frac{\partial}{\partial t}\int\frac{\rho_0 V_n}{R}|\nabla f|\delta(f)\delta(g)d\vec{y}\,d\tau \\ &-\frac{\partial}{\partial x_i}\int\frac{l_i}{R}|\nabla f|\delta(f)\delta(g)d\vec{y}\,d\tau + \frac{\partial^2}{\partial x_i\partial x_j}\int\frac{T_{ij}}{R}\delta(g)d\vec{y}\,d\tau,\end{aligned} \tag{2.64}$$

Due to the property of the Dirac function, the domain of integration can have one dimension reduced, which makes two terms reduced to be the surface integrals, called "the surface source term", whereas one term would still be the volume integral, thus called "the volume source term".

2.3.3 The fundamental integral formulas of the surface sources in time domain

In the solution for Eq. (2.63) to the inhomogeneous wave equation (2.62), the integrand contains the Dirac function $\delta(f)$ and $\delta(g)$. Thus, to further simplify the integral expression, a relation between $d\vec{y}\,d\tau$ and $df\cdot dg$ is required.

For this purpose, we will correspondingly derive three fundamental integration formulas in this section.

2.3.3.1 Surface integral formula

Now the concern is to solve for the sound field in the time domain, i.e., to determine the sound pressure at an arbitrary observation point $\vec{x}$ in three-dimensional space at time t. Apparently, the sound waves emitted at a certain instant of time τ_0 from all the sources located on the blade surface will not be received at the same moment t_1 by the observer located at x_1. In other words, the sources of all the sound waves, received by a specific observer at the time t_1 and emitted at the same time τ_0, are not the collection of all the original sources located on the blade surface. Therefore, we need to utilize a different integral surface rather than the blade surface.

Consider two surfaces $f=0$ and $g=0$. $f=0$ denotes the blade surface whereas $g=0$ represents the spherical surface $R=c_0(t-\tau)$, an assemblage of all the locations of those potential sound sources emitted at time τ and the sound waves arriving at the observation point x at time t. The intersection curve Γ of these two surfaces forms a surface F with time going on ($\tau \in (-\infty, t)$), which can be written as:

$$F(\vec{y}, \vec{x}, t) = f\left(\vec{y}, \tau = t - \frac{R}{c_0}\right). \tag{2.65}$$

This equation represents the sound waves emitted at different instants by the sources on the blade surface that are received at the same moment for the same observation point. Therefore, we can specify a three-dimensional space of $F - r_\perp$ with $d\vec{y} = d\Sigma \cdot dr_\perp$. The $r_\perp$ is defined as the unit normal to the surface F, and $d\Sigma$ is a surface element of the surface $F=0$.

Integrate Eq. (2.63) with respect to τ first. Note that the equation can be formulated as:

$$\begin{aligned} 4\pi\Phi(\vec{x}, t) &= \int \frac{Q(\vec{y}, \tau)}{R} |\nabla f| \delta(f)\delta(g) d\vec{y}\, dg \\ &= \int \left[\frac{Q(\vec{y}, \tau)}{R} |\nabla f| \delta(f)\right]_{ret} d\Sigma dr_\perp \\ &= \int_{F=0} \left[\frac{Q(\vec{y}, \tau)}{R \cdot \Lambda}\right]_{ret} d\Sigma. \end{aligned} \tag{2.66}$$

by employing the relation

$$d\vec{y} = d\Sigma dr_\perp = d\Sigma \frac{dF}{|\nabla F|} = d\Sigma \frac{dF}{|\nabla f| \cdot \Lambda},$$

where

$$\nabla F = \frac{\partial F}{\partial y_i} = \frac{\partial f}{\partial y_i} + \frac{\partial f}{\partial \tau} \cdot \left[-\frac{1}{c_0} \frac{\partial R}{\partial y_i} \right] = \left[n_i - M_n \hat{R}_i \right] \frac{\partial f}{\partial y_i},$$

$$\Lambda = \left| n_i - M_n \hat{R}_i \right| = 1 + M_n^2 - 2M_n \cos\theta,$$

θ denotes the angle between $\mathbf{n}$ and $\hat{\mathbf{R}}$, and M_n cos θ is the Mach number of the blade surface moving relative to the observation point. Λ represents the velocity of the blade surface relative to the sound wave "beams" emitted toward the observation point.

So far, the integral Eq. (2.66) is the source-emitting surface integral we obtained.

2.3.3.2 Collapsing sphere formulation

Consider the emission time τ at which the panel on the blade $f(\vec{y}, \tau) = 0$ and the surface of collapsing sphere $g = \tau - t + R/c_0 = 0$ intersected at curve Γ. That is to say, all the sound signals radiated by the sound sources located on Γ at the time τ arrive at the observation point $\vec{x}$ at the time t. Note that both $f_\perp$ and $g_\perp$ are normal to the curve Γ, respectively (but $f_\perp$ and $g_\perp$ are not necessarily orthogonal). Although we cannot establish a rectangular coordinate system based on $(\Gamma, f_\perp, g_\perp)$, we can still obtain the relation $d\vec{y} = \frac{df_\perp dg_\perp}{\sin\theta} d\Gamma$ and determine the angle θ between $f_\perp$ and $g_\perp$. The relations $df_\perp = \frac{df}{|\nabla f|}$ and $dg_\perp = \frac{dg}{|\nabla g|} = c_0 \cdot dg$ imply the integral of Eq. (2.63) can be rewritten as:

$$\begin{aligned} 4\pi\Phi(\vec{x}, t) &= c_0 \int \frac{Q(\vec{y}, \tau)}{R} |\nabla f| \delta(f) \delta(g) \frac{df}{|\nabla f|} \frac{dg}{\sin\theta} d\Gamma d\tau \\ &= c_0 \int \frac{Q(\vec{y}, \tau)}{R \sin\theta} d\Gamma d\tau, \end{aligned} \tag{2.67}$$

which is the source-emitting line integral, also called the collapsing sphere formulation.

2.3.3.3 Physical surface integral: The retarded time formulation

Establish a coordinate system $\vec{\eta}$ that satisfies the relation $d\vec{\eta} = dSdf_\perp$ and the Jacobian coefficient of the transform from $\vec{y}$ to $\vec{\eta}$ equal to unity. And then the integral of Eq. (2.63) can be transformed into:

$$4\pi\Phi(\vec{x},t)=\int\frac{Q(\vec{y},\tau)}{R}|\nabla f|\delta(f)\delta(g)d\vec{\eta}\,d\tau$$
$$=\int\frac{Q(\vec{y},\tau)}{R}\delta(g)dSd\tau. \tag{2.68}$$

Integration respective to time τ in the coordinate system $\vec{\eta}$ with

$$d\tau=\frac{dg}{\left|\frac{dg}{d\tau}\right|}=\frac{dg}{|1-M_r|},$$

yields

$$4\pi\Phi(\vec{x},t)=\int_{f=0}\left[\frac{Q(\vec{y},\tau)}{R|1-M_r|}\right]_{ret}dS, \tag{2.69}$$

where M_r denotes the Mach number of any fixed point in the system $\vec{\eta}$ that is on the blade surface moving relative to the observation point. The only difference between M_r and $M_n\cos\theta$ in the first integral expression is that M_r is defined for the fixed $\vec{\eta}$. Eq. (2.69) is the physical surface integral expression called the retarded time formulation.

2.3.3.4 The relationship of the three integral formulas

Comparing the three integral formulas discussed earlier and retaining the term containing $\frac{Q}{R}$ implies:

$$\frac{d\Sigma}{\Lambda}=\frac{c_0 d\Gamma d\tau}{\sin\theta}=\frac{dS}{|1-M_r|}. \tag{2.70}$$

This relation can be utilized to transform one integral into another. Among these three integral formulations, Eq. (2.69) has the fastest computation speed; however, there will be integration singularities appearing under the transonic condition, which calls for a switch to other formulas. Therefore, the different formulas may be used in alternation during the actual processes of numerical computation.

2.3.4 The integral expressions of the sound field due to monopoles and dipoles

In the first section of this chapter, we discussed that the dominating sources of propeller noise are the volume displacement due to the blade thickness, and the steady and the unsteady blade forces, which are all of the properties of monopoles and dipoles. Thus, we should mainly consider the first two terms in the sound field solution of Eq. (2.64). Moreover, to eliminate the partial differentiation outside the integral, a set of differential transforms will be introduced. The dipole term resulting from the blade loads is:

$$
\begin{aligned}
4\pi p'(\vec{x}, t) &= -\frac{\partial}{\partial x_i}\int \frac{l_i}{R}|\nabla f|\delta(f)\delta(g)\, d\vec{y}\, d\tau \\
&= -\int l_i|\nabla f|\delta(f)\left[\frac{1}{R}\frac{\partial \delta(g)}{\partial x_i} - \frac{\delta(g)}{R^2}\frac{\partial R}{\partial x_i}\right] d\vec{y}\, d\tau \\
&= -\int l_i|\nabla f|\delta(f)\left[\frac{1}{R}\left(-\frac{\hat{R}_i}{c_0}\frac{\partial \delta(g)}{\partial t}\right) - \frac{\hat{R}_i\delta(g)}{R^2}\right] d\vec{y}\, d\tau.
\end{aligned}
\tag{2.71}
$$

According to the third integral formula, it can be further reduced to:

$$
4\pi p'(\vec{x}, t) = \frac{1}{c_0}\frac{\partial}{\partial t}\int_{f=0}\left[\frac{l_r}{R(1-M_r)}\right]_{ret} dS + \int_{f=0}\left[\frac{l_r}{R^2(1-M_r)}\right]_{ret} dS, \tag{2.72}
$$

where

$$
l_r = l_i\frac{x_i - y_i}{R}, \hat{R}_i = \frac{R_i}{R} = \frac{x_i - y_i}{R},
$$

and

$$
\frac{\partial R}{\partial x_i} = \frac{R_i}{R} = \hat{R}_i,
$$

$$
\frac{\partial \delta(g)}{\partial x_j} = \delta'(g)\frac{\partial g}{\partial x_j} = \delta'(g)\frac{R_i}{c_0 R} = -\frac{\hat{R}_i}{c_0}\frac{\partial \delta(g)}{\partial t}.
$$

Therefore, the total sound field due to the monopole source and the dipoles source are given as:

$$
4\pi p'(\vec{x}, t) = \frac{1}{c_0}\frac{\partial}{\partial t}\int_{f=0}\left[\frac{\rho_0 c_0 V_n + l_r}{R(1-M_r)}\right]_{ret} dS + \int_{f=0}\left[\frac{l_r}{R^2(1-M_r)}\right]_{ret} dS. \tag{2.73}
$$

This formulation can be directly applied to solve for the sound field; however, in view of the fact that the accuracy and validity of numerical solutions largely relies on the accuracy of the given time difference, nowadays the most widely applied solving approach is to take the time derivative of each relevant term appearing in the integrand. Upon using the relation:

$$
\left.\frac{\partial}{\partial t}\right|_x = \frac{\partial}{\partial \tau}\frac{\partial \tau}{\partial t} = \left[\frac{1}{1-M_r}\left.\frac{\partial}{\partial \tau}\right|_x\right]_{ret},
$$

the first term of the integral expression can be derived as:

$$\frac{\partial}{\partial t}\left[\frac{\rho_0 c_0 V_n + l_r}{R(1-M_r)}\right] = \frac{1}{1-M_r}\frac{\partial}{\partial \tau}\left[\frac{\rho_0 c_0 V_n + l_r}{R(1-M_r)}\right]$$

$$= \frac{l_i' \cdot \hat{R}_i}{R(1-M_r)^2} + \frac{l_i}{R(1-M_r)^2}\frac{\partial \hat{R}_i}{\partial \tau} + \frac{\rho_0 c_0}{R(1-M_r)^2}\frac{\partial V_n}{\partial \tau}$$

$$+ \frac{(\rho_0 c_0 V_n + l_r)}{R^2(1-M_r)^3}\left[R_i M_i' - c_0 M_i^2 - V_r\right],$$

where

$$\frac{\partial R}{\partial \tau} = -V_i \cdot \hat{R}_i = -V_r,$$

$$\frac{\partial \hat{R}_i}{\partial \tau} = \frac{\partial}{\partial \tau}\left(\frac{R_i}{R}\right) = \frac{1}{R}\frac{\partial R_i}{\partial \tau} + R_i \frac{\partial}{\partial \tau}\left(\frac{1}{R}\right) = \frac{\hat{R}_i V_r - V_i}{R},$$

$$\frac{\partial M_r}{\partial \tau} = \frac{1}{c_0 R}\left[R_i \frac{\partial V_i}{\partial \tau} + V_r^2 - V_i^2\right].$$

During cruise, the propeller is in steady motion, i.e., $\partial V_n / \partial \tau = 0$. Thus, the sound field due to the dipoles and the monopoles can be expressed as:

(1) The sound generation due to dipoles (caused by the blade loads)

$$4\pi p_L(\vec{x}, t) = \frac{1}{c_0}\int_{f=0}\left[\frac{l_i' \hat{R}_i}{R(1-M_r)^2}\right]_{ret} dS + \int_{f=0}\left[\frac{l_r - l_i M_i}{R^2(1-M_r)^2}\right]_{ret} dS$$
$$+ \int_{f=0}\left[\frac{l_r\left(M_i' R_i + c_0 M_r - c_0 M_i^2\right)}{R^2(1-M_r)^3}\right]_{ret} dS. \tag{2.74}$$

(2) The sound generation due to monopoles (caused by the blade thickness)

$$4\pi p_T'(\vec{x}, t) = \int_{f=0}\left[\frac{\rho_0 V_n\left(R_i M_i' + c_0 M_r - c_0 M_i^2\right)}{R^2(1-M_r)^3}\right]_{ret} dS. \tag{2.75}$$

The summation leads to:

$$p'(\vec{x}, t) = p_L'(\vec{x}, t) + p_T'(\vec{x}, t). \tag{2.76}$$

For the sound field resulting from the monopoles and the dipoles, the integration is apparently conducted on the whole blade surface $f = 0$. For a certain observer at $(\vec{x}, t)$, the value of each "source point" in the integrand should respectively be taken as it is at the corresponding retarded time τ. The physical explain of this integral is as follows. The sound field is formed by superposing all the influences imposed by an infinite number of the "source points" on the blade surface $f = 0$. In other words, the sound pressure $p'(\vec{x}, t)$ is due to a summation of all the sound waves arriving at $(\vec{x}, t)$ emitted by those "sound points" individually at their own retarded times.

The retarded time equation is expressed as $g = \tau - t + |\vec{x} - \vec{y}(\eta, \tau)|/c_0 = 0$. No matter whether for subsonic propeller or transonic propeller, before calculating the sound field we should first solve the retarded time equation. Indeed, the retarded time equation implies for a certain observation point $\vec{x}$, the spherical surface $R = c_0(t - \tau)$ contains all the possible locations of the sources of the sound waves emitted at time τ and received by that observer at time t. With $\tau \to t$, R approaches zero, which derives the name of "collapsing sphere." The origin of the collapsing sphere is located at the observation point, and for any given instant $\tau < t$, the intersection of this sphere and the blade surface is the collection of all sound sources.

After the observation point is specified, for each "source point" the corresponding time τ can be obtained by solving the retarded time equation. The solution can be very different for the subsonic condition and the supersonic condition. For a sound source moving with the velocity of $\mathbf{V}$, if $\mathbf{V}$ is subsonic, the sound field generated in the undisturbed medium would contain a set of nonintersecting wave fronts of spherical waves that propagate forward. Thus, as long as the propagating time is long enough, the sound wave can reach any point in the space. In other words, there is only one retarded time τ for a fixed observer at the observation time t, or we can say the retarded time equation of a source moving with subsonic velocity has only one solution.

However, the case is totally different for the sources moving faster than sound speed. The retarded time equation could have more than one solution. An analysis can be conducted in a similar way as in Chapter 1, so we will not discuss it in detail here.

2.3.5 Introduction to numerical computation methods

(1) The fixed coordinate system and the follow-up coordinate system of propellers

Generally speaking, two sets of coordinate systems are needed for propeller noise prediction. The first one is the fixed coordination relative to the undisturbed medium, which is established as the rectangular coordinates

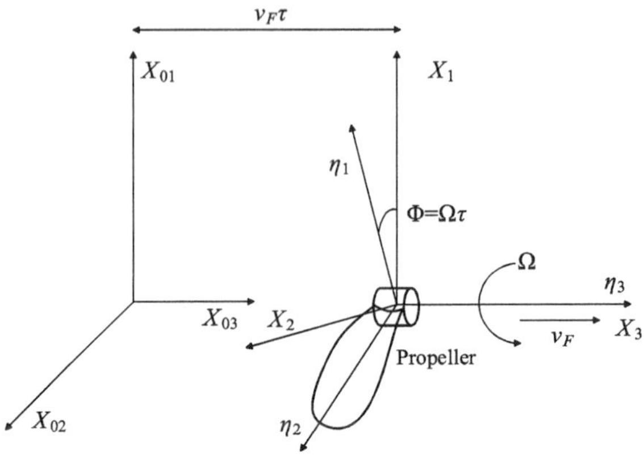

Fig. 2.7 The fixed coordinate system and the follow-up coordinate system of the propeller.

(X_{01}, X_{02}, X_{03}) or cylindrical coordinates (R_0, Φ_0, Z_0). The second one is the follow-up coordinate system (η_1, η_2, η_3) fixed to the moving propeller blades. We use X_1, X_2 and η_1, η_2 to represent the propeller's rotational plane while X_3 andη_3 is parallel to the forward flight direction. Then we introduce the coordinate system (X_1, X_2, X_3) moving with the same flight velocity of the propeller. When $\tau = 0$, coordinates (X_1, X_2, X_3) and (X_{01}, X_{02}, X_{03}) will coincide with each other as shown in Fig. 2.7.

There are many vector dot products and cross products in the computation formula, thus making it convenient to calculate when it comes to the normal vectors or pressure on the blade surface. Note that the sound source's position $\mathbf{Y}$ is always referred to as the coordinates in the coordinate system $\mathbf{X}$, thus $\mathbf{R} = \mathbf{X} - \mathbf{Y}$ and $\hat{R}_i$ should be evaluated in the system $\mathbf{X}$, whereas the vector components need to be calculated in the coordinate system $\boldsymbol{\eta}$. Also, special attention should be paid to the velocity variables. For example, both $M_r = M_i R_i$ and $V_n \cdot \vec{n}$ are evaluated in the coordinates $\mathbf{X}$, thus we need to first determine $V_i = \partial y_i / \partial \tau$ and then use it to calculate $M_r$and V_n.

For the actual propeller blades, the transform from $\mathbf{X}$ to $\boldsymbol{\eta}$ at time τ is formulated as:

$$\begin{bmatrix} X_{01} \\ X_{02} \\ X_{03} \end{bmatrix} = \begin{bmatrix} X_1 \\ X_2 \\ X_3 \end{bmatrix} + \begin{bmatrix} 0 \\ 0 \\ V_F \cdot \tau \end{bmatrix}, \begin{bmatrix} X_1 \\ X_2 \\ X_3 \end{bmatrix} = \begin{bmatrix} \cos\psi & -\sin\psi & 0 \\ \sin\psi & \cos\psi & 0 \\ 00 & 0 & 1 \end{bmatrix} \begin{bmatrix} \eta_1 \\ \eta_2 \\ \eta_3 \end{bmatrix},$$

where $\psi = \Omega\tau + \varphi$ and φ denotes the initial phase angle of the actual blades.

As for the cylindrical coordinates (as shown in Fig. 2.8), it should be:

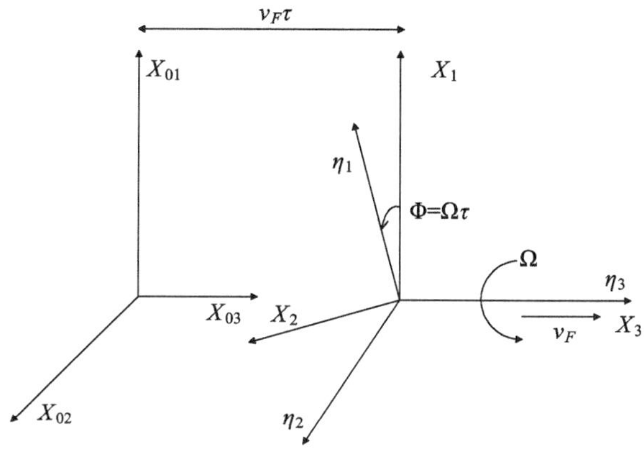

Fig. 2.8 The fixed coordinate system and the follow-up coordinate system of the propeller.

$$\begin{bmatrix} X_{01} \\ X_{02} \\ X_{03} \end{bmatrix} = \begin{bmatrix} R_0 \cos\Phi_0 \\ R_0 \sin\Phi_0 \\ Z_0 \end{bmatrix}, \begin{bmatrix} R_0 \\ \Phi_0 \\ Z_0 \end{bmatrix} = \begin{bmatrix} R_0 \\ \varphi + \Omega\tau \\ Z + V_F \cdot \tau \end{bmatrix}.$$

(2) Integral methods

To solve for the sound field of a moving propeller, the first step is to transform the curve integral into a plane integral, and then resolve it into the line integrals in two directions. The plane is usually chosen as an $X - Y$ plane. Suppose the projected area of the blade surface to the $X - Y$ plane is N, then we have the relation $dS = Ndxdy$. In the blunt region of the blade's leading edge, the curve dS is nearly vertical to Y and $N \to \infty$, which brings about the integral singularity. The consequent expense will either be a reduction in accuracy or an increase of the mesh density around the leading edge. To overcome this difficulty, we can replace coordinate η_2 with the arc length of the blade profile on the $\eta_2 - \eta_3$ plane. However, on different sections $\eta_1 =$ constant, the arc lengths of the blade profile will be different as well. So the dimensionless arc length is scaled with the whole arc length of the blade profile (on the pressure side or the suction side). At last, the lines $\eta_1 =$ constant and $l =$ constant are used to divide the blade surface into grids. Therefore, for an arbitrary point located on the blade surface, its position can be specified by the blade surface equation and the two parameters η_1 and l, with its normal vector determined by:

$$\mathbf{N} = \frac{\partial \mathbf{Q}}{\partial \eta_i} \times \frac{\partial \mathbf{Q}}{\partial l} = (N_1, N_2, N_3), \mathbf{n} = \frac{\mathbf{N}}{|N|}, \quad |N| = \sqrt{N_1^2 + N_2^2 + N_3^2},$$

$$
\begin{cases}
N_1 = \dfrac{\partial \eta_2}{\partial \eta_1} \cdot \dfrac{\partial \eta_3}{\partial l} - \dfrac{\partial \eta_3}{\partial \eta_1} \cdot \dfrac{\partial \eta_2}{\partial l} \\
N_2 = -\dfrac{\partial x_3}{\partial l} \\
N_3 = -\dfrac{\partial x_2}{\partial l}
\end{cases},
$$

as well as the transform relation as $dS = Ndxdl$.

After getting the original integral of S converted into the double integral with respect to η_1 and l, we can move to the investigation of the specific numerical integration methods. Considering the sensibility of acoustic computations to errors, the high-accuracy as well as high-efficiency Gauss-Legendre integral method (*G-L* method) is employed in each direction:

$$
\begin{cases}
\int_a^b f(x)dx = \sum_{i=1}^{m} f(x_x)\omega_i \\
x_i = (b+a)/2 + (b-a)\varsigma_i/2
\end{cases},
$$

where ς_i denotes the locations of the zeros of the Legendre polynomial, and $\varsigma_i \in (-1, 1)$. ω_i is the weighting coefficient, and m is the order of the *G-L* method. The values of the arrays $\{\varsigma_i\}$ and $\{\omega_i\}$ for a given order m can be obtained by look-up table. Finally, $\{\varsigma_i\}$ determines the positions of the discrete sampling points for the integrand. In our problem, all the sample points are located at the mesh points generated on the blade surface with parameters η_1 and l.

Furthermore, in the blade leading-edge region for the integral with respect to l and in the blade-tip region for the integral with respect to η_1, the aerodynamic forces will change drastically in the normal direction. Thus, for these regions, a denser grid is required for the integral, which leads to what is called a subsection integral.

References

[1] D.C. Mikkelson, G.A. Mitchell, L.J. Bober, Summary of recent NASA propeller research, (1983) NASA-TN-83733.
[2] F.J.H. Lynan, The Emission of Sound by Airscrews, Report No. 624 Advisory Committee for Aeroacoustics, Washington, DC, 1919.
[3] G.H. Bryan, H. Lamb, The Acoustics of Moving Sources With Application to Airscrews, Aeronautical Research Committee, London, 1920 R&M No. 684.
[4] L. Gutin, On the sound field of a rotating propeller, (1948) NACA-TM-1195.
[5] H.H. Hubbard, Free-space oscillating pressures near the tips of rotating propellers, (1950) NACA Report 996.

[6] I.E. Garrick, A theoretical study of the effect of forward speed on the free-space sound-pressure field around propellers, (1954) NACA Report 1198.
[7] C.E. Watkins, A method for calculation of free-space sound pressures near a propeller in flight including considerations of the chordwise blade loading, (1956) NACA-TN-3809.
[8] N. Kemp, R.A. Arnoldi, Machine Calculation of Free-Space Sound Pressure Field Around Propellers in Forward Motion, United Aircraft Corporation Research Department Report R-22673-1, East Hartford, CT, 1951.
[9] A. Shashady, N.H. Kemp, A Study of Effective Radius in Propeller Sound Calculation and on Analytical Integration for Sound in the Far-Field, United Aircraft Corporation Research Department Report R-22673-3, East Hartford, CT, 1954.
[10] A. Shashady, Presentation of Calculated Rotational Sound Pressure due to a Propeller in Forward Motion, United Aircraft Corporation Research Department Report R-22673-2, East Hartford, CT, 1954.
[11] W. Ernsthausen, The source of propeller noise, (1937) NACA-TN-825.
[12] A.F. Deming, Noise from propellers with symmetrical sections at zero blade angle, (1937) NACA-TN-605.
[13] W. Ernsthausen, Influence of aerodynamic characteristics on the sound field and radiated power of an airscrew, Akustiche Zeitschrift 6 (1941) 245–261.
[14] W. Ernsthausen, Der rotierende Tragflügel als Strahlungsproblem, Z. Angew. Math. Mech. 31 (1–2) (1951) 20–35.
[15] K.V. Diprose, Some propeller noise calculation showing the effect of thickness and planform, (1955) RAE Technical Note No. MS, 19.
[16] R.A. Arnoldi, Propeller Noise Caused by Blade Thickness, United Aircraft Corporation Research Department Report No. R-0896-1, East Hartford, CT, 1956.
[17] R.A. Arnoldi, Near-Field Computations of Propeller Blade Thickness, United Aircraft Corporation Research Department Report No. R-0896-2, East Hartford, CT, 1956.
[18] C.L. Morfey, The sound field of sources in motion, J. Sound Vib. 23 (1972) 291–295.
[19] A.I.V.D. Vooren, P.J. Zandbergen, Noise field of a rotating propeller in forward flight, AIAA J. 1 (1963) 366–368.
[20] C.L. Morfey, Rotating blades and aerodynamic sound, J. Sound Vib. 28 (1973) 587–617.
[21] L. Prandtl, Remarks on aircraft noise, Z. Tech. Phys. 2 (1921) 244–245.
[22] F.W. Barry, B. Magliozzi, Noise Detectability Prediction Method for Low Tip Speed Propellers, (1971) Technical Report AFAPL-TR-71-37, Wright-Patterson Air Force Base, OH.
[23] M.V. Lowson, J.B. Ollerhead, Studies of helicopter rotor noise (No. WR-68-9), Wyle Labs Huntsville Al Research Staff, 1969.
[24] D.B. Hanson, Measurements of static inlet turbulence, in: 2nd Aeroacoustics Conference, AIAA Paper 75-746, 1975.
[25] B. Magliozzi, The influence of forward flight on propeller noise, (1977) NASA-CR-145105.
[26] R.J. Pegg, F. Farassat, B. Magliozzi, Some measured and calculated effects of forward velocity on propeller noise, in: ASME 1977 International Gas Turbine Conference and Products Show, ASME Paper 77-GT-70, 1977.
[27] B.M. Brooks, Acoustics test and analysis of three turboprop models, (1980) NASA CR-159667.
[28] F. Farassat, Linear acoustic formulas for calculation of rotating blade noise, AIAA J. 19 (1981) 1122–1130.
[29] D.L. Hawkings, M.V. Lowson, Theory of open supersonic rotor noise, J. Sound Vib. 36 (1974) 1–20.
[30] W. Jou, Supersonic propeller noise in a uniform flow, AIAA J. 27 (1989) 385–390.

[31] C. Woan, G. Gregorek, The exact numerical calculation of propeller noise, in: 11th Fluid and Plasma Dynamics Conference, AIAA Paper 78-1122, 1978.
[32] G.P. Succi, Design of quiet efficient propellers, in: SAF Paper 790584, 1979.
[33] P.A. Nystrom, F. Farassat, A numerical technique for calculation of the noise of high-speed propellers with advanced blade geometry, (1980) NASA-TP-1662.
[34] W.E. Zorumski, Aircraft noise prediction program theoretical manual, (1982) NASA-TM-83199.
[35] D.B. Hanson, Influence of propeller design parameters on far-field harmonic noise in forward flight, AIAA J. 18 (11) (1980) 1313–1319.
[36] D.B. Hanson, Compressible helicoidal surface theory for propeller aerodynamics and noise, AIAA J. 21 (1983) 881–889.
[37] D.J.K. Hanson, Helicoidal surface theory for harmonic noise of propellers in the far field, AIAA J. 18 (1980) 1213–1220.
[38] D.B. Hanson, M.R. Fink, The importance of quadrupole sources in prediction of transonic tip speed propeller noise, J. Sound Vib. 62 (1978) 19–38.
[39] D.B. Hanson, Near-field frequency-domain theory for propeller noise, AIAA J. 23 (1985) 499–504.
[40] K.D. Kordan, E. Von Lavante, T.A. White, An alternative method of calculating propeller noise generated at transonic tip speeds including non-linear effects, in: 23rd Aerospace Sciences Meeting, AIAA Paper 85-2, 1985.
[41] F. Farassat, A new aerodynamic integral equation based on an acoustic formula in the time domain, AIAA J. 22 (1984) 1337–1340.
[42] L.N. Long, G.A. Watts, Arbitrary motion aerodynamics using an aeroacoustic approach, AIAA J. 25 (1987) 1442–1448.
[43] A. Das, A unified approach for the aerodynamics and acoustics of propellers in forward motion, (1984) ACARD-CP-366.
[44] L.J. Bober, D.S. Chaussee, P. Kutler, Prediction of high speed propeller flow fields using a three-dimensional Euler analysis, (1983) NASA-TM-83065.
[45] D.B. Hanson, Noise of counter-rotation propellers, J. Aircr. 22 (1985) 609–617.
[46] J.A. Gazzaniga, The acoustic experimental investigation of counterrotating propeller configurations, SAE Paper 871031, 1987.
[47] X. Sun, Z. Hu, S. Zhou, Unsteady loading noise of counter-rotating propeller, J. Aerosp. Power 3 (3) (1988) 227–283.
[48] M.E. Goldstein, Aeroacoustics, McGraw-Hill Book Co, New York, 1976.

CHAPTER 3

Noise prediction in aeroengine

3.1 Noise sources in aeroengine

Aeroengine noise is one main sound source of aircraft noise that cannot be ignored in aircraft design. After World War II, civil aviation received rapid development worldwide, and the most challenging problem at an early stage might be how to reduce jet noise in an aeroengine. However, the application of a turbofan engine leads to more complex noise sources along with increasing the bypass ratio. Now, even if various aeroengine noise control techniques were already employed, the noise generated by the propulsive system is still dominant in aircraft noise, as shown in Fig. 3.1. Therefore, seeking optimal design for various types of aeroengine noise is still a major task in all development strategies and in the planning of aeroacoustics for any countries devoted to developing commercial aviation. It can be seen that no matter how different the main sound source is, the main challenge for every phase of aeroacoustics development is how to effectively reduce sound radiation of the corresponding aeroengine's prominent source. On the other hand, in the following statement, it is noted that with the knowledge accumulation of fluid mechanics and other disciplines, the noise reduction of propulsive system has gone through a series of evolutions.

The turbojet engine was the main type of engine used in the 1950s. Fig. 3.2 depicts the noise components and their proportions in a turbojet

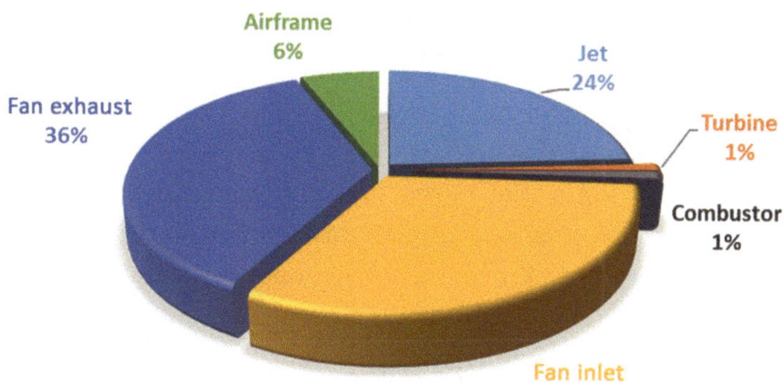

Fig. 3.1 Proportions of various noise sources of aircraft [1].

Fundamentals of Aeroacoustics with Applications to Aeropropulsion Systems
https://doi.org/10.1016/B978-0-12-408069-0.00003-8

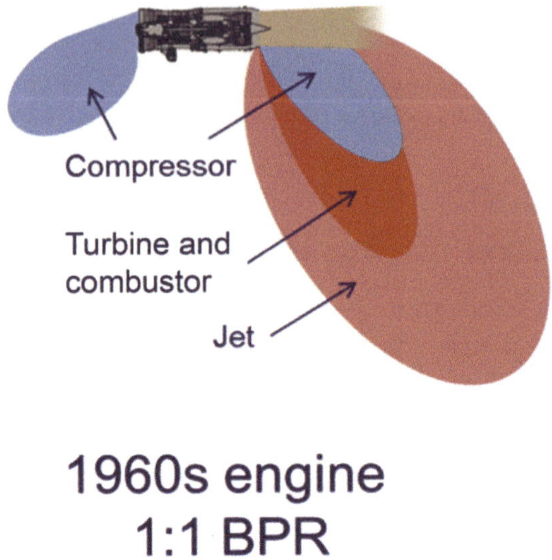

Fig. 3.2 Noise components of a turbojet engine in the early stage [2].

engine. It can be found that jet noise was the major source of noise due to the jet speed of more than 700 m/s, thus obstructing the civil application in aviation [1]. In 1954, the sound analogy theory was established by Lighthill, and an explicit formula of jet noise, also known as the eighth-order law of energy, was reported [3]. This formula explains why the turbojet engine has such strong noise, also providing an effective way to reduce this kind of noise by slowing down the jet speed of the engine. Furthermore, in the 1960s, humankind stepped onto the stage of the turbofan engine, which has higher propulsive efficiency as well as lower jet noise compared with the turbojet engine. Such technological changes led to a reduction of 10 dB in the aeroengine, as shown in Fig. 3.3, and the noise components of turbofan engine in 1990s is shown in Fig. 3.4. It can be seen that, for the turbofan engine, jet noise is not the main proportion anymore, whereas the noise from the fan and compressor became the major component.

After this realization in the 1970s, great attention has been given to the control of fan noise. It took a long time from the establishment and investigation of the theoretical model to their engineering applications. The most significant theoretical basis, established in this phase, might be the concept of "Rotating Mode," proposed by Tyler and Sofrin in 1962 to study the characteristics of noise propagation in a duct [4]. With a series of assumptions, the "cut-off" condition for the propagation of high-order sound modes was

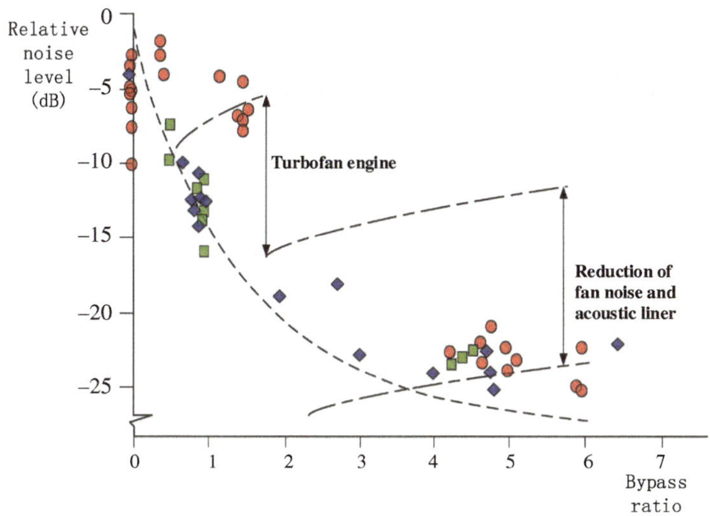

Fig. 3.3 Variation of noise in aeroengine with the bypass ratio [1].

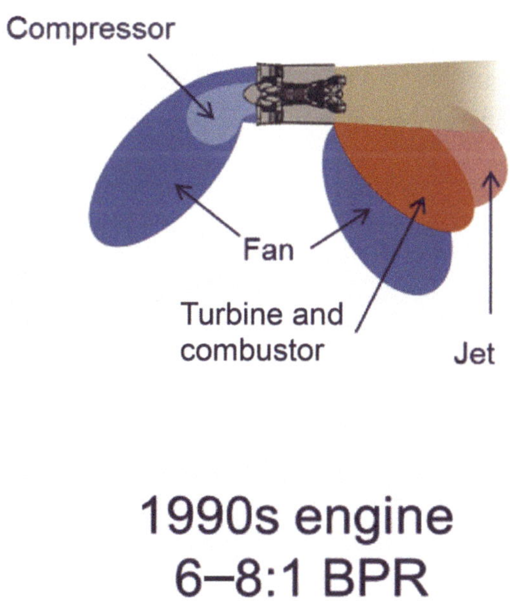

Fig. 3.4 Main noise components of aeroengine in the 1990s [2].

acquired using the eigenvalue approach. The cut-off condition reveals that, when the frequency of fluctuation of the sound source is lower than a critical value, the sound energy of higher-order modes will face exponential attenuation. Based on this theory, the authors also reported on an analysis for the tone noise by rotor/stator interaction and presented an approach for the selection of a blade number so that the soundwave meets the cut-off condition by selecting the proper numbers of blades for the rotor and stator, thus reducing the sound energy. Despite the oversimplification of Tyler and Sofrin's model, as well as ignoring the non-uniform mean flow effect, it is still followed by the control of rotor/stator interaction noise even today.

Afterward, the effect of a moving solid boundary on sound generation has been considered in the study of sound source. The problem of sound generation by a moving body in fluids was solved by Ffowcs Willians and Hawkings [5]. Later, Goldstein [6] investigated sound generation by a moving body in uniform flow using Green's function. Compared with the current trends to pursue a unified solution of fluid and an acoustic field based on a computational approach, past acoustic models were generally set up with two steps, which were, respectively, to obtain acoustic loading either by computational approaches or by experimental data and to calculate the acoustic field by Green's function method. This is also the core of acoustic analogy, transforming a multiscale problem into two distinct scale problems. In spite of some simplifications, the relevant models still constructed a basis for the qualitative analysis and prediction of the sound source generated by a fan/compressor, which were also followed by many models of rotor/stator interaction. More importantly, some effective methods of fan/compressor noise reduction were proposed based on these models, including the design of the blade's geometry, enlarging the gap between the rotor and stator, and applying the backward swept guide vanes, et al., which all succeeded in reducing the noise of the aeroengine and are still used today.

Later in the development of civil aviation, the bypass ratio of large transport aircraft steadily increased. Compared with turbofan engine with low bypass ratio in the early stage whose jet speed reached 500 m/s by mixing the core and bypass, engines for civil usage today have a bypass ratio over 12, with the jet speed lower than 300 m/s or even more [1]. The typical noise from different components of a modern aeroengine is shown in Fig. 3.5.

Due to the large bypass ratio value between 10 and 12, the fan will rotate at a supersonic tip speed, thus resulting in a large contribution to the noise level made by the shockwave during taking off. Spectra analysis to the shockwave noise appears to have a strong characteristic of multiple pure tones. Although the shockwave noise consists of a broad band of spectra, sharp decay occurs easily due to its nonlinearity. Moreover, acoustic liner application is

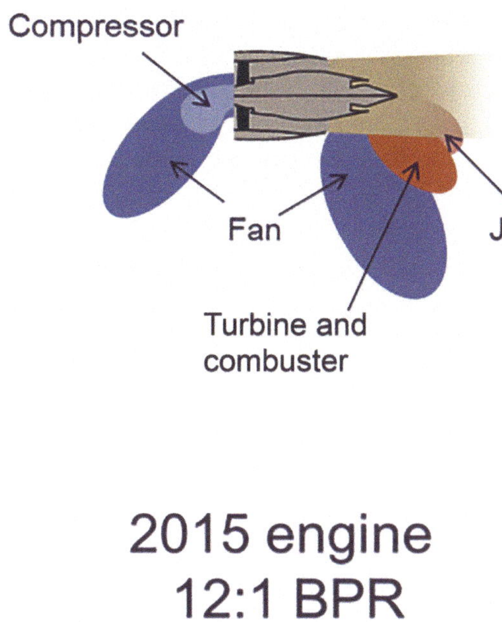

Fig. 3.5 Comparisons of different components in aeroengine noise [2].

effective to suppress this kind of noise because its frequencies are multiple integers to their mode values. Hence, even if the magnitude of shockwave noise is 10 dB higher than that of the rotor/stator interaction tone noise, the latter still makes the most contribution to the noise propagation at the far field, especially during the landing phase, as shown in Fig. 3.6.

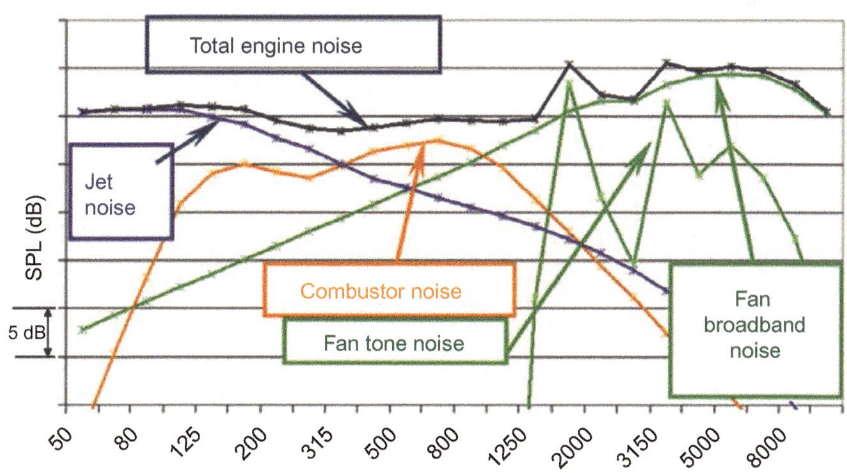

Fig. 3.6 Noise spectra of the turbofan engine with large bypass ratio during landing [7].

With the establishment of advanced theoretical models, as well as the improvement of experimental techniques, more careful analysis has been adapted to every source of aeroengine noise. From Fig. 3.6, the intensities of various components of noise are depicted as a function of frequency, which shows that, with effective suppression to tone noise, the broad-band noise of the fan and the combustion noise in a low-frequency range, as well as the jet noise, become relatively important to the aircraft noise.

Therefore, this chapter will pay more attention to the physical generation mechanism of fan/compressor noise, as well as methods for its prediction, including noise due to the rotor/stator interaction and the shockwave. Meanwhile, a brief introduction to the combustion noise and the jet noise, with totally different properties of sound generation, will also be included in this chapter, with an emphasis on the mechanism of noise generation.

3.2 Tone noise by rotor/stator interaction in fan/compressor

3.2.1 Introduction

Studying and predicting the sound propagation from rotor/stator interaction is significant to the aeroacoustics design and noise reduction of the nacelle in an aeroengine. In the early stage, most of the researches about noise generation focused on the construction of empirical relations between the design parameters and source strength, as well as qualitative proposals for noise reduction, such as the effect from the blade tip speed, pressure ratio, and shaft power. For example, some empirical predicting formulas were reported by Smith and House [8] through experimental measurements for a practical aeroengine, which had a great effect on noise control at the time. In most of these theoretical studies, only an incompressible flow condition was considered. Kemp and Sears [9, 10] investigated the force generated on the blades by the potential flow and viscous wake interaction. In their studies, the force was considered as dipole for sound field calculation. It is obvious that, for the calculation of sound in the far field at incoming flow with low Mach number, using these predicting methods for incompressible condition, the results appeared to be consistent with the experiments. However, if the Mach number becomes higher, an enormous error will occur when these incompressible methods are still applied to compute the unsteady force.

Since the early stage of the 1960s, some new methods were employed by Tyler and Sofrin [4] to study the sound propagation in an aeroengine. They proved that rotating modes occur if blade rows interaction exists, which are going to propagate or decay in the duct. Meanwhile, they also reported that noise reduction can be achieved through the selection of the proper number of blades for rotor and stator to meet the cut-off condition. Moreover, no rotating mode will generate for a single subsonic rotor. The cut-off condition proposed by Tyler and Sofrin is significant in fan noise reduction and was followed by most selections of blade numbers. Nevertheless, these authors did not try to construct methods of noise prediction for blade row interaction.

After these previous works, the calculation of unsteady force and sound field for blade rows in compressible flow condition was highlighted in most of the following studies. Three different aeroacoustics models for rotor/stator interaction noise can be classified: model of free sound field, model of multiple blade rows, and model of sound field in duct.

The model of free sound field assumes that the rotor and stator are all in free space and the effect from casing containment is not considered. Moreover, stationary flow condition is also adapted [11]. This method borrows from the successful aeroacoustics designs of the propeller. Nevertheless, the sound field in duct is constructed on the discrete modes of the soundwave, which has an obvious difference with the propeller noise. Due to this, there exist two inevitable disadvantages in the application of this model to predict rotor/stator noise: the effect from moving media as well as the duct is not included. On the other hand, with the development of unsteady aerodynamics, it does not appear to be much difficulty to calculate the unsteady load on the blade. Therefore, more effective models for rotor/stator interaction are urgently needed.

The model of multiple blade rows uses the two-dimensional infinite flat cascade to model the effect of rotor or stator cascade. Most studies adapting this model pay attention to the sound generation, propagation, and refection by the cascade, whereas different blade row models and mathematical methods are employed in these studies. Some works are widely accepted including those by Kaji and Okazaki [12–14], Mani [15], and Smith [16]. Kaji and Okazaki first studied the sound reflection and transmission by blade rows through the actuator-disk method, and the effect from the gap between blade rows was also included in the application of potential function. Afterward, they reported the prediction for the rotor/stator interaction noise using the no-perturbation boundary condition on blades of both the rotor and stator, which considered the interactive effect from the wake and the secondary potential flow. Mani employed a series of point force to model the blade and acquired an approximate solution for the sound generation of a single blade row. This point force model was verified by Lipstein and Mani [17] through experiments of the interaction between the rotor and the wake from the cylinder bar. Smith followed the model of Kaji and Okazaki and handled the singularity of integral equations in detail. Moreover, a much better mathematical method was used to calculate the load on blades. Meanwhile, the model was also validated through some experiments by Smith. However, compared with the real working condition of the rotor or stator of fan and compressor, the model of the blade row appears to be oversimplified. Hence, this model is always used to explore physical mechanisms instead of practical noise prediction.

It is obvious that the geometrical shape of the annular cascade cannot be considered in a two-dimensional model, especially the shape variation along the radial and spanwise direction, such as the backward swept, as well as these non-uniform characteristics in spanwise such as flow perturbation and blade vibration. Therefore, it was urgent to employ three-dimensional models to replace these two-dimensional ones for accurate prediction, and this appeared to be a chance for the establishment of these aeroacoustics models in the duct.

For these aeroacoustics models in the duct, the rotor or stator is assumed to be placed in a duct with a uniform cross-section area and no reflection. As an approximate method, the casing effect on the sound propagation is also included. The most representative work of this type of model might be the three-dimensional lifting surface theory, and the assumption of linear small perturbation was adapted, which came from the integral formula for thickness noise prediction of rotor blades in compressible flow proposed by McCune in 1958 [18]. After 12 years, Okurounmu and McCune [19] reported formulas for sound field prediction, with a known aerodynamic load on the blade. However, it was not complete because the difference of pressure distribution on the blade surface should be determined through the shape of the surface and the flow condition.

In the 1970s, various models of lifting surface theory, which meet real conditions, were constructed due to the rapid improvement of computational hardware. For the reverse problem of solving the difference of pressure distribution, the solution can be achieved through the integral equation based on the no-perturbation boundary condition on the blade surface, which also constructs the intrinsic connection between the sound field generated by the aerodynamic load and the perturbation of the incoming flow, thus solving the load distribution on the blade. However, the toughest challenge appears to be the complexity of three-dimensionality, which leads to problems of singularity and convergence in the solution. Due to this, the lifting surface theory was proposed by Namba [20–24] and Salaun [25, 26]. Initially, Namba used this theory to study the steady lift on the blade with supersonic and subsonic tip speeds respectively. Salaun employed his method to solve the problem of blade flutter. The problem of noise generated by the interaction between the rotor wake and the stator was then investigated by Namba through the lifting surface theory. Although both of these methods proposed by Namba and Salaun appear to be applications of the lifting surface theory, they dealt with the integral equation in totally different ways. The kernel function was presented in an accurate form in Salsun's method while Namba applied the way of finite radial wave number expansion to replace the kernel function, which made the problem of singularity much easier. Later, another way of radial expansion was employed in the lifting force theory by Schulten [27–31] who further used his method to inlet and outlet guide vanes (OGV) of the fan. Moreover, effects from the backward swept, the incline, as well as the stagger angle and the arched shape, were studied by Schulten. He also acquired consistent results with Namba in calculations for the fourth type of the Benchmark problem [32, 33], that indicated the problem of singularity arose by three-dimensional lifting surface theory had been solved.

Numerical approaches appear to be more superior for solving these problems that cannot be dealt with by classic models. In the prediction of sound source, the complex surface shape of the blade can be considered through numerical studies. However, three-dimensional simulations were still limited to the linearized Euler equation. In 2002, Prasad and Verdon [34, 35] reported three-dimensional codes of the linearized Euler equation, as well as solutions for Benchmark problems. In their studies, the unsteady loads on the blade surface were consistent with those solved through the three-dimensional lifting theory. Elhadidi and Atassi [36] used a three-dimensional numerical method to study effects of the incline as well as the backward swept stator vane, to the noise decay, in 2004. One year later, the numerical method for linearized equations was improved by Cooper and Peake [37], to include the effects of vortices and blade incline. Numerical approaches for linearized governing equations were also more than just covering all problems concerned with theoretical models. Giles [38, 39] worked out two-dimensional codes for the rotor/stator interactions in the time domain, with the nonlinear effect included. Furthermore, these codes were advanced to consider the viscous effect. However, only unsteady loads on the blade surface were reported in Giles' papers. Standing on the basis constructed by Giles' studies, several NASA research groups [40, 41] proposed various rotor/stator interaction models to predict the sound fields of two-dimensional cascades. In spite of this progress, it is believed that a complete three-dimensional numerical simulation for unsteady flow field in turbomachinery will still consume huge computing resources, and a rapid prediction tool for preliminary acoustic design of aeroengine is thus required.

This chapter will expand upon the three-dimensional lifting surface model proposed by Namba in which the prediction for a sound field, generated by steady and unsteady loads on blade surface in the duct, is mainly discussed. The following chapter will also pay attention to the process of solving loads on the blade surface through the lifting surface theory.

3.2.2 Model of sound generation by unsteady aerodynamic load on blade

The computational model, as well as the coordinate, of the rotor/stator interaction for the three-dimensional lifting surface theory is shown in Fig. 3.7. The fundamental assumptions adapted here are:

(1) The internal wall in the duct is solid, and fluids are assumed to be compressible, isentropic, and uniform without viscosity.

(2) Every blade is a flat plate without thickness effect considered.

(3) The angle of attack of incoming flow is 0, and it will not turn its direction.

(4) Flow perturbation is small, and flow governing equations can be linearized.

(5) There is a subsonic flowing condition.

(6) The unsteady loads on blades' trailing edge appear to be finite.

In Fig. 3.7, w_i' represents the periodic wake of upstream rotating blades, which will generate unsteady loads after it interacts with the vane. The periodic and unsteady loads on the stator will generate soundwaves denoted by p'_D and p'_A, which is the cause for the noise of rotor/stator interaction.

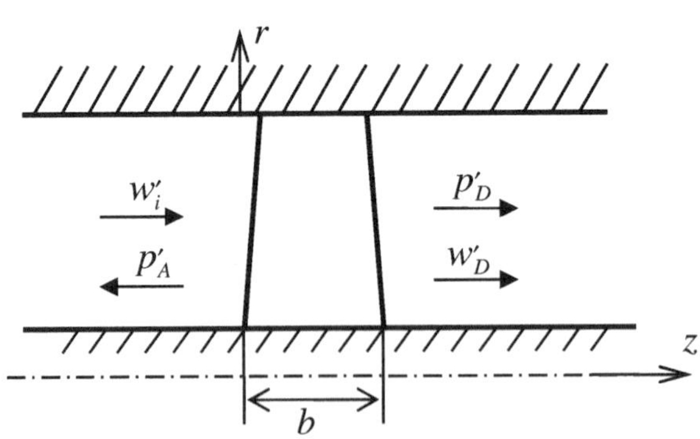

Fig. 3.7 Schematic of noise by rotor wake and stator interaction.

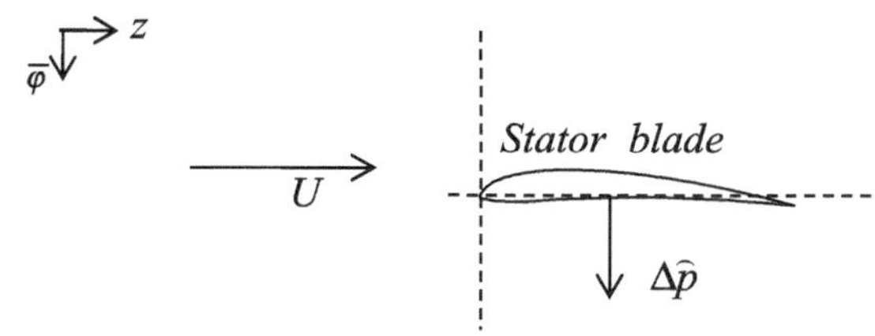

Fig. 3.8 Top view of the stator vane.

3.2.2.1 Governing equations for the problem

For a simple discussion, first, the rotor of the fan or compressor is assumed to be in a uniform flow field with velocity U, and no reflection will occur at two ends of the duct. Angles of backward swept appears to be 0, and vanes are parallel to the z-axis, the top view of which is shown in Fig. 3.8.

Considering the effect of a solid wall on the sound generation of moving media, the formula of sound pressure can be written as Eq. (3.1), based on the General Lighthill's Theory:

$$p'(\vec{x},t) = \int_{-T}^{T}\int_{v(\tau)} \frac{\partial^2 G}{\partial y_i \partial y_j} T'_{ij} d\vec{y}\, d\tau + \int_{-T}^{T}\int_{s(\tau)} \frac{\partial G}{\partial y_i} f_i ds(\vec{y})\, d\tau + \int_{-T}^{T}\int_{s(\tau)} \rho_0 \widetilde{V'}_n \frac{D_0 G}{D\tau} ds(\vec{y})\, d\tau. \tag{3.1}$$

It is obvious that the first item represents the sound generated by the fluctuating viscous stress on the solid wall of the duct, which contributes little to the whole sound field under high Reynolds-number conditions. Meanwhile, because $\widetilde{V'}_n = 0$ on solid walls, they will not contribute to the integration of Eq. (3.1). Moreover, the normal derivative of Green's function can be eliminated, thus leading to the situation that the surface integral in Eq. (3.1) is only needed to be performed on every surface of stator vanes. This means that for practical stator vanes of the fan and compressor, their noise mainly comes from the dipoles that are generated by interactions between these stator vanes and the unsteady wake.

In the discussion of the first chapter, it was mentioned that the second item in Eq. (3.1) represents the sound generated by the fluctuating force on the boundary of flow field, which can be classified as dipole. Hence, in this mathematical problem, Eq. (3.1) can be simplified as:

$$p'(\vec{x},t) = \int_{-T}^{T}\int_{s(\tau)} \frac{\partial G}{\partial y_i} f_i ds(\vec{y})\, d\tau, \tag{3.2}$$

where $s(\vec{y})$ denotes the vane surface and G denotes Green's function in the duct, which meets the equation:

$$\nabla^2 G - \frac{1}{c_0^2} \cdot \frac{D_0^2 G}{D\tau^2} = -\delta(\vec{r} - \vec{r}\,')\delta(t - \tau), \tag{3.3}$$

with its boundary conditions being

$$\begin{cases} G = \dfrac{DG}{D\tau} = 0 \quad (t < \tau) \\ \dfrac{\partial G}{\partial \vec{r}\,'} = 0 \quad \text{on the wall} \end{cases}, \tag{3.4}$$

where $\vec{r}\,'$ and $\vec{r}$ denote the coordinates of the sound source and the observer. To get the solution of the sound field in the duct, which is also the integral result of Eq. (3.2), first, Green's function for a duct with arbitrary shape is needed by solving Eq. (3.3) with its boundary condition as given by Eq. (3.4).

3.2.2.2 Solution for Green's function for a duct with arbitrary shape

As shown in Fig. 3.9, $\vec{r}\,'$ denotes the coordinate in the internal space of a duct with infinite length and solid wall. Its cross-section area is A, and z' represents the axial direction.

Green's function, as well as its Fourier transformation, can be written as:

$$G_\omega = \int_{-\infty}^{\infty} e^{i\omega(\tau - t)} \cdot G \cdot d\tau, \quad G = \frac{1}{2\pi} \int_{-\infty}^{\infty} e^{i\omega(t-\tau)} \cdot G_\omega \cdot d\omega. \tag{3.5}$$

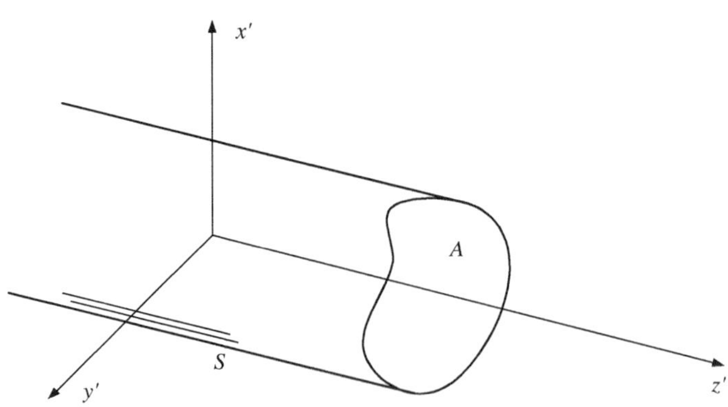

Fig. 3.9 Schematic for the solution of Green's function.

Therefore, we have:

$$\mathbb{F}(\nabla^2 G) = \nabla^2 G_\omega,\ \ \mathbb{F}\left(\frac{\partial G}{\partial \tau}\right) = -i\omega G_\omega,\ \ \mathbb{F}\left(\frac{\partial^2 G}{\partial \tau^2}\right) = -\omega^2 G_\omega, \tag{3.6}$$

$$\nabla^2 G_\omega - \frac{1}{c_0^2}\cdot\left[-\omega^2 G_\omega + 2U_0\frac{\partial}{\partial z'}(-i\omega G_\omega) + U_0^2\frac{\partial^2 G_\omega}{\partial z'^2}\right] = -\delta(\vec{r}-\vec{r}\,'), \tag{3.7}$$

$$\left(\nabla^2 - M_a^2\frac{\partial^2}{\partial z'^2} + 2iM_a k_0\frac{\partial}{\partial z'} + k_0^2\right)G_\omega = -\delta(\vec{r}-\vec{r}\,'). \tag{3.8}$$

Eq. (3.6) can be obtained based on the property of Fourier transformation, and with the transformation performed on the original governing equation of Green's function, Eq. (3.7) is obtained, which can be rearranged as given by Eq. (3.8).

The method of variable separation is employed to solve Eq. (3.8). Let $G_\omega = \sum_m f_m(z')\Phi_m(x', y')$, and Eq. (3.8) can be rewritten as:

$$\begin{aligned}\left(\nabla^2 - M_a^2\frac{\partial^2}{\partial z'^2} + 2iM_a k_0\frac{\partial}{\partial z'} + k_0^2\right)\sum_m f_m(z')\Phi_m(x', y') &= -\delta(\vec{r}-\vec{r}\,')\\ &= -\delta(x-x')\delta(y-y')\delta(z-z').\end{aligned} \tag{3.9}$$

Here, $\Phi(x', y')$ is governed by the two-dimensional Helmholtz equation:

$$\begin{cases}\dfrac{\partial^2\Phi}{\partial x'^2} + \dfrac{\partial^2\Phi}{\partial y'^2} + k^2\cdot\Phi = 0\\ \text{Boundary condition}\quad \dfrac{\partial\Phi}{\partial \vec{r}\,'} = 0\end{cases}. \tag{3.10}$$

In the Helmholtz equation as given by Eq. (3.10), k appears to be the eigenvalue with its solution, $\Phi(x', y')$, being the eigenfunction, which meets the orthogonal condition as given by:

$$\int_A \Phi_n\cdot\Phi_m^* dx'dy' = \begin{cases}0 & m\neq n\\ \Gamma_m & m = n\end{cases}, \tag{3.11}$$

where $\Gamma_m = \int_A |\Phi_m|^2 dx'dy'$. With Fourier transformation to $f_m(z')$, we have:

$$F_\alpha = \int_{-\infty}^{\infty} f_m(z')e^{-i\alpha(z-z')}dz',\ \ f_m(z') = \frac{1}{2\pi}\int_{-\infty}^{\infty} F_\alpha e^{i\alpha(z-z')}d\alpha, \tag{3.12}$$

Performing Fourier transformation again to both sides of Eq. (3.9), and considering Eq. (3.11), we obtain:

$$\sum_m \left(-\beta^2\alpha^2 + 2M_a k_0 \alpha + k_0^2 - k^2\right) F_\alpha \Phi_m(x', y') = -\delta(x - x')\delta(y - y'), \tag{3.13}$$

where $\beta^2 = 1 - M_a^2$. Multiplying this equation by $\Phi_n^*(x', y')$, in which "*" represents the conjugate, and then integrating it over the cross-section, leads to:

$$\int_A \sum_m \left(-\beta^2\alpha^2 + 2M_a k_0 \alpha + k_0^2 - k^2\right) F_\alpha \Phi_n \Phi_m^* dx' dy' = -\Phi_n^*(x, y), \tag{3.14}$$

when $m = n$, with the orthogonality we have:

$$F_\alpha = \frac{\Phi_m^*(x, y)}{\Gamma_m} \frac{1}{\beta^2\alpha^2 - 2M_a k_0 \alpha - k_0^2 + k^2}, \tag{3.15}$$

therefore, it can be obtained that:

$$\begin{aligned} G_\omega &= \sum_m f_m(z')\Phi_m(x', y') \\ &= \frac{1}{2\pi}\sum_m \frac{\Phi_m(x', y')\Phi_m^*(x, y)}{\Gamma_m} \int_{-\infty}^{\infty} \frac{e^{i\alpha(z-z')}}{\beta^2\alpha^2 - 2M_a k_0 \alpha - k_0^2 + k^2} d\alpha. \end{aligned} \tag{3.16}$$

Furthermore, Green's function can be written as:

$$\begin{aligned} G &= \frac{1}{2\pi}\int_{-\infty}^{\infty} e^{i\omega(t-\tau)} \cdot G_\omega \cdot d\omega \\ &= \frac{1}{4\pi^2}\sum_m \frac{\Phi_m(x', y')\Phi_m^*(x, y)}{\Gamma_m} \int_{-\infty}^{\infty}\int_{-\infty}^{\infty} \frac{e^{i\alpha(z-z')}}{\beta^2\alpha^2 - 2M_a k_0 \alpha - k_0^2 + k^2} e^{i\omega(t-\tau)} \cdot d\omega d\alpha. \end{aligned} \tag{3.17}$$

3.2.2.3 Solution of Green's function in an annular duct

For a circular or annular duct, the polar coordinate (r', φ', z'), (r, φ, z) is used to replace the original (x', y', z'), (x, y, z), thus Eq. (3.10) can be rewritten as:

$$\begin{cases} \dfrac{1}{r}\dfrac{\partial}{\partial r}\left(r\dfrac{\partial \Phi}{\partial r}\right) + \dfrac{1}{r^2}\dfrac{\partial^2 \Phi}{\partial \varphi^2} + k^2\Phi = 0 \\ \text{Boundary condition}: \dfrac{\partial \Phi}{\partial r} = 0 \end{cases}. \tag{3.18}$$

In Eq. (3.18), Φ is a function of r and φ, which can be written as $\Phi(r,\varphi)$. Let $\Phi(r,\varphi)=\psi(r)\Psi(\varphi)$, and the equation becomes:

$$\begin{cases} \dfrac{\partial^2\Psi}{\partial\varphi^2}+m^2\Psi=0 \\ \dfrac{1}{r}\dfrac{d}{dr}\left(r\dfrac{\partial\psi}{\partial r}\right)+\left(k^2-\dfrac{m^2}{r^2}\right)\psi=0 \end{cases}. \tag{3.19}$$

A periodic solution can be easily obtained for the first formula in Eq. (3.19) using the eigenvalue approach for the second-order ordinary differential equation, and the second formula appears to be an mth-order Bessel equation, the solution of which comes out to be Eq. (3.20) with the periodic boundary condition as well as the boundary condition on the wall:

$$\begin{cases} \Psi(\varphi)=C\cdot e^{im\varphi} \\ \psi_m(r)=AJ_m(k_{mn}r)+BY_m(k_{mn}r) \end{cases}, \tag{3.20}$$

where $J_m(k_{mn}r)$ and $Y_m(k_{mn}r)$ represent the first and the second type of Bessel function, respectively. When $r=0$, the Bessel function becomes:

$$\begin{cases} J_m=1, \quad m=0 \\ J_m=0, \quad m\neq 0 \\ Y_m=-\infty \end{cases}. \tag{3.21}$$

For a circular duct, the solution must meet the requirement that a finite value of the eigenfunction exists at the center of the circular cross-section, which leads to $B=0$. Therefore, for a circular duct:

$$\begin{cases} \Psi(\varphi)=C\cdot e^{im\varphi} \\ \psi_m(r)=AJ_m(k_{mn}r) \end{cases}. \tag{3.22}$$

For an annular duct, Eq. (3.22) appears not to be the solution. With the boundary condition in Eq. (3.18), we get:

$$\begin{cases} AJ'_m(k_{mn}R_d)+BY'_m(k_{mn}R_d)=0 \\ AJ'_m(k_{mn}R_h)+BY'_m(k_{mn}R_h)=0 \end{cases}. \tag{3.23}$$

A and B cannot be 0 simultaneously, which requires that:

$$\begin{vmatrix} J'_m(k_{mn}R_d) & Y'_m(k_{mn}R_d) \\ J'_m(k_{mn}R_h) & Y'_m(k_{mn}R_h) \end{vmatrix}=0;$$

when $m=0$ and $m\neq 0$, we have:

$$J_1(k_{m,n}R_h)Y_1(k_{m,n}R_d)-J_1(k_{m,n}R_d)Y_1(k_{m,n}R_h)=0,$$

and

$$\begin{gathered}[J_{m-1}(k_{m,n}R_h)-J_{m+1}(k_{m,n}R_h)][Y_{m-1}(k_{m,n}R_d)-Y_{m+1}(k_{m,n}R_d)]- \\ [J_{m-1}(k_{m,n}R_d)-J_{m+1}(k_{m,n}R_d)][Y_{m-1}(k_{m,n}R_h)-Y_{m+1}(k_{m,n}R_h)]=0,\end{gathered}$$

respectively. The eigenvalue, k_{mn}, can be determined from these formulas, in which $n = 1, 2, 3\ldots$. After the eigenvalues are determined, the next task is to find coefficients A and B. Consider the boundary condition as well as variations of A and B, we can let $A = Y'_m(k_{m,n}R_d)$ and $B = -J'_m(k_{m,n}R_d)$, thus obtaining:

$$\psi = Y'_m(k_{m,n}R_d)J_m(k_{m,n}r) - J'_m(k_{m,n}R_d)Y_m(k_{m,n}r). \tag{3.24}$$

It should be noted that there might exist various values for A and B, but at a fixed ratio between these two coefficients. Finally, orthogonal values for these eigenfunctions will be solved. Multiplying both sides of the mth-order Bessel equation by $r^2\psi'_m$ yields:

$$r\psi'_m\frac{d}{dr}(r\psi'_m) + (r^2k^2 - m^2)\psi_m\psi'_m = 0,$$

and then performing integral to this equation with respect to r leads to:

$$\int_{R_h}^{R_d} r\psi'_m d(r\psi'_m) + \int_{R_h}^{R_d}(r^2k^2 - m^2)\psi_m\psi'_m dr = 0.$$

From the boundary condition:

$$\int_{R_h}^{R_d} r\psi'_m d(r\psi'_m) = \frac{1}{2}(r\psi'_m)^2\Big|_{R_h}^{R_d} = 0,$$

therefore

$$\int_{R_h}^{R_d}(r^2k^2 - m^2)\psi_m\psi'_m dr = k^2\int_{R_h}^{R_d} r^2\psi_m d\psi_m - m^2\int_{R_h}^{R_d}\psi_m d\psi_m = 0.$$

Solving the integral equation:

$$k^2\left[r^2\frac{\psi_m^2}{2}\Big|_{R_h}^{R_d} - \int_{R_h}^{R_d} r\psi_m^2 dr\right] - m^2\frac{\psi_m^2}{2}\Big|_{R_h}^{R_d} = 0,$$

and then rearrange it into:

$$\int_{R_h}^{R_d} r\psi_m^2 dr = \frac{1}{2}\left(r^2 - \frac{m^2}{k^2}\right)\psi_m^2\Big|_{R_h}^{R_d},$$

thus leading to:

$$\overline{\Gamma}_{mn} = \int_{R_h}^{R_d} r\psi_m^2 dr = \int_{R_h}^{R_d} r\psi_m^2 dr = \frac{1}{2}\left(r^2 - \frac{m^2}{k^2}\right)\psi_m^2\Big|_{R_h}^{R_d}.$$

where $k = k_{mn}$ represents the eigenvalue of the eigenfunction on the duct. From this derivation, we have:

$$G = \frac{1}{4\pi^2}\sum_{m=-\infty}^{\infty}\sum_{n=1}^{\infty}\frac{\psi_m(k_{mn}r)\psi_m(k_{mn}r')e^{im\varphi}e^{-im\varphi'}}{\Gamma_{mn}} \int_{-\infty}^{\infty}\int_{-\infty}^{\infty}\frac{e^{i\omega(t-\tau)}e^{i\alpha(z-z')}d\alpha d\omega}{\beta^2\alpha^2 - 2M_a k_0\alpha - k_0^2 + k_{mn}^2}. \tag{3.25}$$

The pole point within Green's function appears to be:

$$\alpha = \frac{M_a k_0 \pm \sqrt{k_0^2 - \beta^2 k_{mn}^2}}{\beta^2},$$

in which "+" and "-" represent the upstream and downstream directions of propagation, respectively. Applying the residue theorem with respect to α gives:

$$G = -\frac{i}{4\pi}\sum_{m=-\infty}^{\infty}\sum_{n=1}^{\infty}\frac{\psi_m(r)\psi_m(r')e^{im\varphi}e^{-im\varphi'}}{\Gamma_{mn}}\int_{-\infty}^{\infty}\frac{1}{\kappa_{nm}}e^{i\gamma_{mn}^{\pm}(z-z')}e^{i\omega(t-\tau)}d\omega, \tag{3.26}$$

in which

$$\kappa_{nm} = \begin{cases} \sqrt{k_0^2 - \beta^2 k_{mn}^2} & k_0^2 > \beta^2 k_{mn}^2 \\ -i\sqrt{\beta^2 k_{mn}^2 - k_0^2} & k_0^2 < \beta^2 k_{mn}^2 \end{cases},$$

$$\gamma_{mn}^{\pm}=\begin{cases}\dfrac{M_a k_0}{\beta^2}+\dfrac{\kappa_{nm}}{\beta^2} & \textit{upstream}\\ \dfrac{M_a k_0}{\beta^2}-\dfrac{\kappa_{nm}}{\beta^2} & \textit{downstream}\end{cases},$$

$$\Gamma_{mn}=\int_0^{2\pi}d\varphi\int_{R_h}^{R_d}r\psi_m^2dr=2\pi\int_{R_h}^{R_d}r\psi_m^2dr=\pi\left(r^2-\frac{m^2}{k^2}\right)\psi_m^2\bigg|_{R_h}^{R_d}=2\pi\overline{\Gamma}_{mn}.$$

Let $\phi_{mn}=\psi_{mn}/\sqrt{\Gamma_{mn}}$, we have:

$$\int_{R_h}^{R_d}r\phi_m\phi_n dr=\delta_{mn}=\begin{cases}1 & m=n\\ 0 & m\neq n\end{cases}.$$

Therefore, Green's function can be written as:

$$G=-\frac{i}{4\pi}\sum_{m=-\infty}^{\infty}\sum_{n=1}^{\infty}\frac{\phi_m(k_{mn}r)\phi_m(k_{mn}r')e^{im\varphi}e^{-im\varphi'}}{2\pi}\int_{-\infty}^{\infty}\frac{1}{\kappa_{nm}}e^{i\gamma_{mn}^{\pm}(z-z')}e^{i\omega(t-\tau)}d\omega. \tag{3.27}$$

3.2.2.4 Solution for sound generation by unsteady loads

After Green's function is obtained, we can take it into Eq. (3.2) and find the integral expression of the sound field. For a vane with a specific position, the force acting on it can be decomposed into two components: the axis thrust f_T and the circumferential force f_D. The differential relation can then be written as:

$$f_i\frac{\partial}{\partial y_i}=\frac{f_D}{r'}\frac{\partial}{\partial\varphi'}+f_T\frac{\partial}{\partial z'}. \tag{3.28}$$

With this relation and the expression of Green's function taken into Eq. (3.2), we obtain:

$$p'(\vec{x},t)=-\frac{1}{4\pi}\sum_{m=-\infty}^{\infty}\sum_{n=1}^{\infty}\frac{\psi_m(r)e^{im\varphi}}{\Gamma_{mn}}\int_{-\infty}^{\infty}\frac{1}{\kappa_{nm}}e^{i\gamma_{mn}^{\pm}z}e^{i\omega t}$$
$$\int_{-T}^{T}\int_{s(\tau)}\psi_m(r')e^{-i(m\varphi'+\gamma_{mn}^{\pm}z')}\left(m\frac{f_D}{r'}+\gamma_{mn}^{\pm}f_T\right)e^{-i\omega\tau}ds(\vec{y})d\tau d\omega. \tag{3.29}$$

If the sound generation of the rotor blade is considered, the coordinate $(\overline{r}, \overline{\varphi}, \overline{z})$ that rotates with the rotor can be introduced, and $\overline{r} = r'$, $\overline{\varphi} = \varphi' + \Omega\tau$, $\overline{z} = z'$. Taking them into Eq. (3.29) with the consideration that the limit of the surface integral does not connect with time τ any more after the rotating coordinate is introduced, the fluctuating pressure can thus be written as:

$$p'\left(\vec{x}, t\right) = -\frac{1}{4\pi}\sum_{m=-\infty}^{\infty}\sum_{n=1}^{\infty}\frac{\psi_m(r)e^{im\varphi}}{\Gamma_{mn}}\int_{-\infty}^{\infty}\frac{1}{\kappa_{nm}}e^{i\gamma_{mn}^{\pm}z}e^{i\omega t}$$
$$\int_{-T}^{T}\int_{s(\tau)}\psi_m(r')e^{-i\left(m\varphi' + \gamma_{mn}^{\pm}z'\right)}\left(m\frac{f_D}{r'} + \gamma_{mn}^{\pm}f_T\right)e^{-i(\omega - m\Omega)\tau}ds\left(\vec{y}\right)d\tau d\omega. \tag{3.30}$$

The blade surface facing upward and downward can be denoted by footnotes (1) and (2), respectively. Integrating with respect to the blade surface A in the rotating coordinate of fan/compressor yields:

$$p'\left(\vec{x}, t\right) = -\frac{1}{4\pi}\sum_{m=-\infty}^{\infty}\sum_{n=1}^{\infty}\frac{\psi_m(r)e^{im\varphi}}{\Gamma_{mn}}\int_{-\infty}^{\infty}\frac{1}{\kappa_{nm}}e^{i\gamma_{mn}^{\pm}z}e^{i\omega t}$$
$$\int_{A}\psi_m(r')e^{-im\varphi'}\int_{-T}^{T}\left(\frac{m}{r'}g_D^{\pm} + \gamma_{mn}^{\pm}g_T^{\pm}\right)e^{-i(\omega - m\Omega)\tau}d\tau r' dr' d\varphi' d\omega, \tag{3.31}$$

where

$$g_\alpha^{\pm} = \frac{f_\alpha^{(1)}}{\left|\widetilde{n}_1^{(1)}\right|}e^{-i\gamma_{mn}^{\pm}z'^{(1)}} + \frac{f_\alpha^{(2)}}{\left|\widetilde{n}_1^{(2)}\right|}e^{-i\gamma_{mn}^{\pm}z'^{(2)}}, \alpha = T,\ D,$$

and $\left|\widetilde{n}_1^{(1)}\right|$ denotes the component along z' direction of the unit vector on the surface of fan/compressor rotor blade. If the blade thickness is assumed to be much lower than the chord length, then the effect from the delay time of the rotor blade surface can be ignored, thus leading to:

$$g_\alpha^{\pm} = \widetilde{f}_\alpha e^{-i\gamma_{mn}^{\pm}z'^{(c)}},$$

in which $z'^{(c)}$ denotes the axial component of the blade chord in the rotating coordinate and $\widetilde{f}_\alpha$ represents the blade force, which is one specific point on the blade surface, acting on the unit area of (r', φ') plane. Hence, the fluctuating pressure can be rewritten as:

$$p'(\vec{x},t) = -\frac{1}{4\pi}\sum_{m=-\infty}^{\infty}\sum_{n=1}^{\infty}\frac{\psi_m(r)e^{im\varphi}}{\Gamma_{mn}}\int_{-\infty}^{\infty}\frac{1}{\kappa_{nm}}e^{i\gamma_{mn}^{\pm}z}e^{i\omega t}$$
$$\int_A \psi_m(r')e^{-i\left(m\varphi'+\gamma_{mn}^{\pm}z'^{(c)}\right)}\int_{-T}^{T}\left(\frac{m}{r'}\widetilde{f}_D+\gamma_{mn}^{\pm}\widetilde{f}_T\right)e^{-i(\omega-m\Omega)\tau}d\tau r'dr'd\varphi'd\omega.$$
(3.32)

It should be noted that Eq. (3.32) only considers sound generated by the dipole of fan/compressor, as well as its propagation in a straight annular duct, in which the positive and the negative symbols denote the upward and the downward propagation, respectively.

3.2.3 Prediction for tone noise by rotor/stator interaction

3.2.3.1 Sound pressure expression for stator/rotor interaction

The rotor of a fan/compressor is assumed to rotate at an angular speed of Ω, and stator vanes of number V are placed upstream of the rotor cascade. The unsteady force acting on these rotor blades, by the stator wake, can be written in a Fourier expansion as:

$$\widetilde{f}_\alpha = \sum_{-\infty}^{\infty}F_p^\alpha e^{ipV\Omega\tau}, \quad (\alpha = T, D) \tag{3.33}$$

and their coefficients are:

$$F_p^\alpha = \frac{\Omega}{2\pi}\int_0^{2\pi/\Omega}\widetilde{f}_\alpha\cdot e^{-ipV\Omega\tau}d\tau. \tag{3.34}$$

Based on these two formulas and with attention to the arbitrary order component in the wake, we can obtain:

$$\left(\frac{m}{r'}\widetilde{f}_D+\gamma_{mn}^{\pm}\widetilde{f}_T\right)e^{-i(\omega-m\Omega)\tau} = \left(\frac{m}{r'}F_p^D+\gamma_{mn}^{\pm}F_p^T\right)e^{-i[\omega-(m+pV)\Omega]\tau}.$$

Taking it into Eq. (3.32), the fluctuating sound pressure might be expressed as:

$$p'(\vec{x},t) = -\frac{1}{4\pi}\sum_{m=-\infty}^{\infty}\sum_{n=1}^{\infty}\frac{\psi_m(r)e^{im\varphi}}{\Gamma_{mn}}\int_{-\infty}^{\infty}\frac{1}{\kappa_{nm}}e^{i\gamma_{mn}^{\pm}z}e^{i\omega t}$$
$$\int_A \psi_m(r')e^{-i\left(m\varphi'+\gamma_{mn}^{\pm}z'^{(c)}\right)}\int_{-T}^{T}\left(\frac{m}{r'}F_p^D+\gamma_{mn}^{\pm}F_p^T\right)e^{-i[\omega-(m+pV)\Omega]\tau}d\tau r'dr'd\varphi'd\omega.$$
(3.35)

With the relation of $\lim_{T\to\infty}\int_{-T}^{T} e^{-i[\omega-(m+pV)\Omega]\tau}d\tau=2\pi\delta$ $[\omega-(m+pV)\Omega]$ taken into account in Eq. (3.35), and by using the integral characteristic of the Dirac function, it can be further rewritten as:

$$p'(\vec{x},t)=-\frac{1}{2}\sum_{m=-\infty}^{\infty}\sum_{n=1}^{\infty}\frac{\psi_m(r)e^{im\varphi}}{\Gamma_{mn}}\frac{e^{i\gamma_{mn}^{\pm}z}e^{i\omega_0 t}}{\kappa_{nm}}$$
$$\int_A \psi_m(r')e^{-i\left(m\varphi'+\gamma_{mn}^{\pm}z'^{(c)}\right)}\left(mF_p^D+r'\gamma_{mn}^{\pm}F_p^T\right)dr'd\varphi', \tag{3.36}$$

in which $\omega_0=(m+pV)\Omega$ and both κ_{nm} , $\gamma_{mn}^{\pm}$ are functions of ω_0.

Eq. (3.36) appears to be the sound field expression for a single rotor blade. For a cascade with the same blades of number B, the incoming unsteady pressures have the same amplitude F_p^{α} for all rotor blades but with phase between them. The phase angle between sounds generated by unsteady fluctuating forces on the kth and the first blade, comes out to be:

$$F_p^{\alpha}\left[r',\varphi'+\frac{2\pi}{B}(k-1),\tau-\frac{2\pi}{B\Omega}(k-1)\right].$$

After phase angles of $\tau-\frac{2\pi}{B\Omega}(k-1)$ and $\varphi'+\frac{2\pi}{B}(k-1)$ are taken into Eq. (3.33) and Eq. (3.36), respectively, the sound field generated by a cascade, which contains the same blades of number B, can be obtained through a similar approach employed before as:

$$p'(\vec{x},t)=-\frac{1}{2}\sum_{m=-\infty}^{\infty}\sum_{n=1}^{\infty}\frac{\psi_m(r)e^{im\varphi}}{\Gamma_{mn}}\frac{e^{i\gamma_{mn}^{\pm}z}e^{i\omega_0 t}}{\kappa_{nm}}$$
$$\int_A \psi_m(r')e^{-i\left(m\varphi'+\gamma_{mn}^{\pm}z'^{(c)}\right)}\sum_{k=1}^{B}e^{-i(k-1)\frac{2\pi}{B}[m+pV]}\left(\frac{m}{r'}F_p^D+\gamma_{mn}^{\pm}F_p^T\right)r'dr'd\varphi', \tag{3.37}$$

in which $\omega_0=(m+pV)\Omega$. Moreover, we have:

$$\sum_{k=1}^{B}e^{-i(m+pV)\frac{2\pi}{B}(k-1)}=\begin{cases}B & pV+m=sB\\ 0 & pV+m\neq sB\end{cases}\quad\Rightarrow\quad m=sB-pV.$$

Hence, only when $m=sB-pV$, the sound generated by the interaction between the upstream stator with V vanes and the rotor with B blades appears to be non-zero. Taking this formula into Eq. (3.37), the interaction noise by the rotor blades can be written as:

$$p'(\vec{x},t)=-\frac{B}{2}\sum_{m=-\infty}^{\infty}\sum_{n=1}^{\infty}\frac{\psi_m(r)e^{im\varphi}}{\Gamma_{mn}}\frac{e^{i\gamma_{mn}^{\pm}z}e^{i\omega_0 t}}{\kappa_{nm}}$$
$$\int_A \psi_m(r')e^{-i\left(m\varphi'+\gamma_{mn}^{\pm}z'^{(c)}\right)}\left(\frac{m}{r'}F_p^D+\gamma_{mn}^{\pm}F_p^T\right)r'dr'd\varphi', \tag{3.38}$$

with $\omega_0 = (m + pV)\Omega = sB\Omega$. The total sound field generated by the rotor/stator interaction in the duct thus can be expressed as:

$$p'\left(\vec{x}, t\right) = \sum_{s=-\infty}^{\infty} p_{SB}\left(\vec{x}\right) \cdot e^{isB\Omega t}, \tag{3.39}$$

in which

$$p_{sB}\left(\vec{x}\right) = -\frac{B}{2} \sum_{m=-\infty}^{\infty} \sum_{n=1}^{\infty} \frac{\psi_m(r) e^{im\varphi}}{\Gamma_{mn}} \frac{e^{i\gamma_{mn}^{\pm} z}}{\kappa_{nm}} \cdot \left[m D_{m,n}^{\pm} + \gamma_{mn}^{\pm} T_{m,n}^{\pm} \right], \tag{3.40}$$

$$\begin{cases} D_{m,n}^{\pm} = \int_A \psi_m(r') e^{-i\left(m\varphi' + \gamma_{mn}^{\pm} z'^{(c)}\right)} F_p^D dr' d\varphi' \\ T_{m,n}^{\pm} = \int_A \psi_m(r') e^{-i\left(m\varphi' + \gamma_{mn}^{\pm} z'^{(c)}\right)} F_p^T r' dr' d\varphi' \end{cases}, \tag{3.41}$$

$$\kappa_{nm} = \sqrt{\left(\frac{sB\Omega}{c_0}\right)^2 - \beta^2 k_{mn}^2}, \quad \gamma_{mn}^{\pm} = \frac{M_a k_0}{\beta^2} \pm \frac{\kappa_{nm}}{\beta^2}. \tag{3.42}$$

Based on the derivation for the solution of sound field, it can be found that some specific properties exist for the sound generated by these periodic unsteady loads on rotor blades, which will be discussed in detail in the following content.

Frequency characteristic

From the expression of the angular frequency ω_0 in Eq. (3.38), it is obvious that the frequency variable of the sound field appears to be:

$$f = \frac{sB\Omega}{2\pi}, \quad (s = 0, \pm 1, \pm 2, \ldots)$$

which is the blade pass frequency as well as its harmonics. This indicates that the noise of rotor/stator interaction in the fan/compressor distributes at the blade pass frequency and its harmonics while the noise at other frequencies appears to be 0.

Propagation characteristic

During the derivation for Green's function in a circular or annular duct, the expression of κ_{nm} has already been obtained, which will directly determine the axial wave number $\gamma_{mn}^{\pm}$, which is the propagation characteristic of the soundwave along the axial direction. κ_{nm} appears to be a complex number, and the state when it becomes a pure imaginary number is first discussed, which requires:

$$k_0^2 < \beta^2 k_{mn}^2 \Rightarrow \left(\frac{sB\Omega}{c_0}\right)^2 < \beta^2 k_{mn}^2. \tag{3.43}$$

The expression of κ_{nm} can be written as $\kappa_{nm} = -i\sqrt{\beta^2 k_{mn}^2 - k_0^2}$, with these corresponding axial wave numbers being:

$$\gamma_{mn}^{\pm} = \begin{cases} \dfrac{M_a k_0}{\beta^2} - \dfrac{i\sqrt{\beta^2 k_{mn}^2 - k_0^2}}{\beta^2} & \text{upstream wave} \\ \dfrac{M_a k_0}{\beta^2} + \dfrac{i\sqrt{\beta^2 k_{mn}^2 - k_0^2}}{\beta^2} & \text{downstream wave} \end{cases}. \tag{3.44}$$

For the downstream propagation wave, the observer is located downstream of the sound source, which means $z > z'$. The axial sound propagation can be expressed as $e^{i\gamma_{mn}^{\pm}(z-z')}$, and it can be found that the attenuation factor $e^{-\frac{\sqrt{\beta^2 k_{mn}^2 - k_0^2}}{\beta^2}(z-z')}$ will occur along the downstream propagation, which approaches 0 with the increase of z. Hence, when the condition of pure imaginary eigenvalue is met, the sound will decay exponentially in the duct. On the other hand, A similar rule also exists for the upstream traveling soundwave. For this reason, Eq. (3.43) is called the cut-off condition.

The expression of the cut-off condition includes the eigenvalue k_{mn}, which needs to be determined based on the duct shape and boundary condition. The case of a circular duct with a solid wall will be discussed as a sample for detail properties of the cut-off condition. For a circular duct, the eigenvalue k_{mn} is determined by $J_m'(k_{mn}R_d) = 0$, which is defined by:

$$J_m'(k_{mn}R_d) = \left.\frac{dJ_m(x)}{dx}\right|_{x=k_{mn}R_d} = 0 \tag{3.45}$$

Based on the property of the Bessel function, the smallest root value of Eq. (3.45) at $n = 1$ meets the condition $k_{m1} \approx \frac{m}{R_d}$. Furthermore, the cut-off condition can be expressed as:

$$\left(\frac{sB\Omega}{c_0}\right)^2 < \beta^2 \left(\frac{sB - qV}{R_d}\right)^2, \tag{3.46}$$

which is also

$$\frac{M_t}{\sqrt{1 - M_a^2}} < \left|\frac{sB - qV}{sB}\right|. \tag{3.47}$$

In Eq. (3.47), M_t denotes the circumferential Mach number of the rotor blade tip. Considering this equation, it can be found that different combinations of numbers of blades and vanes help to generate a soundwave that will propagate or decay. However, for a fixed frequency such as the first-order blade pass frequency, the cut-off condition can be written as:

$$\frac{M_t}{\sqrt{1-M_a^2}} < \left|1 - \frac{qV}{B}\right|. \tag{3.48}$$

When the relative Mach number of the blade tip appears to be subsonic, the left side of inequality Eq. (3.48) must be lower than 1. Meanwhile, if the ratio of the blade and vane numbers is larger than 2, its right side becomes larger than 1, which indicates that, in the subsonic condition, when the vane number is at less than 2 times of the blade number, the first-order blade-pass-frequency component in the soundwave will be cut off. This cut-off condition plays a significant role in the modern aeroacoustics design of a fan/compressor because appropriate selection of numbers of blade and vane can contribute to considerable benefits in noise reduction.

Let's return to consider the noise generation by the steady load on the blades. Under this situation, $q = 0$, thus $m = sB$. Eq. (3.47) can be simplified as:

$$\frac{M_t}{\sqrt{1-M_a^2}} < 1.$$

From this analysis, it can be found that this condition is equal to that at subsonic tip speed. Hence, for turbomachinery with subsonic rotating speed, all sound generated by the steady load will be cut off. Furthermore, only when the relative speed of the blade tip is supersonic will the soundwave generated by the steady load propagate without attenuation. Under this circumstance, the connection between the noise generated by a steady load and those by other sources, such as shockwave and rotor/stator interaction, should be analyzed thoroughly to seek the most effective ways for noise reduction.

3.2.3.2 Sound pressure expression for rotor/stator interaction

For the noise generated by the interaction between the upstream blades of number B and downstream vanes of number V, the unsteady load on vane surfaces caused by the wake of the upstream rotor, can be expressed as a form of Fourier series:

$$\widetilde{f}_{\alpha}=\sum_{-\infty}^{\infty}F_{s}^{\alpha}e^{isB\Omega\tau}.\quad(\alpha=T,D) \tag{3.49}$$

in which

$$F_{s}^{\alpha}=\frac{\Omega}{2\pi}\int_{0}^{2\pi/\Omega}\widetilde{f}_{\alpha}\cdot e^{-isB\Omega\tau}d\tau. \tag{3.50}$$

The unsteady force acting on the first vane is now named as $\widetilde{f}_{\alpha}^{0}(r',\varphi',\tau)$, and the sound generated by the kth vane at the moment τ is equal to that of the first one at the position of $\varphi'\sim\varphi'+\frac{2\pi}{V}(k-1)$ and moment $\tau\sim\tau+\frac{2\pi}{V\Omega}(k-1)$, which can be written as:

$$\widetilde{f}_{\alpha}^{0}\left[r',\varphi'+\frac{2\pi}{V}(k-1),\tau+\frac{2\pi}{V\Omega}(k-1)\right].$$

Taking this expression into the sound field expression in the stationary coordinate corresponding to Eq. (3.32), we obtain:

$$\begin{aligned}p'(\vec{x},t)=&-\frac{1}{4\pi}\sum_{m=-\infty}^{\infty}\sum_{n=1}^{\infty}\frac{\psi_m(r)e^{im\varphi}}{\Gamma_{mn}}\int_{-\infty}^{\infty}\frac{1}{\kappa_{nm}}e^{i\gamma_{mn}^{\pm}z}e^{i\omega t}\\&\int_A\psi_m(r')e^{-i\left(m\varphi'+\gamma_{mn}^{\pm}z'^{(c)}\right)}e^{i\frac{2\pi}{V}(sB-m)(k-1)}\\&\int_{-T}^{T}\left(\frac{m}{r'}F_s^D+\gamma_{mn}^{\pm}F_s^T\right)e^{-i(\omega-sB\Omega)\tau}d\tau r'dr'd\varphi'd\omega,\end{aligned} \tag{3.51}$$

and with the formula $\lim_{T\to\infty}\int_{-T}^{T}e^{-i[\omega-sB\Omega]\tau}d\tau=2\pi\delta[\omega-sB\Omega]$, as well as the integral property of the Dirac function, Eq. (3.51) can be further simplified to:

$$\begin{aligned}p'(\vec{x},t)=&-\frac{1}{2}\sum_{m=-\infty}^{\infty}\sum_{n=1}^{\infty}\frac{\psi_m(r)e^{im\varphi}}{\Gamma_{mn}}\frac{1}{\kappa_{nm}}e^{i\gamma_{mn}^{\pm}z}e^{i\omega_0 t}\\&\int_A\psi_m(r')e^{-i\left(m\varphi'+\gamma_{mn}^{\pm}z'^{(c)}\right)}\sum_{k=1}^{V}e^{i\frac{2\pi}{V}(sB-m)(k-1)}\left(mF_s^D+r'\gamma_{mn}^{\pm}F_s^T\right)dr'd\varphi',\end{aligned} \tag{3.52}$$

in which $\omega_0 = sB\Omega$. With the consideration to:

$$\sum_{k=1}^{V} e^{i(sB-m)\frac{2\pi}{V}(k-1)} = \begin{cases} V & sB-m=pV \\ 0 & sB-m \neq pV \end{cases} \Rightarrow m = sB - pV,$$

it can be found that, when the mode of sound in the duct appears to be $m = sB - pV$, the corresponding sound frequency generated by the sth-order component in the wake becomes $\omega_0 = sB\Omega$, and the sound field can be further expressed as:

$$\begin{aligned} p'(\vec{x}, t) = -\frac{V}{2} \sum_{m=-\infty}^{\infty} \sum_{n=1}^{\infty} \frac{\psi_m(r) e^{im\varphi}}{\Gamma_{mn}} \frac{1}{\kappa_{nm}} e^{i\gamma_{mn}^{\pm} z} e^{i\omega_0 t} \\ \int_A \psi_m(r') e^{-i\left(m\varphi' + \gamma_{mn}^{\pm} z'^{(c)}\right)} \left(mF_s^D + r' \gamma_{mn}^{\pm} F_s^T\right) dr' d\varphi'. \end{aligned} \tag{3.53}$$

If all blade wakes are taken into account, the sound pressure should be:

$$p'(\vec{x}, t) = \sum_{s=-\infty}^{\infty} p_{sB}(\vec{x}) \cdot e^{isB\Omega t}, \tag{3.54}$$

in which

$$p_{sB}(\vec{x}) = -\frac{V}{2} \sum_{m=-\infty}^{\infty} \sum_{n=1}^{\infty} \frac{\psi_m(r) e^{im\varphi}}{\Gamma_{mn}} \frac{1}{\kappa_{nm}} e^{i\gamma_{mn}^{\pm} z} \left(mD_{m,n}^{\pm} + \gamma_{mn}^{\pm} T_{m,n}^{\pm}\right), \tag{3.55}$$

$$\begin{cases} D_{m,n}^{\pm} = \int_A \psi_m(r') e^{-i\left(m\varphi' + \gamma_{mn}^{\pm} z'^{(c)}\right)} F_s^D dr' d\varphi' \\ T_{m,n}^{\pm} = \int_A \psi_m(r') e^{-i\left(m\varphi' + \gamma_{mn}^{\pm} z'^{(c)}\right)} F_s^T r' dr' d\varphi' \end{cases}, \tag{3.56}$$

$$\kappa_{nm} = \sqrt{\left(\frac{sB\Omega}{c_0}\right)^2 - \beta^2 k_{mn}^2}, \quad \gamma_{mn}^{\pm} = \frac{M_a k_0}{\beta^2} \pm \frac{\kappa_{nm}}{\beta^2}. \tag{3.57}$$

Compare Eq. (3.53) with Eq. (3.38), and it is easy to find that, no matter the sound generated by rotor/stator or stator/rotor interaction, the sound field distributes at the blade pass frequency as well as its harmonics, thus giving them the same characteristics of frequency and propagation mode.

3.3 Shockwave noise in fan/compressor

3.3.1 Physical mechanism of shockwave noise in fan/compressor

When the rotor of fan/compressor rotates at supersonic speed, the relative Mach number of the blade tip appears larger than 1. Discrete spectra

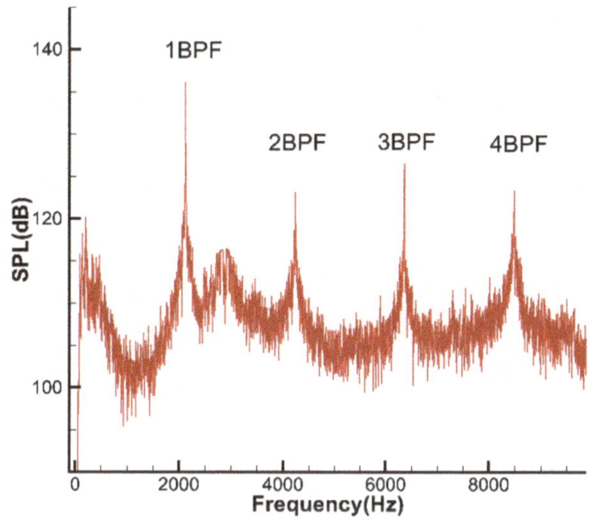

Fig. 3.10 Spectrum of a subsonic fan.

obtained through a sound test show an obvious difference with the frequency characteristic of a subsonic rotor. For a subsonic rotor, components at the blade pass frequency and its harmonics make the main contribution to its noise. However, as shown in Fig. 3.10, for a supersonic rotor, besides these components in a subsonic rotor, the axis pass frequency as well as its harmonics also takes a place as shown in Fig. 3.11, which is called multiple pure tone noise, combination tone noise, or buzz noise based on this frequency characteristic of supersonic rotor. These components in noise are essentially caused by the shockwave at the leading edge of the supersonic rotor blade, which can be simply called shockwave noise.

Shockwave noise is caused by the propagation and evolution of the discontinuous surface caused by shockwave. The shockwave that expands out of the passage generates on the tip's leading edge of every blade, with several expansion waves, thus constructing a complex system of shockwaves on the leading edge of the rotor, which spirals forward in the duct during the rotation. Standing on a rotating coordinate, it can be found that every shockwave propagates stably forward. However, due to the chase and merge of initial shockwaves, strong nonlinearity is observed during practical propagation.

Fig. 3.12 shows the flow field for a two-dimensional cascade with ideal blades. It is assumed that every blade is the same, the distance between two near blades keeps a constant value, and the incoming flow is uniform. When

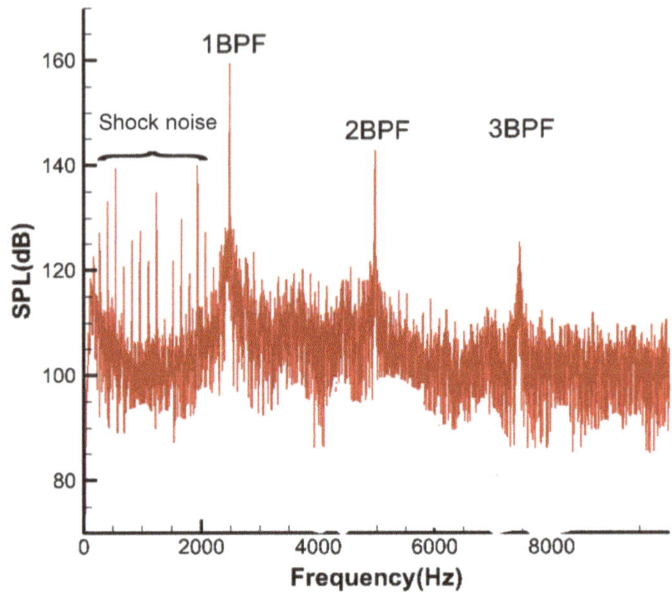

Fig. 3.11 Spectrum of a supersonic fan.

the rotor rotates at a supersonic speed, a shockwave expands out of the passage on the blade's leading edge. As illustrated in Fig. 3.12(A), The shockwave over the range of A and B almost covers the whole blade passage, which appear to be of strong strength, while the shockwave between B and C is the expanding part with much weaker strength. The weak shockwave is parallel with the Mach cone of incoming flow. The sound field is formed by this part of the shockwave propagating forward. Meanwhile, an expansion wave exists between the two near weak shockwaves, which can be attributed to the Prandtl-Meyer stream flowing around the blade. These expansion waves meet these weak shockwaves in a close range upstream, thus contributing to the attenuation of the weak shockwave. This assumption of identical blades results in uniform pressure wave pattern as shown in Fig. 3.12(B).

In practice, no matter how accurate the manufacturer is, slight differences between two blades are inevitable in the tolerance. Meanwhile, error also exists during installation, and the strength of shockwave is highly sensitive to these errors. Fig. 3.13(A) shows the real state of the flow field around the two-dimensional cascade. When the rotor, with blade number of B, rotates at a supersonic speed, expanding shockwaves of number B will also be formed. Due to these manufacturer errors and the random incoming fluctuation, the strength and direction of these expanding shockwaves

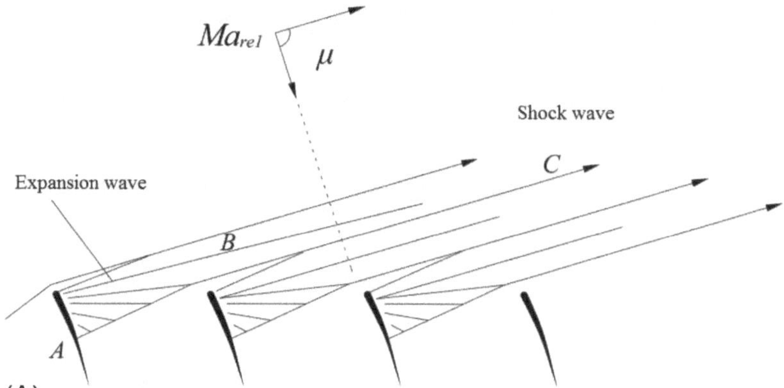

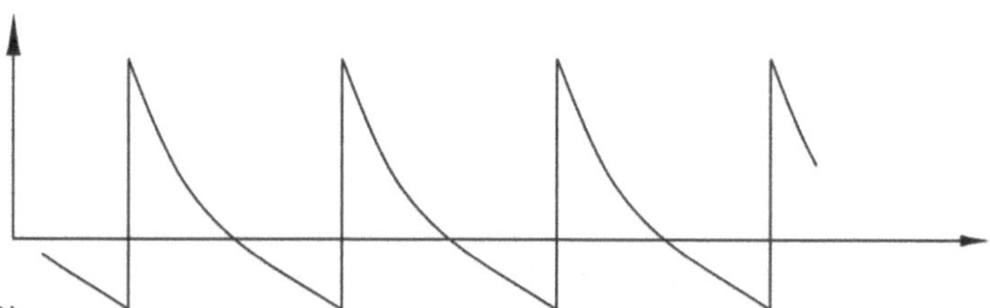

Fig. 3.12 Propagation model of shockwave in ideal supersonic cascade [42]. (A) shock wave (B) pressure wave pattern.

appear to be slightly different but still approximately the same at the beginning, and these components at the blade pass frequency and its harmonics take the main place in the spectra. Within these shockwaves spiraling forward, the slight difference between them at the initial point is also being amplified, as shown in Fig. 3.13 (B), and phenomena of chasing and merging are observed, which contributes to the gradual decrease of the number of shockwaves, as well as changes of spectra and pressure's time trace. Meanwhile, these shockwaves are strengthened through merging, and the attenuation is delayed, with the extreme situation that all shockwaves merge into a single one. Hence, measurements outside the inlet will obtain spectra of multiple pure tone noise, as shown in Fig. 3.11.

The explanation for shockwave noise is not completely rigorous, especially if the variation of shockwave strength along the radial direction is not considered. However, it is still able to quantitatively explain the generation and propagation of shockwave noise. In reality, factors like the geometry of blade, the rotor pitch, as well as the stagger angle have significant effect on noise generation and propagation.

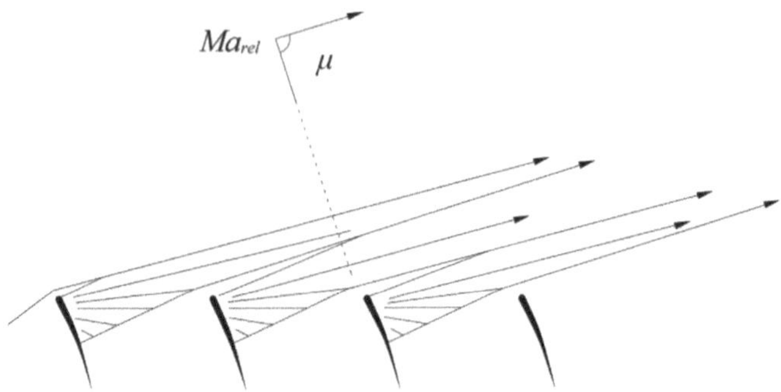

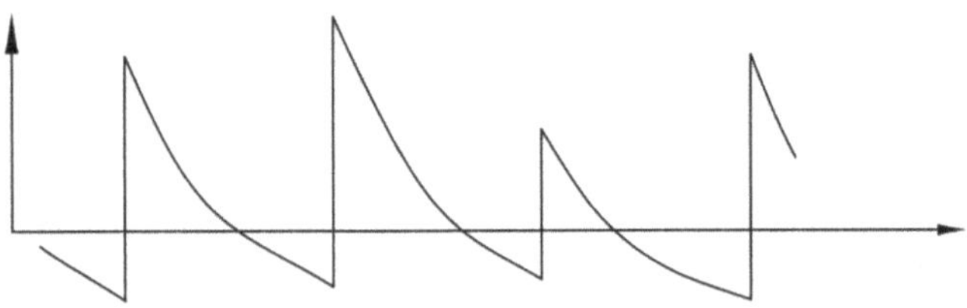

Fig. 3.13 Propagation model of shockwave in practical supersonic cascade [42]. (A) shock wave (B) pressure wave pattern.

Shockwave noise occurs repeatedly for a supersonic rotor, and shockwaves will generate on these surfaces of every blade tip, thus being an ordered process. However, the frequency distribution of the main component in shockwave noise appears to be random in a range around the axial frequency and its harmonics, which is dominated by the nonlinear characteristic of the shockwave's propagation. Although measurement can be adapted close to the cross-section of the rotor, it is still impossible for the mode of shockwave at the blade pass frequency to repeat accurately. When shockwaves radiate from the rotor, the randomness of these modes of shockwave noise is intensified. The slight irregularity of shockwaves at the initial point will be amplified due to its nonlinearity, and with the effect of non-uniform incoming flow, it becomes more challenging to predict shockwave noise accurately.

3.3.2 Shockwave noise prediction method

Generation of shockwave noise and its propagation is a typical problem of nonlinearity. In the early stage of the 1970s, Hawkings [42], Morfey [43], Kurosake [44], et al., proposed several classic one-dimensional models for shockwave noise propagation by employing different approaches, which explained the generation of the multiple pure tones in shockwave noise observed in experiments to some extent. Actually, this type of method appeared to be the expansion of the propagation theory of a sonic boom, proposed by Whitham [45] in the 1950s, and it could take practical boundary conditions into account at the inlet and outlet of the transonic rotor. Due to this defect, Hawkings [46] reported a shockwave-aeroacoustics model with the geometry of an inlet boundary taken into consideration, by using the theory of radiative sound. In 1979, following Hawking's studies, Tsai [47] included the condition with non-uniform media in his investigation. Nevertheless, due to the adoption of the symmetry assumption, both methods of Hawking and Tsai can only consider the effect of a quasi-three-dimensional boundary. In 1984, Vaidya [48] explained the problem of energy transfer between the multiple pure tones and the blade pass frequency using the theory of nonlinear resonance. However, it still lacked accurate descriptions to the essential problem of discontinuity for a shockwave. Hence, deeper discussion is still urgent for shockwave noise.

In the following content of this section, using a one-dimensional model of propagation to predict shockwave noise for transonic rotor will be introduced, to depict the nonlinear characteristic of propagation of shockwave noise.

3.3.2.1 Analytic model

Generation of shockwave noise is well related to the shockwave on the leading edge of the blades, as shown in Fig. 3.14,which contains obvious nonlinear effect. Does nonlinearity play a significant role during its propagation? In this chapter, the propagation process will be explained in mechanism through an analytic model.

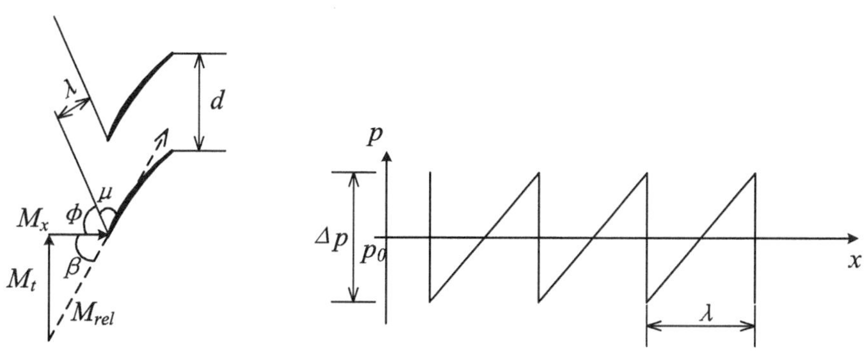

Fig. 3.14 Shockwave model for the two-dimensional cascade [43].

An analytic approach models the evolution of a complex shockwave system of practical fan/compressor as one-dimensional propagation of a detached shockwave on the leading edge of a two-dimensional cascade. The ideal gas assumption is adopted, with these thermal and aerodynamic variables being $p_0,\ \rho_0,\ T_0,\ v_0$ upstream of the shockwave and $p_1,\ \rho_1,\ T_1,\ v_1$ downstream. In a stationary coordinate, the upstream incoming Mach number is $M = v_0/c_0$, which appears to be the Mach number of the shockwave in a coordinate with a constant speed v_0. Take the parameters' relation between the two sides of the normal shockwave into account, we have:

$$\frac{p_1}{p_0} = \frac{2\gamma}{\gamma+1}M^2 - \frac{\gamma-1}{\gamma+1}, \quad \frac{\rho_1}{\rho_0} = \frac{(\gamma+1)M^2}{2+(\gamma-1)M^2}, \tag{3.58}$$

and the strength of the shockwave can be given by:

$$Z = \frac{p_1 - p_0}{p_0} = \frac{p_1}{p_0} - 1, \tag{3.59}$$

which can be also written as:

$$Z = \frac{2\gamma}{\gamma+1}\left(M^2 - 1\right), \tag{3.60}$$

Furthermore, it can be obtained that:

$$\frac{\rho_1}{\rho_0} = \frac{1 + \dfrac{\gamma+1}{2\gamma}Z}{1 + \dfrac{\gamma-1}{2\gamma}Z}. \tag{3.61}$$

The process of gas flowing through the normal shockwave will make a contribution to entropy production, and for the ideal gas with constant specific heat, the entropy production can be written as:

$$\Delta S = S_1 - S_0 = c_p \ln\left(\frac{T_1}{T_0}\right) - R \ln\left(\frac{p_1}{p_0}\right). \tag{3.62}$$

With the ideal gas equation:

$$p = \rho R T, \quad \frac{T_1}{T_0} = \frac{p_1 \rho_0}{p_0 \rho_1}, \tag{3.63}$$

and

$$c_p = \gamma c_v, \quad R = (\gamma - 1)c_v, \tag{3.64}$$

taken into the formula of entropy production, we can obtain that:

$$\frac{\Delta S}{c_v} = \ln\left[\left(\frac{p_1}{p_0}\right)\left(\frac{p_0}{p_1}\right)^{\gamma}\right]. \tag{3.65}$$

Taking Eq. (3.61) and Eq. (3.64) into Eq. (3.65), the entropy production can be rewritten as:

$$\frac{\Delta S}{c_v} = \ln\left\{(1+Z)\left[\frac{2\gamma+(\gamma-1)Z}{2\gamma+(\gamma+1)Z}\right]^{\gamma}\right\}. \tag{3.66}$$

If the shockwave is weak, which means $Z \ll 1$, the series expansion can be applied to Eq. (3.66):

$$\frac{\Delta S}{c_v} = \frac{\gamma^2-1}{12\gamma^2}Z^3 - \frac{\gamma^2-1}{8\gamma^2}Z^4 + \cdots \approx \frac{\gamma^2-1}{12\gamma^2}Z^3, \tag{3.67}$$

thus it can be obtained that:

$$\Delta S = \frac{R(\gamma+1)}{12\gamma^2}Z^3, \tag{3.68}$$

which shows the entropy production of unit fluid mass after it flows through the shockwave. In the one-dimensional duct with same cross-section area, the entropy production of fluids on the unit area in the time of t, after they past the shockwave, comes out to be:

$$\Delta S = \frac{R(\gamma+1)}{12\gamma^2}Z^3\rho_0 v_0 t. \tag{3.69}$$

For the undisturbed media in front of the shockwave, the speed of sound propagation is c_0, and $v_0 = c_0$. During the time of dt, the entropy production, dS, can be calculated as:

$$\frac{dS}{dt} = \rho_0 c_0 R\frac{(\gamma+1)}{12\gamma^2}Z^3, \tag{3.70}$$

yet on the other hand,

$$dS = \frac{de + pdv}{T}. \tag{3.71}$$

For the undisturbed media, $dq = de$, thus $dS = -de/T$. Consider this relation in Eq. (3.70); the entropy production of fluids passing the unit area of the shockwave, in the time of dt, can be expressed as:

$$\frac{dS}{dt}=\rho_0 c_0 R\frac{(\gamma+1)}{12\gamma^2}Z^3=-\frac{1}{T_0}\frac{dE}{dt}. \tag{3.72}$$

By applying the external energy formula of a sawtooth wave:

$$E=p_0 Z^2\lambda/12\gamma,\quad \frac{dE}{dt}=p_0\frac{\lambda}{12\lambda}2Z\frac{dZ}{dt}. \tag{3.73}$$

Substituting Eq. (3.73) into Eq. (3.72) leads to:

$$\frac{dZ}{dt}=-\frac{\gamma+1}{2\gamma}\frac{c_0}{\lambda}Z^2, \tag{3.74}$$

and it is also:

$$\frac{d}{dt}\left(\frac{1}{Z}\right)=\frac{\gamma+1}{2\gamma}\frac{c_0}{\lambda}. \tag{3.75}$$

Assume the initial strength of the shockwave is Z_s, and at the moment t, it can be expressed as:

$$\frac{1}{Z}-\frac{1}{Z_s}=\frac{c_0}{\lambda}t\frac{\gamma+1}{2\gamma}. \tag{3.76}$$

The axial location of the shockwave is denoted by x, at the moment t. From Eq. (3.76), the shockwave strength can be written as:

$$\frac{\Delta p}{p_0}=Z=\left[\frac{1}{Z_s}+\frac{c_0 t}{\lambda}\frac{\gamma+1}{2\gamma}\right]^{-1}. \tag{3.77}$$

The shockwave travels toward its normal direction by the sound speed, which varies with time and the shockwave strength, whereas the Mach number of this speed keeps constant as 1. When the shockwave travels to the location x upstream of the rotor, its axial Mach number becomes:

$$M_u=\sin\varphi-M_x. \tag{3.78}$$

As shown in Fig. 3.14, With the geometrical relation $\varphi+\beta+\mu=\pi$, in which φ denotes the angle formed by the shockwave; the axial direction, β denotes the relative flowing angle; and μ denotes the Mach angle, it can be written as:

$$\cos\beta=M_x/M_{rel}\quad,\ \sin\mu=1/M_{rel}. \tag{3.79}$$

By using the relation of trigonometric functions, it can be solved that:

$$\begin{cases}\sin\varphi=\dfrac{1}{M_{rel}^2}\left[M_x+M_t\left(M_{rel}^2-1\right)^{1/2}\right]\\[2mm] \cos\varphi=\dfrac{1}{M_{rel}^2}\left[M_t+M_x\left(M_{rel}^2-1\right)^{1/2}\right]\end{cases}. \tag{3.80}$$

Taking these two formulas into Eq. (3.78), we can obtain:

$$M_u = \frac{M_t}{M_{rel}}\left(1 - \frac{1}{M_{rel}^2}\right)^{1/2} - M_x\left(1 - \frac{1}{M_{rel}^2}\right), \tag{3.81}$$

in which $M_t = (M_{rel}^2 - M_x^2)^{1/2}$. At the location away from the rotor by x, $\lambda/(c_0t) = \lambda M_u/x$ and $\lambda = d\cos\varphi$. Hence, we can write:

$$\frac{\lambda}{c_0 t} = \frac{d}{x}\cos\varphi(\sin\varphi - M_x) = \frac{d}{x}F(M_{rel}, M_x), \tag{3.82}$$

and

$$F(M_{rel}, M_x) = \frac{\left(M_{rel}^2 - 1\right)^{1/2}}{M_{rel}^4}\left[M_x\left(M_{rel}^2 - 1\right)^{1/2} - M_t\right]^2, \tag{3.83}$$

in which $F(M_{rel},M_x)$ is called the propagation function of the shockwave. Take Eq. (3.82) into Eq. (3.77), thus it can be written as:

$$\left(\frac{\gamma + 1}{2\gamma}\right)Z = \frac{Fd}{x}\bigg/\left(1 + \frac{2\gamma}{\gamma + 1}\frac{Fd}{xZ_s}\right), \tag{3.84}$$

and it reveals the nonlinear propagation of the shockwave. When x becomes larger, the propagation equation reads:

$$\left(\frac{\gamma + 1}{2\gamma}\right)Z \approx \frac{Fd}{x}. \tag{3.85}$$

It can be found that, when the shockwave travels a far distance as denoted by x, the strength will decay at a ratio of inverse x, also becoming irrelevant with the initial strength. This derivation, the simplified approach for the problem of shockwave noise, appears to be as following:

(1) First, the shockwave is assumed to only generate around the blade tip. Then one circumferential cross-section can be designated, thus the rotor is modeled at an annular cascade.

(2) With circumferential expansion, the model can be further simplified as a two-dimensional cascade.

(3) Consider the problem at a rotating coordinate, and it becomes a steady flowing condition.

(4) By this step, an analytic solution is still impossible if the steady flowing problem around a flat cascade is considered. Further simplification is needed. The shockwave system, generated on the leading edge of the rotor, will decay gradually during its upward propagation, which leads to an analogy between this problem and the condition of a shockwave moving in a one-dimensional duct with constant cross-section area. The one-dimensional coordinate is constructed following the normal direction of the shockwave, thus the problem can be modeled as one-dimensional and unsteady. It should be noted here that this analogy is not a completely rigorous simplification, because for a physical process in practice, the shockwave appears to be stationary with respect to the rotor, and no propagation exists in a rotating frame, whereas these analytic methods describe the process as a condition where the shockwave travels upward in the rotating frame, which it is consistent with the real process. The motion of a shockwave in an one-dimensional duct with constant cross-section area comes out to be a typical problem, and various proposals, including the SOD shockwave tube and the Riemann problem, appear to be descriptions for this problem [49]. Moreover, it is also widely regarded as a Benchmark problem to check the ability of shockwave capture for a differential scheme.

(5) It becomes much easier to model this problem after the simplification to a one-dimensional case. Meanwhile, some other assumptions, such as the duct having the same cross-section area, ideal gas conditions, and weak shockwaves, can be understood easily, which we will not pay much attention to.

Although such a simplified model is adopted, some basic rules can still be obtained through it. The shockwave noise will decay gradually during its upward propagation with a ratio of inverse x, which denotes the distance it travels. Later numerical investigations for the shockwave noise reported consistent results with these by analytic methods. These studies by Prasad, et al. [50], of Pratt & Whitney showed that the computational fluid dynamics (CFD) results for the two-dimensional flat cascade matched well with the analytic solution; such was also for the three-dimensional case.

Regular sawtooth is highlighted in this model, with the initial wave (also the initial shockwave strength in this problem) known. The effect of linear is not included. The ability of accurate prediction for this model is limited due to neglects of the variation of the cross-section area, the non-uniformity of the background flow, and its three-dimensionality. Since the 1970s, not much dramatic advancement has been achieved for the model of shockwave noise. However, in recent years, Fisher and McAlpine, et al. proposed another new analytic model [51–55] that attracted much attention. Compared with the previous analytic methods, the new model is able to include the irregularity of a sawtooth wave, thus multiple pure tones can be predicted by it, as well as the interaction between acoustic liner and the shockwave noise. However, the initial wave information still needs to be obtained through experiments. These studies were well supported by Rolls-Royce.

The model, proposed by Fisher and McAlpine, first considers the problem in the time domain. In their model, the moving velocity of the shockwave, v_s, is calculated by mean pressures on two sides of the shockwave as:

$$\begin{cases} v(p) = a_0 + \dfrac{\gamma + 1}{2\rho_0 a_0} p \\ v_s(p_m) = a_0 + \dfrac{\gamma + 1}{2\rho_0 a_0} p_m \end{cases} . \tag{3.86}$$

For a regular shockwave system, each shockwave holds the same and constant velocity, and all of them keep their initial regular shape of sawtooth wave thereafter. However, for an irregular system, traveling velocities vary for these irregular shockwaves, leading to chasing and merging between them. The method shown in Fig. 3.15 is adopted to deal with merging shockwaves in this new model. Hence, after processes of shockwave propagation and merging are descripted, the analytic model in the time domain becomes complete.

Fisher and McAlphine, et al. paid much attention to the frequency-domain model. When the sound field is decomposed in a frequency domain with various modes obtained, it becomes easier to model the interaction between the liner and different sound modes, as well as the cut-off characteristic of each mode. Consider the problem in a circular-cylinder duct, where there exists uniform background flow of Mach number M_a, as shown in Fig. 3.16.

Because the flow field appears to be steady in the rotating coordinate, we can draw the conclusion that, if one specific tone component in the shockwave noise coincides to be the mth harmonic of the axial frequency, the number of its circumferential modes appears to be m. Meanwhile, because the relative Mach number of blade tip is only slightly larger than 1, these second and higher circumferential modes will be cut off. Therefore, it can be assumed that the $(m, 1)$ mode, the lowest circumferential one,

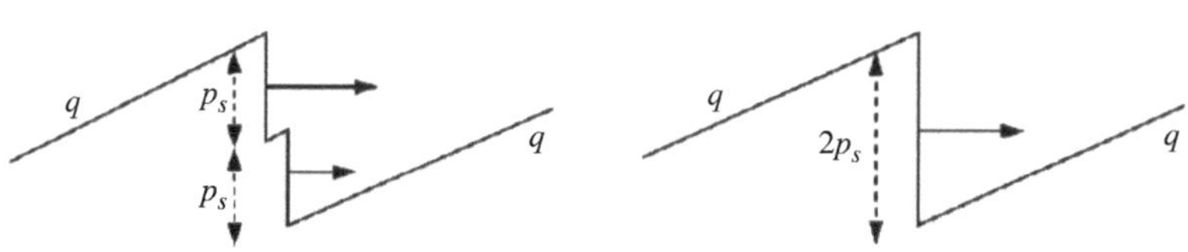

Fig. 3.15 Process for shockwave merging [54].

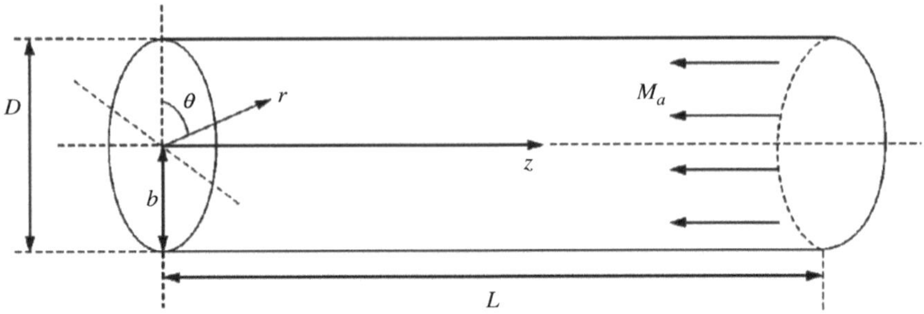

Fig. 3.16 Schematic of the geometry for the model in frequency domain [54].

contains the most sound energy. Due to the nonlinearity during the propagation of the shockwave noise, the nonlinear wave equation should be applied to model this problem:

$$\frac{\partial p}{\partial t}+\frac{\gamma+1}{2\rho_0 a_0}p\frac{\partial p}{\partial x}=0. \tag{3.87}$$

Following non-dimensional parameters are adopted:

$$\begin{cases} T=\dfrac{a_0 t}{\lambda} \\ X=\dfrac{2\pi x}{B\lambda} \\ P=\left(\dfrac{\gamma+1}{2\gamma}\right)\dfrac{p}{P_0} \end{cases}, \tag{3.88}$$

and the nonlinear wave equation can be written as:

$$\frac{\partial P}{\partial T}+\frac{2\pi}{B}P\frac{\partial P}{\partial X}=0. \tag{3.89}$$

With the irregular sawtooth wave expanded as a Fourier series:

$$P(X,T)=\sum_{m=-\infty}^{\infty} C_m(T)e^{\mathbf{i}mX}. \tag{3.90}$$

Eq. (3.89) can be rewritten as:

$$\frac{dC_m}{dT}=-\frac{\mathbf{i}m\pi}{B}\left(\sum_{l=1}^{m-1} C_{m-l}C_l+2\sum_{l=m+1}^{\infty} C_l\widetilde{C}_{l-m}\right). \tag{3.91}$$

It should be noted that the propagation loss is not included in Eq. (3.91). The nonlinear items in the governing equation can present the energy transfer process between different modes, but the sound decay during propagation is not included. To overcome this disadvantage, a dissipative item is added into Eq. (3.89), and the new governing equation is obtained as Burgers equation:

$$\frac{\partial P}{\partial T}+\frac{2\pi}{B}P\frac{\partial P}{\partial X}=\frac{\varepsilon}{B^2}\frac{\partial^2 P}{\partial X^2}. \tag{3.92}$$

The coefficient of viscosity ε added into the equation is impossible to be written out through the governing equation or the physical mechanism, but rather by numerical simulations. With the analytic solution regarded as the accurate solution, results by the frequency-domain method can be consistent with these through calculations in the time domain.

3.3.2.2 Numerical model

The advantage of analytic methods is that an accurate propagation characteristic can be obtained through less computational cost. Nevertheless, obvious defects exist in all these analytic models. Therefore, it is urgent to construct an analytic model with as many as possible practical physical conditions considered.

On the other hand, due to the fast advancement of computer hardware, numerical approaches have been employed for the problem of shockwave noise propagation. By solving the complete governing equation through numerical methods, the propagation process of shockwave noise can be depicted thoroughly without limitation of various assumptions in analytic models. If we only consider the multiple pure tones after merging without taking the difference between various shockwaves into account, results can be achieved by solving these steady flow equations in one blade passage. With the CFD techniques stepping into a mature stage, studying the characteristic of shockwave noise through numerical approaches is highly achievable, which is widely accepted in engineering design of the industry.

Because the shockwave system on the leading edge of fan appears to be steady in the rotating frame, it becomes a rare sample of aeroacoustics problems that can be simulated by the steady CFD technique, which is much more mature than the unsteady approach. Therefore, it has already been a regular analysis tool applied widely in aeroengine design.

Governing equation

Obvious three-dimensionality and viscosity are observed in flow fields of turbomachinery. Reynolds Average Navier-Stokes (RANS) is mostly applied in the steady CFD simulation. Even through solving inviscid Euler equation, shockwave still can be well simulated, thus the viscous effect is ignored during the construction of these analytic models for shockwave noise. The generation of shockwave noise, as well as its propagation, is treated as inviscid processes. Investigations by G.E.(General Electric) and P.W.(Pratt & Whitney) show that the practical mechanism can be captured through inviscid calculations, and the results are almost the same of those by viscous simulations. Fig. 3.17 is cited from the literature [50], which shows both two-dimensional results by the inviscid and the viscous calculations, respectively, as well as the analytic solution. The dashed lines and the solid lines represent the decay process of the shockwave with respect to the traveling distance, predicted by the analytic approach and CFD, respectively. If these two curves are tangential to each other, as shown in Fig. 3.17, we can say the results by the different approaches are consistent. It can also be found that curves predicted by the viscous and the inviscid approaches match closely, further indicating that viscosity plays a nonsignificant role in the main physical mechanism of shockwave noise.

Fig. 3.18 shows the results of three-dimensional viscous and inviscid cases. It can be found that shockwave decay in three dimensions also present a similar trend with the quasi-one-dimensional analytic solution. However, the obvious difference is observed between the present and the two-dimensional case, in the initial strength of shockwave and the decay process, which suggests that three-dimensionality has a great effect on the characteristic of shockwave noise. Therefore, numerical calculations for engineering applications are all performed under the three-dimensional condition [56–58].

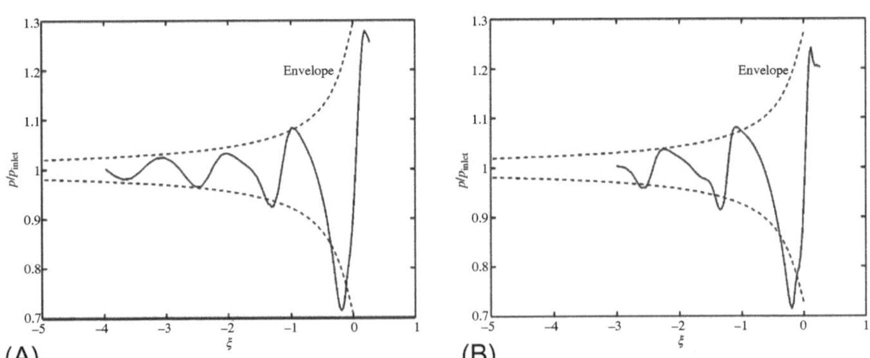

Fig. 3.17 Results by two-dimensional viscous and inviscid computation [50]. (A) 2D inviscid, (B) 2D viscous.

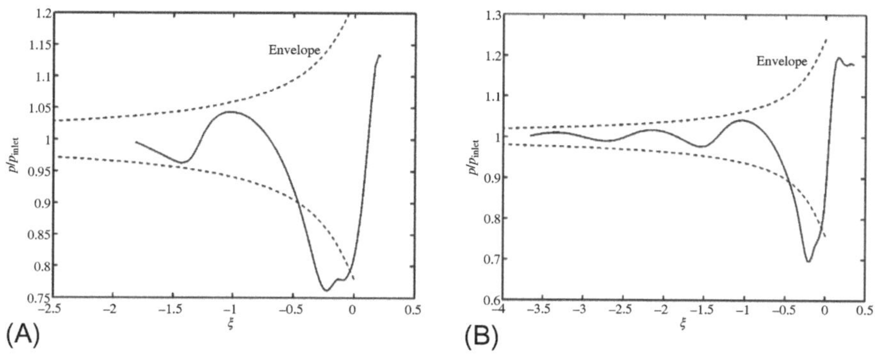

Fig. 3.18 Results by three-dimensional viscous and inviscid computation [50]. (A) 3D viscous, (B) 3D inviscid.

Because viscosity is a secondary factor in shockwave noise, both the complete and inviscid governing equations are accepted in the practical application. It should be noted that, in the inviscid calculation, due to the lack of blocking by viscous mechanism, the rate of flow solved out to be higher than the practical value. Therefore, the back pressure needs to be adjusted to assure that the working condition in the calculation meets the real working line. Similar adjustment for the back pressure is also necessary for the viscous calculation, which appears to be slighter than that in the inviscid case. After the working line is appropriately adjusted, the results by the inviscid and viscous approaches closely match each other.

Differential scheme

Two essential problems are involved in these differential schemes concerning shockwave noise: the first one is how to capture the discontinuous phenomenon, or shockwave, and the latter one appears to be how to suppress the dissipation and the dispersion of differential scheme with the fluctuation of shockwave noise.

It is unexpected that, seen from these published literatures, the second-order central differential scheme with the second-order and the fourth-order artificial viscosity is widely accepted by most investigators, although various total variation diminishing (TVD) schemes can be used for shockwave capture, and computational aeroacoustics (CAA) schemes are effective for the suppression of numerical dissipation and dispersion. As can be found in Figs. 3.17 and 3.18, the widely applied numerical scheme is not qualified for the resolution of shockwave but provides consistent results with analytic methods.

Computational domains, grids, and boundary conditions

Results in published literature, by numerical simulations and experiments for fan, show that the OGV of the fan has little effect on shockwave noise. Furthermore, the literature [56] reported that the effect from the blade tip clearance as well as the shunt ring can also be ignored through the comparison between prototypes with different geometries. Therefore, it is possible for calculations of the fan to only contain a single rotor fan. To give a combinative estimation of the shockwave noise and the aerodynamic performance for the fan, these ignored geometrical factors can be again included in the computational domain.

The propagation process of the shockwave noise appears to be that, after its generation on leading edges of these blades, it will travel upward and past the nacelle lip before it is radiated to the far field. In numerical studies, for different needs, the geometrical model for the propagation path of shockwave noise also varies with each other.

One type is to place a length of duct with a constant cross-section area upward of the fan, to simulate the propagation of the shockwave noise in the duct to reveal its propagation mechanism, which is similar with practical aeroengine experiments on land. It can be found by the comparison between the results of experiments and numerical predictions that some other factors can also be excluded, such as the variation of the incoming flow Mach number, in the aeroacoustics estimation for a fan. Studies in the literature [50, 56, 59] applied this type of computational domain, and Fig. 3.19 shows the computation domain employed in the literature [54].

Another type is to depict the detailed geometry information of the nacelle to take the aerodynamic effect of the inlet into account. This setting is more close to the real installation state of an aeroengine because the shockwave noise under the state of practical installation can be assessed

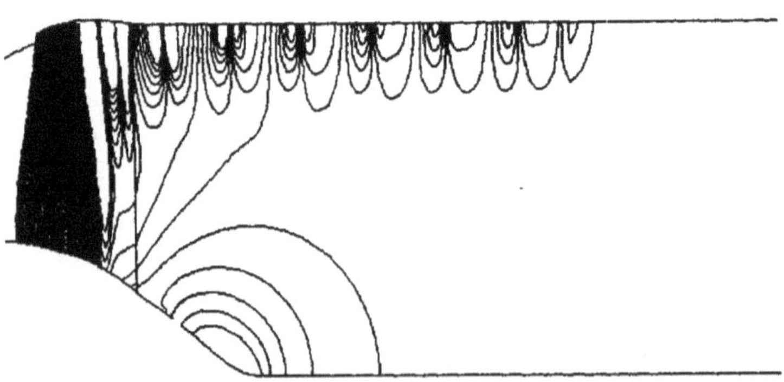

Fig. 3.19 Computational domain with duct of a constant cross-section area [59].

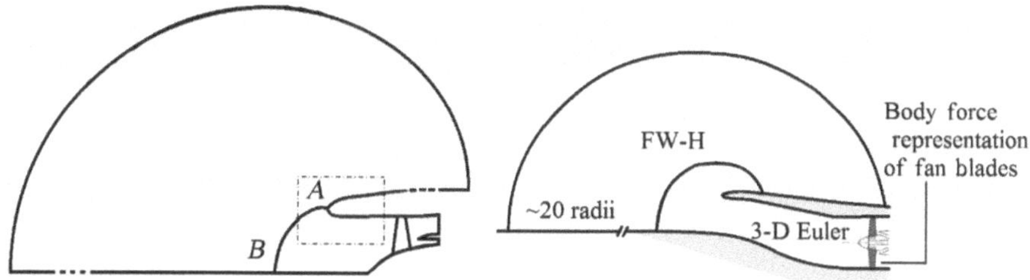

Fig. 3.20 Computational domain with the shape of inlet considered [60].

quantitatively when seeking the integral design of the aircraft and its engine, as well as the following optimization. This approach was applied in the literature [60], the computational domains of which are shown in Fig. 3.20.

There appears to be no essential variation between the numerical simulations for the shockwave noise and traditional aerodynamics in turbomachinery, therefore the principles for grids generation did not receive much modification. Most studies pay less attention to the grid topology and the criterion of quality, instead of the grid density. For the second-order central differential scheme, low dispersion needs to be achieved by setting 25 grid points per wavelength [60]. Shockwave noise does not come out to be a classic wave problem. By numerical studies, it was found that to precisely predict the shockwave noise, the grid points per wavelength could not be less than 18 [50]. The density criterion accepted by the literature [57] is 30–40 grid points per axial wavelength.

Boundary condition plays a significant role in aeroacoustics computations, in which a practical soundwave is radiated freely through the boundaries of the computational domain. Therefore, a non-reflecting boundary condition is needed in the computation. Universal and applicative boundary condition for various flowing states, such as viscous flow and non-uniform background flow, is almost impossible. For the shockwave noise, incoming flow at upward positions far from the rotor can be considered as uniform. Based on this assumption, the non-reflecting boundary condition proposed by Giles could be employed [61]. The computational domain used in the literature [57] is shown in the left section of Fig. 3.15, in which Giles' boundary condition was also applied.

3.3.3 Power computation of shockwave noise

If the shockwave shape is regular, numerical calculation can be performed in one blade passage, whereas multiple passages are required in the computational domain if the shape shows to be irregular. In this subchapter, how to estimate the strength of shockwave noise after the CFD result in one blade passage is obtained will be introduced. For multiple blade passages and the whole annulus, only the parameter of blades number is modified.

The vector of sound strength in non-uniform flow is:

$$\mathbf{I} = \left[\overline{\mathbf{u}} \cdot \mathbf{u}' + \frac{p'}{\overline{\rho}}\right] \left[\overline{\rho}\mathbf{u}' + \frac{\overline{\mathbf{u}}\overline{\rho}}{\gamma \overline{p}} p'\right], \tag{3.93}$$

and the sound power at one specific cross-section can be written as:

$$P = \int_S \overline{\mathbf{I}} \cdot d\mathbf{S}, \tag{3.94}$$

in which

$$\overline{\mathbf{I}} = \frac{1}{T_0} \int_{t_0}^{t_0 + T_0} \mathbf{I} dt, \tag{3.95}$$

becomes the time-average value of sound strength. The coordinate is set to rotate with the rotor, and we have the coordinate transformation as follows:

$$d\theta = \Omega dt, \quad T_0 = \frac{2\pi}{B\Omega}. \tag{3.96}$$

In this transformation, Ω denotes the angular frequency of the blade and B denotes the blade number. Take them into Eq. (3.95), which defines the time-average sound strength, and it can be obtained that:

$$\overline{\mathbf{I}} = \frac{B\Omega}{2\pi} \int_{\theta_0}^{\theta_0 + 2\pi/B} \mathbf{I} \frac{d\theta}{\Omega} = \frac{B}{2\pi} \int_{\theta_0}^{\theta_0 + 2\pi/B} \mathbf{I} d\theta. \tag{3.97}$$

Consider this expression in Eq. (3.94) and assume that the cross-section S for the sound power calculation is perpendicular to the x-axis, we can obtain the sound power expressed as:

$$\begin{aligned} P &= \int_S \overline{\mathbf{I}} \cdot d\mathbf{S} \\ &= \frac{B}{2\pi} \int_S \int_{\theta_0}^{\theta_0 + 2\pi/B} \left[\overline{\mathbf{u}} \cdot \mathbf{u}' + \frac{p'}{\overline{\rho}}\right] \left[\overline{\rho} u_x{}' + \frac{\overline{u}_x \overline{\rho}}{\gamma \overline{p}} p'\right] d\theta dS \\ &= \frac{B}{2\pi} 2\pi \int_r r dr \int_{\theta_0}^{\theta_0 + 2\pi/B} \left[\overline{\mathbf{u}} \cdot \mathbf{u}' + \frac{p'}{\overline{\rho}}\right] \left[\overline{\rho} u_x{}' + \frac{\overline{u}_x \overline{\rho}}{\gamma \overline{p}} p'\right] d\theta \\ &= B \int_r r dr \int_{\theta_0}^{\theta_0 + 2\pi/B} \left[\overline{\mathbf{u}} \cdot \mathbf{u}' + \frac{p'}{\overline{\rho}}\right] \left[\overline{\rho} u_x{}' + \frac{\overline{u}_x \overline{\rho}}{\gamma \overline{p}} p'\right] d\theta, \end{aligned} \tag{3.98}$$

in which the time-average value of each parameter can be computed by:

$$\overline{\phi} = \frac{B}{2\pi}\int_{\theta_0}^{\theta_0 + 2\pi/B} \phi d\theta. \tag{3.99}$$

By integrating flowing variables in one blade passage using Eqs. (3.98) and (3.99), the sound power of shockwave noise can be thus obtained.

3.4 Combustion noise

Combustion noise can be classified into two categories, i.e., "direct noise" that results from combustion processes associated with unsteady heat release, and "indirect noise" that originates from temperature or entropy fluctuation passing through regions of rapidly accelerated mean flow [7, 62–64].

The classical investigation of direct combustion noise was conducted by Thomas and Williams [65]. They filled soap bubbles with a homogeneous reactive mixture and recorded the pressure field radiated by the burning of each isolated bubble, as shown in Fig. 3.21. The mixture was ignited by

Fig. 3.21 Instantaneous snapshot of spherically expanding flame after ignition in time [7].

spark at the center. Because the bubble size is much smaller than the acoustic wave length emitted by the flame, the ignited mixture can be seen as a monopole source of sound. They showed that the pressure fluctuation at a distance of d from the source can be given by the expression:

$$p' = \frac{\rho_\infty}{4\pi d}\frac{d}{dt}\left(\frac{dV}{dt}\right), \tag{3.100}$$

where p' denotes the pressure fluctuation induced by the combustion, ρ_∞ the density of medium through which acoustic wave propagates, and V the volume of the bubble. Thus, the sound was produced by volumetric expansion effect induced by combustion according to the previous equation.

Based on Eq. (3.100), turbulent flame can be assumed equivalently as distributed monopole sound sources with different phases [66]. Therefore, the pressure fluctuation in the soundwaves that emanate from the ith in the far field can be expressed as:

$$p_i'(t) = (\rho_\infty/4\pi d)(E-1)[dq_i/dt]_{t-\tau}, \tag{3.101}$$

where d is the distance from the source, ρ_∞ the density of surrounding medium, and q_i the volume rate of consumption of mixture in the ith flame element, whereas τ represents the retarded time taken for the soundwave to propagate from the source to the observation point. For combustion at constant pressure, E denotes the mean density ratio of unburnt and burnt gases. This equation assumes that the distance from the source is much larger than the wavelength, therefore the noise generated by the flame can be calculated by summing up all the noises emitted by different flame elements.

Combustion theory is further improved by Strahle [67], Dowling, and Ffowcs Williams and Crighton [68, 69] by using Lighthill's analogy [70]. The far field noise then can be expressed as:

$$p'(\vec{x}, t) = \frac{\gamma - 1}{4\pi c_0^2 |\vec{x}|}\int_v \ddot{q}(\vec{y}, t - |\vec{x}|/c_0)\, d^3\vec{y}, \tag{3.102}$$

where $\vec{x}$ is considered far field, v denotes the volume that encloses the combustion region, and the combustion source is assumed to be compact. This equation clearly explained that the direct combustion noise is associated with the temporal change of heat release, which is closely related to the dynamics of combustion.

When the combustion processes take place in a confined environment, more phenomenon occur associated with interactions between the entropy wave and acoustic wave, the acoustic wave and flames, and so forth.

And when a feedback is established between acoustic waves emitted by combustion and unsteady combustion, the noise radiation is tuned onto one of the resonance frequencies. Otherwise, the noise radiation is broadband.

Usually, in practical applications, the disturbances in a combustor can be described by an acoustic disturbance, an entropic disturbance, and a vertical disturbance independent of each other. With no heat source, the acoustic disturbance can be obtained by solving the wave equation:

$$\left(\frac{1}{c_0^2}\frac{D^2}{Dt^2}-\nabla^2\right)p'=0, \tag{3.103}$$

where D/Dt is the convective derivative. The entropic disturbance and vertical disturbance convect with the mean flow and can be described by:

$$\frac{D\vec{\xi'}}{Dt}=0. \tag{3.104}$$

The disturbances upstream and downstream of the heat source can be related by conservation laws, based on which the investigation about the effect of combustion noise on thermoacoustic instability and the ratio between indirect noise and direct noise can be carried out [7, 71, 72].

3.5 Jet noise

When rapid mixture occurs between the high-speed flow and stationary media, turbulence with violent fluctuation will be formed, thus contributing to so-called jet noise, which is widely met not only in jet aircraft and rockets with sound power higher than 100 million watts, but also in compression systems, like the discharged steam of a boiler and the exhaust of an internal combustion engine. Moreover, noise from high-speed flow in a gas pipe, water pipe, and valve also belongs to this type. Therefore, generation mechanism of jet noise, as well as its reduction approach, have always been one of the most attractive branches since aero-acoustics was born.

Lighthill's sound analogy along with the classic sound theory constructed the fundamental of aeroacoustics [4, 74]. Meanwhile, through the application of Lighthill's theory, the physical and mathematical models were established for high-speed jet noise, which indicates that jet noise arose by turbulence and shows a strong characteristic of quadrupole, the strength of which is proportional to the eighth order of jet speed. These analyses have provided us with a basic understanding of Lighthill's theory [74].

In the early stage of the 1970s, more rigorous and systematic experimental studies were performed for the investigation of jet noise, with enormous experimental data published [75–77] that verified the validity of Lighthill's theory. However, some phenomena found in experiments could not be predicted by the theory. For example, the total sound power predicted by Lighthill's model was consistent with these experimental results, whereas an obvious gap existed for the directivities obtained by these two approaches, especially when the observer was located within 0–30 degrees with respect to the jet central axis. What leads to this gap between the experiments and Lighthill's theory thus became a popular issue later. It is generally explained that, in theoretical prediction, although the sound source is moving, the media is still assumed to be stationary. This naturally excludes the interaction between a soundwave and background flow in the jet range, which contributes to the gap. Therefore, many studies [78–80] tried to establish a more general theory of sound analogy to include the effect of non-uniform or shear background flow using some treatments. Among all these models, a study in the literature [7] achieved a better result, which is called Lilley's equation today. However, all of these models appeared to be quite complex for a complete analytical solution with direct physical explanations; for example, Lilley's equation only obtained the asymptotic solutions at low and high frequencies. In recent years, new approaches have been employed for more general cases [81].

On the other hand, with the advancement of computational fluid dynamics, directly solving jet noise has become possible. In fact, Musafir [81] applied the method of a renormalization group to compute the turbulent flow and obtained the far-field solution by the Kirchhoff Method. In the literature [82], the RDT(Rapid Distortion Theory) method was employed to calculate the turbulent near-field flow, and the sound field was then solved through matching with the flow at far field. It can be seen that more factors are considered in these studies, thus deepening the physical understanding to the jet noise. More recently, more computational approaches can be found in many references [83–86].

This subchapter will pay much attention to the physical mechanism of jet noise, and some fundamental numerical approaches will be introduced.

3.5.1 Solution of Lighthill's equation

As can be known from Eq. (1.102), when no solid boundary exists, the second and the third items will be eliminated. Therefore, the Lighthill' equation can be written as

$$\rho' = \frac{1}{c_0^2} \int\limits_{v(\tau)} T_{ij}(y, \tau) \frac{\partial^2 G}{\partial y_i \partial y_j} dy, \tag{3.105}$$

where

$$G = \frac{\delta(\tau - t + R/c_0)}{4\pi R}.$$

3.5.1.1 Stationary media

Due to:

$$\frac{\partial G}{\partial x_i} = -\frac{\partial G}{\partial y_i},$$

Eq. (3.105) can be rewritten as:

$$\begin{aligned} \rho' &= \frac{1}{\partial x_i \partial x_j} \int\limits_{v(\tau)} T_{ij}(y, \tau) \frac{\partial \delta(\tau - t + R/c_0)}{4\pi R c_0^2} dy \\ &= \frac{1}{\partial x_i \partial x_j} \int\limits_{v(\tau)} \frac{T_{ij}(y, t - R/c_0)}{4\pi R c_0^2} dy. \end{aligned} \tag{3.106}$$

Eq. (3.106) is the fundamental formula we need. It can be seen that there is no spatial differential of the sound source, which simplifies further derivation. Actually, using Eq. (3.106) for the calculation still appears to be extremely complex. Therefore, we consider the mathematical relation:

$$\frac{\partial}{\partial x_i} \frac{T_{ij}}{R} = -\frac{R_i}{R^3} \left[T_{ij} \right] + \frac{R_i}{c_0^2 R} \frac{\partial T_{ij}}{\partial t},$$

and with the high-order item of $1/R$ ignored, we can obtain:

$$\rho' = \frac{1}{4\pi c_0^2} \frac{\partial^2}{\partial x_i \partial x_j} \int\limits_{v(\tau)} \frac{T_{ij}}{R} dy = \frac{1}{4\pi c_0^2} \int \frac{R_i R_j}{R^3 c_0^2} \frac{\partial^2 T_{ij}}{\partial t^2} dy. \tag{3.107}$$

For the far field, Eq. (3.107) can be further simplified as:

$$\rho'(x, t) = \frac{1}{4\pi c_0^2} \frac{x_i x_j}{x^3} \int\limits_{v(\tau)} \frac{\partial^2 T_{ij}}{\partial t^2} dy, \tag{3.108}$$

in which $x = |x_i|$. By applying this equation, the model of jet noise can be easily constructed.

3.5.1.2 Effect of convective velocity

Assume the sound source moves downstream at a specific velocity, and we can write out the solution of Eq. (3.105) as:

$$\begin{aligned} \rho' &= \frac{\partial^2}{\partial x_i \partial x_j} \int\limits_{v(\tau)} T_{ij}(y,\tau) \frac{\partial \delta(\tau - t + R/c_0)}{4\pi R c_0^2} dy \\ &= \frac{\partial^2}{\partial x_i \partial x_j} \int\limits_{v(\tau)} \frac{T_{ij}[y(\tau), t - R/c_0]}{4\pi R c_0^2 |1 - M_c \cos\theta|} dy. \end{aligned} \tag{3.109}$$

where $M_c = \frac{dy}{c_0 d\tau}$ and θ denotes the angle formed between the directions of the source's motion and observer.

If M_c keeps a constant, it can be found from formulas in Chapter 1 that Eq. (3.109) can be expressed as:

$$\rho' = \int\limits_{v(\tau)} \frac{(x_i - y_i)(x_j - y_j)}{4\pi R^3 c_0^4 (1 - M_c \cos\theta)^3} \frac{\partial^2 T_{ij}}{\partial \tau^2} dy.$$

If we let $x_i - y_i \approx x_i$, $x_j - y_j \approx x_j$ and take it into the previous formula, then it can be rewritten as:

$$\rho' = \frac{x_i x_j}{4\pi R^3 c_0^4 (1 - M_c \cos\theta)^3} \int\limits_{v(\tau)} \frac{\partial^2 T_{ij}}{\partial \tau^2} dy. \tag{3.110}$$

Comparing Eq. (3.110) with its original form, the Doppler factor due to the source's motion is added. If the convective velocity M_c appears to be of low values ($M_c \approx 0$), the result of Eq. (3.110) is consistent with that of Eq. (3.108).

3.5.2 Prediction of jet noise

As can be found from Eq. (3.110) or Eq. (3.108), if we want to predict jet noise, Lighthill's stress tensor must be known first. Therefore, deep understanding of flow structures is required.

3.5.2.1 A simplified jet structure

The structure of a real jet is quite complicated with multiscale characteristics. As a typical analysis approach, some simplification to jet structure is necessary. As shown in Fig. 3.22, the jet formed by a circular duct with diameter D is described as three zones, which can be characterized by the fully

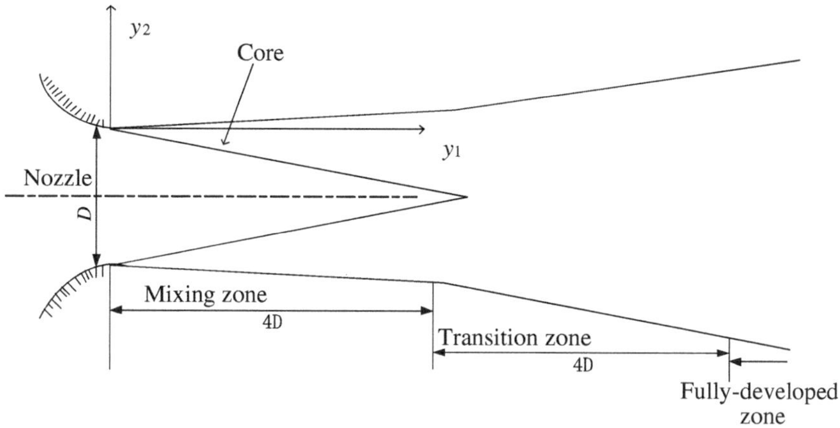

Fig. 3.22 Schematic of the jet from the circular outlet to stationary air [86].

developed jet: the mixing zone, the transition zone, and the fully developed zone. Fig. 3.22 shows the case when fluids are injected into stationary air. The mixing zone only contains one limited part of turbulence, and the velocity in the core keeps a constant, equaling the outflowing velocity. The length of the core appears to be almost 4.5D. With the cross-section area of jet becoming larger, it is different to determine its boundary, and half of the expanding angle comes out to be almost 8 degrees (the angle between the jet boundary and the central axis is almost 8 degrees). The mixing zone is followed by the transition zone. Turbulent flow is everywhere in the transition zone, with the mean velocity decreasing along the positive direction of x_1. The end of the transition zone almost extends to the position at 10 times that of the diameter ($x_1/D = 10$), which is also the starting location of the fully developed zone.

Along the axial direction, the jet velocity keeps a constant in the mixing zone, decreases rapidly in the transition zone, and becomes inversely proportional to the distance away from the outlet (x_1). The variation of the radial velocity is related to the axial distance. In the mixing zone, the radial velocity almost stays a constant until the jet boundary and then starts to decrease rapidly. In the downstream area, the profile of velocity gets wider with lower velocity gradient. After it flows into the fully developed zone, velocities do not vary almost along the radial direction. The turbulence intensity, represented by the ratio of the fluctuation of axial velocity and the jet velocity at the outlet, is much related to the velocity gradient. Among all these cross-sections of x_1 axis, the cross-section around

$x_2/D = 0.5$appears to have the strongest turbulence intensity. In the fully developed zone, the velocity gradient is the lowest, thus the turbulence intensity also gets weaker. Noise measurements have verified that jet noise mostly generates by the turbulence in the mixing zone and the transition zone.

On the other hand, due to the randomness of the turbulent fluctuation, the far-field sound also fluctuates randomly, as can be found from Eq. (3.110) or (3.108). It is natural that time-average properties of jet noise attract much attention. Therefore, the average value of the sound strength can be obtained through the self-relevant function of the sound pressure, as follows:

$$\Gamma(x,\tau) = \frac{\overline{[p(x,t+\tau)-p_0][p(x,t)-p_0]}}{\rho_0 c_0}. \tag{3.111}$$

It can be known from Eq. (3.108) that:

$$\Gamma(x,\tau) = \frac{1}{16\pi^2 c_0^5 \rho_0} \frac{x_i x_j x_k x_l}{x^6} \iint \overline{\frac{\partial^2 T_{ij}(y',t')}{\partial t^2} \frac{\partial^2 T_{kl}(y'',t'')}{\partial t^2}} dy' dy''. \tag{3.112}$$

It has been proven [86] that:

$$\overline{\frac{\partial^2 T_{ij}(y',t')}{\partial t^2} \frac{\partial^2 T_{kl}(y'',t'')}{\partial t^2}} = \frac{\partial^4}{\partial \tau^4} \iint \overline{T_{ij}(y',t') T_{kl}(y'',t'')}.$$

Substituting this formula into Eq. (3.112) leads to:

$$\Gamma(x,\tau) = \frac{1}{16\pi^2 c_0^5 \rho_0} \frac{x_i x_j x_k x_l}{x^6} \frac{\partial^4}{\partial \tau^4} \iint \overline{T_{ij}(y',t') T_{kl}(y'',\tau_0)} dy' dy'', \tag{3.113}$$

where $\tau_0 = t + \tau + \frac{\mathrm{x}}{xc_0}(y'' - y')$. Let $\eta = y'' - y'$ and transform variables y' and y'' to y' and η, Eq. (3.113) can be expressed as:

$$\Gamma(x,\tau) = \frac{\rho_0}{16\pi^2 c_0^5} \frac{x_i x_j x_k x_l}{x^6} \frac{\partial^4}{\partial \tau^4} \iint R_{ijkl}\left(y, \eta, \tau + \frac{\eta}{c_0} \cdot \frac{\mathrm{x}}{\mathbf{x}}\right) dy d\eta, \tag{3.114}$$

where

$$R_{ijkl}(y,\eta,\tau) = \frac{\overline{T_{ij}(y',t) T_{kl}(y'',t+\tau)} - \varphi_{ijkl}(y,\eta)}{\rho_0^2},$$

And φ_{ijkl} represents a time-irrelevant variable, which will be used to simplify equations, and:

$$\begin{cases} t' = t - \dfrac{|x - y'|}{c_0} \\ t' = t + \tau - \dfrac{|x - y''|}{c_0} \end{cases}.$$

Moreover, because the self-relevant function of a steady function has no connection with time translation, we have:

$$\overline{T_{ij}(y',t)T_{kl}(y'',t'')}=\overline{T_{ij}(y',t)T_{kl}\left(y'',t+\tau+\frac{|x-y'|-|x-y''|}{c_0}\right)}.$$

with $|x-y'|=x-\frac{x}{x}\cdot y'$. For large value of x, we can write out that $\frac{|x-y'|-|x-y''|}{c_0}\sim\frac{x}{\mathbf{x}}\cdot\frac{(y''-y')}{c_0}$. Performing Fourier transformation to Eq. (3.114), the spectrum function of the sound strength can be expressed as:

$$\overline{I}_\omega(\mathrm{x})=\frac{\omega^4\rho_0}{32\pi^3c_0^5}\frac{x_ix_jx_kx_l}{x^6}\int_{-\infty}^{\infty}\iint e^{i\omega\left[\tau-\frac{\mathrm{x}}{x}\cdot\frac{\eta}{c_0}\right]}R_{ijkl}(y,\eta,\tau)dyd\eta d\tau. \tag{3.115}$$

Basically, Eq. (3.115) can be used for the computation of a sound field generated by turbulent fluctuation.

3.5.2.2 Dimensionless analysis

For a subsonic jet, compared with the Reynolds stress ρu_iu_j, the viscous stress can be ignored, thus the Lighthill's stress tensor can by simplified as:

$$T_{ij}\approx\rho u_iu_j.$$

With a vortex of length L considered, we have $\omega\approx\frac{|u|}{L}$ and

$$R_{ijkl}=\frac{\rho^2}{\rho_0{}^2}\overline{u_iu_ju_ku_l}\sim\frac{\rho^2u^4}{\rho_0{}^2}.$$

The average jet velocity at outlet is assumed to be U. From experiments, it is known that the maximum fluctuating velocity of turbulence can reach:

$$\overline{u}_{\max}\sim0.16U.$$

Therefore, it is noted from the magnitude that:

$$\omega\sim\frac{U}{l},\quad R_{ijkl}\sim\frac{\rho^2U^4}{\rho_0^2}.$$

Hence, the sound strength generated by the vortex can be approximately written as:

$$\overline{I}_\omega(\mathrm{x})\sim\frac{\rho_0}{32\pi^3c_0^5x^2}\left(\frac{U}{L}\right)^4\frac{\rho^2U^4}{\rho_0^2}\left(L^3\right)^2d\tau\approx\frac{\rho_0U^3L^2}{32\pi^3c_0^5x^2}\frac{\rho^2}{\rho_0^2}. \tag{3.116}$$

From Eq. (3.116), it can be found that the quadrupole noise arising by turbulence is proportional to the eighth order of the outflowing velocity, which is the famous Lighthill's eighth-order law.

3.5.2.3 Effect of the convective velocity

The axial convective velocity of jet is denoted by M_c, and let

$$\eta = \xi + ic_0 M_c \tau,$$

we can obtain the equation as:

$$\overline{I}_\omega(\mathrm{x}) = \frac{\omega^4 \rho_0}{32\pi^3 c_0^5} \frac{x_i x_j x_k x_l}{x^6} \iiint e^{i\omega\left[(1-M_c\cos\theta)\tau - \frac{\mathrm{x}}{x}\cdot\frac{\eta}{c_0}\right]} R_{ijkl}(y, \xi, \tau)\, dy\, d\xi\, d\tau \tag{3.117}$$

where

$$\cos\theta = \frac{x_1}{x}.$$

With the inverse Fourier transformation to Eq. (3.117), it can be written as:

$$\begin{aligned}
\Gamma(x,\tau) &= \frac{\rho_0}{16\pi^2 c_0^5} \frac{x_i x_j x_k x_l}{x^6} \int \left(\frac{1}{1 - M_c\cos\theta}\right) \frac{\partial^4}{\partial t^4} \\
&\quad \int R_{ijkl}\left(y', \xi, \frac{t + \frac{\mathrm{x}}{x}\cdot\frac{\xi}{c_0}}{1 - M_c\cos\theta}\right) d\xi\, dy' \\
&= \frac{\rho_0}{16\pi^2 c_0^5} \frac{x_i x_j x_k x_l}{x^6} \int \frac{1}{(1 - M_c\cos\theta)^5} \\
&\quad \times \left\{\frac{\partial^4}{\partial \tau^4} \int R_{ijkl}\left[y', \xi, \tau + \frac{\mathrm{x}}{x}\cdot\frac{\xi}{c_0(1 - M_c\cos\theta)}\right]\right\}_\tau d\xi\, dy,
\end{aligned} \tag{3.118}$$

where

$$\tau = \frac{t}{1 - M_c\cos\theta}.$$

Finding out the exact solution for the sound field from Eq. (3.118) is almost impossible because all these fourth-order variables relevant to turbulence are required to be known. Therefore, further simplification is inevitable. In general, the formulas for jet noise prediction today take these errors of simplification into account through the modification by empirical coefficients obtained from experiments. It can be known from Eq. (3.116) that the sound power of radiation for unit volume is proportional to:

$$\frac{\rho_0 U^8}{Lc_0^5}\left(\frac{\rho}{\rho_0}\right)^2.$$

It is known that the mixing zone of turbulence generates the strongest jet noise, and its length appears to be four times that of the outlet diameter almost (4D). With the cross-section area of the jet in the mixing zone being:

$$A(y_1) \approx \pi D \times (\text{thickness of mixing area}) \approx \frac{\pi D}{4} y_1,$$

the sound power of radiation by the turbulence in the mixing zone can be expressed as:

$$W_M = k\frac{\rho_0 U^8 D^2}{Lc_0^5}\left(\frac{\rho}{\rho_0}\right)^2. \tag{3.119}$$

Furthermore, if the effect of convective velocity is included, the sound power becomes:

$$W_M \approx 2.5 \times 10^{-6} K \frac{\rho_0 U^8 D^2}{Lc_0^5} \frac{1+M_c^2}{\left(1-M_c^2\right)^4}. \tag{3.120}$$

In most cases, $M_c \approx 0.62\frac{U}{c_0}$ and K is a non-dimensional coefficient. Generally, the sound power from the mixing zone takes half of the total sound power in jet noise. Therefore, the total sound power of jet noise can be written as:

$$W_T \approx 5.0 \times 10^{-6} K \frac{\rho_0 U^8 D^2}{c_0^5} \frac{1+M_c^2}{\left(1-M_c^2\right)^4}. \tag{3.121}$$

The results of some experiments show that the variation of sound power in jet noise is consistent with the eighth-order law. However for different cases, these coefficients for modification, which are also called Lighthill's parameters, are different, and they need to be determined through experiments. Fig. 3.23 shows the result of total sound power measured in an experiment [87], as well as the theoretical prediction. Fig. 3.24 gives the comparison of the total sound pressure level in different locations between the measurement and the theory [75].

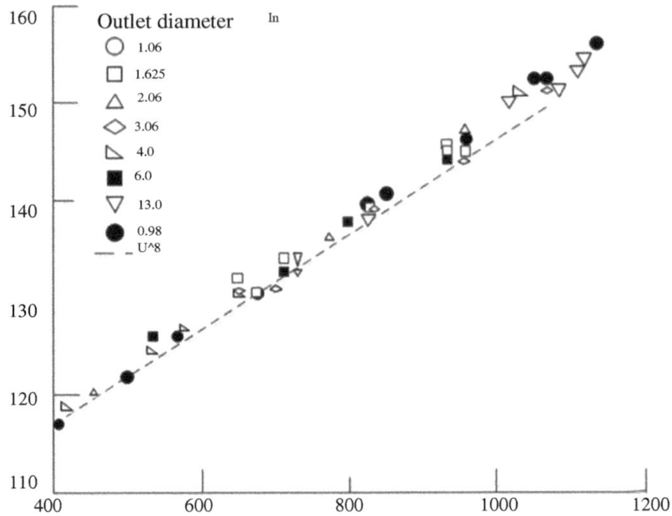

Fig. 3.23 Variation of the total sound power with jet velocity [87].

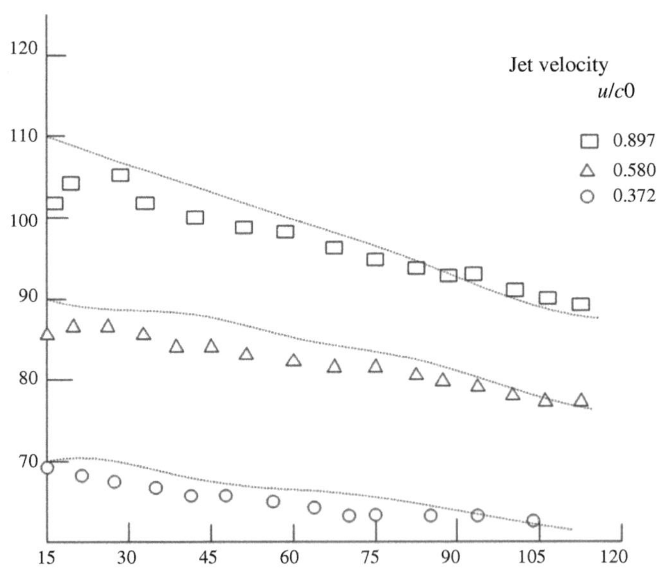

Fig. 3.24 Variations of the total sound pressure level with azimuth angle [75].

3.5.3 Effect of non-uniform flow—Lilley's equation

The continuity and the momentum equations can be written as:

$$\frac{1}{\rho}\frac{Dp}{D\tau}+\frac{\partial v_j}{\partial y_j}=0, \tag{3.122}$$

$$\frac{Dv_i}{D\tau}=-\frac{1}{\rho}\frac{\partial p}{\partial y_i}+\frac{1}{\rho}\frac{\partial e_{ij}}{\partial y_j}, \tag{3.123}$$

where

$$\frac{D}{D\tau}=\frac{\partial}{\partial \tau}+v_j\frac{\partial}{\partial y_j}.$$

For ideal gas:

$$\frac{d\rho}{\rho}=\frac{1}{k}\frac{dp}{p}-\frac{dS}{c_p}, \tag{3.124}$$

where $k=\frac{c_p}{c_v}$ denotes the adiabatic coefficient. Take Eq. (3.124) into Eq. (3.122) and we can obtain:

$$\frac{D\Pi}{D\tau}+\frac{\partial v_i}{\partial y_i}=\frac{1}{c_p}\frac{DS}{D\tau}, \tag{3.125}$$

where

$$\Pi=\frac{1}{k}\ln\frac{p}{p_0}.$$

Furthermore, the momentum equation can be rewritten as:

$$\frac{Dv_i}{D\tau}=-c^2\frac{\partial \Pi}{\partial y_i}+\frac{1}{\rho}\frac{\partial e_{ij}}{\partial y_j}. \tag{3.126}$$

After some transformation performed on Eq. (3.125) and Eq. (3.126), it can be obtained that:

$$\frac{D^2\Pi}{D\tau^2}-\frac{\partial}{\partial y_i}c^2\frac{\partial \Pi}{\partial y_i}=\frac{\partial v_j}{\partial y_i}\frac{\partial v_i}{\partial y_j}-\frac{\partial}{\partial y_i}\frac{1}{\rho}\frac{\partial e_{ij}}{\partial y_j}+\frac{D}{D\tau}\frac{DS}{D\tau}, \tag{3.127}$$

and with its time differential taken, Eq. (3.127) can be expressed as:

$$\frac{D}{D\tau}\left(\frac{D^2\Pi}{D\tau^2}-\frac{\partial}{\partial y_i}c^2\frac{\partial \Pi}{\partial y_i}\right)=2\frac{\partial v_j}{\partial y_i}\frac{D}{D\tau}\frac{\partial v_i}{\partial y_j}-\frac{D}{D\tau}\frac{\partial}{\partial y_i}\frac{1}{\rho}\frac{\partial e_{ij}}{\partial y_j}+\frac{D^2}{D\tau^2}\frac{1}{c_p}\frac{DS}{D\tau}, \tag{3.128}$$

$$\frac{D}{D\tau}\left(\frac{D^2\Pi}{D\tau^2}-\frac{\partial}{\partial y_i}c^2\frac{\partial \Pi}{\partial y_i}\right)=2\frac{\partial v_j}{\partial y_i}\frac{D}{D\tau}\frac{\partial v_i}{\partial y_j}-\frac{D}{D\tau}\frac{\partial}{\partial y_i}\frac{1}{\rho}\frac{\partial e_{ij}}{\partial y_j}+\frac{D^2}{D\tau^2}\frac{1}{c_p}\frac{DS}{D\tau}. \tag{3.129}$$

Furthermore,

$$\frac{D}{D\tau}\left(\frac{D^2\Pi}{D\tau^2}-\frac{\partial}{\partial y_i}c^2\frac{\partial \Pi}{\partial y_i}\right)+2\frac{\partial v_j}{\partial y_i}\frac{\partial}{\partial y_j}c^2\frac{\partial \Pi}{\partial y_i}=-2\frac{\partial v_j}{\partial y_i}\frac{\partial v_k}{\partial y_j}\frac{\partial v_i}{\partial y_k}+\phi. \tag{3.130}$$

Eq. (3.130) is the so-called Lilley's equation, in which:

$$\phi=2\frac{\partial v_j}{\partial y_i}\frac{\partial}{\partial y_j}\frac{1}{\rho}\frac{\partial e_{ik}}{\partial y_k}-\frac{D}{D\tau}\frac{\partial}{\partial y_i}\frac{1}{\rho}\frac{\partial e_{ij}}{\partial y_j}+\frac{D^2}{D\tau^2}\frac{1}{c_p}\frac{DS}{D\tau},$$

and ϕ represents these effects of flow viscosity and entropy fluctuation on the sound generation.

Eq. (3.130) includes the convective and the reflective effect of the gradient of mean flow. On the other hand, a much more complex solution will be encountered compared with the solution of Lighthill's equation. Moreover, with linearization to Eq. (3.130) along with shear flow assumption, the formula will degenerate into the expression of wave equation under non-uniform condition as seen in Chapter 1. By the application of Eq. (3.130), the directivity of jet noise can be well presented [88].

References

[1] M.J.T. Smith, Aircraft Noise, Cambridge Aerospace Series, Cambridge University, 1989.
[2] D. V. Zante, D. Nark and H. Fernandez, Propulsion noise reduction research in the NASA advanced air transport technology project, International Society for Air Breathing Engines 2017, 22697.
[3] M.J. Lighthill, On sound generated aerodynamically. II. Turbulence as a source of sound, Proc. Roy. Soc. A 222 (1148) (1954) 1–32.
[4] J.M. Tyler, T.G. Sofrin, Axial flow compressor noise studies, Soc. Automot. Eng. Trans. 70 (1962) 309–332.
[5] J.E. Ffowcs Williams, D.L. Hawkings, Sound generation by turbulence and surfaces in arbitrary motion, Phil. Trans. R. Soc. A 264A (1969) 321–342.
[6] M.E. Goldstein, Unified approach to aerodynamic sound generation in the presence of sound boundaries, J. Acoust. Soc. Am. 56 (1974) 497–509.
[7] A.P. Dowling, Y. Mahmoudi, Combustion noise, Proc. Combust. Inst. 35 (2015) 65–100.
[8] M.J.T. Smith, M.E. House, Internally generated noise from gas turbine engines: measurement and prediction, ASME J. Eng. Power 89 (1) (1967) 177–190.
[9] N.H. Kemp, W.R. Sears, Aerodynamic interference between moving blade rows, J. Aeronaut. Sci. 20 (1952) 585–597.
[10] N.H. Kemp, W.R. Sears, The unsteady forces due to viscous waves in turbomachines, J. Aeronaut. Sci. 22 (1955) 478–483.
[11] L. Gutin, On the Sound Field of a Rotating Propeller, (1948) NASA-TM-1195.

[12] S. Kaji, T. Okazaki, Propagation of sound waves through a blade row I. Analisis based on the semi-actuator disk theory, J. Sound Vib. 11 (3) (1970) 339–353.
[13] S. Kaji, T. Okazaki, Propagation of sound waves through a blade row II. Analisis based on the acceleration potential method, J. Sound Vib. 11 (3) (1970) 355–375.
[14] S. Kaji, T. Okazaki, Generation of sound by rotor-stator interaction, J. Sound Vib. 13 (1970) 281–307.
[15] R. Mani, Noise due to interaction of inlet turbulence with isolated stators and rotors, J. Sound Vib. 17 (1971) 251–560.
[16] S.N. Smith, Discrete frequency sound generation in axial flow turbomachines, (1972) Cambridge No. 3709.
[17] N.J. Lipstein, R. Mani, Experimental investigation of discrete frequency noise generated by unsteady blade forces, J. Basic Eng. Trans. ASME 92 (1970) 155–164.
[18] J.E. McCune, A three-dimensional theory of axial compressor blade rows—application in subsonic and supersonic flows, J. Aero Space Sci. 25 (9) (1958) 544–560.
[19] O. Okurounmu, J.E. McCune, Three-dimensional vortex theory of axial compressor blade rows at subsonic and transonic speeds, AIAA J. 8 (7) (1970) 1275–1283.
[20] M. Namba, Three-dimensional analysis of blade force and sound generation for an annular cascade in distorted flows, J. Sound Vib. 50 (4) (1977) 479–508.
[21] M. Namba, Three-Dimensional Flow, (1987) AFARD-AG-298.
[22] M. Namba, A. Ishikawa, Three-dimensional aerodynamic characteristics of oscillating supersonic and thransonic annular cascades, ASME J. Eng. Power 105 (1983) 138–146.
[23] H. Kodama, M. Namba, Unsteady lifting surface theory for a rotating cascade of swept blades, J. Turbomach. 112 (1990) 411–417.
[24] M. Namba, Lifting surface theory for a rotating subsonic or transonic blade row, (1972) British Reports & Memoranda No. 3740.
[25] P. Salaun, Flutter instabilities in an annular cascade, in: Proceedings of IUTAM Symposium on Aeroelasticity in Turbomachines, 1976, pp. 35–38.
[26] P. Salaun, Flutter instabilities in an annular cascade with supersonic relative flow, (1979) ONERA. Publication No. 1979-3.
[27] J.B.H.M. Schulten, Sound generated by rotor wakes interacting with a leaned vane stator, AIAA J. 20 (1982) 1352–1358.
[28] J.B.H.M. Schulten, Vane stagger angle and camber effects in fan noise generation, AIAA J. 22 (8) (1984) 1071–1079.
[29] J.B.H.M. Schulten, Sound Generation by Ducted Fans and Propellers as a Lifting Surface Problem, Univ. of Twente, Netherlands, 1993.
[30] J.B.H.M. Schulten, Transmission of Sound through a Rotor, DGLR/AIAA Paper 92-02-085, (1992).
[31] J.B.H.M. Schulten, Vane sweep effects on rotor/stator interaction noise, AIAA J. 35 (6) (1997) 945–951.
[32] M. Dahl, Third Computational Aeroacoustic Workshop on Benchmark Problems, NASA CP-2000-209790, (1999).
[33] M. Namba, J.B.H.M. Schulten, Category 4 - fan Stator with Harmonic Excitation by Rotor Wake: Numerical Results of Lifting Surface Theory, NASA CP-2000-209790, (2000).
[34] J.M. Verdon, Linerized unsteady aerodynamic analysis of the acoustic response to Wake/blade-row interaction, (2001) NASA CR-2001-210713.
[35] D. Prasad, J.M. Verdon, Validation of a Three-Dimensional Linearized Euler Analysis for Classical Wake/Stator Interactions, GT-2002-30400, (2002).
[36] B. Elhadidi, H.M. Atassi, Passive noise control by blade lean and sweep, in: AIAA, 2004, pp. 2004–2999.
[37] A.J. Cooper, N. Peake, Rotor-stator interaction noise in swirling flow: stator sweep and lean effects, AIAA J. 44 (5) (2006) 981–991.

[38] M.B. Giles, Calculation of unsteady wake/rotor interaction, J. Propulsion 4 (4) (1988) 356–362.
[39] M. Giles, R. Haimes, Validation of numerical method for unsteady flow calculations, J. Turbomach. 115 (1993) 110–117.
[40] S. Sawyer, M. Nallasamy, R. Hixon, R.W. Dyson, A computational aeroacoustic prediction of discrete-frequency rotor-stator interaction noise-a linear theory analysis, Int. J. Aeroacoust. 3 (1) (2004) 67–86.
[41] X.Y. Wang, A. Himansu, S.C. Chang, et al., Computation on Wake/Stator Interaction in a 2D Cascade, Springer, Berlin, Heidelberg, 2004.
[42] D. Hawkings, Multiple tone generation by transonic compressors, J. Sound Vib. 17 (2) (1971) 241–250.
[43] C.L. Morfey, M.J. Fisher, Shock-wave radiation from a supersonic ducted rotor, Aeronaut. J. 74 (715) (1970) 579–585.
[44] M. Kurosaka, A note on multiple pure tone noise, J. Sound Vib. 19 (4) (1971) 453–462.
[45] G.B. Witham, Linear and Non-linear Waves, John Wiley, New York, 1974.
[46] D.L. Hawkings, The effects of inlet conditions on supersonic cascade noise, J. Sound Vib. 33 (3) (1974) 353–369.
[47] M.S. Tsai, Propagation of Buzz-Saw noise in a non-uniform medium, in: AIAA-1979-1022, 1979.
[48] P.G. Vaidya, Multiple pure tone generation in aeroengine fans at subsonic and supersonic relative tip speeds, AIAA J. 22 (10) (1984) 1366–1373.
[49] G.A. Sod, A survey of several finite difference methods for systems of nonlinear hyperbolic conservation laws, J. Comput. Phys. 27 (1) (1978) 1–31.
[50] A. Prasad, Evolution of upstream propagating shock waves from a transonic compressor rotor, Physica C 125 (1) (2003) 273–277.
[51] A. McAlpine, P.J.G. Schwaller, M.J. Fisher, et al., "Buzz-saw" noise: prediction of the rotor-alone pressure field, J. Sound Vib. 331 (22) (2012) 4901–4918.
[52] A. Mcalpine, M.J. Fisher, B.J. Tester, "Buzz-saw" noise: a comparison of modal measurements with an improved prediction method, J. Sound Vib. 306 (3–5) (2007) 419–443.
[53] A. McAlpine, M.J. Fisher, B.J. Tester, "Buzz-saw" noise: a comparison of measurement with prediction, J. Sound Vib. 290 (3–5) (2006) 1202–1233.
[54] J. Fisher, A. Mcalpine, On the prediction of "buzz-saw" noise in aero-engine inlet ducts, J. Sound Vib. 248 (1) (2001) 123–149.
[55] A. Mcalpine, M.J. Fisher, On the prediction of "buzz-saw" noise in acoustically lined aero-engine inlet ducts, J. Sound Vib. 265 (1) (2003) 175–200.
[56] P. Gliebe, R. Mani, H. Shin, B. Mitchell, G. Ashford, S. Salamah, … D. Huff, Aeroacoustic prediction codes, (2000).
[57] D. Prasad, Propagation and decay of shock waves in turbofan engine inlets, J. Turbomach. 127 (1) (2005) 118–127.
[58] F. Han, A. Sharma, U. Paliath, et al., Multiple pure tone noise prediction, J. Sound Vib. 333 (2013) 6942–6959.
[59] A. Breard, A. Sayma, M. Imregun, et al., A CFD-based non-linear model for the prediction of tone noise in lined ducts, in: AIAA-2001-2176, 2001.
[60] J. Defoe, A. Narkaj, Z. Spakovszky, A body-force based methodology for predicting multiple-pure-tone noise: validation, in: AIAA Paper 2010-3747, 16th AIAA/CEAS Aeroacoustics Conference, 2010.
[61] M.B. Giles, Non-reflecting boundary conditions for Euler equation calculations, AIAA J. 28 (12) (2012) 2050–2058.
[62] W.C. Strahle, Combustion noise, Prog. Energy Combust. Sci. 4 (3) (1978) 157–176.
[63] S. Candel, D. Durox, T. Schuller, P. Palies, J.F. Bourgouin, J.P. Moeck, Progress and challenges in swirling flame dynamics, C.R. Mec. 340 (11–12) (2012) 758–768.

[64] F.E. Marble, S.M. Candel, Acoustic disturbance from gas non-uniformities convected through a nozzle, J. Sound Vib. 55 (2) (1977) 225–243.
[65] A. Thomas, G.T. Williams, Flame noise – sound emission from spark-ignited bubbles of combustible gas, Proc. Roy. Soc. A 294 (1966) 449–466.
[66] R.B. Price, I.R. Hurle, T.M. Sugden, Optical studies of the generation of noise in turbulent flames, Symp. Combust. 12 (1) (1969) 1093–1102.
[67] W. Strahle, On combustion generated noise, J. Fluid Mech. 49 (2) (1971) 399–414.
[68] A.P. Dowling, J.E. Ffowcs Williams, W.M. Wright, Sound and Sources of Sound, Ellis Horwood Limited, 1983.
[69] D.G. Crighton, A.P. Dowling, J.E. Ffowcs Williams, et al., Modern Methods in Analytical Acoustics, Springer, 1992.
[70] M.J. Lighthill, Sound generated aerodynamically, Proc. Roy. Soc. A 267 (1329) (1962) 147–182.
[71] M. Leyko, F. Nicoud, T. Poinsot, Comparison of direct and indirect combustion noise mechanisms in a model combustor, AIAA J. 47 (11) (2013) 2709–2716.
[72] I. Duran, S. Moreau, F. Nicoud, T. Livebardon, E. Bouty, T. Poinsot, Combustion noise in modern aero-engines, AerospaceLab 7 (2014) 1.
[73] M.J. Lighthill, On sound generated aerodynamically. I. General theory, Proc. R. Soc. Lond. 267 (1329) (1962) 564–587.
[74] E.J. Ffowcs Williams, The noise from turbulence convected at high speed, Philos. Trans. R. Soc., B 255 (1061) (1963) 469–503.
[75] P.A. Lush, Measurements of subsonic jet noise and comparison with theory, J. Fluid Mech. 46 (3) (1971) 477–500.
[76] H.K. Tanna, An experimental study of jet noise part I: Turbulent mixing noise, J. Sound Vib. 50 (3) (1977) 405–428.
[77] B.J. Tester, C.L. Morfey, Developments in jet noise modelling—theoretical predictions and comparisons with measured data, J. Sound Vib. 46 (1) (1976) 79–103.
[78] G. Lilley, P. Morris, B. Tester, On the theory of jet noise and its applications, in: Aeroacoustics Conference, 1973, p. 987.
[79] C.L. Morfey, Amplification of aerodynamic noise by convected flow inhomogeneities, J. Sound Vib. 31 (4) (1973) 391–397.
[80] M.S. Howe, Application of energy conservation to the solution of radiation problems involving uniformly convected source distributions, J. Sound Vib. 43 (1) (1975) 77–86.
[81] R.E. Musafir, On the solution of Lilley's equation, in: Proceedings of ICA 1992, 1992, p. 27.
[82] C.H. Berman, J.I. Ramos, Computation of turbulence noise, in: D. Lee, A. Cakmak, R. Vichevetsky (Eds.), Computational Acoustics, Vol. 2, 1990, p. 241.
[83] T. Suzuki, A review of diagnostic studies on jet-noise sources and generation mechanisms of subsonically convecting jets, Fluid Dyn. Res. 42 (2010) 014001 (30pp).
[84] G. Bres, J. Nichols, S. Lele, F. Ham, Unstructured large eddy simulation of a hot supersonic over-expanded jet with chevrons, in: AIAA Paper, AIAA 2012-2213, 18th AIAA/CEAS Aeroacoustics Conference, 2012.
[85] S. Buhler, L. Kleiser, Simulation of subsonic turbulent nozzle jet flow and its near-field sound, AIAA J. 52 (8) (2014) 1653–1669.
[86] M.E. Goldstein, Aeroacoustics, McGraw-Hill International Book Company, New York, 1976.
[87] W. Olsen, O. Gutierrez, R. Dorsch, The effect of nozzle inlet shape, lip thickness, and exit shape and size on subsonic jet noise, in: AIAA-73-187, 1973.
[88] R. Mani, The influence of jet flow on jet noise, part 1: the noise of unheated jets, J. Fluid Mech. 73 (4) (1976) 753–778.

CHAPTER 4

Linearized unsteady aerodynamics for aeroacoustic applications

4.1 Introduction

It is generally believed that unsteady aerodynamic computation undertakes an indispensable role in aeroelastic analysis of moving objects as well as in aeroacoustic prediction. Moreover, aeroacoustic prediction, compared with the former, is more concerned with the calculation of the unsteady forces caused by the disturbances of incoming flow, also called a gust response problem. Currently, the mainstream methods for these types of problems can be classified into two groups: numerical methods and analytical methods. In most cases, with a linearized approximation, numerical methods first deal with the solution of the steady mean flow field to determine the relevant coefficients in the unsteady governing equations, and then various fluctuating quantities can be obtained accordingly by solving these equations. This class of methods have been in widespread use for aeroelastic stability analysis so far [1].

However, for a gust response problem corresponding to fan/compressor cascade, the calculation of unsteady blade loading becomes more complicated than that for aeroelastic stability analysis [2]. The reason is that such calculation requires more accurate far-field and surface conditions. As we know, the velocity decomposing theorem is widely used in various linearized calculations. For its application, the relationship between the fluctuating quantities on the body surfaces with those in the region upstream or far upstream needs to be known in advance. In this way, the boundary conditions on the body surfaces for the governing equations of unsteady flows can only be determined. Actually, it is quite difficult to realize such a numerical simulation for any linearized problem. The common approach is thus to assume that the thickness of an object is negligible together with a zero angle of attack as well as a stationary medium. For example, we usually assume the airfoil or blade is thin and uncambered, which means the magnitude and direction of the perturbations propagating from the oncoming flow will

Fundamentals of Aeroacoustics with Applications to Aeropropulsion Systems
https://doi.org/10.1016/B978-0-12-408069-0.00004-X

remain unchanged. It has been proven that, in these approximations, the disturbed velocity field can be decomposed into an acoustic term associated with pressure and the vortical term independent of pressure [3]. Then by imposing body surface boundary conditions, we can naturally build up the relation between the acoustic velocity on the body surfaces and the incoming disturbance velocity and then eventually solve the equations. In fact, such treatment for unsteady flow problems discussed previously played an essential role in the study of both aeroelastic and aeroacoustic problems before the mid-1980s, which were based on the solution of some integral equations for two- or three-dimensional unsteady flow analysis of turbomachinery. Up to now, the fast algorithm for fan/compressor noise prediction is still the basis of such an approach.

On the other hand, more generalized methods were also sought to include the influence of more relevant parameters, still accompanied by the linearized assumptions. As mentioned previously, the key point is to establish the relation between the fluctuations at an arbitrary space or body surface position and the incoming disturbances from far field. In 1978, Goldstein [4] obtained some very valuable results based on the assumption of potential mean flow. In the past decades, the general approach developed by Goldstein and Atassi [4, 5] has been extended to the computations for turbulent flows and the various unsteady problems mentioned earlier [6–10].

The velocity decomposing theorem for uniform and non-uniform flows, as an application of linearized unsteady aerodynamics, will be introduced in the following section. Their further application can be found in a study by Atassi [5]. Aimed at fast prediction of the fan/compressor aerodynamic noise in the design phase, the linearized calculation of unsteady blade loading has played a crucial role in such an estimation, along with uniform mean flow assumptions. Therefore, in the last two sections of this chapter, the computation of unsteady blade forces for both subsonic and supersonic flow conditions will be introduced in detail with an emphasis on the solution of the relevant integral equation [11].

4.2 Basic linearized unsteady aerodynamic equations

4.2.1 Velocity decomposing theorem for uniform flows

Assume the velocity of the incoming flow is a constant as U_∞, and then the linearized continuity equation and momentum equation are formulated as:

$$\frac{1}{\rho_0 c_0^2}\frac{D_0 p}{D\tau} = -\nabla \cdot \mathbf{u}, \tag{4.1}$$

$$\rho_0 \frac{D_0 \mathbf{u}}{D\tau} = -\nabla p, \tag{4.2}$$

where

$$\frac{D_0}{D\tau} = \frac{\partial}{\partial \tau} + U_\infty \frac{\partial}{\partial y_1}.$$

Then we are going to prove that the velocity can be decomposed into two parts by:

$$\mathbf{u} = \mathbf{u}_1 + \mathbf{u}_2, \tag{4.3}$$

where

$$\nabla \times \mathbf{u}_1 = \nabla \cdot \mathbf{u}_2 = 0, \tag{4.4}$$

$$\rho_0 \frac{D\mathbf{u}_1}{D\tau} = -\nabla p, \tag{4.5}$$

$$\frac{1}{\rho_0 c_0^2} \frac{D_0 p}{D\tau} = -\nabla \cdot \mathbf{u}_1, \tag{4.6}$$

$$\frac{D_0 \mathbf{u}_2}{D\tau} = 0. \tag{4.7}$$

Consider the fact that any vector field can be divided into an irrotational field and a solenoidal (zero divergence) field, i.e.,

$$\mathbf{u} = \mathbf{u}_1' + \mathbf{u}_2', \tag{4.8}$$

and

$$\nabla \times \mathbf{u}_1' = \nabla \cdot \mathbf{u}_2' = 0. \tag{4.9}$$

Let:

$$A = \rho_0 \frac{D_0 \mathbf{u}_1'}{D\tau} + \nabla p = -\rho_0 \frac{D_0 \mathbf{u}_2'}{D\tau}. \tag{4.10}$$

Assume there is an arbitrary vector $\mathbf{u}_0$ that satisfies the relation:

$$A = \rho_0 \frac{D_0 \mathbf{u}_0}{D\tau}, \tag{4.11}$$

and

$$\mathbf{u}_0(y, \tau) = \frac{1}{\rho_0} \int_0^\tau \mathbf{A}[\mathbf{y} + \mathbf{i}(t - \tau) U_\infty, \tau] dt, \tag{4.12}$$

Inserting Eq. (4.11) into Eq. (4.10) yields:

$$\frac{D_0(\mathbf{u}_1' - \mathbf{u}_0)}{D\tau} + \nabla p = 0, \tag{4.13}$$

$$\frac{D_0(\mathbf{u}_2' + \mathbf{u}_0)}{D\tau} = 0. \tag{4.14}$$

With a comparison between Eq. (4.5) and Eq. (4.6), we can obtain:

$$\mathbf{u}_1 = \mathbf{u}_1' - \mathbf{u}_0, \tag{4.15}$$

$$\mathbf{u}_2 = \mathbf{u}_2' - \mathbf{u}_0. \tag{4.16}$$

Eq. (4.5) implies that, because the irrotational quantity $\mathbf{u}_1$ is associated with pressure fluctuations, it is called acoustic velocity, whereas $\mathbf{u}_2$ is called vortical velocity correspondingly, and

$$\omega = \nabla \times \mathbf{u} = \nabla \times \mathbf{u}_2. \tag{4.17}$$

The decomposing theorem indicates, in uniform flow fields, acoustic velocity and vortical velocity will not interact with each other in a linearized approximation; in other words, there is no energy exchange in the flow field. Indeed, we can further deduce that nonlinear interactions would only be produced in the second order or higher-order terms.

Also, when there are solid boundaries existing in the flow field, the disturbance velocity **u** will interact with the solid boundaries such that the condition should be satisfied that it must vanish on the solid boundary surfaces:

$$\mathbf{u} \cdot \mathbf{n} = 0. \tag{4.18}$$

Then we have:

$$\mathbf{u}_2 \cdot \mathbf{n} = \mathbf{u}_1 \cdot \mathbf{n}. \tag{4.19}$$

Now it is clear that, with the existence of solid boundaries in uniform flow fields, acoustic velocity and vortical velocity are completely coupled with each other and thus will have an energy exchange. This condition is a most commonly applied boundary condition in linearized unsteady aerodynamics, and this effect exactly accounts for the noise generation by the interaction of solid boundaries and a fluctuating quantity independent of pressure.

Moreover, the velocity decomposing theorem, for uniform flows and the boundary conditions on object surfaces together, lays the theoretical foundation for predicting unsteady blade loads and the sound field due to rotor-blade interactions by employing analytical methods. With these basic assumptions, we will discuss how to determine the unsteady aerodynamic forces for two- and three-dimensional blades in turbomachinery. The physical problems will involve the fields of both aeroelastics and aeroacoustics.

4.2.2 Disturbance velocity decomposition in non-uniform flow fields: Goldstein's equation

4.2.2.1 Basic equations

For ideal fluids, the continuity equation, momentum equation, and energy equation can be written as follows:

$$\begin{cases} \rho \dfrac{D\mathbf{V}}{Dt} = -\nabla p \\ \dfrac{D\rho}{Dt} + \rho \nabla \cdot \mathbf{V} = 0, \\ \dfrac{DS}{Dt} = 0 \end{cases} \tag{4.20}$$

where

$$\frac{D}{Dt} = \frac{\partial}{\partial t} + \mathbf{v} \cdot \nabla,$$

and

$$\begin{cases} p(x,t) = p_0(x) + p(x,t) \\ \mathbf{V}(x,t) = \mathbf{U}(x) + \mathbf{u}(x,t). \\ \rho_t(x,t) = \rho_0(x) + \rho(x,t) \end{cases}$$

The vector $\mathbf{U} = (U_1, U_2, U_3)$, p_0, and ρ_0 denote the velocity, pressure, and density of the mean flow, respectively. Besides, when $x_1 \to -\infty$,

$$\mathbf{V} = \mathbf{i}U_{-\infty} + \mathbf{U}_{-\infty}(x_1 - U_{-\infty}t, x_2, x_3), \tag{4.21}$$

$$\nabla \cdot \mathbf{u}_{-\infty} = 0, \tag{4.22}$$

$$s = S_{-\infty}(x_1 - U_{-\infty}t, x_2, x_3), \tag{4.23}$$

In view of the linearized approximation we can obtain:

(1) for steady flow:

$$\rho_0 \mathbf{U} \cdot \nabla \mathbf{U} = -\nabla p_0, \tag{4.24}$$

$$\mathbf{U} \cdot \nabla \rho_0 + \rho_0 \nabla \cdot \mathbf{U} = 0, \tag{4.25}$$

(2) for unsteady flow:

$$\rho_0\left(\frac{D_0\mathbf{u}}{Dt}+\mathbf{u}\cdot\nabla\mathbf{U}\right)+\rho'\mathbf{U}\cdot\nabla\mathbf{U}=-\nabla p', \tag{4.26}$$

$$\frac{D_0\rho'}{Dt}+\rho'\nabla\cdot\mathbf{U}+\nabla\cdot(\rho_0\mathbf{u})=0, \tag{4.27}$$

$$\frac{D_0 s'}{Dt}=0, \tag{4.28}$$

where

$$\frac{D_0}{Dt}=\frac{\partial}{\partial t}+\mathbf{U}\cdot\nabla,$$

$$S_1-S_2=c_p\ln\frac{p_1}{p_2}-c_v\ln\frac{\rho_1}{\rho_2}, \tag{4.29}$$

$$s'=-c_p\frac{\rho'}{\rho_0}-c_v\frac{p'}{p_0}, \tag{4.30}$$

and

$$\frac{p_0}{\rho_0^k}=\text{const.}$$

Therefore:

$$\frac{D_0\mathbf{u}}{Dt}+\mathbf{u}\cdot\nabla\mathbf{U}-\frac{s'}{c_p}\mathbf{U}\cdot\nabla\mathbf{U}=\frac{p'}{k\rho_0 p_0}\nabla p_0-\frac{1}{\rho_0}\nabla p'=-\nabla\left(\frac{p'}{\rho_0}\right).$$

Further, the previous equation can be simplified as:

$$\frac{D_0\mathbf{u}^*}{Dt}+\mathbf{u}^*\cdot\nabla\mathbf{U}=-\nabla\left(\frac{p'}{\rho_0}\right), \tag{4.31}$$

with

$$\mathbf{u}^*=\mathbf{u}-\frac{s'}{2c_p}\mathbf{U}. \tag{4.32}$$

Then the continuity equation can be rewritten in the form of:

$$\frac{D_0}{Dt}\left(\frac{\rho'}{\rho_0}\right)+\frac{1}{\rho_0}\nabla\cdot(\rho_0\mathbf{u})=0. \tag{4.33}$$

Considering Eq. (4.30) leads to:

$$\frac{\rho'}{\rho_0} = \frac{p'}{c_0^2 \rho_0} - \frac{1}{c_p} s' \quad \left(c_0^2 = \frac{c_p p_0}{c_v \rho_0} \right). \tag{4.34}$$

Then, with Eq. (4.28), the continuity equation is transformed into:

$$\frac{D_0}{Dt}\left(\frac{p'}{c_0^2 \rho_0} \right) + \frac{1}{\rho_0} \nabla \cdot (\rho_0 \mathbf{u}) = 0. \tag{4.35}$$

4.2.2.2 Goldstein's equation

We have:

$$\frac{dx}{U_x} = \frac{dy}{U_y} = \frac{dz}{U_z} = dt. \tag{4.36}$$

This equation determines the streamline equations as follows:

$$y = y_s(x, y, z), \quad z = z_s(x, y, z).$$

Then introducing Lighthill's drift function:

$$\Delta(x, y, z) = \frac{x}{U_{-\infty}} + \int_{-\infty}^{x} \left[\frac{1}{U_x(x', y_s, z_s)} - \frac{1}{U_{-\infty}} \right] dx'. \tag{4.37}$$

Define:

$$\mathbf{X} = (X_1, X_2, X_3),$$

with its components given by

$$X_1 = U_{-\infty}\Delta, \quad X_2 = Y, \quad X_3 = Z,$$

and the condition

$$\frac{D_0}{Dt}(\mathbf{X} - \mathbf{i} U_\infty t) = 0. \tag{4.38}$$

If we assume:

$$p' = -\rho_0 \frac{D\varphi}{Dt}, \quad u^* = \nabla \varphi, \tag{4.39}$$

then its general solution can be determined by solving the following equation:

$$\frac{D_0 \mathbf{u}^{(H)}}{Dt} + \mathbf{u}^{(H)} \cdot \nabla \mathbf{U} = 0. \tag{4.40}$$

Because we assume the mean flow is irrotational, we can obtain the relation (noting that we will let $x_1 = x$, $x_2 = y$, $x_3 = z$):

$$\frac{\partial U_i}{\partial x_j} - \frac{\partial U_j}{\partial x_i} = 0. \tag{4.41}$$

Thus, with Eq. (4.38), we can derive the solution of Eq. (4.40) as:

$$\mathbf{u}^{(H)} = \mathbf{A}(\mathbf{X} - \mathbf{i}U_{-\infty}t)\frac{\partial \mathbf{X}}{\partial x_i}. \tag{4.42}$$

So the general solution of Eq. (4.31) is in the form of:

$$\mathbf{u}^* = \nabla\varphi + \mathbf{u}^{(H)}. \tag{4.43}$$

Further, we can get:

$$\mathbf{u} = \nabla\varphi + \mathbf{u}^{(I)}, \tag{4.44}$$

where

$$u_i^{(I)} = \frac{s'}{2c_p}U_i + \mathbf{A}(\mathbf{X} - \mathbf{i}U_{-\infty}t)\frac{\partial \mathbf{X}}{\partial x_i}, \quad (i = 1, 2, 3) \tag{4.45}$$

Inserting the first equation of Eqs. (4.39) and (4.43) into Eq. (4.35) yields:

$$\frac{D_0}{Dt}\left(\frac{1}{c_0^2}\frac{D_0\varphi}{Dt}\right) - \frac{1}{\rho_0}\nabla\cdot(\rho_0\nabla\varphi) = \frac{1}{\rho_0}\nabla\cdot\left(\rho_0\mathbf{u}^{(I)}\right), \tag{4.46}$$

which is the basic equation of nonlocal flows proposed by Goldstein [12]. And this equation indicates that, under the non-uniform flow condition, the perturbations from the oncoming flows will not only interact with acoustic disturbances on the boundaries but also couple with acoustic fluctuations during the propagation process, giving rise to the source term on the equation's right side.

4.2.2.3 Relation between the fluctuations at an arbitrary point and the incoming disturbances in far field

We have:

$$\nabla\rho_0\mathbf{u}^{(I)} = 0, \tag{4.47}$$

$$\mathbf{A}(\mathbf{X} - \mathbf{i}U_{-\infty}t) = u_{-\infty}(\mathbf{X} - \mathbf{i}U_{-\infty}t) - \mathbf{i}\frac{U_{-\infty}}{2c_p}\hat{S}(\mathbf{X} - \mathbf{i}U_{-\infty}t), \tag{4.48}$$

where

$$\hat{S}(\mathbf{X}-\mathbf{i}U_{-\infty}t) = s_{-\infty}(\mathbf{X}-\mathbf{i}U_{-\infty}t). \tag{4.49}$$

Finally, we get:

$$s' = s_{-\infty}(\mathbf{X}-\mathbf{i}U_{-\infty}t), \tag{4.50}$$

$$\begin{aligned} u_i^{(I)} &= u_{-\infty}(\mathbf{X}-\mathbf{i}U_{-\infty}t)\frac{\partial \mathbf{X}}{\partial x_i} \\ &+ \frac{1}{2c_p}s_{-\infty}(\mathbf{X}-\mathbf{i}U_{-\infty}t)\left(U_i - U_{-\infty}^2\frac{\partial \Delta}{\partial x_i}\right), \quad (i=1,2,3) \end{aligned} \tag{4.51}$$

By using the equations presented here, we can solve the unsteady aerodynamic forces on the body surfaces due to the oncoming disturbances and the sound field resulting from these.

As for the application of Goldstein's equation, more examples can be found in Atassi [5].

4.3 Unsteady loading for two-dimensional supersonic cascades with subsonic leading-edge locus

Under supersonic conditions, the calculation of unsteady blade forces is much more complex due to the need to include the effect of Mach wave reflections. Similar to the case of subsonic cascades, the analytical methods for supersonic cascade problems also employ the mathematical model of the pressure wave equation with inviscid flow and small disturbance approximations. For this equation, various solving approaches have been developed. In 1957, by employing Laplace transform, Lane [13] obtained the analytical solution of a cascade with an infinite number of blades and zero stagger angle. Although this kind of cascade does not exist in the practical cases, it was still a pioneering research work for that time. These thoughts left a significant effect on deeper investigations and later studies of this problem. In the middle of the 1960s, Gorelov [14] conducted an analysis of the unsteady computation for a supersonic cascade with subsonic axial flow and arbitrary angle of stagger. However, it still did not reach the stage of unsteady blade force predictions for supersonic cascades with subsonic inlet axial flow that makes practical sense. In this case, the perturbation wave was propagated to the whole flow field, so the mathematical treatment of the latter is more difficult than that of the axial supersonic flow. In the early 1970s, Kurosaka [15, 16] obtained the analytical solution of unsteady blade loading with subsonic leading-edge locus under a specific reduced frequency condition. However, it is difficult to extend this method to be suitable for an arbitrary frequency. In 1973, Verdon [17] obtained the solution of a finite cascade through the combined use of closed-form and numerical procedures. By using the method of velocity potential, Verdon [18–20] further developed his work to the calculation of an unsteady force of a supersonic cascade with arbitrary frequency parameters and axial subsonic speed. Meanwhile, Nagashima and Whitehead [21], Goldstein and Adamczyke [22], and Platzer [23] et al. independently developed their own methods of unsteady blade force predictions for supersonic cascades with subsonic inlet axial flow. Also, Ni [24] , Bendiksen [25] , Butenko [26] , and Namba [27] made some deeper investigations on these problems. It has been proven through the fundamental calculating examples that the results obtained by the various methods previously mentioned are coincident with each other. However, a common feature of all these methods is that the final expressions for numerical computation are all in very complicated forms and require considerable effort to be applied toward the treatments of generalized functions and singularities. Thus, there is great difficulty in raising the convergence rate and improving the computation accuracy. Goldstein [28] tried to develop a simpler approach to obtain the solution. In this work, he first took the Fourier transforms of the basic equation to derive an integral expression for disturbance velocity or pressure, and then, by analyzing the properties of the integrand, the residual theorem was applied to further simplify the integral. The final equation used for numerical calculation is the second kind of nonsingular Fredholm integral equation. So far, the kernel function derived by this method is simpler than those given by any other existing methods. However, the solution he gave didn't include the effect of Mach wave reflections, therefore it is still not applied to practical computations. According to Goldstein [28], there was not any numerical calculation examples presented.

According to Sun [29], Goldstein's solution [28] was further developed to consider the reflection coefficients of Mach waves and depict the pressure fluctuations on the blade surfaces by using a continuous function. By using the collocation method, the numerical results of the second kind of nonsingular Fredholm integral equation were finally obtained. The practical calculations manifested the validity of this method. Indeed, the method presented here provides a simple and effective approach to calculate the unsteady blade loads of supersonic cascades with subsonic leading-edge locus, and meanwhile, it can be also used to predict the unstalled flutter of supersonic cascades.

4.3.1 Physical and mathematical models

At the beginning, the following assumptions are made for the flow model we will discuss:

(1) The fluid is inviscid.

(2) The Mach number of incoming mean flow is greater than unity ($Ma > 1$).

(3) The fluctuating quantities are small relative to the steady background flow.

(Note: The first three assumptions imply the governing equations for the flow field should be the linearized Euler's equations.)

(4) In the linear approximation, the angle of attack of the oncoming flow, the blade camber, and thickness affect the steady blade forces only while their effects on unsteady forces are negligible.

With these assumptions, we can first derive the linearized two-dimensional Euler's equations:

$$\begin{cases} \dfrac{\partial \rho'}{\partial t} + U\dfrac{\partial \rho'}{\partial X} + \rho_0\left(\dfrac{\partial u_1}{\partial X} + \dfrac{\partial u_2}{\partial Y}\right) = 0 \\ \rho_0\left(\dfrac{\partial u_1}{\partial t} + U\dfrac{\partial u_1}{\partial X}\right) + \dfrac{\partial p'}{\partial X} = 0 \\ \rho_0\left(\dfrac{\partial u_2}{\partial t} + U\dfrac{\partial u_2}{\partial X}\right) + \dfrac{\partial p'}{\partial Y} = 0 \end{cases}, \tag{4.52}$$

The subscript "0" denotes the quantities of steady mean flow, and the superscript "′" represents fluctuating quantities. In the following contents, we will omit the notation "′" for simplicity. u_i is defined as the projected component of $\mathbf{u}$ along the positive direction of the axis x_i.

With an approach similar to the unified solution of flow field and sound field, we can derive from the previously presented momentum equation as well as the isentropic condition that:

$$\left(1 - Ma^2\right)\frac{\partial^2 p}{\partial X^2} + \frac{\partial^2 p}{\partial Y^2} - \frac{2Ma_0}{c_0}\frac{\partial^2 p}{\partial X \partial t} - \frac{1}{c_0^2}\frac{\partial^2 p}{\partial t^2} = 0, \tag{4.53}$$

where we have introduced the Mach number $Ma = U/c_0$.

Assume the fluctuating pressure p can be expended into a Fourier series with respect to time, and then we can take an arbitrary term out of the whole series as p for analysis purposes, i.e., let $p = \overline{p}e^{-i\omega t}$. Then Eq. (4.53) can be rewritten as:

$$\frac{\partial \Delta_+}{\partial \alpha} = \frac{1}{2}\left(-\frac{s^+}{b} + \frac{s\beta_r}{b}\frac{\alpha}{\sqrt{\alpha^2 - K^2}}\right) = \frac{1}{2\gamma}\left(-\frac{\gamma s^+}{b} + \frac{s\beta_r}{b}\alpha\right). \tag{4.54}$$

To simplify the equation, we can introduce the dimensionless Prandtl-Glauert coordinates: as shown in Fig. 4.1

$$\begin{cases} \zeta = \dfrac{X}{b} \\ \eta = \dfrac{Y\beta_r}{b} \end{cases}. \tag{4.55}$$

with a transform relation given by:

$$\overline{p} = \psi(\zeta, \eta) e^{iMaK\zeta}, \tag{4.56}$$

where $\beta_r = \sqrt{Ma^2 - 1}$, the wave number $K = \frac{\omega b}{c_0 \beta_r^2}$, and b is defined as the half-chord length of the blades.

Therefore, Eq. (4.54) can be further developed into the form of:

$$\frac{\partial^2 \psi}{\partial \eta^2} - \frac{\partial^2 \psi}{\partial \zeta^2} - K^2 \psi = 0. \tag{4.57}$$

Taking the Fourier transform of this equation with respect to ζ:

$$\begin{aligned} \overline{\psi}(\alpha, \eta) &= \int_{-\infty}^{+\infty} \psi(\zeta, \eta) e^{-i\alpha\zeta} d\zeta \\ \psi(\zeta, \eta) &= \frac{1}{2\pi} \int_{-\infty}^{+\infty} \overline{\psi}(\zeta, \eta) e^{i\alpha\zeta} d\alpha, \end{aligned} \tag{4.58}$$

yields

$$\frac{\partial^2 \overline{\psi}}{\partial \eta^2} + \left(\alpha^2 - K^2\right) \overline{\psi} = 0. \tag{4.59}$$

The solution of this equation is in the form of:

$$\overline{\psi}(\alpha, \eta) = \frac{g_1(\alpha)}{2} \mathrm{e}^{i\gamma\eta} + \frac{g_2(\alpha)}{2} \mathrm{e}^{-i\gamma\eta}, \tag{4.60}$$

where $\gamma = \sqrt{\alpha^2 - K^2}$. The value of γ is assigned according to the criterion that its imaginary part must be negative.

As is shown in Fig. 4.1, m coordinate systems are fixed to m rotor blades, respectively. Let the dimensionless coordinates in the coordinate system fixed to the 0th blade be denoted by $(\zeta\ ,\eta)$, and the dimensionless coordinates in the coordinate system fixed to the mth blade be denoted by $(\zeta_m\ ,\eta_m)$, thus they are related by:

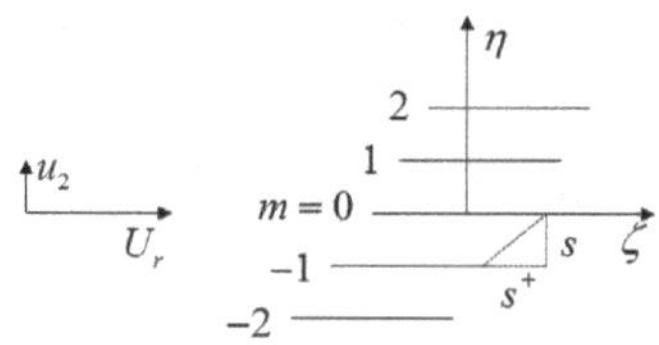

Fig. 4.1 The coordinates of rotor blade row.

$$\begin{cases} \zeta_m = (x_1 - ms^+)\dfrac{1}{b} = \zeta - \dfrac{ms^+}{b} \\ \eta_m = (x_2 - ms)\dfrac{\beta_r}{b} = \eta - \dfrac{ms}{b}\beta_r \end{cases}, \tag{4.61}$$

where m is the integer that represents the order of the blades, and s and s^+ are respectively defined as the distance along the direction vertical to the blade chord and the chordwise distance between two adjacent blades.

Then we have:

$$\overline{\psi}_m(\alpha, \eta) = \frac{g_{1m}(\alpha)}{2}\mathrm{e}^{i\gamma\eta_m} + \frac{g_{2m}(\alpha)}{2}\mathrm{e}^{-i\gamma\eta_m}, \tag{4.62}$$

$$\psi_m(\zeta, \eta) = \frac{1}{2\pi}\int_{-\infty}^{+\infty}\left(\frac{g_{1m}(\alpha)}{2}\mathrm{e}^{i(\alpha\zeta_m + \gamma\eta_m)} + \frac{g_{2m}(\alpha)}{2}\mathrm{e}^{i(\alpha\zeta_m - \gamma\eta_m)}\right)d\alpha. \tag{4.63}$$

In practical cases, outgoing waves have the following property: the pressure due to the perturbations coming from a certain blade will decrease with the distance along the direction vertical to increasing blade chord. Because γ is of a negative imaginary part, we should take $-|\eta_m|$ in the exponent. Furthermore, to ensure the pressure above the blade is positive and the pressure below the blade is negative, the quantity should be multiplied by $\mathrm{sgn}(\eta_m)$. Then it is guaranteed that the coefficient functions of the solution are assigned with the values of physical significance. Therefore, a solution of Eq. (4.57) with respect to the mth blade is given by:

$$\psi_m(\zeta, \eta) = \frac{\mathrm{sgn}(\eta_m)}{2\pi}\int_{-\infty}^{+\infty}\frac{f_m(\alpha)}{2}\mathrm{e}^{i(\alpha\zeta_m + \gamma|\eta_m|)}d\alpha \tag{4.64}$$

where sgn is the sign function:

$$\mathrm{sgn}(\eta_m) = \begin{cases} 1 & (\eta_m > 0) \\ -1 & (\eta_m > 0) \end{cases}. \tag{4.65}$$

For each given value of m, this solution shows a jump discontinuity cross the line where the mth blade is, i.e., the line $\eta = \frac{ms}{b}\beta_r$:

$$\Delta\psi_m(\zeta) = \frac{1}{2\pi}\int_{-\infty}^{+\infty} f_m(\alpha)\mathrm{e}^{i\alpha\zeta_m}\,d\alpha. \tag{4.66}$$

In view of $\overline{p} = \psi(\zeta, \eta)e^{iMaK\zeta}$, the jump discontinuity of pressure can be described by:

$$\Delta\overline{p}_m(\zeta) = \frac{1}{2\pi}\int_{-\infty}^{+\infty} f_m(\alpha)\mathrm{e}^{i(\alpha + MaK)\zeta_m}\,d\alpha. \tag{4.67}$$

If the solution is just derived out by considering a single specific value of m, then the solution would appear to be continuous in the space where the nth ($n \neq m$) blade is. Obviously, it does not coincide with the physical truth. Only when the pressure p turns out to be discontinuous in every region occupied by every blade will the solution be of any physical meaning. Therefore, we seek the solution in the form of the superposition, i.e.,

$$\psi(\zeta, \eta) = \sum_{m=-\infty}^{+\infty} \psi_m(\zeta, \eta). \tag{4.68}$$

Thus, there will be a pressure discontinuity across each blade. The pressure jump $\Delta\overline{p}_m(\zeta)$ can be uniquely determined by:

$$\Delta\overline{p}(\zeta)\big|_m = \Delta\overline{p}_m(\zeta) = \frac{1}{2\pi}\int_{-\infty}^{+\infty} f_m(\alpha)\mathrm{e}^{i(\alpha + Ma_0K)\zeta_m}\,d\alpha. \tag{4.69}$$

Because the pressure is in the form of $p = \overline{p}e^{-i\omega t}$, we can assume $u_2 = \overline{u}_2 e^{-i\omega t}$. Inserting it into the third equation of the linearized Euler equations leads to:

$$-i\omega\overline{u}_2 + U\frac{\partial\overline{u}_2}{\partial X} = -\frac{1}{\rho_0}\frac{\partial\overline{p}}{\partial Y} \tag{4.70}$$

Introduce the Prandtl-Glauert dimensionless coordinates and substitute it into the expression of $\overline{p}$, then we can obtain:

$$-\frac{ib\omega}{U}\overline{u}_2 + \frac{\partial\overline{u}_2}{\partial\zeta} = -\frac{\beta_r}{\rho_0 U}e^{iMaK\zeta}\frac{\partial\psi}{\partial\eta}. \tag{4.71}$$

Based on $-\frac{ib\omega}{U}\overline{u}_2+\frac{\partial\overline{u}_2}{\partial\zeta}=\mathrm{e}^{i\frac{\omega}{bU}\zeta}\frac{\partial}{\partial\zeta}\left(\overline{u}_2\mathrm{e}^{-i\frac{\omega}{bU}\zeta}\right)$, Eq. (4.71) can be transformed into:

$$\frac{\partial}{\partial\zeta}\left(\overline{u}_2e^{-i\frac{\omega}{bU}\zeta}\right)=-\frac{\beta_r}{\rho_0U}e^{-i\left(\frac{\omega}{bU}-MaK\right)\zeta}\frac{\partial\psi}{\partial\eta}. \tag{4.72}$$

Then we can directly conduct an integral decomposition:

$$\begin{aligned}\overline{u}_2&=-e^{i\frac{\omega b}{U}\zeta}\int_{-\infty}^{\zeta}\frac{\beta_r}{\rho_0U}e^{-i\left(\frac{\omega b}{U}-MaK\right)\zeta}\frac{\partial\psi}{\partial\eta}d\zeta\\&=-e^{i\frac{\omega b}{U}\zeta}\int_{-\infty}^{\zeta}\frac{\beta_r}{\rho_0U}e^{-i\left(\frac{\omega b}{U}-MaK\right)\zeta}\\&\quad\times\frac{\partial}{\partial\eta}\left[\sum_{m=-\infty}^{+\infty}\frac{\mathrm{sgn}(\eta_m)}{2\pi}\int_{-\infty}^{+\infty}\frac{f_m(\alpha)}{2}\mathrm{e}^{i(\alpha\zeta_m+\gamma|\eta_m|)}d\alpha\right]d\zeta\\&=-\frac{\beta_r}{4\pi\rho_0U}e^{i\frac{K\beta_r^2}{Ma}\zeta}\frac{\partial}{\partial\eta}\\&\quad\left\{\int_{-\infty}^{+\infty}\left[\int_{-\infty}^{\zeta}e^{i\frac{K}{Ma}\zeta}\cdot\sum_{m=-\infty}^{+\infty}\mathrm{sgn}(\eta_m)f_m(\alpha)\mathrm{e}^{i(\alpha\zeta_m+\gamma|\eta_m|)}d\zeta\right]d\alpha\right\}\\&=-\frac{\beta_r}{4\pi i\rho_0U}\frac{\partial}{\partial\eta}\int_{-\infty}^{+\infty}\frac{e^{iKMa\zeta}}{\left(\alpha+\frac{K}{Ma}\right)}\cdot\sum_{m=-\infty}^{+\infty}\mathrm{sgn}(\eta_m)f_m(\alpha)\mathrm{e}^{i(\alpha\zeta_m+\gamma|\eta_m|)}d\alpha.\end{aligned} \tag{4.73}$$

The flow is subjected to the boundary conditions on the body surfaces, i.e., the normal velocity must vanish on the body surfaces, which can be expressed by $(u_2+u_{\infty,x2})|_{\eta=\frac{ms}{b}\beta_r}=0$. $u_{\infty,\,x2}$ represents the component of vortical velocity $\boldsymbol{u}_\infty$ (which is independent of pressure) in the direction normal to the blade chord. After the reasonable expression of $\boldsymbol{u}_\infty$ has been derived, we can further deduce that, on the upper surfaces of two adjacent blades, the component $u_2|_{\eta=\frac{ms}{b}\beta_r}$ of acoustic velocity in the normal direction of blade chord would have a constant phase difference σ, which can be described by:

$$\overline{u}_2\left(\zeta+\frac{ms^+}{b},\frac{ms}{b}\beta_r\right)=-e^{im\sigma}\overline{u}_2(\zeta,0). \tag{4.74}$$

It can be proved with Eq. (4.73) that, only if the following equation holds, the boundary conditions on the body surfaces mentioned earlier can be guaranteed.

$$f_m(\alpha) = e^{im\Gamma} f_0(\alpha), \tag{4.75}$$

where

$$\Gamma = \sigma - \frac{MaKs^+}{b}, \tag{4.76}$$

Therefore:

$$\overline{u}_2(\zeta,\eta) = -\frac{\beta_r}{4\pi i \rho_0 U} \frac{\partial}{\partial \eta} \int_{-\infty}^{+\infty} \frac{f_0(\alpha)}{\left(\alpha + \dfrac{K}{Ma}\right)} \sum_{m=-\infty}^{+\infty} \operatorname{sgn}(\eta_m) e^{i[m\sigma + (KMa+\alpha)\zeta_m + \gamma|\eta_m|]} d\alpha. \tag{4.77}$$

Furthermore:

$$\sum_{m=-\infty}^{+\infty} syn(\eta_m) e^{i\left[m\sigma - \frac{ms^+}{b}(KMa+\alpha) + \gamma|\eta_m|\right]} = \frac{1}{2i}\left[\frac{e^{i(\Delta_- + \eta\gamma)}}{\sinh\Delta_-} + \frac{e^{i(\Delta_+ - \eta\gamma)}}{\sinh\Delta_+}\right], \tag{4.78}$$

where the following relation has been used:

$$\sum_{m=0}^{+\infty} z^m = \frac{1}{1-z} \quad (|z|<1) \tag{4.79}$$

$$\Delta_\pm = \frac{1}{2}\left(\Gamma - \frac{\alpha s^+}{b} \pm \frac{s\beta_r}{b}\gamma\right), \tag{4.80}$$

and we already set $\gamma = \sqrt{\alpha^2 - K^2}$. Then, Eq. (4.77) becomes:

$$\overline{u}_2(\zeta,\eta) = \frac{\beta_r}{8\pi \rho_0 U} \frac{\partial}{\partial \eta} \int_{-\infty}^{+\infty} \frac{f_0(\alpha) e^{i(KMa+\alpha)\zeta}}{\left(\alpha + \dfrac{K}{Ma}\right)} \left[\frac{e^{i(\Delta_- + \eta\gamma)}}{\sinh\Delta_-} + \frac{e^{i(\Delta_+ - \eta\gamma)}}{\sinh\Delta_+}\right] d\alpha. \tag{4.81}$$

And from Eq. (4.69) we can obtain:

$$\Delta\overline{p}(\zeta)|_0 = \Delta\overline{p}_0(\zeta) = \frac{1}{2\pi} \int_{-\infty}^{+\infty} f_0(\alpha) e^{i(\alpha + MaK)\zeta} d\alpha. \tag{4.82}$$

The inverse Fourier transform leads to:

$$f_0(\alpha) = \int_{-1}^{1} \Delta\overline{p}_0(\zeta) \mathrm{e}^{-i(\alpha + MaK)\zeta} d\zeta, \tag{4.83}$$

where the integral interval is defined to be $[-1,1]$. Consider the actual case: there would be no pressure difference on the axis of ζ except for the region where the blade is, i.e., $\Delta\overline{p}_0(\zeta) = 0$ for $\zeta \in (-\infty, 1) \cup (1, +\infty)$.

Then we know:

$$\overline{u}_2(\zeta, \eta) = \int_{-1}^{1} \frac{\Delta\overline{p}_0(\zeta_0)}{\rho_0 U} \frac{\beta_r}{8\pi} \frac{\partial}{\partial\eta} \int_{-\infty}^{+\infty} \frac{\mathrm{e}^{i(KMa + \alpha)(\zeta - \zeta_0)}}{\left(\alpha + \dfrac{K}{Ma}\right)} \left[\frac{\mathrm{e}^{i(\Delta_- + \eta\gamma)}}{\sinh\Delta_-} + \frac{\mathrm{e}^{i(\Delta_+ - \eta\gamma)}}{\sinh\Delta_+} \right] d\alpha d\zeta_0.$$

By defining the following quantities:

$$F_0(\zeta_0) = \frac{\Delta\overline{p}_0(\zeta_0)}{\rho_0 U} \tag{4.84}$$

$$K(\zeta, \eta) = \frac{\beta_r}{8\pi} \frac{\partial}{\partial\eta} \int_{-\infty}^{+\infty} \frac{\mathrm{e}^{i(KMa + \alpha)\zeta}}{\left(\alpha + \dfrac{K}{Ma}\right)} \left[\frac{\mathrm{e}^{i(\Delta_- + \eta\gamma)}}{\sinh\Delta_-} + \frac{\mathrm{e}^{i(\Delta_+ - \eta\gamma)}}{\sinh\Delta_+} \right] d\alpha, \tag{4.85}$$

we can obtain:

$$\overline{u}_2(\zeta, \eta) = \int_{-1}^{1} F_0(\zeta_0) K(\zeta - \zeta_0) d\zeta_0. \tag{4.86}$$

When our concern is the velocity field on the blade surfaces, we can take the limit:

$$\begin{aligned} K(\zeta, \eta) &= \lim_{\eta \to 0} K(\zeta, \eta) \\ &= \frac{\beta_r}{8\pi} \lim_{\eta \to 0} \frac{\partial}{\partial\eta} \int_{-\infty}^{+\infty} \frac{\mathrm{e}^{i(KMa + \alpha)\zeta}}{\left(\alpha + \dfrac{K}{Ma}\right)} \left[\frac{\mathrm{e}^{i(\Delta_- + \eta\gamma)}}{\sinh\Delta_-} + \frac{\mathrm{e}^{i(\Delta_+ - \eta\gamma)}}{\sinh\Delta_+} \right] d\alpha, \end{aligned} \tag{4.87}$$

This integral can be solved by applying the residual theorem for complex variable functions. It is easy to figure out that the expression of $\overline{u}_2$ has two kinds of first-order poles:

(1) $\alpha = -\frac{K}{Ma}$

(2) $\Delta_{\pm} = n\pi \quad (n = 0, \pm 1, \pm 2, \cdots)$ (According to this, an infinite number of first-order poles can be determined.)

Inserting $\Delta_{\pm} = n\pi$ into Eq. (4.80) yields:

$$\frac{1}{2}\left(\Gamma - \frac{\alpha_n^{\pm} s^+}{b} \pm \frac{s\beta_r}{b}\sqrt{\left(\alpha_n^{\pm}\right)^2 - K^2}\right) = n\pi \tag{4.88}$$

where $\alpha_n^{\pm}$ represents the values of α that satisfy the condition $\Delta_{\pm} = n\pi$, and the superscript "+" and "−" respectively denote propagating upstream and downstream.

The solution comes out to be:

$$\alpha_n^{\pm} = \frac{\Gamma_n s^+}{b(d^+)^2} \mp \frac{s\beta_r}{bd^+}\sqrt{\left(\frac{\Gamma_n}{d^+}\right)^2 - K^2} \quad (\Gamma_n \leq 0) \tag{4.89}$$

where

$$d^+ = \frac{\sqrt{(s^+)^2 + s^2\beta_r^2}}{b} \tag{4.90}$$

$$\Gamma_n = \Gamma - 2n\pi \tag{4.91}$$

For $\zeta > 0$ and $\zeta < 0$, the integral $K_c(\zeta, \eta)$ in the complex plane are different. For $\zeta > 0$, $K_c(\zeta, \eta)$ should be the opposite of the sum of all the residues in the upper half complex plane, whereas for $\zeta < 0$, $K_c(\zeta, \eta)$ equals the sums of all the residues in the lower half plane, i.e.,

$$K(\zeta, \eta) = -2\pi i \frac{\partial}{\partial \eta} \sum_{n=-\infty}^{+\infty} \mathrm{Res}\left[F_c(\alpha), \alpha_n^-\right] \quad (\zeta > 0) \tag{4.92}$$

$$K(\zeta, \eta) = 2\pi i \frac{\partial}{\partial \eta} \sum_{n=-\infty}^{+\infty} \mathrm{Res}\left[F_c(\alpha), \alpha_n^+\right] \quad (\zeta < 0) \tag{4.93}$$

where

$$\begin{cases} F_c(\alpha) = \dfrac{P_c(\alpha)}{Q_c(\alpha)} \\ P_c(\alpha^{\pm}) = \dfrac{\beta_r}{8\pi} \mathrm{e}^{i(KMa + \alpha)\zeta} \mathrm{e}^{i(\Delta_{\pm} \mp \eta\gamma)} \\ Q_c(\alpha^{\pm}) = \sinh \Delta_{\pm} \end{cases} \tag{4.94}$$

Due the property of the first-order poles, the residue $Res[F_c(\alpha), \alpha_n^{\pm}]$ can be calculated by the following formula:

$$\mathrm{Res}\left[F_c(\alpha), \alpha_n^{\pm}\right] = \frac{P_c\left(\alpha_n^{\pm}\right)}{Q_c'\left(\alpha_n^{\pm}\right)}, \tag{4.95}$$

because

$$\frac{\partial \Delta_+}{\partial \alpha} = \frac{1}{2}\left(-\frac{s^+}{b} + \frac{s\beta_r}{b}\frac{\alpha}{\sqrt{\alpha^2 - K^2}}\right) = \frac{1}{2\gamma}\left(-\frac{\gamma s^+}{b} + \frac{s\beta_r}{b}\alpha\right), \tag{4.96}$$

Eq. (4.88) implies:

$$\gamma\left(\alpha_n^+\right) = -\frac{b}{s\beta_r}\left(\Gamma_n - \frac{\alpha_n^+ s^+}{b}\right), \tag{4.97}$$

Thus:

$$\left.\frac{\partial \Delta_+}{\partial \alpha}\right|_{\alpha=\alpha_n^+} = \frac{\left[\left(\Gamma_n - \frac{\alpha_n^+ s^+}{b}\right)\frac{s^+}{b} + \frac{s^2\beta_r^2}{b^2}\alpha_n^+\right]}{-2\left(\Gamma_n - \frac{\alpha_n^+ s^+}{b}\right)} = \frac{s\beta_r}{b}\frac{\left(\Gamma_n\frac{s^+}{b} + d^{+2}\alpha_n^+\right)}{2\gamma\left(\alpha_n^+\right)} \tag{4.98}$$

In a similar fashion, we get:

$$\left.\frac{\partial \Delta_-}{\partial \alpha}\right|_{\alpha=\alpha_n^-} = \frac{s\beta_r}{b}\frac{\left(\Gamma_n\frac{s^+}{b} - d^{+2}\alpha_n^-\right)}{2\gamma\left(\alpha_n^-\right)} \tag{4.99}$$

To sum up, we obtain:

(1) $X < 0$

$$\begin{aligned} K(X^+) &= \frac{s\beta_r^2}{2b}\sum_{n=-\infty}^{+\infty}\frac{\left(\alpha_n^{+2} - K^2\right)e^{i\left(KMa + \alpha_n^+\right)\zeta}}{\left(\alpha_n^+ + \frac{K}{Ma}\right)\left(\Gamma_n\frac{s^+}{b} - d^{+2}\alpha_n^+\right)} \\ &= -\frac{s\beta_r^2}{2bd^{+2}}\sum_{n=-\infty}^{+\infty}\frac{\left(\alpha_n^{+2} - K^2\right)e^{i\left(KMa + \alpha_n^+\right)\zeta}}{\left(\alpha_n^+ + \frac{K}{Ma}\right)\left(\alpha_n^+ - \Gamma_n\frac{s^+}{bd^{+2}}\right)}, \end{aligned} \tag{4.100}$$

(2) $X > 0$

$$K_1(X^-) = \frac{s\beta_r^2}{2bd^{+2}} \sum_{n=-\infty}^{+\infty} \frac{\left(\alpha_n^{-2} - K^2\right) e^{i\left(KMa + \alpha_n^-\right)\zeta}}{\left(\alpha_n^- + \dfrac{K}{Ma}\right)\left(\alpha_n^- - \Gamma_n \dfrac{s^+}{bd^{+2}}\right)}. \tag{4.101}$$

For $\alpha = -\frac{K}{Ma}$, because:

$$\Delta_+ = \frac{1}{2}\left(\Gamma - \frac{\alpha s^+}{b} + \frac{s\beta_r}{b}\gamma\right)\Bigg|_{\alpha = -\dfrac{K}{Ma}} = \frac{1}{2}\left(\sigma + \frac{K\beta_r^2 s}{Mab} i\right), \tag{4.102}$$

$$\Delta_- = \frac{1}{2}\left(\sigma - \frac{K\beta_r^2 s}{Mab} i\right), \tag{4.103}$$

then we have

$$K_2(X^-) = \frac{\beta_r^2 K}{2Ma} \frac{\sinh\left(\dfrac{\beta_r^2 Ks}{Mab}\right) \exp\left(i\dfrac{\beta_r^2 K}{Ma}\zeta\right)}{\cosh\left(\dfrac{\beta_r^2 Ks}{Mab}\right) - \cos\left(\Gamma + \dfrac{Ks^+}{Mab}\right)}. \tag{4.104}$$

Obviously, for $X > 0$:

$$\begin{aligned} K(X^-) &= K_1(X^-) + K_2(X^-) \\ &= \frac{s\beta_r^2}{2bd^{+2}} \sum_{n=-\infty}^{+\infty} \frac{\left(\alpha_n^{-2} - K^2\right) e^{i\left(KMa + \alpha_n^-\right)\zeta}}{\left(\alpha_n^- + \dfrac{K}{Ma}\right)\left(\alpha_n^- - \Gamma_n \dfrac{s^+}{bd^{+2}}\right)} \\ &\quad + \frac{\beta_r^2 K}{2Ma} \frac{\sinh\left(\dfrac{\beta_r^2 Ks}{Mab}\right) \exp\left(i\dfrac{\beta_r^2 K}{Ma}\zeta\right)}{\cosh\left(\dfrac{\beta_r^2 Ks}{Mab}\right) - \cos\left(\Gamma + \dfrac{Ks^+}{Mab}\right)}. \end{aligned} \tag{4.105}$$

4.3.2 Discussion concerning the convergence of the kernel function

A deeper investigation on the kernel function expressions Eqs. (4.100) and (4.105) indicates that, when $n \to \infty$, the series is divergent, which is just opposite to the case of subsonic conditions. We are now going to discuss the convergence of the kernel functions and then propose a method that can ensure numerical convergence.

According to the expression of $\alpha_n^{\pm}$, it is known that, when $n \to \infty$:

$$\alpha_n^+ = \frac{\Gamma_n s^+}{bd^{+2}} + \frac{s\beta_r \Gamma_n}{bd^+}\left(1 - \frac{d^{+2}K^2}{2\Gamma_n^2}\right) = \frac{b\Gamma_n}{s^+ - s\beta_r} - \frac{s\beta_r K^2}{2b\Gamma_n} + o\left(n^{-2}\right), \quad (4.106)$$

$$\alpha_n^- = \frac{b\Gamma_n}{s^+ + s\beta_r} - \frac{s\beta_r K^2}{2b\Gamma_n} + o\left(n^{-2}\right). \quad (4.107)$$

Substituting these two formulas into Eqs. (4.100) and (4.105) leads to:

$$\begin{aligned} K_s^+ &= \frac{-\beta_r}{2(s^+ - \beta_r s)} e^{iKMa\zeta} \sum_{n=-\infty}^{+\infty} e^{i\frac{\Gamma_n}{s^+ - \beta_r s}\zeta} \\ &= \frac{-\beta_r}{2(s^+ - \beta_r s)} e^{iKMa\zeta} e^{i\frac{\sigma - MaKs^+}{s^+ - \beta_r s}\zeta} \sum_{n=-\infty}^{+\infty} e^{-i\frac{2\pi n}{s^+ - \beta_r s}\zeta} \quad (\zeta < 0), \end{aligned} \quad (4.108)$$

In view of:

$$\sum_{n=-\infty}^{+\infty} e^{-i\frac{2\pi n}{s^+ - \beta_r s}\zeta} = -(s^+ - \beta_r s) \sum_{n=-\infty}^{+\infty} \delta[\zeta - n(s^+ - \beta_r s)], \quad (4.109)$$

Eq. (4.108) can be further simplified to:

$$\begin{aligned} K_s^+ &= \frac{\beta_r}{2} e^{iKMa\zeta} e^{i\frac{\sigma - MaKs^+}{s^+ - \beta_r s}\zeta} \sum_{n=-\infty}^{+\infty} \delta(\zeta_n + \beta_r sn) \\ &= \frac{\beta_r}{2} e^{iKMa\zeta} \sum_{n=-\infty}^{+\infty} e^{in(\sigma - MaKs^+)} \delta(\zeta_n + \beta_r sn) \\ &= \frac{\beta_r}{2} \sum_{n=-\infty}^{+\infty} e^{i(n\sigma + MaKsn)} \delta(\zeta_n + \beta_r sn) \\ &= \frac{\beta_r}{2} \sum_{n=-\infty}^{+\infty} e^{in(\sigma - MaK\beta_r s/b)} \delta(\zeta - nd_2). \end{aligned} \quad (4.110)$$

In a similar fashion we can obtain:

$$K_s^- = \frac{\beta_r}{2} \sum_{n=-\infty}^{+\infty} e^{in(\sigma + MaK\beta_r s/b)} \delta(\zeta - nd_1) \quad (\zeta > 0), \quad (4.111)$$

where

$$\begin{cases} d_2 = s^+ - \beta_r s \\ d_1 = s^+ + \beta_r s \end{cases}. \quad (4.112)$$

d1 and d2 are indicated in Fig. 4.2.

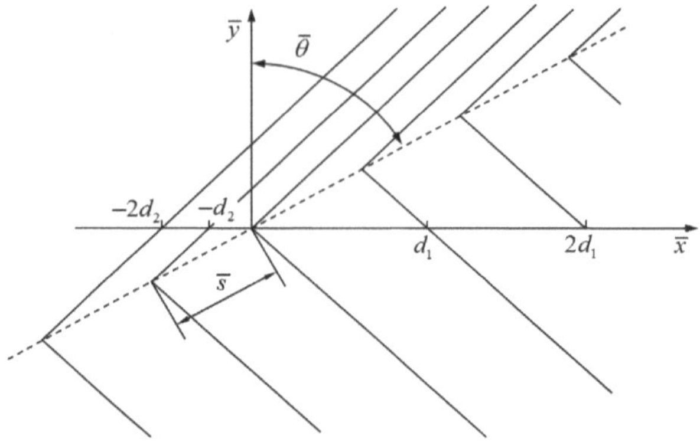

Fig. 4.2 The Mach waves for a supersonic cascade with subsonic axial flow velocity.

With the coordinate transformation, d_1 and d_2 can be depicted as shown in Fig. 4.2.

Eqs. (4.110) and (4.111) reveal that the kernel function would turn out to be a step function when n approaches infinity. When $K = 0$, i.e., the blades are in steady condition, the only term left in the kernel function is K_s^+ or K_s^-, which means we can derive the steady solution due to the Mach wave by inserting K_s^+ and K_s^- into the integral equation. Hence, the kernel function can be resolved into the steady part and the unsteady part to ensure the convergence.

(1) $\zeta > 0$:

$$K(X^-) = \frac{2}{bs}\sum_{n=-\infty}^{+\infty}\left\{ \frac{\left(\Gamma_n - \frac{s^+}{b}\alpha_n^-\right)^2 e^{i\left(\alpha_n^- + MaK\right)\zeta}}{\left(\alpha_n^- + \frac{K}{Ma}\right)\left(\Gamma_n \frac{s^+}{b} - d^{+2}\alpha_n^-\right)} - \frac{\beta_r s}{s^+ + \beta_r s} e^{i\frac{b\Gamma_n}{s^+ + \beta_r s}\zeta}\right\}$$
$$+ \frac{\beta_r}{2}\sum_{n=-\infty}^{+\infty} e^{in(\sigma + MaK\beta_r s/b)}\delta(\zeta - nd_1)$$
$$+ \frac{K\beta_r^2}{2Ma}\frac{\sinh\left(\frac{\beta_r^2 Ks}{Mab}\right) e^{i\frac{\beta_r^2 K}{Ma}\zeta}}{\cosh\left(\frac{\beta_r^2 Ks}{Mab}\right) - \cos\left(\Gamma + \frac{Ks^+}{Mab}\right)}, \tag{4.113}$$

(2) $\zeta < 0$:

$$K(X^+) = \frac{2}{bs}\sum_{n=-\infty}^{+\infty}\left\{\frac{\left(\Gamma_n - \frac{s^+}{b}\alpha_n^+\right)^2 e^{i\left(\alpha_n^+ + MaK\right)\zeta}}{\left(\alpha_n^+ + \frac{K}{Ma}\right)\left(\Gamma_n\frac{s^+}{b} - d^{+2}\alpha_n^+\right)} - \frac{\beta_r s}{s^+ - \beta_r s}e^{i\frac{b\Gamma_n}{s^+ - \beta_r s}\zeta}\right\}$$
$$+\frac{\beta_r}{2}\sum_{n=-\infty}^{+\infty} e^{in(\sigma - MaK\beta_r s/b)}\delta(\zeta - nd_2). \tag{4.114}$$

By conducting such transforms, the series in Eqs. (4.113) and (4.114) become convergent. Nevertheless, the relevant references indicate the convergence speed of the computations based on a direct application of Eqs. (4.113) and (4.114) is very slow. To improve the numerical convergence speed, we can retain the $\frac{1}{n}$ term of α_n^+ and α_n^- when $n \to \infty$.

Now inserting Eq. (4.107) into the nth term of Eq. (4.101) gives:

$$\frac{\left(\Gamma_n - \frac{\alpha_n^- s^+}{b}\right)^2 e^{i\alpha_n^-\zeta}}{\left(\alpha_n^- + \frac{K}{Ma}\right)\left(\Gamma_n\frac{s^+}{b} - d^{+2}\alpha_n^-\right)}$$
$$\approx \frac{\left(1 - \frac{s^+}{s^+ + \beta_r s}\right)^2 e^{i\left(\frac{\Gamma_n b}{s^+ + \beta_r s} + \frac{K^2 s\beta_r}{\Gamma_n b}\right)\zeta}}{\left(\frac{b}{s^+ + \beta_r s} + \frac{K}{Ma\Gamma_n}\right)\left[s^+ - (s^+ - \beta_r s)\right]\frac{1}{b}}$$
$$\approx \frac{\beta_r s}{s^+ + \beta_r s}\left[1 - \frac{K(s^+ + \beta_r s)}{Ma\Gamma_n b}\right]\left(1 + i\frac{K^2 s\beta_r}{2b\Gamma_n}\zeta\right)\exp\left(i\frac{\Gamma_n b}{s^+ + \beta_r s}\zeta\right)$$
$$\approx \frac{\beta_r s}{s^+ + \beta_r s}\left\{1 + \frac{K}{2\pi nbMa}\left[s^+ + s\beta_r\left(1 - i\frac{MaK}{2}\zeta\right)\right]\right\}\exp\left(i\frac{\Gamma_n b}{s^+ + \beta_r s}\zeta\right). \tag{4.115}$$

Similarly, putting Eq. (4.106) into Eq. (4.100) leads to:

$$-\frac{\left(\Gamma_n - \frac{\alpha_n^+ s^+}{b}\right)^2 e^{-i\alpha_n^+\zeta}}{\left(\alpha_n^+ + \frac{K}{Ma}\right)\left(\Gamma_n\frac{s^+}{b} - d^{+2}\alpha_n^+\right)}$$
$$\approx \frac{\beta_r s}{s^+ - \beta_r s}\left\{1 + \frac{K}{2\pi nbMa}\left[s^+ - s\beta_r\left(1 - i\frac{MaK}{2}\zeta\right)\right]\right\}\exp\left(i\frac{\Gamma_n b}{s^+ - \beta_r s}\zeta\right). \tag{4.116}$$

Upon using Eq. (4.115), Eqs. (4.116), (4.113), and (4.114) can be further deduced,for $\zeta > 0$:

$$K(\zeta^-) = \frac{K\beta_r^2}{2Ma}\frac{\sinh\left(\dfrac{\beta_r^2 Ks}{Mab}\right)e^{i\frac{\beta_r^2 K}{Ma}\zeta}}{\cosh\left(\dfrac{\beta_r^2 Ks}{Mab}\right) - \cos\left(\Gamma + \dfrac{Ks^+}{Mab}\right)}$$
$$+\frac{2}{bs}\sum_{n=-\infty}^{+\infty}\left\{\frac{\left(\Gamma_n - \dfrac{s^+}{b}\alpha_n^-\right)^2 e^{i(\alpha_n^- + MaK)\zeta}}{\left(\alpha_n^- + \dfrac{K}{Ma}\right)\left(\Gamma_n\dfrac{s^+}{b} - d^{+2}\alpha_n^-\right)}\right.$$
$$\left.-\frac{\beta_r s e^{iMaK\zeta}}{s^+ + \beta_r s}\left[1 + \frac{K}{2\pi nbMa}\left[s^+ + s\beta\left(1 - \frac{iMaK\zeta}{2}\right)\right]\right]\exp\left(i\frac{\Gamma_n b}{s^+ + \beta_r s}\zeta\right)\right\}$$
$$+\frac{\beta_r}{2}\sum_{n=-\infty}^{+\infty} e^{in(\sigma + MaK\beta_r s/b)}\delta(\zeta - nd_1)$$
$$+\sum_{n=-\infty}^{+\infty}\frac{K\beta_r e^{iMaK\zeta}}{4\pi nMa(s^+ + \beta_r s)}\left[s^+ + s\beta_r\left(1 - \frac{iMaK\zeta}{2}\right)\right]\exp\left(i\frac{\Gamma_n b}{s^+ + \beta_r s}\zeta\right), \tag{4.117}$$

for$\zeta < 0$:

$$K(\zeta^-) = -\frac{2}{bs}\sum_{n=-\infty}^{+\infty}\left\{\frac{\left(\Gamma_n - \dfrac{s^+}{b}\alpha_n^+\right)^2 e^{i(\alpha_n^+ + MaK)\zeta}}{\left(\alpha_n^+ + \dfrac{K}{Ma}\right)\left(\Gamma_n\dfrac{s^+}{b} - d^{+2}\alpha_n^+\right)}\right.$$
$$\left.+\frac{\beta_r s e^{iMaK\zeta}}{s^+ - \beta_r s}\left[1 + \frac{K}{2\pi nbMa}\left[s^+ - s\beta\left(1 - \frac{iMaK\zeta}{2}\right)\right]\right]\exp\left(i\frac{\Gamma_n b}{s^+ - \beta_r s}\zeta\right)\right\}$$
$$+\frac{\beta_r}{2}\sum_{n=-\infty}^{+\infty} e^{in(\sigma - MaK\beta_r s/b)}\delta(\zeta - nd_2)$$
$$+\sum_{n=-\infty}^{+\infty}\frac{K\beta_r e^{iMaK\zeta}}{4\pi nMa(s^+ - \beta_r s)}\left[s^+ - s\beta_r\left(1 - \frac{iMaK\zeta}{2}\right)\right]\exp\left(i\frac{\Gamma_n b}{s^+ - \beta_r s}\zeta\right). \tag{4.118}$$

For the two previous equations, introducing the following variables:

$$K_W(\zeta^-) = \sum_{n=-\infty}^{+\infty}\frac{K\beta_r e^{iMaK\zeta}}{4\pi nMa(s^+ + \beta_r s)}\left[s^+ + s\beta_r\left(1 - \frac{iMaK\zeta}{2}\right)\right]\exp\left(i\frac{\Gamma_n b}{s^+ + \beta_r s}\zeta\right), \tag{4.119}$$

$$K_W(\zeta^+) = \sum_{n=-\infty}^{+\infty}\frac{K\beta_r e^{iMaK\zeta}}{4\pi nMa(s^+ - \beta_r s)}\left[s^+ - s\beta_r\left(1 - \frac{iMaK\zeta}{2}\right)\right]\exp\left(i\frac{\Gamma_n b}{s^+ - \beta_r s}\zeta\right), \tag{4.120}$$

leads to:

$$
\begin{aligned}
K_W(\zeta^-) = &\frac{K\beta_r}{4\pi Ma}\left[\frac{s^+ + s\beta_r\left(1-\frac{iMaK\zeta}{2}\right)}{s^+ + \beta_r s}\right]\\
&\cdot \exp\left(i\frac{\sigma+\beta_r MaK\frac{s}{b}}{s^+ + \beta_r s}\zeta\right)\cdot \sum_{n=-\infty}^{+\infty}\frac{(1-\delta_{n,0})}{n}e^{-\frac{2\pi nb}{s^+ + \beta_r s}\zeta},
\end{aligned}
\tag{4.121}
$$

$$
\begin{aligned}
K_W(\zeta^+) = &\frac{K\beta_r}{4\pi Ma}\left[\frac{s^+ - s\beta_r\left(1-\frac{iMaK\zeta}{2}\right)}{s^+ - \beta_r s}\right]\\
&\cdot \exp\left(i\frac{\sigma-\beta_r MaK\frac{s}{b}}{s^+ - \beta_r s}\zeta\right)\cdot \sum_{n=-\infty}^{+\infty}\frac{(1-\delta_{n,0})}{n}e^{-\frac{2\pi nb}{s^+ - \beta_r s}\zeta},
\end{aligned}
\tag{4.122}
$$

where

$$
\delta_{n,0} = \begin{cases} 1 & (n=0) \\ 0 & (n\neq 0) \end{cases}
\tag{4.123}
$$

Because:

$$
\begin{aligned}
\sum_{n=-\infty}^{+\infty}\frac{(1-\delta_{n,0})}{n}e^{-\frac{2\pi nb}{s^+ + \beta_r s}\zeta} &= \sum_{n=1}^{+\infty}-\frac{1}{n}\left[\cos\left(\frac{2\pi b\zeta}{s^+ + \beta_r s}n\right)+i\sin\left(\frac{2\pi b\zeta}{s^+ + \beta_r s}n\right)\right]\\
&+\sum_{n=1}^{+\infty}\frac{1}{n}\left[\cos\left(\frac{2\pi b\zeta}{s^+ + \beta_r s}n\right)-i\sin\left(\frac{2\pi b\zeta}{s^+ + \beta_r s}n\right)\right]\\
&= -2i\sum_{n=1}^{+\infty}\frac{\sin\left(\frac{2\pi b\zeta}{s^+ + \beta_r s}n\right)}{n}\\
&= -2i\sum_{n=1}^{+\infty}\frac{\sin 2\pi\left(\frac{b\zeta}{s^+ + \beta_r s}-LI\right)n}{n} \quad \left(LI=\frac{b\zeta}{s^+ + \beta_r s}\text{ takes an integer value}\right)\\
&= -2i\frac{\pi-2\pi\left(\frac{b\zeta}{s^+ + \beta_r s}-LI\right)}{2}\\
&= -\pi i\left(1-\frac{2b\zeta}{s^+ + \beta_r s}+2LI\right),
\end{aligned}
\tag{4.124}
$$

with a similar approach, we obtain:

$$\sum_{n=-\infty}^{+\infty} \frac{(1-\delta_{n,0})}{n} e^{-\frac{2\pi nb}{s^+-\beta_r s}\zeta} = \pi i \left(1 + \frac{2b\zeta}{s^+-\beta_r s} - 2LJ\right), \tag{4.125}$$

where $LJ = \frac{b\zeta}{s^+-\beta_r s}$ is assigned with an integer value.

With the analysis previously discussed, an expression for the kernel function that is of improved convergence speed can be formulated as follows: for $\zeta > 0$:

$$\begin{aligned}
K(\zeta^-) = {} & \frac{K\beta_r^2}{2Ma} \frac{\sinh\left(\frac{\beta_r^2 Ks}{Mab}\right) e^{i\frac{\beta_r^2 K}{Ma}\zeta}}{\cosh\left(\frac{\beta_r^2 Ks}{Mab}\right) - \cos\left(\Gamma + \frac{Ks^+}{Mab}\right)} \\
& + \frac{2}{bs} \sum_{n=-\infty}^{+\infty} \left\{ \frac{\left(\Gamma_n - \frac{s^+}{b}\alpha_n^-\right)^2 e^{i(\alpha_n^- + MaK)\zeta}}{\left(\alpha_n^- + \frac{K}{Ma}\right)\left(\Gamma_n \frac{s^+}{b} - d^{+2}\alpha_n^-\right)} \right. \\
& \left. - \frac{\beta_r s}{s^+ + \beta_r s} e^{iMaK\zeta + i\frac{\Gamma_n b}{s^+ + \beta_r s}\zeta} \left[1 + \frac{K}{2\pi nbMa}\left[s^+ + s\beta\left(1 - \frac{iMaK\zeta}{2}\right)\right]\right] \right\} \\
& + \frac{\beta_r}{2} \sum_{n=-\infty}^{+\infty} e^{in(\sigma + MaK\beta_r s/b)} \delta(\zeta - nd_1) - \frac{iK\beta_r}{4Ma} \left[\frac{s^+ + s\beta_r\left(1 - \frac{iMaK\zeta}{2}\right)}{s^+ + s\beta_r}\right] \\
& \cdot \exp\left(i\frac{b\sigma + \beta_r MaKs}{s^+ + \beta_r s}\zeta\right)\left(1 - \frac{2b\zeta}{s^+ + \beta_r s} + 2LI\right),
\end{aligned} \tag{4.126}$$

for $\zeta < 0$:

$$\begin{aligned}
K(\zeta^-) = {} & -\frac{2}{bs} \sum_{n=-\infty}^{+\infty} \left\{ \frac{\left(\Gamma_n - \frac{s^+}{b}\alpha_n^+\right)^2 e^{i(\alpha_n^+ + MaK)\zeta}}{\left(\alpha_n^+ + \frac{K}{Ma}\right)\left(\Gamma_n \frac{s^+}{b} - d^{+2}\alpha_n^+\right)} \right. \\
& \left. + \frac{\beta_r s}{s^+ - \beta_r s} e^{iMaK\zeta + i\frac{\Gamma_n b}{s^+ - \beta_r s}\zeta} \left[1 + \frac{K}{2\pi nbMa}\left[s^+ - s\beta\left(1 - \frac{iMaK\zeta}{2}\right)\right]\right] \right\} \\
& + \frac{\beta_r}{2} \sum_{n=-\infty}^{+\infty} e^{in(\sigma - MaK\beta_r s/b)} \delta(\zeta - nd_2) + \frac{iK\beta_r}{4Ma} \left[\frac{s^+ - s\beta_r\left(1 - \frac{iMaK\zeta}{2}\right)}{s^+ - s\beta_r}\right] \\
& \cdot \exp\left(i\frac{b\sigma - \beta_r MaKs}{s^+ - \beta_r s}\zeta\right)\left(1 + \frac{2b\zeta}{s^+ + \beta_r s} - 2LJ\right).
\end{aligned} \tag{4.127}$$

4.3.3 Reflection coefficients of Mach waves and the solution of the integral equation

In Eq. (4.87), taking the limit of $\eta \to 0$ leads to:

$$\frac{\overline{u}_2(\zeta)}{U} = \int_{-1}^{1} F_p(\zeta_0) K(\zeta - \zeta_0) d\zeta_0, \tag{4.128}$$

where $F_p(\zeta_0) = \frac{\Delta \overline{p}_0(\zeta_0)}{\rho_0 U^2}$ and $K(\zeta - \zeta_0)$ are defined by Eqs. (4.126) and (4.127), respectively.

As earlier analyzed, the emergence of the step function in the kernel function of Eqs. (4.126) and (4.127) should be attributed to the reflections of Mach waves in the cascade. At the same time, it is obvious that, when we conduct the integral by inserting Eqs. (4.126) and (4.127) into the integral Eq. (4.128), the term in the form of $F_p(\zeta + nd_{1,\,2})$ would appear, which corresponds to the value at the pressure discontinuity point $(\zeta + nd_{1,\,2})$ due to the reflections of Mach waves. These discontinuities set up huge obstacles in the numerical computation process of the integral equation. To overcome this difficulty, it is suggested that the discontinuous pressure function F_p could be decomposed into a continuous function plus the additions relevant to the Mach wave reflections [21]. The decomposition is formulated by:

$$F_p(\zeta) = g(\zeta) + \sum_{N_r} F_\alpha \left[g(\zeta + d_p) \right]_{\zeta_1}^{\zeta_2}, \tag{4.129}$$

where F_α is the reflectance of Mach wave, N_r is the number of reflection times, and (ζ_1, ζ_2) denotes the influence domain of the Mach wave reflectance.

Finally, combined with the solution addressed in this book in a similar fashion as that given by Nagashima and Whitehead [21], the Mach wave reflectance and the region of its domain of influence are obtained as given in Table 4.1.

Table 4.1 The Mach wave reflectance and the region of its domain of influence.

Starting point of waves	Reflection times	F_α	d_p	ζ_1	ζ_2
Upstream	1	$-E_2$	d_2	-0.5	$0.5 - d_2$
Downstream	1	$-E_1$	$-d_1$	$-0.5 + d_1$	0.5
Upstream	2	E_1E_2	$-d_1 + d_2$	$-0.5 + d_1$	0.5
Downstream	2	E_1E_2	$-d_1 + d_2$	$-0.5 + d_1 - d_2$	$0.5 - d_2$
Upstream	3	$-E_1E_2$	$-d_1 + 2d_2$	$-0.5 + d_1 - d_2$	$0.5 - d_2$

where

$$\begin{cases} E_1 = \exp\left(i\sigma + iMaK\beta_r\frac{s}{b}\right) \\ E_2 = \exp\left(-i\sigma + iMaK\beta_r\frac{s}{b}\right). \\ d_1 = s^+ + s\beta_r \\ d_2 = s^+ - s\beta_r \end{cases}$$

Fig. 4.3 depict several common shapes of the Mach waves. By inserting Eq. (4.129) into Eq. (4.128), we can deduce that:

$$\frac{\overline{u}_2(\zeta)}{U} = \int_{-1}^{1} g(\zeta)\left\{K(\zeta-\zeta_0) + \sum_{N_r}\left[F_\alpha K(\zeta-\zeta_0+d_p)\right]_{\zeta_1+d_p}^{\zeta_2+d_p}\right\}d\zeta_0. \quad (4.130)$$

It can be demonstrated, upon substituting the kernel function into this equation, that in the curly brace, all the functions other than $\delta(\zeta - \zeta_0)$ will cancel each other eventually. Thus, there will be no essential difficulty any longer in the solving procedure of the integral equation. In fact, in view of the property of the only remaining term $\delta(\zeta - \zeta_0)$, a rearrangement of this expression yields [30]:

$$\frac{\overline{u}_2(\zeta)}{U} = \frac{\beta_r}{2}g_N(\zeta_i) + {\sum_{j=0}^{N}}' W_j K_q(\zeta_i, \zeta_{0i})g_N(\zeta_i) \quad (i=0, 1, \ldots, N), \quad (4.131)$$

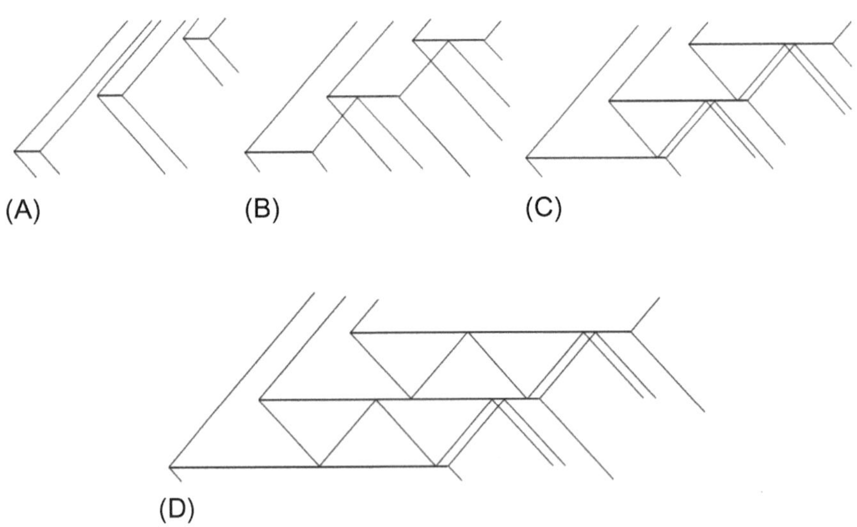

Fig. 4.3 The Mach waves. (A) $d_2 > 1$, (B) $d_2 < 1$, $d_1 > 1$, (C) $d_1 < 1$, $(2d_1 - d_2) > 1$, (D) $(2d_1 - d_2) < 1$, $(3d_1 - 2d_2) > 1$.

where

$$W_j = \frac{4}{N} \sum_{j=0}^{N} \frac{1}{1-r^2} \cos\left(\frac{rj\pi}{N}\right) \quad (r \text{ is assigned with the even numbers}), \tag{4.132}$$

$$\zeta_j = \zeta_{0j} = \cos\left(\frac{j\pi}{N}\right) \quad (j = 0, 1, \cdots, N), \tag{4.133}$$

and $\sum_{j=0}^{N}{}'$ denotes the summation with the first term, and the last term in the summation takes their half values. Also:

$$K_q(\zeta, \zeta_0) = K_q(\zeta - \zeta_0) + \sum_{N_r} \left[F_\alpha K(\zeta - \zeta_0 + d_p)\right]_{\zeta_1 + d_p}^{\zeta_2 + d_p}. \tag{4.134}$$

Therefore, the pressure distribution function $g_N(\zeta)$ can be finally determined by solving the algebraic equation (4.131).

4.3.4 Comparison of numerical solutions for unsteady blade loading

Table 4.2 shows a comparison of the calculation results obtained by solving the earlier integral equation with the data from Nagashima and Whitehead [21], along with the definitions of lift and moment coefficients for a bending vibration:

$$C_{Fq} = \frac{1}{\rho_0 U_r \overline{v}_b H b} \int_0^1 \int_0^\pi \Delta\overline{p}(\xi_0, \varsigma_0) d\xi_0 d\varsigma_0 \tag{4.135}$$

$$C_{Mq} = \frac{1}{\rho_0 U_r \overline{v}_b H b^2} \int_0^1 \int_0^\pi \xi_0 \Delta\overline{p}(\xi_0, \varsigma_0) d\xi_0 d\varsigma_0 \tag{4.136}$$

Table 4.2 Lift and moment coefficients.

	Bending		Torsion	
	C_{Fq}	C_{Mq}	$C_{F\alpha}$	$C_{M\alpha}$
Nagashima and Whitehead study [21]	−0.1592 + i0.3822	−0.0241 + i0.0648	−0.1396 + i0.3894	−0.0181 + i0.0417
Present	−0.1592 + i0.3822	−0.0239 + i0.0649	−0.1397 + i0.3894	−0.0179 + i0.0418

$\omega/U = 0.301$, $M = 1.3454$, $s'/2b = 0.7889$, $\theta = 59.53$, $\sigma = -30$

and for torsional vibration

$$C_{F\alpha} = \frac{1}{\rho_0 U_r^2 \alpha_t H b} \int_0^1 \int_0^{\pi} \Delta \overline{p}(\xi_0, \varsigma_0) d\xi_0 d\varsigma_0 \tag{4.137}$$

$$C_{M\alpha} = \frac{1}{\rho_0 U_r^2 \alpha_t H b^2} \int_0^1 \int_0^{\pi} \xi_0 \Delta \overline{p}(\xi_0, \varsigma_0) d\xi_0 d\varsigma_0 \tag{4.138}$$

As indicated in Table 4.2, a very good agreement can be found for two different methods. Fig. 4.4 gives the unsteady pressure distribution of blade torsional vibration for two methods [21, 29]. Again, the results fit with each other very well.

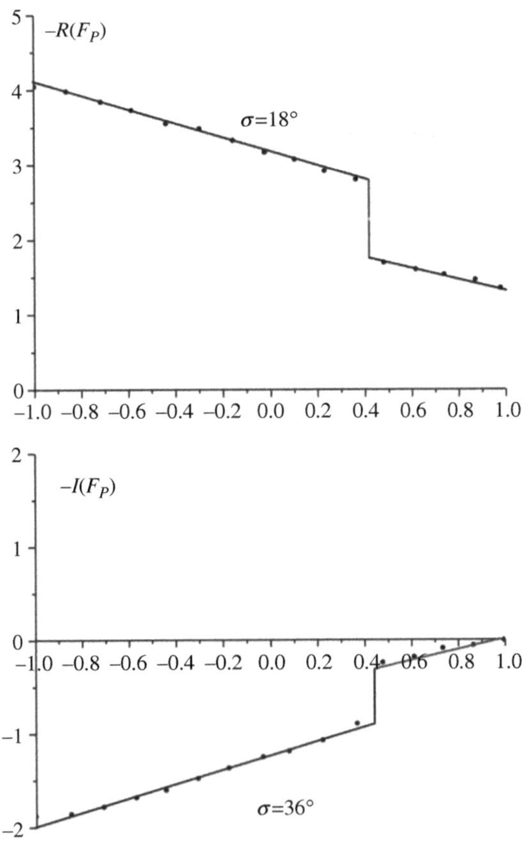

Fig. 4.4 Unsteady pressure distribution of blade torsional vibration [29]. $\omega/U = 0.301$, $M = 1.3454$, $s'/2b = 0.7889$, $\theta = 59.53$, $\sigma = -30$.

Table 4.3 Moment coefficient $C_{M\alpha}$.

σ	Present analysis	Adamcyk's analysis [22]	Verdon's analysis [18, 19]
−180	$0.5259 - i0.3718$	$0.5212 - i0.3641$	$0.51 - i0.37$
−150	$0.3330 - i0.5107$	$-0.3309 - i0.5030$	$-0.34 - i0.5$
−120	$0.0848 - i0.5225$	$0.08566 - i0.5153$	$0.11 - i0.57$
−90	$-0.1819 - i0.3969$	$-0.1790 - i0.3910$	$0.175 - i0.408$
−60	$0.1534 - i0.2917$	$0.1617 - i2895$	$0.160 - i0.290$
−30	$-0.190 - i0.2840$	$-0.1872 - i0.2824$	$-0.190 - i0.28$
0	$-0.2375 + i0.0529$	$-0.2354 + i0.5372$	$-0.24 + i0.04$
30	$-0.0552 + i0.2603$	$-0.05435 + i0.2602$	$-0.04 + i0.26$
60	$0.1859 + i0.3257$	$0.1851 + i0.3257$	$0.19 + i0.30$
90	$0.4161 + i0.2560$	$0.4123 + i0.2567$	$0.405 + i0.24$
120	$0.5725 + i0.0782$	$0.5674 + i0.08128$	$0.58 + i0.05$
150	$0.6134 - i0.1534$	$0.6073 - i0.1479$	$0.60 - i0.16$
180	0.5256–0.3716	0.5212–0.3641	0.51–0.37

$\omega/U = 0.602$, $M = 1.3454$, $s'/2b = 0.7889$, $\theta = 59.53$

Table 4.3 gives comparison results for a different parametric setup. Especially for different inter-blade phase angles, the numerical results given by the present method show good agreement with the existing predictions [18, 19, 22] for the moment coefficient.

4.4 Lifting surface theory for unsteady analysis of fan/compressor cascade

The most representative work in predicting the noise generated by rotor-stator interactions may be the three-dimensional lifting surface theory. The origin of this theory based on the linearized small disturbance assumption can be traced back to the integral expression for the rotor blade thickness noise in a compressible flow field proposed by McCune [31] in 1958. Twelve years later, Okurounmu and McCune [32] established an expression of the sound field generated by a given aerodynamic load distribution on the rotor blade surfaces. However, it still cannot be called a complete solving method because the pressure difference distribution on blade surfaces should have been determined by the actual blade geometry and flow conditions.

After that, with a rapid development of computation capability, a lot of researches concerning the lifting surface theory in a real sense sprung up in the1970s. For the off-design problems about solving the pressure difference on blade surfaces, first an integral equation was established on the basis of the impermeable boundary conditions (i.e., the normal velocity must vanish on the solid body surfaces). In other words, a relation between the disturbed flow and the sound field generated by blade forces was built up, and then the blade load distribution can be determined by solving this equation. However, the greatest difficulty confronting this approach is the complexity due to its three dimensions, which brings about problems concerning the singularity and convergence of the integral equation. To overcome this, Namba [27, 33–36] and Salaun [37, 38] further developed a lifting surface theory, respectively. At the beginning, Nambe concentrated on the steady lift problems under the conditions of subsonic and supersonic relative velocity at the blade tip, whereas Salaun devoted efforts to study blade flutter problems by using his own method. Then Namba applied his theory to solve the problems of noise generation due to the interactions between the wakes from the front stator and the rear rotor. Although the work conducted by both of them all belong to the applications of lifting surface theory, they have different treatments for the integral equation. Salaun adopted an exact form of the kernel function in his own method, whereas Namba resolved the kernel functions into modal components and then approximated them with the finite series expansions in terms of low-order radial modes. Comparably, Namba's method is more efficient to overcome the singularity problem, and it can be extended to the transonic cascade conditions. Subsequently, Schulten [39–43] developed the lifting surface theory by using another kind of radial mode expansion and then applied it to problems involving a compressor's inlet guide vanes. Meanwhile, he carried on studies for the effects of back-sweep blades, lean blades, stagger angle, and blade camber on noise variations. The computation results obtained by Namba and Schulten [44] reveal a satisfying coincidence for the fourth kind of benchmark problems presented by NASA in 2000 [45]. Therefore, it can be concluded that the difficulty in solving the integral equation of the three-dimensional lifting surface theory can be overcome in different ways.

4.4.1 A unified framework for acoustic field and unsteady flow

For a long time, the calculation of unsteady blade forces for aeroacoustic applications was based on two-dimensional methods. Nonetheless, for the compressors and the outlet guide vanes of civil aeroengines, the two-dimensional assumption, and even the strip theory established on the basis of it, seems unreasonable or inadaptable any longer. The reason for this lies in the fact that the radial interactions of unsteady blade forces are not negligible for blades with a large aspect ratio, especially for those with sweep or lean geometry. Hence, for this class of cascades, the three-dimensional effects should be considered by the lifting surface theory.

In this section, the analysis will also be based on the unified solution of flow field and sound field, and velocity decomposition theorem for uniform flows. In the linearized approximation, the solution of unsteady blade forces of three-dimensional cascade are going to be derived, and then the expression for the corresponding sound field will be formulated.

To begin, a computation model for three-dimensional lifting surface method focused on the interactions between oncoming unsteady disturbances and rotor blades as well as the coordinate system we will use are shown in Fig. 4.5.

For the discussion in this section, the following necessary assumptions are made:

(1) The annular duct has a rigid inner wall, and the flow within the duct is compressible, inviscid, isentropic, and uniform.
(2) The blades can be viewed as flat plates with zero thickness.
(3) The incident angle of flow equals zero, and the uniform flow would pass the blades in a fixed direction.
(4) All the disturbances in the flow are small relative to the background flow field, therefore, the governing equations can be linearized.
(5) The flow is subsonic.
(6) The unsteady blade forces at the trailing edge are of finite value.

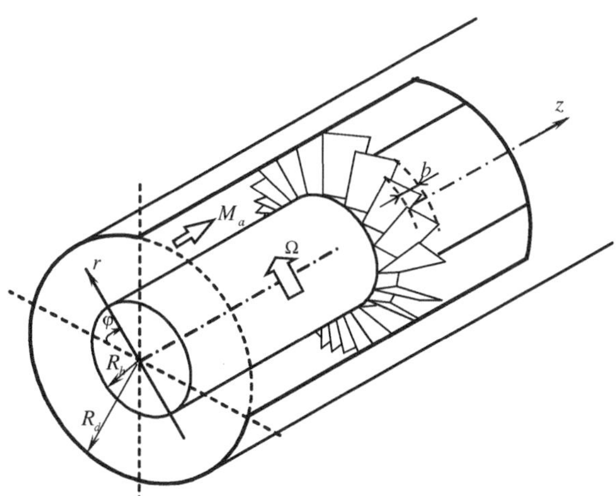

Fig. 4.5 The rotor blade row of the computation model.

Because the oncoming flow is assumed to be uniform with zero incident angle, and meanwhile if we assume the exit flow from the blade's trailing edge is in the axial direction and stays uniform as well, the only possibility is that the blades are twisted along the span. Therefore, they cannot be modeled as flat plates parallel to the duct axis like the vane blades.

In Fig. 4.6, we specify the twist angles at the blade hub, the middle span, and the blade tip, respectively. For convenience, the case considered here is that the leading edge of the rotor blade is vertical to the duct axis, and the axial projection of the blade chord is at the constant length denoted by b.

The key problem concerning the noise generation due to the interactions of oncoming disturbances and rotor blades is to calculate the unsteady loads on the blade surfaces and then determine the sound field in the duct. As shown in Fig. 4.6C, the pressure difference on the blade surface is denoted by $\Delta\widehat{p}$. Considering the effect of solid boundaries on the moving medium and the generalized Lighthill's theory [12], we can deduce that the unsteady blade loads as sound sources have the property of dipoles:

$$p'(\vec{x},t)=\int_{-T}^{T}\int_{s(\tau)}\Delta\widehat{p}_i\frac{\partial G}{\partial y_i}ds(\vec{y})\,d\tau, \tag{4.139}$$

where

$$\begin{aligned}\Delta\widehat{p}_i\frac{\partial}{\partial y_i}&=\Delta p_\varphi\frac{\partial}{r'\partial\varphi'}-\Delta p_z\frac{\partial}{\partial z'}=\Delta p_\varphi\left(\frac{\partial}{r'\partial\varphi'}-\tan\theta\cdot\frac{\partial}{\partial z'}\right)\\&=\Delta\widetilde{p}\left(\frac{\partial}{r'\partial\varphi'}-\frac{\Omega r'}{U}\frac{\partial}{\partial z'}\right),\end{aligned} \tag{4.140}$$

To establish a unified expression, here we substitute Δp for Δp_φ to represent the circumferential projection of the pressure difference on the blade surfaces. Then by inserting it into the sound field expression, we can obtain:

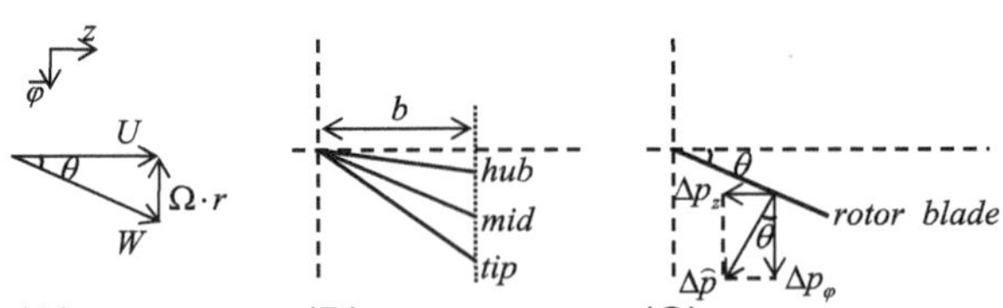

Fig. 4.6 The rotor blades. (A) velocity triangle (B) projection of rotor blades (C) blade force triangle

$$p'(\vec{x},t)=\frac{i}{4\pi^2}\sum_{m=-\infty}^{\infty}\sum_{n=1}^{\infty}\frac{\psi_m(r)e^{im\overline{\varphi}}}{\Gamma_m}\int_{s(\tau)}\psi_m(r')e^{-im\overline{\varphi'}}$$
$$\int_{-\infty}^{\infty}\int_{-\infty}^{\infty}\left(\alpha\frac{\Omega r'}{U}-\frac{m}{r'}\right)\frac{e^{i\alpha(z-z')}}{\beta^2\alpha^2-2Mk_0\alpha-k_0^2+k^2}\int_{-T}^{T}\Delta\widetilde{p}\cdot e^{i\omega(t-\tau)}d\tau d\omega d\alpha ds(\vec{y}).$$

(4.141)

The expression of the sound field is obtained in the duct-fixed coordinates and thus can be directly applied to the stator. As for the rotor, because it is the rotational component, the integral for the rotor blades in the fixed coordinate system must be time-dependent, which means it cannot be directly used to derive an integral solution. Therefore, the coordinate transformation is in need. To make the integral independent of time, we introduce the rotating coordinate system fixed to the rotor:

$$\overline{r'}=r',\quad \overline{\varphi'}=\varphi'\pm\Omega\tau,\quad \overline{z'}=z', \tag{4.142}$$

where $\overline{\varphi}'$ represents the fixed coordinate and φ' represents the rotating coordinate, and "+" and "−" respectively denote that Ω and φ' are in the same direction, and that they are in the opposite directions. Here, we consider the case when Ω and φ' are in opposite directions. Inserting $\overline{\varphi}'=\varphi'-\Omega\tau$ into the sound pressure expression leads to:

$$p'(\vec{x},t)=\frac{i}{4\pi^2}\sum_{m=-\infty}^{\infty}\sum_{n=1}^{\infty}\frac{\psi_m(r)e^{im\overline{\varphi}}}{\Gamma_m}\int_{s(\tau)}\psi_m(r')e^{-im\varphi'}$$
$$\int_{-\infty}^{\infty}\int_{-\infty}^{\infty}\left(\alpha\frac{\Omega r'}{U}-\frac{m}{r'}\right)\frac{e^{i\alpha(z-z')}}{\beta^2\alpha^2-2Mk_0\alpha-k_0^2+k^2} \tag{4.143}$$
$$\int_{-T}^{T}\Delta\widetilde{p}\cdot e^{i\omega(t-\tau)}e^{im\Omega\tau}d\tau d\omega d\alpha ds(\vec{y}).$$

When the perturbations in the oncoming disturbed flows interact with the rotor blades, the rotor blade forces should have the same time-frequency character with the disturbed flow, which can be expressed by:

$$\Delta\widetilde{p}=\Delta p\cdot e^{i\lambda\tau}, \tag{4.144}$$

where λ is the fluctuating frequency in the rotating coordinates fixed to the rotor. If we consider the stator blade wakes of the front stage interact with the rear rotor blades, then the unsteady rotor blade forces due to the wakes of the upstream V vanes can be expanded into a Fourier series in the form of:

$$\Delta\widetilde{p} = \sum_{p=-\infty}^{\infty} \Delta p_p \cdot e^{ipV\Omega\tau}. \tag{4.145}$$

The unsteady load of an arbitrary order can also respectively defined by $\Delta p \cdot e^{i\lambda\tau}$. Thus, for the convenience of discussion, we use the form of perturbations in the incoming flow as a starting point and then insert the expression of $\Delta\widetilde{p}$ into Eq. (4.143):

$$p'(\vec{x}, t) = \frac{i}{4\pi^2} \sum_{m=-\infty}^{\infty} \sum_{n=1}^{\infty} \frac{\psi_m(r) e^{im\overline{\varphi}}}{\Gamma_m} \int_{s(\tau)} \psi_m(r') e^{-im\varphi'} \int_{-\infty}^{\infty} \int_{-\infty}^{\infty} \left(\alpha \frac{\Omega r'}{U} - \frac{m}{r'}\right) \frac{e^{i\alpha(z-z')}}{\beta^2\alpha^2 - 2Mk_0\alpha - k_0^2 + k^2} \int_{-T}^{T} \Delta p \cdot e^{i\lambda\tau} e^{i\omega(t-\tau)} e^{im\Omega\tau} d\tau d\omega d\alpha ds(\vec{y}). \tag{4.146}$$

The discussion given here is only focused on the noise generated by a single rotor blade. Now we will consider the rotor with B identical equally spaced blades. As shown in Fig. 4.7, when Ω and φ' are in opposite directions, for the kth rotor blade, the disturbances acting on it at time τ is exactly the same with the disturbances experienced by the first blade at time $\tau - \frac{2\pi}{B\Omega}(k-1)$. The sound field generated by the kth rotor blade has nothing different from that generated by the first blade but a constant phase difference $\varphi' + \frac{2\pi}{B}(k-1)$. Therefore, for the different rotor blades, $\Delta p_{(r,z)}$ is of the same amplitude but with a phase difference, and consequently the sound fields resulting from different blades share the same magnitude of sound pressure but have corresponding phase differences as well.

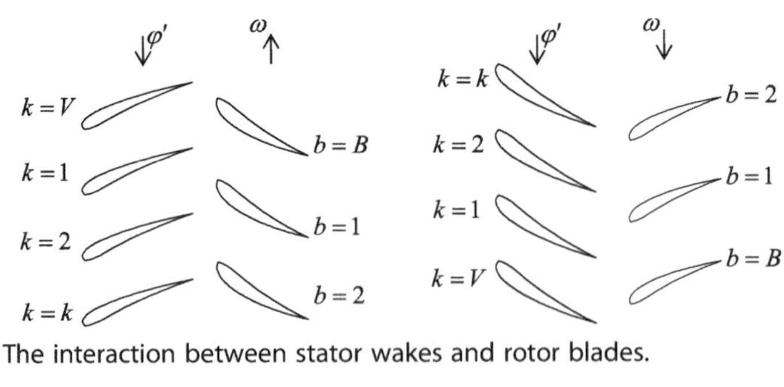

Fig. 4.7 The interaction between stator wakes and rotor blades.

$$\varphi' \sim \varphi' + \frac{2\pi}{B}(k-1), \quad \tau \sim \tau - \frac{2\pi}{B\Omega}(k-1). \tag{4.147}$$

Add the time factor to the series expansion of the unsteady blade forces:

$$\Delta\widetilde{p} = \Delta p \cdot e^{i\lambda\left[\tau - \frac{2\pi}{B\Omega}(k-1)\right]}. \tag{4.148}$$

Accordingly, the pressure field caused by the B rotor blades can be formulated by:

$$p_q'(\vec{x}, t) = \frac{-i}{4\pi^2}\sum_{m=-\infty}^{\infty}\sum_{n=1}^{\infty}\frac{\psi_m(r)e^{im\overline{\varphi}}}{\Gamma_m}\int_{s(\tau)}\sum_{k=1}^{B}\Delta p\psi_m(r')e^{-im\varphi'}e^{-i(\lambda+m\Omega)\frac{2\pi}{B\Omega}(k-1)}$$
$$\int_{-\infty}^{\infty}e^{i\omega t}\int_{-\infty}^{\infty}\left(\frac{m}{r'} - \alpha\frac{\Omega r'}{U}\right)\frac{e^{i\alpha(z-z')}}{\beta^2\alpha^2 - 2Mk_0\alpha - k_0^2 + k^2}$$
$$\int_{-T}^{T}e^{i(-\omega+m\Omega+\lambda)\tau}d\tau d\omega d\alpha ds(\vec{y}). \tag{4.149}$$

Consider the property $\lim_{T\to\infty}\int_{-T}^{T}e^{i(-\omega+m\Omega+\lambda)\tau}d\tau = 2\pi\delta(\omega - m\Omega - \lambda)$. When $\omega = \lambda + m\Omega$, we have:

$$p_q'(\vec{x}, t) = \frac{-ie^{i\omega t}}{2\pi}\sum_{m=-\infty}^{\infty}\sum_{n=1}^{\infty}\frac{\psi_m(r)e^{im\overline{\varphi}}}{\Gamma_m}\int_{s(\tau)}\sum_{k=1}^{B}\Delta p\psi_m(r')e^{-im\varphi'}e^{-i(\lambda+m\Omega)\frac{2\pi}{B\Omega}(k-1)}$$
$$\int_{-\infty}^{\infty}\left(\frac{m}{r'} - \alpha\frac{\Omega r'}{U}\right)\frac{e^{i\alpha(z-z')}}{\beta^2\alpha^2 - 2Mk_0\alpha - k_0^2 + k^2}d\alpha ds(\vec{y}). \tag{4.150}$$

Putting the relation:

$$\sum_{k=1}^{B}e^{-i\left(\frac{\lambda}{\Omega}+m\right)\frac{2\pi}{B}(k-1)} = \begin{cases} B & \frac{\lambda}{\Omega} + m = sB \\ 0 & \frac{\lambda}{\Omega} + m \neq sB \end{cases} \Rightarrow m = sB - \frac{\lambda}{\Omega}, \tag{4.151}$$

into the expression for the sound field, we can obtain:

$$p_q'(\vec{x}, t) = -\frac{iBe^{i\omega t}}{2\pi}\sum_{s=-\infty}^{\infty}\sum_{n=1}^{\infty}\frac{\psi_m(r)e^{im\overline{\varphi}}}{\Gamma_m}\int_{s(\tau)}\Delta p\psi_m(r')e^{-im\varphi'}$$
$$\int_{-\infty}^{\infty}\left(\frac{m}{r'} - \alpha\frac{\Omega r'}{U}\right)\frac{e^{i\alpha(z-z')}}{\beta^2\alpha^2 - 2Mk_0\alpha - k_0^2 + k^2}d\alpha ds(\vec{y}). \tag{4.152}$$

Then, with the relations $\phi_m(k_{mn}r) = \frac{\psi_m(k_{mn}r)}{\sqrt{\overline{\Gamma}_{mn}}}$ and $\Gamma_{mn} = 2\pi\overline{\Gamma}_{mn}$, the equation can be further rearranged into the simpler form:

$$p'_q(\vec{x}, t) = -\frac{iBe^{i\omega t}}{4\pi^2} \sum_{s=-\infty}^{\infty} \sum_{n=1}^{\infty} \phi_m(r) e^{im\overline{\varphi}} \int_{s(\tau)} \Delta p \phi_m(r') e^{-im\varphi'}$$
$$\int_{-\infty}^{\infty} \left(\frac{m}{r'} - \alpha \frac{\Omega r'}{U} \right) \frac{e^{i\alpha(z-z')}}{\beta^2\alpha^2 - 2Mk_0\alpha - k_0^2 + k^2} d\alpha ds(\vec{y}). \quad (4.153)$$

For the noise generation due to the stator wake-rotor interaction, we have $\lambda = qV\Omega$ and $m = sB - qV$. And then applying the residue theorem with respect to α leads to:

$$p'_q(\vec{x}, t) = \frac{-Be^{-i\omega t}}{4\pi} \sum_{s=-\infty}^{\infty} \sum_{n=1}^{\infty} \frac{\phi_{mn}(k_{mn}r) e^{im\varphi}}{\kappa_{n,m}} \int_{s(\tau)} \Delta p \phi_{mn}(k_{mn}r') e^{-im\varphi'}$$
$$\left\{ H(z-z') \left(\frac{m}{r'} - \alpha_1 \frac{\Omega r'}{U} \right) e^{i\alpha_1(z-z')} + H(z'-z) \left(\frac{m}{r'} - \alpha_2 \frac{\Omega r'}{U} \right) e^{i\alpha_2(z-z')} \right\} ds(\vec{y}), \quad (4.154)$$

where

$$\alpha_1 = \frac{Mk_0 - \sqrt{k_0^2 - \beta^2 k^2}}{\beta^2} \text{ (downstream)}, \quad (4.155)$$

$$\alpha_2 = \frac{Mk_0 + \sqrt{k_0^2 - \beta^2 k^2}}{\beta^2} \text{ (upstream)} \quad (4.156)$$

Eq. (4.154) is the expression for the sound field produced by the unsteady loads on the rotor blade surfaces. In most cases, the unsteady blade force distribution is not easily known as a priori. Thus, it is usually obtained by solving an integral equation established according to the impermeable boundary conditions on the blade surfaces as well as the oncoming disturbed flow conditions. Now we are going to present a detailed discussion for the relevant problems in the following contents.

4.4.2 Integral equation for the solution of unsteady blade load

In the previous contents, we have derived the sound pressure expression in the fixed coordinates. However, according to the velocity decomposing theorem, to analyze the sound generation due to the vortical perturbations that come from the incoming flow and pass through the blade row, it is still needed to consider the impermeable boundary conditions on the rigid blade surfaces, i.e., $\mathbf{u}\cdot\mathbf{n}=0$. To eliminate the time dependence of the sound field expression due to the rotor's rotational motion, we should first introduce one rotating coordinate system fixed to the rotor by the coordinate transformation given by:

$$\overline{r}=r,\quad \overline{\varphi}=\varphi-\Omega t,\quad \overline{z}=z,$$

and another co-moving frame (r'', φ'', z'') with one axis parallel to the streamline as shown in Fig. 4.8.

The coordinate transformation between these two co-moving coordinates satisfies the following relations:

$$r=r'', \tag{4.157}$$

$$r\varphi''=r\varphi\cos\theta-z\sin\theta,\quad z''=z\cos\theta+r\varphi\sin\theta, \tag{4.158}$$

$$r\varphi=r\varphi''\cos\theta+z''\sin\theta,\quad z=z''\cos\theta-r\varphi''\sin\theta, \tag{4.159}$$

Furthermore:

$$\frac{\partial\varphi}{\partial\varphi''}=\cos\theta,\quad \frac{\partial z}{r\partial\varphi''}=-\sin\theta. \tag{4.160}$$

Thus, the boundary condition on the body surfaces can be rewritten as:

$$v'_{\varphi''}+w'_{i,\varphi''}=0, \tag{4.161}$$

where $w'_{i,\varphi''}$ represents the normal velocity component of the external disturbances from the incoming flow, on the blade surface, which can be either the pressure perturbations or the vortical perturbations, and $v'_{\varphi''}$ denotes the normal velocity component of the unsteady fluctuations generated on the rotor blade surfaces, including both the pressure fluctuations and the vortical fluctuations. For a specific problem, the expression of $w'_{i,\varphi''}$ would be inserted into the equation as a priori, which will be discussed later. Now we will determine the expression of $v'_{\varphi''}$.

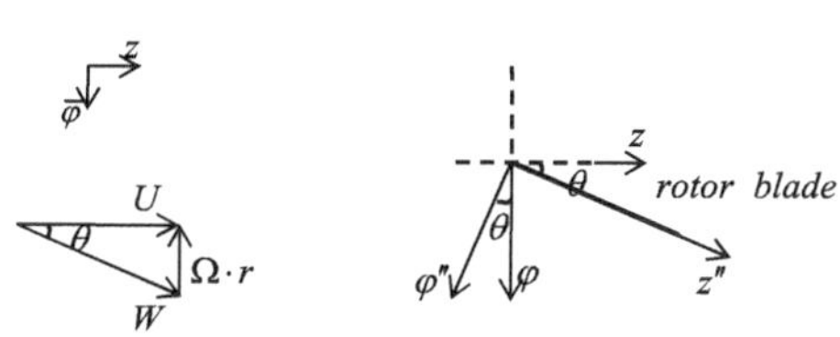

Fig. 4.8 Two co-moving coordinate systems.

First, we need to transform the expression for fluctuating pressure waves to the rotating coordinates. In the co-moving coordinate system fixed to the rotor, the fluctuating pressure can be described by:

$$p'_q(\vec{x},t) = -\frac{iBe^{i\lambda t}}{2\pi}\sum_{s=-\infty}^{\infty}\sum_{n=1}^{\infty}\frac{\psi_m(r)e^{im\overline{\varphi}}}{\Gamma_m}\int_{s(\tau)}\Delta p\psi_m(r')e^{-im\varphi'} \int_{-\infty}^{\infty}\left(\frac{m}{r'}-\alpha\frac{\Omega r'}{U}\right)\frac{e^{i\alpha(z-z')}}{\beta^2\alpha^2-2Mk_0\alpha-k_0^2+k^2}d\alpha ds(\vec{\gamma}). \tag{4.162}$$

Establish the momentum equation in the coordinates (r'', φ'', z'') with the acoustic fluctuating velocity normal to the blade surfaces defined by $v_{\varphi''}$, and then we can obtain:

$$\frac{\partial\widetilde{v}_{\varphi''}}{\partial t}+W\frac{\partial\widetilde{v}_{\varphi''}}{\partial z''}=-\frac{1}{\rho}\frac{\partial\widetilde{p}}{r\partial\varphi''}. \tag{4.163}$$

Thus, the corresponding homogenous equation is:

$$\frac{\partial\widetilde{v}_{\varphi''}}{\partial t}+W\frac{\partial\widetilde{v}_{\varphi''}}{\partial z''}=0. \tag{4.164}$$

Note that the time characteristic of the fluctuating velocity is also in the form of $\widetilde{v}_{\varphi''}=v'_{\varphi''}\cdot e^{i\lambda t}$. Therefore, the solution to the homogeneous equation should be $v'_{\varphi''}=ce^{-i\frac{\lambda}{W}z''}$. Apply the constant variation method and assume $v'_{\varphi''}=v^1_{\varphi''}e^{-i\frac{\lambda}{W}z''}$. Substituting it into the momentum equation leads to:

$$i\lambda v^1_{\varphi''}e^{-i\frac{\omega}{W}z''}+W\left(\frac{\partial v^1_{\varphi''}}{\partial z''}e^{-i\frac{\lambda}{W}z''}-i\frac{\lambda}{W}v^1_{\varphi''}e^{-i\frac{\lambda}{W}z''}\right)=-\frac{1}{\rho}\frac{\partial p'}{r\partial\varphi''}, \tag{4.165}$$

which yields

$$\frac{\partial v^1_{\varphi''}}{\partial z''}=-\frac{1}{\rho W}\frac{\partial p}{r\partial\varphi''}e^{i\frac{\lambda}{W}z''},$$

$$v^1_{\varphi''}=-\frac{1}{\rho W}\int\frac{\partial p}{r\partial\varphi''}e^{i\frac{\lambda}{W}z''}dz''.$$

Thus, we get:

$$v'_{\varphi''}=-\frac{1}{\rho W}e^{-i\frac{\lambda}{W}z''}\int\frac{\partial p}{r\partial\varphi''}e^{i\frac{\lambda}{W}z''}dz''. \tag{4.166}$$

According to the velocity triangle relationship $\frac{z''}{W}=\frac{z}{U}$, it can be further rewritten as:

$$v'_{\varphi''}=-\frac{1}{\rho W}e^{-i\frac{\lambda}{W}z''}\int\frac{\partial p}{r\partial\varphi''}e^{i\frac{\lambda}{W}z''}dz''=-\frac{1}{\rho U}e^{-i\frac{\lambda}{U}z}\int\frac{\partial p'}{r\partial\varphi''}e^{i\frac{\lambda}{U}z}dz. \quad (4.167)$$

Moreover, the partial differentiation relation in Eq. (4.160) implies:

$$\frac{\partial}{r\partial\varphi''}=\frac{\partial}{r\partial\varphi}\frac{\partial\varphi}{\partial\varphi''}+\frac{\partial}{\partial z}\frac{\partial z}{r\partial\varphi''}=\cos\theta\frac{\partial}{r\partial\varphi}-\sin\theta\frac{\partial}{\partial z}. \quad (4.168)$$

By putting this into the expression of the normal component of acoustic velocity, and meanwhile noting that $\cos\theta=\frac{U}{W}$, $\tan\theta=\frac{\Omega r}{U}$, we can obtain:

$$v'_{\varphi''}=-\frac{1}{\rho W}e^{-i\frac{\lambda}{U}z}\int\left[\frac{\partial}{r\partial\varphi}-\frac{\Omega r}{U}\frac{\partial}{\partial z}\right]p'\Bigg|_{\varphi=\varphi_0+\frac{\Omega}{U}z,\ \varphi'=\varphi'_0+\frac{\Omega}{U}z'}e^{i\frac{\lambda}{U}z}dz. \quad (4.169)$$

Upon the sound pressure expression and an integral with respect to z, the normal fluctuating velocity on the blade surfaces can be formulated by:

$$v'_{\varphi''}=\frac{iBe^{i\lambda t}}{2\pi\rho W}\sum_{s=-\infty}^{\infty}\sum_{n=1}^{\infty}\frac{\psi_m(r)e^{im\varphi_0}}{\Gamma_m}\int_{s(\tau)}\Delta p\psi_m(r')e^{-im\varphi'_0}$$
$$\int_{-\infty}^{\infty}\left(\frac{m}{r}-\alpha\frac{\Omega r}{U}\right)\left(\frac{m}{r'}-\alpha\frac{\Omega r'}{U}\right) \quad (4.170)$$
$$\frac{e^{im\frac{\Omega}{U}(z-z')}e^{i\alpha(z-z')}}{\left(\alpha+\frac{\lambda}{U}+m\frac{\Omega}{U}\right)\left(\beta^2\alpha^2-2Mk_0\alpha-k_0^2+k^2\right)}d\alpha ds\left(\vec{\gamma}\right).$$

Then by using the relations $\phi_m(k_{mn}r)=\frac{\psi_m(k_{mn}r)}{\sqrt{\overline{\Gamma}_{mn}}}$ and $\Gamma_{mn}=2\pi\overline{\Gamma}_{mn}$, it can be simplified to:

$$v'_{\varphi''}=\frac{iBe^{i\lambda t}}{4\pi^2\rho W}\sum_{s=-\infty}^{\infty}\sum_{n=1}^{\infty}\phi_m(r)e^{im\varphi_0}\int_{s(\tau)}\Delta p\phi_m(r')e^{-im\varphi'_0}$$
$$\int_{-\infty}^{\infty}\left(\frac{m}{r}-\alpha\frac{\Omega r}{U}\right)\left(\frac{m}{r'}-\alpha\frac{\Omega r'}{U}\right)$$
$$\frac{e^{im\frac{\Omega}{U}(z-z')}e^{i\alpha(z-z')}}{\left(\alpha+\frac{\lambda}{U}+m\frac{\Omega}{U}\right)\left(\beta^2\alpha^2-2Mk_0\alpha-k_0^2+k^2\right)}d\alpha ds\left(\vec{\gamma}\right).$$

Furthermore, applying the residue theorem with respect to α leads to:

$$v'_{\varphi''} = \frac{-Be^{i\lambda t}}{2\pi\rho W}\sum_{s=-\infty}^{\infty}\sum_{n=1}^{\infty}\phi_{mn}(k_{mn}r)e^{im\varphi_0}\int_{s(\tau)}\Delta p\phi_{mn}(k_{mn}r')e^{-im\varphi'_0}$$

$$\left\{H(z-z')\left[\frac{\beta^2 Me^{im\frac{\Omega}{U}(z-z')}e^{i\alpha_1(z-z')}}{2\kappa_{n,m}(M\kappa_{n,m}-k_0)}\left(\frac{m}{r'}-\alpha_1\frac{\Omega r'}{U}\right)\left(\frac{m}{r}-\alpha_1\frac{\Omega r}{U}\right)\right.\right.$$

$$\left.+\frac{M^2 e^{im\frac{\Omega}{U}(z-z')}e^{i\alpha_3(z-z')}}{k_0^2+M^2k_{m,n}^2}\left(\frac{m}{r'}-\alpha_3\frac{\Omega r'}{U}\right)\left(\frac{m}{r}-\alpha_3\frac{\Omega r}{U}\right)\right]$$

$$\left.-H(z'-z)\frac{\beta^2 Me^{im\frac{\Omega}{U}(z-z')}e^{i\alpha_2(z-z')}}{2\kappa_{n,m}(k_0+M\kappa_{n,m})}\left(\frac{m}{r'}-\alpha_2\frac{\Omega r'}{U}\right)\left(\frac{m}{r}-\alpha_2\frac{\Omega r}{U}\right)\right\}ds(\vec{y}). \tag{4.171}$$

where

$$\alpha_1 = \frac{Mk_0-\sqrt{k_0^2-\beta^2k^2}}{\beta^2}\ \text{(downstream)}, \tag{4.172}$$

$$\alpha_2 = \frac{Mk_0+\sqrt{k_0^2-\beta^2k^2}}{\beta^2}\ \text{(upstream)}, \tag{4.173}$$

$$\alpha_3 = -\frac{\omega}{U}\ \text{(downward)} \tag{4.174}$$

Eq. (4.171) is the integral equation containing unknown unsteady blade load $\Delta p(r', z')$. For its solution, we still need to know how to give the upwash velocity on an airfoil.

4.4.3 Upwash velocity for three different incoming conditions

In the introduction of this chapter, we claimed that the unsteady aerodynamic force calculation is a most essential part of the aeroelastic analysis as well as the aeroacoustic prediction for moving objects. From the earlier discussions, it can be easily extrapolated that the key step in the computation process from the unsteady disturbances in the disturbed flow to the unsteady blade forces and then to the sound field is the linearized analysis of the unsteady blade forces. In the previous section, we obtained the integral equation for solving these unsteady aerodynamic loads, and meanwhile the unsteady blade loads are the only unknown functions within it. Thus, in this section, we will discuss how to determine the known term on the right-hand side of the integral equation, i.e., the external disturbances from the oncoming flow.

As early as 1972, based on the velocity decomposing theorem for uniform flows and the unified solution of both flow field and sound field, Smith [46] presented the form of the solution to the sound field resulting from the interactions between the incoming perturbations and a two-dimensional blade row. When handling the three-dimensional cascades problem, the basic mechanism of the interactions won't change. Therefore, upon the analysis conducted by Smith, we can divide the interaction problems between the disturbances from oncoming flows and the blade rows into three classes:

(1) The interactions between viscous wakes and blade rows. The noise due to interactions between the viscous wakes produced by the front and the rear blade rows, commonly referred as stator-rotor interaction noise or rotor-stator interaction noise, are the most dominating components of the compressor noise in aeroengines.

(2) The interactions between potential flows and blade rows. These problems are concerned with the transmission and reflection of sound waves in the duct when passing through the rotor or stator blade rows. After generated by the unsteady blade forces, the sound waves would propagate upstream and downstream, going through the reflections and transmissions when passing through other blade rows and the open ends of the duct, and finally radiate from the inlet and outlet of the duct. During the reflections and transmissions among the blade rows, it is possible for the sound waves to develop the acoustic resonance characteristic of a certain mode. Hence, the interactions with potential flows take an important place in the acoustic computation.

(3) Cascade flutter problems. During the process of vibration, the blades would interact with the acoustic perturbations generated by themselves, i.e., the linearized potential flow interactions, and might fall into the unstable condition, i.e., self-oscillations in the linearized scope. This kind of interaction can not only result in blade instabilities but also produce the sound with certain characteristics of its frequencies and the modes, which is totally different from the basic aerodynamic sound of engine with the blade passing frequency (BPF) or shaft frequency, as is fundamental frequency.

It is noteworthy that all discussions for the interactions mentioned here are in the linearized approximation. Then we will talk about how to determine the disturbance velocity in the incoming flows aimed at these three situations, respectively.

4.4.3.1 The viscous wakes-rotor interaction

Here we take the viscous wakes-rotor interaction. For example, we will study the first kind of interference problems, i.e., the prediction of the unsteady blade forces and the resultant sound field due to the interactions between viscous wakes and the rotor blade row in the linearized approximation.

Assume the incoming disturbances upstream of the rotor are a certain kind of viscous wakes resulting from obstructions, denoted by w_i'. Without generality, we assume the viscous wakes of the disturbed flow can rotate as well. Their rotational angular velocity is Ω_w, and the number of those obstacles far upstream is given as N_w. Therefore, we can expand the axial component of the external disturbance velocity referred to the duct-fixed coordinates into a Fourier series in the form of:

$$\overline{w'}_{i,z}(r, \overline{\varphi}, z, t) = \sum_{q=-\infty}^{\infty} A_q(r) \cdot e^{imN_w[\Omega_w t - \overline{\varphi} + \Delta\varphi(r) - \Omega_w z/U]}, \tag{4.175}$$

Ω_w is assigned to be negative when the rotation of the oncoming disturbances is in the same direction as the rotor. The circumferential phase angle $\Delta\varphi(r)$ represents the phase variation in the radial direction. With a coordination transformation to the rotor-fixed coordinate system and if the fundamental modal component ($q = 1$) is only considered, the previous expression can be rewritten as:

$$w'_{i,z}(r, \varphi, z, t) = A_q(r) \cdot e^{iN_w[(\Omega_w + \Omega)t - \varphi + \Delta\varphi(r) - \Omega_w z/U]}. \tag{4.176}$$

The corresponding circumferential component is:

$$w'_{i,\varphi}(r, \varphi, z, t) = -\frac{\Omega_w r}{U} \cdot A_q(r) \cdot e^{iN_w[(\Omega_w + \Omega)t - \varphi + \Delta\varphi(r) - \Omega_w z/U]}. \tag{4.177}$$

Furthermore, switching to the co-moving frame where one of whose axes is parallel to the streamline, and by using the coordinate transformation relation $w'_{i,\varphi''} = w'_{i,\varphi} \cos\theta - w'_{i,z} \sin\theta$, we can obtain:

$$w'_{i,\varphi''} = -\frac{A_q(r) \cdot (\Omega_w + \Omega) \cdot r}{W} \cdot e^{iN_w[(\Omega_w + \Omega)t - \varphi + \Delta\varphi(r) - \Omega_w z/U]}. \tag{4.178}$$

This is the upwash component on the blade surface, referred to the rotor-fixed coordinate system, of the external disturbance velocity. Then, based on the impermeable boundary condition on the rigid blade surfaces, we can establish a relation between $w'_{i,\varphi''}$ and $v'_{\varphi''}$ to solve for the unsteady blade forces.

A comparison of the fluctuating frequency of the disturbed flow and the disturbance velocity on the blade surfaces implies:

$$\lambda = N_w(\Omega_w + \Omega). \tag{4.179}$$

Consider the case that the perturbations in the disturbed flows are also in the rational motion, i.e., $\Omega_w \neq 0$, which can be viewed as the wakes interference problem for antidirection rotors. Then some necessary adjustments should be made to the three-dimensional lifting surface theory we previously derived, and the series expansion of the unsteady blade forces in Eq. (4.148) should be modified by:

$$\Delta\widetilde{p} = \Delta p \cdot e^{i\lambda\left[\tau - \frac{2\pi}{B(\Omega+\Omega_w)}(k-1)\right]}. \tag{4.180}$$

Correspondingly, the modal characteristic becomes:

$$m = sB - \frac{\lambda}{(\Omega + \Omega_w)} = sB - N_w, \tag{4.181}$$

with other properties unchanged. Furthermore, the condition that the upwash component of the resultant velocity must vanish on the blade surfaces, in other words, the upwash component of the external disturbance velocity $w'_{i,\varphi''}$ must be cancelled by the induced upwash velocity $v'_{\varphi''}$. Thus, it can be formulated by:

$$v'_{\varphi''} + w'_{i,\varphi''} = 0. \tag{4.182}$$

In this integral equation, there is only one unknown quantity, Δp, which is the circumferential component of the pressure difference on the blade surfaces. Thus, the equation can be solved. By putting the obtained pressure difference function $\Delta p(r', z')$ into Eq. (4.154), we can finally determine the sound field in the duct.

According to the discussion for the rotor-stator interaction noise and the stator-rotor interaction noise we made in the last chapter, and the lifting surface theory developed in this chapter, we can easily come to the conclusion that the wake interference problems are a kind of special cases of the previous discussion, i.e., the cases of $\Omega_w = 0$. Consequently, the corresponding frequency and modal characteristics of the sound fields due to the wake interference effect are also in the special forms of the expressions previously developed.

4.4.3.2 Linearized analysis for the interaction between potential flow and blade row

In the previous content, we investigated the noise generation due to wake-rotor blade interactions in detail, and we also took into consideration the rational motion of the external disturbances when discussing the unsteady oncoming flows. Indeed, the interactions between the sound waves and the blades are similar to those between the rotating external disturbances and the blades, just with a different expression of the right-hand term. Consider the sound wave with a certain frequency and a certain mode propagating downstream with reference to the duct-fixed coordinates:

$$\tilde{p}'_{A,1} = e^{i\omega t} \sum_{m=-\infty}^{\infty} \sum_{\mu=1}^{\infty} A_{m\mu}^{1p} \cdot \phi_m\left(k_{m\mu} r\right) \cdot e^{i\alpha_2(z-l)} \cdot e^{im\overline{\varphi}}. \tag{4.183}$$

Correspondingly, the induced unsteady fluctuations on the rotor blade surfaces have the normal velocity components in the form of:

$$\widetilde{v}_{\varphi'',A,1} = -\frac{1}{\rho W} \sum_{q=-\infty}^{\infty} \sum_{\mu=1}^{\infty} A_{m\mu}^{1p} \cdot \phi_m\left(k_{m\mu} r\right) \cdot \frac{\dfrac{m}{r} - \alpha_2 \dfrac{\Omega r}{U}}{\alpha_2 + m\dfrac{\Omega}{U} + \dfrac{\lambda_1}{U}} e^{i\alpha_2(z-l)} e^{im\frac{\Omega}{U}z} e^{i\lambda_1 t}. \tag{4.184}$$

The lifting surface theory, developed in the previous section, indicates that the interactions between potential flow and blade row would generate the sound waves of a certain mode and frequency as well.

For the sound waves that have passed through the rotor blade row, the frequencies and the modes of the sound field would be:

$$m' = sB + m, \quad \omega' = sB\Omega + \omega, \quad s = 0, \pm 1, \pm 2\cdots$$

Similarly, the frequencies and the acoustic modes of the sound field that have passed through a stator would be:

$$m' = sV + m, \quad \omega' = \omega, \quad s = 0, \pm 1, \pm 2\cdots$$

It can be easily seen that there will be acoustic scatterings affecting both the frequencies and the modes when the sound waves pass through the rotor, and they are related to each other in one-to-one correspondence. But when it comes to the stator blade row, there will be only the acoustic scatterings for the modes while the frequency characteristic will remain unchanged.

4.4.3.3 Linearized unsteady analysis for cascade flutter

To establish the integral equation for cascade flutter problems, we need to know the normal projection of the blade vibration velocity. Here we assume the blade displacement is $\alpha(r, z)$. The final expression will be given at the beginning and then we will discuss its derivation process.

The rotor blade vibrations can be classified into two main groups: bending vibrations and torsional vibrations. The blade vibration displacement function can be given by:

$$\alpha(r,z) = H \cdot C_a \cdot h_\alpha(r) \Big/ \sqrt{1+\left(\frac{\Omega r}{U}\right)^2} + \Theta \cdot \theta_\beta(r) \cdot (z - z_e) \cdot \sqrt{1+\left(\frac{\Omega r}{U}\right)^2}, \tag{4.185}$$

where the first and the second term represent the bending vibration and the torsional vibration, respectively. H and Θ denote the amplitudes of the bending and the torsional vibration, respectively, while $h_\alpha(r)$ and $\theta_\beta(r)$ denote the mode shapes. As shown in Fig. 4.9, for the bending vibration:

$$h_\alpha(r) = \frac{\eta}{\eta_{\max}} = \frac{\eta(r)}{\eta(R_d)}. \tag{4.186}$$

Consider the first-order bending vibration, i.e., $\alpha = 1$, and the mode shape of the bending vibration is given by:

$$\eta(r) = 0.707[\cosh(y) - \cos(y)] - 0.518[\sinh(y) - \sin(y)], \tag{4.187}$$

where

$$y(r) = 0.597\pi \frac{r - R_h}{R_d - R_h}.$$

Thus, as shown in Fig. 4.10, the bending displacement is:

$$\alpha_B(r,z) = H \cdot b \cdot h_\alpha(r) \Big/ \sqrt{1+\left(\frac{\Omega r}{U}\right)^2}. \tag{4.188}$$

For the torsional vibration, the fundamental mode is shown in Fig. 4.11. The expression for $\theta_\beta(r)$ of the blade torsional vibration is defined by:

$$\theta_\beta(r) = \sin\left(\frac{\beta\pi}{2} \cdot \frac{r - R_h}{R_d - R_h}\right). \tag{4.189}$$

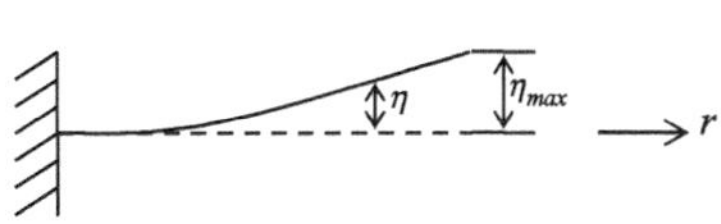

Fig. 4.9 The bending vibration mode.

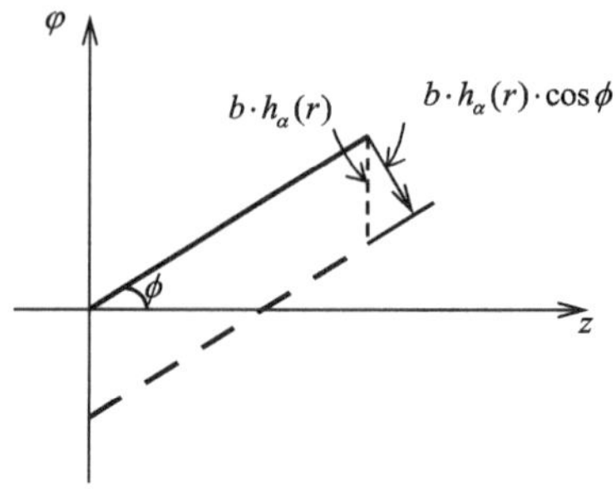

Fig. 4.10 The normal projection of the blade bending vibration mode.

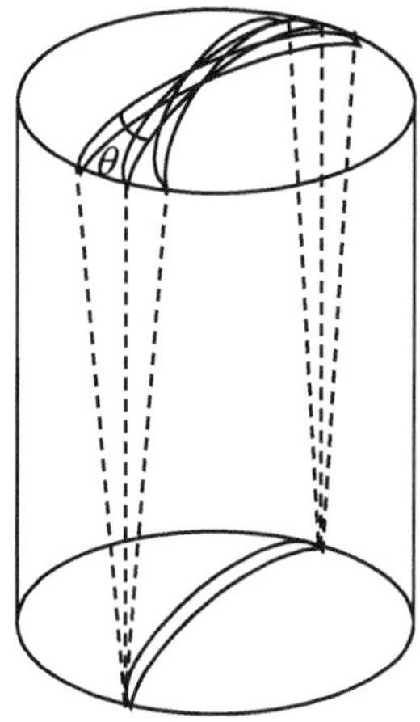

Fig. 4.11 The blade torsional vibration mode.

Also consider the first-order mode, i.e., $\beta = 1$, and then the torsional angle should be:

$$\theta_\beta(r) = \sin\left(\frac{\pi}{2} \cdot \frac{r - R_h}{R_d - R_h}\right). \tag{4.190}$$

As shown in Fig. 4.12, the corresponding torsional displacement is:

$$\alpha_T(r, z) = \Theta \cdot \theta_\beta(r) \cdot (z - z_e) \cdot \sqrt{1 + \left(\frac{\Omega r}{U}\right)^2}$$

Then, by superposing the displacement of the bending vibration and that of the torsional vibration, we can obtain Eq. (4.185). Therefore, the normal projection of the blade vibration velocity should be:

$$w(r, z) = \frac{\partial \alpha}{\partial t} + W\frac{\partial \alpha}{\partial z''} = \frac{\partial \alpha}{\partial t} + U\frac{\partial \alpha}{\partial z}. \tag{4.191}$$

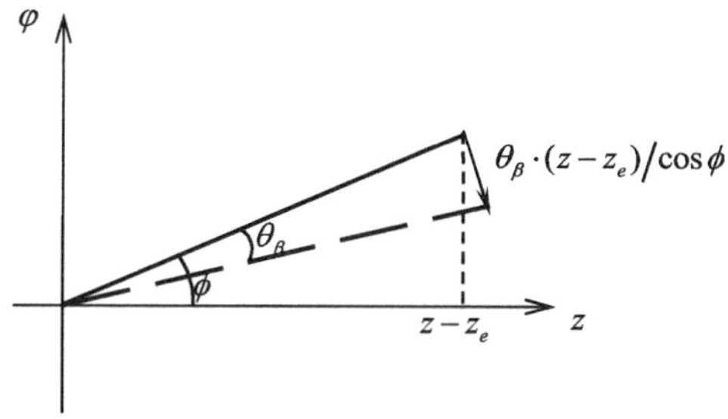

Fig. 4.12 The normal projection of the blade torsional vibration mode.

Assume the natural vibration frequency of the rotor blades is Ω_s. Then a rearrangement leads to:

$$
\begin{aligned}
w(r,z) = \frac{\partial \alpha}{\partial t} + U\frac{\partial \alpha}{\partial z} &= i\Omega_s\left[H \cdot b \cdot h_\alpha(r) \Big/ \sqrt{1+\left(\frac{\Omega r}{U}\right)^2}\right] \\
&+ i\Omega_s\left[\Theta \cdot \theta_\beta(r) \cdot (z - z_e) \cdot \sqrt{1+\left(\frac{\Omega r}{U}\right)^2}\right] \\
&+ U\left[\Theta \cdot \theta_\beta(r) \cdot \sqrt{1+\left(\frac{\Omega r}{U}\right)^2}\right].
\end{aligned}
\tag{4.192}
$$

According to the analysis we conducted for establishing the integral equation in the last section, the characteristics of the sound field due to this kind of interaction are:

$$m = sB - V, \tag{4.193}$$

$$\omega = \Omega_s + (sB - V)\Omega, \tag{4.194}$$

where V is the interblade phase angle coefficient of the rotor blade row vibration defined by $V = \sigma/2\pi$, where σ is the interblade phase angle. Thus, we can conclude that the sound generated by the cascade vibrations is not just discretely distributed at the shaft frequency and its harmonics. Indeed the frequencies of these sound waves also include the natural vibration frequency of the rotor blades, which is called non-synchronous vibrating phenomenon.

In general cases, the blade vibrations of an annular cascade would couple with the disk vibration. Hence, the interblade phase angle σ should be an integer multiple of 2π, and the acoustic modes obtained from the calculation should be a corresponding integer. However, in actual situations, σ will not be an integer multiple of 2π due to the factors such as installation error, and then the sound field in the duct would appear to be a result of the superposition of adjacent modes. For detailed discussions, one can refer to existing research results [47].

4.4.4 Solution to the integral equation

In the previous content, we have studied in detail the interaction problems that can be solved by the applications of the unified solution of flow and sound fields. Also, for various conditions, we have formulated the expressions for the external disturbance term on the right-hand side of the integral equation. In this section, we will discuss the solving approach of the integral equation.

Actually among various solving methods for the integral equation (4.182), the most commonly used and easiest one is the collocation method. However, the collocation method reveals a poor convergence performance for the application of lifting method theory to the rotor problems. In addition, even with considerable collocation points taken in the axial and radial directions, it is still hard to achieve satisfying accuracy. The reason behind this is that, compared with the process of calculating the unsteady loads on the stator blade surfaces, there is an increased complexity for the rotor problems due to the additional torsional character and the spanwise variation of the relative flow velocity. Both the stagger angle and the relative flow velocity are largest at the blade tip, so it is hard to obtain the solution to the integral equation with a simple arrangement of the collocation points. However, if the number of collocation points is too large, the computation speed will be seriously limited. Therefore, in the solving process of the integral equation on the rotor blade surfaces, we will employ another approach, which is to expand the unsteady blade forces into the sinusoidal series.

Before introducing any specific method, we will first discuss the singularity and the convergence problems [34]. Both of these problems are generated due to the superposition of the circumferential modes. Thus, we should conduct a treatment for the Bessel function of order m and, specifically, approximate it by a finite series expansion in terms of the zero-order Bessel functions. Based on the completeness of the functions themselves, we can assume:

$$\phi_m(k_{mn}r) = \sum_{l=1}^{L} B^m_{n,l}\phi_0(k_{0l}r). \tag{4.195}$$

For the convenience of the discussion, there is another expansion formula that needs to be introduced, which is:

$$\frac{\phi_{mn}(k_{mn}r)}{r^2} = \sum_{l=1}^{L} C_{n,l}^{m}\phi_0(k_{0l}r). \tag{4.196}$$

Inserting Eq. (4.195) into Eq. (4.196) leads to:

$$\frac{\sum_{j=1}^{L} B_{n,j}^{m}\phi_0(k_{0j}r)}{r^2} = \sum_{l=1}^{L} C_{n,l}^{m}\phi_0(k_{0l}r). \tag{4.197}$$

With both sides multiplied by $r\phi_0(k_{0i}r)$, we can obtain:

$$\sum_{j=1}^{L} B_{n,j}^{m}\int\frac{\phi_0(k_{0j}r)\phi_0(k_{0l}r)}{r}dr = C_{n,l}^{m}. \tag{4.198}$$

Furthermore, according to the derivation of the Bessel equation,

$$B_{n,k}^{m}\left(\frac{k_{m,n}^2}{m^2} - \frac{k_{0,k}^2}{m^2}\right) = \sum_{l=1}^{L} B_{n,l}^{m}\int_{r_h}^{r_d}\frac{\psi_0(k_{0,k}r)\psi_0(k_{0,l}r)}{r}dr. \tag{4.199}$$

Putting it into Eq. (4.198) yields:

$$C_{n,l}^{m} = B_{n,l}^{m}\left(\frac{k_{m,n}^2}{m^2} - \frac{k_{0,l}^2}{m^2}\right) = B_{n,l}^{m}\left(\mathrm{K}_{m,n}^2 - \frac{k_{0,l}^2}{m^2}\right). \tag{4.200}$$

Other properties of the Bessel function are the same with what we discussed in the last chapter, so we will not describe them again here. Express the disturbance velocity of Eq. (4.171) in terms of the kernel function:

$$v'_{\varphi''} = \int_{s(\tau)} \Delta p \cdot K\, ds(\vec{y}), \tag{4.201}$$

and assume

$$K = K_1 + K_2 + K_3, \tag{4.202}$$

where

$$K_1 = \frac{-B}{2\pi\rho W}\sum_{s=-\infty}^{\infty}\sum_{n=1}^{\infty}\phi_{mn}(k_{mn}r)\phi_{mn}(k_{mn}r')e^{im\varphi_0}e^{-im\varphi'_0}$$
$$H(z-z')\frac{\beta^2 M e^{im\frac{\Omega}{U}(z-z')}e^{i\alpha_1(z-z')}}{2\kappa_{n,m}(M\kappa_{n,m}-k_0)}\left(\frac{m}{r'} - \alpha_1\frac{\Omega r'}{U}\right)\left(\frac{m}{r} - \alpha_1\frac{\Omega r}{U}\right), \tag{4.203}$$

represents the pressure wave propagating downstream,

$$K_2 = \frac{B}{2\pi\rho W}\sum_{s=-\infty}^{\infty}\sum_{n=1}^{\infty}\phi_{mn}(k_{mn}r)\phi_{mn}(k_{mn}r')e^{im\varphi_0}e^{-im\varphi_0'}$$
$$H(z'-z)\frac{\beta^2 M e^{im\frac{\Omega}{U}(z-z')}e^{i\alpha_2(z-z')}}{2\kappa_{n,m}(k_0+M\kappa_{n,m})}\left(\frac{m}{r'}-\alpha_2\frac{\Omega r'}{U}\right)\left(\frac{m}{r}-\alpha_2\frac{\Omega r}{U}\right), \tag{4.116b}$$

represents the pressure wave propagating upstream, and

$$K_3 = \frac{-B}{2\pi\rho W}\sum_{s=-\infty}^{\infty}\sum_{n=1}^{\infty}\phi_{mn}(k_{mn}r)\phi_{mn}(k_{mn}r')e^{im\varphi_0}e^{-im\varphi_0'}$$
$$H(z-z')\frac{M^2 e^{im\frac{\Omega}{U}(z-z')}e^{i\alpha_3(z-z')}}{k_0^2+M^2k_{m,n}^2}\left(\frac{m}{r'}-\alpha_3\frac{\Omega r'}{U}\right)\left(\frac{m}{r}-\alpha_3\frac{\Omega r}{U}\right), \tag{4.116c}$$

represents the vorticity wave propagating downstream.

With the expansions of Eqs. (4.195) and (4.196), the rearrangements of the earlier expressions leads to:

$$K_1 = \frac{-B}{2\pi\rho W}\sum_{l=1}^{L}r\phi_0(k_{0l}r)\sum_{l'=1}^{L}r'\phi_0(k_{0l'}r')\sum_{s=-\infty}^{\infty}\sum_{n=1}^{\infty}e^{im\varphi_0}e^{-im\varphi_0'}$$
$$H(z-z')\frac{\beta^2 M e^{im\frac{\Omega}{U}(z-z')}e^{i\alpha_1(z-z')}}{2\kappa_{n,m}(M\kappa_{n,m}-k_0)}\left(mC_{n,l}^m-\alpha_1\frac{\Omega}{U}B_{n,l}^m\right)\left(mC_{n,l'}^m-\alpha_1\frac{\Omega}{U}B_{n,l'}^m\right), \tag{4.204}$$

$$K_2 = \frac{B}{2\pi\rho W}\sum_{l=1}^{L}r\phi_0(k_{0l}r)\sum_{l'=1}^{L}r'\phi_0(k_{0l'}r')\sum_{s=-\infty}^{\infty}\sum_{n=1}^{\infty}e^{im\varphi_0}e^{-im\varphi_0'}$$
$$H(z'-z)\frac{\beta^2 M e^{im\frac{\Omega}{U}(z-z')}e^{i\alpha_2(z-z')}}{2\kappa_{n,m}(k_0+M\kappa_{n,m})}\left(mC_{n,l}^m-\alpha_2\frac{\Omega}{U}B_{n,l}^m\right)\left(mC_{n,l'}^m-\alpha_2\frac{\Omega}{U}B_{n,l'}^m\right), \tag{4.117b}$$

$$K_3 = \frac{-B}{2\pi\rho W}\sum_{l=1}^{L}r\phi_0(k_{0l}r)\sum_{l'=1}^{L}r'\phi_0(k_{0l'}r')\sum_{s=-\infty}^{\infty}\sum_{n=1}^{\infty}e^{im\varphi_0}e^{-im\varphi_0'}$$
$$H(z-z')\frac{M^2 e^{im\frac{\Omega}{U}(z-z')}e^{i\alpha_3(z-z')}}{k_0^2+M^2k_{m,n}^2}\left(mC_{n,l}^m-\alpha_3\frac{\Omega}{U}B_{n,l}^m\right)\left(mC_{n,l'}^m-\alpha_3\frac{\Omega}{U}B_{n,l'}^m\right). \tag{4.117c}$$

4.4.4.1 The convergence of the vorticity wave

By considering $\alpha_3 = -\omega/U$, the kernel function of the vorticity wave can therefore be further simplified to:

$$K_3 = \frac{-B}{2\pi\rho W} e^{-i\frac{\lambda}{U}(z-z')} \sum_{l=1}^{L} r\phi_0(k_{0l}r) \sum_{l'=1}^{L} r'\phi_0(k_{0l'}r') \sum_{s=-\infty}^{\infty} \sum_{n=1}^{\infty} e^{im\varphi_0} e^{-im\varphi'_0}$$
$$H(z-z') \frac{M^2}{k_0^2 + M^2 k_{m,n}^2} \left(mC_{n,l}^m - B_{n,l}^m \alpha_3 \frac{\Omega}{U} \right) \left(mC_{n,l'}^m - \alpha_3 \frac{\Omega}{U} B_{n,l'}^m \right). \tag{4.205}$$

With Eq. (4.200) inserted into it, we have:

$$K_3 = \frac{-B}{2\pi\rho W} e^{-i\frac{\lambda}{U}(z-z')} \sum_{l=1}^{L} r\phi_0(k_{0l}r) \sum_{l'=1}^{L} r'\phi_0(k_{0l'}r') \sum_{s=-\infty}^{\infty} \sum_{n=1}^{\infty} B_{n,l}^m B_{n,l'}^m e^{im\varphi_0} e^{-im\varphi'_0}$$
$$H(z-z') \frac{M^2}{k_0^2 + M^2 k_{m,n}^2} \left(m\mathrm{K}_{m,n}^2 - m\frac{k_{0,l}^2}{m^2} - \alpha_3 \frac{\Omega}{U} \right) \left(m\mathrm{K}_{m,n}^2 - m\frac{k_{0,l'}^2}{m^2} - \alpha_3 \frac{\Omega}{U} \right). \tag{4.206}$$

Consider the term involving the mode of order m, and then insert $\alpha_3 = -\frac{\lambda}{U} - \frac{m\Omega}{U}$ and $k_0 = \frac{m\Omega}{c_0} + \frac{\lambda}{c_0}$ into it:

$$K_3 = \frac{-B}{2\pi\rho W} e^{-i\frac{\lambda}{U}(z-z')} \sum_{l=1}^{L} r\phi_0(k_{0l}r) \sum_{l'=1}^{L} r'\phi_0(k_{0l'}r') \sum_{s=-\infty}^{\infty} \sum_{n=1}^{\infty} B_{n,l}^m B_{n,l'}^m e^{im\varphi_0} e^{-im\varphi'_0}$$
$$\frac{H(z-z') c_0^2 M^2 m^2}{(m\Omega + \lambda)^2 + c_0^2 M^2 k_{m,n}^2} \left(\mathrm{K}_{m,n}^2 + \frac{\Omega^2}{U^2} + \frac{\lambda\Omega}{mU^2} - \frac{k_{0,l}^2}{m^2} \right)$$
$$\left(\mathrm{K}_{m,n}^2 + \frac{\Omega^2}{U^2} + \frac{\lambda\Omega}{mU^2} - \frac{k_{0,l'}^2}{m^2} \right). \tag{4.207}$$

Now make an investigation of the part relevant to m:

$$\frac{c_0^2 M^2 m^2}{m^2\Omega^2 + 2m\Omega\lambda + \lambda^2 + c_0^2 M^2 k_{m,n}^2} \left(\mathrm{K}_{m,n}^2 + \frac{\Omega^2}{U^2} + \frac{\lambda\Omega}{mU^2} - \frac{k_{0,l}^2}{m^2} \right)$$
$$\left(\mathrm{K}_{m,n}^2 + \frac{\Omega^2}{U^2} + \frac{\lambda\Omega}{mU^2} - \frac{k_{0,l'}^2}{m^2} \right) \tag{4.208}$$

The simplification yields:

$$\frac{c_0^2M^2}{\Omega^2+\dfrac{2\Omega\lambda}{m}+\dfrac{\lambda^2}{m^2}+c_0^2M^2\mathrm{K}_{m,n}^2}\left(\mathrm{K}_{m,n}^2+\frac{\Omega^2}{U^2}\right)\left(\mathrm{K}_{m,n}^2+\frac{\Omega^2}{U^2}\right)+\mathrm{O}\left(\frac{1}{m}\right) \tag{4.209}$$

Take the limit of $m\rightarrow\infty$, and we can further obtain:

$$\frac{c_0^2M^2}{\Omega^2+\dfrac{2\Omega\lambda}{m}+\dfrac{\lambda^2}{m^2}+c_0^2M^2\mathrm{K}_{m,n}^2}\left(\mathrm{K}_{m,n}^2+\frac{\Omega^2}{U^2}\right)\left(\mathrm{K}_{m,n}^2+\frac{\Omega^2}{U^2}\right)$$
$$+\mathrm{O}\left(\frac{1}{m}\right)\overset{m\rightarrow\infty}{\longrightarrow}\left(\mathrm{K}_{m,n}^2+\frac{\Omega^2}{U^2}\right). \tag{4.210}$$

Then decompose the kernel function into the regular part and the singular part, i.e., $K_3=K_3^{Rw}+K_3^{Sw}$. The discussion in the last chapter implies the singular part doesn't make sense. Therefore, the kernel function can be further simplified to:

$$K_3=\frac{B}{2\pi\rho W}e^{-i\frac{\lambda}{U}(z-z')}\sum_{l=1}^{L}r\phi_0(k_{0l}r)\sum_{l'=1}^{L}r'\phi_0(k_{0l'}r')\sum_{s=-\infty}^{\infty}\sum_{n=1}^{\infty}\left[\frac{B_{n,l}^mB_{n,l'}^mM^2m^2}{k_0^2+M^2k_{m,n}^2}\right.$$
$$\left(\mathrm{K}_{m,n}^2+\frac{\Omega^2}{U^2}+\frac{\lambda\Omega}{mU^2}-\frac{k_{0,l}^2}{m^2}\right)\left(\mathrm{K}_{m,n}^2+\frac{\Omega^2}{U^2}+\frac{\lambda\Omega}{mU^2}-\frac{k_{0,l'}^2}{m^2}\right)$$
$$\left.-B_{n,l}^{\infty}B_{n,l'}^{\infty}\left(\mathrm{K}_{m,n}^2+\frac{\Omega^2}{U^2}\right)\right], \tag{4.211}$$

where $m=sB-N_w,\quad s=0,\pm1,\pm2\cdots$. It can be proven that the vorticity wave expression after this treatment has good convergence performance and thus can be directly employed for solving the integral equation.

4.4.4.2 The singularity of pressure wave

The kernel functions concerning the pressure waves can be written in the unified form of:

$$K_{1,2}=\frac{-B}{2\pi\rho W}\sum_{l=1}^{L}r\phi_0(k_{0l}r)\sum_{l'=1}^{L}r'\phi_0(k_{0l'}r')\sum_{s=-\infty}^{\infty}\sum_{n=1}^{\infty}e^{im\varphi_0}e^{-im\varphi_0'}$$
$$\frac{\beta^2Me^{im\frac{\Omega}{U}(z-z')}e^{i\alpha_{1,2}(z-z')}}{2\kappa_{n,m}[M\kappa_{n,m}\mathrm{sgn}(z-z')-k_0]}\left(mC_{n,l}^m-\alpha_{1,2}\frac{\Omega}{U}B_{n,l}^m\right)\left(mC_{n,l'}^m-\alpha_{1,2}\frac{\Omega}{U}B_{n,l'}^m\right). \tag{4.212}$$

Taking Eq. (4.200) into it leads to:

$$K_{1,2}=\frac{-B}{2\pi\rho W}\sum_{l=1}^{L}r\phi_0(k_{0l}r)\sum_{l'=1}^{L}r'\phi_0(k_{0l'}r')\sum_{s=-\infty}^{\infty}\sum_{n=1}^{\infty}B_{n,l}^{m}B_{n,l'}^{m}e^{im\varphi_0}e^{-im\varphi_0'}$$
$$\frac{\beta^2 M e^{im\frac{\Omega}{U}(z-z')}e^{i\alpha_{1,2}(z-z')}}{2\kappa_{n,m}[M\kappa_{n,m}\,\mathrm{sgn}(z-z')-k_0]}\left(m\mathrm{K}_{m,n}^2-m\frac{k_{0,l}^2}{m^2}-\alpha_{1,2}\frac{\Omega}{U}\right)$$
$$\left(m\mathrm{K}_{m,n}^2-m\frac{k_{0,l'}^2}{m^2}-\alpha_{1,2}\frac{\Omega}{U}\right). \tag{4.213}$$

By putting $\alpha_{1,2}=\frac{M_a}{\beta^2}\frac{sB\Omega}{c_0}-\frac{\kappa_{n,m}}{\beta^2}\mathrm{sgn}(z-z')$ into the previous expression, we can obtain:

$$K_{1,2}=\frac{-B}{2\pi\rho W}e^{-i\frac{\lambda}{U}(z-z')}\sum_{l=1}^{L}r\phi_0(k_{0l}r)\sum_{l'=1}^{L}r'\phi_0(k_{0l'}r')\sum_{s=-\infty}^{\infty}\sum_{n=1}^{\infty}B_{n,l}^{m}B_{n,l'}^{m}e^{im\varphi_0}e^{-im\varphi_0'}$$
$$\cdot\frac{\beta^2 M_a e^{i\frac{sB\Omega}{U\beta^2}(z-z')}e^{-i\frac{\kappa_{n,m}}{\beta^2}|z-z'|}}{2\kappa_{n,m}[M\kappa_{n,m}\,\mathrm{sgn}(z-z')-k_0]}$$
$$\cdot\left[m\mathrm{K}_{m,n}^2-m\frac{k_{0,l}^2}{m^2}-\frac{M_a}{\beta^2}\frac{(m\Omega+\lambda)}{c_0}\frac{\Omega}{U}+\frac{\kappa_{n,m}}{\beta^2}\mathrm{sgn}(z-z')\frac{\Omega}{U}\right]$$
$$\cdot\left[m\mathrm{K}_{m,n}^2-m\frac{k_{0,l'}^2}{m^2}-\frac{M_a}{\beta^2}\frac{(m\Omega+\lambda)}{c_0}\frac{\Omega}{U}+\frac{\kappa_{n,m}}{\beta^2}\mathrm{sgn}(z-z')\frac{\Omega}{U}\right]. \tag{4.214}$$

Take the limit of $m\rightarrow\infty$ and make a discussion for the following two cases:

$$\mathrm{K}_{m,n}\xrightarrow{m\rightarrow\infty}\mathrm{K}_{\infty,n}, \tag{4.215}$$

$$\kappa_{n,m}=-i\sqrt{\beta^2k_{mn}^2-\left(\frac{m\Omega}{c_0}+\frac{\lambda}{c_0}\right)^2}=-i\left[\widehat{\mathrm{K}}_{m,n}|m|-\frac{\Omega\lambda}{c_0^2\widehat{\mathrm{K}}_{m,n}}\mathrm{sgn}(m)+\mathrm{O}(m^{-1})\right]$$
$$\xrightarrow{m\rightarrow\infty}-i\left[\widehat{\mathrm{K}}_{\infty,n}|m|-\frac{\Omega\lambda}{c_0^2\widehat{\mathrm{K}}_{\infty,n}}\mathrm{sgn}(m)+\mathrm{O}(m^{-1})\right]. \tag{4.216}$$

The term in the kernel function relevant to the mode of order m and worth discussion is:

$$\frac{\beta^2 M_a}{2\kappa_{n,m}[\mathrm{sgn}(z-z')M\kappa_{n,m}-k_0]}$$
$$\cdot\left[m\mathrm{K}^2_{m,n}-m\frac{k^2_{0,l}}{m^2}-\frac{M_a}{\beta^2}\frac{(m\Omega+\lambda)}{c_0}\frac{\Omega}{U}+\frac{\kappa_{n,m}}{\beta^2}\mathrm{sgn}(z-z')\frac{\Omega}{U}\right]$$
$$\cdot\left[m\mathrm{K}^2_{m,n}-m\frac{k^2_{0,l'}}{m^2}-\frac{M_a}{\beta^2}\frac{(m\Omega+\lambda)}{c_0}\frac{\Omega}{U}+\frac{\kappa_{n,m}}{\beta^2}\mathrm{sgn}(z-z')\frac{\Omega}{U}\right].$$

To simplify this term, we assume:

$$\frac{A\cdot m^2+B\cdot|m|+\mathrm{O}(m^0)}{P\cdot m^2+Q\cdot|m|+\mathrm{O}(m^0)}=a+\frac{b}{|m|}+\mathrm{O}\left(m^{-2}\right), \tag{4.217}$$

where

$$A=\left[\mathrm{K}^2_{m,n}-\frac{\Omega^2}{\beta^2c_0^2}-\frac{i\Omega\widehat{\mathrm{K}}_{m,n}}{\beta^2U}\mathrm{sgn}(m)\,\mathrm{sgn}(z-z')\right]^2\beta^2M_a,$$

$$B=2\beta^2M_a\left[-\frac{\lambda\Omega\widehat{\mathrm{K}}^2_{m,n}}{\beta^4c_0^2}\mathrm{sgn}(m)+\frac{\Omega^3\lambda}{\beta^4U^2c_0^2}\mathrm{sgn}(m)+2i\frac{\lambda\Omega^2\widehat{\mathrm{K}}_{m,n}}{\beta^4Uc_0^2}\mathrm{sgn}(z-z')\right],$$

$$P=2i\frac{\Omega}{c_0}\widehat{\mathrm{K}}_{m,n}\mathrm{sgn}(m)-2M\widehat{\mathrm{K}}^2_{m,n}\mathrm{sgn}(z-z'),$$

$$Q=4M\frac{\Omega\lambda}{c_0^2}\mathrm{sgn}(z-z')\,\mathrm{sgn}(m)+2i\frac{\lambda}{c_0}\widehat{\mathrm{K}}_{m,n}-2i\frac{\Omega^2\lambda}{c_0^3\widehat{\mathrm{K}}_{m,n}}.$$

By solving this equation, we can get:

$$a=\frac{A}{P}=\frac{\left[\mathrm{K}^2_{m,n}-\frac{\Omega^2}{\beta^2c_0^2}-\frac{i\Omega\widehat{\mathrm{K}}_{m,n}}{\beta^2U}\mathrm{sgn}(m)\,\mathrm{sgn}(z-z')\right]^2\beta^2M_a}{2i\frac{\Omega\widehat{\mathrm{K}}_{m,n}}{c_0}\mathrm{sgn}(m)-2M\widehat{\mathrm{K}}^2_{m,n}\mathrm{sgn}(z-z')}.$$

Inserting $\widehat{\mathrm{K}}^2_{m,n}=\beta^2\mathrm{K}^2_{m,n}-\frac{\Omega^2}{c_0^2}$ into this expression and then rearranging it leads to:

$$a = -\frac{1}{2\beta^2}\left[\widehat{\mathrm{K}}_{m,n}^2 \operatorname{sgn}(z-z') - \frac{i\Omega}{U}\widehat{\mathrm{K}}_{m,n}\operatorname{sgn}(m)\right]. \tag{4.218}$$

Then we can determine that the coefficient of $\frac{1}{|m|}$ is:

$$b = \frac{B}{P} - \frac{aQ}{P} = \frac{i\frac{\lambda}{\beta^2 c_0}\widehat{\mathrm{K}}_{m,n}\operatorname{sgn}(z-z')\beta^2\mathrm{K}_{m,n}^2 + \frac{\Omega\lambda}{Uc_0}\mathrm{K}_{m,n}^2\operatorname{sgn}(m)}{2i\frac{\Omega}{c_0}\widehat{\mathrm{K}}_{m,n}\operatorname{sgn}(m) - 2M\widehat{\mathrm{K}}_{m,n}^2\operatorname{sgn}(z-z')},$$

which can be further simplified to:

$$b = -\frac{i}{2}\frac{\lambda \mathrm{K}_{m,n}^2}{U\widehat{\mathrm{K}}_{m,n}}. \tag{4.219}$$

Hence, when $m \to \infty$, we have:

$$\frac{\beta^2 M_a}{2\kappa_{n,m}[\operatorname{sgn}(z-z')M\kappa_{n,m} - k_0]}.$$
$$\left[m\mathrm{K}_{m,n}^2 - m\frac{k_{0,l}^2}{m^2} - \frac{M_a}{\beta^2}\frac{(m\Omega+\lambda)}{c_0}\frac{\Omega}{U} + \frac{\kappa_{n,m}}{\beta^2}\operatorname{sgn}(z-z')\frac{\Omega}{U}\right].$$
$$\left[m\mathrm{K}_{m,n}^2 - m\frac{k_{0,l'}^2}{m^2} - \frac{M_a}{\beta^2}\frac{(m\Omega+\lambda)}{c_0}\frac{\Omega}{U} + \frac{\kappa_{n,m}}{\beta^2}\operatorname{sgn}(z-z')\frac{\Omega}{U}\right]$$
$$\xrightarrow{m\to\infty} -\frac{1}{2\beta^2}\left[\widehat{\mathrm{K}}_{\infty,n}^2\operatorname{sgn}(z-z') - i\frac{\Omega}{U}\widehat{\mathrm{K}}_{\infty,n}\operatorname{sgn}(m)\right] - \frac{i}{2}\frac{\lambda\mathrm{K}_{\infty,n}^2}{U\widehat{\mathrm{K}}_{\infty,n}}\frac{1}{|m|} + \mathrm{O}(m^{-2}), \tag{4.220}$$

where

$$\widehat{\mathrm{K}}_{\infty,n} = \sqrt{\beta^2\mathrm{K}_{\infty,n}^2 - \frac{\Omega^2}{c_0^2}}. \tag{4.221}$$

Decompose the kernel function for the pressure waves into the regular part and the singular part, i.e.,

$$K_{1,2} = K_{1,2}^{(Rp)} + K_{1,2}^{(Sp)}. \tag{4.222}$$

where

$$K_{1,2}^{(Rp)} = \frac{-B}{2\pi\rho W} e^{-i\frac{\lambda}{U}(z-z')} \sum_{l=1}^{L} r\phi_0(k_{0l}r) \sum_{l'=1}^{L} r'\phi_0(k_{0l'}r')$$
$$\sum_{s=-\infty}^{\infty} \sum_{n=1}^{\infty} \Big\{ e^{im\varphi_0} e^{i\frac{sB\Omega}{U\beta^2}(z-z')} e^{-i\frac{\kappa_{n,m}}{\beta^2}|z-z'|}$$
$$\frac{\left(\beta^2 M m C_{n,l}^m - \alpha_{1,2}\frac{\Omega}{U} B_{n,l}^m\right)\left(m C_{n,l'}^m - \alpha_{1,2}\frac{\Omega}{U} B_{n,l'}^m\right)}{2\kappa_{n,m}\left[M\kappa_{n,m}\,\mathrm{sgn}(z-z') - k_0\right]} \tag{4.223}$$
$$- e^{i\frac{sB\Omega}{U\beta^2}(z-z')} e^{-\frac{\left[\widehat{K}_{\infty,n}|m| - \frac{\Omega\lambda}{c_0^2 \widehat{K}_{\infty,n}}\mathrm{sgn}(m)\right]}{\beta^2}|z-z'|} B_{n,l}^{\infty} B_{n,l'}^{\infty}$$
$$\cdot \left[a + b\frac{1}{|m|} + O\left(m^{-2}\right)\right]\Big\}.$$

$$K_{1,2}^{(Sp)} = \frac{-B}{2\pi\rho W} e^{-i\frac{\lambda}{U}(z-z')} \sum_{l=1}^{L} r\phi_0(k_{0l}r) \sum_{l'=1}^{L} r'\phi_0(k_{0l'}r') \sum_{s=-\infty}^{\infty} \sum_{n=1}^{\infty} B_{n,l}^{\infty} B_{n,l'}^{\infty} e^{im\varphi_0}$$
$$\cdot e^{i\frac{sB\Omega}{U\beta^2}(z-z')} e^{-\frac{\left[\widehat{K}_{\infty,n}|m| - \frac{\Omega\lambda}{c_0^2 \widehat{K}_{\infty,n}}\mathrm{sgn}(m)\right]}{\beta^2}|z-z'|} \cdot \left[a + b\frac{1}{|m|} + O\left(m^{-2}\right)\right]. \tag{4.224}$$

Then by conducting a treatment for the singular part, $K_{1,2}^{(Sp)}$, one can derive that, with $|z-z'| \to 0$, the expression for the singular part turns out to be:

$$K_{1,2}^{(Sp)} = \frac{-B}{2\pi\rho W} \sum_{l=1}^{L} r\phi_0(k_{0l}r) \sum_{l'=1}^{L} r'\phi_0(k_{0l'}r')$$
$$\sum_{n=1}^{\infty} B_{n,l}^{\infty} B_{n,l'}^{\infty} \left[-\frac{\widehat{K}_{m,n}}{B\beta^2} \frac{1}{(z-z')} - \frac{i\lambda K_{m,n}^2}{BU\widehat{K}_{m,n}} \log|z-z'| \right]. \tag{4.225}$$

Indeed, during the integration process of the kernel function, the first term of the singular part, which contains a factor of $1/(z-z')$, can be evaluated directly by applying the trapezoidal integration rule. However, the second term, involving $\log|z-z'|$, is still hard to be directly integrated, thus requiring certain extra treatments that will be discussed in the following contents.

4.4.4.3 Expansion method for solving the integral equation

As mentioned before, although the collocation method has achieved a satisfying application in problems concerning the rotor wakes-stator interactions, in the three-dimensional lifting surface method for rotors, due to a series of problems resulting from the rotor's rotational motion, the collocation method reveals a very poor convergence performance and thus cannot be directly used for the solving process. Therefore, to solve the integral equation on the rotor blade surfaces, we adopt another approach that employs the expansion of Δp [34].

Consider the integral equation (4.201) and assume:

$$\Delta p = \sum_{j'=1}^{J} \phi_{0j'}(r') \left[A_{1j'} \cot\frac{\vartheta'}{2} + \sum_{i'=2}^{I} A_{i'j'} \sin(i'-1)\vartheta' \right], \tag{4.226}$$

where

$$z' = \frac{b}{2}(1-\cos\vartheta'); \quad z = \frac{b}{2}(1-\cos\vartheta), \quad z, z' \in (0,b); \quad \vartheta, \vartheta \in (0,\pi),$$

and

$$dz' = \frac{b}{2}\sin\vartheta' \, d\vartheta'.$$

Insert this expansion into the expression for the induced fluctuating velocity on the blade surface, and: we can obtain

$$v'_{\varphi''} = \int_{s(\tau)} \sum_{j'=1}^{J} \phi_{0j'}(r') \left[A_{1j'} \cot\frac{\vartheta'}{2} + \sum_{i'=2}^{I} A_{i'j'} \sin(i'-1)\vartheta' \right] \cdot K(r|r', \varphi, z-z') \, ds(\vec{\gamma}). \tag{4.227}$$

By investigating the expression of the kernel function, one can easily find that, due to the orthogonality of the Bessel functions, the convergence performance in the radial integration process can be greatly improved, which is one of the main reasons why the expansion method is more advantageous than the collocation method. Eq. (4.227) can then be further simplified.

Assume that:

$$v'_{\varphi''} = \int_{0}^{\pi} \sum_{l'=1}^{J} \left[A_{1l'} \cot\frac{\vartheta'}{2} + \sum_{i'=2}^{I} A_{i'l'} \sin(i'-1)\vartheta' \right] \cdot K^{\Delta}(r|r', \varphi, z-z') \frac{b}{2}\sin\vartheta' \, d\vartheta', j' = l' \tag{4.228}$$

where

$$K^{\Delta} = K_1^{\Delta} + K_2^{\Delta} + K_3^{\Delta}, \tag{4.229}$$

$$K_1^{\Delta} = \frac{-B}{2\pi\rho W}\sum_{l=1}^{L} r\phi_0(k_{0l}r)\sum_{s=-\infty}^{\infty}\sum_{n=1}^{\infty} e^{im\varphi_0}e^{-im\varphi_0'}$$
$$H(z-z')\frac{\beta^2 M e^{im\frac{\Omega}{U}(z-z')}e^{i\alpha_1(z-z')}}{2\kappa_{n,m}(M\kappa_{n,m}-k_0)}\left(mC_{n,l}^m-\alpha_1\frac{\Omega}{U}B_{n,l}^m\right)\left(mC_{n,l'}^m-\alpha_1\frac{\Omega}{U}B_{n,l'}^m\right), \tag{4.230}$$

$$K_2^{\Delta} = \frac{B}{2\pi\rho W}\sum_{l=1}^{L} r\phi_0(k_{0l}r)\sum_{s=-\infty}^{\infty}\sum_{n=1}^{\infty} e^{im\varphi_0}e^{-im\varphi_0'}$$
$$H(z'-z)\frac{\beta^2 M e^{im\frac{\Omega}{U}(z-z')}e^{i\alpha_2(z-z')}}{2\kappa_{n,m}(k_0+M\kappa_{n,m})}\left(mC_{n,l}^m-\alpha_2\frac{\Omega}{U}B_{n,l}^m\right)\left(mC_{n,l'}^m-\alpha_2\frac{\Omega}{U}B_{n,l'}^m\right), \tag{4.143b}$$

$$K_3^{\Delta} = \frac{B}{2\pi\rho W}e^{-i\frac{\lambda}{U}(z-z')}\sum_{l=1}^{L} r\phi_0(k_{0l}r)\sum_{s=-\infty}^{\infty}\sum_{n=1}^{\infty}\left[\frac{B_{n,l}^m B_{n,l'}^m M^2 m^2}{k_0^2+M^2k_{m,n}^2}\right.$$
$$\left(\mathrm{K}_{m,n}^2+\frac{\Omega^2}{U^2}+\frac{\lambda\Omega}{mU^2}-\frac{k_{0,l}^2}{m^2}\right)\left(\mathrm{K}_{m,n}^2+\frac{\Omega^2}{U^2}+\frac{\lambda\Omega}{mU^2}-\frac{k_{0,l'}^2}{m^2}\right) \tag{4.143c}$$
$$\left.-B_{n,l}^{\infty}B_{n,l'}^{\infty}\left(\mathrm{K}_{\infty,n}^2+\frac{\Omega^2}{U^2}\right)\right].$$

Meanwhile, considering that the impermeable boundary condition should be satisfied at various locations on the blade surfaces, we assume:

$$r_j = R_h + (R_d - R_h)\cdot\frac{j-1}{J-1};\; j=1,2,\ldots,J, \theta_i = \frac{i\pi}{I};\; i=1,2,\ldots,I.$$

Then the integral equation can be converted into the form of an algebraic equation:

$$w_{ij}' = \sum_{i'=1}^{I}\sum_{j'=1}^{J} A_{i'j'}\cdot \Im_{v(ij,i'j')},\; j'=l' \tag{4.231}$$

For the singular term of the pressure waves that contains $\log|z-z'|$ and therefore cannot be directly integrated, it is in the form of:

$$v'^{Sp}_{\varphi'',ij} = \frac{B}{2\pi\rho W}\frac{b}{2}\sum_{l'=1}^{J}\sum_{i'=1}^{I} A_{i'l'}\cdot\sum_{l=1}^{L} r_j\phi_0(k_{0l}r_j)\sum_{n=1}^{\infty} B_{n,l}^{\infty}B_{n,l'}^{\infty}\frac{i\lambda \mathrm{K}_{m,n}^2}{BU\widehat{\mathrm{K}}_{m,n}}$$
$$\int_0^{\pi} f_{i'}'(\vartheta')\log|z-z'|d\vartheta', \tag{4.232}$$

where

$$f'_{i'}(\vartheta')=\begin{cases}2\cos^2\dfrac{\vartheta'}{2} & i'=1\\ \sin(i'-1)\vartheta'\cdot\sin\vartheta' & i'\geq 2\end{cases}. \tag{4.233}$$

In the previous chapter, we introduced the standard trapezoidal integration method for the case including the logarithmic term proposed by Whitehead. The singular part is given by:

$$\int_0^{\pi} f'_{i'}(\vartheta')\log|\cos\vartheta-\cos\vartheta'|d\vartheta'=\sum_{k'=1}^{K'_{num}}\frac{\pi}{K'_{num}}S_{i,k'}, \tag{4.234}$$

where

$$\vartheta'_{k'}=\frac{k'\pi}{K'_{num}};\ k'=1,2,\cdots,K'_{num}, \tag{4.235}$$

$$S_{i,k'}=-\left\{\log 2+2\sum_{q=1}^{L}\frac{1}{q}\left(\cos q\theta'_{k'}\right)\left(\cos q\theta_i\right)\right\}. \tag{4.236}$$

Thus, the singular part can be rewritten as:

$$v'^{Sp}_{\varphi'',ij}=\frac{B}{2\pi\rho W}\frac{b}{2}\sum_{l'=1}^{J}\sum_{i'=1}^{I}A_{i'l'}\cdot\sum_{l=1}^{L}r_j\phi_0\left(k_{0l}r_j\right)\sum_{n=1}^{\infty}B^{\infty}_{n,l}B^{\infty}_{n,l'}\frac{i\lambda\mathrm{K}^2_{m,n}}{BU\widehat{\mathrm{K}}_{m,n}}\sum_{k'=1}^{K'_{num}}\frac{\pi}{K'_{num}}S_{i,k'}. \tag{4.237}$$

By considering the treatment for the singularity of the logarithmic term, the coefficient matrix of the original algebraic equation (4.231) can be given by:

$$\begin{aligned}\mathfrak{I}_{v(ij,i'j')}=-\Big\{&K^{\nabla}_{ij,i'j'}\left(r_j,\ \cos\vartheta_i\right)\\&+\sum_{l=1}^{L}r_j\phi_0\left(k_{0l}r_j\right)\cdot Q_{l,l'}\cdot\sum_{k'=1}^{K'_{num}}\frac{\pi}{K'_{num}}\left[S_{i,k'}-\log\left|\cos\vartheta'_{k'}-\cos\vartheta_i\right|\right]\Big\},\end{aligned} \tag{4.238}$$

where

$$Q_{l,l'}=\frac{B}{2\pi\rho W}\frac{b}{2}\sum_{n=1}^{\infty}B^{\infty}_{n,l}B^{\infty}_{n,l'}\frac{i\lambda\mathrm{K}^2_{m,n}}{BU\widehat{\mathrm{K}}_{m,n}}, \tag{4.239}$$

$$K^{\nabla}_{ij,i'j'}\left(r_j,\ \cos\vartheta_i\right)=\int_0^{\pi}K^{\Delta}_{ij,j'}\cdot f'_{i'}(\vartheta')\frac{b}{2}d\vartheta'. \tag{4.240}$$

So far, the equation has been ready to be solved by a direct application of the trapezoidal integration method. It is still noteworthy that we need to discuss the propagating properties of the sound waves and the vorticity waves during the integration process, i.e., to discuss the two cases that $\vartheta > \vartheta'$ and $\vartheta < \vartheta'$. Also, it can be found that K_{num}' is an independent parameter for the integration process. Because there is no requirement that the parameter value at the source point ϑ' should coincide with that at the observation point, as there is in the collocation method, the convergence speed is significantly improved, which is one of the main reasons why the expansion method is superior to the collocation method in this case.

4.4.5 Numerical validation of unsteady blade loading

To clearly describe the solving procedure of the unified flow-acoustic field solution, two illustrative examples will be discussed in the following text.

The first example is based on the classical test case in Third Computational Aeroacoustics (CAA) Workshop on Benchmark Problems: Category 4——Fan Stator with Harmonic Excitation by Rotor Wake. A comparison is made between the present results with the data obtained by Namba and Schulten [44]. As depicted in Fig. 4.13, consider the sound generation by the interaction between the rotor wakes and the downstream stator in an infinitely long annular duct, where the stator vanes of zero thickness and zero chamber are aligned with the uniform axial mean flow. The case configurations are as follows: the number of rotor blades $B = 16$; the number of stator vanes $V = 24$; the boss ratio $R_h/R_d = 0.5$; at the blade tip, the stator vane spacing is equal to its chord length b, i.e., $2\pi R_d/V = b$; and the mean

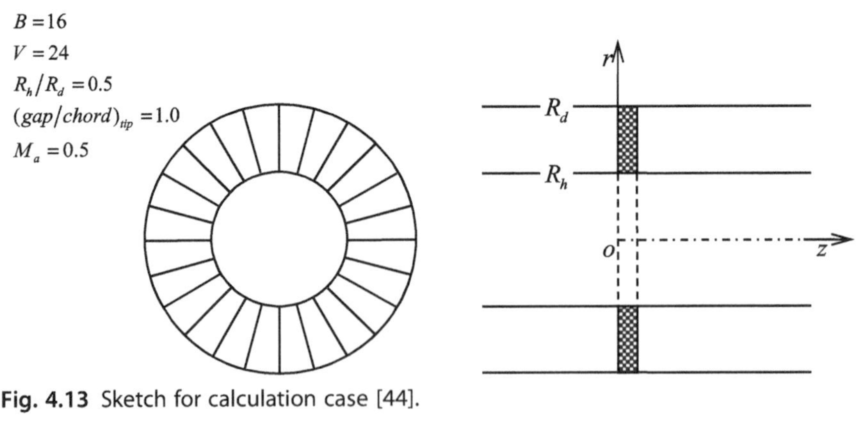

Fig. 4.13 Sketch for calculation case [44].

flow Mach number $M_a = 0.5$. Taking the cylindrical coordinates (z, r, φ), one can expand the rotor wake velocity in the harmonic series:

$$w_i' = U \sum_{s=0}^{\infty} W_s \cdot e^{isB[\Omega z/U + \varphi - \vartheta(r) - \Omega t]} \tag{4.241}$$

where U is the mean flow velocity and Ω is the angular speed of rotor rotation. It should be noted that the wake representation and other definitions given in the CAA Workshop are based on the harmonic time dependence $e^{-i\omega t}$, but here, in consistence with the present liner model, the time dependence is assumed to be $e^{i\omega t}$. There is no fundamental distinction between these two forms of the harmonic time dependence assumption, but the imaginary parts of the corresponding prediction results will differ by a minus sign. Such effect has been taken into account in the comparison; in other words, we have made the corresponding phase adjustment to the results obtained by the present approach to allow an effective comparison to the published results. In this case, only the fundamental harmonic perturbation, i.e., $s = 1$, at the blade passing frequency is considered, and correspondingly, the amplitude of the upwash perturbation is $W_s = 0.1$. In Eq. (4.241), the term related to the radial variation is given by:

$$\vartheta(r) = -\frac{2\pi\imath}{B}\frac{r - R_h}{R_d - R_h} \tag{4.242}$$

where $\vartheta(R_d) = -2\pi\imath/B$ denotes the phase variation in the spanwise direction, and when $\imath = 0$, the excitations are in-phase from the blade root to the blade tip.

First of all, the comparison is made for the existing two-dimensional results. The present case is degenerated into a two-dimensional one by setting $R_h/R_d = 0.98$ and discarding the radial variation term, i.e., letting $\vartheta(r) = 0$. Four different flow conditions are investigated, where the blade tip Mach number $M_t = 0.3897$, 0.4330, 0.4763, and 0.6495, respectively. For the convenience of comparing the computational results, we introduce the complex coefficient A_m, which denotes the modal amplitude of pressure fluctuations:

$$p'(z, \varphi, t) = p_0 \sum_{q=-\infty}^{\infty} A_m(z) \cdot e^{i(m\varphi - sB\Omega t)} \tag{4.243}$$

Consider the axial stations one chord length upstream and downstream of the blade edges, i.e., $z = -b$ and $z = +2b$. Moreover, for this two-dimensional case, the unsteady lift coefficient with respect to the middle span is defined by:

$$C_L = \frac{\int_0^b \Delta p(r, z)dz}{b \cdot W_1} \tag{4.244}$$

Table 4.4 gives the calculation results of the unsteady lift coefficients for different blade tip Mach numbers. The unsteady lift coefficient as a function of the reduced frequency $\frac{2\pi B}{V} \cdot \frac{M_t}{M_a}$ is shown in Fig. 4.14, where the results obtained by the present method are compared with the two-dimensional results of a study by Namba [44].

Table 4.4 Comparison of the lift coefficient predictions.

	2D		TEM	
M_t	Real part	Image part	Real part	Image part
0.3897	−2.336E−01	1.971E−01	−2.209E−01	1.837E−01
0.4330	−4.030E−02	1.039E−01	−4.046E−02	1.132E−01
0.4763	1.816E−02	1.413E−01	1.368E−02	1.491E−01
0.6495	9.709E−02	1.586E−01	9.499E−02	1.787E−01

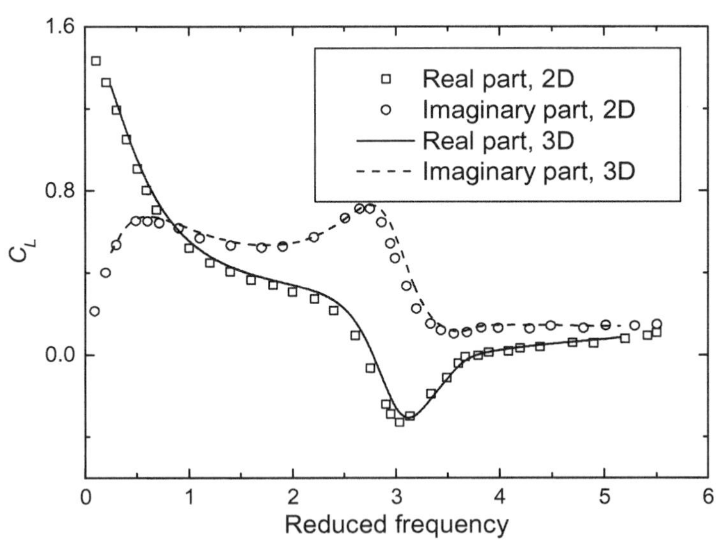

Fig. 4.14 The unsteady lift coefficient as a function of the reduced frequency (see Namba [44] for 2D result).

After the validation has been made for the two-dimensional case, more attention is paid to the three-dimensional case. Figs. 4.15 and 4.16 show the unsteady aerodynamic load distribution along the spanwise and the chordwise directions, where the present results predicted by TEM are compared to those given by Namba [44] in 2000 and by Prasad and Verdon [48] in 2002.

In general, the predictions of the unsteady blade loading distribution by the present model are in satisfactory agreement with Namba's results, whereas some discrepancies are observed when compared with Verdon's results, which is considered to be a consequence of the use of different approaches. Furthermore, the effect of the radial non-uniformness of the excitation disturbances is investigated. With the radial variation of the oncoming flow in the form of Eq. (4.242) taken into consideration, the prediction results of the unsteady aerodynamic loads on the stator vane are given in Fig. 4.17.

With the effect of the radial non-uniform wake perturbations included, the unsteady blade loading distribution predicted by the present model is still in close agreement with Namba's prediction, except that some deviations are observed near the blade leading edge. These comparatively evident deviations are due to the fact that the infinite unsteady aerodynamic load at the leading edge may take an arbitrarily large value in numerical computations and the evident annular duct effect near the blade tip (where the value of the Bessel function is large).

In the second case, we investigate the unsteady aerodynamic forces generated by the interaction between the rotor blades and the unsteady oncoming flow, and a comparison will be made with Nambe's results [33]. Consider the case where the number of rotor blades $B = 40$, the axial length of blade chord $b = 0.054978$, the angular speed of rotor rotation $\Omega = 294.4536$, the hub-to-tip ratio $R_h/R_d = 0.4$, and the Mach number of the uniform mean flow $M_a = 0.35$. The oncoming disturbances with the circumferential periodicity $N_w = 35$ are spinning with the angular velocity $\Omega_w = 210.347$ in the same direction with the rotor rotation. In the rotor-fixed frame of reference fixed to the rotor, the angular frequency of the upwash disturbances is thus $\lambda = N_w(\Omega + \Omega_w) = 2943.72$. The representation of the oncoming disturbances has been given previously in the duct-fixed frame of reference. Here, consider that $A_q(r) = \varepsilon \cdot U$ where ε is a small quantity. The dimensionless lift coefficient along the span is defined by:

$$C_L(r) = \frac{\int_{-b/2}^{b/2} \Delta p dz'}{\pi \bar{\rho} b W \cdot Q(r)} \tag{4.245}$$

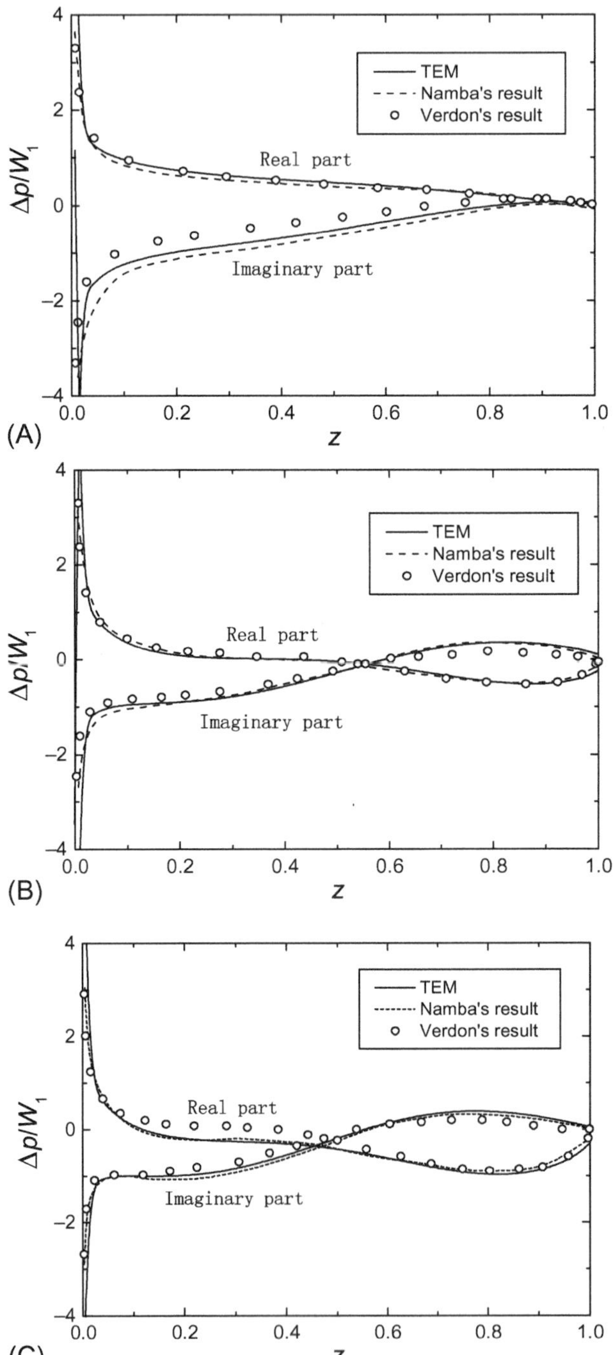

Fig. 4.15 Unsteady blade loading distribution along the chordwise direction with $M_T = 0.783$ (see [44] for Namba's result, and [48] for Verdon's result). (A) Tip, (B) mid, (C) hub.

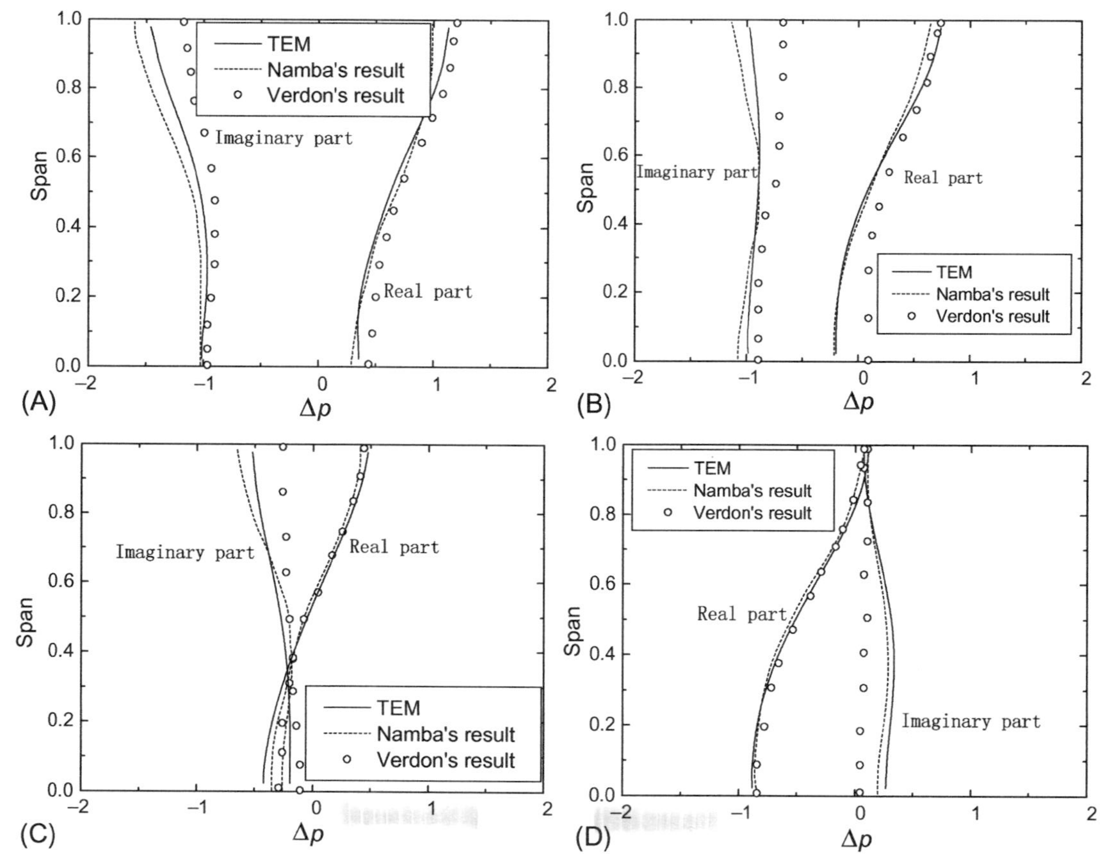

Fig. 4.16 Unsteady blade loading distribution along the spanwise direction with $M_T = 0.783$ (see [44] for Namba's result, and [48] for Verdon's result). (A) $x = 0.06$, (B) $x = 0.2$, (C) $x = 0.5$, (D) $x = 0.9$.

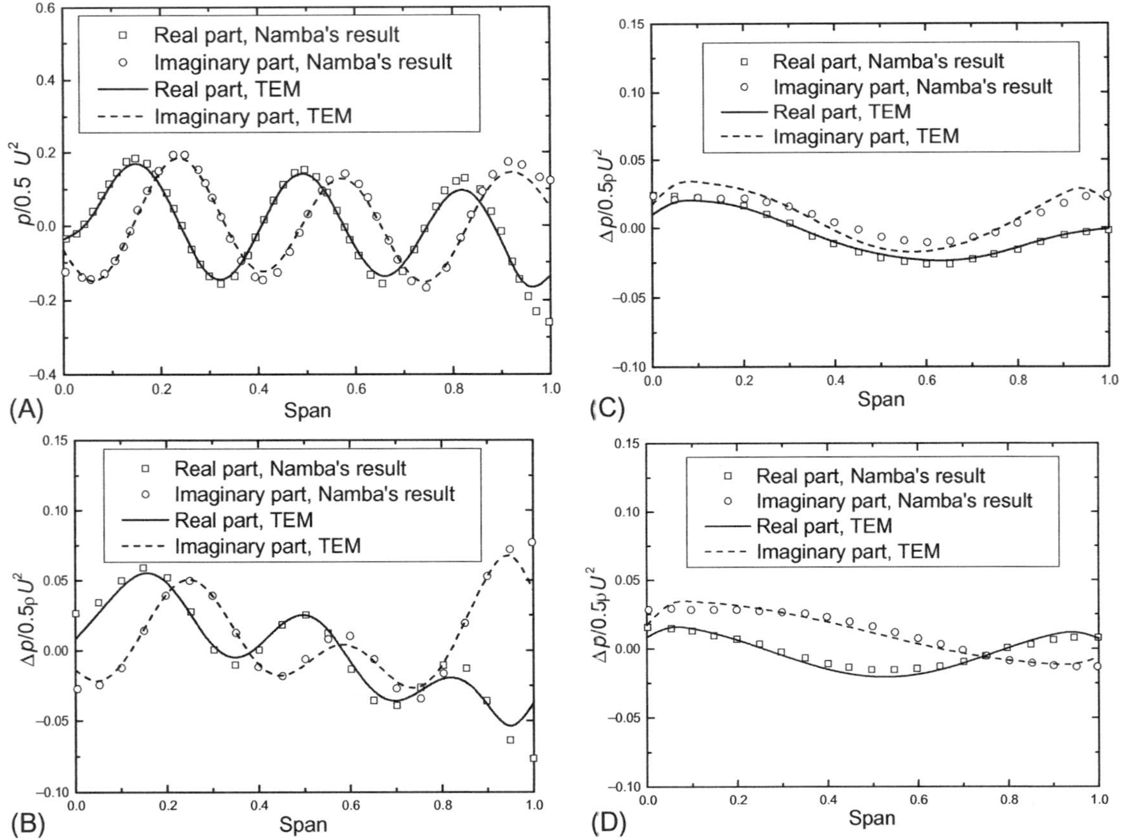

Fig. 4.17 Unsteady blade loading distribution along the spanwise direction with $\iota = 3$, $M_T = 0.783$ (see [44] for Namba's result). (A) $x = 0.06$, (B) $x = 0.2$, (C) $x = 0.5$, (D) $x = 0.9$.

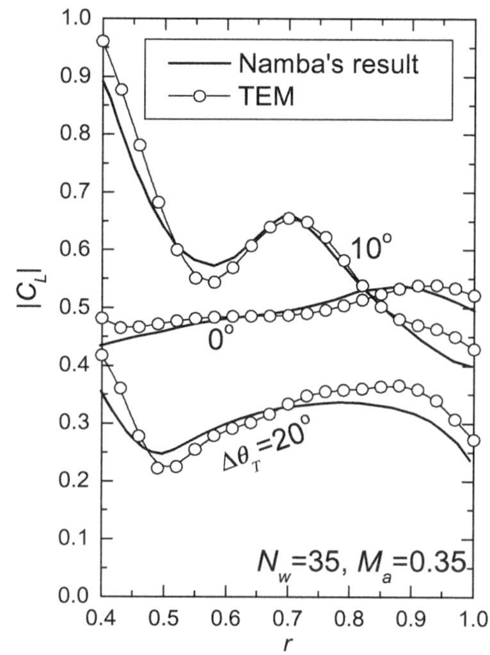

Fig. 4.18 Lift coefficient on the radial distribution for different oncoming disturbances (see [33] for Namba's result).

where the amplitude of the viscous wake $Q(r) = \frac{A_q(r) \cdot (\Omega_w + \Omega) \cdot r}{W}$. The prediction results of the lift coefficients for different radial distributions of the oncoming disturbances are presented in Fig. 4.18, where $\Delta\varphi(r) = \Delta\theta_T (r - R_h)/(R_d - R_h)$. It is found that, even though both the present model and Namba's calculations are based on the lifting surface theory, the different expansion forms adopted lead to small discrepancies in the predictions, especially near the blade tip and hub regions. But the overall trends predicted by the two approaches are identical.

References

[1] Z. Sheng, An Introduction to Aeroelasticity of Turbomachinery, National Defence Industry Press, 1989.

[2] K.C. Hall, E.F. Crawley, Calculation of unsteady flows in turbomachinery using the linearized Euler equations, AIAA J. 27 (6) (1989) 777–787.

[3] B. Chu, L.S.G. Kovasznay, Non-linear interactions in a viscous heat-conducting compressible gas, J. Fluid Mech. 3 (05) (1958) 494–514.

[4] M.E. Goldstein, Unsteady vortical and entropic distortions of potential flows round arbitrary obstacles, J. Fluid Mech. 89 (03) (1978) 433–468.

[5] H.M. Atassi, Unsteady aerodynamics of vortical flows: Early and recent developments, in: K.-Y. Fung (Ed.), Aerodynamics and Aeroacoustics, World Scientific, 1994, pp. 121–172.
[6] M.E. Goldstein, Turbulence generated by the interaction of entropy fluctuations with non-uniform mean flows, J. Fluid Mech. 93 (02) (1979) 209–224.
[7] H. Atassi, M. Goldstein, Unsteady Lift Forces on Highly Cambered Airfoils Moving Through a Gust 12th Aerospace Sciences Meeting, American Institute of Aeronautics and Astronautics, 1974.
[8] M. Goldstein, H.M. Atassi, A complete second-order theory for the unsteady flow about an airfoil due to a periodic gust, J. Fluid Mech. 74 (04) (1976) 741–765.
[9] H.M. Atassi, J. Grzedzinski, Unsteady disturbances of streaming motions around bodies, J. Fluid Mech. 209 (1989) 385–403.
[10] H.M. Atassi, A.A. Ali, O.V. Atassi, I.V. Vinogradov, Scattering of incident disturbances by an annular cascade in a swirling flow, J. Fluid Mech. 499 (2004) 111–138.
[11] L.M. Delves, J.L. Mohamed, Computational Methods for Integral Equation, Cambridge University Press, 1956.
[12] M.E. Goldstein, Aeroacoustics, McGraw-Hill International Book Company, New York, 1976.
[13] F. Lane, Supersonic flow past an oscillating cascade with supersonic leading-edge locus, J. Aeronaut. Sci. 24 (1957) 65–66.
[14] D.N. Gorelov, Cascade of plates in an unsteady supersonic flow, Fluid Dyn. 1 (4) (1966) 34–39.
[15] M. Kurosaka, On the unsteady supersoinc cascade with a subsonic leading edge—an exact first order theory—part I, J. Eng. Power Trans. ASME 96 (1974) 13–21.
[16] M. Kurosaka, On the unsteady supersoinc cascade with a subsonic leading edge—an exact first order theory—part II, J. Eng. Power Trans. ASME 96 (1974) 22–31.
[17] J.M. Verdon, Unsteady aerodynamics of a finite supersonic cascade with subsonic axial flow, J. Appl. Mech. Trans. ASME 40 (1973) 667–671.
[18] J.M. Verdon, J.E. Mccune, Unsteady supersonic cascade in subsonic axial flow, AIAA J. 13 (1975) 193–201.
[19] J.M. Verdon, Further developments in the aerodynamic analysis of unsteady supersonic cascades, part I: the unsteady pressure field, J. Eng. Power Trans. ASME 99 (1977) 509–516.
[20] J.M. Verdon, Further developments in the aerodynamic analysis of unsteady supersonic cascades, part II: aerodynamic response predictions, J. Eng. Power Trans. ASME 99 (1977) 517–525.
[21] T. Nagashima, D.S. Whitehead, Linearized Supersonic Unsteady Flow in Cascades, Aeronautical Research Council, 1978 Reports and Menoranda, No. 3811.
[22] J.J. Adamczyk, M.E. Goldstein, Unsteady flow in a supersonic cascade with subsonic leading-edge locus, AIAA J. 16 (1978) 248–254.
[23] W.B. Christian, M.F. Platzer, Theoretical investigation of supersonic flow past oscillating cascades with subsonic leading-edge locus, in: 12th Aerospace Sciences Meeting, AIAA-74-14, 1974.
[24] R.H. Ni, A rational analysis of periodic flow perturbation in supersonic two-dimensional cascade, J. Eng. Power Trans. ASME 101 (1979) 431–439.
[25] O.O. Bendiksen, P.P. Friedmann, Coupled bending-torsion flutter in a supersonic cascade, AIAA J. 19 (1981) 774–781.
[26] K.K. Butenko, A.A. Osipov, Calculation of unsteady supersonic flow past a two-dimensional cascade of plates ween vorticity inhomogeneities of the flow act on it, Fluid Dyn. 21 (1986) 466–473.
[27] M. Namba, A. Ishikawa, Three-dimensional aerodynamic characteristics of oscillating supersonic and transonic annular cascades, ASME J. Eng. Power 105 (1983) 138–146.

[28] M.E. Goldstein, Cascade with subsonic leading-edge locus, AIAA J. 13 (1975) 1117–1119.
[29] X. Sun, Investigation of aeroacoustic problems about aviation power, Doctoral thesis-Beihang University, 1988.
[30] L.M. Delves, J.L. Mohamed, Computational Methods for Integral Equation, Cambridge University Press, 1985, pp. 105–106.
[31] J.E. McCune, A three-dimensional theory of axial compressor blade rows - application in subsonic and supersonic flows, J. Aero/Space Sci. 25 (9) (1958) 544–560.
[32] O. Okurounmu, J.E. McCune, Three-dimensional vortex theory of axial compressor blade rows at subsonic and transonic speeds, AIAA J. 8 (7) (1970) 1275–1283.
[33] M. Namba, Three-dimensional analysis of blade force and sound generation for an annular cascade in distorted flows, J. Sound Vib. 50 (4) (1977) 479–508.
[34] M. Namba, Three-Dimensional Flow, (1987) AFARD-AG-298.
[35] H. Kodama, M. Namba, Unsteady lifting surface theory for a rotating cascade of swept blades, J. Turbomach. 112 (1990) 411–417.
[36] M. Namba, Lifting Surface Theory for a Rotating Subsonic or Transonic Blade Row, (1972) British Reports & Memoranda, No. 3740.
[37] P. Salaun, Flutter instabilities in an annular cascade, in: Proceeding of IUTAM Symposium on Aeroelasticity in Turbomachines, Paris, 1976.
[38] P. Salaun, Flutter Instabilities in an Annular Cascade With Supersonic Relative Flow, ONERA, 1979 Publication, No. 1979-3.
[39] J.B.H.M. Schulten, Sound generated by rotor wakes interacting with a leaned vane stator, AIAA J. 20 (1982) 1352–1358.
[40] J.B.H.M. Schulten, Vane stagger angle and camber effects in fan noise generation, AIAA J. 22 (8) (1984) 1071–1079.
[41] J.B.H.M. Schulten, Sound Generation by Ducted Fans and Propellers as a Lifting Surface Problem, (Ph.D) University of Twente, Netherland, 1993.
[42] J.B.H.M. Schulten, Transmission of sound through a rotor, in: 14th DGLR/AIAA Aeroacoustics Conference, DGLR/AIAA Paper 92-02-085, 1992.
[43] J.B.H.M. Schulten, Vane sweep effects on rotor/stator interaction noise, AIAA J. 35 (6) (1997) 945–951.
[44] M. Namba, J.B.H.M. Schulten, Category 4 - Fan Stator With Harmonic Excitation by Rotor Wake: Numerical Results of Lifting Surface Theory, NASA CP-2000-209790 (2000) pp. 73–85.
[45] M. Dahl, Third Computational Aeroacoustic Workshop on Benchmark Problems, (1999), NASA CP-2000-209790.
[46] M.J.T. Smith, Aircraft Noise, Cambridge University Press, 1989.
[47] D. Zhou, X. Wang, J. Chen, X. Jing, X. Sun, Sound generation by non-synchronously oscillating rotor blades in turbomachinery, J. Sound Vib. 355 (2015) 150–171.
[48] D. Prasad, J.M. Verdon, Validation of a three-dimensional linearized Euler analysis for classical wake/stator interactions, in: Proceedings of ASME TURBO EXPO 2002, Amsterdam, 2002.

CHAPTER 5

Vortex sound theory

5.1 Introduction to sound generation induced by vortex flow

As is well known, the most important hypothesis of Lighthill's theory is that the source of a sound induced by flow is already known, so then the flow and sound fields can be solved separately through experimental or numerical methods. Although this key hypothesis has greatly promoted extensive use of Lighthill's theory in various engineering applications, the following two aspects of limitations cannot be ignored. First, the hypothesis of a pre-known source of sound is basically reasonable for the sound generation in a free space where the effect of a solid boundary can be ignored, such as jet noise. However, how to more accurately identify the real sound sources in a turbulent shear layer has been an unsolved problem because such complicated flow contains too many scales to separate one from another. Second, because the governing equations of flow and sound fields are identical, the two kind of fields can be theoretically solved simultaneously. However, the flow and acoustic fields are manually separated in Lighthill's analogy, which would consequently become useless in describing the wave-vortex interaction and energy transfer between them.

Aiming at the exploration of the underlying mechanism of sound generation induced by flow and the interaction between sound and turbulence flow, continuing developments and improvements have been made by Powell [1], Doak [2, 3], Howe [4, 5], Yates [6, 7], Obermeier [8], Goldstein [9], and so forth since the 1960s. Almost without exception, their works proved that generation of sound was closely related to the vortex motion, and the energy transfer was implemented through the nonlinear interaction between the sound and flow field. These investigations have greatly motivated researchers to further study the vortex sound problem. Powell [1] first derived the vortex sound equation, which can be seen as an approximation to Lighthill's equation at the condition of low Mach number flow. Howe [4] further developed Powell's theory to evaluate the vortex sound in more complicated cases, such as vortex motion with heat transfer, by introducing

Fundamentals of Aeroacoustics with Applications to Aeropropulsion Systems
https://doi.org/10.1016/B978-0-12-408069-0.00005-1

the total enthalpy of the isentropic flow as the acoustic variable. Powell's and Howe's theories are usually considered to be the origination of vortex sound research, which is a complementary addition to Lighthill's theory. And vortex sound theory has already become an important branch of the modern discipline of aeroacoustics, which plays an increasingly significant role in depicting the underlying mechanism of various vortex sound interaction phenomena.

In the first section of this chapter, the derivation of basic equations of vortex sound theory is shown in detail. Several applications of vortex sound theory are also illustrated in the succeeding sections, including the model of trailing edge noise of a flat plate and impedance model of bias flow acoustic liner, along with special attention on the effect of grazing flow and nonlinearity on the vortex sound interaction of perforated plates or resonator.

5.2 Basic equations of vortex sound

5.2.1 Powell's equation

For an inviscid and isentropic flow, the continuity and momentum equations can be written as:

$$\frac{\partial \rho}{\partial t} + \nabla \cdot (\rho \mathbf{u}) = 0, \tag{5.1}$$

$$\frac{\partial \mathbf{u}}{\partial t} + (\mathbf{u} \cdot \nabla)\mathbf{u} + \frac{1}{\rho}\nabla p = 0. \tag{5.2}$$

Applying the following vector invariants:

$$\nabla \cdot (\rho \mathbf{u}) = \nabla \rho \cdot \mathbf{u} + \rho \nabla \cdot \mathbf{u},$$

$$\nabla\left(\frac{1}{2}\mathbf{u} \cdot \mathbf{u}\right) = (\mathbf{u} \cdot \nabla)\mathbf{u} - (\nabla \times \mathbf{u}) \times \mathbf{u} = (\mathbf{u} \cdot \nabla)\mathbf{u} - \boldsymbol{\omega} \times \mathbf{u},$$

where $\boldsymbol{\omega} = \nabla \times \mathbf{u}$ is the vorticity, the continuity and momentum equations can be derived as:

$$\frac{\partial \rho}{\partial t} + \nabla \rho \cdot \mathbf{u} + \rho \nabla \cdot \mathbf{u} = 0, \tag{5.3}$$

$$\rho \frac{\partial \mathbf{u}}{\partial t} + \rho(\boldsymbol{\omega} \times \mathbf{u}) + \nabla\left(\frac{\rho u^2}{2}\right) = -\nabla p. \tag{5.4}$$

Multiplying $\mathbf{u}$ to Eq. (5.1) and ρ to Eq. (5.2), then adding these two equations, one obtains:

$$\frac{\partial(\rho\mathbf{u})}{\partial t}+\nabla\cdot(\rho\mathbf{u}\mathbf{u})+\nabla p=0. \tag{5.5}$$

Applying $\partial/\partial t$ to Eq. (5.1), and imposing ∇ to Eq. (5.5) yields:

$$\nabla^2 p-\frac{\partial^2\rho}{\partial t^2}=-\nabla\cdot[\nabla\cdot(\rho\mathbf{u}\mathbf{u})]. \tag{5.6}$$

Considering $\nabla p\approx c_0^2\,\nabla\rho$, and substituting Eq. (5.5) into Eq. (5.6), we have:

$$\nabla^2 p-\frac{1}{c_0^2}\frac{\partial^2 p}{\partial t^2}=\nabla\cdot\nabla p+\nabla\cdot\frac{\partial(\rho\mathbf{u})}{\partial t}. \tag{5.7}$$

Expanding the last term of Eq. (5.6), and then substituting Eq. (5.4) into this equation, we find:

$$\nabla^2 p-\frac{1}{c_0^2}\frac{\partial^2 p}{\partial t^2}=-\nabla\cdot\left\{\rho(\boldsymbol{\omega}\times\mathbf{u})+\nabla\left(\frac{\rho u^2}{2}\right)-\mathbf{u}\frac{\partial\rho}{\partial t}-\frac{u^2}{2}\nabla\rho\right\}. \tag{5.8}$$

When the characteristic Mach number is low, this equation can be simplified by ignoring the small terms of higher order, and the following equation can be obtained:

$$\nabla^2 p-\frac{1}{c_0^2}\frac{\partial^2 p}{\partial t^2}=-\nabla\cdot\{\rho(\boldsymbol{\omega}\times\mathbf{u})\}. \tag{5.9}$$

Eq. (5.8) is the vortex sound equation derived by Powell [1] in the 1960s. Compared with Lighthill's equation, the relationship between vortex and sound is illustrated in detail in this equation. Especially when the Mach number is low, the sources of sound exist only at the places where the moving vortices present in the flow field, which can be easily concluded from Eq. (5.9).

5.2.2 Howe's acoustic analogy

5.2.2.1 Sound propagation in the irrotational mean flow

The average velocity potential of the mean flow is assumed to be $\psi_0(\mathbf{x})$, and the relevant fluctuation is $\psi_1(\mathbf{x})$. Therefore, the whole velocity potential can be expressed as:

$$\psi(\mathbf{x})=\psi_0(\mathbf{x})+\psi_1(\mathbf{x}). \tag{5.10}$$

For an inviscid and isentropic flow, the integral form of momentum equation is written as:

$$\frac{\partial \psi}{\partial t}+\frac{1}{2}\left(\frac{\partial \psi}{\partial x_i}\right)^2+\int \frac{dp}{\rho}=\text{Constant}. \tag{5.11}$$

And the continuity equation becomes:

$$\frac{1}{\rho}\frac{D\rho}{Dt}+\nabla^2\psi=0, \tag{5.12}$$

where $D/Dt=\partial/\partial t+\partial\psi/\partial x_j(\partial/\partial x_j)$. Considering that the effects of viscosity and heat conduction are negligible, the density and pressure variations can be related by:

$$\frac{Dp}{Dt}=c_0^2\frac{D\rho}{Dt}. \tag{5.13}$$

Applying D/Dt to Eq. (5.11) and substituting Eq. (5.13) into this equation yields:

$$\frac{D}{Dt}\left\{\frac{\partial \psi}{\partial t}+\frac{1}{2}\left(\frac{\partial \psi}{\partial x_i}\right)^2\right\}+\frac{c_0^2}{\rho}\frac{D\rho}{Dt}=0. \tag{5.14}$$

Eliminating the $(D\rho/Dt)/\rho$ term in Eq. (5.12) and Eq. (5.14), the resulting equation becomes:

$$\left(\frac{\partial}{\partial t}+\frac{\partial \psi}{\partial x_j}\frac{\partial}{\partial x_j}\right)\left\{\frac{\partial}{\partial t}+\frac{1}{2}\frac{\partial \psi}{\partial x_i}\frac{\partial}{\partial x_i}\right\}\psi-c_0^2\nabla^2\psi=0. \tag{5.15}$$

If the speed of sound is independent of time, and considering the assumption that $\partial\psi/\partial t$ is small, the previous equation can be derived as:

$$\left(\frac{D^2}{Dt^2}+\frac{D\mathbf{v}}{Dt}\cdot\frac{\partial}{\partial \mathbf{x}}-c_0^2\nabla^2\right)\psi=0, \tag{5.16}$$

where $\frac{D}{Dt}=\frac{\partial}{\partial t}+\mathbf{v}\cdot\frac{\partial}{\partial \mathbf{x}}$, $\mathbf{v}=\frac{\partial \psi}{\partial \mathbf{x}}=\frac{\partial \psi_0}{\partial \mathbf{x}}+\frac{\partial \psi_1}{\partial \mathbf{x}}$. Furthermore, this equation can be written as the following expression by defining $\mathbf{U}=\nabla\psi_0$:

$$\left[\left(\frac{\partial}{\partial t}+\mathbf{U}\cdot\frac{\partial}{\partial \mathbf{x}}\right)^2+\frac{\partial}{\partial \mathbf{x}}\left(\frac{1}{2}U^2\right)\cdot\frac{\partial}{\partial \mathbf{x}}-c_0^2\nabla^2\right]\psi=0. \tag{5.17}$$

If the flow Mach number is low, i.e., $U^2 \ll c_0^2$, Eq. (5.17) can be simplified by omitting the second term in the bracket:

$$\left[\left(\frac{\partial}{\partial t}+\mathbf{U}\cdot\frac{\partial}{\partial \mathbf{x}}\right)^2-c_0^2\nabla^2\right]\psi=0. \tag{5.18}$$

This is the wave equation derived by Howe [4] to illustrate the sound propagation in the non-uniform mean flow.

5.2.2.2 Howe's equation

It can be shown from the previous equations that the variables of pressure and density do not satisfy the classic wave equation in non-uniform mean flow; therefore, the fluctuation velocity potential is chosen instead, which is more convenient. Furthermore, Howe defined another independent acoustic variable from the integration of Bernoulli equation (Eq. (5.11)) to study the character of sound wave propagation, which is written as:

$$B = \int \frac{dp}{\rho} + \frac{1}{2}\left(\frac{\partial \psi}{\partial x_i}\right)^2. \tag{5.19}$$

In fact, this is the stagnation enthalpy per unit mass of an isentropic flow. Eq. (5.19) denotes that B becomes constant in steady irrotational flow without sound pressure.

Introduce the specific heat function w, which is related to the pressure, density, temperature, and specific entropy, to widen the definition of B:

$$dw = \frac{1}{p} dP + TdS.$$

The specific stagnation enthalpy of an arbitrary flow can be defined as:

$$B(\mathbf{x}, t) = w + \frac{1}{2} v^2. \tag{5.20}$$

Then Crocco's form of the momentum equation becomes:

$$\frac{\partial \mathbf{v}}{\partial t} + \nabla B = -\boldsymbol{\omega} \times \mathbf{v} + T \nabla S. \tag{5.21}$$

If the viscous dissipation and heat conduction are negligible, the continuity equation is:

$$\frac{1}{\rho c_0^2} \frac{Dp}{Dt} + \nabla \cdot \mathbf{v} = 0. \tag{5.22}$$

Taking the divergence of Eq. (5.21) and differentiating Eq. (5.22) with respect to time, then subtracting term $\nabla \cdot (\partial \mathbf{v}/\partial t)$ of these two equations yields:

$$\frac{\partial}{\partial t}\left(\frac{1}{\rho c_0^2} \frac{Dp}{Dt}\right) - \nabla^2 B = \nabla \cdot (\boldsymbol{\omega} \times \mathbf{v} - T \nabla S)$$

Similar to Eq. (5.16), this equation can be rearranged as:

$$\begin{aligned}\frac{1}{c_0^2}\frac{D^2B}{Dt^2}+\frac{1}{c_0^2}\frac{D\mathbf{v}}{Dt}\cdot\frac{\partial B}{\partial\mathbf{x}}-\nabla^2B&=\nabla\cdot(\boldsymbol{\omega}\times\mathbf{v}-T\nabla S)\\&+\left[\frac{1}{c_0^2}\frac{D^2B}{Dt^2}+\frac{1}{c_0^2}\frac{D\mathbf{v}}{Dt}\cdot\frac{\partial B}{\partial\mathbf{x}}-\frac{\partial}{\partial t}\left(\frac{1}{\rho c_0^2}\frac{Dp}{Dt}\right)\right].\end{aligned}\tag{5.23}$$

Noticing that $p=\rho RT$, and according to Eq. (5.21), we can obtain:

$$\frac{D\mathbf{v}}{Dt}\cdot\frac{DB}{D\mathbf{x}}=-\frac{\partial\mathbf{v}}{\partial t}\cdot\frac{D\mathbf{v}}{Dt}-\frac{D\mathbf{v}}{Dt}\cdot(\boldsymbol{\omega}\times\mathbf{v}-T\nabla S).$$

Because $DS/Dt=0$, the right-hand side of Eq. (5.23) can be written as:

$$\nabla\cdot(\boldsymbol{\omega}\times\mathbf{v}-T\nabla S)-\frac{1}{c_0^2}\frac{D\mathbf{v}}{Dt}\cdot(\boldsymbol{\omega}\times\mathbf{v}-T\nabla S)-\frac{1}{\rho}\frac{\partial p}{\partial t}\cdot\frac{D}{Dt}\left(\frac{1}{c_0^2}\right).$$

Also,

$$\frac{1}{\rho}\frac{\partial p}{\partial t}=\frac{1}{\rho}\frac{Dp}{Dt}-\frac{1}{\rho}\mathbf{v}\cdot\frac{\partial p}{\partial x}=\frac{Dw}{Dt}+\frac{D}{Dt}\left(\frac{1}{2}v^2\right)=\frac{DB}{Dt},$$

$$\frac{1}{\rho}\frac{\partial p}{\partial t}\frac{D}{Dt}\left(\frac{1}{c_0^2}\right)=\frac{D}{Dt}\left(\frac{1}{c_0^2}\right)\frac{DB}{Dt}.$$

Substituting these relations into Eq. (5.23), the final expression is written as:

$$\begin{aligned}\left[\frac{D}{Dt}\left(\frac{1}{c_0^2}\frac{D}{Dt}\right)+\frac{1}{c_0^2}\frac{D\mathbf{v}}{Dt}\cdot\frac{\partial}{\partial\mathbf{x}}-\nabla^2\right]B&=\nabla\cdot(\boldsymbol{\omega}\times\mathbf{v}-T\nabla S)\\&-\frac{1}{c_0^2}\frac{D\mathbf{v}}{Dt}\cdot(\boldsymbol{\omega}\times\mathbf{v}-T\nabla S).\end{aligned}\tag{5.24}$$

For the ideal gas without viscous and heat conduction, the previous equation is strictly derived, as the term D/Dt contains the actual velocity of the fluid, not merely that of the mean flow.

If the heat conduction cannot be ignored, i.e.,$DS/Dt\neq 0$, the continuity equation is:

$$\frac{1}{\rho c_0^2}\frac{Dp}{Dt}+\nabla\cdot\mathbf{v}=\frac{1}{c_p}\frac{DS}{Dt},$$

where c_p is the specific heat at constant pressure. Consequently, the basic equation is derived as:

$$\left[\frac{D}{Dt}\left(\frac{1}{c_0^2}\frac{D}{Dt}\right)+\frac{1}{c_0^2}\frac{D\mathbf{v}}{Dt}\cdot\frac{\partial}{\partial\mathbf{x}}-\nabla^2\right]B$$
$$=\nabla\cdot(\boldsymbol{\omega}\times\mathbf{v}-T\nabla S)-\frac{1}{c_0^2}\frac{D\mathbf{v}}{Dt}\cdot(\boldsymbol{\omega}\times\mathbf{v}-T\nabla S)+\frac{D}{Dt}\left(\frac{T}{c_0^2}\frac{DS}{Dt}\right)+\frac{\partial}{\partial t}\left(\frac{1}{c_p}\frac{DS}{Dt}\right). \tag{5.25}$$

Eqs. (5.24) and (5.25) are called Howe's acoustic analogy equations [4], and it can be easily derived that the equations reduce to Powell's equation (5.9) in an isentropic flow with low Mach number. However, Howe's equation contains more parameters than Powell's, which can be used to describe more complicated problems, such as interaction between sound and flow or sound emission from fluctuating heat source, and so forth.

5.2.2.3 Solution of Howe's equation

Strictly speaking, it is extremely difficult to solve the equations like Eqs. (5.8) and (5.24). Therefore, only the solutions with some special restrictions are discussed in this section.

When the Mach number $M \ll 1$, the flow can be considered to be isentropic, i.e., $T\nabla S=0$, and term $1/c_0^2(D\mathbf{v}/Dt)$ can be ignored, Eq. (5.24) reduces to:

$$\left[\frac{D}{Dt}\left(\frac{1}{c_0^2}\frac{D}{Dt}\right)-\nabla^2\right]B=\nabla\cdot(\boldsymbol{\omega}\times\mathbf{v}). \tag{5.26}$$

Assume that the speed of sound is constant, the prior equation can thus be approximated to:

$$\frac{1}{c_0^2}\frac{D^2B}{Dt^2}-\nabla^2B=\nabla\cdot(\boldsymbol{\omega}\times\mathbf{v}). \tag{5.27}$$

According to the definition of Eq. (5.19), the sound pressure can be given by linearized approximation:

$$p=\frac{B}{\rho_0},$$

where ρ_0 is the uniform value of density. Then the sound pressure generated in the far field is governed by the following equation:

$$\frac{1}{c_0^2}\frac{D^2p}{Dt^2}-\nabla^2p=\nabla\cdot(\boldsymbol{\omega}\times\mathbf{v}). \tag{5.28}$$

This is exactly Powell's equation as shown as Eq. (5.9).

Obviously, the general solution of Eq. (5.28) in free space is written as:

$$p(\mathbf{x},t)=\int G(\mathbf{x},t;\mathbf{y},\tau)q(\mathbf{y},t)d^3\mathbf{y}dt, \tag{5.29}$$

where $G(\mathbf{x},t;\mathbf{y},\tau)$ is the free space Green's function, $q(\mathbf{x},t)=\rho_0\nabla\cdot(\boldsymbol{\omega}\times\mathbf{u})=\rho_0\partial L_i/\partial y_i$, $L_i=(\boldsymbol{\omega}\times\mathbf{u})_i$. The solution of Eq. (5.29) can be given by partial integration:

$$p(\mathbf{x},t)=-\rho_0\int\frac{\partial G}{\partial y_i}(\mathbf{x},t;\mathbf{y},\tau)\cdot L_i(\mathbf{y},t)d^3\mathbf{y}dt. \tag{5.30}$$

Introducing a vector into Green's function, one obtains:

$$\nabla\mathbf{G}=\nabla\times\mathbf{G}. \tag{5.31}$$

Eq. (5.30) can be written as:

$$p(\mathbf{x},t)=-\rho_0\int\mathbf{G}\cdot(\nabla\times\mathbf{L})d^3\mathbf{y}dt.$$

According to Helmholtz equation, variables $\boldsymbol{\omega}$ and $\mathbf{L}$ are related by:

$$\frac{\partial\boldsymbol{\omega}}{\partial t}+\nabla\times\mathbf{L}=0.$$

Therefore, the solution of vortex sound equation becomes:

$$\begin{aligned}p(\mathbf{x},t)&=\rho_0\int\mathbf{G}(\mathbf{x},t;\mathbf{y},\tau)\frac{\partial\boldsymbol{\omega}}{\partial\tau}d^3\mathbf{y}dt\\&=\rho_0\frac{\partial}{\partial\tau}\int\mathbf{G}(\mathbf{x},t;\mathbf{y},\tau)\cdot\boldsymbol{\omega}d^3\mathbf{y}dt.\end{aligned} \tag{5.32}$$

It can be seen from this equation that the sound wave can only be generated in the regions that contain a time-dependent vortex, and it is only necessary to calculate the unsteady variation instead of the induced velocity of the vortex to obtain the sound pressure. Unfortunately, as a general rule, the vector of Green's equation does not exist, because G is a harmonic function, i.e., $\nabla G=0$ concluded from Eq. (5.31), and G has to satisfy the following equation:

$$\nabla^2G=\frac{1}{c_0^2}\frac{\partial^2G}{\partial\tau^2}-\delta(\mathbf{x}-\mathbf{y})\delta(t-\tau). \tag{5.33}$$

When the Mach number is low, the second term on the right of Eq. (5.33) can be ignored in the far field, and then such a vector Green's function can be found. Several researches have been done using Eq. (5.32) to study the sound generation of vortex pair [10], vortex ring [11], and vortex-surface interaction [12, 13].

5.2.2.4 Energy transfer between sound and vortex

Vortex is recognized as the source or sink of sound generation through the energy transformation between the sound and flow fields. Let the velocity induced by the vortex be $\mathbf{v}(\mathbf{x}, t)$, and the velocity of the irrotational flow relevant to the sound field is $\mathbf{u}(\mathbf{x}, t) = \nabla \psi$. Therefore, the total velocity of the flow is $\mathbf{u} + \mathbf{v}$, and the rate of energy transformed from the sound field to flow field is:

$$\prod_T = \rho_0 \frac{\partial}{\partial t} \int \frac{1}{2} v^2 d^3\mathbf{y} \prod_T = \rho_0 \frac{\partial}{\partial t} \int \frac{1}{2} v^2 d^3\mathbf{y} \tag{5.34}$$

For an inviscid and isentropic flow, the momentum equation is:

$$\frac{\partial \mathbf{v}}{\partial t} + \nabla \left[\int \frac{dp}{\rho_0} + \frac{1}{2}(\mathbf{v} + \mathbf{u})^2 + \frac{\partial \psi}{\partial t} \right] = -\boldsymbol{\omega} \times \mathbf{v} - \boldsymbol{\omega} \times \mathbf{u}. \tag{5.35}$$

Taking the dot product of this equation with $\mathbf{v}$, we have:

$$\mathbf{v} \cdot \frac{\partial \mathbf{v}}{\partial t} + \mathbf{v} \cdot \nabla \left[\int \frac{dp}{\rho_0} + \frac{1}{2}(\mathbf{v} + \mathbf{u})^2 + \frac{\partial \psi}{\partial t} \right] = \mathbf{v} \cdot [-\boldsymbol{\omega} \times \mathbf{v} - \boldsymbol{\omega} \times \mathbf{u}].$$

Noticing that $\nabla \cdot \mathbf{v} = 0$ and $-\mathbf{v} \cdot (\boldsymbol{\omega} \times \mathbf{u}) = (\boldsymbol{\omega} \times \mathbf{v}) \cdot \mathbf{u}$ imply:

$$\frac{\partial}{\partial t}\left(\frac{1}{2} v^2\right) + \nabla \left\{ \mathbf{v} \left[\int \frac{\mathrm{d}p}{\rho_0} + \frac{1}{2}(\mathbf{v} + \mathbf{u})^2 + \frac{\partial \psi}{\partial t} \right] \right\} = (\boldsymbol{\omega} \times \mathbf{v}) \cdot \mathbf{u}.$$

Integrating this equation over the bounded space, it can be proved that the second term on the left-hand side would vanish when the integral surface tends to be infinite. The resulting equation becomes:

$$\frac{\partial}{\partial t} \int \frac{1}{2} v^2 d^3\mathbf{y} = \int (\boldsymbol{\omega} \times \mathbf{v}) \cdot \mathbf{u} d^3\mathbf{y}, \tag{5.36}$$

or

$$\Pi_T = \int \rho_0 (\boldsymbol{\omega} \times \mathbf{v}) \cdot \mathbf{u} d^3\mathbf{y}. \tag{5.37}$$

This equation illustrates the energy transfer between the sound and vortex. The term $\rho_0(\boldsymbol{\omega} \times \mathbf{v})$ can also be recognized to be the unsteady vortical lifting force, then Π_T is the rate of working of the vortical lifting force due to the acoustic field $\mathbf{u}$, and this relation has been widely used to study the interaction between vortex and sound by many researchers [14–18].

5.2.3 The equivalence of Curle's equation and Howe's equation

To compare Howe's equation with Curle's equation, sound radiated from the interaction between vortex and stationary surface is discussed in this section. Assuming that the flow Mach number is low, the general solution of Howe's equation, as shown in Eq. (5.30), is:

$$p(\mathbf{x}, t) = -\rho_0 \int (\boldsymbol{\omega} \times \mathbf{v})(\mathbf{y}, \tau) \cdot \nabla G(\mathbf{x}, \mathbf{y}, t-\tau) d^3\mathbf{y} dt, \tag{5.38}$$

where G is the free space Green's function in three dimensions:

$$G(\mathbf{x}, \mathbf{y}, t-\tau) = \frac{1}{4\pi|\mathbf{x}-\mathbf{y}|} \delta\left(t - \tau - \frac{|\mathbf{x}-\mathbf{y}|}{c}\right). \tag{5.39}$$

Expanding Green's function to the first order in the retarded time yields:

$$G(\mathbf{x}, \mathbf{y}, t-\tau) \approx \frac{1}{4\pi|\mathbf{x}|} \delta\left(t - \tau - \frac{|\mathbf{x}|}{c}\right) + \frac{x_j Y_j}{4\pi c|\mathbf{x}|^2} \delta'\left(t - \tau - \frac{|\mathbf{x}|}{c}\right), \quad |\mathbf{x}| \to \infty \tag{5.40}$$

where Y_j is the jth component of the Krichhoff vector for the surface, defined as the velocity potential of an incompressible flow past the surface that has unit speed in the j direction at far field. Substituting Eq. (5.40) into Eq. (5.38), one obtains:

$$p(\mathbf{x}, t) = -\frac{\rho_0 x_j}{4\pi c|\mathbf{x}|^2} \frac{\partial}{\partial t} \left[\int (\boldsymbol{\omega} \times \mathbf{v})(\mathbf{y}, \tau) \cdot \nabla Y_j(\mathbf{y}) \right] d^3\mathbf{y}, \tag{5.41}$$

where the brackets ([ ]) denote that the enclosed quantity is to be evaluated at the retarded time $\tau = t - |\mathbf{x}|/c$.

According to the Biot-Savart law [19], the force exerted on the solid surface by the fluid is written as:

$$F_j = -\int (\boldsymbol{\omega} \times \mathbf{v})(\mathbf{y}, \tau) \cdot \nabla Y_j(\mathbf{y}) d^3\mathbf{y}. \tag{5.42}$$

Substituting Eq. (5.42) into Eq. (5.41) yields:

$$p(\mathbf{x}, t) = -\frac{\rho_0 x_j}{4\pi c|\mathbf{x}|^2} \frac{\partial}{\partial t} F_j(\mathbf{y}, t - |\mathbf{x}|/c). \tag{5.43}$$

This is exactly Curle's equation. It indicates that the Curle's equation and the Howe's equation are equivalent when determining the far field sound pressure radiated from the interaction between vortex and stationary surface in a low Mach number flow.

5.3 Vortex sound model of trailing edge noise

In the beginning of this section, we should emphasize that most of the studies on trailing edge noise generated from flow past bluff bodies are based on Curle's acoustic analogy, in which the surface dipole is considered to be the dominant source due to the unsteady loading on the body. However, the dipole source can be essentially considered a passive source that is simply a reaction to the force exerted by fluid on the body [20]. It is reasonable to calculate the sound pressure by directly solving the flow parameters instead of the surface forces. Actually, the equivalence of Curle's equation and Howe's equation in low Mach number flow has been proven as shown in Section 5.2.3. And plenty of works relating to trailing edge noise issues have illustrated the accuracy and efficiency of the vortex sound theory [12, 20–22]. Here, the general solution of a vortex sound equation (5.30) is applied to study the noise of trailing edge flow from a two-dimensional flat plate.

In this section, the discrete vortex method is used to simulate the plate wake flow, which was first established by Clements [23], and only the main steps are reviewed here for brevity. A two-dimensional rigid plate with an airfoil leading edge and a square trailing edge is placed in the incompressible uniform flow. The thickness of the plate is $2s_0$, and the velocity of the incoming flow is $\mathbf{V}$ with an amplitude of U_g. The incident of the plate is zero, and emphasis is only placed on the vortices shed from the square trailing edge; therefore, the plate can be recognized as semi-infinite in the upstream direction. The incoming flow is assumed to be irrotational except at the places where a point vortex locates, and in addition, it is inviscid except at the two corners of the trailing edge where the vortices separate from the plate.

At the start of each time step, a nascent point vortex is shed from each corner of the trailing edge, and its trajectory can be calculated by the following ordinary differential equation:

$$\frac{d\mathbf{z}_n}{dt} = \boldsymbol{W}(\mathbf{z}_n), \tag{5.44}$$

where $\mathbf{z}_n$ is the complex coordinate of the nth point vortex and $\boldsymbol{W}(\mathbf{z}_n)$represents the complex velocity.

To simplify the modeling process, a Schwartz-Christoffel conformal mapping is employed, as shown in Fig. 5.1.

$$\mathbf{z} = -(2s_0/\pi)\left[\ln\left(\mathbf{i}\lambda + \sqrt{1-\lambda^2}\right) + \mathbf{i}\lambda\sqrt{1-\lambda^2}\right]. \tag{5.45}$$

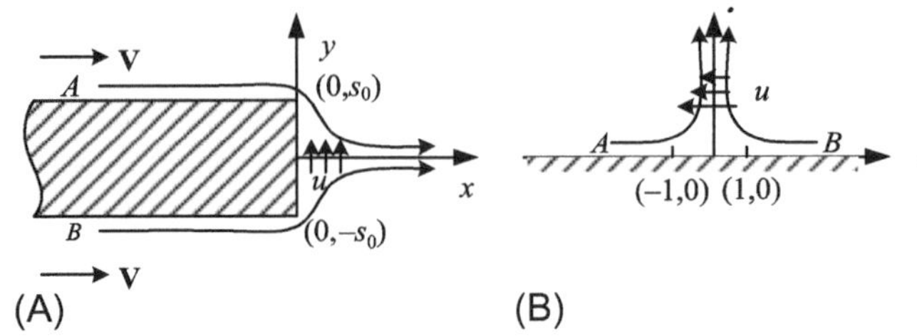

Fig. 5.1 Conformal mapping between the physical and computational domains: (A) physical domain, (B) computational domain.

The complex velocity potential of any points in the transform domain is the sum of the irrational flow around the plate, and the velocity potential induced by the vortices and their images. The irrational flow field includes two components; one is the incoming uniform flow grazing the plate surface, $\boldsymbol{V} \rightarrow (U_g, 0)$, and the other is the circulatory flow around the trailing edge, which represents the initial disturbance to motivate the vortex shedding or the feedback acoustic particle velocity, $z = 0$, $\boldsymbol{u} = (0, -k_0 U_g)$, $|z| \rightarrow \infty$, and $\boldsymbol{u} \rightarrow 0$, where k_0 is a coefficient relating to the circulatory flow.

The complex velocity potential ($\boldsymbol{W}$) in the transform domain can be written as:

$$\boldsymbol{W}(\boldsymbol{\lambda}) = -\frac{2s_0}{\pi} U_g \boldsymbol{\lambda}^2 + \left(4s_0 k_0 U_g/\pi\right)\boldsymbol{\lambda} + \sum_{j=1}^{\infty} \frac{\Gamma_j}{2\pi \mathbf{i}} \left[\ln\left(\boldsymbol{\lambda} - \boldsymbol{\lambda}_j\right) - \ln\left(\boldsymbol{\lambda} - \boldsymbol{\lambda}_j^*\right) \right], \tag{5.46}$$

where Γ_j and $\boldsymbol{\lambda}_j$ are the strength and position of the jth point vortex, respectively, and the asterisk denotes the complex conjugate.

Substituting Eqs. (5.45) and (5.46) into Eq. (5.44), the velocity of the kth point vortex in physical domain can be expressed as:

$$u(\mathbf{z}_k) - iv(\mathbf{z}_k) = \frac{d\boldsymbol{W}_{\mathbf{z}_k}(\mathbf{z})}{d\mathbf{z}} = -\left(1 - \boldsymbol{\lambda}_k^2\right)^{-\frac{1}{2}} \left[\mathbf{i} U_g \boldsymbol{\lambda}_k - \mathbf{i} k_0 U_g - \sum_{\substack{j=1 \\ j \neq k}}^{\infty} \frac{\Gamma_j}{8s_0} \frac{1}{\boldsymbol{\lambda}_k - \boldsymbol{\lambda}_j} + \sum_{j=1}^{\infty} \frac{\Gamma_j}{8s_0} \frac{1}{\boldsymbol{\lambda}_k - \boldsymbol{\lambda}_j^*} - \frac{\Gamma_k}{16s_0} \boldsymbol{\lambda}_k \left(1 - \boldsymbol{\lambda}_k^2\right)^{-1} \right]. \tag{5.47}$$

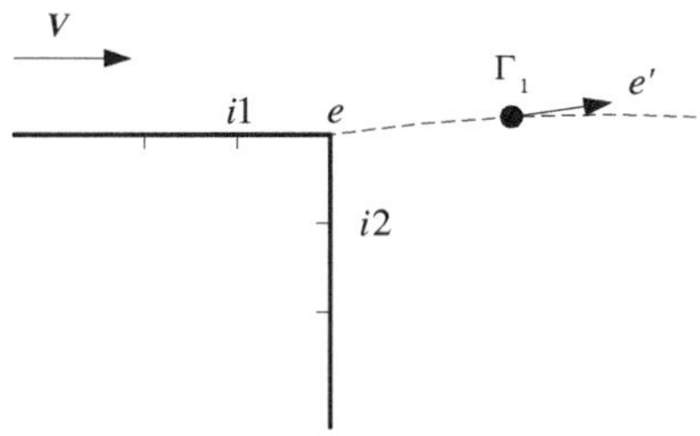

Fig. 5.2 The sketch of nascent vortex connecting with the corner.

In the present model, only one nascent vortex is shed from the upper and lower corner of the plate trailing edge, respectively, during each time step. As shown in Fig. 5.2, the local velocity of the nascent vortex can be written as:

$$\mathbf{z}_1 = \mathbf{z}_e + \frac{1}{2}(\boldsymbol{u}_{i1} + \boldsymbol{u}_{i2})\Delta t, \tag{5.48}$$

where $\mathbf{z}_e$ is the position of the corner, and $\boldsymbol{u}_{i1}$ and $\boldsymbol{u}_{i2}$ are the velocities of the positions on the wall next to the corner.

The strength of nascent vortices is determined by the velocity U_s at the outer edge of the boundary layer over the plate surface, shown as:

$$\Gamma_1 = \begin{cases} -0.5U_s^2\Delta t & \text{for the upper corner} \\ 0.5U_s^2\Delta t & \text{for the lower corner} \end{cases}. \tag{5.49}$$

In a low Mach number, isentropic and inviscid flow, the sound wave equation can be expressed as:

$$\left(\frac{1}{c_0^2}\frac{\partial^2}{\partial t^2} - \nabla^2\right)\frac{p}{\rho_0} = \nabla \cdot (\boldsymbol{\omega} \times \mathbf{v}). \tag{5.50}$$

What should be pointed out here is that the acoustic source term of Eq. (5.50) consists of the vorticity and velocity of the vortices, and these two hydrodynamic parameters both contain a fluctuating component and a time-invariant component. Therefore, the pressure results obtained by directly solving this equation contain both the acoustic pressure and a time-invariant pressure component, which is preferred to be subtracted from the results. This time-mean value can be obtained by averaging the pressure results over an acoustic period.

In the two-dimensional free space, the sound pressure induced by the nth vortex can be written as:

$$\frac{p_n}{\rho_0} = \int_0^t \int_\Omega \nabla \cdot (\boldsymbol{\omega}_n \times \mathbf{v}_n) G d\Omega d\tau = -\int_0^t \int_\Omega (\boldsymbol{\omega}_n \times \mathbf{v}_n) \cdot \nabla G d\Omega d\tau, \tag{5.51}$$

where $\boldsymbol{\omega}_n$ is the vorticity of the nth vortex, $\boldsymbol{\omega}_n = \mathbf{k}\Gamma_n\delta(x - x_n(\tau))\delta(y - y_n(\tau))$, $\mathbf{v}_n$ is the convection velocity of the nth vortex, $\mathbf{v}_n = v_{nx} + \mathbf{i}v_{ny}$, G is the two-dimensional Green's function, and Ω represents the finite two-dimensional region.

Here, we define that the coordinate of observation point is $\mathbf{x}(x_1, x_2)$, the source point is $\mathbf{y}(y_1, y_2)$, and the position of the vortex is $\mathbf{s}(x_n, y_n)$, then we have:

$$\begin{aligned}\frac{p}{\rho_0} &= -\int_0^t \int_\Omega (\boldsymbol{\omega}_n \times \mathbf{v}_n) \cdot \nabla G d\Omega d\tau \\ &= \int_0^t \Gamma_n v_{ny} \frac{\partial G}{\partial y_1}\bigg|_{(x_n(\tau),\, y_n(\tau),\, \tau)} d\tau - \int_0^t \Gamma_n v_{nx} \frac{\partial G}{\partial y_2}\bigg|_{(x_n(\tau),\, y_n(\tau),\, \tau)} d\tau.\end{aligned} \tag{5.52}$$

In the two-dimensional free space, the Green's function is expressed as:

$$G(\mathbf{x}, t/\mathbf{y}, \tau) = \frac{c_0 H[c_0(t-\tau) - R]}{2\pi\sqrt{c_0^2(t-\tau)^2 - R^2}}, \tag{5.53}$$

where R is the distance between the source and observation points, and H is the Heaviside function.

Substituting Eq. (5.53) into Eq. (5.52), the sound pressure radiated by the nth vortex can be expressed as [24]:

$$\begin{aligned}\frac{p_n}{\rho_0} &= \int_0^t \Gamma_n \left(v_{nx}\frac{\partial R}{\partial y_2} - v_{ny}\frac{\partial R}{\partial y_1} \right) \frac{c_0[c_0(t-\tau) - R]H[c_0(t-\tau) - R]}{2\pi\sqrt{[c_0^2(t-\tau)^2 - R^2]^3}} d\tau \\ &+ \int_0^t \frac{\partial\left[\Gamma_n \left(v_{nx}\frac{\partial R}{\partial y_2} - v_{ny}\frac{\partial R}{\partial y_1} \right)\right]}{\partial\tau} \frac{H[c_0(t-\tau) - R]}{2\pi\sqrt{c_0^2(t-\tau)^2 - R^2}} d\tau,\end{aligned} \tag{5.54}$$

Finally, for all of the vortices in the flow field, the total sound pressure is:

$$p = \sum_{n=1}^{N} p_n, \tag{5.55}$$

where N is the total number of the vortices.

To calculate the sound pressure of the trailing edge, we apply the free space Green's function to solve Eq. (5.52), which means the reflection of the plate surface should be taken into consideration to obtain the accurate results. Here, the two-dimensional time-domain boundary element method is introduced.

The integration expression of sound wave equation in two-dimension space without initial sound field can be written as [25]:

$$4\pi c(Q)p(Q,t)=\int_0^t\int_\Pi\left(G\frac{\partial p}{\partial n}-p\frac{\partial G}{\partial n}\right)d\Pi d\tau+\int_0^t\int_\Omega\nabla\cdot(\boldsymbol{\omega}_n\times\mathbf{v}_n)Gd\Omega d\tau,\tag{5.56}$$

where Π is the boundary of the two-dimensional region, $\mathbf{n}$ is outward vector normal to the boundary, and $c(Q)$ is a constant number related to the position of field point Q, that is:

$$c(Q)=\begin{cases}1 & Q\in\Omega\\ 1/2 & Q\in\Pi.\\ 0 & Q\notin\Omega\end{cases}$$

Substituting Eq. (5.53) into Eq. (5.56), and after a series of complicated operations, the following expression is obtained:

$$c(Q)p(Q,t)=\int_0^t\int_\Pi b(Q,t)Gd\Pi d\tau+\int_0^t\int_\Pi\frac{\partial R}{\partial\mathbf{n}}\left(pB+\frac{\partial p}{\partial\tau}\frac{1}{c_0}G\right)d\Pi d\tau+S,\tag{5.57}$$

where

$$b(Q,t)=\frac{\partial p(Q,t)}{\partial\mathbf{n}},$$

$$B=\frac{[c_0(t-\tau)-R]H[c_0(t-\tau)-R]}{2\pi\sqrt{\left[c_0^2(t-\tau)^2-R^2\right]^3}},$$

$$S=\int_0^t\int_\Omega\nabla\cdot(\boldsymbol{\omega}_n\times\mathbf{v}_n)Gd\Omega d\tau.$$

Actually, S is the sound source term of the vortex shedding we have obtained in Eq. (5.54).

The vortex street shed from the plate trailing edge is shown in Fig. 5.3A, and the relevant near field sound pressure of the vortex is plotted on the right. The far field sound pressure radiated from the plate trailing edge and the relevant directivity are plotted in Fig. 5.3C–D.

The results indicate that, based on the proper description of the plate wake flow, the dipole character of the trailing edge noise is well depicted by the vortex sound theory. As a meshless model, this model has great advantages in calculation efficiency compared with traditional numerical methods, which can be readily extended to other engineering issues.

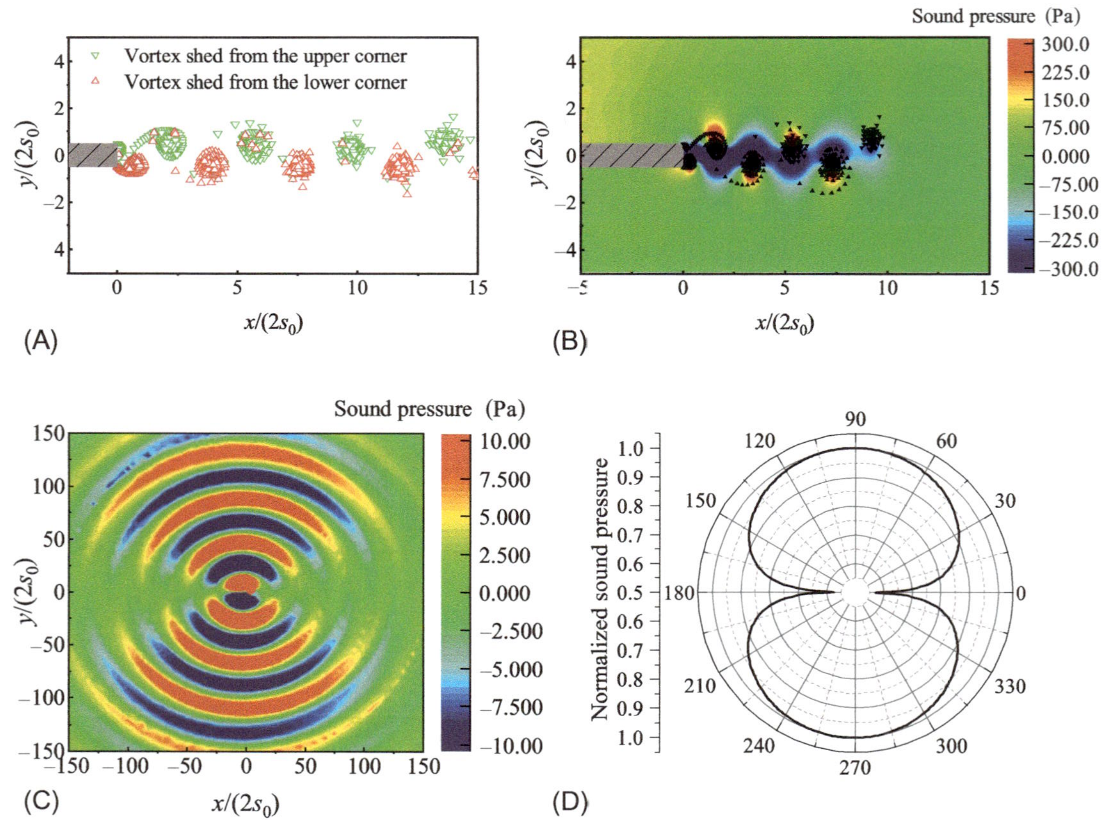

Fig. 5.3 Flow and sound field results: (A) Vortex street behind the plate, (B) the near field sound pressure, (C) the far field sound pressure, and (D) directivity of the far-field sound pressure.

5.4 Vortex sound model of liner impedance

According to vortex sound theory, the vortex acts as the source or sink of the sound wave, which implies the energy can transfer between the acoustic and flow field through vortex motion. In this chapter, the vortex sound theory is applied to study the character of liner impedance, that is, nonlinear properties of the liner with high-intensity sound pressure [26].

Early works [27] showed that the nonlinear acoustic impedance of an orifice in a plate, which is in close connection with the resonators, strongly depends on the amplitude of the applied sound wave. Flow separation and the formation of vortex rings were observed in flow visualizations [28–31], when a high-amplitude sound wave is incident on an orifice. It has been concluded that the absorption mechanism is the conversion of acoustic energy to vortical energy at a sharp edge, in other words, the vortex shedding process transfers acoustic energy into kinetic energy associated with the rotational motion of the vortices. In this section, a discrete vortex model based on axisymmetric potential flow is employed to study the nonlinear process of vortex shedding occurring at an orifice in a plate.

As shown in Fig. 5.4, an orifice is located in an infinitely thin, rigid plate. The radius of the orifice is R. When a sound pressure difference is applied across the plate, there will be vortex shedding at the sharp edge of the orifice due to the effect of viscosity. Consider that the sound pressure is uniformly distributed in the far field and has a sinusoidal time dependency; that is:

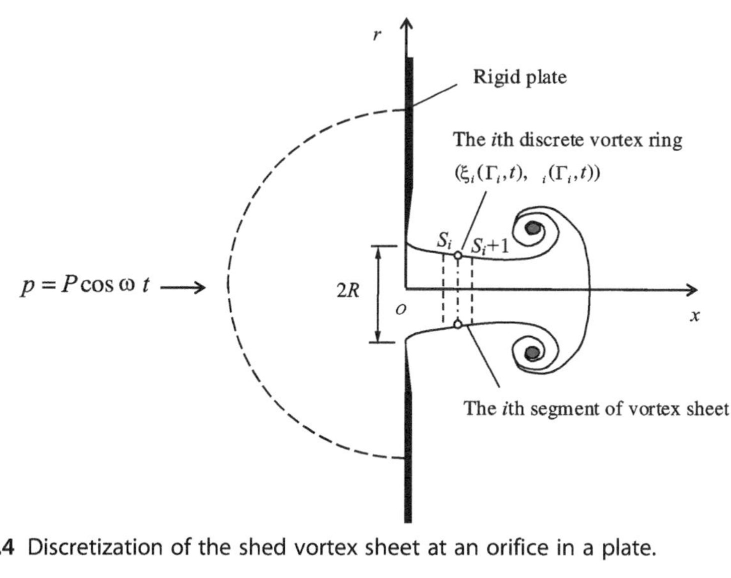

Fig. 5.4 Discretization of the shed vortex sheet at an orifice in a plate.

$$p = \begin{cases} P\cos(\omega t) & (x < 0, x^2 + r^2 \to \infty) \\ 0 & (x > 0, x^2 + r^2 \to \infty) \end{cases}. \tag{5.58}$$

At low frequency, the sound wave length greatly exceeds the radius of the orifice, and so the flow is assumed to be incompressible in the vicinity of the orifice. The flow is also assumed to be inviscid, and the viscous effect is only taken into consideration by applying the Kutta condition at the orifice edge. At high Reynolds numbers, the free shear layer is very thin, which is represented by an array of discrete vortex rings $[\xi_i(\Gamma_i, t), \rho_i(\Gamma_i, t)]$. According to the Kelvin theorem, the differential equations that govern the motion of the ith vortex ring with circulation Γ_i are as follows:

$$\frac{d\xi_i}{dt} = u(\xi_i, \rho_i) + u_I(\xi_i, \rho_i), \quad \frac{d\rho_i}{dt} = v(\xi_i, \rho_i), \tag{5.59}$$

where ξ_i and ρ_i are the axial and radial coordinates of the ith discrete vortex ring, respectively; $u$is the local axial velocity; u_I is the self-induced velocity of discrete vortex ring; and v is the local radial velocity. The variables are normalized as:

$$\begin{aligned} &\bar{x} = x/R, \ \bar{r} = r/R, \ \bar{\phi} = \phi/\omega R^2, \ \bar{\Gamma} = \Gamma/\omega R^2, \\ &\bar{t} = \omega t, \ \bar{u} = u/\omega R, \ \bar{v} = v/\omega R, \ \bar{p} = p/\rho\omega^2 R^2. \end{aligned}$$

In this section, the bars are omitted from the nondimensional variables for simplicity. Then the expression of the velocity potential ϕ_i is:

$$\begin{aligned} \phi_i(x, r; \xi_i, \rho_i) = H(x\xi_i)E_i(x, r; \xi_i, \rho_i) + H(x\xi_i)\hat{E}_i(x, r; \xi_i, \rho_i) \\ + F_i(x, r; \xi_i, \rho_i), \end{aligned} \tag{5.60}$$

where

$$E_i(x, r; \xi_i, \rho_i) = -\frac{1}{2}\operatorname{sgn}(x - \xi_i)\Gamma_i\rho_i \int_0^\infty \exp(-k|x - \xi_i|)J_0(kr)J_1(k\rho_i)dk,$$

$$\hat{E}_i(x, r; \xi_i, \rho_i) = \frac{1}{2}\operatorname{sgn}(x + \xi_i)\Gamma_i\rho_i \int_0^\infty \exp(-k|x + \xi_i|)J_0(kr)J_1(k\rho_i)dk,$$

$$\begin{aligned} F_i(x, r; \xi_i, \rho_i) = -\frac{1}{\pi}\operatorname{sgn}(x)\Gamma_i\rho_i \int_0^\infty \int_0^\infty \int_0^1 \exp(-k|x| - \lambda|\xi_i|) \\ \times J_0(kr)J_1(k\rho_i)\cos ks \cos \lambda s ds d\lambda dk, \end{aligned}$$

where $H(y)$ is the Heaviside function. $E_i(x, r; \xi_i, \rho_i)$ represents the velocity potential induced by the ith discrete vortex ring in free space, $\hat{E}_i(x, r; \xi_i, \rho_i)$ expresses the velocity potential induced by its image in free space, and $F_i(x, r; \xi_i, \rho_i)$ is the contribution of the orifice. The vortex sheet is represented by N discrete vortex rings, and ϕ_i has to satisfy the following boundary condition:

$$\phi_i(x, r; \xi_i, \rho_i) \to 0, \quad x^2 + r^2 \to \infty,$$

Thus, the finial expression of the velocity potential is written as:

$$\phi(x, r) = \sum_{i=1}^{N} \phi_i(x, r; \xi_i, \rho_i) + C(t)G(x, r), \tag{5.61}$$

$$G(x, r) = \begin{cases} 1 - \dfrac{1}{\pi}\displaystyle\int_0^\infty k^{-1} \sin k \exp(kx) J_0(kr)\, dk & (x < 0) \\ \dfrac{1}{\pi}\displaystyle\int_0^\infty k^{-1} \sin k \exp(-kx) J_0(kr)\, dk & (x > 0) \end{cases},$$

$$C(t) = -P \sin t + \Gamma(t),$$

and Γ is the total amount of circulation that has shed into the flow field from the orifice edge at time t. The axial and radial velocity are given by the derivatives of the velocity potential:

$$u = \frac{\partial \phi}{\partial x}, \quad v = \frac{\partial \phi}{\partial r}.$$

Then:

$$\begin{aligned} u(x, r) = {} & \sum_{i=1}^{N} H(x\xi_i) U_i^V(x, r; \xi_i, \rho_i) + \sum_{i=1}^{N} H(x\xi_i) \hat{U}_i^V(x, r; \xi_i, \rho_i) \\ & + \sum_{i=1}^{N} U_i^A(x, r; \xi_i, \rho_i) + C(t) U^P(x, r), \end{aligned} \tag{5.62}$$

$$\begin{aligned} v(x, r) = {} & \sum_{i=1}^{N} H(x\xi_i) V_i^V(x, r; \xi_i, \rho_i) + \sum_{i=1}^{N} H(x\xi_i) \hat{V}_i^V(x, r; \xi_i, \rho_i) \\ & + \sum_{i=1}^{N} V_i^A(x, r; \xi_i, \rho_i) + C(t) V^P(x, r), \end{aligned} \tag{5.63}$$

where

$$\begin{aligned} U_i^V(x, r; \xi_i, \rho_i) = -\frac{\Gamma_i}{2\pi r} \Bigg\{ & \left(\frac{r - \rho_i}{R_1} + \frac{r + \rho_i}{R_2} \right) \left[K(\lambda) - \frac{1}{1 - \lambda^2} E(\lambda) \right] \\ & - \left(\frac{\lambda}{1 - \lambda^2} \right) \left(\frac{r - \rho_i}{R_1} - \frac{r + \rho_i}{R_2} \right) E(\lambda) \Bigg\}, \end{aligned}$$

$$\hat{U}_i^V(x,r;\xi_i,\rho_i)=\frac{\Gamma_i}{2\pi r}\left\{\left(\frac{r-\rho_i}{\hat{R}_1}+\frac{r+\rho_i}{\hat{R}_2}\right)\left[K(\hat{\lambda})-\frac{1}{1-\hat{\lambda}^2}E(\hat{\lambda})\right]\right.$$
$$\left.-\left(\frac{\hat{\lambda}}{1-\hat{\lambda}^2}\right)\left(\frac{r-\rho_i}{R_1}-\frac{r+\rho_i}{\hat{R}_2}\right)E(\hat{\lambda})\right\},$$

$$V_i^V(x,r;\xi_i,\rho_i)=-\frac{\Gamma_i}{2\pi}\frac{(x-\xi_i)}{r}\left(\frac{1}{R_1}+\frac{1}{R_2}\right)\left[K(\lambda)-\frac{1+\lambda^2}{1-\lambda^2}E(\lambda)\right],$$

$$\hat{V}_i^V(x,r;\xi_i,\rho_i)=\frac{\Gamma_i}{2\pi}\frac{(x-\xi_i)}{r}\left(\frac{1}{\hat{R}_1}+\frac{1}{\hat{R}_2}\right)\left[K(\hat{\lambda})-\frac{1+\hat{\lambda}^2}{1-\hat{\lambda}^2}E(\hat{\lambda})\right],$$

$$U_i^A(x,r;\xi_i,\rho_i)=\frac{1}{\pi}\Gamma_i\rho_i\int_0^\infty\int_0^\infty\int_0^1\exp(-k|x|-\lambda|\xi_i|)$$
$$\times J_0(kr)J_1(\lambda\rho_i)k\cos ks\cos\lambda s\,ds\,d\lambda\,dk,$$

$$V_i^A(x,r;\xi_i,\rho_i)=\frac{1}{\pi}\operatorname{sgn}(x)\Gamma_i\rho_i\int_0^\infty\int_0^\infty\int_0^1\exp(-k|x|-\lambda|\xi_i|)$$
$$\times J_1(kr)J_1(\lambda\rho_i)k\cos ks\cos\lambda s\,ds\,d\lambda\,dk,$$

$$U^P(x,r)=-\frac{1}{\pi}\int_0^\infty\sin k\exp(-k|x|)J_0(kr)\,dk,$$

$$V^P(x,r)=-\operatorname{sgn}(x)\frac{1}{\pi}\int_0^\infty\sin k\exp(-k|x|)J_1(kr)\,dk.$$

In these equations, $R_{1,2}^2=(r-\rho_i)^2+(x\mp\xi_i)^2$, $\hat{R}_{1,2}^2=(r-\rho_i)^2+(x\pm\xi_i)^2$, $\lambda=(R_2-R_1)/(R_2+R_1)$, $\hat{\lambda}=(\hat{R}_2-\hat{R}_1)/(\hat{R}_2+\hat{R}_1)$, and $K(y)$ and $E(y)$ are the complete elliptic integrals of the first and second kind, respectively.

Note that when Eqs. (5.62) and (5.63) are used to determine the convection velocity of the jth discrete vortex ring, the induced velocity by itself is not considered; that is:

$$u(x_j,r_j)=\sum_{\substack{i=1\\i\neq j}}^{N}H(x_j\xi_i)U_i^V(x_j,r_j;\xi_i,\rho_i)+\sum_{i=1}^{N}H(x_j\xi_i)\hat{U}_i^V(x_j,r_j;\xi_i,\rho_i)$$
$$+\sum_{i=1}^{N}U_i^A(x_j,r_j;\xi_i,\rho_i)+C(t)U^P(x_j,r_j),\tag{5.64}$$

$$v(x_j, r_j) = \sum_{\substack{i=1 \\ i \neq j}}^{N} H(x_j \xi_i) V_i^V (x_j, r_j; \xi_i, \rho_i) + \sum_{i=1}^{N} H(x_j \xi_i) \hat{V}_i^V (x_j, r_j; \xi_i, \rho_i)$$
$$+ \sum_{i=1}^{N} V_i^A (x_j, r_j; \xi_i, \rho_i) + C(t) V^P (x_j, r_j), \tag{5.65}$$

It can be proven that the self-induced velocity of the *j*th discrete vortex ring u_I is written as [32]:

$$u_I(\xi_j, \rho_j) = \frac{\Gamma_j}{4\pi\rho_j}\left(\ln\frac{8\rho_j}{\sigma_j} - \frac{1}{4}\right), \tag{5.66}$$

where σ_j is the core radius of the *j*th discrete vortex ring, which is calculated from the following relation [32]:

$$\sigma_j \approx 0.25\Delta_j, \tag{5.67}$$

where Δ_j is the length of the *j*th vortex sheet segment, which can be determined by the Kutta condition at the sharp edge of the orifice.

As shown in Fig. 5.5, the nascent vortex ring represents a small segment of the vortex sheet *EP* that starts from the sharp edge *E*. In the present model, the starting vortex sheet is chosen from a small segment of the steady jet. To obtain smooth flow near the sharp edge, the starting vortex sheet is subdivided into N_1 subpanels of equal length. Each subpanel is

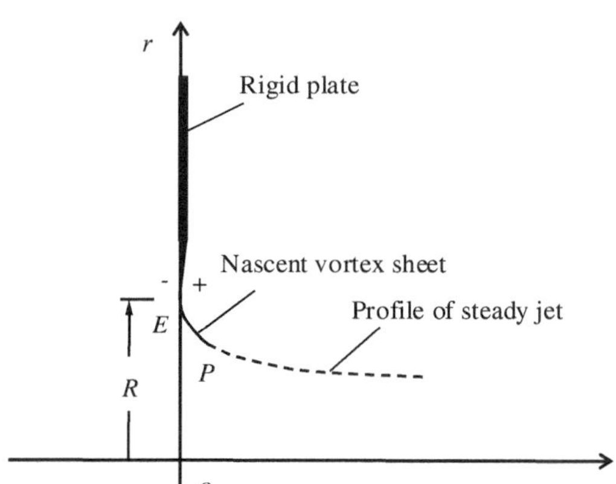

Fig. 5.5 Geometry of the starting vortex sheet *EP*.

represented by a subvortex ring (ξ_{1j}, ρ_{1j}) with circulation Γ_1/N_1. The circulation of the nascent vortex ring is determined by the Kutta condition, which requires:

$$u(0, 1^-) = 0. \tag{5.68}$$

Based on this condition, Γ_1 is given by the following relation:

$$\frac{\Gamma_1}{N_1} = \left[C(t) - \sum_{i=2}^{N} \Gamma_i K_i(\xi_i, \rho_i)\right] / \sum_{j=1}^{N} K_j\left(\xi_{1j}, \rho_{1j}\right), \tag{5.69}$$

where

$$K_i(\xi_i, \rho_i) = \rho_i \int_0^{\infty} \exp\left(-k|\xi_i|\right) J_1(k\rho_i) \cos k dk.$$

The circulation shedding rate at the orifice edge is:

$$\frac{d\Gamma}{dt} = \frac{1}{2}\left(v_-^2 - v_+^2\right), \tag{5.70}$$

where v_- and v_+ are slip velocities at the two sides of the orifice edge. To calculate these two velocities, the contribution of the first discrete vortex ring Γ_1 will be substituted by that of the N_1 subdiscrete vortex rings with circulation Γ_1/N_1. Then, Eq. (5.63) can be written as:

$$\begin{aligned} v(x, r) = &\sum_{j=1}^{N_1} H(x\xi_{1j})\left[V_j^V\left(x, r; \xi_{1j}, \rho_j\right) + \hat{V}_j^V\left(x, r; \xi_{1j}, \rho_j\right)\right] \\ &+ \sum_{j=1}^{N_1} V_j^A\left(x, r; \xi_{1j}, \rho_j\right) + \sum_{i=2}^{N} H(x\xi_j)\left[V_i^V(x, r; \xi_i, \rho_i)\right. \\ &\left.+ \hat{V}_i^V(x, r; \xi_i, \rho_i)\right] + \sum_{i=2}^{N} V_i^A(x, r; \xi_i, \rho_i) + C(t) V^p(x, r), \end{aligned} \tag{5.71}$$

The slip velocities v_- and v_+ are calculated from Eq. (5.71):

$$v_- = v(-\varepsilon_x, 1 + \varepsilon_r), \quad v_+ = v(\varepsilon_x, 1 + \varepsilon_r), \tag{5.72}$$

where $\varepsilon_x \sim O(10^{-2})$ and $\varepsilon_r \sim O(10^{-2})$. The circulation of the nascent vortex and the length of the starting vortex sheet are calculated through an iteration procedure. First of all, a value of Δ_1 is chosen, and the coordinates of the subvortex rings (ξ_{1j}, ρ_{1j}) are determined. Next, the circulation of the nascent vortex Γ_1^K is calculated by Eq. (5.69). After the calculation of slip velocities from Eq. (5.72), the circulation of the nascent vortex Γ_1^V can also be calculated according to Eq. (5.70):

$$\Gamma_1^V = \frac{1}{2}\left(v_-^2 - v_+^2\right)\Delta t. \tag{5.73}$$

The Γ_1^V and Δ_1 are employed as the required value until the following condition is fulfilled:

$$\left|\Gamma_1^K - \Gamma_1^V\right| / \left|\Gamma_1^V\right| < \varepsilon, \tag{5.74}$$

where ε is a control error.

The motion of discrete vortex rings governed by Eq. (5.59) is solved by the second-order Runge-Kutta scheme, and then the volume flux through the orifice is given by:

$$Q(t) = 2\pi \int_0^1 \frac{\partial \phi}{\partial x}(0^+, r) r dr. \tag{5.75}$$

Integrating the previous equation analytically yields:

$$Q(t) = 2\sum_{i=1}^{N} \Gamma_i \left\{1 + \left[\left(\xi_i^2 + \rho_i^2 - 1\right)^2 + 4\xi_i^2\right]^{\frac{1}{4}} \sin(\alpha/2)\right\} - 2C(t), \tag{5.76}$$

and

$$\alpha = \begin{cases} \arctan\left[-2|\xi_i| / \left(\rho_i^2 + \xi_i^2 - 1\right)\right] & \left(\rho_i^2 + \xi_i^2 > 1\right) \\ -\pi/2 & \left(\rho_i^2 + \xi_i^2 = 1\right) \\ -\pi + \arctan\left[-2|\xi_i| / \left(\rho_i^2 + \xi_i^2 - 1\right)\right] & \left(\rho_i^2 + \xi_i^2 < 1\right) \end{cases}.$$

The vortex shedding process for $\overline{P} = 5.0$ and $\Delta t = 0.0261$ is shown in Fig. 5.6. The evolution of the shed vortex sheet is calculated qualitatively, and the rolling-up nature of the shed vortex sheet, including the forming process of the vena contracta, is clearly demonstrated by the numerical simulation.

The normalized special acoustic impedance of an orifice is given by the following relation:

$$z_r = \overline{P}\cos[\phi(1)] / \overline{U}(1), z_x = \overline{P}\sin[\phi(1)] / \overline{U}(1). \tag{5.77}$$

where $\overline{U}(1)$ is the first harmonic and $\phi(1)$ is the relevant phase angle of the average velocity through the orifice. As shown in Fig. 5.7, the normalized specific acoustic resistance increases markedly while the normalized specific acoustic reactance decreases with the increase of the nondimensional amplitude of the applied sound pressure.

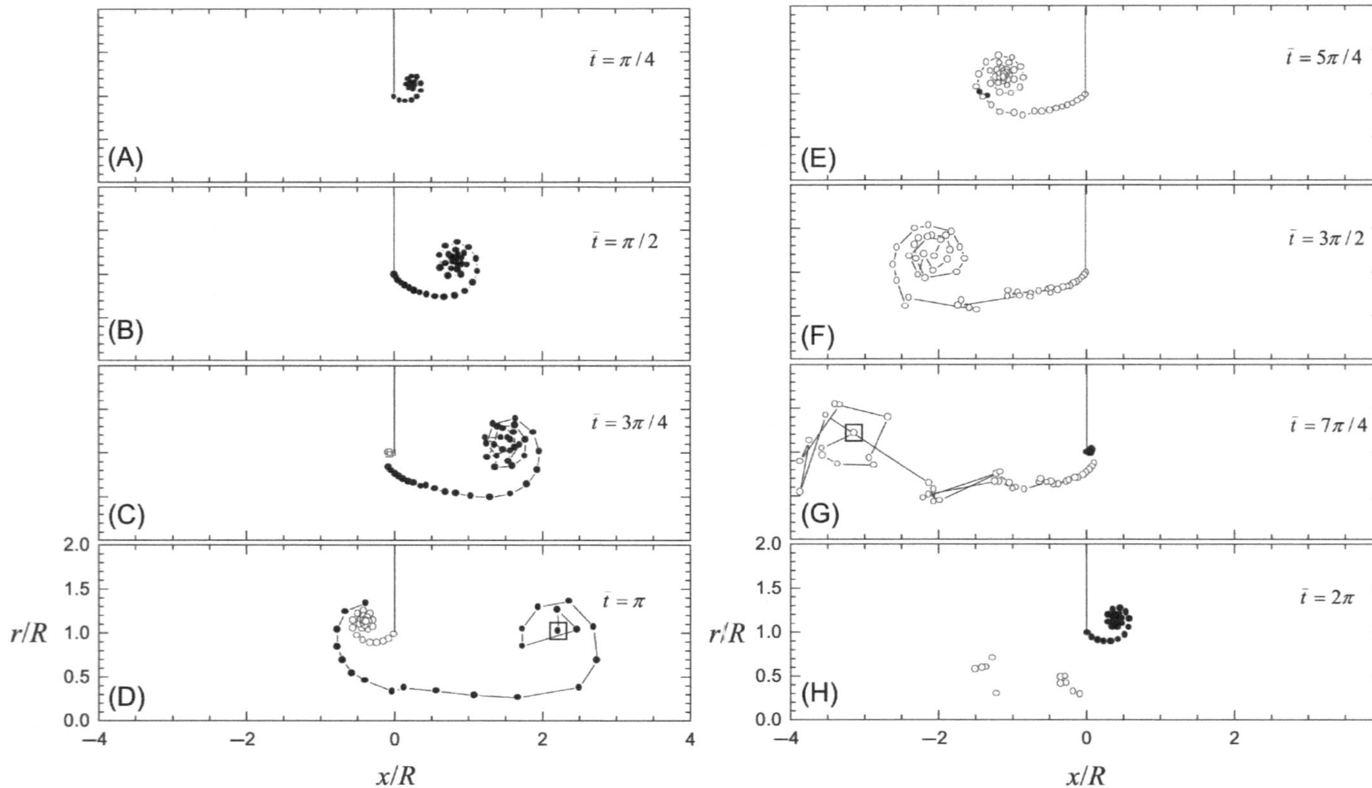

Fig. 5.6 Computed vortex shedding process at the edge of an orifice: (A) to (H) are the instant results at every 1/8 cycle of a full period. The solid and open circles indicate respectively the vortex strength is positive and negative, and the open squares mark the positions of the amalgamation centers..

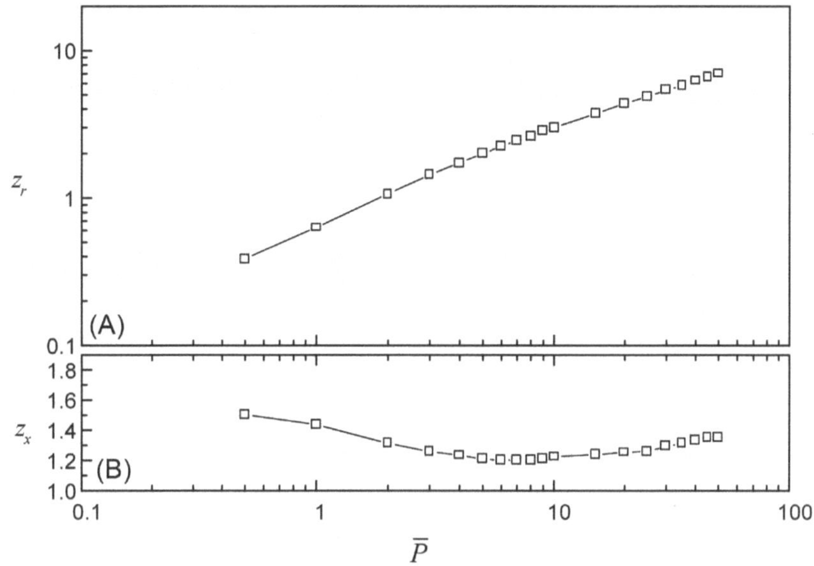

Fig. 5.7 Normalized special acoustic impedance as a function of the amplitude of the applied sound pressure: (A) normalized special acoustic resistance and (B) normalized special acoustic reactance.

As $\overline{P}$ increases, the normalized special acoustic reactance decreases to 0.71 of its linear value at its lowest point. This result is in excellent agreement with the ratio 0.7 obtained by Thurston et al. [30]. Because the viscosity is of little importance in the present model, the numerical results reconfirm the conclusion that the absorption mechanism is the conversion of acoustic energy into vortical energy.

5.5 Effect of grazing flow on vortex sound interaction of perforated plates

It has been well known that grazing flow markedly changes the acoustic impedance of a perforated liner due to the strong interaction between the vortical and acoustic flow at an orifice in the presence of grazing flow. The models based on the mechanism of vortex sound interaction have achieved great success in predicting the effect of grazing/bias flow on the acoustic impedance of an orifice, and two typical models are introduced in this section.

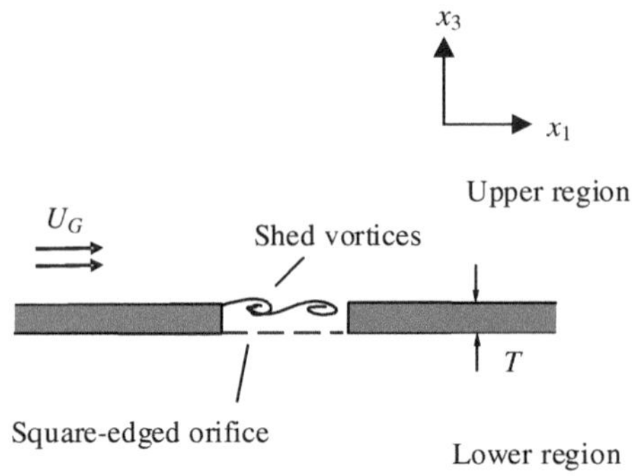

Fig. 5.8 Uniform grazing flow of velocity U_G past a squared-edged orifice of arbitrary shape in a rigid plate of small, but finite, thickness.

5.5.1 Effect of grazing flow on the acoustic impedance of perforated plates

In this section, a linearized potential flow model is developed on the mechanism of sound-vortex interaction to illustrate the influence of grazing flow on the acoustic behaviors of an orifice [33].

As shown in Fig. 5.8, there is an incompressible mean flow U_G over the upper opening of a square-edged orifice in a rigid plate of small, but finite, thickness. As a general case, the orifice is of arbitrary shape. A low-frequency harmonic sound wave incident on the plate produces a uniform pressure difference, $\Delta p = (p_+ - p_-) \exp(-i\omega t)$, between the two sides of the plate in the vicinity of the orifice. The unsteady flow through the orifice forced by this pressure difference will be considered.

A rectangular coordinate system (x_1, x_2, x_3) is introduced. The plane $x_3 = 0$ of the coordinate system is taken to coincide with the midplane of the plate, the x_1 axis is parallel to the mean flow, and the coordinate origin is at the geometrical center of the orifice. On the condition that the plate thickness T is far smaller compared with the sound wavelength, it is assumed that the velocity of the fluid confined in the orifice only has a vertical component, and its value is independent of x_3. Let $w(x_1, x_2)$ be the fluctuating velocity of the fluid particles in the orifice. In accordance with the particle velocity continuity boundary condition rather than the displacement, the vertical component of the fluctuating velocity of fluid particles just above the upper orifice opening and below the lower orifice opening is equal to $w(x_1, x_2)$. In the approximations of low-frequency incident sound and small plate thickness, the sound pressure above and below the plate can be formulated as:

$$
\begin{aligned}
p &= p_+ - i\rho_0\left(\omega + iU_G\frac{\partial}{\partial x_1}\right)\int_S \frac{w(y_1,y_2)}{2\pi|\mathbf{x}-\mathbf{y}|}dy_1dy_2 \quad \left(x_3 > \frac{T}{2}\right) \\
&= p_- + i\rho_0\omega\int_S \frac{w(y_1,y_2)}{2\pi|\mathbf{x}-\mathbf{y}|}dy_1dy_2 \quad \left(x_3 < -\frac{T}{2}\right),
\end{aligned}
\tag{5.78}
$$

where p_+ and p_- are, respectively, the sound pressure in the far distance of the region where $x_3 > 0$ and $x_3 < 0$, ω is angular frequency of the incident sound. The motion of the fluid in the orifice is governed by:

$$
-\rho_0 T\frac{dw}{dt} = i\rho_0\omega Tw = [p] \quad \left(|x_3| < \frac{T}{2},\ \sqrt{x_1^2 + x_2^2} < R\right), \tag{5.79}
$$

where R is the radius of a circular orifice, and $[p]$ is the fluctuating pressure difference between the upper and lower orifice opening. Using the pressure continuity condition, the following differential equation can be deduced from Eqs. (5.78) and (5.79):

$$
\frac{p_+ - p_-}{\rho_0} = \left(2i\omega - U_G\frac{\partial}{\partial x_1}\right)\int_S \frac{w(y_1,y_2)}{2\pi\sqrt{(x_1-y_1)^2 + (x_2-y_2)^2}}dy_1dy_2 + i\omega Tw. \tag{5.80}
$$

Integrate Eq. (5.80) with respect to x_1, and introduce the following nondimensionalization:

$$
X = x/B,\ Y = y/B,\ Sr = \omega B/U_G,\ Z = i\rho_0 SrU_G w/\pi(p_+ - p_-).
$$

where B is characteristic orifice length, equal to the radius of a circular orifice or the half length of the side of a rectangular orifice parallel to grazing flow, and Sr is the Strouhal number.

Then we obtain the following integral equation:

$$
\begin{aligned}
&\int_S \frac{ZdY_1dY_2}{\sqrt{(X_1-Y_1)^2 + (X_2-Y_2)^2}} - 2i\pi Sr(T/B)\int_{LE}^{X_1} Ze^{2iSr(X_1-Y_1)}dY_1 \\
&= 1 + \alpha e^{2iSrX_1},
\end{aligned}
\tag{5.81}
$$

where LE is the leading edge of the orifice and α is the unknown function of X_2 that must be determined to ensure that the Kutta condition is satisfied at the LE of the orifice. Eq. (5.81) can be solved numerically to obtain the acoustic impedance of the orifice.

The fluctuating volume flux through the orifice is given by:

$$Q = \int_S w(x_1, x_2) dx_1 dx_2. \tag{5.82}$$

Then the acoustic impedance of the orifice can be obtained from its definition as:

$$z = \frac{p_+ - p_-}{\rho_0 c u_{av}}, \tag{5.83}$$

where the average velocity through the orifice is given by $u_{av} = Q/S$.

It is obvious from the preceding basic equations that the present model has no such limitation that the method is applicable only for a rectangular orifice with large value of W/L, and W is the length of the side of a rectangular orifice transverse to grazing flow. For a circular orifice, the normalized special acoustic impedance given by Eq. (5.83) is the function of three nondimensional parameters: grazing flow Mach number M, Helmholtz number kR, and normalized plate thickness T/R. In the case of a rectangular orifice, four nondimensional parameters are needed: grazing flow Mach number M, Helmholtz number $kL/2$, normalized plate thickness $2T/L$, and normalized orifice width $2W/L$, where k is the sound wave number.

Figs. 5.9 and 5.10 show the variation of acoustic impedance with the increasing grazing flow speed of perforated plates with rectangular and circular orifice, respectively, where both the experimental and theoretical results are given. It should be emphasized that two kinds of theoretical results are given here: the first one is calculated through an existing model based on the particle displacement match (PDM), and the second one is obtained by a present model based on the particle velocity match (PVM).

It can be seen from the two figures that there are large discrepancies between the theoretical results based on PDM [34] and the experimental data [33], especially for the acoustic reactance. The computed acoustic reactance based on PDM increases too fast compared with the experimental results as the grazing flow speed is increased. By comparison, the present calculations using PVM are in generally good agreement with the experimental data.

The present model demonstrates that, in terms of the variation of the acoustic resistance, there are two different regions: sinking and linearly increasing. The acoustic resistance increases slowly or even decreases with the increase of grazing flow speed in the sinking region, whereas it goes up in proportion to grazing flow speed in the linearly increasing region.

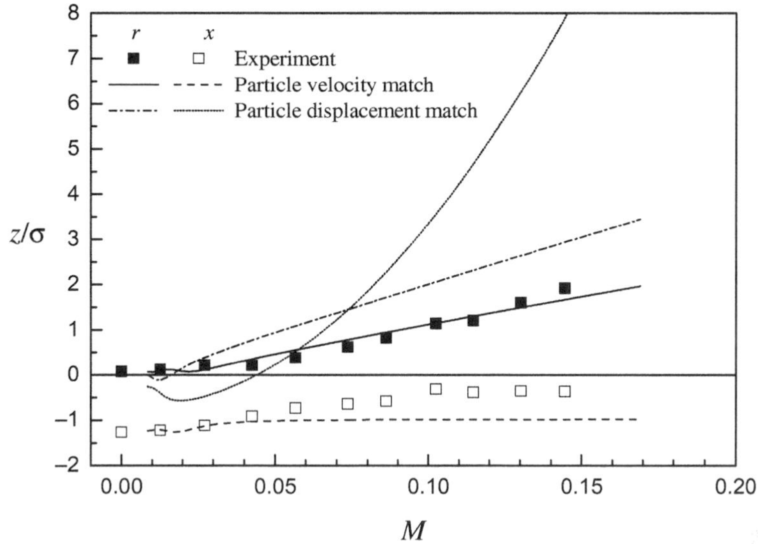

Fig. 5.9 Comparison between the calculations based on PVM and PDM in the case of a circular orifice; normalized special acoustic impedance plotted as a function of the grazing flow Mach number at 500.0 Hz; $T/R = 0.14$; and $kR = 3.23 \times 10^{-2}$.

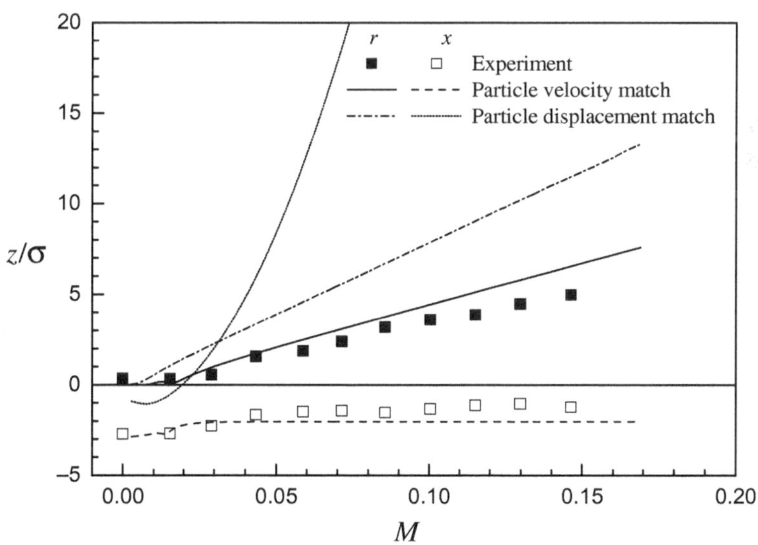

Fig. 5.10 Comparison between the calculations based on PVM and PDM in the case of a rectangular orifice; normalized special acoustic impedance plotted as a function of the grazing flow Mach number at 500.0 Hz; $kL = 1.85 \times 10^{-2}$; $T/L = 0.5$; and $W/L = 0.5$.

It is further shown that the presence of this sinking region is due to the influence of finite plate thickness, which will be discussed in more detail in the next section. For the acoustic reactance, increasing plate thickness leads to an increase of its absolute value. It can also be concluded that, compared with the particle displacement continuity boundary condition, the use of the particle velocity continuity boundary condition results in a great improvement of the theoretical predictions both for the acoustic resistance and reactance.

5.5.2 Effect of plate thickness on impedance of perforated plates with bias flow

In this section, a numerical model based on the vortex sound theory is presented of the Rayleigh conductivity of a square-edged orifice in the presence of finite plate thickness and bias flow [35]. Particular attention is paid to the evaluation of the effect of finite plate thickness on its acoustic impedance.

As shown in Fig. 5.11, an orifice is located in a plate of small, but finite thickness, and there is a low Mach number and high Reynolds number steady flow through the orifice. The flow separates from the inlet edge of the orifice and forms a jet. A cylindrical coordinate system (x_1, θ, r) is introduced. Consider a uniform sinusoidal sound pressure perturbation due to a low-frequency sound wave incident on the plate in the region $x_1 < 0$:

$$p = P\exp(-i\omega t), \tag{5.84}$$

where p is the applied sound pressure with an amplitude of P and an angular frequency of ω.

The mean flow is assumed to be incompressible and inviscid, and the viscous effect is only taken into consideration by applying the Kutta condition

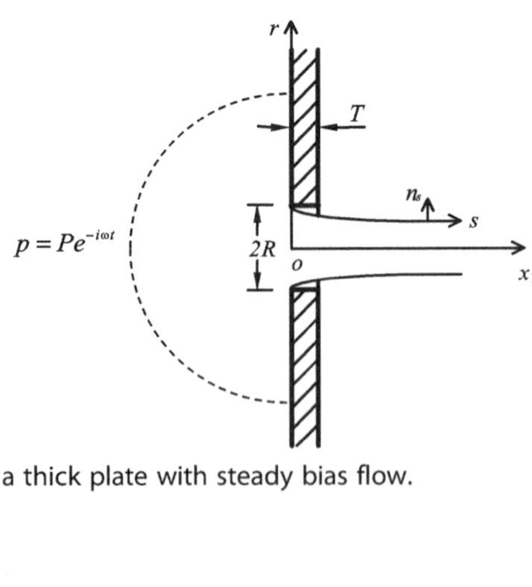

Fig. 5.11 Orifice in a thick plate with steady bias flow.

at the inlet edge. A hypothesis is introduced that the convection of the unsteady vortex plays a dominant role and the perturbation of the mean shear layer is ignored, thus the governing equations can be rewritten as follows:

$$\nabla^2 h = -\nabla \cdot (\varpi \times \mathbf{U}), \tag{5.85}$$

where ϖ is the strength of shed vortices, $\mathbf{U}$ is the mean flow velocity, and h is the perturbation Bernoulli enthalpy defined by the following relation:

$$h = \frac{p}{\rho_0} + \mathbf{U} \cdot \mathbf{u} = -\frac{\partial \phi}{\partial t} + cc, \tag{5.86}$$

where ρ_0 is the air density, $\mathbf{u}$ is the perturbation velocity, ϕ is the perturbation velocity potential, and cc is a constant that takes different values in the two regions of potential flow separated by the free shear layer of the bias flow.

On the condition of high Reynolds number based on the mean flow speed and orifice radius, the shear layer is very thin. The strength of the unsteady vortices can be written as:

$$\varpi = \gamma_e \exp\left[-i\omega(t - s/U_c)\right]\delta(n_s)\mathbf{k}, \tag{5.87}$$

where γ_e is the strength of shed vortex sheet at orifice inlet edge, s is the tangential coordinate along free streamline of bias flow, and U_c is the convected speed of shed vortices.

Substituting Eq. (5.87) into Eq. (5.85) and suppressing the time-dependent part, the following axisymmetric Possion equation can be obtained:

$$\frac{1}{r}\frac{\partial}{\partial r}\left(r\frac{\partial h}{\partial r}\right) + \frac{\partial^2 h}{\partial x_1^2} = -\nabla \cdot \left[\gamma_e U_c \exp\left(i\omega s/U_c\right)\delta(n_s)\mathbf{n}_s\right], \tag{5.88}$$

where n_s is the normal coordinate of free streamline of bias flow and $\mathbf{n}_s$ is the unit vector.

By the application of the half-space Green's function, Eq. (5.88) can be converted into integral equations. As shown in Fig. 5.12, the integration space is divided into two domains separated by boundary Σ_0. The surfaces in Fig. 5.13 are explained as follows: Σ_0 is where $x_1 = 0$ and $r < R$; Σ_1 is where $0 < x_1 < T$ and $r = R$; Σ_2 is where $x_1 = T$ and $r > R$; Σ_3 is the free-stream surface; Σ_4 is where $x_1 < 0$ and $x_1^2 + r^2 \rightarrow \infty$; Σ_5 is where $x_1 > 0$ and $x_1^2 + r^2 \rightarrow \infty$; and Σ_6 is where $x_1 = 0$ and $r > R$.

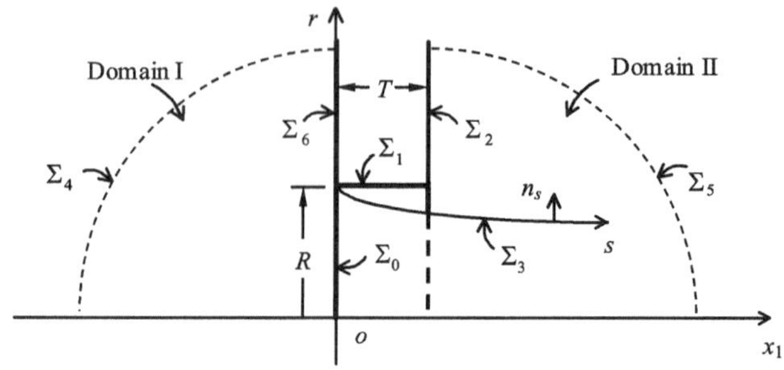

Fig. 5.12 Arrangement of the integration domains and the integration surfaces.

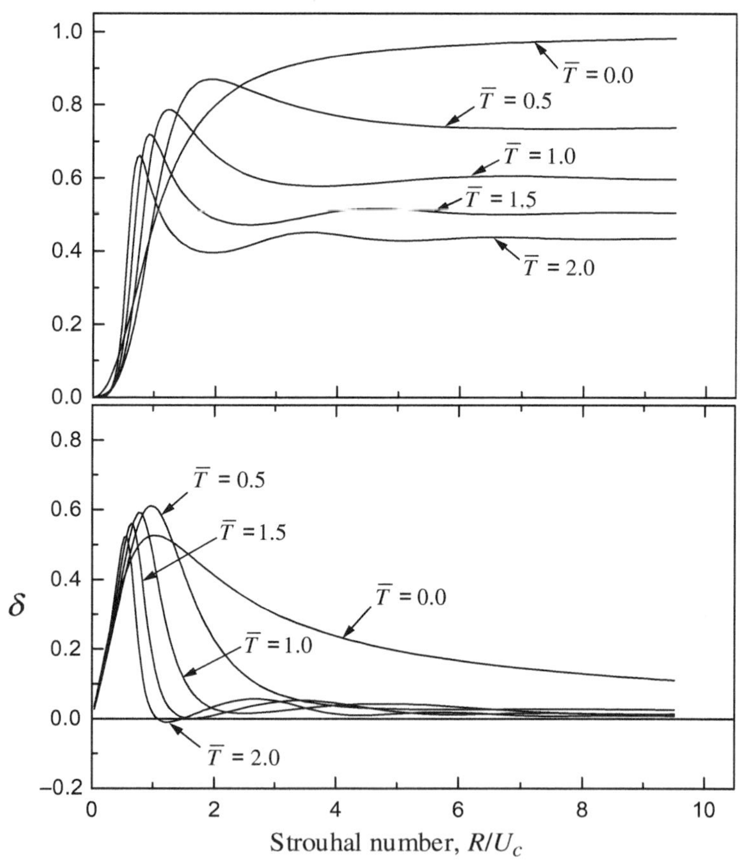

Fig. 5.13 Normalized Rayleigh conductivity as a function of Strouhal number for different nondimentional plate thickness.

To solve Eq. (5.88), the following boundary conditions are applied:

$$\begin{cases} \dfrac{\partial h}{\partial n} = 0, & \text{on the solid wall} \\ h = \dfrac{P}{\rho_0}, \quad \dfrac{\partial h}{\partial n} = 0, & \text{on } \Sigma_4 \\ h = 0, \quad \dfrac{\partial h}{\partial n} = 0, & \text{on } \Sigma_5 \end{cases}.$$

In a cylindrical coordinate system, the half-space Green's function G_h is expressed as:

$$G_h(x_1, \theta, r; y_1, \varphi, \rho) = -\frac{1}{4\pi}\left[\frac{1}{\sqrt{(x_1 - y_1)^2 + r^2 + \rho^2 - 2\rho r\cos(\theta - \varphi)}} + \frac{1}{\sqrt{(x_1 + y_1)^2 + r^2 + \rho^2 - 2\rho r\cos(\theta - \varphi)}}\right]. \tag{5.89}$$

Finally, the expressions of h are obtained. In the domain $x_1 < 0$:

$$h = \frac{P}{\rho_0} - \int_{\Sigma_0} \frac{\partial h}{\partial y_1} G_h \mathrm{d}S. \tag{5.90}$$

In the domain $x_1 > 0$:

$$\begin{aligned} ch = & \int_{\Sigma_0} \frac{\partial h}{\partial y_1} G_h dS + \int_{\Sigma_1} h \frac{\partial G_h}{\partial \rho} dS - \int_{\Sigma_2} h \frac{\partial G_h}{\partial y_1} dS \\ & - \int \nabla \cdot \left[\gamma_e U_c \exp\left(\frac{i\omega s}{U_c}\right) \delta(n_s) \mathbf{n}_s\right] G_h dV, \end{aligned} \tag{5.91}$$

where $c = 1/2$ on the boundary and $c = 1$ in the domain. After integrating the last integral in Eq. (5.91) by parts and carrying out the integration of δ function, the volume integral can be converted to a surface integral, so that Eq. (5.91) is simplified as:

$$\begin{aligned} ch = & \int_{\Sigma_0} \frac{\partial h}{\partial y_1} G_h dS + \int_{\Sigma_1} h \frac{\partial G_h}{\partial \rho} dS - \int_{\Sigma_2} h \frac{\partial G_h}{\partial y_1} dS \\ & - \int_{\Sigma_3} \gamma_e U_c \exp(i\omega s / U_c) \frac{\partial G_h}{\partial n_s} dS. \end{aligned} \tag{5.92}$$

In the inlet cross-section of the orifice, the perturbation Bernoulli enthalpy and its normal derivative are continuous:

$$\begin{cases} \dfrac{\partial h}{\partial x_1}(x_1=0^+,r)=\dfrac{\partial h}{\partial x_1}(x_1=0^-,r) \\ h(x_1=0^+,r)=h(x_1=0^-,r) \end{cases}. \tag{5.93}$$

Eqs. (5.90) and (5.92) are related to each other by the continuous condition of Eq. (5.93), and so the following integral equations can be derived:

$$\begin{aligned} &\frac{P}{\rho_0}-2\int_{\Sigma_0}\frac{\partial h}{\partial y_1}G_h dS-\int_{\Sigma_1}h\frac{\partial G_h}{\partial \rho}dS+\int_{\Sigma_2}h\frac{\partial G_h}{\partial y_1}dS \\ &\quad+\int_{\Sigma_3}\gamma_e U_c \exp\left(\frac{isSr}{R}\right)\frac{\partial G_h}{\partial n}dS=0, \quad (\text{on } \Sigma_0) \end{aligned} \tag{5.94}$$

$$\begin{aligned} &\frac{1}{2}h-\int_{\Sigma_0}\frac{\partial h}{\partial y_1}G_h dS-\int_{\Sigma_1}h\frac{\partial G_h}{\partial \rho}dS+\int_{\Sigma_2}h\frac{\partial G_h}{\partial y_1}dS \\ &\quad+\int_{\Sigma_3}\gamma_e U_c \exp\left(\frac{isSr}{R}\right)\frac{\partial G_h}{\partial n}dS=0, \quad (\text{on } \Sigma_1 \text{ or } \Sigma_2) \end{aligned} \tag{5.95}$$

where $Sr=\omega R/U_c$ is the Strouhal number. The strength of the shedding vortex sheet γ_e is determined by applying the Kutta condition, which requires the normal derivative of the Bernoulli enthalpy to be zero at the inlet edge of the orifice; that is:

$$\frac{\partial h}{\partial x_1}(x_1=0, r=R^-)=0. \tag{5.96}$$

It is equivalent to requiring the normal velocity to be zero. The Rayleigh conductivity C is defined by the relation:

$$C=\frac{i\omega\rho_0 Q}{p^+-p^-}, \tag{5.97}$$

where p^+-p^- is the applied pressure difference across the orifice and the volume flux Q is given by:

$$Q=\frac{1}{i\omega}\int_{\Sigma_0}\frac{\partial h}{\partial x_1}dS. \tag{5.98}$$

The Rayleigh conductivity of an orifice in a thin plate is equal to $2R$ when there is no mean flow through it. Thus, C and Q are normalized as:

$$\overline{C}=C/2R, \quad \overline{Q}=Q/\pi\omega R^3.$$

Then:

$$\overline{C} = \gamma - i\delta = -i\pi\overline{Q}/2\overline{H}, \tag{5.99}$$

where γ and δ are the functions of the Strouhal number Sr and the nondimensional thickness T, respectively.

In Fig. 5.13, the normalized Rayleigh conductivity is plotted as a function of the Strouhal number for different nondimensional thickness. When $\overline{T} \to 0$ and the shape of the vortex sheet is selected to be cylindrical, the present numerical results of the normalized Rayleigh conductivity are compared with the analytical results of Howe [36]. When $\overline{T} = 2.0$, $\delta$is negative in the range of $1.08 < Sr < 1.48$. Further numerical tests show that, when $\overline{T} = 1.5$, there exists a certain range of Strouhal number where δ is negative, which is contrary to the result that δ is always positive in the situation of an infinitely thin plate.

It is shown that plate thickness, which is characteristic of the real structure, has marked influence on both the acoustic resistance and the acoustic reactance from the following aspects:

(1) When the bias flow speed is near zero, the acoustic reactance approaches the sum of the values due to the plate thickness and end correction. As the flow speed increases, the acoustic reactance decreases. As a result, all of the value due to the plate thickness is lost at high flow speed.

(2) As the bias flow speed increases, the increasing tendency of the acoustic resistance is generally unchanged. However, within a certain range of the bias flow speed that is near zero, the acoustic resistance decreases with the increase of the bias flow speed. Furthermore, the acoustic resistance can even decrease to negative values provided that the ratio between the plate thickness and orifice radius is larger than 1.5.

(3) At high flow speed, the effect of plate thickness on the acoustic resistance and acoustic reactance gradually diminishes. Therefore, the acoustic properties of a thick perforated plate with bias flow are similar to the corresponding thin one.

(4) The separation of the mean stream at the inlet edge of the orifice accounts for the described acoustic properties of a perforated plate with bias flow, which are by no means explained by the simple combination of the values due to the plate thickness with the end correction.

5.6 Nonlinear model of vortex sound interaction

Acoustic resonance, a common phenomenon in various engineering applications, is induced by the interaction of vortex and sound within an enclosed space. In general, to predict the finite amplitude of the resonance, nonlinear models are necessary. In this section, a nonlinear time-domain model based on dynamic balance between the flow and sound fields is presented [37], and, at the beginning, the nonlinear model of vortex sound interaction occurring at a slit is given as a foundation [37, 38].

5.6.1 The nonlinear model of vortex sound interaction occurring at a slit

In Fig. 5.14, a slit in an infinitely large thin plate subjected to the excitation of a high-intensity sound wave is shown, and the width and length of the slit are $2s$ and b, respectively. The acoustic pressure drives the oscillatory flow through the slit, resulting in the unsteady vortex shedding at the slit edges. Thus, by the mechanism of dynamical exchange between acoustic and the vortical motions, the acoustic behavior of the slit is highly dependent on the amplitude of the incident sound pressure.

If the sound wavelength is much larger than the size of the slit, the unsteady flow around the slit can be approximately treated as incompressible. For this condition, the incident sound wave is considered as the fluctuating pressure uniformly distributed in the far field:

$$p_i = \begin{cases} P\cos\omega t & \left(x_2 > 0,\, x_1^2 + x_2^2 + x_3^2 \to \infty\right) \\ 0 & \left(x_2 < 0,\, x_1^2 + x_2^2 + x_3^2 \to \infty\right) \end{cases}, \tag{5.100}$$

where p_i is the incident sound pressure with an amplitude of P, and ω is the angular frequency.

When the Reynolds number is high, the viscous effect can be ignored everywhere except for the slit edges where vortex shedding takes place. Therefore, the discrete vortex model is employed to describe the vortex sound interaction phenomenon.

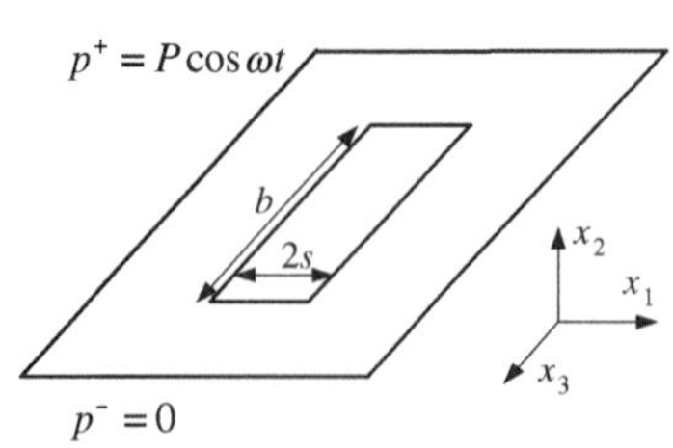

Fig. 5.14 A slit in a thin plate subjected to the high-intensity sound wave.

The 2D incompressible assumption allows the near-field flow to be solved by means of the complex variable theory. The Schwartz-Christoffel transformation is written as:

$$z = \frac{1}{2}\left(\zeta + \frac{s^2}{\zeta}\right). \tag{5.101}$$

In the ζ plane, the complex potential, W, is expressed as:

$$W = \frac{Q}{\pi}\ln\zeta + \sum_{n=1}\frac{\Gamma_n}{2\pi i}\left[\ln(\zeta-\zeta_n) - \ln\left(\zeta-\zeta_n^*\right)\right]. \tag{5.102}$$

Note that the asterisk denotes the complex conjugate. The resulting complex velocity is given by the derivative of Eq. (5.102):

$$w^*(z) = \frac{dW}{dz} = \frac{dW}{d\zeta}\frac{d\zeta}{dz}. \tag{5.103}$$

When this equation is used to compute the convection velocity of each discrete vortex, special care must be taken such that the induced velocity due to the vortex itself is excluded. Then, for the nth point vortex:

$$w^*(z_n) = \lim_{z\to z_n}\frac{d}{dz}\left[W + \frac{\Gamma_n}{2\pi i}\ln(z-z_n)\right] = \left.\frac{d\widetilde{W}}{dz}\right|_{z_n} + \frac{\Gamma_n\zeta''(z_n)}{4\pi i\zeta'(z_n)}, \tag{5.104}$$

where $\widetilde{W}(z)$ denotes the complex potential excluding the contribution of the nth discrete vortex. The second term on the right-hand side of Eq. (5.104) is the Routh's correction. Substituting Eq. (5.102) into Eq. (5.104) yields:

$$w^*(z_n) = \left(\frac{2\zeta_n^2}{\zeta_n^2 - s^2}\right)\left[\frac{Q}{\pi\zeta_n} + \sum_{\substack{n'=1\\ n'\neq n}}\frac{\Gamma_{n'}}{2\pi i}\frac{1}{\zeta_n-\zeta_{n'}} - \sum_{n'}\frac{\Gamma_{n'}}{2\pi i}\frac{1}{\zeta_n-\zeta_{n'}^*}\right] + \frac{i\Gamma_n}{\pi}\frac{\zeta_n s^2}{\left(\zeta_n^2-s^2\right)^2}. \tag{5.105}$$

The mathematical form of the Kelvin theorem is:

$$\frac{dz_n}{dt} = w(z_n), \tag{5.106}$$

$$\Gamma_n = \text{const.} \tag{5.107}$$

Eq. (5.106) is numerically integrated by means of the second-order Adams-Bashforth scheme:

$$z_n\big|_{t+\Delta t} = z_n\big|_t + \frac{1}{2}\left(3w_n\big|_t - w_n\big|_{t-\Delta t}\right)\Delta t. \tag{5.108}$$

It is clear from this derivation that, once Γ_n and ζ_n are given at the initial time t_0, the motion of the vortices and thus the distribution of the circulation in the flow field can be determined at subsequent times.

According to Brown and Michael [39], the shed vortices are represented by a concentrated point vortex connected to the slit edge by a vortex-free feeding sheet, as shown in Fig. 5.15. At each time step, the strength and location and of the nascent vortices from the slit edges are determined by:

$$\Gamma_1 = 2\pi i W'(-s)\frac{(s+\zeta_1)(s+\zeta_1^*)}{\zeta_1^* - \zeta_1}, \tag{5.109}$$

$$\zeta_1 = \sqrt{\frac{4}{5}\left\{\frac{Q}{\pi} + \sum_{n=2}^{N}\frac{s^2\Gamma_n}{2\pi i}\left[\frac{1}{(\zeta_n+s)^2} - \frac{1}{(\zeta_n^*+s)^2}\right]\right\}\Delta t} - s. \tag{5.110}$$

As shown in Fig. 5.16, the Bernoulli equation is applied along the path ACB to obtain the far-field pressure difference across the slit plate.

$$\frac{p_A - p_B}{\rho_0} = -\frac{d\Gamma_t}{dt} + \frac{\partial}{\partial t}\int_A^B \vec{u}_\Gamma \cdot d\vec{l} + \frac{\partial}{\partial t}\int_A^B \vec{u}_Q \cdot d\vec{l}. \tag{5.111}$$

Fig. 5.15 Schematic of Brown and Michael's model of vortex shedding from the edge of a thin plate.

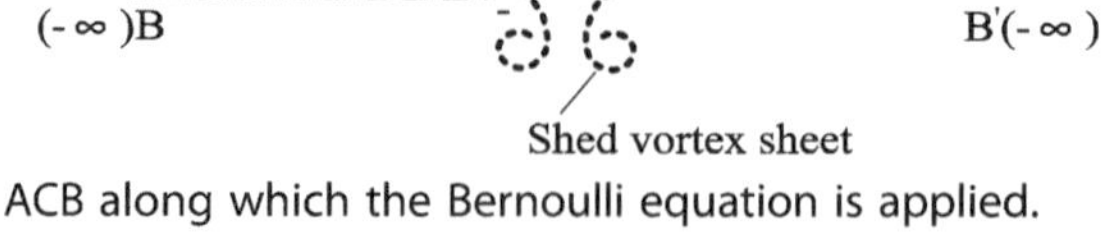

Fig. 5.16 Path ACB along which the Bernoulli equation is applied.

Note that there are two integral terms on the right of Eq. (5.111), and the path integral of $\vec{u}_\Gamma$ is straightforwardly written as:

$$\int_A^B \vec{u}_\Gamma \cdot d\vec{l} = \sum_{n=1}^{N} \frac{\Gamma_n}{\pi}\left[\frac{\pi}{2} - \arctan\left(\frac{\xi_{1n}}{\xi_{2n}}\right)\right]. \tag{5.112}$$

However, the path integral of $\vec{u}_Q$ becomes divergent as the end points, A and B, go into the far field, extending to infinity, if the flow is assumed to be 2D everywhere. Therefore, a quasi-3D strategy is introduced to deal with the problem. The basic idea is that the far-field difference of the velocity potential due to the fluctuating flow through the slit is given by a quasi-3D potential function, Φ_Q, instead of the complete 2D one, W_Q. By means of half-space Green's function of the Laplace equation for the incompressible potential flow, we can write:

$$\Phi_Q(x_1, x_2, x_3) = \Phi_{Q,\pm\infty} \mp \frac{1}{2\pi}\int_{-b/2}^{b/2}\int_{-1}^{1} \frac{\partial \Phi_Q}{\partial x_2'} \frac{dx_1' dx_3'}{\sqrt{(x_1 - x_1')^2 + (x_3 - x_3')^2 + x_2^2}}. \tag{5.113}$$

Then, because Φ_Q and $\partial\Phi_Q/\partial x_2$ are continuous across the slit, we derive from this equation that:

$$\int_{-\infty}^{+\infty} \vec{u}_Q \cdot d\vec{l} = \Phi_{Q,+\infty} - \Phi_{Q,-\infty} = \frac{1}{\pi}\int_{-b/2}^{b/2}\int_{-1}^{1} \frac{\partial \Phi_Q}{\partial x_2'} \frac{dx_1' dx_3'}{\sqrt{(x_1 - x_1')^2 + (x_3 - x_3')^2}}. \tag{5.114}$$

Based on the assumption of quasi-3D flow, Φ_Q can be well approximated by W_Q in the vicinity of the slit:

$$\frac{\partial \Phi}{\partial x_2} \approx \frac{dW_Q}{dz} = \frac{Q(t)}{\pi\sqrt{1 - x_1^2}}. \tag{5.115}$$

Substituting this equation into Eq. (5.114), and performing a spanwise averaging in the way that both sides of the equation are integrated over $-b/2 \leq x_3 \leq b/2$, yields:

$$\int_{-\infty}^{+\infty} \vec{u}_Q \cdot d\vec{l} = Q(t)C(b), \tag{5.116}$$

where

$$C(b) = \frac{1}{\pi^2 b}\int_{-1}^{1} \frac{F(x_1, x_1')}{\sqrt{1 - x_1'^2}} dx_1',$$

$$F(x_1, x_1') = \int_{-b/2}^{b/2}\int_{-b/2}^{b/2} \frac{dx_3' dx_3}{\sqrt{(x_1 - x_1')^2 + (x_3 - x_3')^2}}$$

$$= -2b\left\{ \ln|x_1 - x_1'| - \ln\left[b + \sqrt{b^2 + (x_1 - x_1')^2}\right] + \sqrt{1 + (x_1 - x_1')^2/b^2} - |x_1 - x_1'|/b \right\}.$$

Then, substituting Eqs. (5.112) and (5.116) into Eq. (5.111) yields:

$$\frac{p_A - p_B}{\rho_0} = -\frac{d\Gamma_t}{dt} + \frac{\partial}{\partial t}\left[\sum_{n=1}^{N} \frac{\Gamma_n}{\pi}\left(\frac{\pi}{2} - \arctan\frac{\xi_{1n}}{\xi_{2n}}\right)\right] + C(b)\frac{\partial Q(t)}{\partial t}. \quad (5.117)$$

The unsteady vortical and velocity flow field of the slit under the sinusoidal pressure excitation, $p_i = P\cos\omega t$, is computed in the time domain. A typical time step of the time-marching approach is performed as follows:

(1) Evaluate $Q(t)$ from Eq. (5.117). The rate of the vortex circulation shed from the left slit edge is given by:

$$\frac{d\Gamma_t}{dt} = \frac{\Gamma_1}{dt}. \quad (5.118)$$

Note that the time marching begins at the initial time t_0 of a small value, and the initial condition is approximately $\Gamma_1(t_0) = 0$.

(2) Calculate the coordinates of the nascent vortices according to Eq. (5.110), and then update the positions of all the point vortices by solving Eq. (5.106).

(3) Release the two nascent vortices by assigning them the circulations given by Eq. (5.109), and introduce a small cutoff radius for the nascent vortices as an artificial smoothing parameter. Assuming the vortex blob is a Rankine type, the induced velocity of a point vortex with a cutoff radius is given by the following equation:

$$w^* = \frac{\Gamma_{n'}}{2\pi i}\frac{1}{\zeta_n - \zeta_{n'}}\begin{cases} 1 & (|\zeta_n - \zeta_{n'}| > r_c) \\ r_c^{-2}|\zeta_n - \zeta_{n'}|^2 & (|\zeta_n - \zeta_{n'}| \le r_c) \end{cases}, \quad (5.119)$$

where the cutoff radius $r_c = \kappa d$ in which κ is a selected constant smaller than unity.

(4) Shift the indices of the vortices as follows:

$$\Gamma_{n+2,new} = \Gamma_{n,previous}, \zeta_{n+2,new} = \zeta_{n,previous}, w_{n+2,new} = w_{n,previous}. \tag{5.120}$$

(5) Increment the time by the interval:

$$t_m = t_{m-1} + \Delta t. \tag{5.121}$$

To limit the computation time, each vortex is gradually "damped," i.e., its circulation is decreased as it is convected away from the slit, so it is eventually not involved in the computation when its circulation becomes zero.

Due to the occurrence of the nonlinear distortion, the average velocity through the slit, v_{av}, is periodic but deviates somewhat from being sinusoidal. So, Fourier decomposition is employed to get the spectrum of the average velocity as a function of time. Provided that the first harmonic component is dominant in the spectrum of the average velocity, the nonlinear acoustic impedance of the slit can be defined as:

$$r = ks\frac{P\cos[\phi(1)]}{v_{av}(1)}, \quad x = -ks\frac{P\sin[\phi(1)]}{v_{av}(1)}. \tag{5.122}$$

The acoustic impedance is normalized by $\rho_0 c_0$.

It has long been noticed that the unsteady vortices are shed when high-intensity sound impinges upon the rim of an orifice. This vortex-shedding mechanism is important because it accounts for the nonlinear acoustic impedance of the orifice. It is assumed that the fluctuating flow of the slit is dominated by the vortex shedding occurring at the long edges of the slit with a large aspect ratio (AR). In Fig. 5.17, the shedding and evolution process of the vortices are displayed at each successive time interval of $T/4$ during the first two periods of time.

From the qualitative point of view, the simulation of the formation, evolution, and interaction of the vortex pair at the slit shows marked resemblance with the previous PIV flow visualization [40].

It can be seen from the Fig. 5.18 that the time-history curve of v_{av} is nearly sinusoidal at relatively low *SPL* of 128.5 dB, whereas it exhibits obvious nonlinear distortion when *SPL* is increased to 148.5 dB.

The nonlinear acoustic impedance is calculated according to Eq. (5.122), and the results for the three slits with different ARs are presented in

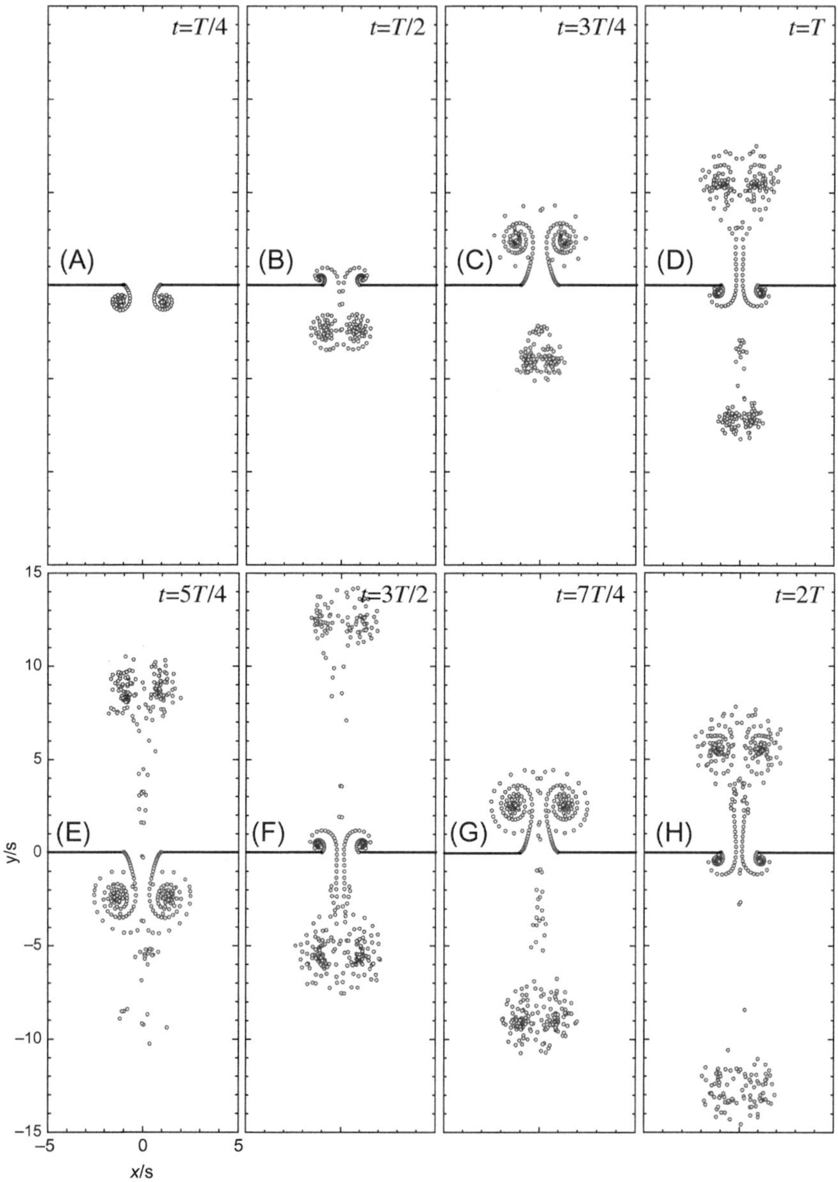

Fig. 5.17 The vortex-shedding process and the evolution of the vortices at a slit subjected to high-intensity sound: (A) to (H) are the instant results at every 1/8 cycle of a full period.

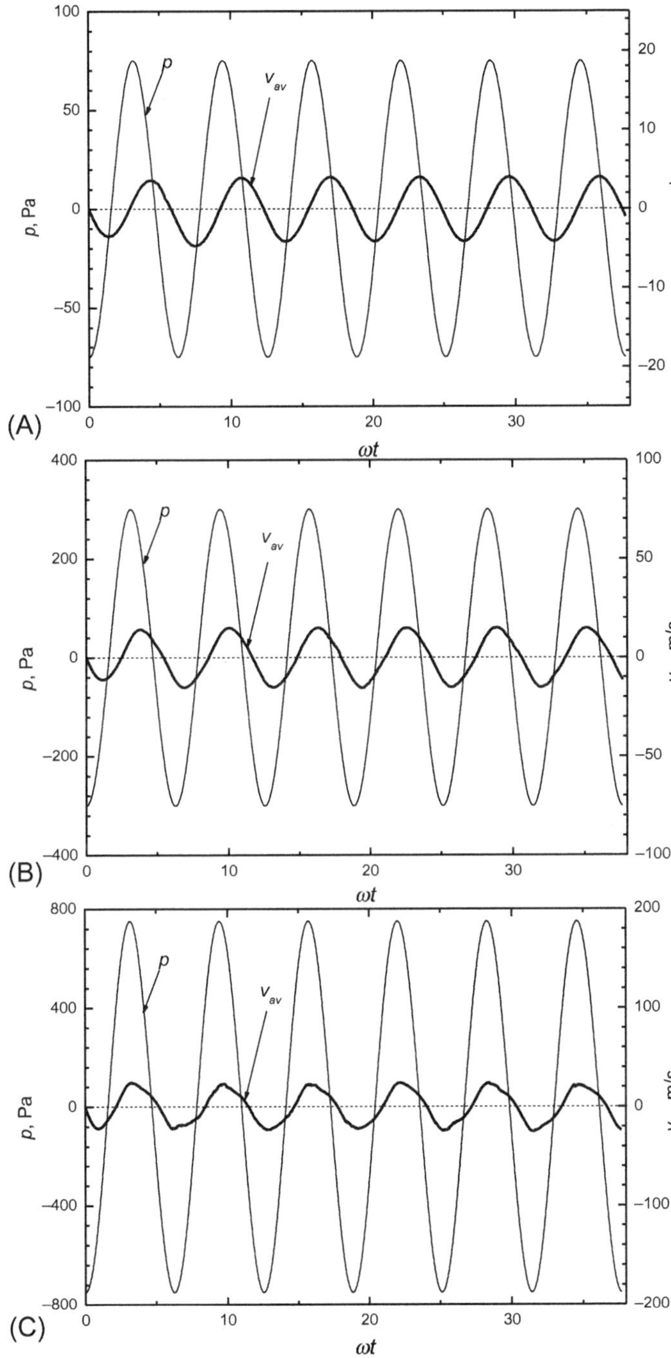

Fig. 5.18 The average velocity through the slit for different cases: (A) $P = 128.5$ dB, (B) $P = 140.5$ dB, and (C) $P = 148.5$ dB.

Fig. 5.19. It is shown that the acoustic resistance increases markedly with the sound pressure level, whereas the acoustic reactance decreases to some extent. As the AR increases from 5 to 15, the nonlinear acoustic resistance slightly increases, whereas the nonlinear acoustic reactance decreases. Obviously, the acoustic reactance changes with AR more prominently than the acoustic resistance. The influence of AR on the acoustic impedance tends to be less important when the SPL exceeds a certain high value (around 145 dB in the present case).

5.6.2 Flow-excited acoustic resonance of a Helmholtz resonator

The flow-excited resonance in a Helmholtz resonator results from the coupling of the shear layer instabilities over the opening with the acoustic mode inside the cavity via a feedback loop. And accurate amplitude prediction is the very open challenge that motivates research in this area. As we know, the linear models are restricted to the prediction of the onset condition and the frequencies of the self-excited instabilities. For a model to predict the oscillation amplitude, it must be nonlinear in nature, because it is the nonlinear effects that prevent the initial disturbance from growing unbounded with time, which includes the nonlinear development of shear layer and the amplitude-dependent dissipations. In this section, a nonlinear time-domain model for describing the flow-excited oscillations in a Helmholtz resonator is illustrated, and some results are discussed.

Consider a Helmholtz resonator with a rectangular opening in a thin but rigid facing plate subjected to a low Mach number flow U_g grazingly passing it, as shown in Fig. 5.20. The incoming flow is simplified to be uniform, ignoring the boundary layer effects. When the Reynolds number is high, viscosity is restricted to only causing flow separation at the edges of the opening. Note that vortex shedding at the downstream and side edges is ignored. It is also assumed that the shed vorticity is confined in a thin shear layer, i.e., vortex sheet. Because the streamwise variation of the shear layer motion in the opening region is dominant over that in the spanwise direction, a two-dimensional simplification is made for the near-field flow around the opening. Provided that frequency is sufficiently low, the flow in the vicinity of the opening that is acoustically compact is treated as incompressible.

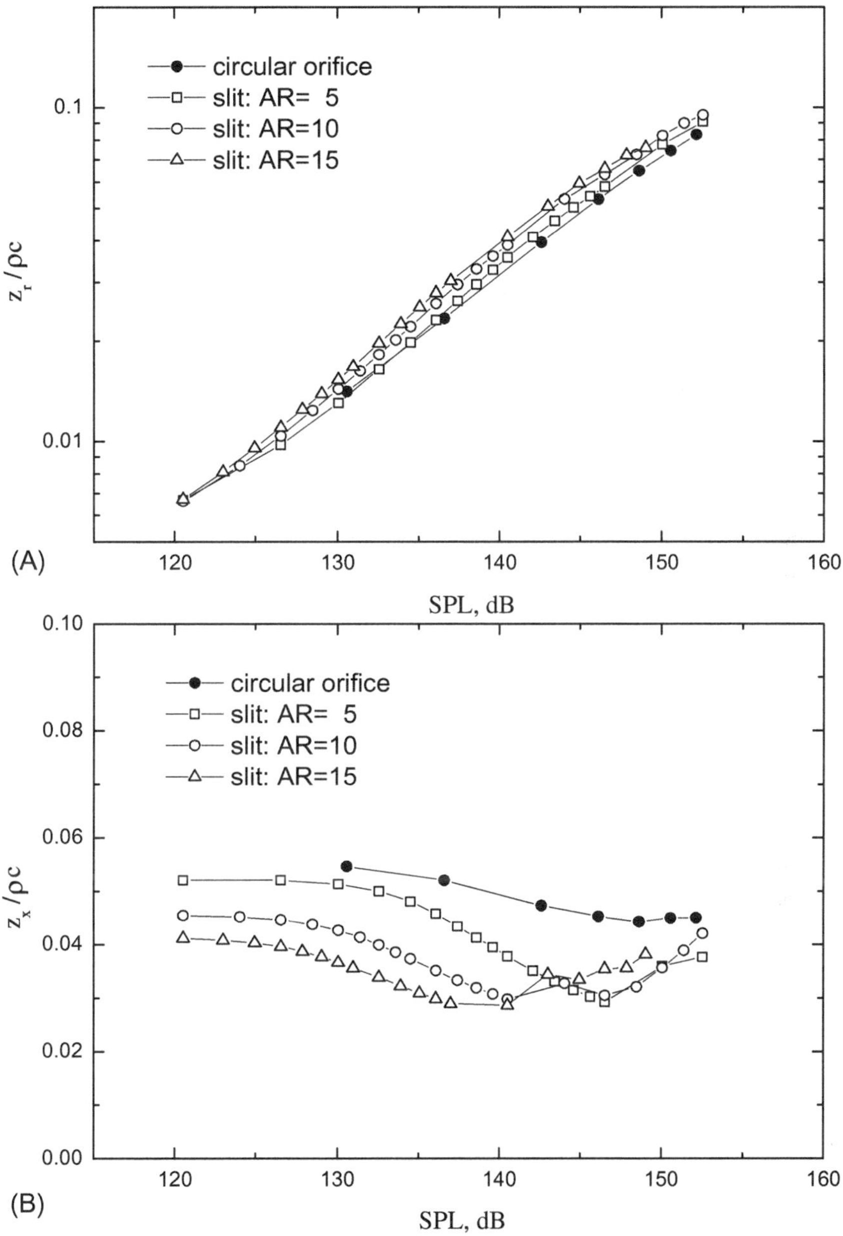

Fig. 5.19 Normalized specific acoustic impedance as a function of the applied SPL for different aspect ratios: (A) acoustic resistance and (B) acoustic reactance.

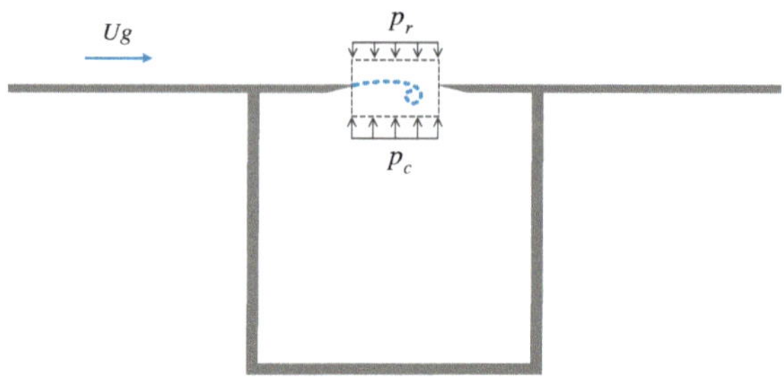

Fig. 5.20 Sketch of the model of flow-excited acoustic resonance in a cavity.

As the geometry of the resonator's opening is similar to the slit in Section 5.6.1, therefore, the same discrete vortex method is employed to simulate the evolution of the vortex sheet over the opening. Note that the grazing flow effect is not considered in the model of Section 5.6.1, which should be added to the velocity potential of the flow, and the velocity of the nth discrete vortex is expressed as:

$$w^*(z_n) = \left(\frac{2\zeta_n^2}{\zeta_n^2 - s^2}\right)\left[\frac{U_g}{2} + \frac{Q}{\pi\zeta_n} + \sum_{\substack{n'=1 \\ n'\neq n}} \frac{\Gamma_{n'}}{2\pi i}\frac{1}{\zeta_n - \zeta_{n'}} - \sum_{n'=1} \frac{\Gamma_{n'}}{2\pi i}\frac{1}{\zeta_n - \zeta_{n'}^*}\right] + \frac{i\Gamma_n}{\pi}\frac{\zeta_n s^2}{\left(\zeta_n^2 - s^2\right)^2}. \tag{5.123}$$

The oscillating fluid slug in the opening of the Helmholtz resonator behaving like an "air piston" receives pressure from the fluids on both sides, p_r and p_c, as shown in Fig. 5.20. Additionally, there is a viscous force, p_v, to resist the oscillation of the "air piston" in the opening region. Thus, from Eq. (5.117), the equation that relates the acoustic mode to the vortical motion in the opening region is obtained:

$$\frac{\Delta p}{\rho} = \frac{p_r(t)}{\rho} + \frac{p_v(t)}{\rho} - \frac{p_c(t)}{\rho} = -\frac{d\Gamma_t}{dt} - \frac{U_g^2}{2} + \frac{\partial}{\partial t}\left[\sum_{n=1}^{N} \frac{\Gamma_n}{\pi}\left(\frac{\pi}{2} - \arctan\frac{\xi_{1n}}{\xi_{2n}}\right)\right] + C(b)2s\frac{\partial v_{av}(t)}{\partial t}, \tag{5.124}$$

where $v_{av}(t)$ is the area-averaged velocity through the opening, $v_{av}(t) = Q(t)/2s$. It can be seen that the acoustic mode is related to the shear layer motion through Eq. (5.124), which is essentially the force-balancing relation between the two sides of the cavity opening.

Assuming that sound pressure is uniform over the cross-section of the cavity, the cavity pressure is given by the one-dimensional acoustic propagation model inside the cavity with the hardwall boundary condition at the bottom in the frequency domain:

$$P_c(\omega) = -i\rho_0 c \frac{A_o}{A_c} \cot\left(\frac{\omega D}{c}\right) V_{av}(\omega), \tag{5.125}$$

where $P_c(\omega)$ and $V_{av}(\omega)$ are the respective Fourier transform of the acoustic pressure at the open end of the chamber $p_c(t)$ and the area-averaged velocity through the opening $v_{av}(t)$, A_o is the opening area, A_c is the cross-section area of the cavity, and D is the depth of the cavity. The acoustic pressures due to the radiation and viscous effects can also be related to $V_{av}(\omega)$ in the frequency domain, which can be expressed as:

$$P_r(\omega) = \frac{\rho_0 A_o \omega^2}{2\pi c} V_{av}(\omega), \tag{5.126}$$

$$P_v(\omega) = \rho_0 \sqrt{8\nu\omega}\, V_{av}(\omega), \tag{5.127}$$

where $P_r(\omega)$ and $P_v(\omega)$ are the respective Fourier transform of $p_r(t)$ and $p_v(t)$, and ν is the dynamic viscosity coefficient. The present time-domain model can deal with the situation that there is more than one resonant peak in the spectrum of the flow-excited resonance. This is done by translating the frequency-domain impedance models in Eqs. (5.125), (5.126), and (5.127) into their time-domain equivalents.

Based on the discrete vortex method coupled to the cavity mode model, the unsteady vorticity motion around the opening and the excited acoustic pressure inside the cavity are computed in the time domain. The time-marching approach is similar to that in Section 5.6.1, and finally the frequency and amplitude characteristics of the resonance sound within the cavity are obtained. Besides, to validate the accuracy of present model, an experiment is conducted which can be inferred to Ref. [37] for more details.

The predicted and measured spectra of the oscillation pressures are presented in Fig. 5.21. Generally speaking, the present model reasonably predicts both the frequency and amplitude of the resonance without resorting to empirical parameters or adjustments. At low flow velocity, there are two harmonic oscillations with similar amplitudes at two frequencies: one is below and the other is above the natural frequency of the Helmholtz resonator. The dual-peak pattern results from the co-existence of the first and second hydrodynamic modes (H1 and H2). With the increase of U_g, the frequency of H1 is close to the natural frequency of the resonator (f_{HR}).

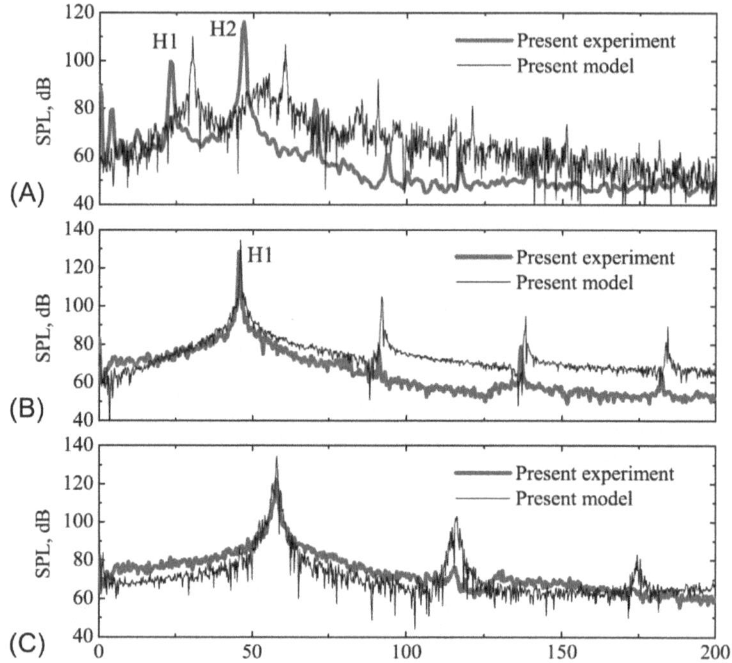

Fig. 5.21 Comparison of the present numerical and experimental results of the sound pressure spectrum at base of the cavity: (A) $U_g = 7.3\ \text{ms}^{-1}$, (B) $U_g = 15.28\ \text{ms}^{-1}$, and (C) $U_g = 26.96\ \text{ms}^{-1}$.

Correspondingly, there is a single sharp peak emerging near f_{HR} in the spectrum. As U_g further increases, the frequencies of H1 and H2 considerably exceed f_{HR}. Consequently, the coupling between the shear layer and the acoustic mode of the cavity declines, and the mode amplitude decreases to a lower level.

For the typical oscillation states, vortex shedding and evolution patterns at the opening of the cavity are shown in Fig. 5.22. When the grazing flow has a relatively low speed, the vortical flow at the opening exhibits the second hydrodynamic mode with the flow feature of a dual-vortex pattern. Then the two vortices evolve differently in the flow, which is the reason for the co-existence of H1 and H2 in the spectrum. At $U_g = 20\,\text{ms}^{-1}$, strong resonance occurs near the natural frequency of the resonator as the cavity acoustic mode is coupled with the first hydrodynamic mode characterized by a single vortex structure flapping over the opening. In other words, when the cavity resonance is strong, the vortical flow is controlled by both the large-scale coherent vortex structure and the flapping shear layer. As U_g further increases, the excitation frequency of the hydrodynamic modes far

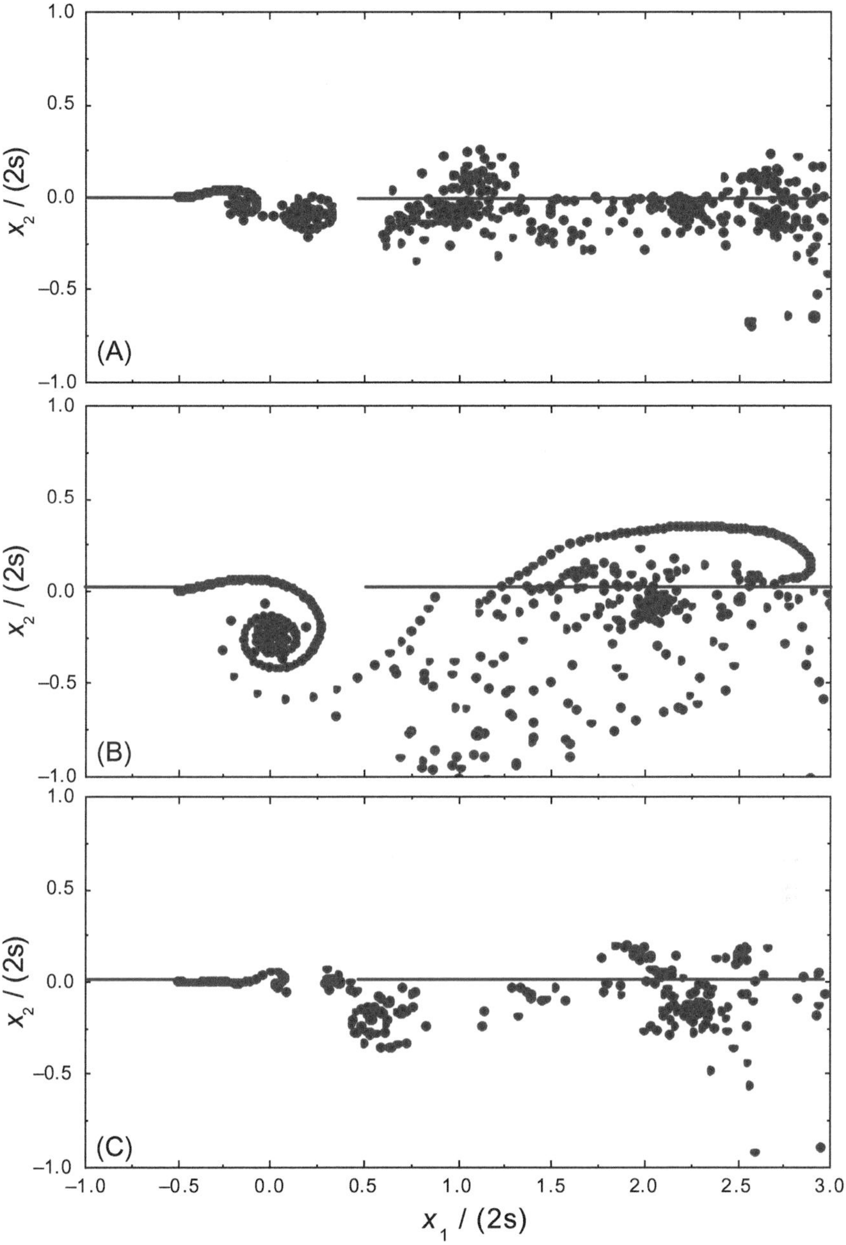

Fig. 5.22 Typical pattern of the vortex sheet over the opening: (A) $U_g = 6\ \mathrm{ms}^{-1}$, (B) $U_g = 20\ \mathrm{ms}^{-1}$, and (C) $U_g = 34\ \mathrm{ms}^{-1}$.

exceeds the natural frequency of the resonator; thus, the vortical-acoustic coupling is much weaker compared with the strong resonant case. The vortical flow shows significantly less unsteadiness, and there is no clear sign of concentrated vortex structure.

It has been recognized that not only the vortex evolution but also the vorticity shed from the separation point is also related to the cavity acoustic mode. To illustrate the effects of the acoustic mode on the vorticity shedding rate, the circulation of the nascent point vortex released from the upstream edge, Γ_1, is plotted as a function of time in Fig. 5.23. It is shown that Γ_1 has a mean value of $-\Delta t U_g^2/2$, resulting from the grazing flow, on which is superimposed an unsteady component caused by the acoustic mode. At $U_g = 6\,\text{ms}^{-1}$ and $U_g = 34\,\text{ms}^{-1}$, $\Gamma_1$fluctuates in relatively low amplitude compared with the mean value, indicating the predominant effect of grazing flow on the vorticity shedding. At $U_g = 20\,\text{ms}^{-1}$ where resonance occurs, however, the time-varying amplitude of Γ_1 is of the same order with the mean value. Thus, the nonlinear acoustic effect on the vorticity shedding is so significant that it should be taken into account.

The flow-excited oscillations can be detected over the range of U_g from around 2 ms^{-1} to over 30 ms^{-1} in both the present numerical and experimental data acquired at the cavity bottom, and their frequencies and amplitudes are plotted as a function of U_g in Fig. 5.24. It is shown that first and second hydrodynamic modes (H1 and H2) are simultaneously present when

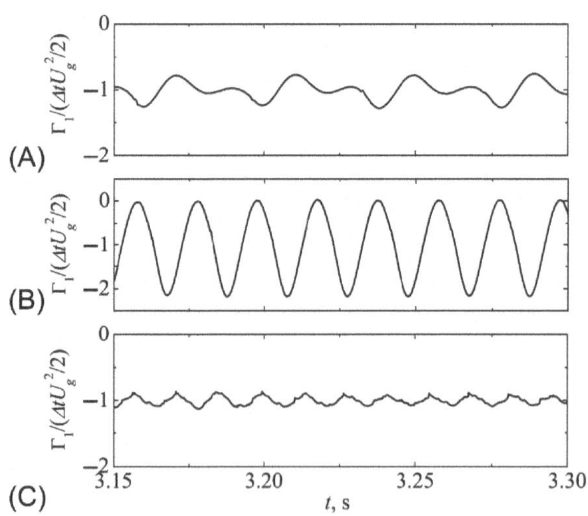

Fig. 5.23 Time history of the nascent vortex strength normalized by $\Delta t U_g^2/2$: (A) $U_g = 6\,\text{ms}^{-1}$, (B) $U_g = 20\,\text{ms}^{-1}$, and (C) $U_g = 34\,\text{ms}^{-1}$.

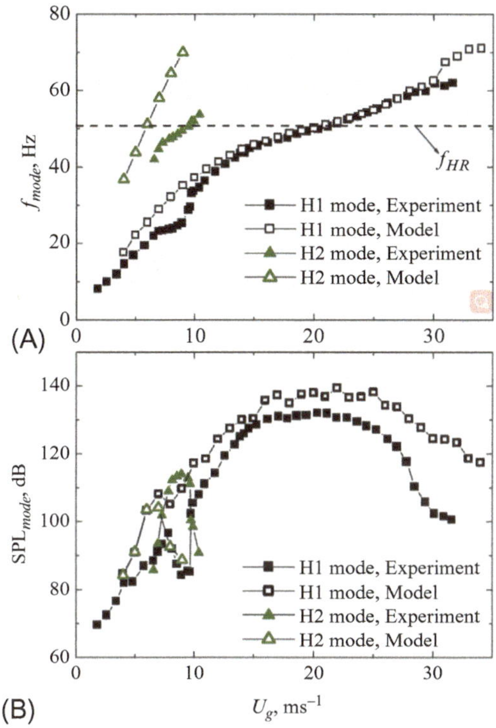

Fig. 5.24 Comparison of the present numerical and experimental results of the mode frequencies and amplitudes as a function of U_g: (A) frequency and (B) amplitude.

U_g is low and in the narrow range of about 4–10 ms^{-1}. However, above $U_g = 10$ ms^{-1}, only H1 is active. In this situation, the predicted frequency of the pressure oscillations agrees fairly well with the experimental data. Generally, the computed and measured oscillation frequencies increase linearly with U_g, except near the natural frequency of the resonator where the lock-in effect obviously reduces the slope of the increase. As for H2, the model overestimates the frequency and underestimates the amplitude. The difficulty in H2 prediction might be due to the higher order modes cause relatively weak oscillations and they are subjected to the influence of H1.

References

[1] A. Powell, Theory of vortex sound, J. Acoust. Soc. Am. 36 (1964) 177–195.

[2] P.E. Doak, Analysis of internally generated sound in continuous materials: 2. A critical review of the conceptual adequacy and physical scope of existing theories of

aerodynamic noise, with special reference to supersonic jet noise, J. Sound Vib. 25 (2) (1972) 263–335.
[3] P.E. Doak, Analysis of internally generated sound in continuous materials: 3. The momentum potential field description of fluctuating fluid motion as a basis for a unified theory of internally generated sound, J. Sound Vib. 26 (1) (1973) 91–120.
[4] M.S. Howe, Contributions to the theory of aerodynamic sound with application to excess jet noise and to theory of the flute, J. Fluid Mech. 71 (1975) 625–673.
[5] M.S. Howe, On the absorption of sound by turbulence and other hydrodynamic flows, IMA J. Appl. Math. 32 (1–3) (1980) 187–209.
[6] J. Yates, G. Sandri, Bernoulli enthalpy—a fundamental concept in the theory of sound, in: 2nd Aeroacoustics Conference, AIAA Paper 1975-439, 1975.
[7] J. Yates, Interaction with and production of sound by vortex flows, in: 4th Aeroacoustics Conference, AIAA Paper 1977-1352, 1977.
[8] F. Obermeier, On a new representation of aeroacoustic source distribution II. Two-dimensional model flows, Acta Acust. United Ac. 42 (1) (1979) 62–71.
[9] M.E. Goldstein, Turbulence generated by the interaction of entropy fluctuations with non-uniform mean flows, J. Fluid Mech. 93 (02) (1979) 209–224.
[10] W. Möhring, On vortex sound at low Mach number, J. Fluid Mech. 85 (1978) 685–691.
[11] T. Kamber, T. Minota, Acoustic wave radiated by head-on collision of two vortex rings, Proc. Roy. Soc. A 286 (1983) 277–308.
[12] M.S. Howe, Trailing edge noise at low Mach numbers, J. Sound Vib. 225 (2) (1999) 211–238.
[13] M.S. Howe, Sound generated by fluid-structure interactions, Comput. Struct. 65 (3) (1997) 433–446.
[14] M.C. Welsh, A.N. Stokes, R. Parker, Flow-resonant sound interaction in a duct containing a plate, part I: semi-circular leading edge, J. Sound Vib. 95 (3) (1984) 305–323.
[15] A. Mohany, S. Ziada, Numerical simulation of the flow-sound interaction mechanisms of a single and two-tandem cylinders in cross-flow, J. Press. Vessel. Technol. 131 (3) (2009) 135–144.
[16] B.T. Tan, M.C. Thompson, K. Hourigan, Sources of acoustic resonance generated by flow around a long rectangular plate in a duct, J. Fluids Struct. 18 (6) (2003) 729–740.
[17] A. Mohany, D. Arthurs, M. Bolduc, M. Hassan, S. Ziada, Numerical and experimental investigation of flow-acoustic resonance of side-by-side cylinders in a duct, J. Fluids Struct. 48 (2014) 316–331.
[18] S.A.T. Stoneman, K. Hourigan, A.N. Stokes, M.C. Welsh, Resonant sound caused by flow past two plates in tandem in a duct, J. Fluid Mech. 192 (1988) 455–484.
[19] M.S. Howe, Theory of Vortex Sound, Cambridge University Press, 2003.
[20] C. Cheong, P. Joseph, Y. Park, S. Lee, Computation of aeolian tone from a circular cylinder using source models, Appl. Acoust. 69 (2) (2008) 110–126.
[21] Y.S.K. Liow, B.T. Tan, M.C. Thompson, K. Hourigan, Sound generated in laminar flow past a two-dimensional rectangular cylinder, J. Sound Vib. 295 (1–2) (2006) 407–427.
[22] M.A. Langthjem, M. Nakano, A numerical study of the hole-tone phenomenon subjected to non-axisymmetric shape perturbations of the jet nozzle, Theor. Comput. Fluid Dyn. 29 (3) (2015) 127–153.
[23] R.R. Clements, An inviscid model of two-dimensional vortex shedding, J. Fluid Mech. 57 (02) (1973) 321–336.
[24] Z. Hong, X. Wang, X. Jing, X. Sun, Frequency lock-in mechanism in flow-induced acoustic resonance of a cylinder in a flow duct, J. Fluid Mech. 884 (2020) A42
[25] J. Dominguez, Boundary Elements in Dynamics, Wit Press, 1993.

[26] X. Jing, X. Sun, Discrete vortex simulation on the acoustic nonlinearity of an orifice, AIAA J. 38 (9) (2000) 1565–1572.
[27] U. Ingard, H. Ising, Acoustic nonlinearity of an orifice, J. Acoust. Soc. Am. 42 (1) (1967) 6–17.
[28] U. Ingard, S. Labate, Acoustic circulation effects and the nonlinear impedance of orifices, J. Acoust. Soc. Am. 22 (2) (1950) 211–218.
[29] U. Ingard, Nonlinear distortion of sound transmitted through an orifice, J. Acoust. Soc. Am. 48 (1) (1970) 32–33.
[30] G.B. Thurston, L.E. Hargrove, B.D. Cook, Nonlinear properties of circular orifices, J. Acoust. Soc. Am. 29 (9) (1957) 992–1001.
[31] M. Salikuddin, K.K. Ahuja, Acoustic power dissipation on radiation through duct terminations: experiments, J. Sound Vib. 91 (4) (1983) 479–502.
[32] I. Sugiaka, S.E. Widnall, A panel method study of vortex sheets with special emphasis on sheets of axisymmetric geometry, (1985)NASA CR-177365.
[33] X. Jing, X. Sun, Effect of grazing flow on the acoustic impedance of an orifice, AIAA J. 39 (8) (2000) 1478–1484.
[34] M.S. Howe, M.I. Scott, S.R. Sipcic, The influence of tangential mean flow on the Rayleigh conductivity of an aperture, Proc. R. Soc. Lond. A Math. Phys. Sci. 452 (1996) 2303–2317.
[35] X. Jing, X. Sun, Effect of plate thickness on impedance of perforated plates with bias flow, AIAA J. 38 (9) (2000) 1573–1578.
[36] M.S. Howe, On the theory of unsteady high Reynolds number flow through a circular aperture, Proc. R. Soc. Lond. A Math. Phys. Sci. 366 (1979) 205–223.
[37] X. Dai, X. Jing, X. Sun, Flow-excited acoustic resonance of a Helmholtz resonator: discrete vortex model compared to experiments, Phys. Fluids 27 (5) (2015) 057102.
[38] X. Dai, X. Jing, X. Sun, Vortex shedding and its nonlinear acoustic effect occurring at a slit, AIAA J. 49 (12) (2011) 2684–2694.
[39] C.E. Brown, W.H. Michael, Effect of leading edge separation on the lift of a delta wing, J. Aeronaut. Sci. 21 (1954) 690–706.
[40] N. Forestier, L. Jacquin, P. Geffroy, The mixing layer over a deep cavity at high-subsonic speed, J. Fluid Mech. 475 (2003) 101–145.

CHAPTER 6

Sound generation, propagation, and radiation in/from an aeroengine nacelle

6.1 Introduction

As the main noise source of modern commercial aircraft, the acoustic design of the aeroengine has received more and more attention along with increasingly stringent requirements for environmental protection [1]. Therefore, how to further reduce aeroengine noise by all means has long been a problem of great importance. It is well known that the best way is to control the noise sources themselves [2, 3] at the design phase of the aeroengine and suppress its propagation in the ducts [4–6]. In fact, it is noted that aeroengine noise is generated by either an internal or external source. The externally generated noise is due mainly to jet exhaust. The internally generated noise is due primarily to the rotating turbomachinery blades and the combustion process. To achieve lower specific fuel consumption, future aircraft engines are expected to have bypass ratios in the range of 12–15 [1]. This results in a reduction of the jet noise to relatively insignificant proportions but leads to an increase of the fan noise. If there is no acoustic treatment to control the noise propagation, some tests show that the noise level of such engines is greater than the noise level of low bypass engines and consists of a broadband spectrum and high-pitched discrete frequency components [7]. The internal flow field past the blades is generally unsteady due to the presence of upstream blades and vanes, and viscous wakes. The interaction of this unsteady flow and the moving blades produces upstream and downstream traveling pressure waves (acoustic modes) with blade-passing frequency (BPF) as their frequency characteristics. Some of these modes decay naturally (cut-off modes), whereas the others (cut-on modes) need to be suppressed by acoustically treating the engine ducts [8]. The latter is just the main objective of the acoustic design of an aeroengine nacelle. To achieve noise reductions in the nacelle as much as possible, the acoustic treatment must be optimized, and in addition, how to include the effect of the interaction

Fundamentals of Aeroacoustics with Applications to Aeropropulsion Systems
https://doi.org/10.1016/B978-0-12-408069-0.00006-3

between the sound source and acoustic treatment on noise attenuation is also a problem of concern.

It is noted that considerable work has been conducted for the prediction of sound radiation from a lined duct over the last decade, including the boundary integral equation method (BIEM) or boundary element methods (BEMs) [9–13], the finite element methods (FEMs) [14–18], and the numerical simulation method based on computational aeroacousitc (CAA) technique [19–22]. These methods are playing diverse roles at different stages of the acoustic design. For the determination of final design parameters, it is necessary to use the numerical simulation tools as accurately as possible to check the results, provided that the relevant computing cost is affordable. However, for the purpose of preliminary design considerations, an acoustic engineer must conduct a large number of parametric studies. In this situation, it is required to provide fast, useful, and reliable tools to acoustic engineers.

The purpose of this part is to introduce the fundamental knowledge regarding the methods of determining the sound attenuation, source-acoustic treatment interaction, and sound radiation from the inlet and exhaust of an aeroengine. Especially, emphasis will be placed on how to use a kind of new strategy, the transfer element method (TEM), to realize the prescribed noise reduction objective at the preliminary phase of the acoustic design of an aeroengine nacelle.

6.2 Basic theory of sound propagation in ducts

In the theory of fundamental acoustics, the sound field in a duct can be expressed as the superposition of multiple modal acoustic waves, and each modal acoustic wave has a particular cut-off frequency, which is the characteristic of the duct acoustics. In aeroacoustics, the theory of sound propagation in ducts is much more complicated than that in fundamental acoustics because of the inclusion of flow and the consideration of the characteristics of the sound source in the engine. For this reason, we first consider the basic law of sound propagation inside the engine. In this part, we first discuss the sound propagation in ducts with flow and introduce the concept of high-order mode on the basis of the theory of duct acoustics in the fundamental acoustics.

As shown in Fig. 6.1, a straight duct with an arbitrary cross-section shape is considered. The duct wall is rigid, and there is a uniform flow in the direction of the z-axis. Assuming no sound source, the sound field in the duct satisfies the wave equation

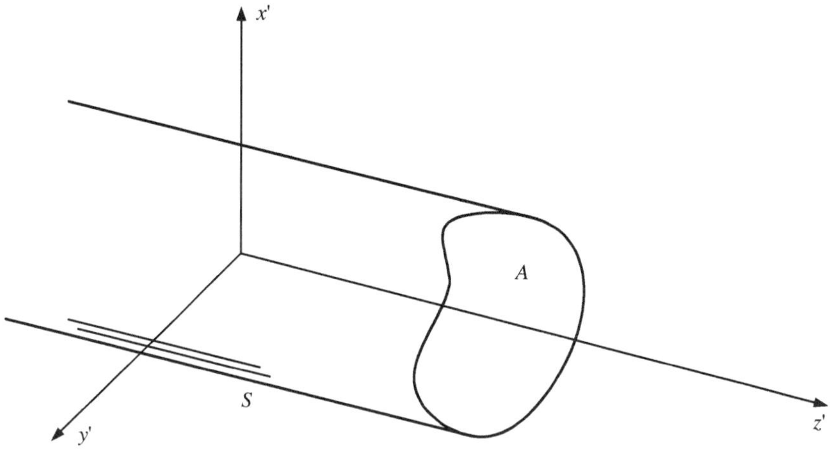

Fig. 6.1 A straight duct with an arbitrary section shape.

$$\nabla^2\widetilde{p'} - \frac{1}{c_0^2}\frac{D_0^2\widetilde{p'}}{Dt^2} = 0, \tag{6.1}$$

where

$$\frac{D_0}{Dt} = \frac{\partial}{\partial t} + U\frac{\partial}{\partial z}. \tag{6.2}$$

The wall boundary condition is:

$$\frac{\partial\widetilde{p'}}{\partial n} = 0, \tag{6.3}$$

where n represents outer normal direction of wall. Considering a single frequency, i.e., $\widetilde{p'} = p' \cdot e^{i\omega t}$, Eq. (6.1) is changed into:

$$\nabla^2 p' + \left(k_0^2 p' - 2iMk_0\frac{\partial p'}{\partial z} - M^2\frac{\partial^2 p'}{\partial z^2}\right) = 0, \tag{6.4}$$

where $k_0 = \frac{\omega}{c_0}$, $M = \frac{U}{c_0}$. Assuming

$$p'(x, y, z) = X(x) \cdot Y(y) \cdot Z(z), \tag{6.5}$$

we obtain three ordinary differential equations (ODEs) of independent coordinates as follows

$$\begin{cases} \dfrac{d^2X(x)}{dx^2} + k_x^2 \cdot X(x) = 0 \\ \dfrac{d^2Y(y)}{dy^2} + k_y^2 \cdot Y(y) = 0 \\ (1 - M^2)\dfrac{d^2Z(z)}{dz^2} - 2iMk_0\dfrac{dZ(z)}{dz} + k_z^2 Z(z) = 0 \end{cases}, \tag{6.6}$$

where k_x, k_y, k_z are three undetermined coefficients, which satisfy:

$$k_x^2 + k_y^2 + k_z^2 = k_0^2. \tag{6.7}$$

The solutions of ODEs in x and y directions depend on the cross-section shape of the duct. Considering the general shape of cross-sections of the aeroengine duct, the circular duct will be discussed here as an example. With the introduction of cylindrical coordinates (r, φ, z), Eq. (6.6) changes into:

$$\begin{cases} \dfrac{d^2R(r)}{dr^2} + \dfrac{1}{r}\dfrac{dR(r)}{dr} + \left(k_r^2 - \dfrac{m^2}{r^2}\right)\cdot R(r) = 0 \\ \dfrac{d^2\Phi(\varphi)}{d\varphi^2} + m^2\cdot\Phi(\varphi) = 0 \\ (1-M^2)\dfrac{d^2Z(z)}{dz^2} - 2iMk_0\dfrac{dZ(z)}{dz} + k_z^2 Z(z) = 0 \end{cases}, \tag{6.8}$$

where $k_z^2 = k_0^2 - k_r^2$. Because the equation associated with function Φ is required to meet the circumferential periodicity condition $\Phi(\varphi) = \Phi(\varphi + 2\pi)$, the parameter m must be an integer. According to the equation, we obtain:

$$\Phi(\varphi) = A_\varphi e^{im\varphi}. \tag{6.9}$$

The equation related to coordinate r is a typical m-order Bessel equation, and the general solution can be written as:

$$R_m(k_r r) = A_r \cdot J_m(k_r r) + B_r \cdot N_m(k_r r), \tag{6.10}$$

where $J_m(k_r r)$ and $N_m(k_r r)$ are Bessel functions of the first and second kinds. This solution is suitable for ducts of circular or annular cross-section. For a circular duct, $B_r = 0$ as the finite-value boundary condition at $r = 0$.

Assuming that we have an infinitely long duct without axial reflection, the solution of the equation in z direction is:

$$Z(z) = A_z \cdot e^{i\gamma_z z}. \tag{6.11}$$

Based on the previous three equations, the general expression of a sound field in circular and annular ducts can be obtained as:

$$\begin{aligned} \widetilde{p'}(r, \varphi, z) &= AJ_m(k_r r)e^{im\varphi}e^{i\gamma_z z}\cdot e^{i\omega t} \quad \textit{for circular duct}, \\ \widetilde{p'}(r, \varphi, z) &= AR_m(k_r r)e^{im\varphi}e^{i\gamma_z z}\cdot e^{i\omega t} \quad \textit{for annular duct}, \end{aligned} \tag{6.12}$$

where k_r, m, and γ_z are radial, circumferential, and axial wavenumbers, respectively. (Note that γ_z is the same quantity as k_z.) The relationship between k_r and γ_z is:

$$\gamma_z = \frac{Mk_0 \pm \sqrt{M^2k_0^2 + \beta^2k_r^2}}{\beta^2}. \tag{6.13}$$

The expression of Eq. (6.12) is then the general solution of a convective wave equation in a straight duct. Note that the wave equation does not account for the source term, so the expression gives only the propagation characteristics of sound waves in a duct. Considering the boundary condition of a rigid wall in Eq. (6.3), we obtain:

$$\frac{d}{dr}R_m(k_r r) = R'_m(k_r r) = 0. \tag{6.14}$$

According to the characteristics of Bessel function, there are infinite number of zeros of its derivative. Denote these zeros and the corresponding axial wavenumbers as k_{mn} and $\gamma_{mn}^{\pm}$, respectively. Here, the superscript "+" means the propagating direction of sound wave is consistent with the positive direction of the z-axis whereas "−" means the opposite. Eq. (6.13) is then expressed as:

$$\gamma_{mn}^{\pm} = \frac{Mk_0 \pm \sqrt{k_0^2 - \beta^2k_{mn}^2}}{\beta^2}. \tag{6.15}$$

6.3 Computational approaches for duct acoustics

6.3.1 Sound propagation in an aeroengine nacelle

As shown in Fig. 6.2, to efficiently control the noise propagation and radiation from an aeroengine nacelle, various acoustic treatments have to be installed in all available places, such as the inlet and outlet of the nacelle, and even the duct surface between the rotor and stator. In the past several decades, many sound propagation models have been developed to calculate noise attenuation. In general, different acoustic treatments are used to realize an optimal attenuation for various sound sources with broad frequency ranges. Obviously, different liners will interact with each other due to discontinuous impedance distribution, and the presence of the rotor and stator as a sound source or the reflection of the nozzle could also interact with the acoustic liner. In fact, a good model for the acoustic design of the aeroengine nacelle is to include such interactions as much as possible to realize a more accurate description to the special sound environment as shown in Fig. 6.2. This is indeed not an easy task. Fig. 6.3 just shows the complex process toward this objective.

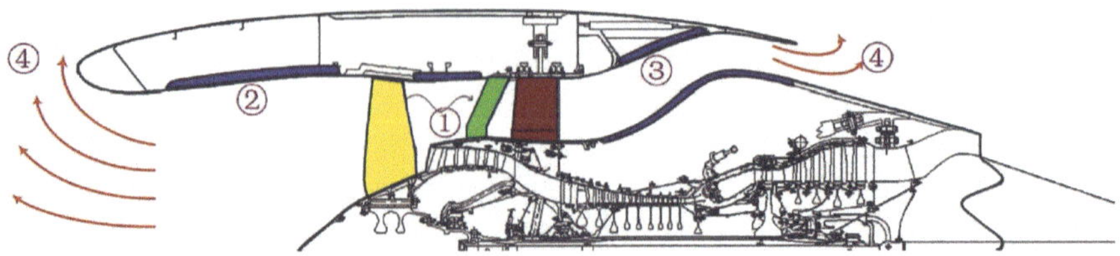

Fig. 6.2 Schematic of simplified aeroengine acoustic model: ① rotor/stator interaction noise; ② inlet liner; ③ outlet liner; ④ noise radiation.

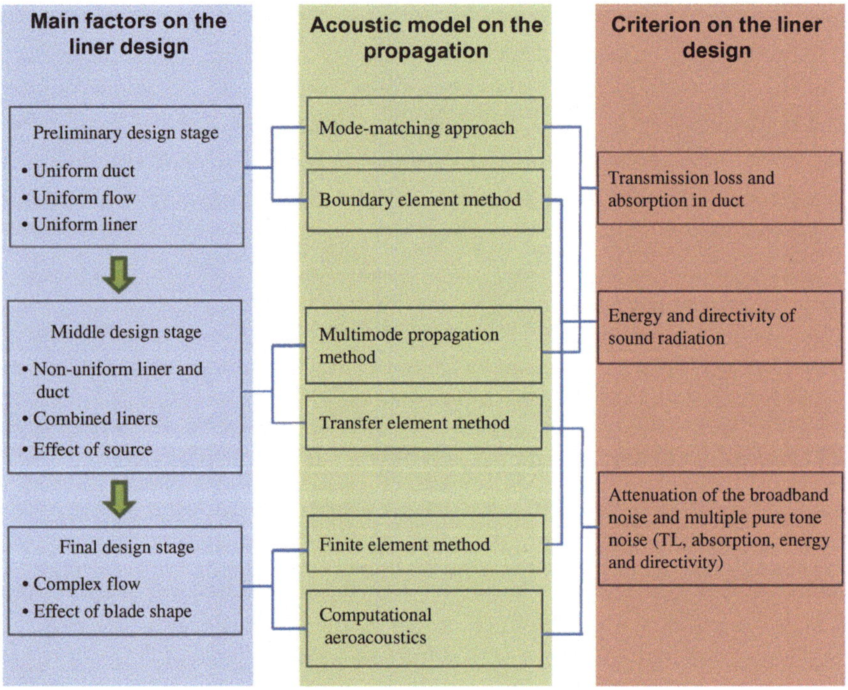

Fig. 6.3 Block diagram of requirements of liner design.

At the preliminary stage of acoustic liner design, to obtain the structure parameters of a liner as soon as possible, it is necessary to introduce a simplified calculation model with the assumptions of mean flow, infinite duct, and strength-fixed sound sources. At this stage, some famous simplified models, such as the Mode-Matching Approach (MMA) and BIEM can be used to estimate sound attenuation for different liner combinations. However, these design models can still not satisfy the increasing requirement of acoustic treatment design for the modern turbofan aeroengine. Therefore, it is indeed required to have a kind of new model to consider the effect of non-uniform duct, complex flow, and the interaction between sound source and various liners on sound attenuation. On the other hand, a more elaborate numerical calculation method like CAA is also quite important due to its capability to include any complex factors. However, because a huge amount of time and computational costs are inevitable in this regard, it is very difficult to use it to conduct a parametric study with the requirement of fast implementation. So, from a practical point of view, a fast method that can be used to optimize all design parameters of acoustic treatment with fewer restrictions is still required.

In addition, the impedance model is also an important factor for liner design, but in this chapter, we only assume that the impedance description is available, such as Ingard's model [23] and Jing et al.'s model [24–27]. So, an emphasis will be placed on the various propagation models in ducts and especially on the connection between the acoustic attenuation and liner structure parameters.

As discussed earlier, to satisfy the requirement of acoustic liner design for a modern turbofan nacelle, we need a new model to solve the following problems: (a) the model can be used in the preliminary design stage for calculating the sound propagation in the duct and sound radiation from the inlet and exhaust; (b) the model must include more effects of liner design such as non-uniform flow and varying cross-sectional duct; (c) the model considers the effect of the interaction between the sound sources and acoustic treatments. For these three purposes, we will first briefly introduce several famous existing models like the MMA and Boundary Integral Method, and then focus on expounding how the TEM was developed to solve the sound propagation problem in an aeroengine nacelle with these three main features.

6.3.1.1 Mode-matching approach (MMA)

Acoustic engineers only designed the acoustic treatment via calculating the eigenvalue of an infinitely long acoustic liner before MMA was published [28–30]. In 1974, Zorumski [31] developed the MMA to consider the sound propagation in multisectioned ducts that are interconnected, which was committed to solve all acoustic problems including sound generation in, transmission through, and radiation from a duct, as shown in Fig. 6.4. However, the calculation of sound propagation based on the eigenvalue method always meets with difficulties associated with the convergence of eigenvalue

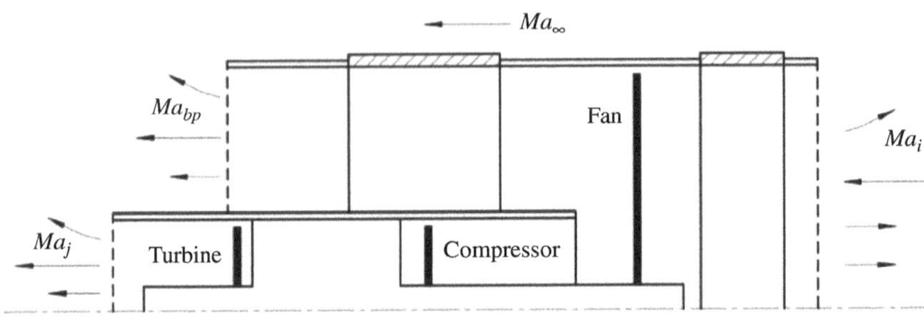

Fig. 6.4 Schematic of engine acoustic model.

iteration due to the solution of a complex transcendental equation, whereas this error will severely affect the computational accuracy of sound propagation. In addition, the sound source on the blades was regarded as the spinning point source without considering the interaction between the sound field and the unsteady blade loading. In spite of this, compared with the computational model in an infinite duct, MMA provided many advantages over the previous sound propagation models.

Aimed at establishing a matrix equation for sound propagation, this method divides the acoustic pressure into a multimode sum. As shown in Fig. 6.5, the infinite duct solution must be generalized to account for the effects of some finite sections, and relationships must be developed to account for the acoustic coupling between these sections. First, to define the acoustic field in terms of wave amplitudes, we can write sound pressure by the following equation:

$$p'_m(r,z) = \sum_{\mu=1}^{\infty} \left[A^{+j}_{m\mu} \psi^{+j}_m (k_{m\mu} r) e^{i\gamma^{+j}_{m\mu}(z-z^j)} - A^{-j}_{m\mu} \psi^{-j}_m (k_{m\mu} r) e^{i\gamma^{-j}_{m\mu}(z-z^j)} \right], \quad (6.16)$$

where $A^{+j}_{m\mu}$ and $A^{-j}_{m\mu}$ are the amplitudes of the acoustic modes at z^j. $\psi_{m\mu}(r)$ denotes the eigenfunction of the cross-section in the duct and $k_{m\mu}$ is the

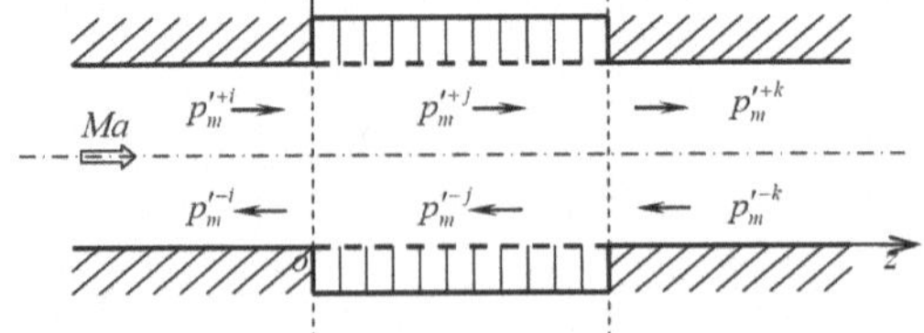

Fig. 6.5 Transmission wave and reflection wave in duct by utilizing the Mode-Matching Approach.

eigenvalue. $\gamma_{m\mu}$ denotes the axial wavenumber. Therefore, the wave generation, transmission, and reflection effects at the interface $j-k$ can be represented by matrix equations as:

$$\begin{aligned}\left\{A_{m\mu}^{+k}\right\}&=\left[T_{m\mu\nu}^{+k+j}\right]\left\{A_{m\nu}^{+j}\right\}+\left[R_{m\mu\nu}^{+k-k}\right]\left\{A_{m\nu}^{-k}\right\}+\left\{Q_{m\mu}^{+k}\right\},\\ \left\{A_{m\mu}^{-j}\right\}&=\left[T_{m\mu\nu}^{-j-k}\right]\left\{A_{m\nu}^{-k}\right\}+\left[R_{m\mu\nu}^{-j+j}\right]\left\{A_{m\nu}^{+j}\right\}+\left\{Q_{m\mu}^{-j}\right\},\end{aligned} \tag{6.17}$$

where the factors R and T are called the reflection coefficient and transmission coefficient of interface, respectively. $\{Q_{m\mu}^{+k}\}$ indicates the source terms for the forward-travelling waves in section k. Utilizing the matching condition on the interface $j-k$, we can obtain:

$$\begin{aligned}\left[T_{m\mu\nu}^{+k+j}\right]&=\left[W_{m\mu\nu}^{-j+k}\right]^{-1}\left[W_{m\mu\nu}^{-j+j}\right],\\ \left[R_{m\mu\nu}^{+k-k}\right]&=\left[W_{m\mu\nu}^{-j+k}\right]^{-1}\left[W_{m\mu\nu}^{-j-k}\right],\end{aligned} \tag{6.18}$$

where $[W_{m\mu\nu}^{-j+k}]$ is the function of $[I_{m\mu\nu}^{-j+k}]$, and

$$I_{m\mu\nu}^{-j+k}=\int_a^b r\psi_{m\mu}^{j}(r)\psi_{m\nu}^{k}(r)\,dr. \tag{6.19}$$

Obviously, to obtain the transmission and reflection coefficient, the most important thing is to calculate the eigenvalue and eigenfunction of these two adjacent sections. It is easy to obtain the eigenvalue only if there is a hard wall in the duct. However, if the impedance boundary conditions are different between these two elements, it becomes very difficult to solve the eigenvalue in Eq. (6.19) in which the modes are not orthogonal. The reason is that we have described the wall boundary condition as a transcendental equation with the help of the concept of acoustic impedance instead of directly considering the sound propagation inside the liner. For example, in the rectangular duct with a one-side liner, we can set up the impedance equation as:

$$\tan(k_x l)=\frac{-ik_0\beta}{k_x}\left(1-Ma\frac{k_z}{k_0}\right)^2, \tag{6.20}$$

where k_x and k_z refer to the wavenumbers for each duct propagation mode in x and z directions. $k_0=\omega/c_0$ and $\beta^2=1-Ma^2$, where ω is angular frequency, c_0 is local sound speed, and Ma is the Mach number of uniform flow. Unfortunately, there are many numerical instabilities and mode jump phenomena under these kind of equations. This eigenvalue equation derived from the impedance condition is usually solved by Eversman's integration method [32, 33], in which a nonlinear ODE is adopted via introducing a parameter perturbation to the transcendental algebraic equation. There is singularity at the lowest-order mode, i.e., the plane wave, in the ODE derived by Eversman's method for both rectangular and circular ducts. This problem has been solved by a homotopy method developed by Sun et al. [34] until 2007. So far, MMA is still one of the most efficient methods of liner design.

As mentioned earlier, in Zorumski's model, he hopes to solve all the acoustic problems involved in sound generation in, transmission through, and radiation from duct systems. In this model, however, it is really difficult to give a unified equation including sound source and liner because the eigenvalue and eigenfunction of a soft-walled duct have been used. On the other hand, there are some inextricable problems to calculate the sound radiation from the duct as it is required to input the mode reflection coefficient of the engine inlet or exhaust duct termination planes via both theoretical method and experimental method [35, 36]. Moreover, the MMA only considers very simple conditions, i.e., uniform mean flow and uniform duct. As shown in Fig. 6.3, using this model, we cannot calculate the sound radiation from the duct without mode sound reflection coefficient at the end of the duct. To overcome these difficulties, it is required to set up a combined equation including both sound propagation in ducts and sound radiation from the inlet and outlet of ducts to simultaneously obtain the acoustic radiation field.

6.3.1.2 Boundary integral equation method (BIEM)

To better understand the effect of aeroengine noise on the acoustic environment in the cabin and airport, acoustic engineers are eager to predict the far-field acoustic characteristics including both amplitude and directivity of radiation.

At first, the Wiener-Hopf method was used to obtain an analytical solution by assuming that the duct has a prescribed length-diameter ratio [37, 38]. However, such approximation is too far away from a practical application. In 1995, Myers [9, 10] presented the boundary integral equation of sound radiation from a duct with a spinning point source and liner by applying Ffowcs Williams-Hawkings equation, which assumed that length of a liner is equal to the duct. Therefore, Myers's method may also be considered to a kind of approximation to the practical problem. Later, Dunn et al. [11, 39] developed different kinds of BEM models, which considered the effect of the multisegment liner with arbitrary impedance distribution in the duct. The boundary integral equation is a fast method because it establishes a unified equation including both duct sound propagation and sound radiation. In this method, it is unnecessary to calculate the eigenvalue of impedance wall, so BEM has become one of the most effective and powerful tools of the acoustic design of an aeroengine nacelle.

After that, Yang and Wang [40] developed a direct boundary element, which could calculate the effect of non-uniform impedance in circumferential direction and the scarf inlet engine. This method overcomes some calculating difficulties due to the complex singularity treatment on the impedance wall and is easily used to analyze the effect of various geometrical and aerodynamic parameters on the sound field.

In this model, the acoustic fields are split into two parts: one is inside of the duct, the other one is outside of the duct, and these two parts are connected with the continuity condition on the duct terminations. As shown in Fig. 6.6, the non-compact source and non-uniform liner can be included.

The governing equation without mean flow in frequency domain can be written as:

$$\left[k_0^2+\frac{1}{r}\frac{\partial}{\partial r}\left(r\frac{\partial}{\partial r}\right)+\frac{1}{r^2}\frac{\partial^2}{\partial \varphi^2}+\frac{\partial^2}{\partial z^2}\right]p'=0. \tag{6.21}$$

The Green function of the three-dimensional Helmholtz equation is:

$$G=\frac{e^{-ikR}}{4\pi R}, \tag{6.22}$$

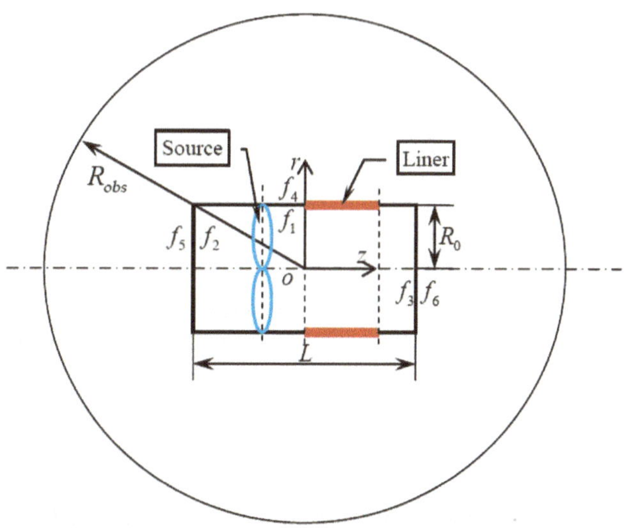

Fig. 6.6 Schematic of sound radiation from a duct.

where

$$R=\sqrt{r^2-r'^2-2rr'\cos(\varphi-\varphi')+(z-z')^2}. \tag{6.23}$$

So we can obtain the total acoustic pressure:

$$p_q'=\int\left[p_q(\vec{y})\frac{\partial G(\vec{x},\vec{y})}{\partial n}-G(\vec{x},\vec{y})\frac{\partial p_q(\vec{y})}{\partial n}\right]dS_q+p_{i,q}', \tag{6.24}$$

where $q=1$ denotes the acoustic field inside the duct, and the integral surfaces are composed of duct interior surface S_1 including $f_1=f_2=f_3=0$. The $p_{i,1}'=p_i'$ expresses the sound source inside the duct. $q=2$ denotes the acoustic field outside the duct, and the integral surfaces are composed of duct exterior surface S_2 including $f_4=f_5=f_6=0$. For the impedance wall and the hard wall, the boundary condition can be described as:

$$\frac{\partial p}{\partial n}=0\ ,\ \frac{\partial p}{\partial n}=\frac{ik}{Z_{\text{im}}}p, \tag{6.25}$$

respectively. Here, Z_{im} is the impedance on the duct wall. According to the continuity condition of pressure and its derivative on the inlet and outlet terminations, the integral equation can be given by the following forms:

$$\begin{cases} p_t'\big|_{(f_2=0)}=p_t'\big|_{(f_5=0)} \\ p_t'\big|_{(f_3=0)}=p_t'\big|_{(f_6=0)} \\ \left.\dfrac{\partial p_t'}{\partial n}\right|_{(f_2=0)}=\left.\dfrac{\partial p_t'}{\partial n}\right|_{(f_5=0)} \\ \left.\dfrac{\partial p_t'}{\partial n}\right|_{(f_3=0)}=\left.\dfrac{\partial p_t'}{\partial n}\right|_{(f_6=0)} \end{cases}, \tag{6.26}$$

We can obtain the source items on the interior and exterior surfaces of the duct by solving these integral equations and then obtain the acoustic field by using the integral expression Eq. (6.24). For the situation with mean flow inside and outside the duct, we assume that the medium is at rest and the duct is moving in the axial direction with Mach number $-M_a$.

It is easy to see that BIEM essentially serves to solve the scattering problem of a finite-length duct in the free field. As a fast method, BIEM has been used for practical liner design because it can simultaneously obtain both the acoustic attenuation and sound radiation results, as shown in Fig. 6.3. As mentioned earlier, however, this model cannot deal with more complex boundary impedance, such as a nonlocally reacting liner, which has been indeed considered for reducing broadband noise, as the acoustic treatment has been regarded as a fixed derivative condition. Besides, it is unable to investigate the interactions between different kinds of liners, such as local and nonlocal reaction liners, and between sound sources and acoustic treatments, as shown in Fig. 6.29. To avoid this trouble, it is necessary to solve the sound propagation inside the duct by another method, such as TEM or FEM.

6.3.1.3 Finite element method (FEM)

Prediction and control of sound propagation in ducts of varying cross-sectional area has been an important topic in acoustics for a long time. There are many situations occurring in practice including the liner design in the aeroengine nacelle where sound propagates in such non-uniform ducts. Both MMA and BIEM mentioned before cannot deal with this problem without making certain approximations. However, many numerical methods have been developed to handle these difficult problems, such as the FEM [16, 17, 41, 42] and CAA [19–22] methods. The advantages of these two methods depend on what the simulation issues are. Especially, for a sound propagation problem in a duct with non-uniform flow, the linearization of the basic flow equation is generally inevitable.

Now, considering the sound propagation in a duct, the governing equation can be a linearized Euler equation, which can be written as:

$$\frac{\partial \widetilde{\mathbf{q}}}{\partial t} + \widetilde{A}\frac{\partial \widetilde{\mathbf{q}}}{\partial z} + \widetilde{B}\frac{\partial \widetilde{\mathbf{q}}}{\partial r} + \widetilde{C}\frac{\partial \widetilde{\mathbf{q}}}{\partial \varphi} + \widetilde{D}\widetilde{\mathbf{q}} = \mathbf{0}, \tag{6.27}$$

where $\widetilde{\mathbf{q}} = \left[\widetilde{\rho'}, \widetilde{u'}, \widetilde{v'}, \widetilde{w'}, \widetilde{p'}\right]^{\mathrm{T}}$ is the acoustic perturbation vector. It is noted that Eq. (6.27) has been used to simulate a duct acoustic problem in a time domain with complex flow [22].

If we only want to know the features of an acoustic field in a frequency domain, it can be expressed as an elliptic equation by harmonic component assumption $\widetilde{\mathbf{q}} = \mathbf{q} \cdot \mathrm{e}^{\mathrm{i}(\omega t + m\varphi)}$. For example,

$$A\frac{\partial \mathbf{q}}{\partial z} + B\frac{\partial \mathbf{q}}{\partial r} + C\mathbf{q} = \mathbf{0}. \tag{6.28}$$

This kind of equation need not be solved by time-advancement. So, FEM is more suitable to solve the sound propagation in a frequency domain, because the FEM needs to set up the unified equation of the whole calculation domain by using the principle of minimum potential energy. However, when we need to consider an accurate variation of the acoustic field with time, one of the best ways is to solve an equation like Eq. (6.27) by CAA techniques [21, 43, 44].

Although FEM assumes that the flow is potential flow, it is more easily implemented than CAA in terms of both computation time and storage. The reason is that the former is only used to solve the linear Euler equation in a frequency domain, which has the advantage of time-saving computation over CAA for the same approximation. Therefore, our discussion for numerical methods in duct acoustics will only focus on the application of FEM.

As a kind of numerical method, FEM can handle the issue including complex boundary and non-uniform flow along with the potential flow assumption. As shown in Fig. 6.7, Eversman [18] used this method to calculate the sound propagation in a non-uniform duct like an aeroengine nacelle.

The governing equations are:

$$\nabla\cdot(\rho_0\mathbf{V}_0)=0,\ \frac{1}{2}|\mathbf{V}_0|^2+\frac{C}{\gamma-1}=E,\ C^2=\gamma\frac{p_0}{\rho_0}=\rho_0^{\gamma-1}\text{ for mean flow field}, \tag{6.29}$$

and

$$i\omega\rho'+\nabla\cdot(\rho_0\nabla\phi+\rho'\mathbf{V}_0)=0,\ i\omega\phi+\mathbf{V}_0\cdot\nabla\phi+\frac{p'}{\rho_0}=0, \quad p'=C^2\rho'\text{ for acoustic field}, \tag{6.30}$$

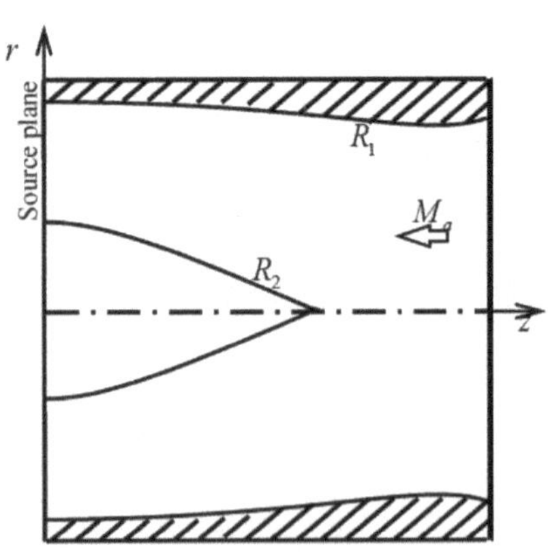

Fig. 6.7 Sound propagation in a non-uniform duct [18].

where ϕ is the velocity potential and $\widetilde{\phi} = \Phi + \phi \cdot e^{i\omega t}$. ρ_0, V_0, P, and C are the density, particle velocity, pressure, and sound speed of mean flow, respectively.

For the mean flow, given the boundary condition on the inlet and outlet terminal surfaces $\partial\Phi/\partial n = Ma_0$, $\Phi = 0$, we can write the Galerkin weak formulation as:

$$\int_v \nabla N \cdot (\rho_0 \nabla \Phi) dv = \int_S N \cdot (\rho_0 \nabla \Phi) \cdot ndS. \tag{6.31}$$

Weighting function N is from the class of continuous functions on volume v of the duct bounded by the duct surface, S, which includes the duct walls, source, and terminal planes. Assuming acoustic perturbations as harmonic in time with frequency ω and harmonic in the angular coordinate θ, the weak formulation on which a finite-element model for acoustic propagation was based is:

$$\begin{aligned}
&\int_v \frac{\rho_0}{C^2}\{C^2 \nabla N \cdot \nabla \phi - (\boldsymbol{V}_0 \cdot \nabla N)(\boldsymbol{V}_0 \cdot \nabla \phi) \\
&\qquad + i\omega[N(\boldsymbol{V}_0 \cdot \nabla \phi) - (\boldsymbol{V}_0 \cdot \nabla N)\phi] - \omega^2 N\phi\} dv \\
&= \int_S \frac{\rho_0}{C^2}\{C^2 N \nabla \phi - \boldsymbol{V}_0 N(\boldsymbol{V}_0 \cdot \nabla \phi) - i\omega \mathbf{V}_0 N\phi\} \cdot ndS\,.
\end{aligned} \tag{6.32}$$

The surface integral on the right-hand side of Eq. (6.32) is the natural boundary condition. On the terminal planes, it provides the boundary condition with the nonreflecting boundary conditions. On the lining walls, the natural boundary condition is used to introduce the Myers boundary condition [45]. Ostensibly, there is no trouble in this FEM, and we can obtain the sound field as long as the noise source is determined and the computational grid is convergent. In the practice of aeroengine acoustic design, however, there is a lot of acoustic waves of higher-order modes and high frequencies. This means that we need a huge number of computational grids to simulate the sound propagation. The cost is unacceptable for a non-uniform liner in circumferential direction even in the optimization of locally reacting liner. That is why it is still difficult to use FEM to calculate the radiation of sound from a duct, as shown in Fig. 6.3. So far, this method has been used to solve the fan noise radiation from a semi-infinite duct [46] without the interaction between inlet and outlet terminal planes. On the other hand, although it can consider more factors of sound generation in the duct, there are still certain difficulties to include the effect of the interaction between the second sound reflection by rotor, stator, the acoustic treatments, and the end of the duct.

In summary, all of the analysis and numerical methods previously discussed are useful for calculating sound propagation in a lined duct. It is true that these methods have their own advantages. However, if aimed at considering the effect of the various interactions inside an aeroengine nacelle as shown in Fig. 6.3, it is indeed indispensable to seek a new model to suffice for the optimal acoustic design of a modern turbofan nacelle. Considering the advantages and disadvantages of the existing models and several requirements for the study of various interactions, any new approach must be able to overcome the following difficulties:

1. avoiding the difficulty of solving eigenvalues of the soft-walled duct;
2. maintaining the orthogonality of duct eigenfunctions for computational convergence;
3. enabling acoustic design to consider the interaction between all kinds of acoustic liners, or sound sources, or even between acoustic liner and sound source;
4. studying the sound radiation with the effect of incident and reflection waves from the inlet and outlet of a finite-length duct.

6.3.2 Fundamental idea of the transfer element method

As shown in Fig. 6.8, considering the circular or annular straight duct, we take this as an example to discuss the establishment of the transfer matrix. If two interfaces, namely $A-A$ and $D-D$, are artificially established in the infinite duct, then the sound field and vortex field on these two interfaces should satisfy the continuity of pressure and velocity, that is:

$$p^{+}=p^{-},\ v^{+}=v^{-},\ w^{+}=w^{-}. \tag{6.33}$$

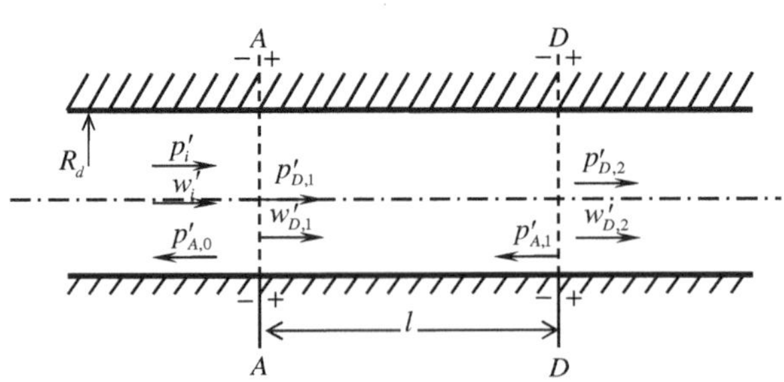

Fig. 6.8 Interface matching schematic diagram in a straight duct.

The whole duct is divided by the interfaces into three parts. The middle part of the interfaces is called a transfer element. There are two sets of sound waves and one set of vortex waves when no acoustic treatments are included in this part:

$$\widetilde{p'}_{A,1}=e^{i\omega t}\sum_{m=-\infty}^{\infty}\sum_{\mu=1}^{\infty}A_{m\mu}^{1p}\cdot\phi_m^1\left(k_{m\mu}^1 r\right)\cdot e^{i\gamma_{m\mu}^{1+}(z-l)}\cdot e^{im\overline{\varphi}},\tag{6.34}$$

$$\widetilde{p'}_{D,1}=e^{i\omega t}\sum_{m=-\infty}^{\infty}\sum_{\mu=1}^{\infty}D_{m\mu}^{1p}\cdot\phi_m^1\left(k_{m\mu}^1 r\right)\cdot e^{i\gamma_{m\mu}^{1-}z}\cdot e^{im\overline{\varphi}},\tag{6.35}$$

$$\widetilde{w'}_{\overline{\varphi},D,1}=e^{i\omega t}\sum_{m=-\infty}^{\infty}\sum_{\mu=1}^{\infty}D_{m\mu}^{1w}\cdot\frac{\phi_m\left(k_{m\mu}r\right)}{r}\cdot e^{i\alpha^1 z}\cdot e^{im\overline{\varphi}}.\tag{6.36}$$

Because both acoustic and vortex waves satisfy the superposition principle, we can calculate the sound field in the frequency domain, namely, assuming $\widetilde{p'}=p'\cdot e^{i\omega t}$ and $\widetilde{w'}=w'\cdot e^{i\omega t}$. There are three waves, the incident sound wave, the incident vortex wave, and the reflected sound wave, on the left side of the interface $A-A$:

$$p_i'=\sum_m\sum_\mu p_{im\mu}\phi_{m\mu}^0\left(k_{m\mu}^0 r\right)e^{i\gamma_{m\mu}^{0-}z}\cdot e^{im\overline{\varphi}},\tag{6.37}$$

$$p_{A,0}'=\sum_m\sum_\mu A_{m\mu}^{0p}\phi_m^0\left(k_{m\mu}^0 r\right)e^{i\gamma_{m\mu}^{0+}z}\cdot e^{im\overline{\varphi}},\tag{6.38}$$

$$w_i'=\sum_m\sum_\mu A_{m\mu}^{0w}\cdot\frac{\phi_m^0\left(k_{m\mu}^0 r\right)}{r}\cdot e^{i\alpha^0 z}\cdot e^{im\overline{\varphi}}.\tag{6.39}$$

There are two waves, the transmitted sound wave and the transmitted vortex wave, on the right side of the interface $D-D$:

$$p_{D,2}'=\sum_m\sum_\mu D_{m\mu}^{2p}\phi_{m\mu}^2\left(k_{m\mu}^2 r\right)e^{i\gamma_{m\mu}^{2-}(z-l)},\tag{6.40}$$

$$w_{D,2}'=\sum_m\sum_\mu D_{m\mu}^{2w}\cdot\frac{\phi_m^2\left(k_{m\mu}^2 r\right)}{r}\cdot e^{i\alpha^2(z-l)}\cdot e^{im\overline{\varphi}},\tag{6.41}$$

where $m=-M,\cdots,0,\cdots M;\quad n=1,2,\ldots,N$. The sound particle velocity corresponding to these sound waves can be determined by the momentum equation:

$$\frac{\partial u'}{\partial t}+U\frac{\partial u'}{\partial z}=-\frac{1}{\rho_0}\frac{\partial p'}{\partial z}. \tag{6.42}$$

Substituting these expressions into the matching conditions on interfaces $A-A$ and $D-D$, we can obtain the matching equations as follows:

$$\begin{aligned}&\sum_m\sum_\mu A^{0p}_{m\mu}\phi^0_m\left(k^0_{m\mu}r\right)-\sum_{m=-\infty}^{\infty}\sum_{\mu=1}^{\infty}D^{1p}_{m\mu}\cdot\phi^1_m\left(k^1_{m\mu}r\right)\\&-\sum_{m=-\infty}^{\infty}\sum_{\mu=1}^{\infty}A^{1p}_{m\mu}\cdot\phi^1_m\left(k^1_{m\mu}r\right)\cdot e^{-i\gamma^{1+}_{m\mu}l}\\&=-\sum_m\sum_\mu p_{im\mu}\phi^0_{m\mu}\left(k^0_{m\mu}r\right),\end{aligned} \tag{6.43}$$

$$\begin{aligned}&\sum_m\sum_\mu A^{0p}_{m\mu}\phi^0_m\left(k^0_{m\mu}r\right)\frac{\gamma^{0+}_{mn}}{U_0\gamma^{0+}_{mn}+\omega}-\sum_{m=-\infty}^{\infty}\sum_{\mu=1}^{\infty}D^{1p}_{m\mu}\cdot\phi^1_m\left(k^1_{m\mu}r\right)\frac{\gamma^{1-}_{mn}}{U_1\gamma^{1-}_{mn}+\omega}\\&-\sum_{m=-\infty}^{\infty}\sum_{\mu=1}^{\infty}A^{1p}_{m\mu}\cdot\phi^1_m\left(k^1_{m\mu}r\right)\cdot\frac{\gamma^{1+}_{mn}e^{-i\gamma^{1+}_{m\mu}l}}{U_1\gamma^{1+}_{mn}+\omega}\\&=-\sum_m\sum_\mu p_{im\mu}\phi^0_{m\mu}\left(k^0_{m\mu}r\right)\frac{\gamma^{0-}_{mn}}{U_0\gamma^{0-}_{mn}+\omega},\end{aligned} \tag{6.44}$$

$$\sum_{m=-\infty}^{\infty}\sum_{\mu=1}^{\infty}D^{1w}_{m\mu}\cdot\frac{\phi_m\left(k_{m\mu}r\right)}{r}=\sum_m\sum_\mu A^{0w}_{m\mu}\cdot\frac{\phi^0_m\left(k^0_{m\mu}r\right)}{r}, \tag{6.45}$$

$$\begin{aligned}&\sum_{m=-\infty}^{\infty}\sum_{\mu=1}^{\infty}D^{1p}_{m\mu}\cdot\phi^1_m\left(k^1_{m\mu}r\right)\cdot e^{i\gamma^{1-}_{m\mu}l}+\sum_{m=-\infty}^{\infty}\sum_{\mu=1}^{\infty}A^{1p}_{m\mu}\cdot\phi^1_m\left(k^1_{m\mu}r\right)\\&-\sum_m\sum_\mu D^{2p}_{m\mu}\phi^2_{m\mu}\left(k^2_{m\mu}r\right)=0,\end{aligned} \tag{6.46}$$

$$\begin{aligned}&\sum_{m=-\infty}^{\infty}\sum_{\mu=1}^{\infty}D^{1p}_{m\mu}\cdot\phi^1_m\left(k^1_{m\mu}r\right)\cdot e^{i\gamma^{1-}_{m\mu}l}\frac{\gamma^{1-}_{mn}}{U_1\gamma^{1-}_{mn}+\omega}\\&+\sum_{m=-\infty}^{\infty}\sum_{\mu=1}^{\infty}A^{1p}_{m\mu}\cdot\phi^1_m\left(k^1_{m\mu}r\right)\frac{\gamma^{1+}_{mn}}{U_1\gamma^{1+}_{mn}+\omega}\\&-\sum_m\sum_\mu D^{2p}_{m\mu}\phi^2_{m\mu}\left(k^2_{m\mu}r\right)\frac{\gamma^{2-}_{mn}}{U_2\gamma^{2-}_{mn}+\omega}=0,\end{aligned} \tag{6.47}$$

$$\sum_{m=-\infty}^{\infty}\sum_{\mu=1}^{\infty} D_{m\mu}^{1w}\cdot\frac{\phi_m\left(k_{m\mu}r\right)}{r}\cdot e^{i\alpha^1 l}-\sum_m\sum_\mu D_{m\mu}^{2w}\cdot\frac{\phi_m^2\left(k_{m\mu}^2 r\right)}{r}=0\ . \qquad (6.48)$$

For these six equations, the common solving methods are the point allocation method and the orthogonal method. In the following chapters, we will see the point allocation method is difficult to achieve the required precision, especially when the rotor blade is included in the varying cross-section duct. So, in the process of solving the equations, the orthogonal solution method should be adopted. In the equations of continuity of pressure and sound particle velocity, after multiplying both sides of equations of pressure and sound particle velocity by $\phi_m(k_{mn}r)\cdot r$ (only considering the straight duct here; the varying cross-section duct will be discussed in the following sections), and multiplying both sides of the equation of vortex particle velocity by $\phi_m(k_{mn}r)\cdot r^2$, we can simplify the equations by using the orthogonality of Bessel function:

$$\begin{cases} A_{m\mu}^{0p}-D_{m\mu}^{1p}-A_{m\mu}^{1p}=-p_{im\mu} \\ A_{m\mu}^{0p}\dfrac{\gamma_{mn}^{0+}}{U_0\gamma_{mn}^{0+}+\omega}-D_{m\mu}^{1p}\dfrac{\gamma_{mn}^{1-}}{U_1\gamma_{mn}^{1-}+\omega}-A_{m\mu}^{1p}\dfrac{\gamma_{mn}^{1+}}{U_1\gamma_{mn}^{1+}+\omega}e^{-i\gamma_{m\mu}^{1+}l}=-p_{im\mu}\dfrac{\gamma_{mn}^{0-}}{U_0\gamma_{mn}^{0-}+\omega} \\ D_{m\mu}^{1w}=A_{m\mu}^{0w} \\ D_{m\mu}^{1p}e^{i\gamma_{m\mu}^{1-}l}+A_{m\mu}^{1p}-D_{m\mu}^{2p}=0 \\ D_{m\mu}^{1p}\cdot e^{i\gamma_{m\mu}^{1-}l}\dfrac{\gamma_{mn}^{1-}}{U_1\gamma_{mn}^{1-}+\omega}+A_{m\mu}^{1p}\dfrac{\gamma_{mn}^{1+}}{U_1\gamma_{mn}^{1+}+\omega}-D_{m\mu}^{2p}\dfrac{\gamma_{mn}^{2-}}{U_2\gamma_{mn}^{2-}+\omega}=0 \\ D_{m\mu}^{1w}\cdot e^{i\alpha^1 l}-D_{m\mu}^{2w}=0 \end{cases}, \qquad (6.49)$$

The equations mentioned here are the matching equations of the sound propagation of a straight duct shown by Fig. 6.8. There are $6\times 2M\times N$ unknowns, which are A^{0p}, D^{1p}, A^{1p}, D^{1w}, D^{2p}, and D^{2w}, respectively, and the number of the equations is $6\times 2M\times N$ in total as each equation set contains $2M\times N$ equations. We can write the equations in a matrix product form as:

$$\begin{pmatrix} ss_{A_{m\mu}^{0p}}^{p_{a-a}} & ss_{D_{m\mu}^{1p}}^{p_{a-a}} & ss_{A_{m\mu}^{1p}}^{p_{a-a}} & 0 & 0 & 0 \\ ss_{A_{m\mu}^{0p}}^{u_{a-a}} & ss_{D_{m\mu}^{1p}}^{u_{a-a}} & ss_{A_{m\mu}^{1p}}^{u_{a-a}} & 0 & 0 & 0 \\ 0 & 0 & 0 & ss_{D_{m\mu}^{1w}}^{w_{a-a}} & 0 & 0 \\ 0 & ss_{D_{m\mu}^{1p}}^{p_{d-d}} & ss_{A_{m\mu}^{1p}}^{p_{d-d}} & 0 & ss_{D_{m\mu}^{2p}}^{p_{d-d}} & 0 \\ 0 & ss_{D_{m\mu}^{1p}}^{u_{d-d}} & ss_{A_{m\mu}^{1p}}^{u_{d-d}} & 0 & ss_{D_{m\mu}^{2p}}^{u_{d-d}} & 0 \\ 0 & 0 & 0 & ss_{D_{m\mu}^{1w}}^{w_{d-d}} & 0 & ss_{D_{m\mu}^{2w}}^{w_{d-d}} \end{pmatrix}\begin{bmatrix} A_{m\mu}^{0p} \\ D_{m\mu}^{1p} \\ A_{m\mu}^{1p} \\ D_{m\mu}^{1w} \\ D_{m\mu}^{2p} \\ D_{m\mu}^{2w} \end{bmatrix}=\begin{bmatrix} p_{im\mu} \\ v_{im\mu} \\ A_{m\mu}^{0w} \\ 0 \\ 0 \\ 0 \end{bmatrix}, \qquad (6.50)$$

where ss denotes a $2M \times N$ matrix. The coefficient matrix inside the transfer element is defined as:

$$\begin{bmatrix} ss^{p_{a-a}}_{D^{1p}_{m\mu}} & ss^{p_{a-a}}_{A^{1p}_{m\mu}} & 0 \\ ss^{u_{a-a}}_{D^{1p}_{m\mu}} & ss^{u_{a-a}}_{A^{1p}_{m\mu}} & 0 \\ 0 & 0 & ss^{w_{a-a}}_{D^{1w}_{m\mu}} \\ ss^{p_{d-d}}_{D^{1p}_{m\mu}} & ss^{p_{d-d}}_{A^{1p}_{m\mu}} & 0 \\ ss^{u_{d-d}}_{D^{1p}_{m\mu}} & ss^{u_{d-d}}_{A^{1p}_{m\mu}} & 0 \\ 0 & 0 & ss^{w_{d-d}}_{D^{1w}_{m\mu}} \end{bmatrix}_{(6\cdot 2M\cdot N)\times(3\cdot 2M\cdot N)} . \tag{6.51}$$

This matrix is a transfer matrix of an independent element. That is to say, it is independent of external sound waves. Different solutions can be obtained under different incident and reflection conditions. When the transfer element contains something that plays a role in the sound field or the vortex field, the coefficients will become more complex, and partial 0 matrices in the equations will also contain the information of the scattering field. In addition, note that in the process of establishing the transfer matrix, the equations are strictly closed. If inside the transfer element there exist some elements that interact with the sound field or the vortex field, no new unknown variable can be added, otherwise the equation group will no longer be closed. This problem will be discussed in detail in a latter chapter.

Now we extend this method to ducts containing multiple transfer elements, as shown in Fig. 6.9. Assuming that the duct contains nd different independent transfer elements, the matching equations can be established in each transfer element as described before. We can also write the final equations in a matrix product form.

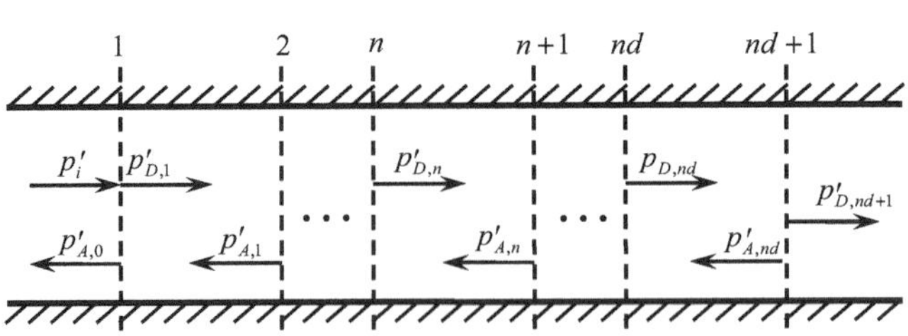

Fig. 6.9 A duct containing multiple transfer elements.

The coefficient matrix is:

$$\begin{bmatrix}
ss^{p_{a-a}}_{A^{0p}_{m\mu}} & ss^{p_{a-a}}_{D^{1p}_{m\mu}} & ss^{p_{a-a}}_{A^{1p}_{m\mu}} & 0 & & & & & \\
ss^{u_{a-a}}_{A^{0p}_{m\mu}} & ss^{u_{a-a}}_{D^{1p}_{m\mu}} & ss^{u_{a-a}}_{A^{1p}_{m\mu}} & 0 & & & & & \\
 & 0 & 0 & ss^{w_{a-a}}_{D^{1w}_{m\mu}} & & & & & \\
 & ss^{p_{d-d}}_{D^{1p}_{m\mu}} & ss^{p_{d-d}}_{A^{1p}_{m\mu}} & 0 & \cdots & \cdots & \cdots & & \\
 & ss^{u_{d-d}}_{D^{1p}_{m\mu}} & ss^{u_{d-d}}_{A^{1p}_{m\mu}} & 0 & \cdots & \cdots & \cdots & & \\
 & 0 & 0 & ss^{w_{d-d}}_{D^{1w}_{m\mu}} & \cdots & \cdots & \cdots & & \\
 & & & & \cdots & \cdots & \cdots & ss^{p_{d-d}}_{D^{n_d p}_{m\mu}} & \\
 & & & & \cdots & \cdots & \cdots & ss^{u_{d-d}}_{D^{n_d p}_{m\mu}} & \\
 & & & & \cdots & \cdots & \cdots & & ss^{w_{d-d}}_{D^{n_d w}_{m\mu}}
\end{bmatrix}_{nz\times nz}, \quad (6.52)$$

where the matrix order is $nz = 3 \cdot (nd + 1) \cdot 2M \cdot N$.

6.3.3 Construction of transfer element for a locally reacting lined duct

To avoid solving soft-walled duct eigenvalues, the equivalent surface source method proposed by Namba and Fukushige [47] is extended to handle the effect of an acoustic liner. In this method, the acoustic impedance boundary is regarded as an equivalent distribution of monopoles with unknown strengthes, which effectively avoids the process of eigenvalue calculation in essence. To obtain the sound field in the lined duct, Namba and Fukushige proposed a way to deal with the complex singularity problem, which are the key links of solving the singular integral equation. We found that the treatment of a singularity could be avoided by using a sinusoidal transformation, thus directly evaluating the noise reduction of sound waves.

The key idea of the equivalent surface source is not to consider the perforated plate as the boundary condition of the soft wall, instead, to regard it as the equivalent monopole distribution source. The sound field in the lined duct is computed by the superposition of two secondary sound fields: one is the propagating sound field with a hard wall boundary condition and the other is the sound field generated by the equivalent surface source. As shown in Fig. 6.10, there are three kinds of waves (expressed in the form of sound pressure) on the transfer element interface: incident and reflected

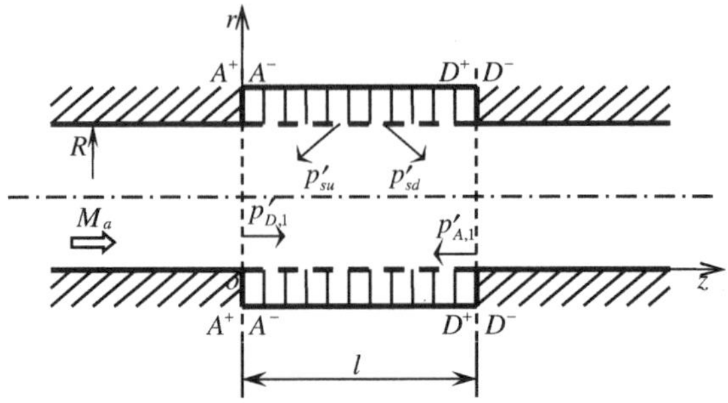

Fig. 6.10 TEM for local liner.

waves with hard wall boundary condition $p'_{D,1}$ and $p'_{A,1}$, and scattering waves emitted by the equivalent monopole distribution source p'_s.

In the last section, we mentioned that, to apply the TEM, we must retain the orthogonal characteristic of the acoustic expression and establish the explicit relation between the scattering field p'_s and the interface pressure waves $p'_{D,1}$ and $p'_{A,1}$, and then transfer all the unknown relations to the interface. The sound field in a hard wall duct can be described as follows:

$$p'_i = p'_{D,1} + p'_{A,1} = \sum_m \sum_\mu A^1_{m\mu} \phi^1_{m\mu}\left(k^1_{m\mu} r\right) e^{i\gamma^{1+}_{m\mu}(z-l)} + \sum_m \sum_\mu D^1_{m\mu} \phi^1_{m\mu}\left(k^1_{m\mu} r\right) e^{i\gamma^{1-}_{m\mu} z}, \tag{6.53}$$

where $A^1_{m\mu}$ and $D^1_{m\mu}$ are the amplitudes of the forward and backward propagating waves, respectively. m and μ represent the circumferential and radial modes, respectively. $\gamma^{1+}_{m\mu}$ and $\gamma^{1-}_{m\mu}$ are the forward and backward axial wavenumbers, respectively. The number "1" in the superscripts or subscripts represents the first section of the transfer element, which is prepared for the establishment of the subsequent multisection transfer element. $\phi^1_{m\mu}$ indicates the radial eigenfunction, whereas $k^1_{m\mu}$ represents the eigenvalue. Due to the linear superposition characteristics of the acoustic circumferential mode, only a fixed circumferential mode is considered to facilitate the discussion of the problem. When $m = m_0$, the upper formula could be simplified to:

$$p'_i = p'_{D,1} + p'_{A,1} = \sum_\mu A^1_{m\mu} \phi^1_{m\mu}\left(k^1_{m\mu} r\right) e^{i\gamma^{1+}_{m\mu}(z-l)} + \sum_\mu D^1_{m\mu} \phi^1_{m\mu}\left(k^1_{m\mu} r\right) e^{i\gamma^{1-}_{m\mu} z} \tag{6.54}$$

This expression is similar to that of the sound wave given in the previous section. To establish the transfer element of the locally reacting acoustic liner, it is important to give the expression of the scattering sound wave p_s' and transfer it to the interface. At the same time, we should establish the relation between the scattering sound wave and the interface sound waves with hard wall boundary conditions to maintain the number of unknowns.

According to the basic idea of the equivalent surface source method, the locally reacting liner can be regarded as an equivalent monopole distribution source along the soft wall, as shown in Fig. 6.11. The surface of an inner wall with axial coordinate among $0 \sim l$ is lined with a locally reacting liner. The original incident wave in the duct (the sound waves p_i' with the hard wall boundary condition) will introduce the vibration of the inner air column in the acoustic liner. The vibration of the inner air column can be regarded as an equivalent monopole source distributed along the wall, which excites a secondary sound field in the duct. This secondary sound field is called the scattering sound field. The superposition of the original incident sound field and the scattering sound field excited by the vibration of the air column inside the acoustic liner is the total sound field when there is an acoustic liner inside the duct. Note that the precondition for the application of the equivalent surface source method is to assume that the vibration of the air column in the acoustic liner is uniform and continuous along the wall (Fig. 6.11).

According to this hypothesis, the sound field generated by the distributed source can be expressed as:

$$\widetilde{p'}\left(\vec{x},t\right) = -\int_{-T}^{T}\int_{s(\tau)} \bar{\rho}\widetilde{V'}_n \frac{D_0 G}{D\tau} ds\left(\vec{y}\right) d\tau, \tag{6.55}$$

where "−" indicates that the outer normal direction of the fluid is positive, $\widetilde{V'}_n$ represents the vibrating velocity of the fluid element at the soft wall, and G is the Green's function in the duct, the full expression of which can be seen in the appendix. The total derivative of the Green's function is:

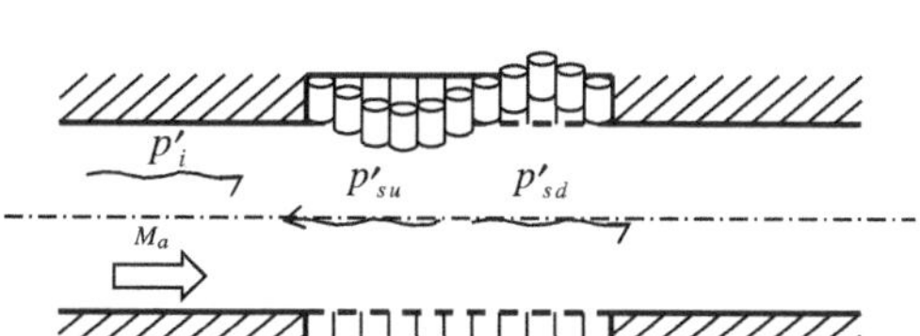

Fig. 6.11 The principle of the equivalent surface source method.

$$\frac{D_0 G}{D\tau}=\frac{\partial G}{\partial \tau}+U\frac{\partial G}{\partial z'}$$
$$=\frac{-1}{4\pi}\sum_{m=-\infty}^{\infty}\sum_{n=1}^{\infty}\frac{\Phi^*_{m,n}(x,y)\Phi_{m,n}(x',y')}{\Gamma_{m,n}}\int_{-\infty}^{\infty}\frac{\left(\omega+U\gamma^{\pm}_{m,n}\right)}{\kappa_{n,m}}e^{i\omega(t-\tau)+i\gamma^{\pm}_{m,n}(z-z')}d\omega\ . \tag{6.56}$$

Substituting Eq. (6.56) into Eq. (6.55) yields:

$$\widetilde{p'}_s(\vec{x},t)=\frac{\overline{\rho}}{4\pi}\sum_{m=-\infty}^{\infty}\sum_{n=1}^{\infty}\frac{\Phi_{m,n}(x,y)}{\Gamma_{m,n}}\int_{s(\tau)}\Phi^*_{m,n}(x',y')\widetilde{V'}_n \int_{-\infty}^{\infty}\frac{\left[\omega+U\gamma^{\pm}_{m,n}\right]}{\kappa_{n,m}}e^{i\gamma^{\pm}_{m,n}(z-z')}\int_{-T}^{T}e^{i\omega(t-\tau)}d\tau d\omega ds(\vec{y})\ , \tag{6.57}$$

where $\Phi_{m,\ n}(x, y)$ is the eigenfunction on the cross-section.

Assuming that the vibration velocity has the same vibration frequency as that of the incoming sound waves, namely, $\widetilde{V'}_n = V'_n e^{i\omega_0\tau}$, we can obtain:

$$\widetilde{p'}_s(\vec{x},t)=\frac{\overline{\rho}}{4\pi}\sum_{m=-\infty}^{\infty}\sum_{n=1}^{\infty}\frac{\Phi_{m,n}(x,y)}{\Gamma_{m,n}}\int_{s(\tau)}\Phi^*_{m,n}(x',y')V'_n \int_{-\infty}^{\infty}\frac{\left[\omega+U\gamma^{\pm}_{m,n}\right]}{\kappa_{n,m}}e^{i\gamma^{\pm}_{m,n}(z-z')}e^{i\omega t}\int_{-T}^{T}e^{i(\omega_0-\omega)\tau}d\tau d\omega ds(\vec{y})\ . \tag{6.58}$$

According to the relation $\lim\limits_{T\to\infty}\int_{-T}^{T}e^{i(\omega_0-\omega)\tau}d\tau=2\pi\delta(\omega_0-\omega)$, we have:

$$\widetilde{p'}_s(\vec{x},t)=\frac{\overline{\rho}}{2}\sum_{m=-\infty}^{\infty}\sum_{n=1}^{\infty}\frac{\Phi_{m,n}(\vec{x},\vec{y})}{\Gamma_{m,n}}\int_{s(\tau)}\Phi^*_{m,n}(x',y')V'_n \int_{-\infty}^{\infty}\frac{\left[\omega+U\gamma^{\pm}_{m,n}\right]}{\kappa_{n,m}}e^{i\gamma^{\pm}_{m,n}(z-z')}e^{i\omega t}\delta(\omega_0-\omega)d\omega ds(\vec{y})\ . \tag{6.59}$$

When $\omega = \omega_0$, it states:

$$\widetilde{p'}_s(\vec{x},t)=\frac{\overline{\rho}e^{i\omega t}}{2}\sum_{m=-\infty}^{\infty}\sum_{n=1}^{\infty}\frac{\Phi_{m,n}(x,y)}{\Gamma_{m,n}}\int_{s(\tau)}\Phi^*_{m,n}(x',y')V'_n\frac{\left[\omega+U\gamma^{\pm}_{m,n}\right]}{\kappa_{n,m}}e^{i\gamma^{\pm}_{m,n}(z-z')}ds(\vec{y}). \tag{6.60}$$

This integral gives the expression of the scattering sound field produced by the distribution source. Only the vibration velocity distribution function V_n' is unknown, and the other parameters are all functions of known duct geometry or flow parameters. To obtain the total sound field inside the element, the vibration function of the source V_n' must be given. For this reason, we give the impedance boundary condition of the soft wall:

$$\frac{\widetilde{p'}}{-\widetilde{V}_n^f} = Z_{zk} \quad at\ r = R, \tag{6.61}$$

where R is the radius of the internal duct and "-" indicates that the outer normal direction of the fluid is positive. Here, the total sound field is denoted as p', which is the sum of the incident sound field p_i' and the scattering sound field p_s'. So, the equation can be written as:

$$\widetilde{p'}_s + Z_{zk}\widetilde{V}_n^f = -\widetilde{p'}_i \quad at\ r = R, \tag{6.62}$$

where Z_{zk} represents the impedance of the acoustic liner and $\widetilde{V}_n^f$ indicates the local normal particle velocity on the liner surface, that is, the local vibration velocity of the air column inside the liner. According to the continuity condition of particle displacement, the relationship between the local normal particle velocity and the normal particle velocity along with the flow can be established as:

$$V_n' = \frac{\partial \eta}{\partial \tau} + U\frac{\partial \eta}{\partial z'} = V_n^f + \frac{U}{i\omega}\frac{\partial V_n^f}{\partial z'}. \tag{6.63}$$

By replacing Eqs. (6.50) and (6.63) into the Eq. (6.62), an integral equation can be obtained. If the incident sound wave $\widetilde{p'}_i$ is regarded as the known quantity, only the unknown function in this equation is V_n^f, and we can use the sinusoidal transformation to solve the equation.

For a uniform acoustic liner, the amplitude of the normal particle velocity V_n^f at the liner surface is obviously a periodic function of the azimuth angle φ' with a period equal to 2π, so the Fourier series expansion can be carried out. As for the axial coordinates z', we assume that it is a periodic function with a $2l$ period (l is the axial length of the acoustic liner), so that the Fourier series expansion about axial coordinates z' can also be used. For the convenience of calculation in mathematics, the Fourier series expansion about azimuth angle φ' is in the form of a complex exponential series, whereas the Fourier series expansion about the axial coordinate z' are in the form of sinusoidal series:

$$V_n^f = \sum_{k=1}^{\infty} \sum_{\bar{k}=-\infty}^{\infty} V_{k\bar{k}} \sin\frac{k\pi z'}{l} e^{-i\bar{k}\varphi'}, \tag{6.64}$$

where

$$\frac{1}{2\pi} \int_{-\pi}^{\pi} \frac{2}{l} \int_{0}^{l} V_n^f \cdot \sin\frac{j\pi z}{l} \exp\left(i\bar{j}\varphi\right) dz d\varphi = V_{j\bar{j}}, \tag{6.65}$$

is the defined sinusoidal transformation. Substituting the series expansion of V_n^f into the expression of V_n' gives:

$$V_n' = \left(1 + \frac{U}{i\omega}\frac{\partial}{\partial z'}\right) V_n^f = \left(1 + \frac{U}{i\omega}\frac{\partial}{\partial z'}\right) \sum_{k=1}^{\infty} \sum_{\bar{k}=-\infty}^{\infty} V_{k\bar{k}} \sin\frac{k\pi z'}{l} e^{-i\bar{k}\varphi'}. \tag{6.66}$$

Applying the defined sinusoidal transformation in Eq. (6.62) and substituting Eq. (6.65) into Eq. (6.62) lead to:

$$\begin{aligned} &\frac{1}{\pi l} \int_{-\pi}^{\pi} \int_{0}^{l} p_s' \cdot \sin\frac{j\pi z}{l} e^{i\bar{j}\varphi} dz d\varphi + \frac{1}{\pi l} \int_{-\pi}^{\pi} \int_{0}^{l} Z_{zk} V_n^f \cdot \sin\frac{j\pi z}{l} e^{i\bar{j}\varphi} dz d\varphi \\ &\quad = -\frac{1}{\pi l} \int_{-\pi}^{\pi} \int_{0}^{l} p_i' \cdot \sin\frac{j\pi z}{l} e^{i\bar{j}\varphi} dz d\varphi . \end{aligned} \tag{6.67}$$

For the scattering sound wave, the equivalent radiation impedance z_{jk} is introduced:

$$\sum_{k=1}^{\infty} z_{jk} V_{k\bar{j}} = \frac{1}{\pi l} \int_{-\pi}^{\pi} \int_{0}^{l} p_s' \sin\frac{j\pi z}{l} e^{i\bar{j}\varphi} dz d\varphi. \tag{6.68}$$

Then, Eq. (6.67) can be rewritten as:

$$\sum_{k=1}^{\infty} \left(z_{jk} + \delta_{jk} Z_{zk}\right) V_k = -I_j \quad at\ r = R, \tag{6.69}$$

where

$$I_j = -\frac{1}{\pi l} \int_{-\pi}^{\pi} \int_{0}^{l} p_i \cdot \sin\frac{j\pi z}{l} e^{i\bar{j}\varphi} dz d\varphi, \tag{6.70}$$

and

$$\frac{1}{2\pi}\int_{-\pi}^{\pi}\frac{2}{l}\int_{0}^{l} Z_{zk}V_n^f\cdot\sin\frac{j\pi z}{l}\exp\left(i\bar{j}\varphi\right)dzd\varphi = Z_{zk}V_{j\bar{j}}\delta_{jk}\delta_{\bar{j}\bar{k}}. \tag{6.71}$$

From Eq. (6.69), we can see that the integral equations are changed into algebraic equations after applying the defined sinusoidal transformation. This makes the computation procedure much simpler.

For the equivalent radiation impedance, we have:

$$\begin{gathered} z_{jk} = \frac{\overline{\rho}e^{i\omega_0 t}}{\pi l}\sum_{m=-\infty}^{\infty}\sum_{n=1}^{\infty}\frac{\Phi_{m,n}}{\Gamma_{m,n}\kappa_{n,m}} \\ \int_{-\pi}^{\pi}\int_{0}^{l}\int_{s(\tau)}\left[\omega + U\gamma_{m,n}^{\pm}\right]\Phi_{m,n}^{*}e^{i\gamma_{m,n}^{\pm}(z-z')} \\ \left(1+\frac{U}{i\omega}\frac{\partial}{\partial z'}\right)\sin\frac{k\pi z'}{l}ds(\gamma)\sin\frac{j\pi z}{l}e^{i\bar{j}\varphi}dzd\varphi\ . \end{gathered} \tag{6.72}$$

This expression can be applied in a straight duct with an arbitrary cross-section. For the circular or annular duct, $\int_{s(\tau)} f(r', \varphi', z')\, ds = 2\pi R\int_0^l f(r' = R, z')dz'$. According to the orthogonal characteristic of Bessel function,

$$z_{jk} = \frac{\overline{\rho}R}{l}\sum_{m=-\infty}^{\infty}\sum_{n=1}^{\infty}\frac{\phi_m(k_{mn}r)\cdot\phi_m(k_{mn}R)}{\kappa_{n,m}}\mathfrak{R}_{m,n,j,k}, \tag{6.73}$$

where

$$\mathfrak{R}_{m,n,j,k} = \int_0^l \sin\frac{j\pi z}{l}\int_0^l\left(1+\frac{U}{i\omega}\frac{\partial}{\partial z'}\right)\sin\frac{k\pi z'}{l}\left[\omega + U\gamma_{m,n}^{\pm}\right]e^{i\gamma_{m,n}^{\pm}(z-z')}dz'dz. \tag{6.74}$$

Here gives the result of the integral about φ as $\int_{-\pi}^{\pi} e^{i(\bar{j}-m)\varphi}d\varphi = 2\pi\delta(\bar{j}-m)$, and note that $\frac{\Phi_{m,n}}{\sqrt{\Gamma_{m,n}}} = \frac{\phi_m(k_{mn}r)\cdot e^{im\varphi}}{\sqrt{2\pi}}$.

In addition, attention should be paid to the location relationship between z and z' when calculating the integral. Therefore, the relevant result can be expressed as two parts:

$$\mathfrak{R}_{m,n,j,k} = \mathfrak{R}^1_{m,n,j,k} + \frac{k\pi U}{i\omega l}\mathfrak{R}^2_{m,n,j,k}, \tag{6.75}$$

where

$$\begin{aligned}\mathfrak{R}^1_{m,n,j,k} &= \int_0^l\int_0^l \sin k\pi\frac{z'}{l}\sin j\pi\frac{z}{l}\left[\omega_0 + U\gamma^{\pm}_{m,n}\right]e^{i\gamma^{\pm}_{m,n}(z-z')}dz'dz\ ,\\ \mathfrak{R}^2_{m,n,j,k} &= \int_0^l\int_0^l \cos k\pi\frac{z'}{l}\sin j\pi\frac{z}{l}\left[\omega_0 + U\gamma^{\pm}_{m,n}\right]e^{i\gamma^{\pm}_{m,n}(z-z')}dz'dz\ .\end{aligned} \tag{6.76}$$

After rearrangement, we obtain:

$$\begin{aligned}\mathfrak{R}^1_{m,n,j,k} &= \left[M_a c_0\gamma^{-}_{m,n} + \omega_0\right]R^1 + \left[M_a c_0\gamma^{+}_{m,n} + \omega_0\right]Q^1\ ,\\ \mathfrak{R}^2_{m,n,j,k} &= \left[M_a c_0\gamma^{-}_{m,n} + \omega\right]R^2 + \left[M_a c_0\gamma^{+}_{m,n} + \omega\right]Q^2\ .\end{aligned} \tag{6.77}$$

where

$$\left\{\begin{aligned}R^1 &= -\frac{k\pi l\cdot j\pi l\left(e^{i\gamma^{-}_{m,n}l}\cos j\pi - 1\right)}{\left(\gamma^{-}_{m,n}2l^2 - k^2\pi^2\right)\left(\gamma^{-}_{m,n}2l^2 - j^2\pi^2\right)} + \delta_{jk}\frac{i\gamma^{-}_{m,n}l^3}{2\left(\gamma^{-}_{m,n}2l^2 - k^2\pi^2\right)}\\ &\quad + \frac{k\pi l}{\left(\gamma^{-}_{m,n}2l^2 - k^2\pi^2\right)}S\\ Q^1 &= \frac{k\pi l\cdot j\pi l\cos k\pi\left(\cos j\pi - e^{-i\gamma^{+}_{m,n}l}\right)}{\left(\gamma^{+}_{m,n}2l^2 - k^2\pi^2\right)\left(\gamma^{+}_{m,n}2l^2 - j^2\pi^2\right)} - \delta_{jk}\frac{i\gamma^{+}_{m,n}l^3}{2\left(\gamma^{+}_{m,n}2l^2 - k^2\pi^2\right)}\\ &\quad - \frac{k\pi l}{\left(\gamma^{+}_{m,n}2l^2 - k^2\pi^2\right)}S\\ R^2 &= -\frac{i\gamma^{-}_{m,n}l^2\cdot j\pi l\left(e^{i\gamma^{-}_{m,n}l}\cos j\pi - 1\right)}{\left(\gamma^{-}_{m,n}2l^2 - k^2\pi^2\right)\left(\gamma^{-}_{m,n}2l^2 - j^2\pi^2\right)} - \delta_{jk}\frac{k\pi l^2}{2\left(\gamma^{-}_{m,n}2l^2 - k^2\pi^2\right)}\\ &\quad + \frac{i\gamma^{-}_{m,n}l^2}{\left(\gamma^{-}_{m,n}2l^2 - k^2\pi^2\right)}S\\ Q^2 &= \frac{i\gamma^{+}_{m,n}l^2\cdot j\pi l\cos k\pi\left(\cos j\pi - e^{-i\gamma^{+}_{m,n}l}\right)}{\left(\gamma^{+}_{m,n}2l^2 - k^2\pi^2\right)\left(\gamma^{+}_{m,n}2l^2 - j^2\pi^2\right)} + \delta_{jk}\frac{k\pi l^2}{2\left(\gamma^{+}_{m,n}2l^2 - k^2\pi^2\right)}\\ &\quad - \frac{i\gamma^{+}_{m,n}l^2}{\left(\gamma^{+}_{m,n}2l^2 - k^2\pi^2\right)}S\end{aligned}\right. ,$$

and

$$S = \begin{cases}\dfrac{2jl}{(j^2 - k^2)\pi} & k+j \ \text{ is odd}\\ 0 & k+j \ \text{ is even}\end{cases} .$$

Note that the analytic integral expressions for R^1, Q^1, R^2, and Q^2 are only correct in cases where all the denominators are not equal to zero. In other cases, new calculations are needed. For example, when $\gamma_{m,\,n}^{-} l = k\pi = j\pi$, new calculations lead to:

$$R^1 = \frac{l^2}{8} - i\frac{3}{8}\frac{l^2}{k\pi}, \quad R^2 = \frac{l^2}{8k\pi} - i\frac{l^2}{8}.$$

However, these special cases are only encountered at discrete particular frequencies.

After establishing the algebraic relation between the distribution source and the incident sound waves, and avoiding the solution of the eigenvalues of the soft wall, we also need to solve another difficulty, that is, the scattering sound waves must be expressed as the expression of the incident sound waves to keep the total number of unknowns unchanged. The expression of incident sound waves in Eq. (6.54) is substituted into Eq. (6.70):

$$\begin{aligned} I_j = & \frac{2}{l}\sum_{\mu} A_{m\mu}^1 \phi_{m\mu}^1 \left(k_{m\mu}^1 R\right) \int_0^l e^{i\gamma_{m\mu}^{1+}(z-l)} \sin\frac{j\pi z}{l} dz \\ & + \frac{2}{l}\sum_{\mu} D_{m\mu}^1 \phi_{m\mu}^1 \left(k_{m\mu}^1 R\right) \int_0^l e^{i\gamma_{m\mu}^{1-} z} \sin\frac{j\pi z}{l} dz. \end{aligned}$$

Here, the integral related with φ has been conducted. Calculating the remaining integral gives:

$$\begin{aligned} I_j = & \frac{2}{l}\sum_{\mu} A_{m\mu}^1 \phi_{m\mu}^1 \left(k_{m\mu}^1 R\right) \frac{j\pi l\left(\cos j\pi - e^{-i\gamma_{m\mu}^{1+} l}\right)}{\gamma_{m\mu}^{1+2} l^2 - j^2\pi^2} \\ & + \frac{2}{l}\sum_{\mu} D_{m\mu}^1 \phi_{m\mu}^1 \left(k_{m\mu}^1 R\right) \frac{j\pi l\left(e^{i\gamma_{m\mu}^{1-} l} \cos j\pi - 1\right)}{\gamma_{m\mu}^{1-2} l^2 - j^2\pi^2}. \end{aligned} \tag{6.78}$$

Denoting:

$$I_j^{A_{m\mu}^1} = -\frac{2}{l}\frac{j\pi l\left(\cos j\pi - e^{-i\gamma_{m\mu}^{1+} l}\right)}{\gamma_{m\mu}^{1+2} l^2 - j^2\pi^2}, \quad I_j^{D_{m\mu}^1} = -\frac{2}{l}\frac{j\pi l\left(e^{i\gamma_{m\mu}^{1-} l} \cos j\pi - 1\right)}{\gamma_{m\mu}^{1-2} l^2 - j^2\pi^2},$$

we have

$$I_j = -\sum_{\mu} \phi_{m\mu}^1 \left(k_{m\mu}^1 R\right) \left[D_{m\mu}^1 I_j^{D_{m\mu}^1} + A_{m\mu}^1 I_j^{A_{m\mu}^1}\right]. \tag{6.79}$$

Substituting this expression into Eq. (6.69) yields:

$$\sum_{k=1}^{\infty}\left(z_{jk}+\delta_{jk}Z_{zk}\right)V_k=\sum_{\mu}\phi^1_{m\mu}\left(k^1_{m\mu}R\right)\left[D^1_{m\mu}I_j^{D^1_{m\mu}}+A^1_{m\mu}I_j^{A^1_{m\mu}}\right]. \quad (6.80)$$

Then, V_k can be written as the explicit function of the amplitudes $D^1_{m\mu}$ and $A^1_{m\mu}$ of the sound waves with the hard wall condition:

$$V_k^{m,\mu}=\sum_{\mu}\phi^1_{m\mu}\left(k^1_{m\mu}r\right)\sum_j\left[z_{jk}+\delta_{jk}Z_{zk}\right]^{-1}\left[D^1_{m\mu}I_j^{D^1_{m\mu}}+A^1_{m\mu}I_j^{A^1_{m\mu}}\right]. \quad (6.81)$$

So, no more unknowns have been introduced. Here, the superscripts of $V_k^{m,\,\mu}$ mean that this quantity is related to the corresponding mode number. Here, $[\cdot]^{-1}$ represents the inverse matrix.

Substituting this expression into Eq. (6.60), we can obtain the scattering sound field in a circular lined duct as:

$$\begin{aligned}p'_s\left(\vec{x},t\right)&=\sum_n p_{smn}\phi_{mn}(k_{mn}r)\cdot e^{im\varphi}\cdot e^{i\gamma^{\pm}_{m,n}z}\\&=\frac{\overline{\rho}}{2}\sum_{n=1}^{\infty}\frac{\phi_{mn}(k_{mn}r)\cdot e^{im\varphi}\cdot e^{i\gamma^{\pm}_{m,n}z}}{2\pi}\left\{\sum_{\mu}D^1_{m\mu}Q^{D_{m\mu}}_{n\pm}+A^1_{m\mu}Q^{A_{m\mu}}_{n\pm}\right\},\end{aligned} \quad (6.82)$$

where

$$\begin{aligned}Q^{D_{m\mu}}_{n\pm}&=\sum_k\phi^1_{m\mu}\left(k^1_{m\mu}R\right)\sum_j\left[z_{jk}+\delta_{jk}Z_{zk}\right]^{-1}I_j^{D^1_{m\mu}}S_k^{n\pm},\\Q^{A_{m\mu}}_{n\pm}&=\sum_k\phi^1_{m\mu}\left(k^1_{m\mu}R\right)\sum_j\left[z_{jk}+\delta_{jk}Z_{zk}\right]^{-1}I_j^{A^1_{m\mu}}S_k^{n\pm},\end{aligned} \quad (6.83)$$

where $S_k^{n\pm}$ is only related to the expansion order k, and the radial mode number n of scattering acoustic waves is:

$$S_k^{n\pm}=\int_s\phi_{mn}(k_{mn}r')\cdot e^{i(\bar{j}-m)\varphi'}\frac{\left[\omega+U\gamma^{\pm}_{m,n}\right]}{\kappa_{n,m}}e^{-i\gamma^{\pm}_{m,n}z'}\left(1+\frac{U}{i\omega}\frac{\partial}{\partial z'}\right)\sin\frac{k\pi z'}{l}ds\left(\vec{r'}\right). \quad (6.84)$$

The expression of scattering sound waves at the interface is Eq. (6.82). It can be seen from the expression that the scattering sound pressure at the interface is an explicit function of the modal amplitudes $D^1_{m\mu}$ and $A^1_{m\mu}$.

The superposed sound particle velocity and sound pressure at the interface can be expressed as:

$$\begin{aligned} u'_{A^-A^-}(r) &= u'_i\big|_{A^-A^-} + u'_s\big|_{A^-A^-} \\ &= \sum_{\mu} \frac{\phi_{m\mu}(k_{m\mu} r)}{\bar{\rho}} \left[\frac{D_{m\mu}\gamma^-_{m\mu}}{U\gamma^-_{m\mu} + \omega} + \frac{A_{m\mu}\gamma^+_{m\mu} e^{-i\gamma^+_{m\mu} l}}{\gamma^+_{m\mu} U + \omega} \right] \\ &+ \sum_{n=1}^{\infty} \frac{\phi_{mn}(k_{mn} r) \cdot \gamma^+_{mn}}{4\pi(\gamma^+_{mn} U + \omega)} \sum_{\mu} \left\{ D^1_{m\mu} Q^{D_{m\mu}}_{n\pm} + A^1_{m\mu} Q^{A_{m\mu}}_{n\pm} \right\}, \end{aligned} \tag{6.85}$$

$$\begin{aligned} p'_{A^-A^-}(r) &= p'_i\big|_{A^-A^-} + p'_s\big|_{A^-A^-} = \sum_{\mu} D_{m\mu}\phi_{m\mu}(k_{m\mu} r) \\ &+ \sum_{\mu} A_{m\mu}\phi_{m\mu}(k_{m\mu} r) e^{-i\gamma^+_{m\mu} l} \\ &+ \frac{\bar{\rho}}{2} \sum_{n=1}^{\infty} \frac{\phi_{mn}(k_{mn} r)}{2\pi} \left\{ \sum_{\mu} D^1_{m\mu} Q^{D_{m\mu}}_{n\pm} + A^1_{m\mu} Q^{A_{m\mu}}_{n\pm} \right\}, \end{aligned} \tag{6.86}$$

$$\begin{aligned} u'_{D^+D^+}(r) &= u'_i\big|_{D^+D^+} + u'_s\big|_{D^+D^+} \\ &= \sum_{\mu} \frac{\phi_{m\mu}(k_{m\mu} r)}{\bar{\rho}} \left[\frac{D_{m\mu}\gamma^-_{m\mu} e^{i\gamma^-_{m\mu} l}}{U\gamma^-_{m\mu} + \omega} + \frac{A_{m\mu}\gamma^+_{m\mu}}{\gamma^+_{m\mu} U + \omega} \right] \\ &+ \sum_{n=1}^{\infty} \frac{\phi_{mn}(k_{mn} r)\gamma^-_{mn} e^{i\gamma^-_{mn} l}}{4\pi(U\gamma^-_{mn} + \omega)} \sum_{\mu} \left\{ D^1_{m\mu} Q^{D_{m\mu}}_{n\pm} + A^1_{m\mu} Q^{A_{m\mu}}_{n\pm} \right\}, \end{aligned} \tag{6.87}$$

$$\begin{aligned} p'_{D^+D^+}(r) &= p'_i\big|_{D^+D^+} + p'_s\big|_{D^+D^+} = \sum_{\mu} \phi_{m\mu}(k_{m\mu} r)\left[D_{m\mu} e^{i\gamma^-_{m\mu} l} + A_{m\mu}\right] \\ &+ \frac{\bar{\rho}}{2} \sum_{n=1}^{\infty} \frac{\phi_{mn}(k_{mn} r) \cdot e^{i\gamma^-_{mn} l}}{2\pi} \left\{ \sum_{\mu} D^1_{m\mu} Q^{D_{m\mu}}_{n\pm} + A^1_{m\mu} Q^{A_{m\mu}}_{n\pm} \right\}. \end{aligned} \tag{6.88}$$

Merging the calculation items containing the same unknown coefficient into a matrix form yields:

$$\begin{bmatrix} u'_{A^-A^-} \\ p'_{A^-A^-} \\ u'_{D^+D^+} \\ p'_{D^+D^+} \end{bmatrix} = \begin{bmatrix} ss^{u'_{A^-A^-}}_{D_{m\mu}} & ss^{u'_{A^-A^-}}_{A_{m\mu}} \\ ss^{p'_{A^-A^-}}_{D_{m\mu}} & ss^{p'_{A^-A^-}}_{A_{m\mu}} \\ ss^{u'_{D^+D^+}}_{D_{m\mu}} & ss^{u'_{D^+D^+}}_{A_{m\mu}} \\ ss^{p'_{D^+D^+}}_{D_{m\mu}} & ss^{p'_{D^+D^+}}_{A_{m\mu}} \end{bmatrix} \begin{bmatrix} D^1_{m\mu} \\ A^1_{m\mu} \end{bmatrix}. \tag{6.89}$$

The expression has the same form of the acoustic expression on the interface given without any acoustic treatment and special boundary condition within the transfer element, but the coefficient matrices ss are different. The coefficients matrices ss in the expression of Eq. (6.89) can be divided into two parts. For example, coefficients matrix $ss_{D_{m\mu}}^{p'_{A^-}A^-}$ can be expressed as:

$$ss_{D_{m\mu}}^{p'_{A^-}A^-} = sc_{D_{m\mu}}^{p'_{A^-}A^-} + cs_{D_{m\mu}}^{p'_{A^-}A^-}, \tag{6.90}$$

where $sc_{D_{m\mu}}^{p'_{A^-}A^-}$ is consistent with the corresponding term in Eq. (6.51), whereas $cs_{D_{m\mu}}^{p'_{A^-}A^-}$ is the corresponding coefficients matrix in Eq. (6.84) associated with the scattering sound waves.

At this point, we have established the transfer element of the locally reacting lined duct. It can be seen that all the unknown coefficients are transferred to the interface and can be displayed explicitly. According to the previous deduction, we can clearly see that the model of the TEM satisfies all previous requirements. This method avoids solving the eigenvalues of the soft wall, and the expressions of sound pressures and particle velocities at the interface retains the orthogonality of the eigenfunctions $\phi_{m\mu}(k_{m\mu}r)$. In addition, from the essence of the method we can see that the transfer element can still maintain the same form but different expressions of the coefficient matrices when the element contains other different acoustic treatments.

As can be seen from the last section, the transfer matrix can be extended to characterize a duct that contains multisegments lined with different locally reacting liners. Fig. 6.12 gives the calculation results of insertion losses of the axial two segmented acoustic liners.

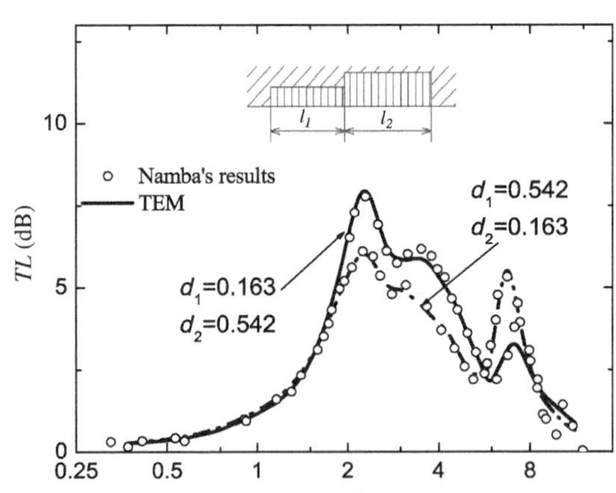

Fig. 6.12 Insertion losses of the axial two segmented acoustic liners for (0,0) mode with different cavity depths (see [47] for Namba's result) $R_{zk1} = 0.8$, $R_{zk2} = 0.6$, $\omega_{01} = \omega_{02} = 25.86$, $L = 4.339$, $L_1/L = 0.5$, $M_a = 0.4$.

6.3.4 Construction of transfer element for a nonlocally reacting lined duct

In the last part, we established the transfer element model for a duct lined with a locally reacting liner that can be applied to the calculation of sound propagation in the lined duct and the calculation of the sound propagation in a varying cross-section duct. Up to now, the locally reacting acoustic liner is widely used in the acoustic treatment design of an aeroengine nacelle. This kind of acoustic liner has a good effect on noise attenuation in the engine. In recent years, the increasing requirements for noise reduction make the design of acoustic treatment more and more difficult. However, the noise attenuation features of the locally reacting acoustic liner determine that such kind of liner is difficult to further apply to reduce the noise in new aeroengines. This is because the noise may be broadband and the sound energy in the low frequency range may be still relatively high, whereas the traditional locally reacting liner is only effective near several discrete resonant frequencies. Therefore, the nonlocally reacting acoustic liner, which may have the potential to overcome the disadvantages of the locally reacting liner, has received increasing attention in recent decades. But unfortunately, there is no specific nonlocally reacting acoustic liner model that can be used directly for the calculation of the sound propagation in the aeroengine duct. So, this section will focus on establishing a transfer element model for a duct lined with a nonlocally reacting liner.

As shown in Fig. 6.13, the difference between the locally reacting liner and the nonlocally reacting liner is whether there is sound wave propagation parallel to the perforated plate in the cavity behind the plate. For a locally reacting liner, only plane waves are propagating in the depth direction of the honeycomb structure due to the cut-off effect, as shown in Fig. 6.13A, whereas there is no constraint of transverse propagation of acoustic waves in the cavities for the nonlocally reacting liner, as shown in Fig. 6.13B. It can be seen that there is an obvious difference between the two kinds of acoustic liners. For locally reacting liners, the impedance model can directly include the effect of the honeycomb structure because

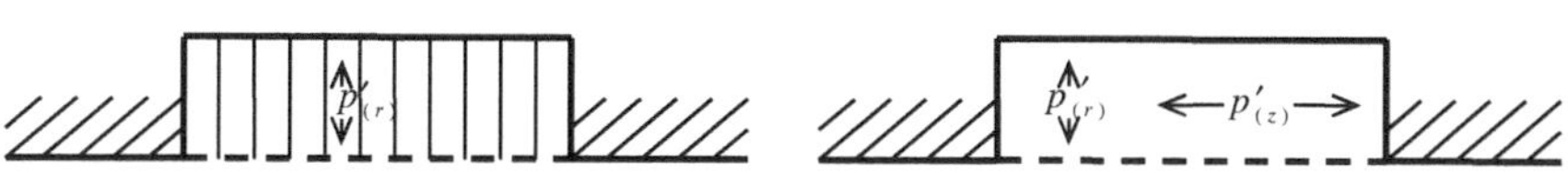

Fig. 6.13 The comparison between the locally reacting liner and the nonlocally reacting liner. (A) locally reacting liner (B) nonlocally reacting liner.

only radial plane waves can propagate in the honeycomb structure. For nonlocally reacting liners, however, the effect of cavities cannot be directly included in the impedance model during the calculation of the sound propagation because there are sound waves in two directions, namely, perpendicular to the perforated plate and parallel to the perforated plate in the back cavity. So, the calculation model needs to be reestablished.

Because the effect of sound pressure p'^{+}_{s} in the back cavity cannot be directly included in the surface impedance, the influence of sound pressure p'^{+}_{s} should be considered when the impedance equation is established. As shown in Fig. 6.14, the impedance equation for the perforated plate is established as

$$\frac{p'^{-} - p'^{+}}{-V_n^f} = Z_{zk}, \tag{6.91}$$

where V_n^f still represents the average vibrating velocity of the air column in the hole on the perforated plate, but Z_{zk} is determined only by the impedance model of the perforated plate, which is independent of the geometric parameters of the back cavity. In the main duct, the sound field is composed of two parts:

$$p'^{-} = p'^{-}_{s} + p'_{i}. \tag{6.92}$$

Based on the basic idea of the equivalent surface source method, p'^{-}_{s} represents the sound pressure field generated by the vibration in the holes of the perforated plate after it is regarded as the sound distribution sources that still belong to the monopole sources. In the back cavity, because there is no external sound source input, the internal sound field contains only the sound field produced by the vibration of the hole on the perforated plate, which can be expressed as:

$$p'^{+} = p'^{+}_{s}, \tag{6.93}$$

Substituting Eqs. (6.92) and (6.93) into the impedance equation gives:

$$p'^{-}_{s} - p'^{+}_{s} + Z_{zk} V_n^f = -p'_{i}. \tag{6.94}$$

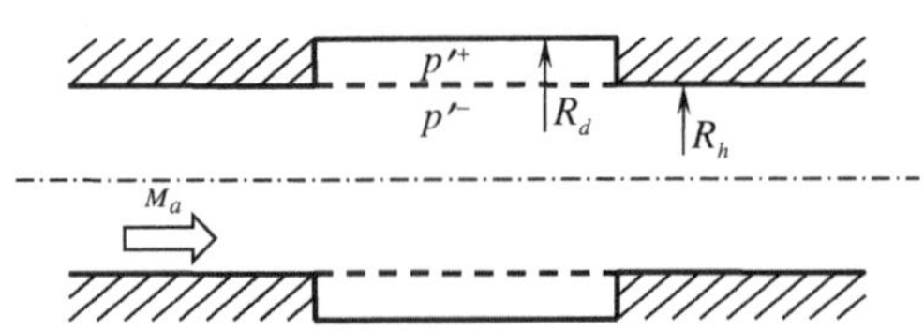

Fig. 6.14 A duct line with a nonlocally reacting liner.

The computation for p'^{-}_{s} has been discussed in detail in the section about the locally reacting liner as:

$$\tilde{p}'(\vec{x}, t) = -\int_{-T}^{T} \int_{s(\tau)} \bar{\rho} \widetilde{V'}_{n} \frac{D_0 G}{D\tau} ds(\vec{y}) d\tau. \tag{6.95}$$

Here, the specific solution is no longer described in detail. For the sound field p'^{+}_{s} in the back cavity, we can derive its specific expression using the knowledge of room acoustics, which will be described later.

In the case of fixed circumferential mode $m = m_0$, it is not necessary to take the circumferential Fourier transform of the impedance equation, but only the sinusoidal transformation in the axial direction is needed. The following narration only considers the case of the fixed circumferential mode.

First, the sinusoidal transformation is made to Eq. (6.94), and then:

$$\frac{2}{l}\int_{0}^{l} \left(p'^{-}_{s} - p'^{+}_{s}\right) \sin\frac{i\pi z}{l} dz + \frac{2}{l}\int_{0}^{l} Z_{zk} V^{f}_{n} \sin\frac{i\pi z}{l} dz = -\frac{2}{l}\int_{0}^{l} p'_{i} \sin\frac{i\pi z}{l} dz. \tag{6.96}$$

The sound field p'^{+}_{s} in the back cavity is formed by the vibration of the air column in the holes of the perforated plate. Therefore, it can also be expressed as a single function of its vibration velocity V^{f}_{n}. To give the unified form, the scattering sound field is still written as an equivalent radiation impedance:

$$\begin{cases} \sum_{k=1}^{\infty} \overset{-}{z}_{jk} V_k = \frac{2}{l}\int_{0}^{l} p'^{-}_{s} \sin\frac{j\pi z}{l} dz \\ \sum_{k=1}^{\infty} \overset{+}{z}_{jk} V_k = \frac{2}{l}\int_{0}^{l} p'^{+}_{s} \sin\frac{j\pi z}{l} dz \end{cases}. \tag{6.97}$$

The original integral equation can be expressed as the form of algebraic equations in the form of:

$$\sum_{k=1}^{\infty} \left(\overset{-}{z}_{jk} - \overset{+}{z}_{jk} + \delta_{jk} Z_{zk}\right) V_k = -I_j. \tag{6.98}$$

The equivalent radiation impedance $\overset{-}{z}_{jk}$ is the same as z_{jk} in the previous chapter for the locally reacting liner. Only the equivalent radiation impedance $\overset{+}{z}_{jk}$ in the back cavity is required to deduce. Eq. (6.98) is an algebraic impedance equation that needs to be solved. The unknown is the expansion coefficient V_k of the vibration velocity of the air column in the orifice of perforated plate.

In Eq. (6.98), I_j is the sinusoidal transformation of the incident sound wave, and Z_{zk} is the impedance of the perforated plate. Because the expression of z_{jk}^{-} has been discussed, only the expression of z_{jk}^{+} needs to be given. According to the generalized Lighthill's theory, the sound pressure in the back cavity can be expressed as:

$$\widetilde{p'}_s^{+}(\vec{x}, t) = \int_{-T}^{T} \int_{s(\tau)} \bar{\rho} \widetilde{V}_n^f \frac{\partial G^+}{\partial \tau} ds(\vec{y}) d\tau. \tag{6.99}$$

The equation is still the expression of the sound field of the monopole distribution source in the duct. G^+ is the Green's function in the cavity, the detailed derivation of which is shown in the appendix. For a circular duct, the expression of G^+ is as follows:

$$G^+ = -\frac{1}{2\pi} \sum_m \sum_n \sum_s \frac{\psi_m(k_{mn}r) e^{im\varphi} \cos\left(\frac{s\pi}{l} z\right) \psi_m(k_{mn}r') e^{im\varphi'} \cos\left(\frac{s\pi}{l} z'\right)}{\Gamma'_{mns}} \int_{-\infty}^{\infty} \frac{e^{i\omega(t-\tau)}}{\left(k_0^2 - k_{mns}^2\right)} d\omega. \tag{6.100}$$

Because there is no flow in the cavity, the total derivative of G^+ is directly degraded into partial derivative of time:

$$\frac{\partial G^+}{\partial \tau} = -\frac{1}{2\pi} \sum_m \sum_n \sum_s \frac{\psi_m(k_{mn}r) e^{im\varphi} \cos\left(\frac{s\pi}{l} z\right) \psi_m(k_{mn}r') e^{-im\varphi'} \cos\left(\frac{s\pi}{l} z'\right)}{\Gamma'_{mns}} \int_{-\infty}^{\infty} \frac{-i\omega \cdot e^{i\omega(t-\tau)}}{\left(k_0^2 - k_{mns}^2\right)} d\omega\ . \tag{6.101}$$

Substituting this equation into the expression of sound pressure in the cavity leads to:

$$\widetilde{p'}_s^{+}(\vec{x}, t) = \frac{i\bar{\rho}}{2\pi} \sum_m \sum_n \sum_s \frac{\psi_m(k_{mn}r) e^{im\varphi} \cos\left(\frac{s\pi}{l} z\right)}{\Gamma'_{mns}} \int_{s(\tau)} V_n^f \psi_m(k_{mn}r') e^{-im\varphi'} \cos\left(\frac{s\pi}{l} z'\right) \int_{-\infty}^{\infty} \frac{\omega}{k_0^2 - k_{mns}^2} \int_{-T}^{T} e^{i\omega_0\tau} e^{i\omega(t-\tau)} d\tau d\omega ds(\vec{y}). \tag{6.102}$$

Because $\lim_{T\to\infty}\int_{-T}^{T} e^{i(\omega_0-\omega)\tau}d\tau = 2\pi\delta(\omega_0-\omega)$, then:

$$\widetilde{p'}_s^{+}(\vec{x},t) = i\overline{\rho}\sum_m\sum_n\sum_s \frac{\psi_m(k_{mn}r)e^{im\varphi}\cos\left(\frac{s\pi}{l}z\right)}{\Gamma'_{mns}}$$
$$\int_{s(\tau)} V_n^f \psi_m(k_{mn}r')e^{-im\varphi'}\cos\left(\frac{s\pi}{l}z'\right)\int_{-\infty}^{\infty}\frac{\omega e^{i\omega t}}{k_0^2-k_{mns}^2}\delta(\omega_0-\omega)\,d\omega ds(\vec{y}) \,. \tag{6.103}$$

When $\omega=\omega_0$, we have:

$$p'^{+}_s(\vec{x},t) = i\overline{\rho}\omega\sum_m\sum_n\sum_s \frac{\psi_m(k_{mn}r)e^{im\varphi}\cos\left(\frac{s\pi}{l}z\right)}{\Gamma'_{mns}}$$
$$\int_{s(\tau)} V_n^f \cos\left(\frac{s\pi}{l}z'\right)\frac{\psi_m(k_{mn}r')e^{-im\varphi'}}{k_0^2-k_{mns}^2}ds(\vec{y}) \,. \tag{6.104}$$

Expanding V_n^f into sine series and substituting the series into Eq. (6.104) gives:

$$p'^{+}_s(\vec{x},t) = i\overline{\rho}\omega\sum_m\sum_n\sum_s \frac{\psi_m(k_{mn}r)e^{im\varphi}\cos\left(\frac{s\pi}{l}z\right)}{\Gamma'_{mns}}$$
$$\left.\int_0^l\int_0^{2\pi}\sum_{k=1}^{\infty}V_k\sin\frac{k\pi z'}{l}\cos\left(\frac{s\pi}{l}z'\right)\frac{\psi_m(k_{mn}r')e^{-im\varphi'}}{k_0^2-k_{mns}^2}dz'd\varphi'\right|_{r'=R_h} \,. \tag{6.105}$$

Making a sinusoid transform to this expression results in:

$$\frac{2}{l}\int_0^l p'^{+}_s\sin\frac{j\pi z}{l}dz = \frac{4i\pi\overline{\rho}\omega R_h}{l}\sum_m\sum_n\sum_s\frac{\psi_m(k_{mn}r)e^{im\varphi}}{\Gamma'_{mns}}$$
$$\sum_{k=1}^{\infty}V_k\frac{\psi_m(k_{mn}R_h)}{k_0^2-k_{mns}^2}\int_0^l\cos\left(\frac{s\pi}{l}z\right)\sin\frac{j\pi z}{l}\int_0^l\sin\frac{k\pi z'}{l}\cos\left(\frac{s\pi}{l}z'\right)dz'dz. \tag{6.106}$$

Using $\phi_m(k_{mn}r) = \frac{\psi_m(k_{mn}r)}{\sqrt{\Gamma_{m,n}}}$, $\Gamma_{mn}=2\pi\overline{\Gamma}_{mn}$, and $\Gamma'_{mns}=\overline{\Gamma}_{mn}\cdot\Gamma_s$, replace the relevant terms in Eq. (6.106). Then, introducing the radiation impedance:

$$z^{+}_{jk} = \frac{2i\overline{\rho}\omega R_h}{l}\sum_m\sum_n\sum_s\frac{\phi_m(k_{mn}R_h)e^{im\varphi}}{\Gamma_s}\frac{\phi_m(k_{mn}R_h)}{k_0^2-k_{mns}^2}\mathfrak{R}_{s,j,k}, \tag{6.107}$$

we have

$$\frac{2}{l}\int_0^l p_s'^{+}\sin\frac{j\pi z}{l}dz=\sum_{k=1}^{\infty}V_k z_{jk}^{+}, \tag{6.108}$$

where

$$\Gamma_s=\begin{cases} l & s=0\\ \dfrac{l}{2} & s\neq 0\end{cases},$$

$$\mathfrak{R}_{s,j,k}=\int_0^l \cos\left(\frac{s\pi}{l}z\right)\sin\frac{j\pi z}{l}\int_0^l \sin\frac{k\pi z'}{l}\cos\left(\frac{s\pi}{l}z'\right)dz'dz$$

$$\int_0^l \cos\frac{k\pi z'}{l}\sin\frac{s\pi z'}{l}dz'=\begin{cases} -\dfrac{l}{s\pi}(\cos s\pi-1) & k=0\\ 0 & k\neq 0, s+k\,\text{is even}\\ \dfrac{2sl}{(s^2-k^2)\pi} & k\neq 0, s+k\,\text{is odd}\end{cases},$$

$$\int_0^l \cos\frac{s\pi z}{l}\sin\frac{j\pi z}{l}dz=\begin{cases} -\dfrac{l}{j\pi}(\cos j\pi-1) & k=0\\ 0 & k\neq 0, j+s\,\text{is even}\\ \dfrac{2jl}{(j^2-s^2)\pi} & k\neq 0, j+s\,\text{is odd}\end{cases}.$$

In this way, we have established the relationship between the sound fields both in the duct and in the back cavity, and the vibration velocity of the air column in the hole of the perforated plate. On this basis, we can develop the TEM by computing the sound field in a duct lined with a nonlocally reacting liner.

As shown in Fig. 6.15, we consider a straight duct for brevity. The transfer matrix for a nonlocally reacting liner and that of locally reacting liner are similar, whereas the expression of $p_s'^{-}$ is much more complicated for a

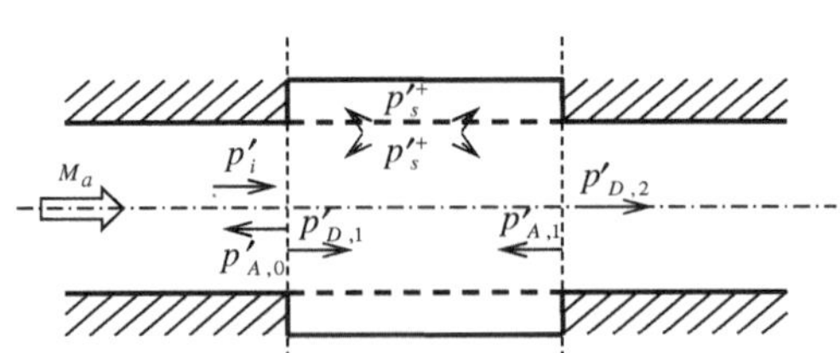

Fig. 6.15 The transfer element for nonlocally reacting liner.

nonlocally reacting liner. For a nonlocally reacting liner, the incident sound pressure wave is:

$$p_i' = p_{D,1}' + p_{A,1}' = \sum_{\mu} A_{m\mu}^1 \phi_{m\mu}^1 \left(k_{m\mu}^1 r\right) e^{i\gamma_{m\mu}^{1+}(z-l)} + \sum_{\mu} D_{m\mu}^1 \phi_{m\mu}^1 \left(k_{m\mu}^1 r\right) e^{i\gamma_{m\mu}^{1-} z}. \tag{6.109}$$

Then, Eq. (6.98) can be written as:

$$\sum_{k=1}^{\infty} \left(z_{jk}^- - z_{jk}^+ + \delta_{jk} Z_{zk}\right) V_k = \sum_{\mu} \phi_{m\mu}^1 \left(k_{m\mu}^1 R\right) \left[D_{m\mu}^1 I_j^{D_{m\mu}^1} + A_{m\mu}^1 I_j^{A_{m\mu}^1}\right], \tag{6.110}$$

where the expressions of $I_j^{D_{m\mu}^1}$ and $I_j^{A_{m\mu}^1}$ are the same as for a locally reacting liner. Solving this equation, the vibration velocity in the form of the matrix product is rewritten as:

$$V_k^{m,\mu} = \sum_{\mu} \phi_{m\mu}^1 \left(k_{m\mu}^1 r\right) \sum_{j} \left[z_{jk}^- - z_{jk}^+ + \delta_{jk} Z_{zk}\right]^{-1} \left[D_{m\mu}^1 I_j^{D_{m\mu}^1} + A_{m\mu}^1 I_j^{A_{m\mu}^1}\right]. \tag{6.111}$$

Note that Z_{zk} here only represents the impedance of the perforated plate. Substitute Eq. (6.111) into the expression of $p_s'^-$, we have:

$$\begin{aligned} p_s'^-\left(\vec{x}, t\right) &= \sum_{n} p_{smn} \phi_{mn}(k_{mn} r) \cdot e^{im\varphi} \cdot e^{i\gamma_{m,n}^{\pm} z} \\ &= \frac{\bar{\rho}}{2} \sum_{n=1}^{\infty} \frac{\phi_{mn}(k_{mn} r) \cdot e^{im\varphi} \cdot e^{i\gamma_{m,n}^{\pm} z}}{2\pi} \left\{\sum_{\mu} D_{m\mu}^1 Q_{n\pm}^{D_{m\mu}} + A_{m\mu}^1 Q_{n\pm}^{A_{m\mu}}\right\}, \end{aligned} \tag{6.112}$$

where the coefficients of $D_{m\mu}^1$ and $A_{m\mu}^1$ are:

$$\begin{cases} Q_{n\pm}^{D_{m\mu}} = \sum_{k} \phi_{m\mu}^1 \left(k_{m\mu}^1 R\right) \sum_{j} \left[z_{jk}^- - z_{jk}^+ + \delta_{jk} Z_{zk}\right]^{-1} I_j^{D_{m\mu}^1} S_k^{n\pm} \\ Q_{n\pm}^{A_{m\mu}} = \sum_{k} \phi_{m\mu}^1 \left(k_{m\mu}^1 R\right) \sum_{j} \left[z_{jk}^- - z_{jk}^+ + \delta_{jk} Z_{zk}\right]^{-1} I_j^{A_{m\mu}^1} S_k^{n\pm} \end{cases}. \tag{6.113}$$

After $p_s'^-$ is expressed as the explicit function of the amplitudes of sound waves with hard wall boundary condition $D_{m\mu}^1$ and $A_{m\mu}^1$, it can be substituted into the transfer element, and the solving process is exactly the same as that of the locally reacting liner. Note that the sound pressure in the back cavity can also be expressed as a function of $D_{m\mu}^1$ and $A_{m\mu}^1$. After solving the algebraic matrix equation of the transfer element to calculate $D_{m\mu}^1$ and $A_{m\mu}^1$, the

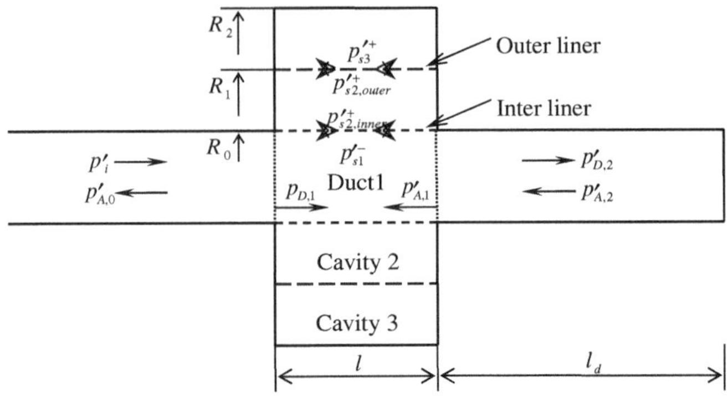

Fig. 6.16 The double-layer nonlocally reacting liner.

expression of the vibration velocity of the air column in the holes of perforated plate can then be obtained. Substituting the result of vibration velocity into Eq. (6.105), the sound field in the back cavity can be obtained.

So far, we have established the transfer element model suitable for sound propagation in ducts lined with a nonlocally reacting liner. Note that the transfer matrix of the element for the nonlocally reacting liner is relatively independent of the external conditions. This means that this method can handle a series of sound problems in the duct, which is the advantage of the method.

When there are multi-nonlocally reacting liners in the duct, the transfer matrix of a new element can also be derived like the procedure described earlier. Here is the case for a two-layer nonlocally reacting liner.

As shown in Fig. 6.16, for the two inner and outer acoustic liners, the acoustic admittance equations can be written as:

$$\begin{cases} p'^{+}_{s2,outer} + p'^{+}_{s2,inner} - p'^{-}_{s1} - p'_i = \dfrac{i\omega\bar{\rho}}{\eta_{inner}} V_n^{f,inner} \\ p'^{+}_{s3,outer} - p'^{+}_{s2,outer} - p'^{+}_{s2,inner} = \dfrac{i\omega\bar{\rho}}{\eta_{outer}} V_n^{f,outer} \end{cases}. \tag{6.114}$$

These two relations can be transformed into algebraic equations by using the sinusoid transformation. $\sin\frac{j\pi z}{l}$ and $\sin\frac{j'\pi z}{l}$ have been used, and the relation between p_s' and V_n^f has been considered. The terms related with scattering sound waves are shown in the form of the equivalent radiation impedances:

$$\begin{cases} \sum\limits_{k'=1}^{\infty} z^{+}_{jk',2} V_{k'}^{outer} + \sum\limits_{k=1}^{\infty} z^{+}_{jk,2} V_{k}^{inner} - \sum\limits_{k=1}^{\infty} z^{-}_{jk,1} V_{k}^{inner} - \dfrac{i\omega\bar{\rho}}{\eta_{inner}} V_{j}^{inner} = I_j \\ \sum\limits_{k'=1}^{\infty} z^{+}_{j'k',3} V_{k'}^{outer} - \sum\limits_{k'=1}^{\infty} z^{+}_{j'k',2} V_{k'}^{outer} - \sum\limits_{k=1}^{\infty} z^{+}_{j'k,2} V_{k}^{inner} - \dfrac{i\omega\bar{\rho}}{\eta_{outer}} V_{j'}^{outer} = 0 \end{cases}.$$

(6.115a,b)

Here, the series expansion of the vibration velocities at outer and inner layers are:

$$V_n^{f,outer} = \sum_{k'=1}^{\infty} V_{k'}^{outer} \cdot \sin\frac{k'\pi z}{l}, \quad V_n^{f,inner} = \sum_{k=1}^{\infty} V_k^{inner} \cdot \sin\frac{k\pi z}{l}. \tag{6.116}$$

In the equivalent radiation impedances, the numbers appearing in subscripts represent Duct 1, Cavity 2, and Cavity 3, respectively, corresponding to the scattering sound fields in different cavities. By solving the admittance of Eq. (6.115b), the velocity of the air column in the outer-layer holes is written as a function of the velocity of the inner-layer holes by:

$$V_{k'}^{outer} = \sum_{j'=1}^{\infty}\left(z_{j'k',3}^{+} - z_{j'k',2}^{+} - \delta_{j'k'}\frac{i\omega\bar{\rho}}{\eta_{outer}}\right)^{-1}\left(\sum_{k=1}^{\infty} z_{j'k,2}^{+} V_k^{inner}\right). \tag{6.117}$$

Substituting this into Eq. (6.115a) and rearranging the equation yields:

$$\begin{aligned}\sum_{k=1}^{\infty}\Bigg[\sum_{j'=1}^{\infty}\sum_{k'=1}^{\infty} z_{jk',2}^{+}\left(z_{j'k',3}^{+} - z_{j'k',2}^{+} - \delta_{j'k'}\frac{i\omega\bar{\rho}}{\eta_{outer}}\right)^{-1} z_{j'k,2}^{+} \\ + z_{jk,2}^{+} - z_{jk,1}^{-} - \delta_{jk}\frac{i\omega\bar{\rho}}{\eta_{inner}}\Bigg] \cdot V_k^{inner} = I_j.\end{aligned} \tag{6.118}$$

Let:

$$\begin{aligned}\mathfrak{I}_{jk} = \sum_{j'=1}^{\infty}\sum_{k'=1}^{\infty} z_{jk',2}^{+}\left(z_{j'k',3}^{+} - z_{j'k',2}^{+} - \delta_{j'k'}\frac{i\omega\bar{\rho}}{\eta_{outer}}\right)^{-1} z_{j'k,2}^{+} + z_{jk,2}^{+} - z_{jk,1}^{-} \\ - \delta_{jk}\frac{i\omega\bar{\rho}}{\eta_{inner}},\end{aligned} \tag{6.119}$$

then, the original equation can be written as:

$$\sum_{k=1}^{\infty} \mathfrak{I}_{jk} \cdot V_k^{inner} = I_j. \tag{6.120}$$

Therefore, the vibration velocity of the air column in the holes of perforated plate can be written as:

$$V_k^{inner} = \sum_{j=1}^{\infty} \mathfrak{I}_{jk}^{-1} I_j. \tag{6.121}$$

When the vibration velocity is written as a function of the incident sound waves, it can be directly applied to the TEM introduced earlier, and the

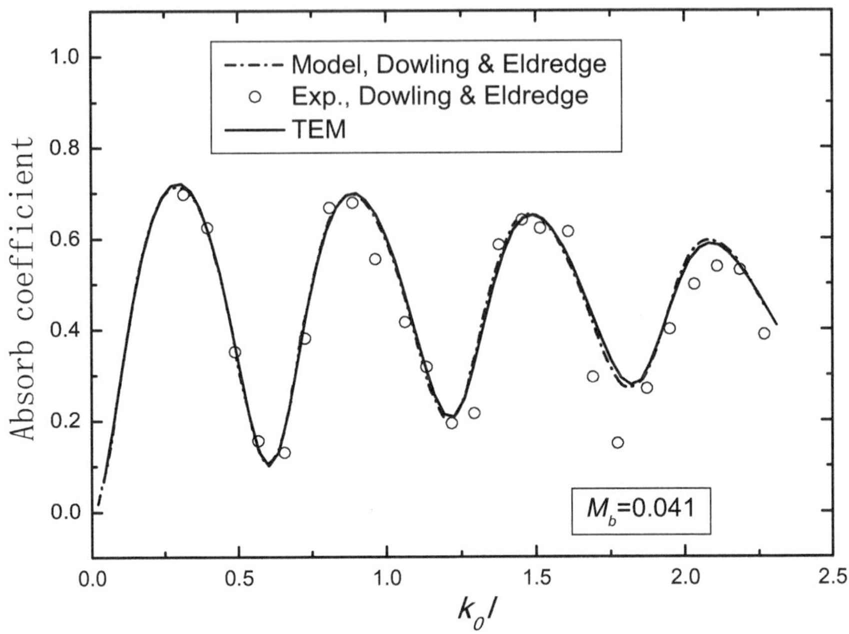

Fig. 6.17 The absorption coefficients of the double layer nonlocally reacting liner (see [48] for Eldredge's results).

problem can be solved directly. The derivation of the formula shows that the model can be applied to the calculation of multilayer acoustic liners. It is important to note that, in the model, the equivalent radiation impedance is different for annular ducts with different sizes. In addition, we need to pay special attention to the summation process of different series when solving $\mathfrak{J}_{jk}$.

Fig. 6.17 gives the calculation result of the double-layer acoustic liner. The right termination of the duct is an open end, and the relationship between the reflected sound wave and the incident sound wave is:

$$\frac{p'_{A,2}}{p'_{D,2}} = -\left[1 - 0.5 \cdot (k_0 \cdot R_0)^2\right] \cdot \exp\left[-i2k_0(l_d + \chi)\right], \tag{6.122}$$

where $\chi \approx 0.6R_0$ is the end correction factor. It can be seen that the reflection coefficient at the open end is a function of the frequency, the radius, and the length of the duct. After the end reflection ratio is given, the number of unknowns in the TEM is still four groups, which does not increase but only includes the total result of the two waves at the ejection end (at the matching condition of the right side of the element). The definition of sound absorption coefficient is:

$$\Delta = 1 - \frac{|p'_{A,0}|^2 + |p'_{D,2}|^2}{|p'_i|^2 + |p'_{A,2}|^2}. \tag{6.123}$$

Note that, in the range of the calculated frequencies, the only cut-on modal waves in the duct are the plane waves. For plane waves, the ratio of the acoustic energies is the same as that of the squared amplitudes of sound pressures. So, this equation represents the sound energy absorbed by the acoustic liner.

6.3.5 Construction of transfer element for a varying cross-section duct

Both the convergent and divergent ducts are varying cross-section ducts, which can be seen in Fig. 6.18A and C.

Assuming that the area change of the cross-section is relatively small, that is, $\Delta A \ll A_1,\ A_2$. Here, we try to establish the matching equation at the interface. The continuity equation and momentum equation are:

$$\rho_1 A_1 U_1 = \rho_2 A_2 U_2, \tag{6.124}$$

$$(\rho_1 U_1)U_1 - (\rho_2 U_2)U_2 = p_2 - p_1, \tag{6.125}$$

where $\rho = \overline{\rho} + \rho'$, $U = U_0 + u'$, $p = \overline{p} + p'$, with $\overline{\rho},\ U,\ \overline{p}$ the averaged variables and $\rho',\ u',\ p'$ the disturbing variables. Linearizing these equations results in:

$$\left(\overline{\rho}_1 u'_1 + \rho'_1 U_{01}\right) A_1 = \left(\overline{\rho}_2 u'_2 + \rho'_2 U_{02}\right) A_2, \tag{6.126}$$

$$p'_1 + \rho'_1 U_1^2 + 2\overline{\rho}_1 U_1 V'_1 = p'_2 + \rho'_2 U_2^2 + 2\overline{\rho}_2 U_1 V'_2 \tag{6.127}$$

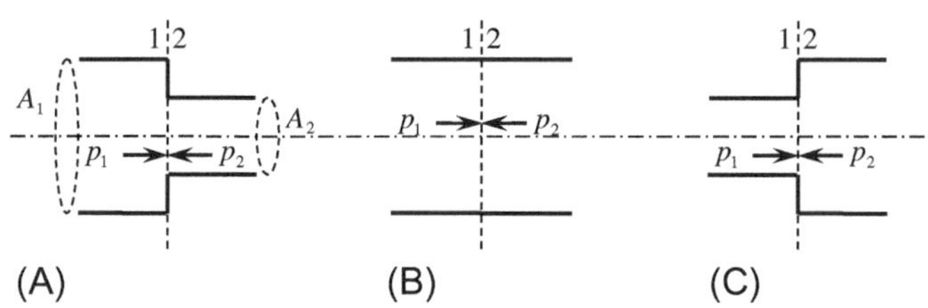

Fig. 6.18 Matching conditions for non-uniform duct. (A) expansive duct (B) uniform duct (C) contractive duct.

Substituting $p' = \rho' \cdot c_0^2$ into the equations leads to:

$$\left(\overline{\rho}_1 V_1' + p_1' \frac{U_1}{c_1^2}\right) A_1 = \left(\overline{\rho}_2 V_2' + p_2' \frac{U_2}{c_2^2}\right) A_2, \tag{6.128}$$

$$p_1' + p_1' \frac{U_1^2}{c_1^2} + 2\overline{\rho}_1 U_1 V_1' = p_2' + p_2' \frac{U_2^2}{c_2^2} + 2\overline{\rho}_2 U_1 V_2'. \tag{6.129}$$

Here, take the divergent duct as an example, namely $A_1 < A_2$. The continuity equation can be written as:

$$\left(\overline{\rho}_1 u_1' + \rho_1' U_{01}\right) A_1 = \left(\overline{\rho}_2 u_2' + \rho_2' U_{02}\right)(A_1 + \Delta A). \tag{6.130}$$

For the points on ΔA,

$$u_2' = 0. \tag{6.131}$$

Therefore:

$$\overline{\rho}_1 u_1' + p_1' \frac{U_{01}}{c_{01}^2} = \overline{\rho}_2 u_2' + p_2' \frac{U_{02}}{c_{02}^2}\left(1 + \frac{\Delta A}{A_1}\right) \tag{6.132}$$

To reduce the flow loss in the case of flow, the area change is assumed to be so small ($\Delta A \ll A_1$) that area change can be negligible. Then,

$$\overline{\rho}_1 u_1' + p_1' \frac{U_{01}}{c_{01}^2} = \overline{\rho}_2 u_2' + p_2' \frac{U_{02}}{c_{02}^2}, \tag{6.133}$$

$$p_1'\left(1 + Ma_1^2\right) + 2\overline{\rho}_1 U_{01} u_1' = p_2'\left(1 + Ma_2^2\right) + 2\overline{\rho}_2 U_{02} u_2'. \tag{6.134}$$

Eqs. (6.133), (6.134), and (6.131) are the matching equations in a variable cross-section duct. For no flow or straight-duct cases, the matching equations degenerate into:

$$u_1' = u_2', \quad p_1' = p_2'. \tag{6.135}$$

According to the matching condition of the varying cross-section duct, it is not difficult to see that the sound propagation problem concerned here is the non-uniform duct with a slightly changed cross-section in the axial direction. The flow field is not strongly influenced by the area change of the cross-section. This is often right in the aeroengine duct. All discussion and analysis here are within the scope of the non-uniform duct with a slightly changed cross-section in the axial direction.

As shown in Fig. 6.19, consider a convergent duct. We take this as an example to discuss the establishment of transfer element equations. There

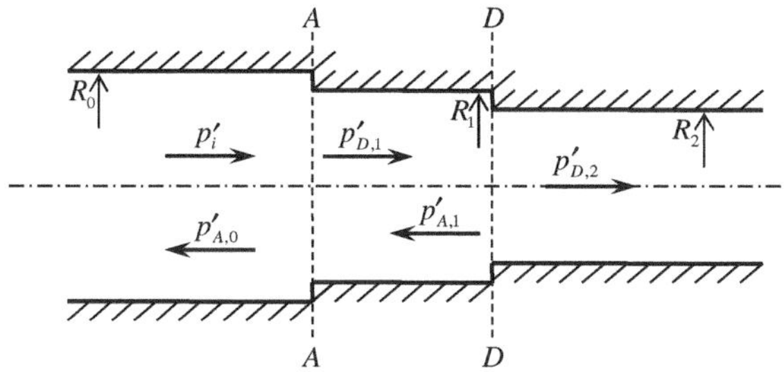

Fig. 6.19 A schematic diagram of the matching interfaces of a convergent duct.

are five groups of sound waves inside the duct. The middle part between interfaces A and D is a typical transfer element. Inside the element, there are two groups of sound waves:

$$p'_{A,1} = \sum_m \sum_n A^1_{mn} \phi^1_{mn}\left(k^1_{mn} r\right) e^{i\gamma^{1+}_{mn}(z-l)}, \tag{6.136}$$

$$p'_{D,1} = \sum_m \sum_n D^1_{mn} \phi^1_{mn}\left(k^1_{mn} r\right) e^{i\gamma^{1-}_{mn} z}. \tag{6.137}$$

There are incident and reflected sound waves at the left side of interface A:

$$p'_i = \sum_m \sum_n p_{imn} \phi^0_{mn}\left(k^0_{mn} r\right) e^{i\gamma^{0-}_{mn} z}, \tag{6.138}$$

$$p'_{A,0} = \sum_m \sum_n A^0_{mn} \phi^0_{mn}\left(k^0_{mn} r\right) e^{i\gamma^{0+}_{mn} z}. \tag{6.139}$$

There is a transmitted sound wave at the right side of interface D:

$$p'_{D,2} = \sum_m \sum_n D^2_{mn} \phi^2_{mn}\left(k^2_{mn} r\right) e^{i\gamma^{2-}_{mn}(z-l)}, \tag{6.140}$$

where $m = -M, \ldots, 0, \ldots M$; $n = 1, 2, \ldots, N$. The particle velocities can be obtained from the momentum equation given by:

$$\frac{\partial u'}{\partial t} + U\frac{\partial u'}{\partial z} = -\frac{1}{\rho_0}\frac{\partial p'}{\partial z}. \tag{6.141}$$

Substituting the expressions both for sound pressures and particle velocities into Eq. (6.135), we can obtain the matching equations at the interfaces A and Ddescribed by:

$$
\begin{aligned}
&\left\{\sum_m\sum_n A^0_{mn}\phi^0_{mn}\left(k^0_{mn}r\right)\frac{U_0}{c_0^2}-\sum_m\sum_n A^0_{mn}\phi^0_{mn}\left(k^0_{mn}r\right)\frac{\gamma^{0+}_{mn}}{U_0\gamma^{0+}_{mn}+\omega}\right\}\\
&-\left\{\sum_m\sum_n D^1_{mn}\phi^1_{mn}\left(k^1_{mn}r\right)\frac{U_1}{c_1^2}-\sum_m\sum_n D^1_{mn}\phi^1_{mn}\left(k^1_{mn}r\right)\frac{\gamma^{1-}_{mn}}{U_1\gamma^{1-}_{mn}+\omega}\right\}\\
&-\left\{\sum_m\sum_n A^1_{mn}\phi^1_{mn}\left(k^1_{mn}r\right)e^{-i\gamma^{1+}_{mn}l}\frac{U_1}{c_1^2}\right.\\
&\left.-\sum_m\sum_n A^1_{mn}\phi^1_{mn}\left(k^1_{mn}r\right)\frac{\gamma^{1+}_{mn}}{U_1\gamma^{1+}_{mn}+\omega}e^{-i\gamma^{1+}_{mn}l}\right\},\quad r\in(0,R_1)\\
&=-\left\{\sum_m\sum_n p_{imn}\phi^0_{mn}\left(k^0_{mn}r\right)\frac{U_0}{c_0^2}-\sum_m\sum_n p_{imn}\phi^0_{mn}\left(k^0_{mn}r\right)\frac{\gamma^{0-}_{mn}}{U_0\gamma^{0-}_{mn}+\omega}\right\}
\end{aligned}
\tag{6.142}
$$

$$
\begin{aligned}
&\left\{\left(1+Ma_0^2\right)\sum_m\sum_n A^0_{mn}\phi^0_{mn}\left(k^0_{mn}r\right)-2U_0\sum_m\sum_n A^0_{mn}\phi^0_{mn}\left(k^0_{mn}r\right)\frac{\gamma^{0+}_{mn}}{U_0\gamma^{0+}_{mn}+\omega}\right\}\\
&-\left\{\left(1+Ma_1^2\right)\sum_m\sum_n D^1_{mn}\phi^1_{mn}\left(k^1_{mn}r\right)-2U_1\sum_m\sum_n D^1_{mn}\phi^1_{mn}\left(k^1_{mn}r\right)\frac{\gamma^{1-}_{mn}}{U_1\gamma^{1-}_{mn}+\omega}\right\}\\
&-\left\{\left(1+Ma_1^2\right)\sum_m\sum_n A^1_{mn}\phi^1_{mn}\left(k^1_{mn}r\right)e^{-i\gamma^{1+}_{mn}l}\right.\\
&\left.-2U_1\sum_m\sum_n A^1_{mn}\phi^1_{mn}\left(k^1_{mn}r\right)\frac{\gamma^{1+}_{mn}}{U_1\gamma^{1+}_{mn}+\omega}e^{-i\gamma^{1+}_{mn}l}\right\},\quad r\in(0,R_1)\\
&=-\left\{\left(1+Ma_0^2\right)\sum_m\sum_n p_{imn}\phi^0_{mn}\left(k^0_{mn}r\right)-2U_0\sum_m\sum_n p_{imn}\phi^0_{mn}\left(k^0_{mn}r\right)\frac{\gamma^{0-}_{mn}}{U_0\gamma^{0-}_{mn}+\omega}\right\}
\end{aligned}
\tag{6.143}
$$

$$
\begin{aligned}
&\sum_m\sum_n p_{imn}\phi^0_{mn}\left(k^0_{mn}r\right)\frac{\gamma^{0-}_{mn}}{U_0\gamma^{0-}_{mn}+\omega}\\
&+\sum_m\sum_n A^0_{mn}\phi^0_{mn}\left(k^0_{mn}r\right)\frac{\gamma^{0+}_{mn}}{U_0\gamma^{0+}_{mn}+\omega}=0,\quad r\in(R_1,R_0),
\end{aligned}
\tag{6.144}
$$

$$\left\{\sum_m\sum_n D^1_{mn}\phi^1_{mn}\left(k^1_{mn}r\right)e^{i\gamma^{1-}_{mn}l}\frac{U_1}{c_1^2}-\sum_m\sum_n D^1_{mn}\phi^1_{mn}\left(k^1_{mn}r\right)\frac{\gamma^{1-}_{mn}}{U_1\gamma^{1-}_{mn}+\omega}e^{i\gamma^{1-}_{mn}l}\right\}$$
$$+\left\{\sum_m\sum_n A^1_{mn}\phi^1_{mn}\left(k^1_{mn}r\right)\frac{U_1}{c_1^2}-\sum_m\sum_n A^1_{mn}\phi^1_{mn}\left(k^1_{mn}r\right)\frac{\gamma^{1+}_{mn}}{U_1\gamma^{1+}_{mn}+\omega}\right\},\quad r\in(0,R_2)$$
$$-\left\{\frac{U_{02}}{c_{02}^2}\sum_m\sum_n D^2_{mn}\phi^2_{mn}\left(k^2_{mn}r\right)-\sum_m\sum_n D^2_{mn}\phi^2_{mn}\left(k^2_{mn}r\right)\frac{\gamma^{2-}_{mn}}{U_2\gamma^{2-}_{mn}+\omega}\right\}=0 \tag{6.145}$$

$$\left\{\left(1+Ma_1^2\right)\sum_m\sum_n D^1_{mn}\phi^1_{mn}\left(k^1_{mn}r\right)e^{i\gamma^{1-}_{mn}l}\right.$$
$$\left.-2U_1\sum_m\sum_n D^1_{mn}\phi^1_{mn}\left(k^1_{mn}r\right)\frac{\gamma^{1-}_{mn}}{U_1\gamma^{1-}_{mn}+\omega}e^{i\gamma^{1-}_{mn}l}\right\}$$
$$\left\{\left(1+Ma_1^2\right)\sum_m\sum_n A^1_{mn}\phi^1_{mn}\left(k^1_{mn}r\right)\right.$$
$$\left.-2U_1\sum_m\sum_n A^1_{mn}\phi^1_{mn}\left(k^1_{mn}r\right)\frac{\gamma^{1+}_{mn}}{U_1\gamma^{1+}_{mn}+\omega}\right\},\quad r\in(0,R_2) \tag{6.146}$$
$$-\left\{\left(1+Ma_2^2\right)\sum_m\sum_n D^2_{mn}\phi^2_{mn}\left(k^2_{mn}r\right)\right.$$
$$\left.-2U_2\sum_m\sum_n D^2_{mn}\phi^2_{mn}\left(k^2_{mn}r\right)\frac{\gamma^{2-}_{mn}}{U_2\gamma^{2-}_{mn}+\omega}\right\}=0$$

$$\sum_m\sum_n A^1_{mn}\phi^1_{mn}\left(k^1_{mn}r\right)\frac{\gamma^{1+}_{mn}}{U_1\gamma^{1+}_{mn}+\omega}$$
$$+\sum_m\sum_n D^1_{mn}\phi^1_{mn}\left(k^1_{mn}r\right)\frac{\gamma^{1-}_{mn}}{U_1\gamma^{1-}_{mn}+\omega}e^{i\gamma^{1-}_{mn}l}=0,\quad r\in(R_2,R_1) \tag{6.147}$$

The common methods to compute these six equations are the allocation point and orthogonal methods, but the allocation point method is difficult to achieve the required precision, especially in the varying cross-section duct. So, in the process of solving the equations, the orthogonal solution method should be adopted.

Multiplying Eq. (6.142) with $\phi^0_{mn}(k^0_{m'n'}r)r$ on both sides, and Eq. (6.143) with $\phi^1_{mn}(k^1_{m'n'}r)r$ on both sides, at the same time, considering the hard wall boundary condition and using the orthogonality of the eigenfunction, we obtain:

$$\left\{\sum_m\sum_n A^0_{mn}\int_0^{R_1}\phi^0_{mn}\left(k^0_{mn}r\right)\phi^0_{mn}\left(k^0_{m'n'}r\right)rdr\frac{U_0}{c_0^2}-\frac{A^0_{m'n'}\gamma^{0+}_{m'n'}}{U_0\gamma^{0+}_{m'n'}+\omega}\right\}$$
$$-\sum_m\sum_n D^1_{mn}\int_0^{R_1}\phi^1_{mn}\left(k^1_{mn}r\right)\phi^0_{mn}\left(k^0_{m'n'}r\right)rdr\left\{\frac{U_1}{c_1^2}-\frac{\gamma^{1-}_{mn}}{U_1\gamma^{1-}_{mn}+\omega}\right\}$$
$$-\sum_m\sum_n A^1_{mn}\int_0^{R_1}\phi^1_{mn}\left(k^1_{mn}r\right)\phi^0_{mn}\left(k^0_{m'n'}r\right)rdre^{-i\gamma^{1+}_{mn}l}\left\{\frac{U_1}{c_1^2}-\frac{\gamma^{1+}_{mn}}{U_1\gamma^{1+}_{mn}+\omega}\right\},$$
$$=-\left\{\sum_m\sum_n p_{imn}\int_0^{R_1}\phi^0_{mn}\left(k^0_{mn}r\right)\phi^0_{mn}\left(k^0_{m'n'}r\right)rdr\frac{U_0}{c_0^2}-\frac{p_{im'n'}\gamma^{0-}_{m'n'}}{U_0\gamma^{0-}_{m'n'}+\omega}\right\}$$
$$m'=-M,\cdots,0,\cdots M \qquad n'=1,2,\cdots,N \tag{6.148}$$

$$\left\{\left(1+Ma_0^2\right)\sum_m\sum_n A^0_{mn}\int_0^{R_1}\phi^0_{mn}\left(k^0_{mn}r\right)\phi^0_{mn}\left(k^0_{m'n'}r\right)rdr-2U_0\frac{A^0_{m'n'}\gamma^{0+}_{m'n'}}{U_0\gamma^{0+}_{m'n'}+\omega}\right\}$$
$$-\sum_m\sum_n D^1_{mn}\int_0^{R_1}\phi^1_{mn}\left(k^1_{mn}r\right)\phi^0_{mn}\left(k^0_{m'n'}r\right)rdr\left\{\left(1+Ma_1^2\right)-\frac{2U_1\gamma^{1-}_{mn}}{U_1\gamma^{1-}_{mn}+\omega}\right\}$$
$$-\sum_m\sum_n A^1_{mn}\int_0^{R_1}\phi^1_{mn}\left(k^1_{mn}r\right)\phi^0_{mn}\left(k^0_{m'n'}r\right)rdre^{-i\gamma^{1+}_{mn}l}\left\{\left(1+Ma_1^2\right)-\frac{2U_1\gamma^{1+}_{mn}}{U_1\gamma^{1+}_{mn}+\omega}\right\},$$
$$=-\left\{\left(1+Ma_0^2\right)\sum_m\sum_n p_{imn}\int_0^{R_1}\phi^0_{mn}\left(k^0_{mn}r\right)\phi^0_{mn}\left(k^0_{m'n'}r\right)rdr-2U_0\frac{p_{im'n'}\gamma^{0-}_{m'n'}}{U_0\gamma^{0-}_{m'n'}+\omega}\right\}$$
$$m'=-M,\cdots,0,\cdots M \qquad n'=1,2,\cdots,N \tag{6.149}$$

$$\left\{\sum_m\sum_n D^1_{mn}\int_0^{R_2}\phi^1_{mn}\left(k^1_{mn}r\right)\phi^1_{mn}\left(k^1_{m'n'}r\right)rdr\frac{U_1}{c_1^2}e^{i\gamma^{1-}_{mn}l}-\frac{D^1_{m'n'}\gamma^{1-}_{m'n'}e^{i\gamma^{1-}_{m'n'}l}}{U_1\gamma^{1-}_{m'n'}+\omega}\right\}$$

$$+\left\{\sum_m\sum_n A^1_{mn}\int_0^{R_2}\phi^1_{mn}\left(k^1_{mn}r\right)\phi^1_{mn}\left(k^1_{m'n'}r\right)rdr\frac{U_1}{c_1^2}-\frac{A^1_{m'n'}\gamma^{1+}_{m'n'}}{U_1\gamma^{1+}_{m'n'}+\omega}\right\},$$

$$-\sum_m\sum_n D^2_{mn}\int_0^{R_2}\phi^2_{mn}\left(k^2_{mn}r\right)\phi^1_{mn}\left(k^1_{m'n'}r\right)rdr\left\{\frac{U_{02}}{c_{02}^2}-\frac{\gamma^{2-}_{mn}}{U_2\gamma^{2-}_{mn}+\omega}\right\}=0$$

$$m'=-M,\cdots,0,\cdots M \qquad n'=1,2,\cdots,N \tag{6.150}$$

$$\left\{\left(1+Ma_1^2\right)\sum_m\sum_n D^1_{mn}\int_0^{R_2}\phi^1_{mn}\left(k^1_{mn}r\right)\phi^1_{mn}\left(k^1_{m'n'}r\right)rdre^{i\gamma^{1-}_{mn}l}\right.$$

$$\left.-2U_1\frac{D^1_{m'n'}\gamma^{1-}_{m'n'}e^{i\gamma^{1-}_{m'n'}l}}{U_1\gamma^{1-}_{m'n'}+\omega}\right\}$$

$$\left\{\left(1+Ma_1^2\right)\sum_m\sum_n A^1_{mn}\int_0^{R_2}\phi^1_{mn}\left(k^1_{mn}r\right)\phi^1_{mn}\left(k^1_{m'n'}r\right)rdr-2U_1\frac{A^1_{m'n'}\gamma^{1+}_{m'n'}}{U_1\gamma^{1+}_{m'n'}+\omega}\right\},$$

$$-\sum_m\sum_n D^2_{mn}\int_0^{R_2}\phi^2_{mn}\left(k^2_{mn}r\right)\phi^1_{mn}\left(k^1_{m'n'}r\right)rdr\left\{\left(1+Ma_2^2\right)-2U_2\frac{\gamma^{2-}_{mn}}{U_2\gamma^{2-}_{mn}+\omega}\right\}=0$$

$$m'=-M,\cdots,0,\cdots M \qquad n'=1,2,\cdots,N \tag{6.151}$$

These four equation sets are then the matching equations of the convergent duct as shown in Fig. 6.19. There is a total of $4\times 2M\times N$ unknowns in these equations, that is, A^0, D^1, A^1, D^2. The number of these equations is also $4\times 2M\times N$(each equation set consists of $2M\times N$ equations). So, the equation sets are closed. Rewriting these equations into the form of a matrix product leads to:

$$\begin{bmatrix} ss^{am}_{A^0} & ss^{am}_{D^1} & ss^{am}_{A^1} & 0 \\ ss^{af}_{A^0} & ss^{af}_{D^1} & ss^{af}_{A^1} & 0 \\ 0 & ss^{dm}_{D^1} & ss^{dm}_{A^1} & ss^{dm}_{D^2} \\ 0 & ss^{df}_{D^1} & ss^{df}_{A^1} & ss^{df}_{D^2} \end{bmatrix}\begin{bmatrix} A^0_{mn} \\ D^1_{mn} \\ A^1_{mn} \\ D^2_{mn}\end{bmatrix}=\begin{bmatrix} p_i^{am} \\ p_i^{af} \\ 0 \\ 0\end{bmatrix}, \tag{6.152}$$

where, ss represents a $2M \times N$ coefficient matrix, in which the superscript a denotes the interface and the subscript corresponds to the unknowns. m and f correspond to the continuous equation and momentum equation, respectively. For example, a specific expression of $ss_{A^0}{}^{am}$ is given by:

$$ss_{A^0}^{am} = \begin{bmatrix} cs_1^1 & \cdots & cs_{2M\times N}^1 \\ & \ddots & \\ \vdots & cs_j^i & \vdots \\ & \ddots & \\ cs_1^{2M\times N} & \cdots & cs_{2M\times N}^{2M\times N} \end{bmatrix} + \begin{bmatrix} sc_1 & 0 & \cdots & 0 \\ 0 & \ddots & & \vdots \\ \vdots & & sc_i & \\ & & \ddots & 0 \\ 0 & \cdots & 0 & sc_{2M\times N} \end{bmatrix}, \tag{6.153}$$

where

$$cs_j^i = \frac{U_0}{c_0^2}\int_0^{R_1} \phi_{mn}^0\left(k_{mn}^0 r\right)\phi_{m'n'}^0\left(k_{m'n'}^0 r\right) r dr$$
$$= \begin{cases} \dfrac{U_0}{c_0^2}\dfrac{R_1^2}{2}\left\{\left[\phi'^0_{m'n'}\left(k_{m'n'}^0 R_1\right)\right]^2 + \left[1 - \dfrac{m^2}{\left(k_{m'n'}^0 R_1\right)^2}\right]\cdot\left[\phi_{m'n'}^0\left(k_{m'n'}^0 R_1\right)\right]^2\right\} \\ m=m', n=n' \\ \dfrac{U_0}{c_0^2}\dfrac{R_1 k_{mn}^0 \phi'^0_{mn}\left(k_{mn}^0 R_1\right)\phi_{m'n'}^0\left(k_{m'n'}^0 R_1\right) - R_1 k_{m'n'}^0 \phi'^0_{m'n'}\left(k_{m'n'}^0 R_1\right)\phi_{mn}^0\left(k_{mn}^0 R_1\right)}{\left(k_{m'n'}^0\right)^2 - \left(k_{mn}^0\right)^2} \\ m=m', n\neq n' \end{cases}, \tag{6.154}$$

and

$$sc_i = \frac{\gamma_{m'n'}^{0+}}{U_0\gamma_{m'n'}^{0+} + \omega}. \tag{6.155}$$

In addition, the integral relation of eigenfunctions in different ducts can be written as follows:

$$\int_0^{R_1} \phi_{mn}^1\left(k_{mn}^1 r\right)\phi_{mn}^0\left(k_{m'n'}^0 r\right) r dr = \frac{R_1 k_{m'n'}^0 \phi'^0_{m'n'}\left(k_{m'n'}^0 R_1\right)\phi_{mn}^1\left(k_{mn}^1 R_1\right)}{\left(k_{mn}^1\right)^2 - \left(k_{m'n'}^0\right)^2}. \tag{6.156}$$

The internal coefficient matrix of the transfer element is:

$$\begin{bmatrix} ss_{D^1}^{am} & ss_{A^1}^{am} \\ ss_{D^1}^{af} & ss_{A^1}^{af} \\ ss_{D^1}^{dm} & ss_{A^1}^{dm} \\ ss_{D^1}^{df} & ss_{A^1}^{df} \end{bmatrix}_{2M\times N}. \tag{6.157}$$

This matrix is a transfer matrix of an independent element. It can be seen that it is independent of the external sound wave. Different solutions can be obtained under different incident and reflection conditions. Note that if the propagating modes contain plane waves, the result of integration will be different.

Two examples are given to illustrate the sound propagation characteristics of variable cross-section ducts. The first is the reflection and transmission of the sound waves passing through a sound hard wall duct. As shown in Fig. 6.20, a variable section can be approximated and treated as the equivalent duct shown in Fig. 6.21. Fig. 6.22 gives the results of the reflection and transmission coefficients of sound waves both in a slowly divergent duct and in a slowly convergent duct. The results of the method here are compared

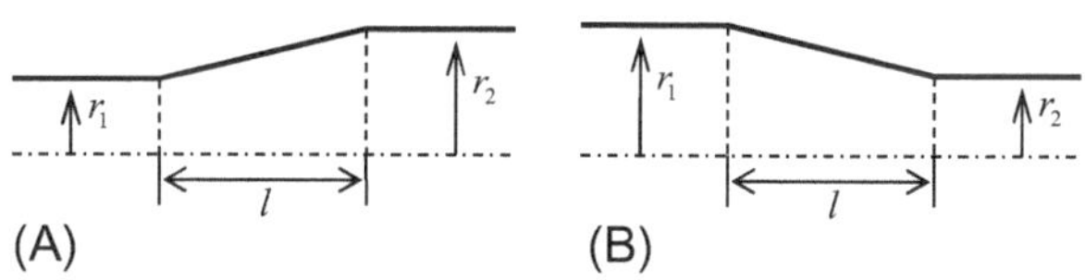

Fig. 6.20 A schematic diagram of a slowly varying cross-section duct. (A) expansive duct (B) contractive duct.

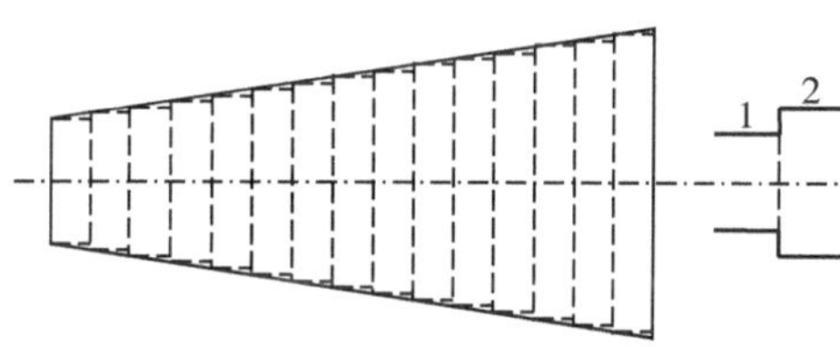

Fig. 6.21 An equivalent duct for a slowly varying cross-section duct.

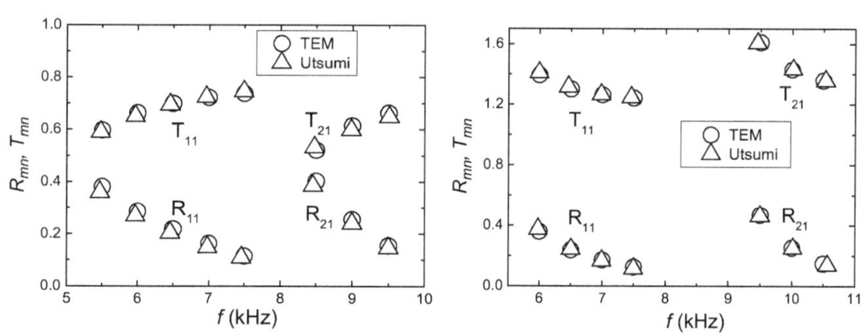

Fig. 6.22 The reflection and transmission coefficients of sound waves both in a slowly divergent duct and in a slowly convergent duct (see [49] for Utsumi's result).

with that of Utsumi's method [49, 50]. The parameters of the divergent duct are: $r_1 = 0.0208$, $r_2 = 0.01785$, $l = 0.0207$; the parameters of the convergent duct are: $r_1 = 0.0208$, $r_2 = 0.01785$, $l = 0.0207$. The reflection coefficient R_{mn} and the transmission coefficient T_{mn} are defined as:

$$R_{mn} = \frac{|A_{mn}^0|}{|p_{imn}|}, \quad T_{mn} = \frac{|D_{mn}^{nd+1}|}{|p_{imn}|}. \tag{6.158}$$

From the calculation results, it can be seen that the acoustic wave through the varying cross-section duct will produce the reflection and scattering effect. For the convergent duct, the phenomenon "close sound", when the transmission coefficient is greater than 1, appears because of these effects.

On this basis, we will discuss a more complex situation in which the nonlocally reacting liner is included in the varying cross-section duct, and the sound absorbing material exists in the back cavity. As shown in Fig. 6.23, the same duct has been discussed by Selame [51] in 2004.

The parameters of perforated plate are: the diameter of hole $d_h = 0.249\,\text{cm}$, perforation rate $\sigma = 8\%$, plate thickness $t_w = 0.09\,\text{cm}$. The cavity behind the perforated plate is filled with porous material, whose characteristics are:

$$\begin{cases} \dfrac{\hat{k}_0}{k_0} = \left[1 + 0.1472\left(\dfrac{f}{R}\right)^{-0.577}\right] - i\left[0.1734\left(\dfrac{f}{R}\right)^{-0.595}\right] \\ \dfrac{\hat{Z}_0}{Z_0} = \left[1 + 0.0855\left(\dfrac{f}{R}\right)^{-0.754}\right] - i\left[0.0765\left(\dfrac{f}{R}\right)^{-0.732}\right] \end{cases}, \tag{6.159}$$

where $Z_0 = \rho c$ and k_0 are the characteristic impedance and wavenumber of the air, and $\hat{Z}_0 = \hat{\rho}\hat{c}$ and $\hat{k}_0$ are the characteristic impedance and wavenumber of porous material, respectively. The flow resistivity of the porous material

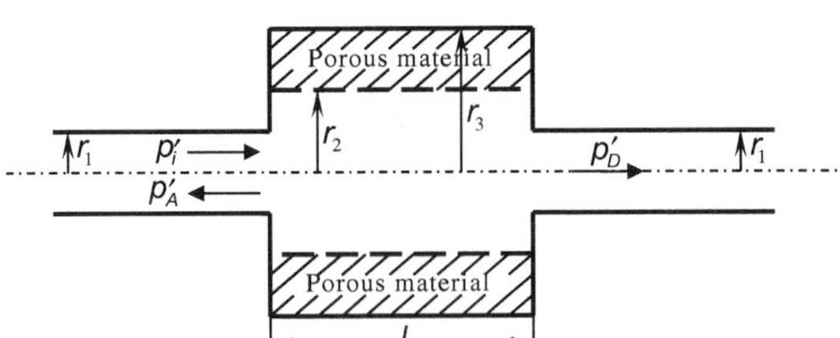

Fig. 6.23 The sound propagation in a varying cross-section duct with a nonlocally reacting liner $r_1 = 2.45\,\text{cm}$, $l = 25.72\,\text{cm}$, $r_3 = 8.22\,\text{cm}$.

$R = 4896$ Rayls/m. The nondimensional impedance of the perforated plate is:

$$Z = \left\{ 0.006 + ik_0 \left[t_w + 0.375 d_h \left(1 + \frac{\hat{Z}_0 \hat{k}_0}{Z_0 k_0} \right) \right] \right\} / \sigma. \tag{6.160}$$

Note that the medium in the cavity is no longer air, so when calculating the equivalent radiation impedance, attention should be paid to the effect of the change in the medium. Fig. 6.24 shows the theoretical and experimental results of the TEM and Selamet's method [51].

On this basis, we will get a more complicated result. As shown in Fig. 6.25, which is a schematic diagram of the aeroengine nacelle's front part, the sound source plane is at the position $z = 0$. The calculation is carried out for a single incident modal wave, and the termination of the duct is assumed to be anechoic. The uniform locally reacting liner is mounted on the outer wall of the duct with radius R_1 and axial range $0 \sim l$. The nondimensional axial length of the liner is $l = 1.86393$, and the nondimensional specific impedance is $Z_{zk} = 2 - i$. Other parameters are:

$$\begin{cases} R_1 = 1 - 0.18453(z/l)^2 + 0.10158 \dfrac{e^{-11(1-z/l)} - e^{-11}}{1 - e^{-11}} & z \in (0,\ 1.86393) \\ R_2 = \max\left[0,\ 0.64212 - (0.04777 + 0.98234 z/l)^{1/2}\right] & z \in (0,\ 1.86393) \end{cases}. \tag{6.161}$$

The inner wall of the annular duct is acoustically hard. In 2001, Rienstra and Eversman [18], respectively, used the perturbation method and the FEM to compute the sound propagation in the varying cross-section duct containing the locally reacting liner, which give the transmission levels of the liner with different parameters. Here, we also calculate the transmission levels by using the TEM, and the comparison is shown in Table 6.1.

Note that the transmission loss is different from the acoustic attenuation definition used previously. The transmission loss is defined as:

$$TLL = -10 \log_{10} \left(\frac{E_D}{E_i - E_A} \right),$$
$$E = \int_s I_s ds = \int_s \left[\left(1 + M_a^2\right) p' v'^* + \frac{M_a}{\bar{\rho} c} p' p'^* + M_a \bar{\rho} c v' v'^* \right] ds. \tag{6.162a,b}$$

Note that the FEM and multiple scales (MS) methods assume that the incident wave is a modal wave in the soft-walled duct, whereas the TEM

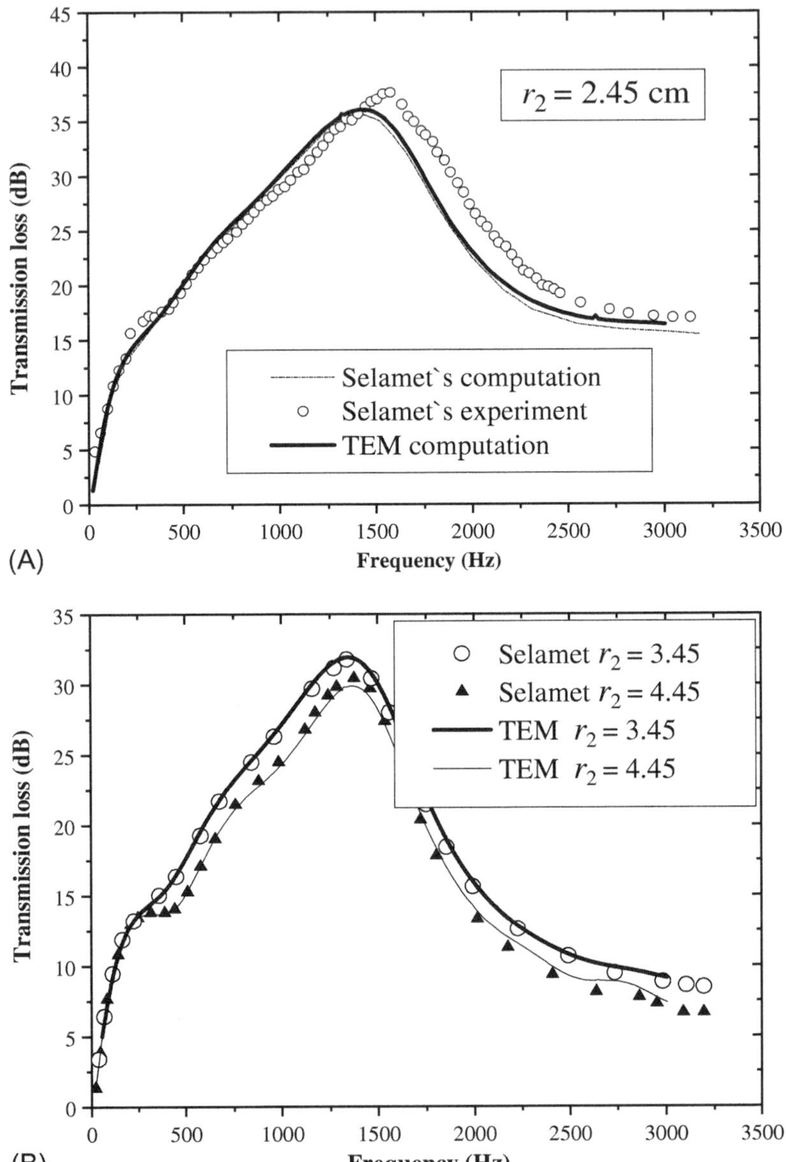

Fig. 6.24 Comparison on the transmission loss in a varying cross-section duct with a nonlocally reacting liner (see [51] for Selamet's results). (A) $r_2=2.45$ with experiment result (B) $r_2=3.45$ and $r_2=4.45$.

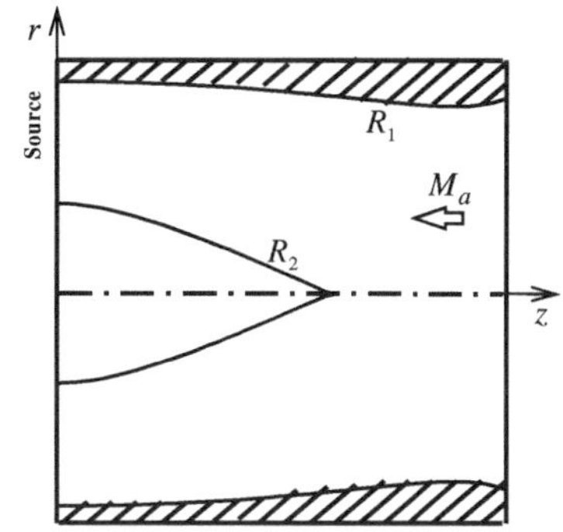

Fig. 6.25 The schematic diagram of aeroengine inlet nacelle lined with locally reacting liner.

Table 6.1 Acoustic attenuation comparison of various methods.

	m	k_0	Ma	FEM [18]	MS [18]	TEM	Cut-on mode	Cut-off ratio
1a	10	16	0	51.6	51.6	50.4	1	1.36
1b	10	16	−0.5	27.2	27.1	28.1	2	1.25
2a	10	50	0	4.7	3.5	5.9	9	3.38
2b	10	50	−0.5	1.5	0.9	2.7	11	3.97
3a	20	50	0	12.5	12.3	13.3	8	1.91
3b	20	50	−0.5	3.9	3.3	4.7	9	2.26
4a	30	50	0	29.0	28.7	28.4	4	1.34
4b	30	50	−0.5	9.7	8.9	10.7	6	1.59
5a	40	50	0	196	210	182.8	2	1.03
5b	40	50	−0.5	28.4	28.6	29.0	3	1.23

method assumes that the incident wave is a modal wave in the hard wall duct.

The results given by Table 6.1 show that, although there are slight differences between the transmission loss obtained by Rienstra and Eversman [18] and that by the TEM, the overall trend is completely consistent.

6.3.6 Calculation of the combined acoustic liner

6.3.6.1 The mechanism of the multiple cavity resonance

In this section, a theoretical model for analyzing the acoustic effect of a multiple cavity resonance liner inside an infinite duct is developed. Concerning a uniform rectangular duct of infinite axial extent, its geometry is depicted in Fig. 6.26. The acoustic liner is installed in the bottom wall of the duct, which

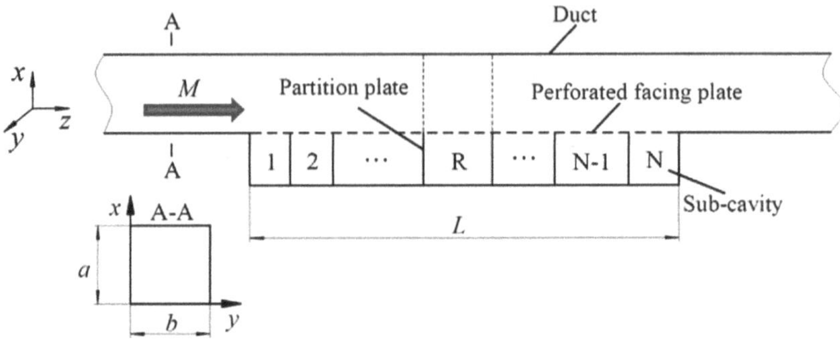

Fig. 6.26 Geometry of an infinite duct with a multiple cavity resonance liner.

means the perforated facing plate locates at the $x = 0$ plane. Moreover, the back cavity of the liner is axially divided by the rigid partition plate, whose thickness is ignored, into N sub-cavities with different axial dimension. The cross-sections of the duct and liner cavity are $a \times b$ and $a_c \times b_c$, respectively. The total length of liner is equal to L. The direction of mean flow in the duct is constantly axial, and the Mach number is M.

As mentioned earlier, sound propagation in a lined duct is currently investigated by employing TEM. The basic idea of the method is to regard the acoustic components as several transfer elements, then establish the matching condition on the interfaces of adjacent elements, and finally, obtain the sound field inside each element by simultaneously solving the matching equations of all element interfaces. Concerning the present duct, the duct segment that corresponds to a sub-cavity can be treated as a transfer element. The element of R-th sub-cavity is illustrated in Fig. 6.27. The total sound pressure inside the element can be decomposed into three parts: the incident sound pressure p_i, the reflected sound pressure p_r, and the disturbance sound pressure $p_e^{\pm}$. The incident and reflected sound pressure are

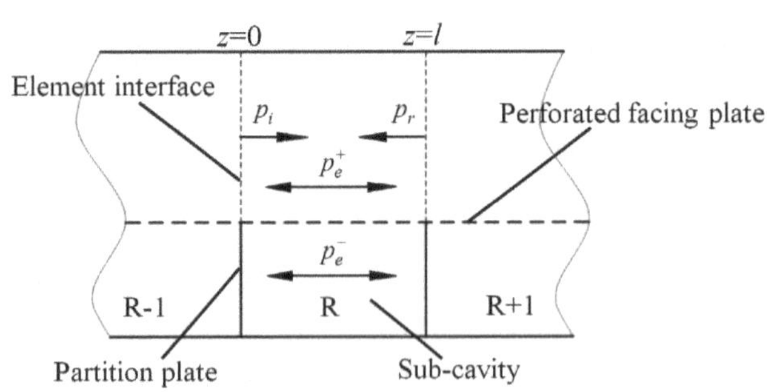

Fig. 6.27 Transfer element of multiple cavity resonance liner.

defined as the pressure that would be realized if every part of the lined walls of the duct were substituted by a solid surface, whereas the disturbance sound pressure is generated by the lined sections of the duct walls, which are considered as a disturbance source.

In the R-th element, the axial coordinate z is referred to as the left end of the element, and it is assumed that the axial length of the duct segment is l. The incident and reflected sound pressures in the element can be written as:

$$p_i = \sum_{m=0}^{\infty}\sum_{n=0}^{\infty} A_{mn}\sqrt{\varepsilon_m}\sqrt{\varepsilon_n}\cos\left(\frac{m\pi x}{a}\right)\cos\left(\frac{n\pi y}{b}\right)\exp\left(i\gamma_{mn}^{+}z\right)\exp\left(i\omega_0 t\right), \tag{6.163}$$

$$p_r = \sum_{m=0}^{\infty}\sum_{n=0}^{\infty} B_{mn}\sqrt{\varepsilon_m}\sqrt{\varepsilon_n}\cos\left(\frac{m\pi x}{a}\right)\cos\left(\frac{n\pi y}{b}\right)\exp\left(i\gamma_{mn}^{-}(z-l)\right)\exp\left(i\omega_0 t\right). \tag{6.164}$$

where A_{mn} and B_{mn} denote the complex amplitude for mode with order (m, n), ω_0 is the sound angular frequency. ε_m and ε_n equal 1 when $m = n = 0$, otherwise they equal 2. $\boldsymbol{r}(x, y, z)$ is the coordinate for an observer, and t is the time associated with the observer. Also, $\gamma_{mn}^{\pm}$ is the axial wavenumber, and based on the sound wavenumber k_0, it can be defined as $\gamma_{mn}^{\pm} = mk_0 \mp \kappa_{mn}/1 - M^2$, where:

$$\kappa_{mn} = \sqrt{k_0^2 - (1-M^2)\left[\left(\frac{m\pi}{a}\right)^2 + \left(\frac{n\pi}{b}\right)^2\right]}. \tag{6.165}$$

The original incident and reflected sound field would induce vibration of the fluid particles inside the perforated facing plate, thus generating an additional acoustic field. According to the equivalent surface source method, the lined surface can be represented by a series of mass source singularities on the surface. The disturbance sound pressure in the element is considered as the generated sound by the mass source. Therefore, on the basis of the definition of the mass source, the disturbance sound pressure inside the duct segment can be obtained in the form of:

$$p_e^{+} = \int_{-T}^{+T}\iint_{S} \bar{\rho}\widetilde{V}_n'\left(\frac{\partial G^{+}}{\partial \tau} + U\frac{\partial G^{+}}{\partial z'}\right)dS(\boldsymbol{r}')d\tau, \tag{6.166}$$

where $\bar{\rho}$ is the mean density, $\widetilde{V}_n'$ is fluid particle velocity of equivalent mass source on the liner surface exposed to the interior of the duct, U is the velocity of mean flow, and G^+ is the Green's function of hard-walled rectangular duct. Additionally, $\boldsymbol{r}'(x', y', z')$ is the coordinate for a source, τ is the time associated with the source, T is a large time interval, and S is the area of liner facing plate.

In consideration of the disturbance sound pressure inside the liner sub-cavity, it can be obtained that:

$$p_e^- = \int_{-T}^{+T} \iint_S \bar{\rho} \widetilde{V}_n^f \frac{\partial G^-}{\partial \tau} dS(\boldsymbol{r}') d\tau, \tag{6.167}$$

where the mean flow is ignored, $\widetilde{V}_n^f$ is the fluid particle velocity on the liner surface exposed to the interior of the sub-cavity, and G^- is the Green's function of the hard-walled rectangular cavity. Furthermore, the minus symbol in this formula denotes that the direction of the fluid velocity is contrary to the positive direction that has been defined.

To reduce the complexity of the present model, it is assumed that $b_c = b$. Then, assuming that the frequency of the equivalent mass source is identical to the sound frequency inside the duct, the preceding fluid particle velocity can be expressed as $\widetilde{V}_n' = V_n' e^{i\omega_0 \tau}$ and $\widetilde{V}_n^f = \widetilde{V}_n^f e^{i\omega_0 \tau}$.

After substituting the Green's functions for duct and cavity described by Zhou et al. [52] into Eqs. (6.166) and (6.167), the disturbance sound pressure $p_e^\pm$ can be rewritten as:

$$p_e^+ = \bar{\rho} e^{i\omega_0 \tau} \sum_m \sum_n \frac{\varepsilon_m \varepsilon_n \cos\left(\frac{m\pi x}{a}\right) \cos\left(\frac{n\pi y}{b}\right)}{2\Gamma_{mn}} \times \int_{S(\tau)} \cos\left(\frac{m\pi x'}{a}\right) \cos\left(\frac{n\pi y'}{b}\right) V_n'\left(\omega_0 + U\gamma_{mn}^\pm\right) \frac{e^{i\gamma_{mn}^\pm (z-z')}}{\kappa_{mn}} dS(\mathbf{y}), \tag{6.168}$$

$$p_e^- = i\bar{\rho}\omega_0 e^{i\omega_0 \tau} \sum_{m_c} \sum_{n_c} \sum_j \frac{\varepsilon_{m_c} \varepsilon_{n_c} \varepsilon_j \cos\left(\frac{m_c \pi x}{a_c}\right) \cos\left(\frac{n_c \pi y}{b_c}\right) \cos\left(\frac{j\pi z}{l}\right)}{\Gamma_{mnj}\left(k_0^2 - \kappa_{mnj}^2\right)} \times \int_{S(\tau)} V_n^f \cos\left(\frac{m_c \pi x'}{a_c}\right) \cos\left(\frac{n_c \pi y'}{b_c}\right) \cos\left(\frac{j\pi z'}{l}\right) dS(\mathbf{y}), \tag{6.169}$$

where $\Gamma_{mn} = ab$, $\Gamma_{mnj} = a_c b_c l$ and $\kappa_{mnj}^2 = (m_c\pi/a_c)^2 + (n_c\pi/b_c)^2 + (j\pi/l)^2$. Also, m_c, n_c, and j denote the orders of cavity mode. The characters of ε_{m_c}, ε_{n_c}, and ε_j are identical to those of ε_m and ε_n.

Because it is assumed that the boundary condition of particle displacement continuity on the lining surface when mean flow exists, the liner surface velocity exposed to the duct interior can be denoted as:

$$V_n' = V_n^f + \frac{U}{i\omega_0}\frac{\partial V_n^f}{\partial z'} \tag{6.170}$$

Next then, the velocity of mass source can be expanded into the form of trigonometric series that:

$$V_n^f = \sum_{k=1}^{\infty}\sum_{k=0}^{\infty} V_{k\bar{k}}(x')\sin\left(\frac{k\pi z'}{l}\right)\cos\left(\frac{\bar{k}\pi y'}{b}\right) \tag{6.171}$$

The effect of perforated facing plate can be represented by the acoustic impedance boundary condition, thus the link between the total sound pressure in the duct and that inside the sub-cavity is constructed and can be illustrated by the following formula:

$$p_e^+ - p_e^- + \widetilde{V}^f_n \bar{\rho} c_0 Z = -(p_i + p_r), \tag{6.172}$$

where Z is the normalized specific acoustic impedance of perforated panel, and c_0 is the sound speed.

Then, carrying out the Fourier transform on both sides of the equal symbol in Eq. (6.172), it can be obtained that:

$$e^{i\omega_0 t}\sum_{k=1}^{\infty}\sum_{\bar{k}=0}^{\infty}\left(Z^+_{\bar{\bar{j}}\bar{j}k\bar{k}} V_{k\bar{k}} - Z^-_{\bar{j}\bar{j}-k\bar{k}} V_{k\bar{k}} + \delta_{jk}\delta_{\bar{j}}\bar{k} Z V_{k\bar{k}}\right) = e^{i\omega_0 t} I_{\bar{\bar{j}}\bar{j}}, \tag{6.173}$$

where $Z^{\pm}_{\bar{\bar{j}}\bar{j}k\bar{k}}$ stands for the equivalent impedance, δ_{jk} and $\delta_{\bar{j}\bar{k}}$ are the Kronecker delta functions, and.

$$e^{i\omega_0 t}\sum_{k=1}^{\infty}\sum_{\bar{k}=0}^{\infty} Z^{\pm}_{\bar{\bar{j}}\bar{j}k\bar{k}} V_{k\bar{k}} = \frac{4}{bl}\int_0^l\int_0^b p_e^{\pm}\sin\left(\frac{j\pi z}{l}\right)\cos\left(\frac{\bar{\bar{j}}\pi y}{b}\right)dydz, \tag{6.174}$$

$$\begin{aligned} e^{i\omega_0 t} I_{\bar{\bar{j}}} - &= -\frac{4}{bl}\int_0^l\int_0^b (p_i + p_r)\sin\left(\frac{\bar{j}\pi z}{l}\right)\cos\left(\frac{\bar{\bar{j}}\pi y}{b}\right)dydz \\ &= e^{i\omega_0 t}\sum_{m=0}^{\infty}\sum_{n=0}^{\infty}\sqrt{\varepsilon_m}\sqrt{\varepsilon_n}\cos\left(\frac{m\pi x}{a}\right)\left(A_{mn} I^{A_{mn}}_{\bar{\bar{j}}\bar{j}} + B_{mn} I^{B_{mn}}_{\bar{\bar{j}}\bar{j}}\right). \end{aligned} \tag{6.175}$$

The term $e^{i\omega_0 t}$ can be removed on both sides of Eq. (6.173). As a further step, solve the equations to acquire $V_{k\bar{k}}$. Then, by substituting the obtained $V_{k\bar{k}}$ into Eq. (6.166), the disturbance sound pressure p_e^+ can be procured.

According to what has been discussed, the adjacent elements are matched by the continuity of the total sound pressure and the axial acoustic particle velocity on the interface between them. As far as the antecedent R-th element is concerned, the following matching condition on the interface $z=0$ and $z=l$ can be established that:

$$\text{The interface of } z=0: \begin{cases} p_i^{(R-1)}+p_r^{(R-1)}+p_{e(+)}^{+(R-1)}=p_i^{(R)}+p_r^{(R)}+p_{e(-)}^{+(R)} \\ u_i^{(R-1)}+u_r^{(R-1)}+u_{e(+)}^{+(R-1)}=u_i^{(R)}+u_r^{(R)}+u_{e(-)}^{+(R)} \end{cases}. \tag{6.176a}$$

$$\text{The interface of } z=l: \begin{cases} p_i^{(R)}+p_r^{(R)}+p_{e(+)}^{+(R)}=p_i^{(R+1)}+p_r^{(R+1)}+p_{e(-)}^{+(R+1)} \\ u_i^{(R)}+u_r^{(R)}+u_{e(+)}^{+(R)}=u_i^{(R+1)}+u_r^{(R+1)}+u_{e(-)}^{+(R+1)} \end{cases}. \tag{6.176b}$$

In the two equations, the subscripts (+) and (−) denote the propagation direction of disturbance sound pressure, the superscripts such as (***R***) and (***R +1***) represent the indices of transfer elements, and u is the axial acoustic particle velocity.

By substituting the incident, reflected, and disturbance sound pressure into Eqs. (6.176a) and (6.176b) and then transforming the equations into the form of a matrix and combining them together, the matching equations of R-th element can be written as:

$$\begin{bmatrix} SS_{A(R-1)}^{(R-1)d} & SS_{B(R-1)}^{(R-1)d} & SS_{A(R)}^{(R)u} & SS_{B(R)}^{(R)u} & 0 & 0 \\ SU_{A(R-1)}^{(R-1)d} & SU_{A(R-1)}^{(R-1)d} & SU_{A(R)}^{(R)u} & SU_{B(R)}^{(R)u} & 0 & 0 \\ 0 & 0 & SS_{A(R)}^{(R)d} & SS_{B(R)}^{(R)d} & SS_{A(R+1)}^{(R+1)u} & SS_{B(R+1)}^{(R+1)u} \\ 0 & 0 & SU_{A(R)}^{(R)d} & SU_{B(R)}^{(R)d} & SU_{A(R+1)}^{(R+1)u} & SU_{B(R+1)}^{(R+1)u} \end{bmatrix} \begin{bmatrix} A_{mn}^{(R-1)} \\ B_{mn}^{(R-1)} \\ A_{mn}^{(R)} \\ B_{mn}^{(R)} \\ A_{mn}^{(R+1)} \\ B_{mn}^{(R+1)} \end{bmatrix} = \begin{bmatrix} 0 \\ 0 \\ 0 \\ 0 \end{bmatrix}. \tag{6.177}$$

where each coefficient is a matrix, each unknown quantity is a vector, and the superscripts u and d represent upstream and downstream, respectively.

Finally, by combining matching equations of all elements, the integrated transfer equations for the amplitude of the incident and reflected sound pressures inside duct can be obtained as:

$$\begin{bmatrix} SS_{B^{(0)}}^{(0)} & SS_{A^{(1)}}^{(1)\mathrm{u}} & SS_{B^{(1)}}^{(1)\mathrm{u}} & & & & \\ SU_{B^{(0)}}^{(0)} & SU_{A^{(1)}}^{(1)\mathrm{u}} & SU_{B^{(1)}}^{(1)\mathrm{u}} & & & & \\ & SS_{A^{(1)}}^{(1)\mathrm{d}} & SS_{B^{(1)}}^{(1)\mathrm{d}} & & & & \\ & & & \cdots\ \cdots & & & \\ & & & \cdots\ \cdots & & & \\ & & & & SU_{A^{(N)}}^{(\mathrm{N})\mathrm{u}} & SU_{B^{(N)}}^{(\mathrm{N})\mathrm{u}} & \\ & & & & SS_{A^{(N)}}^{(\mathrm{N})\mathrm{d}} & SS_{B^{(N)}}^{(\mathrm{N})\mathrm{d}} & SS_{A^{(N+1)}}^{(\mathrm{N}+1)} \\ & & & & SU_{A^{(N)}}^{(\mathrm{N})\mathrm{d}} & SU_{B^{(N)}}^{(\mathrm{N})\mathrm{d}} & SU_{A^{(N+1)}}^{(\mathrm{N}+1)} \end{bmatrix} \begin{bmatrix} B_{mn}^{(0)} \\ A_{mn}^{(1)} \\ B_{mn}^{(1)} \\ \vdots \\ \vdots \\ A_{mn}^{(\mathrm{N})} \\ B_{mn}^{(\mathrm{N})} \\ A_{mn}^{(\mathrm{N}+1)} \end{bmatrix} = \begin{bmatrix} A_{mn}^{(0)} \\ -A_{mn}^{(0)} \dfrac{\gamma_{mn}^{+} \exp\left(i\gamma_{mn}^{+} l\right)}{\omega_0 + U\gamma_{mn}^{+}} \\ 0 \\ \vdots \\ \vdots \\ 0 \\ 0 \\ 0 \end{bmatrix}, \tag{6.178}$$

where it is assumed that the incident sound wave only comes from the upstream hard wall duct section of the liner segment, $A_{mn}^{(0)}$ is the modal amplitude of incident sound wave, and $A_{mn}^{(N+1)}$ is the modal amplitude of transmitted sound wave. It can be seen that the incident sound can involve multiple modes. After solving Eq. (6.178), the modal amplitude of sound wave in each element would be acquired.

Based on the developed TEM model, the coupling effect between PPC perforated plate with cavity system of multiple cavity resonance liner and different incident duct modes are investigated. Here, the basic configurations of a duct and liner are similar to what Fig. 6.26 depicts. The dimension of a duct cross-section is equal to the former experimental one, and the current investigated object is the previous sample liner L2; meanwhile, the thickness of the liner partition plate is ignored. Furthermore, the amplitude of the present incident sound wave is limited to permit linear acoustics.

6.3.6.1.1 Plane wave

A perforated plate together with a back cavity forms an acoustic liner. When the cavity is divided into 10 sub-cavities, the transmission loss of the acoustic liner is given as a function of frequency in Fig. 6.28. There is a single peak on the curve of the transmission loss, and this is evident as the geometry of the sub-cavities only allows the transverse resonance to occur. The peak frequency is close to the theoretical value of 641 Hz, given by the special case of the following expression:

$$f_{n,j,m} = \frac{c}{2\pi}\sqrt{k_{y,n}^2 + \left(\frac{j\pi}{l}\right)^2 + \left(\frac{m\pi}{a_c}\right)^2}, \tag{6.179}$$

when $m = n = 0$. Although the TL peak is up to about 50 dB, the acoustic liner has a very narrow effective bandwidth, which is typically regarded as the locally reacting characteristics.

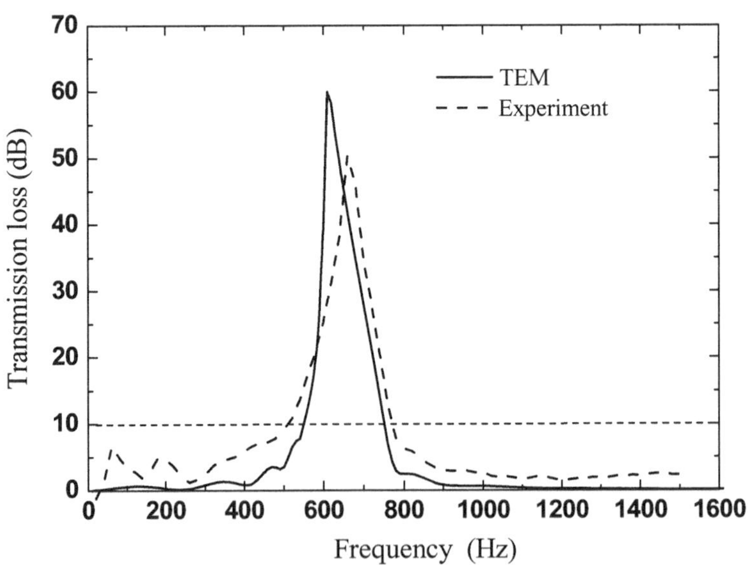

Fig. 6.28 The transmission loss as a function of frequency for the locally reacting acoustic liner.

When all the partition plates are removed from the cavity, more resonance peaks appear on the transmission loss curve, as shown in Fig. 6.29. There is even a flat and high TL band from about 900 Hz to 1100 Hz with the values being over 20 dB on average, thus the nonlocally reacting effect is demonstrated. The TL peaks are related to the multiple resonances due to the couple of the modes cutting on in different directions inside the cavity, and the resonance frequencies can be estimated by Eq. (6.179). Such results are presented and denoted by arrows in Fig. 6.29, showing generally good agreement with those of the experiment and the ESS-TM model. Being of zero order and decoupled from the other two, the x-mode has been omitted in the analysis, and the corresponding index has been dropped from the variable $f_{n,\ j}$ for the theoretical resonance frequencies. It can be seen that, for a nonlocally reacting acoustic liner, the first transverse resonance is not necessarily dominant, as it only produces a small peak at about 640 Hz in this case. The high TL peaks are due to the (1,2) and (1,3) resonances that originate from the couple of the first y-mode with the second and the third modes in z direction, respectively. In this case, the transverse mode and the axial modes have shown an unexpectedly strong coupling effect.

We further consider the situation of a multiresonance acoustic liner inside whose cavity partition plates are purposely arranged. Fig. 6.30 shows

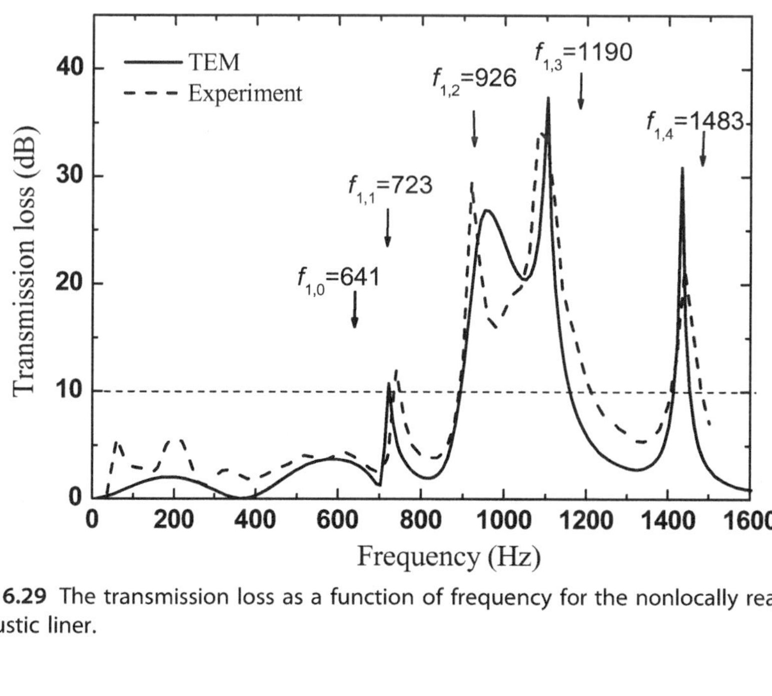

Fig. 6.29 The transmission loss as a function of frequency for the nonlocally reacting acoustic liner.

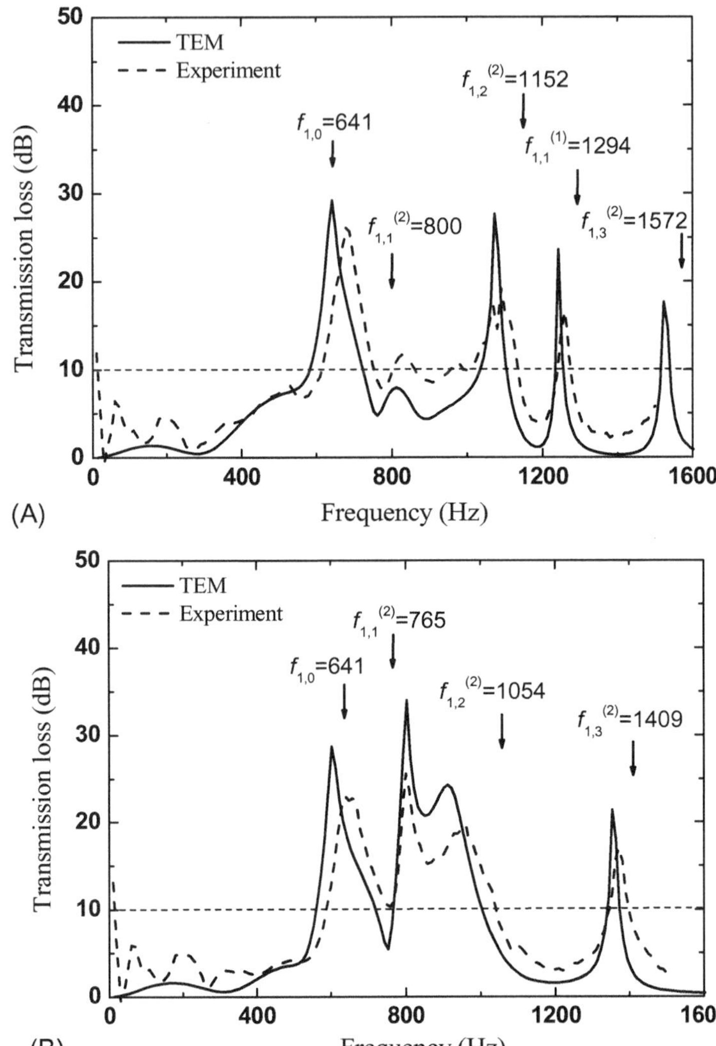

Fig. 6.30 The transmission loss as a function of frequency for the multiresonance liner and a subdivided cavity at the length ratio of (A) 3:7 and (B) 2:8.

the case that the liner cavity is subdivided into two different length ratios. As expected, there occur several peaks on the curve of the transmission loss, and their positions have been considerably changed in comparison with the non-partition-plate case of Fig. 6.29. When the length ratio is 3:7 as shown in Fig. 6.30A, the first transverse resonance peak emerges as the dominant one, and next to it toward the high frequency are (1,1) and (1, 2) resonance peaks at correspondingly 830 Hz and 1080 Hz (measured values), associated with the second sub-cavity. The (1, 1) resonance peak, due to the first sub-cavity, has a higher frequency of 1250 Hz as the cavity has a smaller length. Unfortunately, the (1, 1) resonance peak at 830 Hz is quite a modest one with a value of just above 10 dB. So, although the transmission loss curve is overall raised by the connection of the peaks, the result is not as satisfactory as expected. However, when the length ratio of the sub-cavities is adjusted to 2:8, a very much different result is obtained as shown in Fig. 6.30B. In this case, the first sub-cavity has a length equal to the duct width, so it only generates the transverse resonance that results in the (1,0) TL peak at about 640 Hz, whereas the second sub-cavity produces high TL peaks by (1, 1) and (1, 2) resonances at about 790 Hz and 940 Hz, respectively. We can see from the figure that, as a result of the joint effect of the three peaks, a high TL band appears from 600 Hz to 1000 Hz with the average value being over 15 dB. Compared with the non-partition-plate case in Fig. 6.29, the TL peaks are also moved toward the low frequency, although the total length of the acoustic liner is unchanged.

A perforated plate of different geometrical parameters is also used in the study, and the results are presented in Fig. 6.31. When the length ratio of the

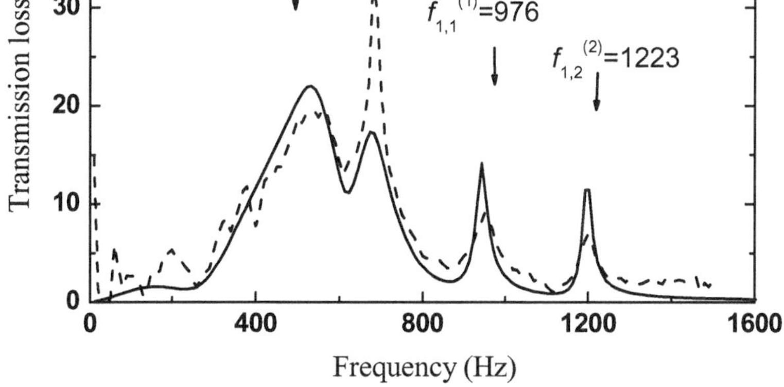

Fig. 6.31 The transmission loss as a function of frequency for the multiresonance liner and a subdivided cavity at the length ratio of 4:6.

sub-cavities is 4:6, it can be seen from the transmission loss curve that the first transverse resonance peak connects with the (1, 1) resonance peak of the second sub-cavity to broaden the bandwidth of the acoustic liner.

Higher-order grazing modes

The properties of multiple cavity resonance liner for absorbing grazing modes are investigated on the foundation of the actual TEM model. Fig. 6.32 presents the absorption coefficient and attenuation of the liner L2 with different grazing modes incidence.

First, it can be seen that the performance of L2 is comparatively poor, especially at high frequency. With regard to the curve of a grazing mode, such as $n = 1$ mode, the absorption coefficient and attenuation reach a comparatively high level within the frequency range near cut-on frequency of the mode, and the magnitude as well as bandwidth of peaks become

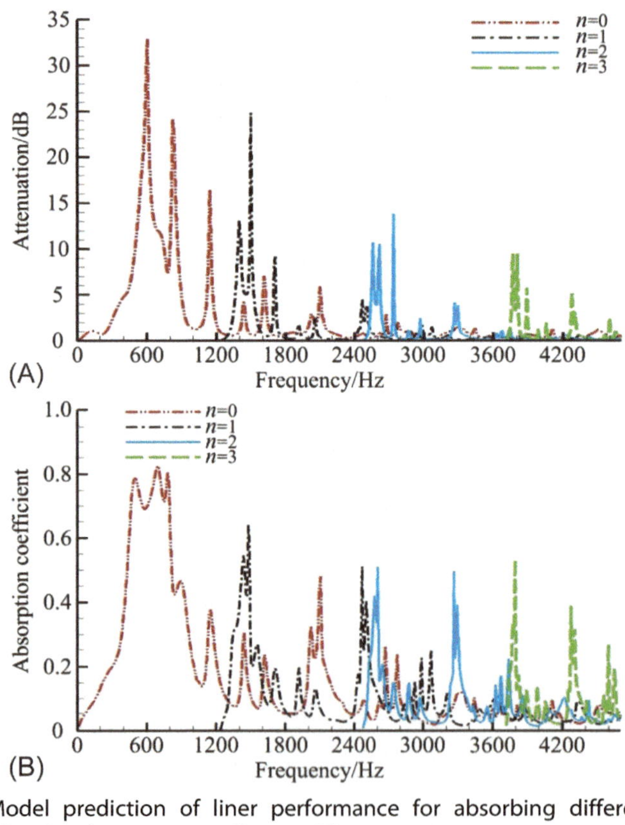

Fig. 6.32 Model prediction of liner performance for absorbing different incident grazing modes. (A) absorption coefficient; (B) attenuation.

greater. When the frequency is increasing, the peaks' magnitude and bandwidth are dropping dramatically. Second, from the comparison of different curves, it can be noted that the absorption peaks become sharper as the order of mode rises. Third, by comparing the aforementioned results with the present model predictions, it can be found that the downstream reflection would alter the magnitude of liner absorption, whereas the positions of resonance peaks are slightly affected.

The sound absorption coefficient of incident modes versus axial wavenumber is exhibited in Fig. 6.33. The absorption curves of different grazing modes generally show comparability that is the so-called self-similarity proposed by Eldredge [48, 53]. At a small axial wavenumber, the agreement among different cases is poor, and it is also known that the discrepancy of the plane wave incidence is the most obvious. As the liner is currently flush-mounted on the duct bottom wall, the propagation state of grazing modes with different order is identical. At a common axial wavenumber, the angle of incidence for the propagating mode is connected with the mode order, and the discrepancy of incidence angle among different modes reduces as the wavenumber increases, thus the liner would show a similar acoustic feature for absorbing these modes. Furthermore, it is necessary to state that the axial wavenumber becomes quite sensitive to the frequency as the axial wavenumber approaches zero, thus the frequency bandwidth of a peak at small axial wavenumber would be much smaller than it appears in Fig. 6.32, particularly for those duct modes with higher order.

The aforementioned Eq. (6.172) describes the link between the acoustic pressure both inside and outside the liner cavity established through the

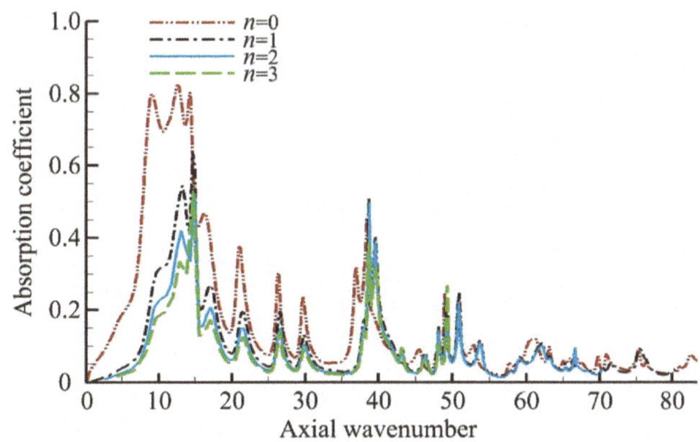

Fig. 6.33 Absorption coefficient of different incident grazing modes versus axial wavenumber.

perforated plate impedance. After converting Eq. (6.172) to a frequency domain, it can be obtained that:

$$ik_0 \iint\limits_S G_d(r',r)u_p(r')dS(r') + ik_0 \iint\limits_S G_c(r',r)u_p(r')dS(r') + Z\overline{\rho}c_0 u_p = -\widetilde{p}_i, \tag{6.180}$$

where $\widetilde{p}_i$ is the incident sound pressure, u_p is the acoustic particle velocity on the perforated plate, G_c is the Green's function for a rectangular cavity, and G_d is the Green's function for a rectangular duct. Furthermore, the mean flow and the reflected sound pressure inside the duct are ignored.

$\widetilde{p}_i$ and u_p can be expanded into the series of cavity eigenfunctions on the perforated plate surface. Which can be expressed by:

$$u_p = \sum_{k=0}^{\infty}\sum_{\overline{k}=0}^{\infty} u_{k\overline{k}}\sqrt{\varepsilon_k}\sqrt{\varepsilon_{\overline{k}}}\cos\left(\frac{k\pi z}{l}\right)\cos\left(\frac{\overline{k}\pi y}{b_c}\right), \tag{6.181a}$$

$$p_i = \sum_{k=0}^{\infty}\sum_{\overline{k}=0}^{\infty} p_{k\overline{k}}\sqrt{\varepsilon_k}\sqrt{\varepsilon_{\overline{k}}}\cos\left(\frac{k\pi z}{l}\right)\cos\left(\frac{\overline{k}\pi y}{b_c}\right), \tag{6.181b}$$

along with the assumption $b = b_c$.

Then, defining the location of perforated panel in x-axis as $x = 0$, substituting the Green's functions of cavity and duct into Eq. (6.180), and multiplying the term $\sqrt{\varepsilon_{\overline{j}}\varepsilon_{\overline{\overline{j}}}}\cos(\overline{j}\pi z/l)\cos(\overline{\overline{j}}\pi y/b_c)$ to both sides and integrating over the surface of the perforated plate lead to

$$\frac{p_{\overline{j}\overline{\overline{j}}}}{\overline{\rho}c_0} = \begin{cases} u_{\overline{j}\overline{\overline{j}}}\left[Z - \dfrac{ik_0}{k_y}\cot\left(k_y a_c\right) - \dfrac{k_0}{2al\pi}\displaystyle\sum_m \frac{\varepsilon_m \varepsilon_{\overline{j}}}{\kappa_{mn}} \times RC_{mn}\right], & j = \overline{j} = k \text{ and } n_c = \overline{\overline{j}} = \overline{k} \\ Zu_{\overline{j}\overline{\overline{j}}} - \dfrac{k_0}{2al\pi}\displaystyle\sum_m \frac{\varepsilon_m}{\kappa_{mn}} \sum_{k=0}^{\infty} u_{\overline{j}\overline{\overline{j}}}\sqrt{\varepsilon_k}\sqrt{\varepsilon_{\overline{j}}} \times RC_{mn}, & j = \overline{j} \neq k \text{ and } n_c = \overline{\overline{j}} = \overline{k} \end{cases}, \tag{6.182}$$

where $k_y = \sqrt{k_0^2 - (j\pi/l)^2 - (n_c\pi/b_c)^2}$, and

$$RC_{m'n'} = \int_0^l \cos\left(\frac{j\pi z}{l}\right)e^{i\gamma^+_{m'n'}z}\left[\int_0^l \cos\left(\frac{k\pi z_0}{l}\right)e^{-i\gamma^+_{m'n'}z'}dz'\right]dz. \tag{6.183}$$

Moreover, the relation $\cot\left(k_y a_c\right) = \left(k_y/a_c\right)\sum_{m_c=0}^{\infty}\varepsilon_{m_c}/\left[k_0^2 - k_{mnj}^2\right]$ has been used to derive Eq. (6.182). From Eq. (6.182), it can be noted that the absolute value of $u_{\overline{j}\overline{j}-}$, which determines the acoustic energy dissipation effect of the perforated panel, is influenced by the perforated panel impedance, liner cavity, and duct comprehensively. The contribution of the liner cavity is reflected in the form of reactance, whereas the contribution of the duct is embodied in not only the form of reactance but also resistance, which means the duct would affect the dissipation effect of the liner. Moreover, the absolute value of $u_{\bar{j}j-}$ would reach extremes as its factors in Eq. (6.182); for example, the terms in the bracket would go to their minimums. This character defines the resonance frequency of the duct-liner system.

After ignoring the effect of the duct, Eq. (6.182) becomes:

$$\frac{-p_{\bar{\bar{j}}j}}{\bar{\rho}c_0} = u_{\bar{\bar{j}}j}\left[z - \frac{ik_0}{k_y}\cot\left(k_y a_c\right)\right] = u_{\bar{\bar{j}}j} \times E(k_0), j = \bar{j} = k \text{ and } n_c = \bar{\bar{j}} = \bar{k}, \tag{6.184}$$

It is evident that the resonance condition is only determined by a liner cavity when the incident sound wave and the parameters of perforated panel are defined. For each combination of j and n_c, the absolute value of the term $E(k_0)$ can reach a sequence of local minimums at a different wavenumber, which correspond to a series of resonance frequencies.

Next, the situation of incoming grazing modes is considered, and the actual incident sound pressure with mode order n can be defined as:

$$\widetilde{p}_i = A_0\sqrt{\varepsilon_n}\cos\left(\frac{n\pi y}{b}\right)e^{ik_z z}, \tag{6.185}$$

where the current axial wavenumber k_z is independent of the spatial location, and A_0 is a constant. Therefore, the term $p_{\bar{j}j-}$ in Eq. (6.184) can be found as:

$$p_{\bar{j}j-} = \frac{iA_0 k_z\sqrt{\varepsilon_{\bar{j}}}\left[1 - \cos\left(\bar{j}\pi\right)e^{ik_z l}\right]}{\left(k_z^2 l^2 - \bar{j}^2\pi^2\right)}, \tag{6.186}$$

where $\bar{\bar{j}} = n_c = n$, otherwise $p_{\bar{j}j-} = 0$. It means the resonance mode order n_c is completely decided by the order of incident grazing mode.

Fig. 6.34 shows one part of the attenuation results in Fig. 6.32B, and the resonance frequencies obtained through calculating the minimums of the term $|E(k_0)|$ in Eq. (6.184) are marked by arrows in the figure. The resonance orders are also denoted above the arrows. The superscript of the marked order expresses the index of the cavity that corresponds to the denoted resonance frequency. If there is no superscript, it means the resonance frequency has connection with all cavities. Here, along the direction

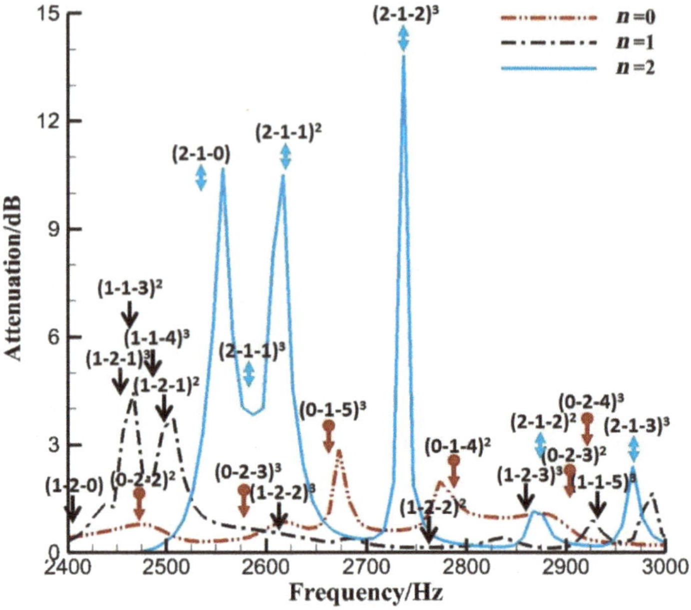

Fig. 6.34 Attenuation of different incident grazing modes within the frequency range of 2400–3000 Hz.

from the duct upstream to downstream, the index of the first liner cavity is defined as 1. Furthermore, the colour of arrows is related to different incident duct modes. It can be seen that the obtained resonance frequencies generally show good agreement with the results of TEM model. Particularly, the comparison results confirm the preceding statement that the resonance order n_c is equal to the order of incident grazing mode. Because the effect of the duct is ignored, the deviation between the two results exists. Besides, the deviation also comes from the interaction among different PPC systems that correspond to different liner sub-cavities.

In addition, from Eq. (6.184), the term $E(k_0)$ actually can be regarded as a kind of normalized specific impedance that involves the contributions of perforated plate impedance and cavity reactance. Consequently, the cavity resonance frequency can be found through calculating the zero point of the imaginary part of $E(k_0)$, which is the combined reactance. For example, concerning the No. 2 cavity of L2, defining $n_c = j = 1$, the profiles of $\mathrm{Im}[E'(f)]$ as well as $|E'(f)|$ are plotted in Fig. 6.35, where $E(k_0)$ has been converted to a new function $E'(f)$ with respect to sound frequency f. It is evident that

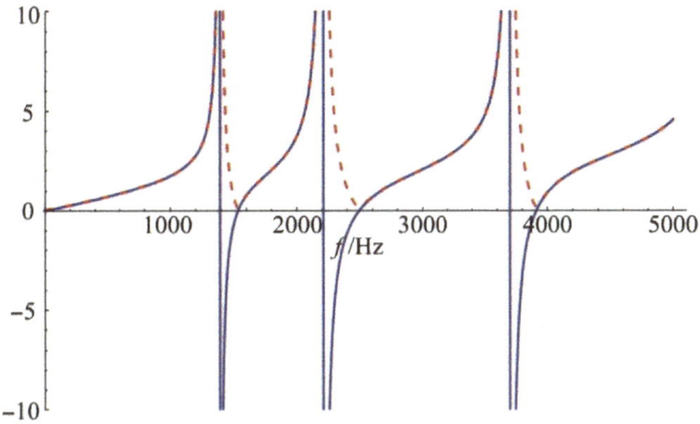

Fig. 6.35 Comparison of Im[$E'(f)$] (solid line) and $|E'(f)|$ *(dashed line).*

the zero-point frequencies of $\mathrm{Im}[E'(f)]$ is equal to that of the $|E'(f)|$ minimum points.

As far as the self-similarity of liner resonance frequency is concerned, it also can be explained according to the analysis of the term $E(k_0)$ in Eq. (6.184). As $k_0^2 = k_z^2 + (n\pi/b)^2$, $E(k_0)$ can be rewritten as a new function with respect to variable k_z:

$$E(k_0) = \overline{E}(k_z) = Z(k_z) - \frac{i\sqrt{k_z^2 + (n\pi/b)^2}}{k_z^2 - (j\pi/l)^2}\cot\left[a_c\sqrt{k_z^2 - (j\pi/l)^2}\right], \quad (6.187)$$

Fig. 6.36 presents the curves of $|\overline{E}(k_z)|$ when different grazing duct modes are incoming, and now $j = 1$. It is apparent that there is no shift

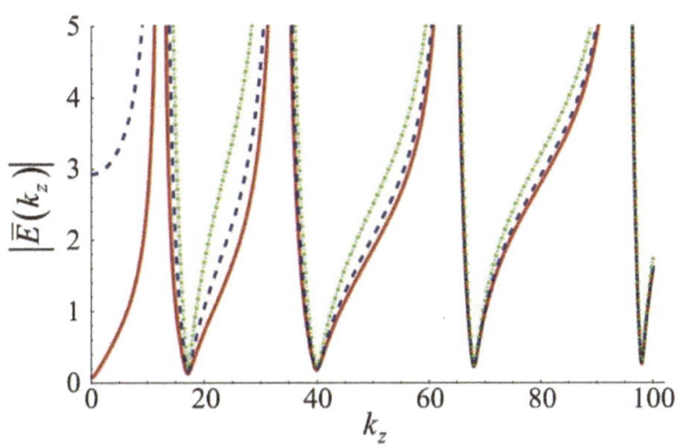

Fig. 6.36 Curves of $|\overline{E}(k_z)|$ for different incident grazing modes; *(red solid line)* $n = 0$; *(blue dashed line)* $n = 1$; *(green dash-dotted line)* $n = 2$.

of the axial wavenumbers that correspond to the local minimums of $|\overline{E}(k_z)|$, and the variation of the profile concentrates on its shape and the part near zero axial wavenumber, which supports the self-similarity feature of liner cavity resonance axial wavenumber.

The sound attenuation mechanism of a perforated liner is a hybrid that contains a dissipative effect and reactive effect. Focusing on the dissipation induced by cavity resonance, its mechanism and properties are presented in the following paragraphs.

As mentioned before, the absolute value of $u_{\bar{j}j-}$ determines the dissipation effect of the PPC system. Similarly, considering the circumstance of grazing mode incidence and ignoring the duct effect, from Eq. (6.184) it can be found that:

$$\frac{u_{\bar{j}j-} = -p_{\bar{j}j-}}{\overline{\rho}c_0 E(k_0).} \tag{6.188}$$

Substituting Eq. (6.186) and Eq. (6.187) into Eq. (6.188), it can be obtained that:

$$u_{\bar{j}j-}(k_z) = \frac{ik_z\sqrt{\varepsilon_{\bar{j}}}\left[\cos(\bar{j}\pi)e^{ik_z l} - 1\right]}{\left(k_z^2 l^2 - \bar{j}^2\pi^2\left\{z(k_z) - \frac{i\sqrt{k_z^2 + (n\pi/b)^2}}{\sqrt{k_z^2 - (j\pi/l)^2}}\cot\left[a_c\sqrt{k_z^2 - (j\pi/l)^2}\right]\right\}\right)}, \quad n_c = n, \tag{6.189}$$

with no loss of generality, it is assumed that $A_0 = \overline{\rho}c_0$. Based on Eq. (6.188), the feature of the liner dissipation corresponding to cavity resonance can be

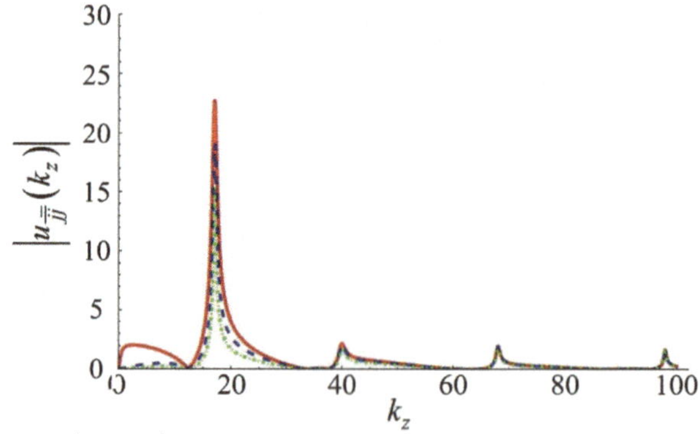

Fig. 6.37 Curves of $|u_{\bar{j}j}(k_z)|$ for different incident grazing modes; *(red solid line)* $n = 0$; *(blue dashed line)* $n = 1$; *(green dash-dotted line)* $n = 2$.

investigated. Fig. 6.37 shows the profiles of $\left|u_{\bar{j}j-} = (k_z)\right|$, where the second cavity of L2 is also considered and j is set as 1. From the comparison with Fig. 6.36, it can be seen that the absolute value of $u_{\bar{j}j-}$ reaches its peaks at the axial wavenumbers of cavity resonance, and the value is diminishing as the axial wavenumber shifts, which shows the influence of liner cavity resonance on the dissipation effect of the perforated plate. Regarding the results of certain mode incidence, such as the curve with respect to $n = 1$, the dissipation effect corresponding to the cavity resonance of order $\widetilde{m}_c = 1$ is the most intense. With the augment of $\widetilde{m}_c$, the intensity reduces. By comparing the results of different incident modes, it can be noticed that there is no shift of resonance peaks, and the variation of dissipation degree among different modes decreases as axial wavenumber increases. The results not only confirm the previous conclusion about the self-similarity of liner cavity resonance axial wavenumber but also support the self-similarity of sound absorption level as shown in Fig. 6.33.

Fig. 6.38 presents the profiles of $\left|u_{\bar{j}j}(k_z)\right|$ with different axial resonance order j, and now $n = 1$. Obviously, the increase of j can shift resonance peaks to a higher wavenumber, and simultaneously, the abatement of the peaks' bandwidth is obvious, which is attributed to the enlargement of perforated plate acoustic reactance to some extent.

By integrating this analysis, it can be concluded that, generally, the dissipation that corresponds to cavity resonance with higher order is comparatively weak, and its range of action is also too limited to provide effective broadband absorption for sound attenuation. Therefore, with respect to the present liner, the high frequency sound is difficult to absorb. Besides, the use

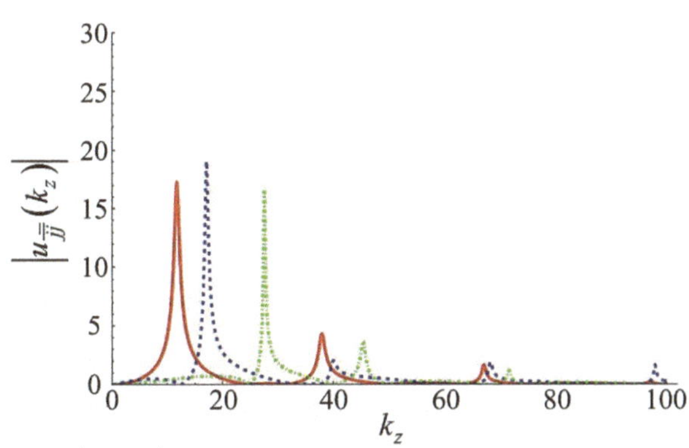

Fig. 6.38 Curves of $\left|u_{\bar{j}j}(k_z)\right|$ for different axial cavity resonance order j; *(red solid line)* $j = 0$; *(blue dashed line)* $j = 1$; *(green dash-dotted line)* $j = 2$.

of low-order cavity resonance should be a prime consideration during nonlocally reacting liner design.

Higher-order obliquely incident modes

Fig. 6.39 presents the TEM model prediction about the absorption coefficient and attenuation of the aforementioned sample liner L2 with different obliquely incident modes. Furthermore, the result of plane wave incidence is also exhibited.

In comparison with the results in Fig. 6.32, it can be seen that the liner performance for absorbing oblique incidence duct modes is much better than that of grazing modes, and similarly the performance degrades as the sound frequency rises. At a certain frequency, the absorption and attenuation of the liner is proportional to the order of an incoming sound wave, and the beneficial connection effect of different absorption peaks, which is the

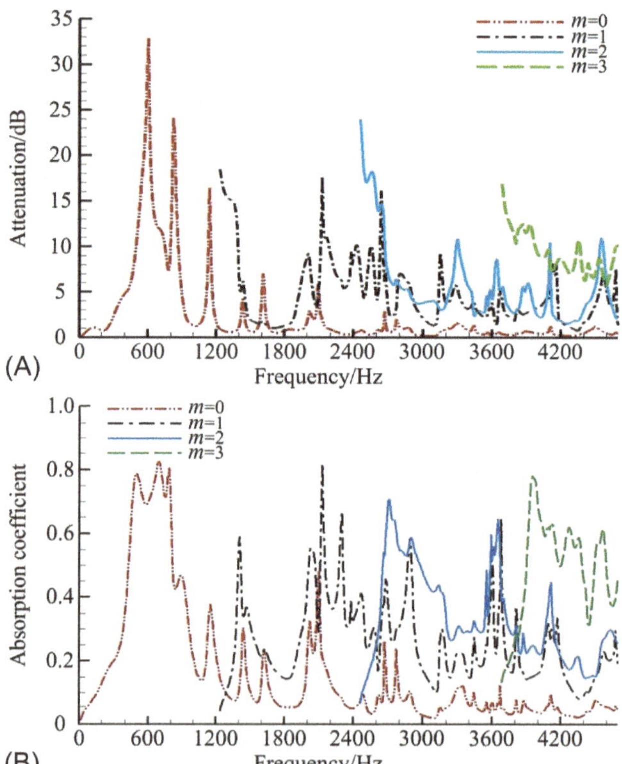

Fig. 6.39 Model prediction of liner performance for absorbing different obliquely incident modes. (A) absorption coefficient; (B) attenuation.

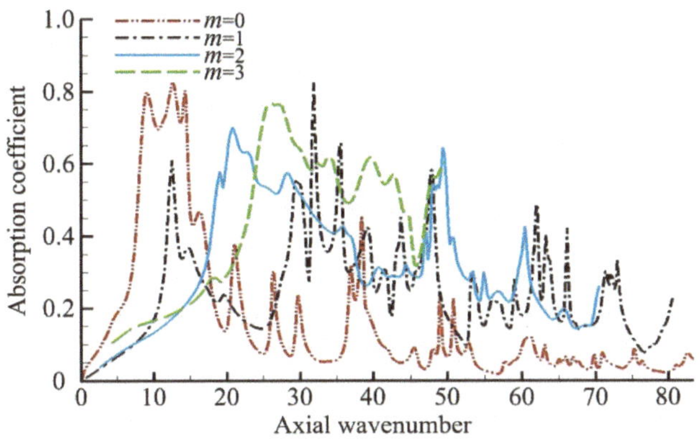

Fig. 6.40 Absorption coefficient of different obliquely incident modes versus axial wavenumber.

so-called parallel absorption mechanism[10], is more obvious for higher-order mode. Moreover, near the cut-on frequency of the incident mode, the liner attenuation is extremely high, which is largely due to the reactive effect of the liner.

The sound absorption coefficient of incident modes versus axial wavenumber is displayed in Fig. 6.40. Currently, the absorption curves of different modes do not show similar behavior. From the view of sound propagation, the incidence direction of propagating mode would tend to be axial as the sound frequency increases, thus an oblique incident mode would behave just like a plane wave to the flush-mounted liner.

On the basis of the preceding approach, the characteristic of liner cavity resonance frequency under the situation of obliquely incident modes can be investigated. Here, it is assumed that the actual incident sound pressure with order m is:

$$\widetilde{p}_i = A(x)e^{ik_z z}, \tag{6.190}$$

where $A(x)$ is a function with respect to the spatial location x. Therefore, the term $p_{\bar{j}\bar{\bar{j}}}$ in Eq. (6.184) can be found as:

$$p_{\bar{j}\bar{\bar{j}}} = \frac{iA(x)k_z\sqrt{\varepsilon_{\bar{j}}}\left[1-\cos\left(\bar{j}\pi\right)e^{ik_z l}\right]}{\left(k_z^2 l^2 - \bar{j}^2\pi^2\right)}, \tag{6.191}$$

where $\bar{\bar{j}} = n_c = 0$, otherwise $p_{\bar{j}\bar{\bar{j}}} = 0$. From the formula, it can be seen that the present cavity resonance mode order is absolutely identical to that of the condition with an incoming plane wave.

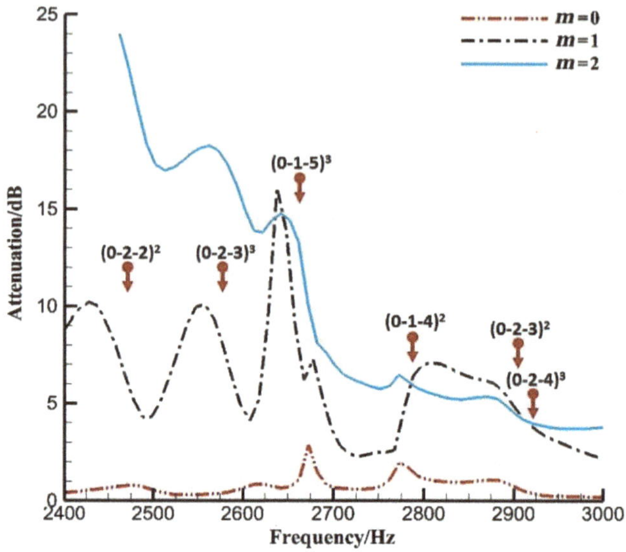

Fig. 6.41 Attenuation of different obliquely incident modes within the frequency range of 2400–3000 Hz.

Fig. 6.41 shows a part of the attenuation results in Fig. 6.39B, and the resonance frequencies obtained through calculating the minimums of the term $|E(k_0)|$ in Eq. (6.184) are marked by arrows in the figure. The orders of resonance frequencies are also denoted above the arrows in the same form of Fig. 6.34. It can be seen that the obtained resonance frequencies generally show good agreement with the results of TEM model, and the resonance frequencies of peaks at different modes are approximately identical to that of plane wave incidence, which confirms the preceding statement. Likewise, because of the duct effect and the interaction among PPC systems, the deviation between the two results of resonance peaks exists. Furthermore, by making comparison with the former results in Fig. 6.34, it can be noticed that currently the deviation is more distinct, which can be ascribed to the enhancement of duct influence to some extent.

Similarly, according to Eq. (6.184) and the relation that $k_0{}^2 = k_z^2 + (m\pi/a)^2$, $\overline{E}(k_z)$ can be written as:

$$\overline{E}(k_z) = Z(k_z) - \frac{i\sqrt{k_z^2 + (m\pi/a)^2}}{\sqrt{k_z^2 + (m\pi/a)^2 - (j\pi/l)^2}} \cot\left[a_c\sqrt{k_z^2 + (m\pi/a)^2 - (j\pi/l)^2}\right]. \tag{6.192}$$

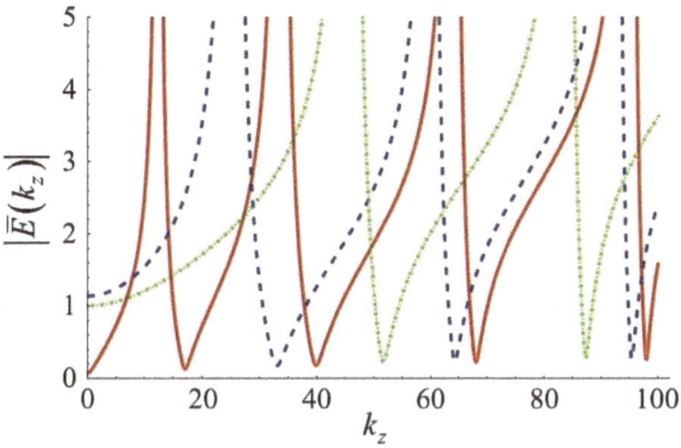

Fig. 6.42 Curves of $|\overline{E}(k_z)|$ for different obliquely incident duct modes; *(red solid line)* $m = 0$; *(blue dashed line)* $m = 1$; *(green dash-dotted line)* $m = 2$.

Fig. 6.42 presents the curves of $|\overline{E}(k_z)|$ with different oblique incidence modes, and now $j = 1$. It is evident that the shift of the axial wavenumbers that correspond to the local minimums of $|\overline{E}(k_z)|$ is pronounced, and the variation of the profile also involves its shape change.

Concerning the dissipation induced by cavity resonance, the former analysis approach based upon Eq. (6.189) is not applicable. Currently the sound scattering effect evoked by perforated panel is noticeable, and from Eq. (6.182) it can be found that the scattering effect focuses on the x direction, which is identical to the distribution direction of oblique incidence mode, thus duct's contribution cannot be ignored.

As mentioned before, at a certain frequency, the liner's performance of absorbing higher order obliquely incident mode is better. While the mode order is enlarging, the axial wavenumber of incoming sound with certain frequency is gradually decreasing, and simultaneously the incidence angle is rising. Accordingly, the dissipation effect of perforated plate is improved so that the parallel absorption mechanism would be constructed.

Moreover, near cut-on frequency, the incident duct modes would behave like a normal incident wave to the perforated panel, thus the contribution of transverse and axial cavity modes would diminish while that of longitudinal mode would be more distinct. In contrast, as the sound frequency increases, the incidence angle of the propagating mode reduces and the mode tends to perform like a plane wave, which leads to the abatement of dissipation and the breakdown of the parallel absorption mechanism.

Based on the developed TEM model, the coupling effect of multiple cavity resonance liner and different incident sound waves was investigated.

Specifically, the features of resonance frequency and dissipation of the liner PPC system under grazing or oblique incidence conditions are revealed. Several conclusions can be made:

(1) The response of PPC system in a nonlocally reacting liner is strongly correlated with incident duct modes. Regarding the design of a nonlocally reacting liner for absorbing multimodal broadband or high frequency noise, it is imperative to notice such effect, and the classical means for plane wave condition might not be applicable.

(2) The comparison of the liner absorption results for different incident grazing modes reveals the self-similarity. On one hand, the axial wavenumber of the PPC system resonance peak does not alter with the change of duct mode order. On the other hand, the dissipation level of the PPC system is also similar at a common axial wavenumber. Also, the order of cavity transverse resonance is totally determined by the incident duct mode. As the cavity resonance order increases, the dissipation effect of PPC system gradually decreases. Hence, during liner design, the utilization of low-order resonance of a PPC system should be majorly considered.

(3) Concerning the situation of obliquely incident duct modes, the self-similarity behavior disappears. Nonetheless, the resonance frequency of the PPC system under different mode incidence shows identity, and the order of the cavity transverse resonance is equal to zero. In addition, at oblique incidence, the liner displays better absorption performance, yet meanwhile the effect of the duct on liner dissipation performance can be more apparent.

6.3.6.2 Combination of the locally reacting liner and the nonlocally reacting liner

The greatest advantage of the TEM introduced in this chapter is that it can consider the interaction between various acoustic liners in the duct and predict the total effect of noise reduction in the duct. In this section, several different combinations of acoustic liners will be discussed. As shown in Fig. 6.43, there are three elements divided in the duct. The first part of acoustic treatments is a locally reacting liner, then a varying cross-section part with a hard wall, and a nonlocally reacting liner mounted behind it. (It should be explained that part B is a duct with a sudden change in its cross-sectional area, which can be seen as the varying cross-section duct or be regarded as a nonlocally reacting acoustic liner whose perforated plate impedance is zero.)

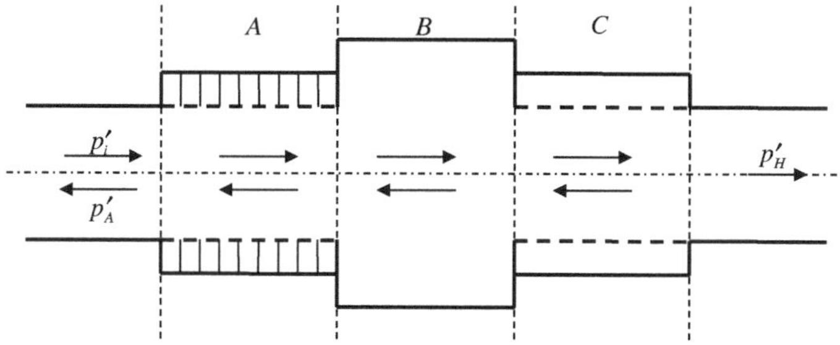

Fig. 6.43 The combination of different acoustic liners without flow.

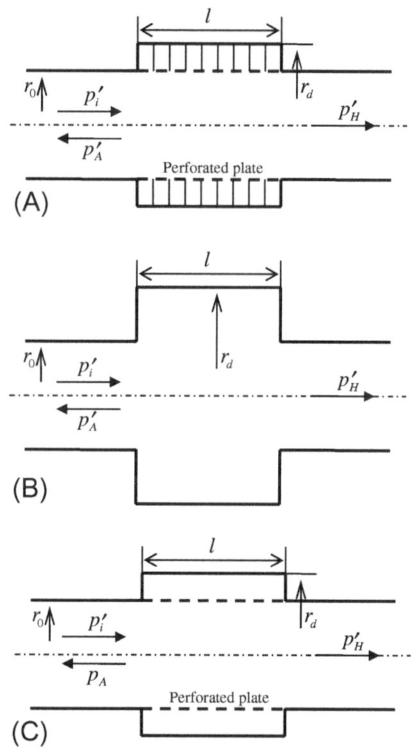

Fig. 6.44 Three kinds of acoustic liners. (A) duct with locally reacting liner (B) duct with expansive cavity (C) duct with nonlocally reacting liner.

Before calculating the total noise attenuation effect of the combined acoustic liner, we first split it into three separate parts to analyze the noise reduction characteristics of each kind of liner, as shown in Fig. 6.44.

Fig. 6.44A is a locally reacting liner, and the duct dimensions are $r_0 = 0.1m$ (corresponding cut-off frequency is $f = 2073Hz$), $r_d = 0.125m$. The geometry parameters of the perforated plate are: liner length

$l = 0.1m$, hole diameter $d_h = 2.5mm$, plate thickness $t_w = 2mm$, and perforation rate $\sigma = 2\%$. The chosen impedance model of the perforated plate is:

$$r = \frac{\sqrt{8\nu\omega}}{\sigma c}\left(1 + \frac{t_w}{d_h}\right), \quad x = \frac{\omega(t_w + 0.85d_h)}{\sigma c} + \frac{\sqrt{8\nu\omega}}{\sigma c}\frac{t_w}{d_h} - \cot\left[\frac{\omega(r_d - r_0)}{c}\right], \tag{6.193}$$

where $\nu = 1.5 \times 10^{-5}$. Fig. 6.44B is an expansion muffler, which can also be treated as a varying cross-section duct. The parameters are: $r_0 = 0.1m$, $r_d = 0.18m$, $l = 0.3m$.

Fig. 6.44C is a nonlocally reacting liner. The parameters are: liner length $l = 0.3m$, hole diameter $d_h = 1.83mm$, plate thickness $t_w = 0.87mm$, and perforation rate $\sigma = 2\%$. The chosen impedance model of the perforated plate is:

$$r = \frac{\sqrt{8\nu\omega}}{\sigma c}\left(1 + \frac{t_w}{d_h}\right), \quad x = \frac{\omega(t_w + 0.85d_h)}{\sigma c} + \frac{\sqrt{8\nu\omega}}{\sigma c}\frac{t_w}{d_h}. \tag{6.194}$$

The transmission loss is calculated as:

$$TL = -20\log_{10}\left|p'_H / p'_i\right|. \tag{6.195}$$

The absorption coefficient of the acoustic liner is defined as:

$$AC = 1 - \frac{|p'_A| + |p'_H|}{|p'_i|}. \tag{6.196}$$

Fig. 6.45 shows the transmission losses and absorption coefficients of all the three acoustic liners.

From Fig. 6.45A, it can be seen that the locally reacting acoustic liner has a single peak characteristic in the range of 0–2000 Hz, and this the same both for sound absorption coefficient and for transmission loss. The noise reduction of expansion duct in Fig. 6.45B satisfies the periodic characteristic in the low frequency range where the plane wave dominates. But as the radius r_d gets larger, higher-order modal waves are cut-on in the duct. So, the noise reduction characteristic is no longer periodic with the frequency in the high frequency range (here >1100 Hz). In addition, because in essence there is no noise dissipation in the expansion duct, the sound absorption coefficient of the whole system is always zero. From Fig. 6.45C, we can see many resonant peaks in the 1000–1500 Hz range for the nonlocally reacting liner, which results in a broadband sound attenuation effect. Also note that for a nonlocally reacting liner, the resonant frequencies of the transmission loss and absorption coefficient are nearly the same.

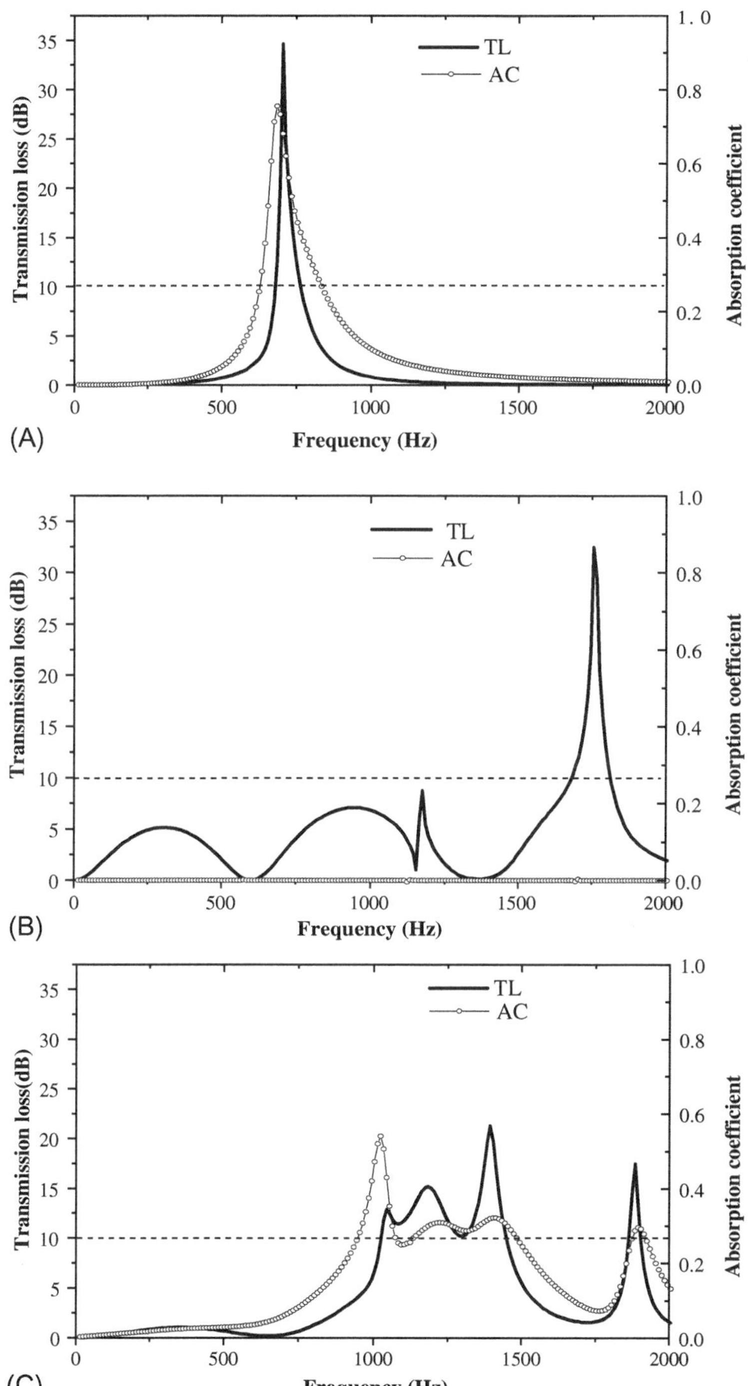

Fig. 6.45 The comparison of transmission losses and absorption coefficients of all three acoustic liners. (A) TL and AC for duct with locally reacting liner (B) TL and AC for duct with expansive cavity (C) TL and AC for duct with nonlocally reacting liner.

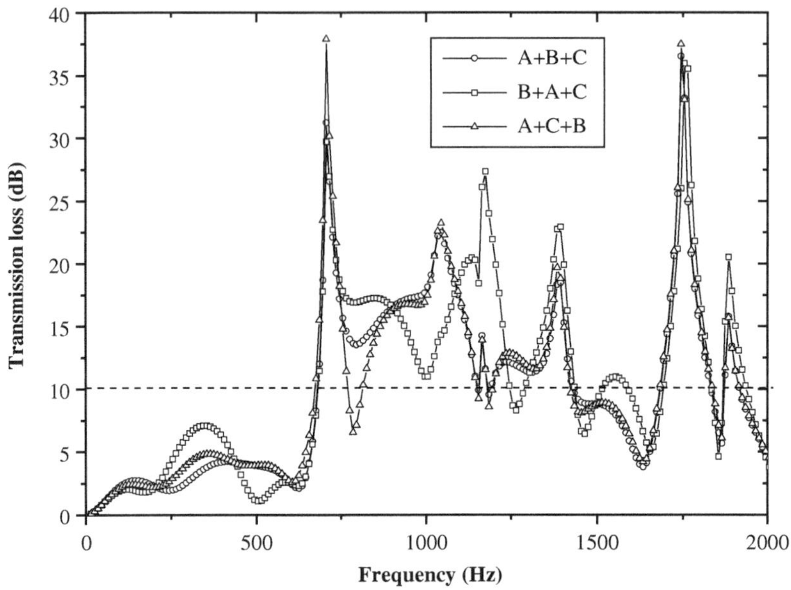

Fig. 6.46 Effect of different combinations on transmission loss.

The purpose of this section is to achieve noise reduction in a wider frequency range by reasonably designing a combination of these three different acoustic liners. As shown in Fig. 6.46, we can also investigate the effects of different combinations of sequences.

Fig. 6.47 shows the noise reduction characteristics of three combinations. The parameters of each independent acoustic liner are the same as before. Although the results of the three combinations are very different in some frequency ranges, in general, there is a good noise reduction effect in the range from 700 to 1400 Hz. Note that the parameters of the acoustic liners are carefully designed. It can be seen that the total noise reduction effect of the combined liner is not equal to a simple superposition of these three acoustic liners, indicating that there is an obvious interaction between the three. Especially, Fig. 6.47A shows the two kinds of combinations have almost the same transmission loss, but Fig. 6.47B indicates that they have obviously different sound absorption coefficient spectra, which is mainly due to the existence of the varying cross-section duct. If the sound absorption coefficient is taken as the evaluation criterion, it is obvious that the combination of C + B + A has a wider bandwidth.

So, it may be expected that a better noise reduction can be achieved by properly designing the combination of three kinds of liners.

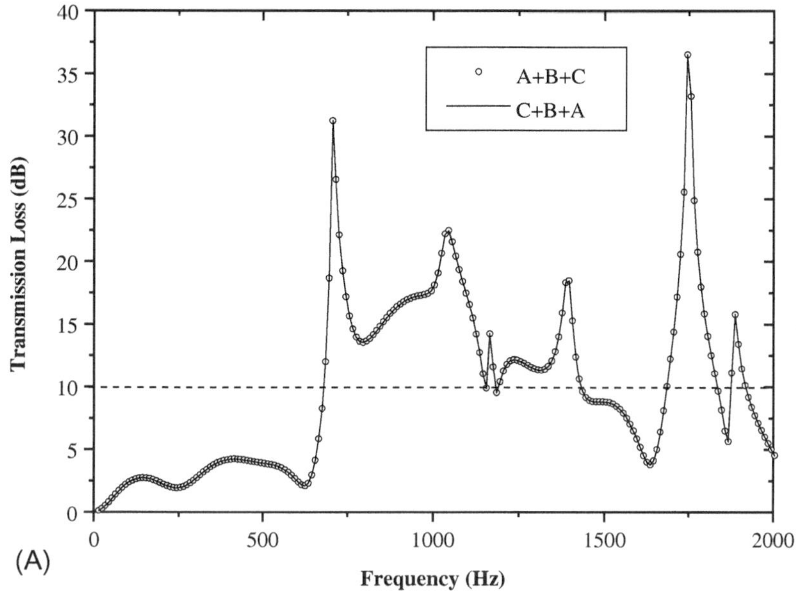

(A)

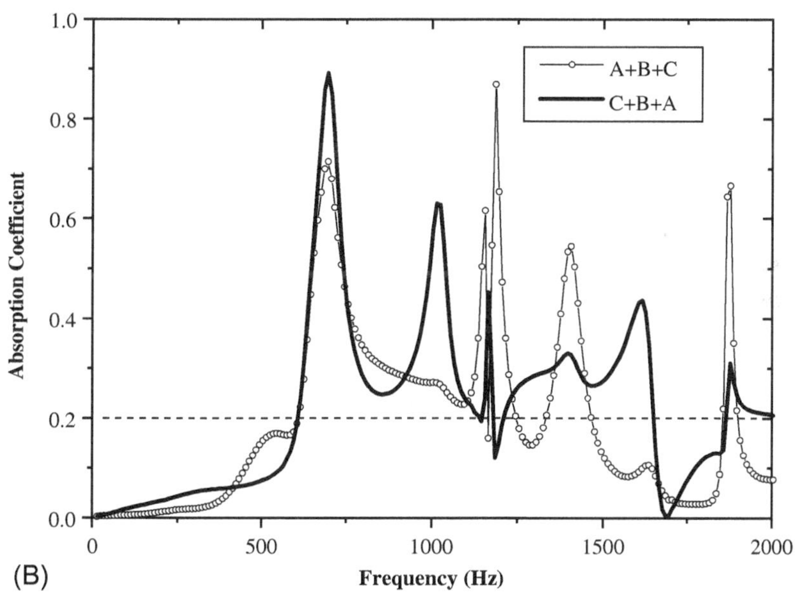

(B)

Fig. 6.47 The comparison between the absorption coefficients and the transmission loss. (A) transmission loss (B) absorption coefficients.

In addition, the application of TEM can make the calculation of noise reduction effect of combined liners simple and feasible. This method also provides an effective and fast tool for acoustic design in varying cross-section ducts.

After discussing the sound propagation in a duct without flow, here we consider a more realistic situation to illustrate the advantages of the combined acoustic liner and the application prospect of the TEM in this kind of problem. Consider the acoustic treatment in an aeroengine nacelle. Here, we replace the original locally reacting liner with the combined acoustic, which includes a nonlocally reacting liner with bias flow. The bias flow can influence the noise reduction effect of the nonlocally reacting liner so that the noise reduction characteristic of the whole system can be controlled. In this example, we are concerned with the sound attenuation effects of the acoustic liner in five cases corresponding to different working states, the descriptions of which are shown in the first three columns of Table 6.2. All cases include the effect of flow. Fig. 6.48 shows the schematic diagram of the front part of the aeroengine nacelle duct. The sound source plane is at position $z = 0$, and modal waves are given on the sound source plane. The duct is terminated with a nonreflective end. There are two kinds of acoustic linings in the varying cross-section duct: one is a nonlocally reacting liner with bias flow, and the other is a locally reacting liner. The axial length of the nonlocally reacting liner is $l_1 = 0.485$m, and the total length of the combined acoustic liner is $l_1 + l_2 = 1.86393$m where l_2 is the axial length of the locally reacting liner. Note that $R_d = 1.2$m is invariant with the axial coordinate, whereas the inner and outer radii are the functions of the axial coordinate:

$$\begin{cases} R_1 = 1 - 0.18453(z/l)^2 + 0.10158\dfrac{e^{-11(1-z/l)} - e^{-11}}{1 - e^{-11}} & z\in(0, 1.86393) \\ R_2 = \max\left[0,\ 0.64212 - (0.04777 + 0.98234z/l)^{1/2}\right] & z\in(0, 1.86393) \end{cases}. \tag{6.197}$$

Table 6.2 The comparison of transmission losses of locally reacting liner and combined liners.

	m	*k*$_0$	*Ma* (*z* = 0)	Locally reacting liner	Combined liner
1b	10	16	−0.5	30.0	22.58
2b	10	50	−0.5	2.5	5.29
3b	20	50	−0.5	4.9	12.74
4b	30	50	−0.5	11.2	16.85
5b	40	50	−0.5	31.4	27.35

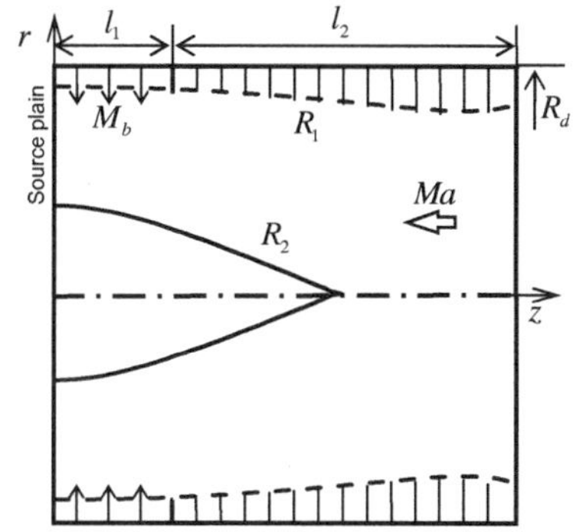

Fig. 6.48 Schematic of combination of locally and nonlocally reacting liners with flow.

Because it is almost impossible for this method to compute nonlocally reacting liners with varying cross-sections, the geometry of the duct needs to be changed a little. The outer radius of the duct section lined with the locally reacting liner is set as invariant with the axial coordinate, that is, $R_1 = 0.995\text{m}$ when $z = 0 \sim l_1$. The Mach number of the axial mean flow is computed by solving the potential flow equation. The flow Mach number versus the axial coordinate is shown in Fig. 6.49.

The parameters of perforated plate for locally and nonlocally reacting liners are the same: plate thickness $t_w = 0.001\text{m}$, hole diameter $d_h = 0.001\text{m}$, and perforation rate $\sigma = 0.02$.

The nondimensional impedance of the perforated plate is $Z_{zk} = \hat{r} + i\hat{x}$. For the nonlocally reacting liner,

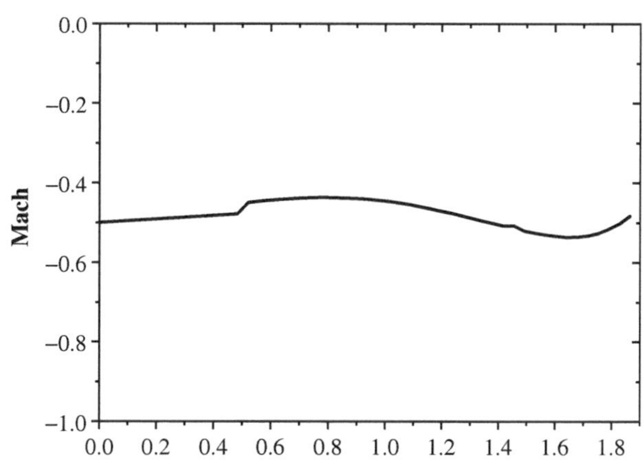

Fig. 6.49 The Mach number of the axial mean flow.

$$\hat{r} = \frac{M_b}{\sigma C_D^2}, \quad \hat{x} = \frac{4k\sigma\gamma}{\pi d_h} + \frac{kt_w}{\sigma},$$
$$C_D^2 = \frac{1}{\left(k_1 + k_2/\lambda + k_3/\lambda^2\right)}, \quad \lambda = \frac{M_b}{M_a}. \tag{6.198}$$

where $k_1 = 2.10$, $k_2 = -0.332$, $k_3 = 0.0556$ (*for outflow*), and $\hat{x}$can be given by the Howe's model [54]. M_b is the bias flow Mach number, which is assumed to be the controllable parameter. For the locally reacting liner, the nondimensional impedance is $Z_{zk} = \hat{r} + i\hat{x}$, where

$$\hat{r} = \frac{\sqrt{8\nu\omega}}{\sigma c}\left(1 + \frac{t_w}{d_h}\right) + 0.3\frac{|M_a|}{\sigma},$$
$$\hat{x} = \frac{\omega(t_w + 0.85d_h)}{\sigma c} + \frac{\sqrt{8\nu\omega}}{\sigma c}\frac{t_w}{d_h} - \cot\left[\frac{\omega(R_d - R_1)}{c}\right]. \tag{6.199}$$

From the last two columns of Table 6.2, we can see that, compared with the locally reacting liner (here, the impedance of the locally reacting liner is set as $Z_{zk} = 2 - i$), the noise reduction of the combined liner is significantly increased in the three cases of 2b, 3b, and 4b, but in two cases of 1b and 5b, the noise reduction is lower but still above 20 dB.

Therefore, the combined acoustic liner shows the advantage that the combined liner is more suitable for achieving a good noise reduction under different working conditions. The combined acoustic liner inherits the goodness of both locally and nonlocally reacting liners. Note that the combined liner can be designed to behave obviously better than a traditional locally reacting liner by properly controlling the bias flow. The effect of bias flow on the transmission loss can be seen in Fig. 6.50.

It can be seen that the optimum value of the bias flow Mach number is different for different working states.

6.4 Fan noise source modeling

In the discussion of Chapter 3, the main sound sources in turbofan engines are introduced. Although for modern commercial aircraft engines, as the bypass ratio increases, fan noise becomes a dominant component. The sound pressure level of "buzz-saw" noise related to shockwaves at the rotor blade tip can be even 160–170 dB in the near field. But because of its nonlinear attenuation properties and distribution characteristics of frequencies and modes, this kind of noise can be easily attenuated. Therefore, the far-field

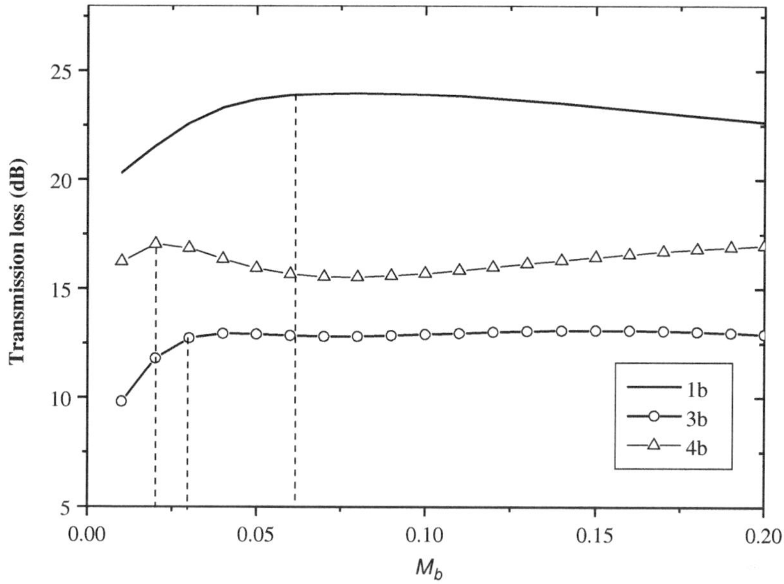

Fig. 6.50 The effect of bias flow on the transmission loss.

frequency spectrum is still dominated by rotor/stator interaction noise. This means that interaction noise is the main sound source to be controlled.

For this kind of noise, the mechanism of its generation has been briefly introduced in Chapter 3. Also, integral methodology to solve the ducted acoustic field generated by the known unsteady surface loading distribution on vanes has been given. In Chapter 4, a three-dimensional lifting surface method is used to solve the unsteady surface loadings within the linear scope based on velocity decomposition and a unified solution of sound and flow fields. So far, the sound field generated by rotor/stator interaction can be solved as long as the incident rotor wakes can be obtained from numerical simulations or empirical models. This will be a part of this section.

In addition, the broadband interaction noise prediction model can be established. The incident disturbance is the stochastic turbulence, which can be described statistically with the assumption of isotropy and homogeneity. The three-dimensional effect can be fully considered by using the three-dimensional lifting surface method to construct the vanes' response function.

As one of the latest developments in nacelle acoustics, the transfer element model for interaction noise prediction will be introduced in this section. The TEM can be directly applied to the analysis of sound propagation and radiation in a finite-length pipe. So, it means that the mutual effect of sound sources and acoustic liners can be discussed by combining the finite-length elements of source and propagation at the same time.

6.4.1 Tonal/broadband interaction noise prediction

6.4.1.1 Wake description

As mentioned previously, the three-dimensional lifting surface method introduced in Chapter 4 can be employed to solve the interaction noise. Once the distribution of rotor wakes is obtained, we can study the source characteristics of tonal interaction noise.

For more specific engineering problems, the velocity distribution of rotor or stator wakes can be calculated by numerical simulation of steady flow and fitted to some curves or formulas. But in most cases, we can generally use the empirical models, such as the Gaussian distribution introduced later.

The flow velocity in the rotor wakes can be regarded as the sum of the mean flow velocity and fluctuating velocity:

$$w = w_\infty + w_t. \tag{6.200}$$

Experimental studies [55] showed that the fluctuating velocity is roughly consistent with the Gaussian distribution law as follows:

$$w_t(y'') = w_c \exp\left[-\delta\left(\frac{y''}{l}\right)^2\right], \tag{6.201}$$

in which w_c is the amplitude, l is the half-width of the wake, and δ is the coefficient of the wake function.

From the definition of the half-width l, we can obtain:

$$w_t = w_c \exp\left(-\delta\left(\frac{1}{2}\right)^2\right) = \frac{1}{2} w_c \quad for\ y'' = l/2, \tag{6.202}$$

and the coefficient can be calculated to be:

$$\delta = 4 \ln 2 \approx 2.7726. \tag{6.203}$$

Coordinate transformation of y'' yields:

$$\frac{w_t}{w_c} = \exp\left\{-\delta\left[(z' \sin\chi + y' \cos\chi)/l\right]^2\right\}. \tag{6.204}$$

For the two-dimensional linear cascade, the wake from the n-th rotor blade can be expressed as:

$$\frac{w_t}{w_c} = \exp\left\{-\delta[(z' \sin\chi + (y' + nd_r)\cos\chi)/l]^2\right\}. \tag{6.205}$$

The whole fluctuating velocity distribution downstream of the rotor cascade can be regarded as the superposition of the infinite number of blade wake flow fields, i.e.,

$$\frac{w_t}{w_c} = \sum_{-\infty}^{\infty} \exp\left\{-\frac{\delta}{l^2}[z' \sin\chi + (y' + nd_r)\cos\chi]^2\right\}. \tag{6.206}$$

Based on the Poisson summation law,

$$\sum_{-\infty}^{\infty} f(n) = \sum_{s=-\infty}^{\infty} \int_{-\infty}^{\infty} f(z) e^{-2\pi isz} dz, \tag{6.207}$$

we use Fourier series expansion and have:

$$\frac{w_t}{w_c} = \sqrt{\frac{\pi}{\delta}} \frac{2l}{d_r \cos\chi} \sum_{s=1}^{\infty} \exp\left[-\left(\frac{s\pi l}{d_r \cos\chi}\right)^2 / \delta\right] \cdot \exp\left[\frac{2\pi is}{d_r}(y' + z' \tan\chi)\right]. \tag{6.208}$$

With the coordinate transformation relations:

$$y = y' - \Omega rte_2, \tag{6.209}$$

$$w_t = \sum w_q \exp i\left(k_0 z + \frac{2\pi sB}{V} n - sB\Omega t\right], \tag{6.210}$$

where

$$w_q = w_c \sqrt{\frac{\pi}{\delta}} \frac{2l}{d_r \cos\chi} \exp\left[-\left(\frac{\pi sl}{d_r \cos\chi}\right)^2 / \delta\right], \tag{6.211}$$

the component of the disturbance velocity in the normal direction of the stator vane can be written as:

$$w_{\varphi'} = w_t \sin(\theta + \chi) = \sum w_q \sin(\theta + \chi) \exp\left[i\left(k_0 z + \frac{2\pi sB}{V} n - sB\Omega t\right)\right]. \tag{6.212}$$

So far, in the integral equation established by the three-dimensional lift surface theory, the flow disturbance term on the right side has been clearly given. The wake depth and half-width of the rotor wake can be fitted to the rotor wake distribution function based on the flow field data obtained by CFD simulations. It can be seen that the tonal interaction noise can be solved completely.

6.4.1.2 Incident turbulence description

The fan discrete noise is distributed at the blade passing frequency and its harmonic frequencies, and the bandwidth is narrow in the spectrum. Therefore, when the fan discrete noise level is no longer significantly higher than the broadband noise component, its proportion in the total noise level is relatively reduced. So, fan broadband noise has begun to receive more and more attention, and the corresponding efficient and effective prediction models are desired to guide the industrial design.

The source of the fan broadband interaction noise is also the unsteady loads on vanes' surfaces. The difference is that, for fan discrete noise, the incoming disturbance is the periodic steady rotor wakes, whereas for broadband interaction noise, the incoming disturbance is random turbulence. This section begins with a general methodology for broadband interaction noise, followed by the analytical process in the three-dimensional annular duct.

The expression of the sound velocity potential in the duct is:

$$\Phi(\vec{x},\omega)=\frac{V}{2pr_0U}\sum_{s=-\infty}^{\infty}\sum_{n=1}^{\infty}\psi(k_{mn}r)e^{im\varphi}\cdot\int_{s0}\psi(k_{mn}r')e^{im\varphi'}\cdot f_{mn}(r')$$

$$\Delta p_0'(r',z',\omega')\cdot\left\{H(z'-z)\frac{\beta^2M}{2k_{nm}(k_0-Mk_{nm})}\cdot e^{i\alpha_1(z-z')}\right. \tag{6.213}$$

$$\left.+H(z-z')\frac{\beta^2M}{2k_{nm}(k_0-Mk_{nm})}\cdot e^{i\alpha_1(z-z')}\right\}ds_0,$$

in which, s_0 is the source surface, i.e., the stator vane surface. Using the Fourier transform, it can be expressed as follows:

$$\phi(\vec{x},t)=\frac{1}{2\pi}\int\Phi(\vec{x},\omega)e^{-i\omega t}d\omega. \tag{6.214}$$

With the definition of acoustic energy in a uniform cross-section duct, when the mean flow is uniform, isentropic, and non-swirling, the expression of the sound flux intensity can be written as:

$$\vec{I}=-\rho_0\frac{\partial\phi^*}{\partial t}\left(\nabla\phi-\frac{\vec{U}}{c_0^2}\frac{D\phi}{Dt}\right). \tag{6.215}$$

Then the acoustic energy propagating upstream and downstream can be obtained by integrating the acoustic flux over the cross-sectional area as follows:

$$\wp^{\pm} = \int \left\langle \vec{I}^{\pm} \right\rangle \cdot \vec{n}\, dS; \ \vec{n}^{\pm} = \pm(0, 0, -1). \tag{6.216}$$

S is the cross-section. Obviously, to determine the acoustic energy in the duct, it is necessary to first obtain the expression of the unsteady surface pressure spectrum $\langle \Delta p_0{}'(r_1', z_1', \omega') \Delta p_0(r_2', z_2', v) \rangle$ on the vane surface.

According to the three-dimensional lifting surface theory, the integral equation can be written into a system of algebraic equations as follows:

$$\Delta p_{i'j'} = \sum_{i_0=1}^{L} \sum_{j_0=1}^{N} \aleph_{i_0 j_0, i' j'} \cdot \omega_{i_0 j_0} \tag{6.217}$$

Based on the linear perturbation assumption, the frequency of the unsteady loadings on the vane surfaces should be the same with the frequency of the upwash velocity spectrum, and the upwash velocity spectrum can be expanded according to the Fourier transform (can be expressed in the form of the Fourier integral):

$$\omega\left(\vec{y}, t\right) = \iint W\left(\vec{K}, \omega\right) e^{i\left(\vec{K} \cdot \vec{y} - \omega t\right)} d\vec{K}\, d\omega. \tag{6.218}$$

Therefore,

$$\Delta p(r', z', \omega) = \int \Im\left(\vec{K}, r', z'\right) W\left(\vec{K}, \omega\right) d\vec{K}, \tag{6.219}$$

where $\Im\left(\vec{K}, r', z'\right)$ represents the unsteady loading on the vanes' surfaces caused by the unit upwash velocity $e^{i\vec{K} \cdot \vec{y}}$ with the wavenumber $\vec{K}$. Thus, the upwash spectrum $\left\langle W^*\left(\vec{K}, \omega\right) \cdot W\left(\vec{k}, v\right) \right\rangle$ can be used to calculate the unsteady surface pressure spectrum.

$$\begin{aligned} \langle \Delta p_0^*(r_1', z_1', \omega) \Delta p_0(r_2', z_2', v) \rangle &= \textstyle\int\int \Im^*\left(\vec{K}, r_1', z_1'\right) \Im\left(\vec{k}, r_2', z_2'\right) \\ &\quad \cdot \langle W^* \cdot W \rangle d\vec{K}\, d\vec{k}, \end{aligned} \tag{6.220}$$

where, $\left\langle W^*\left(\vec{K}, \omega\right) \cdot W\left(\vec{k}, v\right) \right\rangle$ describes the two-point velocity correlation function, which can be solved by spectral analysis of turbulent statistical theory.

The turbulent flow is a random motion that changes drastically in both space and time. Therefore, the exact magnitude and phase of the disturbance at a certain point cannot be obtained, and statistical methods must be introduced to obtain meaningful results. To fully understand a turbulent motion, it is necessary to grasp all possible joint probability distributions of any turbulent random quantities at any time-space points. To simplify the analysis, scientists began to study the simplest isotropic and homogeneous turbulence, and established a statistical theory of turbulence wavenumber spectrum.

The turbulent motion is essentially composed of vortices of different scales. The information regarding the distribution of turbulent energy in vortices of different scales is very important. Through Fourier transform, the isotropic turbulence can be studied in the wavenumber space. In the isotropic turbulent flow field, the velocity correlation function of two unit velocity vectors at two points is:

$$R\left(\vec{r},\ \vec{a},\ \vec{b}\right)=\overline{u'_a u'_b}, \tag{6.221}$$

$\vec{r}$ is the distance vector between these two points.

For the homogeneous turbulence in space, the two-point velocity correlation function is only related to the distance between two points, and the velocity correlation tensor is generally expressed as:

$$R_{ij}\left(\vec{r}\right)=\overline{u'^2}\left[(f-g)\frac{r_i r_j}{r^2}+g\delta_{ij}\right], \tag{6.222}$$

$f(r)$ and $g(r)$ are called the longitudinal correlation coefficient and horizontal correlation coefficient, respectively, and they satisfy the following relation:

$$g=f+\frac{r}{2}\frac{\partial f}{\partial r}. \tag{6.223}$$

In general, when the distance between two points increases to infinity, the velocity correlation tensor R_{ij} quickly approaches zero, so its Fourier transform exists, and turbulence spectral density is defined as:

$$\phi_{ij}\left(\vec{k}\right)=\frac{1}{(2\pi)^3}\iiint R_{ij}(S)\cdot e^{-i\vec{k}\cdot S}dS. \tag{6.224}$$

For the homogeneous turbulence, it can be expressed as:

$$\phi_{ij}=\frac{E\left(\vec{k}\right)}{4\pi k^2}\left[\delta_{ij}-\frac{k_i k_j}{k^2}\right]. \tag{6.225}$$

Assume that the longitudinal correlation coefficient function $f(r)$ is approximately expressed as an exponential form, that is, the Liepmann spectrum [56], in which turbulence wavenumber spectrum can be derived as:

$$E_L(k*) = \frac{8\overline{u'^2}\Lambda}{\pi} \cdot \frac{k*4}{(1+k*2)^3}, \tag{6.226}$$

in which $k* = k\Lambda$. Λ is the turbulent integral scale, and it represents the largest vortex scale.

$$\Lambda = \int_0^\infty f(r)dr. \tag{6.227}$$

Von Karman [57] spectrum can be expressed in the form of:

$$E_V(k*) = \frac{55\Gamma(5/6)}{9\sqrt{\pi}\Gamma(1/3)}\overline{u'^2}\Lambda\frac{k'^4}{(1+k'^2)^{17/6}}, \tag{6.228}$$

in which $k' = \frac{\Gamma(1/3)}{\sqrt{\pi}\Gamma(5/6)}$.

Another form of spectrum assumes that the longitudinal correlation coefficient function follows the normal distribution, i.e., $f(r) = e^{-\frac{\pi r^2}{4\Lambda^2}}$, and the corresponding turbulent wavenumber spectrum is written as:

$$E_G(k*) = \frac{4\overline{\omega^2}\Lambda}{\pi^3} \cdot k*^4 e^{-\frac{k*2}{\pi}}. \tag{6.229}$$

The advantage of the Gaussian spectrum form is that different integration scales can be defined for different directions. This method is often used in numerical simulations of turbulence.

When external turbulence is drawn into the interior of the nacelle, deformation occurs in the axial direction, thus the axial integral scale may be much larger than those of the other two directions. Therefore, this effect needs to be taken into account. Kerschen and Gliebe [58] give the general form of the axisymmetric turbulence spectrum as follows:

$$\begin{aligned}\phi_{ij}\left(\vec{k}\right) &= \left[k^2\delta_{ij} - k_i k_j \delta_{ij}\right]F \\ &+ \left[\left(k^2 - \left(\vec{k}\cdot\vec{\lambda}\right)^2\delta_{ij} - k_i k_j - k^2\lambda_i\lambda_j + \vec{k}\cdot\vec{\lambda}\left(\lambda_i\lambda_j + k_i\lambda_j\right)\right)\right]G,\end{aligned} \tag{6.230}$$

in which $\vec{\lambda}$ is the unit vector perpendicular to the axial direction. Defining $k_t = \sqrt{k_{ty}^2 + k_{tz}^2}$:

$$F = \frac{2\overline{u'^2}}{\pi^2} \frac{\Lambda_a \Lambda_t^4}{\left(1 + \Lambda_a^2 k_a^2 + \Lambda_t^2 k_t^2\right)^3}, \tag{6.231}$$

$$G = \left[2\frac{\overline{u_t'^2}}{\overline{u_a'^2}} - \frac{\Lambda_t^2}{\Lambda_a^2} - 1 \right] F. \tag{6.232}$$

Although isotropic turbulence is a very ideal model, its description is simple and reflects the randomness of turbulence. At present, most of the existing analytical prediction models for fan broadband interaction noise use this assumption as a starting point.

As the mean flow velocity is much larger than the turbulent disturbance velocity, it can be assumed that the turbulence propagates downstream with a frozen pattern. Hanson's derivation [59] shows that the turbulent disturbance can be expressed as follows according to Taylor's hypothesis:

$$\omega\left(\vec{x}, t\right) = \widetilde{\omega}(\widetilde{x}, y, z) = \widetilde{\omega}(x - Ut, y, z), \tag{6.233}$$

where $(\widetilde{x}, y, z)$ is the coordinate system moving with the mean flow velocity, and $(\widetilde{x}, y, z)$ is the stationary coordinate system. By applying the Fourier transform, we have:

$$W\left(\vec{K}, \omega\right) = \frac{\delta(\gamma - \omega/U)}{U} \frac{1}{(2\pi)^3} \iiint \widetilde{\omega}(\widetilde{x}, y, z) e^{-i\left(\gamma\widetilde{x} + \alpha y + kz\right)} d\widetilde{x} dy dz. \tag{6.234}$$

It can be seen that the Taylor's hypothesis is a fairly important hypothesis, which transforms the quadruple integral into a triple integral and greatly simplifies the derivation process. Therefore,

$$\left\langle W^*\left(\vec{K}, \omega\right) \cdot W\left(\vec{k}, \upsilon\right) \right\rangle = \frac{\delta(\gamma - \omega/U)\delta(\gamma' - \upsilon/U)}{U^2} \frac{1}{(2\pi)^6} \iiiint\!\!\iint \langle \widetilde{\omega}(\widetilde{x}, y, z) \widetilde{\omega}(\widetilde{x}', y', z') \rangle \cdot e^{i\left(\gamma\widetilde{x} + \alpha y + kz\right)} e^{-i\left(\gamma'\widetilde{x}' + y' + z'\right)} d\widetilde{x} dy dz d\widetilde{x}' dy' dz' . \tag{6.235}$$

Next, we define $\widetilde{x}' = \widetilde{x} + \Delta x$, $y' = y + \Delta y$, $z' = z + \Delta z$, and rewrite the previous formula as:

$$\left\langle W^*\left(\vec{K},\omega\right)\cdot W\left(\vec{k},\upsilon\right)\right\rangle = \frac{\delta(\gamma-\omega/U)\delta(\gamma'-\upsilon/U)}{U^2}\frac{1}{(2\pi)^6}\iiiint\!\!\iint\langle\widetilde{\omega}(\widetilde{x},y,z)\widetilde{\omega}(\widetilde{x}',y',z')\rangle$$

$$\cdot e^{i(\gamma\widetilde{x}+\alpha y+kz)}e^{-i(\gamma'\widetilde{x}'+y'+z')}d\widetilde{x}dydzd\widetilde{x}'dy'dz'$$

$$=\frac{\delta(\gamma-\omega/U)\delta(\gamma'-\upsilon/U)}{U^2}\frac{1}{(2\pi)^3}\iiint\langle\widetilde{\omega}(\widetilde{x},y,z)\widetilde{\omega}(\widetilde{x}',y',z')\rangle$$

$$\cdot e^{-i(\gamma'\Delta x+\alpha'\Delta y+k'\Delta z)}d\Delta x d\Delta y d\Delta z$$

$$\cdot\frac{1}{2\pi}\int e^{i(\gamma-\gamma')\widetilde{x}}d\widetilde{x}\frac{1}{2\pi}\int e^{i(\alpha-\alpha')y}dy\frac{1}{2\pi}\int e^{i(k-k')z}dz \quad . \tag{6.236}$$

$\langle\widetilde{\omega}(\widetilde{x},y,z)\widetilde{\omega}(\widetilde{x}',y',z')\rangle$ is the two-point velocity correlation function in the turbulent flow. For the homogeneous turbulence, it is independent of the specific coordinate position, so it can be separated from the integral, which is the standard turbulence spectrum:

$$\phi_{ij}\left(\vec{k}\right) = \frac{1}{2\pi^2}\iiint R_{ij}(S)\cdot e^{-i\vec{k}S}dS, \tag{6.237}$$

where $S = (\Delta x, \Delta y, \Delta z)$. Thus, Eq. (6.236) can be finally expressed as:

$$\left\langle W^*\left(\vec{K},\omega\right)\cdot W\left(\vec{k},\upsilon\right)\right\rangle = \delta(\gamma-\omega/U)\cdot\delta(\gamma'-\upsilon/U)\cdot\delta(\omega-\upsilon)$$

$$\cdot\delta(\alpha-\alpha')\cdot\delta(k-k')\frac{1}{U}\phi_{33}. \tag{6.238}$$

Among them, the subscript "33" represents the direction of the upwash velocity component. It can be seen that there is a corresponding relationship between the axial wavenumber of the upwash and the frequency. Considering Delta function $\delta(\omega - \upsilon)$, the correlation of the upwash at different frequencies is zero.

Besides the background turbulence, the rotor wake turbulence is an important component of the incident random disturbance. The distribution of wake turbulence intensity downstream of the rotor is shown in Fig. 6.51. Similar to the rotor wake velocity, the wake turbulence intensity is also assumed to satisfy the Gauss distribution. To describe the velocity correlation function more conveniently, we use the rotor coordinate system (Fig. 6.52),

$$\vec{X} = \vec{x} + \vec{D} + (\Omega r_1 t)\vec{e}_2,\ \vec{Y} = \vec{y} + \vec{D} + (\Omega r_2 \tau)\vec{e}_2, \tag{6.239}$$

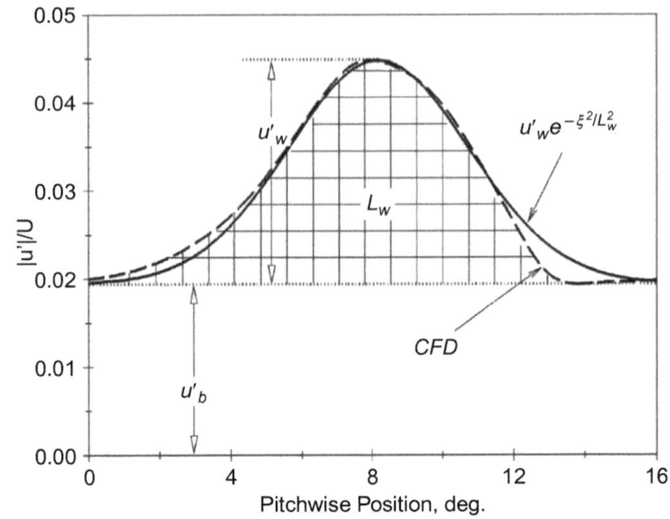

Fig. 6.51 The distribution of rotor wake turbulence intensity [60].

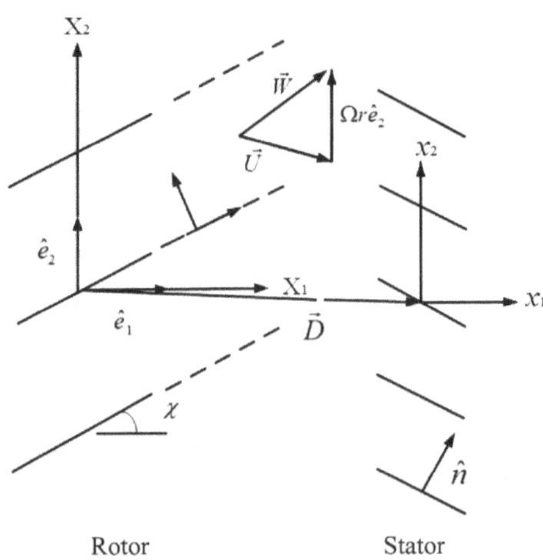

Fig. 6.52 Rotor/stator coordinate system

$$\omega\left(\vec{X}-\vec{W}\,t,r\right)=F\left(\vec{X}\cdot\hat{n},r\right)g\left(\vec{X}-\vec{W}\,t,r\right), \tag{6.240}$$

where $\vec{D}$ is the distance vector between the rotor cascade and the stator cascade.

F is the average turbulence intensity, and g is a statistical random function. According to Nallasamy and Envia [60], F is expressed as:

$$F\left(\vec{X}\cdot\vec{n},\bar{r}\right)=u_b'+\frac{N_B L_\omega u_\omega'}{2\pi\bar{r}}\sum_{m=-\infty}^{\infty}e^{-\frac{1}{4\pi}\left(\frac{mN_B L_\omega}{\bar{r}}\right)^2}\cdot e^{i\frac{mN_B}{\bar{r}\cos\chi}\vec{X}\cdot\vec{n}}. \tag{6.241}$$

So we have:

$$\left\langle W^*\left(\vec{K},\omega,r_1\right)\cdot W\left(\vec{k},\upsilon,r_2\right)\right\rangle=\frac{1}{(2\pi)^6}\delta\left(\omega-\vec{K}\cdot\vec{U}+\Omega r_1 K_3\right)$$

$$\cdot\delta\left(\upsilon-\vec{k}\cdot\vec{U}+\Omega r_2 k_3\right)\cdot e^{-i\left(\vec{K}-\vec{k}\right)\cdot\vec{D}}\cdot\iint F^*\left(\vec{X}\cdot\vec{n},\vec{r}\right)\cdot$$

$$F\left(\vec{Y}\cdot\vec{n},\vec{r}\right)\cdot\hat{\Phi}\left(\vec{\xi},\Delta r\right)\cdot e^{i\vec{K}\cdot\vec{\xi}+i\left(\vec{K}-\vec{k}\right)\cdot\vec{Y}}d\vec{\xi}\,d\vec{Y} \tag{6.242}$$

where $\bar{r}$ represents the average radius of two points, i.e., $\bar{r}=(r_1+r_2)/2$. The correlation function $\Phi\left(\vec{\xi},\Delta r\right)$ is periodic in the circumferential direction, so we have:

$$\Phi\left(\vec{\xi},\Delta r\right)=\frac{1}{2\pi\bar{r}}\sum_{s=-\infty}^{\infty}\phi\left(\vec{\xi},\Delta r\right)e^{is\xi_2/\bar{r}}. \tag{6.243}$$

For an annular duct,

$$\omega\left(\vec{x}_0,t\right)=\sum_{m_g=-\infty}^{+\infty}\sum_{n_g=-\infty}^{+\infty}\iint W_{m_g n_g}\left(k_g,\omega,r\right)\exp\left[i\left(k_g z+m_g\varphi+n_g\bar{r}-\omega t\right)\right]dk_g d\omega. \tag{6.244}$$

The two-point correlation function is written as:

$$\hat{\Phi}\left(\vec{\xi},\Delta r\right)=\frac{\left\langle\omega_i\left(\vec{x}_0-\vec{U}\,t\right)\cdot\omega_i\left(\vec{x}_0'-\vec{U}\,t\right)\right\rangle}{\overline{\omega^2}}, \tag{6.245}$$

Because the turbulence scale is much smaller than the duct size $\Lambda\ll r_d$, two variables are defined by using the multiscale method,

$$\hat{x} = \frac{x}{\Lambda}, \tag{6.246}$$

$$x^* = \frac{x}{r_d}. \tag{6.247}$$

$\hat{x}$ is the fast variable and x^* is the slow variable. For baseline or swept-only stator cascade,

$$k_1 = \frac{\omega}{U}, k_2 = \frac{\pi n_g}{r_d - r_h}, k_3 = \frac{m_g}{r}. \tag{6.248}$$

It is assumed that the transformation of the duct scale does not affect the correlation within the small scale of turbulence, so:

$$\hat{\Phi}\left(\vec{k}, \Delta r\right) = \frac{1}{\omega^2} \frac{\pi}{r_d - r_h} \frac{1}{U} \phi_{ij}, \tag{6.249}$$

$$\omega\left(\vec{x}_0, t\right) = \sum_{m_g=-\infty}^{+\infty} \sum_{n_g=-\infty}^{+\infty} \iint W_{m_g n_g}\left(k_g, \omega, r\right) \exp\left[i\left(k_g z + m_g \varphi + n_g \bar{r} - \omega t\right)\right] dk_g d\omega, \tag{6.250}$$

where

$$\bar{r} = \frac{\pi(r - r_h)}{r_d - r_h}. \tag{6.251}$$

Thus,

$$\begin{aligned}
\left\langle W^*\left(\vec{K}, \omega, r_1\right) \cdot W\left(\vec{k}, \upsilon, r_2\right)\right\rangle &= \left\langle W*\left(\vec{K}, \omega, r_1\right) \cdot W\left(\vec{k}, \upsilon, r_2\right)\right\rangle_{\omega\omega} \\
&+ \left\langle W*\left(\vec{K}, \omega, r_1\right) \cdot W\left(\vec{k}, \upsilon, r_2\right)\right\rangle_{b\omega} \\
&+ \left\langle W*\left(\vec{K}, \omega, r_1\right) \cdot W\left(\vec{k}, \upsilon, r_2\right)\right\rangle_{bb},
\end{aligned}$$

$$\begin{aligned}
\left\langle W*\left(\vec{K}, \omega, r_1\right) \cdot W\left(\vec{k}, \upsilon, r_2\right)\right\rangle_{\omega\omega} &= \left(\frac{N_B L_\omega u'_\omega}{2\pi\bar{r}}\right)^2 \cdot \frac{\pi}{\bar{r} \cdot (r_d - r_h)} \cdot \sum_m \sum_l \sum_{m_g} \sum_{n_g} \\
& e^{-\frac{1}{4\pi}\left(\frac{m N_B L_\omega}{\bar{r}}\right)^2} \cdot e^{-\frac{1}{4\pi}\left(\frac{l N_B L_\omega}{\bar{r}}\right)^2} \cdot e^{-\frac{1}{4\pi}\left(\frac{(m-l) N_B}{\bar{r}\cos\chi}\right)\vec{D} \cdot \vec{n}} \\
& \cdot \delta\left(\omega - \vec{k} \cdot \vec{U}\right) \cdot \delta(\omega' - \omega - \Omega(m-l)N_B) \\
& \cdot \Phi\left(\vec{k} - \frac{m N_B}{\bar{r}\cos\chi} n_1, \Delta r\right) \cdot \delta\left(\vec{K} - \vec{k} + \frac{(m-l)N_B}{\bar{r}\cos\chi}\vec{n}\right) \\
& \cdot \delta\left(k_2 - \frac{\pi n_g}{r_d - r_h}\right) \cdot \delta\left(k_3 - \frac{m N_B + m_g}{\bar{r}}\right)
\end{aligned}$$

$$\left\langle W*\left(\vec{K},\omega,r_1\right)\cdot W\left(\vec{k},v,r_2\right)\right\rangle_{b\omega}=\frac{N_B L_\omega u_b' u_\omega'}{2\pi\bar{r}}\cdot\frac{\pi}{\bar{r}\cdot(r_d-r_h)}\cdot\sum_m\sum_{m_g}\sum_{n_g}$$

$$e^{-\frac{1}{4\pi}\left(\frac{mN_BL_\omega}{\bar{r}}\right)^2}\delta\left(\omega-\vec{k}\cdot\vec{U}\right)\cdot\delta(\omega'-\omega-\Omega mN_B)\cdot$$

$$\left\{e^{i\left(\frac{mN_B}{\bar{r}\cos\chi}\right)\vec{D}\cdot\vec{n}}\cdot\Phi\left(\vec{k}-\frac{mN_B}{\bar{r}\cos\chi}n_1,\Delta r\right)\cdot\delta\left(\vec{K}-\vec{k}+\frac{mN_B}{\bar{r}\cos\chi}n\right)\right.$$

$$\cdot\delta\left(k_2-\frac{\pi n_g}{r_d-r_h}\right)\cdot\delta\left(k_3-\frac{mN_g+m_g}{\bar{r}}\right)$$

$$+e^{-i\left(\frac{mN_B}{\bar{r}\cos\chi}\right)\vec{D}\cdot\vec{n}}\cdot\Phi\left(\vec{k},\Delta r\right)\cdot\delta\left(\vec{K}-\vec{k}+\frac{mN_B}{\bar{r}\cos\chi}n\right)$$

$$\cdot\delta\left(k_2-\frac{\pi n_g}{r_d-r_h}\right)\cdot\delta\left(k_3-\frac{m_g}{\bar{r}}\right)$$

$$\left\langle W*\left(\vec{K},\omega,r_1\right)\cdot W\left(\vec{k},v,r_2\right)\right\rangle_{bb}=u_b'^2\cdot\frac{\pi}{\bar{r}\cdot(r_d-r_h)}\cdot\delta\left(\omega-\vec{k}\cdot\vec{U}\right)\delta(\omega'-\omega)\cdot\Phi\left(\vec{k},\Delta r\right)$$

$$\cdot\delta\left(\vec{K}-\vec{k}\right)\cdot\delta\left(k_2-\frac{\pi n_g}{r_d-r_h}\right)\cdot\delta\left(k_3-\frac{m_g}{\bar{r}}\right).$$

(6.252a,b,c,d)

NASA Source Diagnostic Test (SDT) case has been widely used to verify broadband fan noise models for detailed aerodynamic and aeroacoustic results. The interaction noise caused by turbulence with downstream stator is separated. The background turbulence and rotor wake turbulence are both taken into account by Nallasamy's description. The input parameters are listed in Table 6.3. In the present calculation, the most simplified Liepmann spectrum is used as the input, which is homogeneous without consideration of the spatial variation of the integral length scale and turbulence intensity. The following listed parameters refer to the input of the prediction. The real stagger angle in the SDT case is about 35° near the leading edge, and close to O near the trailing edge. However, in the present prediction, the vanes are assumed to be unstaggered. This effect will be discussed later. As Posson et al. [61] ignored the cross term of the background/wake turbulence, the

Table 6.3 Input parameter of the NASA SDT.

	Λ, *m*	$u_{rms,b}$, *m/s*	$u_{rms,w}$, *m/s*	L_w, *m*
Baseline	0.006	2.4	5.0	0.05
Low-count	0.0055	2.4	5.0	0.03

background turbulence intensity given here is smaller than the value given by Posson et al. The turbulence modal numbers of the incident perturbation are set to be within the range $|m_g| \leq 200$, $|n_g| \leq 50$.

In Posson and Roger's results [61], the upstream acoustic power is in very good agreement with the experiment, but the downstream acoustic power is underestimated, especially in the baseline OGV case. So the upstream and downstream acoustic power levels are close. But according to the experimental data, the downstream acoustic power level is about 5 dB higher than the upstream acoustic power level for both the baseline OGV case and low-count OGV case. This character is well captured by the present model. As Fig. 6.53 shows, the present model gives good predictions in all four cases. Although the stagger angle and swirling effect are out of consideration, and the turbulence model is simply represented by Liepmann spectrum using a constant integral length scale and turbulence intensity value, the level and the shape of the acoustic power spectra are well predicted by the present model. This indicates that the three-dimensional geometry plays quite an important role and should be considered more strictly. The predictions in all the four cases show the similar behavior compared with the experiential data. The acoustic power is underestimated slightly in the medium frequency range and is overestimated a little in the high frequency range. This indicates that the corrections for stagger and swirling mean flow effects are quite necessary for realistic applications.

The baseline OGV has 54 radial vanes, and the low-count OGV has 26 radial vanes. The experimental data shows nearly no difference in the low frequency range. Above 5000 Hz, the variation of the noise level becomes gradually larger as the frequency increases. In Nallasamy and Envia [60], the two-dimensional calculation indicated that the 6.3 dB reduction in acoustic power at high frequencies can be observed if the same stagger angle is applied, and the stagger angle change also contributes to the reduction. But in the low frequency range, the predictions fail to provide any explanation. Fig. 6.54 shows the new predictions of the exhaust acoustic power level at approach condition for 54 and 26 radial vanes by using the present three-dimensional theory. It can be seen that the predicted trends for different stator vane numbers are consistent with the experimental data, although no stagger angle change is considered. As the same turbulence model is used, the trend is assumed to be involved with the propagation of an acoustic wave in an annular duct.

The circumferential mode is expressed as $m = m_g - sV$. Fig. 6.55 plots the exhaust acoustic power as a function of circumferential mode order at

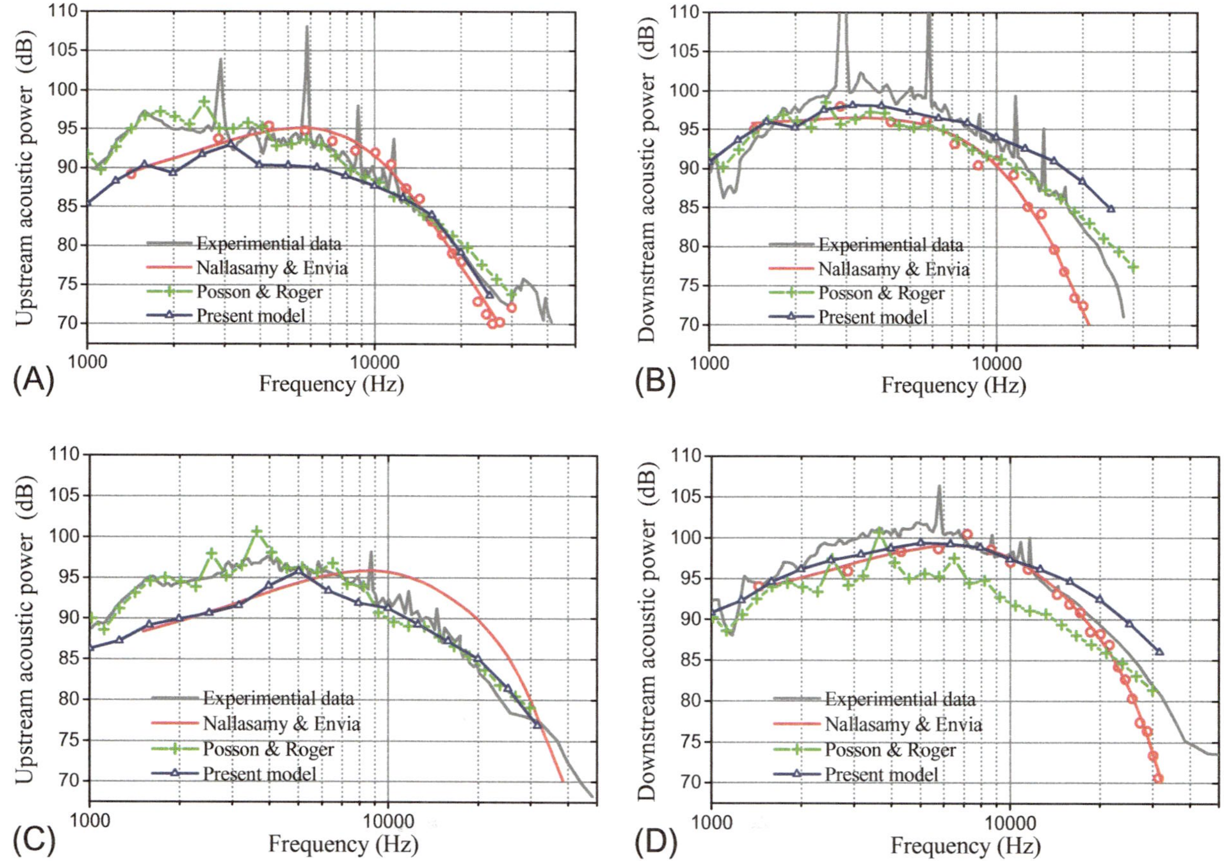

Fig. 6.53 Comparison of the computed and measured spectra for the NASA SDT stator at approach: (A) low count, upstream; (B) low count, downstream; (C) baseline, upstream; and (D) baseline, downstream (see [61] for Nallasamy's result).

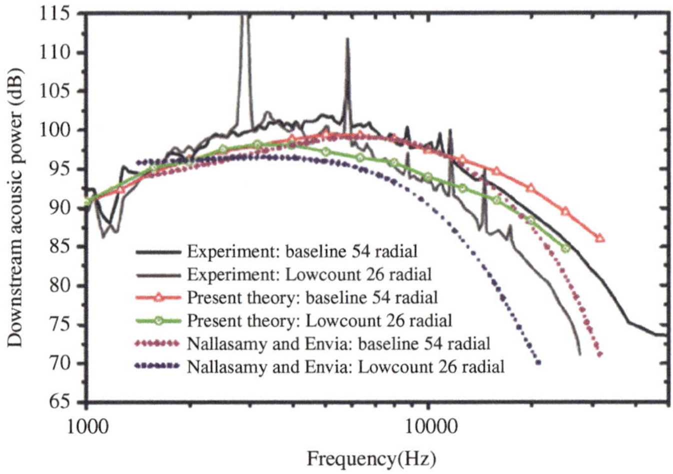

Fig. 6.54 Effect of vane number on exhaust sound power spectra at approach (see [61] for Nallasamy's result).

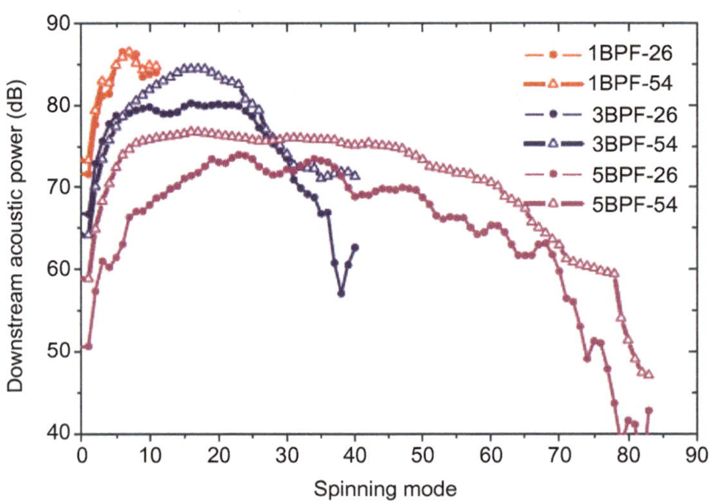

Fig. 6.55 Exhaust sound power distribution as a function of spinning mode order at approach.

three frequencies. For lower frequencies, the number of cut-on modes is small. Moreover, from the shape of the Liepmann spectrum, it can be seen that the main distribution is limited to a relatively small circumferential wavenumber. So the required order for the convergent summation of the series is small or even equal to zero. In other words, vanes number has relatively small effect on the circumferential modes distribution. For an isolated vane, the reduction for decreasing vane number is expected to be 20 log (V_2/V_1) for a constant chord, but the low-count OGV has a larger chord. So ultimately, there is almost no change of acoustic power for all cut-on modes at low frequencies. For higher frequencies, the effective range of value is larger for the predictions for low-count OGV, which means the relationship between the vane number and mode distribution is much more complex. So the authors believe that this effect plays a crucial role on the reduction of broadband noise level for different vane numbers. This still needs further investigation and experimental evidence.

For broadband interaction noise, an extended study has been made by Hanson using the two-dimensional rectilinear cascade model, in which the transformations of several coordinate systems are required to connect the lean/sweep definition with common broadband noise analysis. More understanding on this influence can be obtained when the annular geometry is considered. The fan row can also be written as a transfer element, so the reflection and scattering effects by the fan can be discussed. All these interesting investigations require more experimental and theoretical efforts. Another big issue is to introduce the corrections for stagger and swirling mean flow effects in an annular duct.

6.4.2 The passive control effect of vane sweep and lean

Fig. 6.56 shows the sketch of the stator with swept and leaned vanes. From the vanes' hub to the tip, the vanes sweep backward and the sweep angle is positive; otherwise, the vanes sweep forward and the sweep angle is negative. The lean angle is positive in the fan rotor rotation direction. The disturbance velocity perpendicular to the vane surface is composed of the circumferential velocity u_φ' and radial velocity u_r', which can be deduced

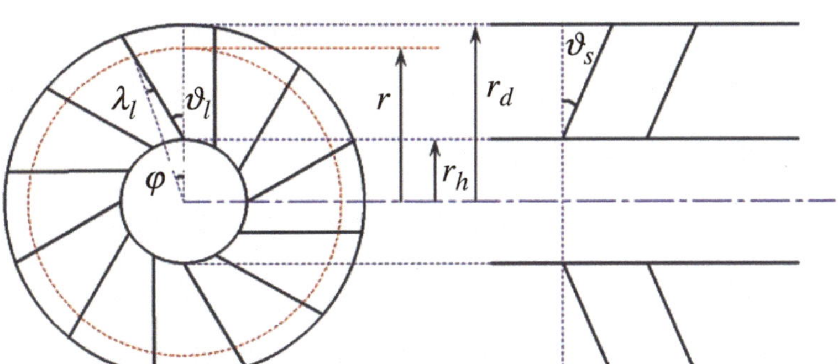

Fig. 6.56 Sketch of swept and leaned vanes.

by the momentum equation. By using the boundary condition on the vane surface, one can obtain an integral equation for the unsteady loading on a single blade in the form of:

$$-w_n = u'_{\varphi} \cdot \cos\lambda_l - u'_r \cdot \sin\lambda_l. \tag{6.253}$$

where w_n is the incident velocity component perpendicular to the vane surface.

We define two functions,

$$\begin{cases} g_{mn}(r) = \dfrac{im}{r}\cos\lambda_l(r) - \dfrac{\psi'(k_{mn}r)}{\psi(k_{mn}r)}\sin\lambda_l(r) \\ f_{mn}(r') = -\dfrac{im}{r'}\cos\lambda_l(r') - \dfrac{\psi'(k_{mn}r')}{\psi(k_{mn}r')}\sin\lambda_l(r') \end{cases}. \tag{6.254}$$

The first term in Eq. (6.254) represents the effect of circumferential acoustic velocity, and the second term represents the effect of radial acoustic velocity, respectively. The integral equation derived by the three-dimensional lifting surface method is:

$$-w_n = \int_{s_0(\tau)} \Delta p_0 \cdot K ds_0(\mathbf{y}). \tag{6.255}$$

in which the kernel function is given by:

$$K = \frac{-V}{2\pi\rho_0 U}\sum_{q=-\infty}^{\infty}\sum_{n=1}^{\infty}\psi(k_{mn}r)e^{im\varphi}\psi(k_{mn}r')e^{-im\varphi'_0}\cdot$$
$$g_{mn}(r)\cdot f_{mn}(r')\cdot$$
$$\left\{H(z-z')\left[\frac{\beta^2 M_a e^{i\alpha_2(z-z')}}{2\kappa_{n,m}(M_a\kappa_{n,m}-k_0)} + \frac{M_a^2 e^{i\alpha_3(z-z')}}{k_0^2 + M_a^2 k_{m,n}^2}\right]\right. \tag{6.256}$$
$$\left. + H(z'-z)\frac{-\beta^2 M_a e^{i\alpha_1(z-z')}}{2\kappa_{n,m}(k_0 + M_a\kappa_{n,m})}\right\}.$$

6.4.2.1 *Vanes sweep effect on the tonal interaction noise*

The NLR low-speed fan experiment was made to validate Schulten's three-dimensional lifting surface method by comparing the computed pressure jump distribution with the experimental data. Based on this low-speed fan configuration, Schulten [62] studied the vanes' sweep effect on the reduction of fan tonal noise. Table 6.4 gives the parameters. The model

Table 6.4 Input parameters of the case proposed by Schulten for vane sweep discussion.

Parameter description	Value
Number of rotating rods	16
Number of stator vanes	18
Hub/tip ratio	0.6
Vane chord length	0.25
Axial Mach number	0.244
Air density/$(\text{kg} \cdot \text{m}^{-3})$	1.225
Sound speed/$(\text{m} \cdot \text{s}^{-1})$	340.43
Nondimensional first harmonic frequency	6.491
Rods circumferential tip Mach number	0.406

consists of a through-flow annular nacelle. The outer duct radius r_d is used for the nondimensionalization of the parameters, and the hub-to-tip ratio is 0.6. The axial Mach number is 0.244. A rotor consisting of 16 cylindrical rods is placed in front of the stator, which has 18 vanes with a constant nondimensional vane chord as 0.25. The rods' circumferential tip Mach number is 0.406. The nondimensional first harmonic frequency is $M_T B$, at which only one acoustic mode (2,0) is cut-on.

In this case, $n_g = 0$. That is to say, the incident disturbance is in phase in the spanwise direction. Fig. 6.57 shows the computed acoustic intensity level as a function of sweep angle at the first harmonic frequency. The

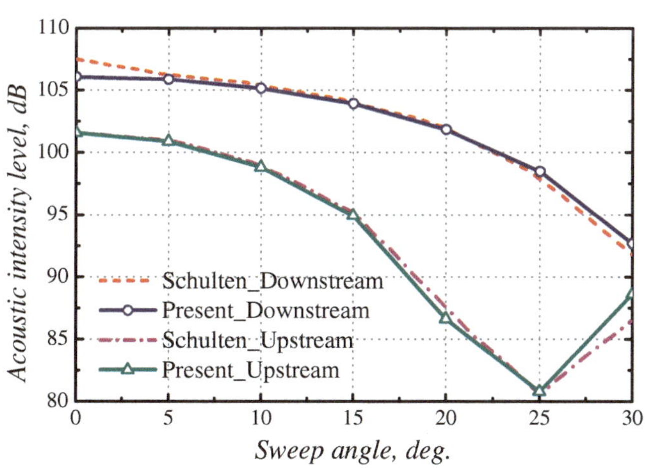

Fig. 6.57 Changes of acoustic intensity level with different sweep angles (see [62] for Schulten's result).

changes of upstream and downstream acoustic intensity levels obtained by using the present formulations are both in close agreement with Schulten's results. It's also validated by Schulten's work that the unsteady pressure jump on the whole vane does not vanish but has a strong spanwise phase variation for a large sweep angle, which is most prominent at the leading edge. Moreover, Schulten found out that sweep does not yield immediate and sizable benefits when he studied the case for the advanced ducted propeller (ADP), which is closer to the real configuration of high bypass turbofans.

Another case is presented here for a high-speed fan configuration. By using the parameters (listed in Table 6.5) in Category 4 Benchmark Problem of the Third CAA Workshop, a classical test for rotor wake/stator interaction, Elhadidi and Atassi [63] studied the sweep effect for a high-speed fan by using the three-dimensional linearized Euler simulation. But their discussion was still limited to the case $n_g = 0$. Good agreement is observed in Fig. 6.58. The calculated results show that a small sweep angle may have a slightly adverse effect on the relative acoustic power changes. With a given sweep

Table 6.5 Input parameters of category 4 benchmark problem of the third CAA workshop.

B	V	R_h/R_d	$Gap_{tip}/Chord$	M_a	M_T
16	24	0.5	1.0	0.5	0.783

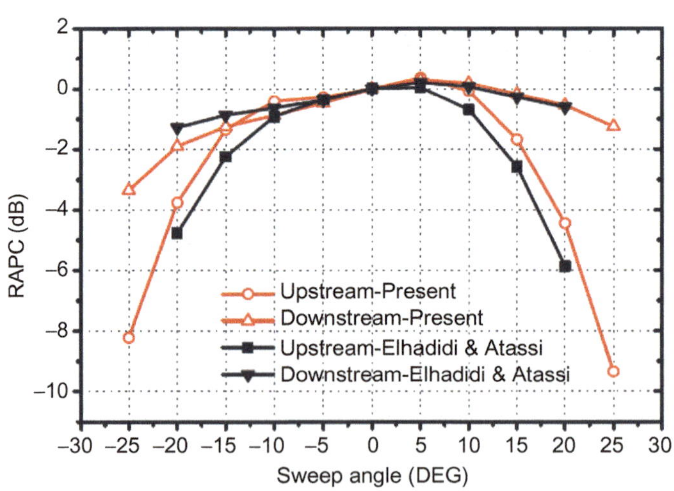

Fig. 6.58 Relative acoustic power changes with different sweep angles, MT = 0.783, ng = 0 (see [63] for Atassi's result).

angle, the reduction for upstream is larger than that for downstream. Schulten [62] and Elhadidi and Atassi [63] reached the same explanation. As indicated by the expressions of the axial acoustic wavenumbers, the upstream-propagating wavelength is smaller than the downstream wavelength for the propagating duct modes. The upstream mode is more sensitive to the axial source variation, which in this chapter is the relative axial position changes of a swept stator at different radial positions, and the noise reduction is induced.

The forward-swept stator seems also beneficial to the reduction of the acoustic power. Such finding is different from the conclusion addressed by two-dimensional strip theory. In the two cases mentioned earlier, the phase variation of the incident gust is not introduced. For $n_g = 0$, the spanwise phase variation of the upwash at the leading edge of the vane is only caused by the sweep. Theoretically, forward-swept and backward-swept stators have the same effect, although in a practical application, forward-swept stators are not efficient for suppressing the fan tonal noise. To explain this conflict, the discussion is extended to include high orders of the radial wavenumbers n_g in the following section.

The aerodynamic response on the vane surface is very sensitive to the radial phase variation of incident disturbance velocity. Also, the distribution of unsteady surface pressure couples with the acoustic field in the annular duct. Atassi et al. [64] claimed that wake scattering into the duct acoustic modes occurs when the radial phase of the incident disturbance is different from that of the duct modes. In other words, it occurs when there is a great difference between n_g and the order of the radial acoustic modes. To take a closer look, a detailed comparison is carried out at two different reduced frequencies. The wake phase variation $n_g = 0,\ \pm 1,\ \pm 2,\ \pm 3,\ \pm 4$, and N_{cuton} denotes the number of the cut-on radial modes. Fig. 6.59A and B show the influence of high orders of n_g on the sound scattering of a radial stator. For the circumferential Mach number at the fan tip $M_T = 0.522$, only one acoustic mode (−8.0) is cut-on, and $N_{cuton} = 1$. The relative acoustic power changes are different for $n_g = -1$ and $n_g = +1$. Two modes, (−8, 0) and (−8, 1) can propagate at $M_T = 0.783$, and $N_{cuton} = 2$. The corresponding range is $|n_g| \leq 2$, in which the sign of n_g has a great effect. This phenomenon is due to the aerodynamic response on the vanes and its coupling with the ducted acoustic field.

Figs. 6.60 and 6.61 show the contours of the real part of the unsteady pressure jump across the vane, scaled with respect to $\rho_0 U_0 W_{m_g ng}$. In the leading edge region of the vane, the radial phase variation of the unsteady surface pressure is consistent with that of the incident disturbance. On the rest of the

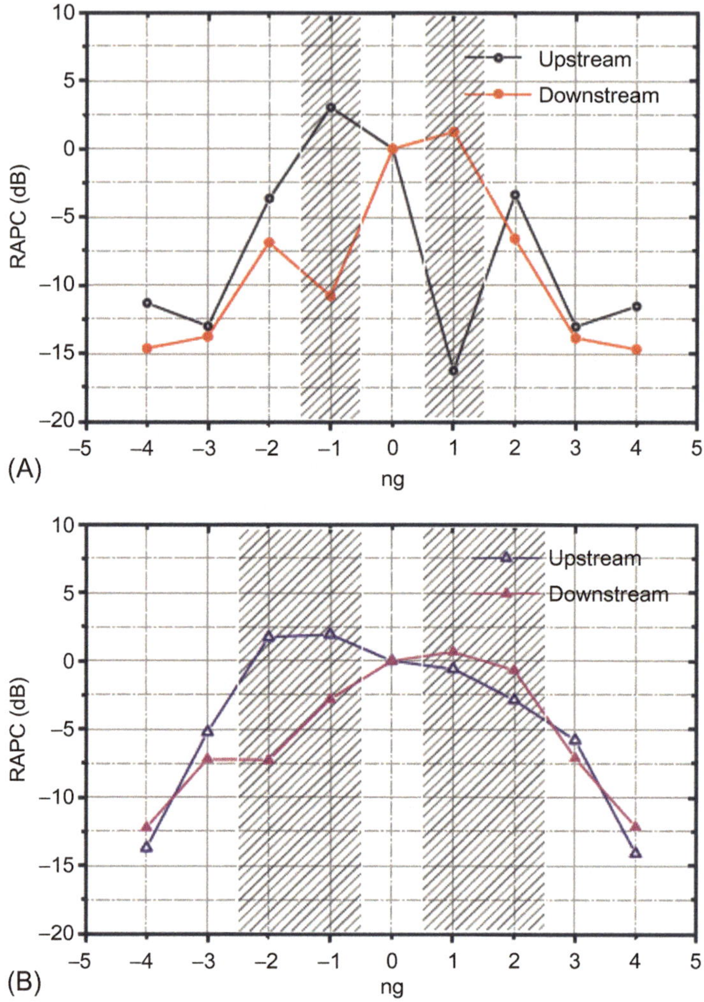

Fig. 6.59 The relative acoustic power changes with different radial phase variation n_g for a radial stator: (A) $M_T = 0.522$ and (B) $M_T = 0.783$.

chord, the distribution also shows notable differences among different n_g, when the absolute value of n_g is less than or equal to the number of the cut-on radial modes N_{cuton} in the annular duct. Meanwhile, the phase pattern along the spanwise direction is more concerned with the cut-on acoustic modes. It is the reason that different relative acoustic power changes are brought by plus or minus n_g, only if $|n_g| \leq N_{cuton}$. The unsteady loadings at the midchord is still comparable in amplitude to the unsteady loadings the leading edge when $|n_g| = N_{cuton} + 1$, but its distributions are nearly

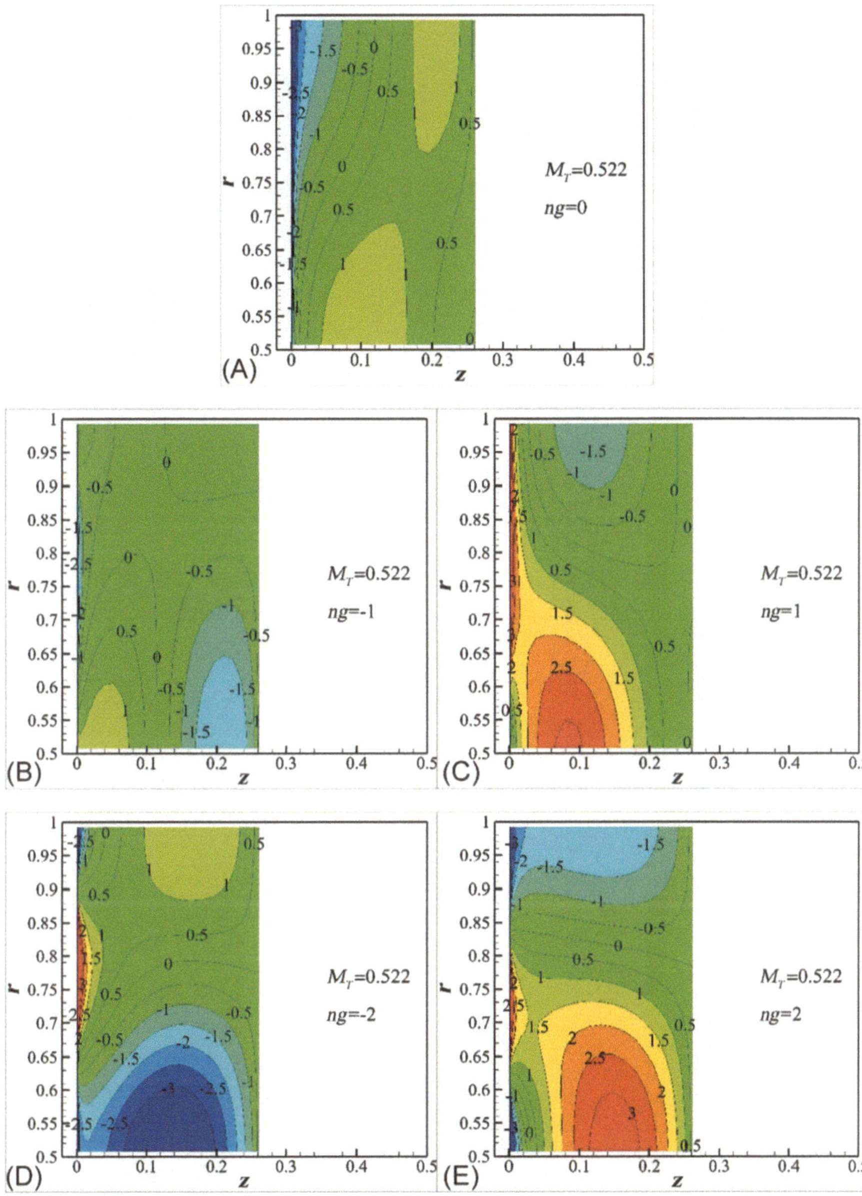

Fig. 6.60 Real part of the unsteady pressure jump across the radial vane with different n_g, $M_T = 0.522$: (A) $n_g = 0$, (B) $n_g = -1$, (C) $n_g = 1$, (D) $n_g = -2$, (E) $n_g = 2$, (F) $n_g = -3$, (G) $n_g = 3$, (H) $n_g = -4$, (I) $n_g = 4$.

(Continued)

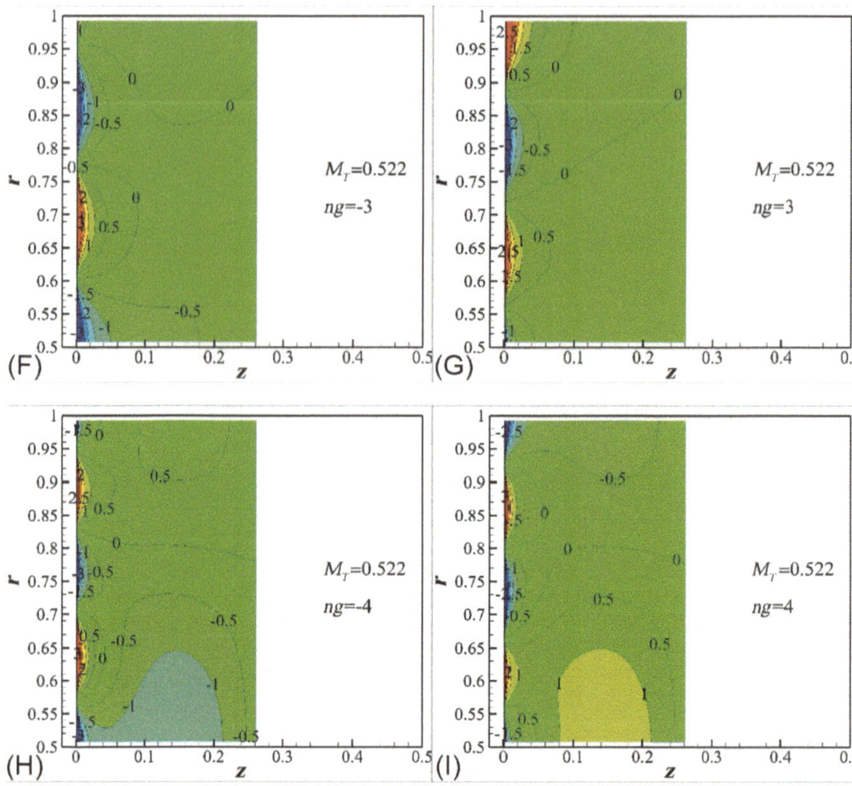

Fig. 6.60, cont'd

symmetric. With further increasing of the absolute value of n_g, the unsteady loadings are then mainly distributed just at the leading edge of the vane. Therefore, the relative acoustic power changes are nearly the same for plus and minus n_g, if $|n_g| > N_{cuton}$.

As mentioned earlier, n_g is positive for the typical wake of a rotor. Fig. 6.62 shows the relative acoustic power changes with different combinations of sweep angle and n_g, under the condition $M_T = 0.522$. The baseline is the case with $n_g = 0$ and $\upsilon_s = 0$. Fig. 6.63 presents the contours of a real part of the unsteady pressure jump with different n_g, across a forward-sweep vane $\upsilon_s = -25^0$ or a back-sweep vane $\upsilon_s = 25^0$. In a uniform flow,

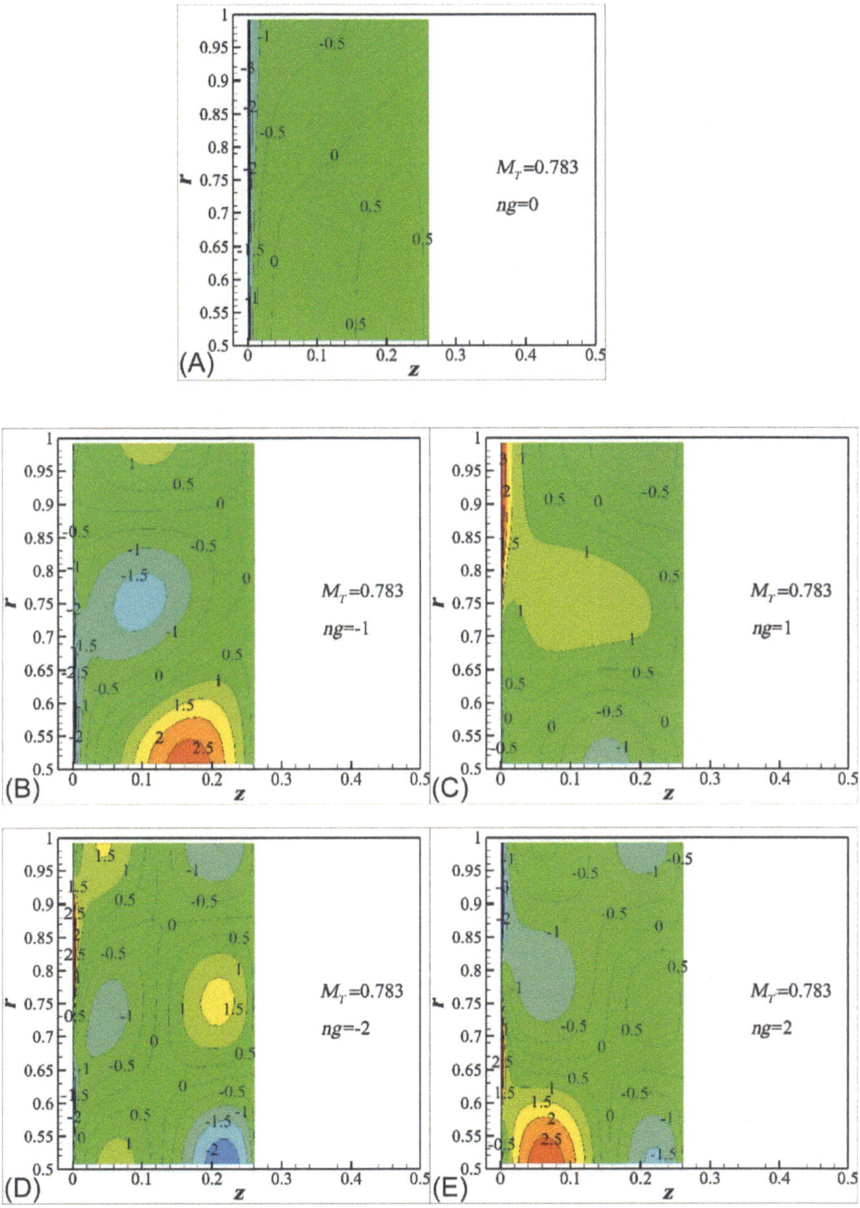

Fig. 6.61 Real part of the unsteady pressure jump across the radial vane with different n_g, $M_T = 0.783$: (A) $n_g = 0$, (B) $n_g = -1$, (C) $n_g = 1$, (D) $n_g = -2$, (E) $n_g = 2$, (F) $n_g = -3$, (G) $n_g = 3$, (H) $n_g = -4$, (I) $n_g = 4$.

(Continued)

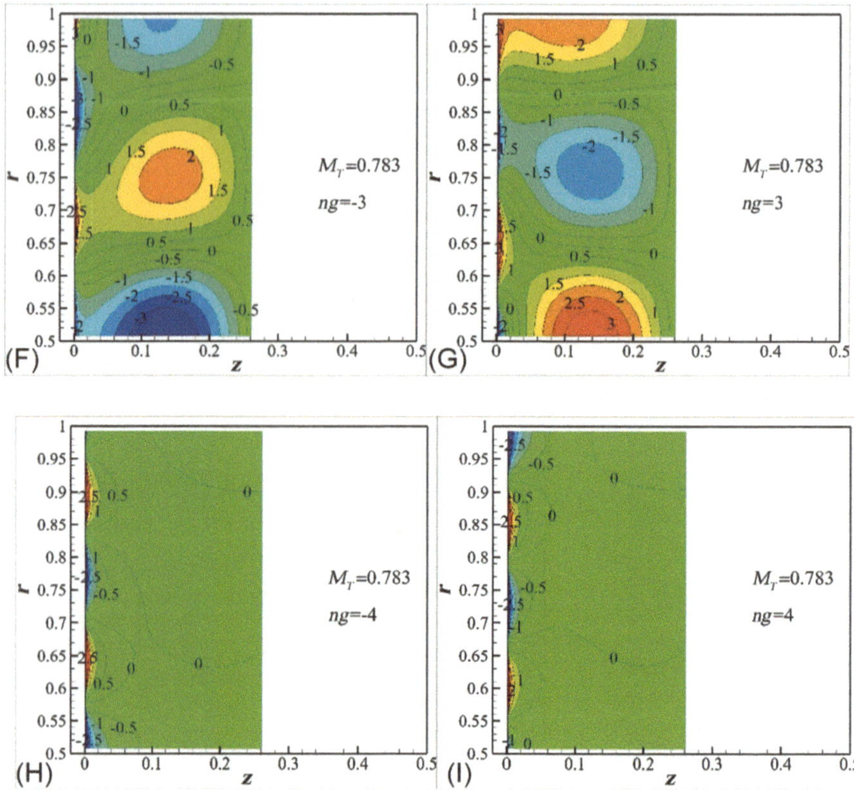

Fig. 6.61, cont'd

the upwash phase variation from vane sweep is proportional to $\frac{\omega}{U}\Delta r \tan\vartheta_s$. Hence, the alternative factor $\widetilde{n}_g$ indicative of the upwash phase variation at the leading edge can then be expressed by:

$$\widetilde{n}_g = n_g + \frac{BM_T(1-h)}{\pi M}\tan\upsilon_s. \tag{6.257}$$

For the comparison purpose, the corresponding results for $M_T = 0.783$ are given in Figs. 6.64 and 6.65.

On one hand, when $\left|\widetilde{n}_g\right| \leq N_{cuton} + 1$, the coupling between the unsteady surface loadings and scattering acoustic modes still play an important role.

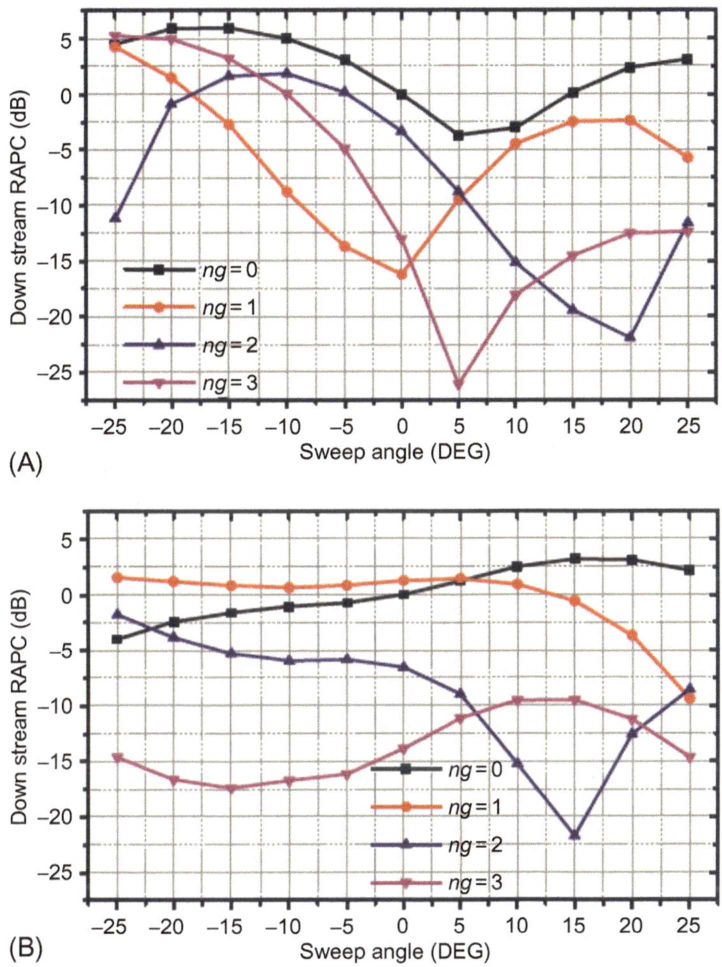

Fig. 6.62 The relative acoustic power changes with different combinations of sweep angle and radial wavenumber n_g, $M_T = 0.522$: (A) upstream, (B) downstream.

For instance, in the case $M_T = 0.522$ and $n_g = 1$, the vane sweep leads to an increase of the acoustic power level in the upstream. On the other hand, If n_g is already larger than $N_{cuton} + 1$, further increase of upwash radial phase variation by the backward sweep of the vanes cannot bring any further significant noise reduction. The best selection of $\widetilde{n}_g$ seems to be that $\widetilde{n}_g$ larger than $N_{cuton} + 1$. When the incident disturbance with high radial mode order n_g is considered, the forward sweep is not always effective. To more clearly illustrate this point, Figs. 6.62 and 6.64 are reproduced to be Figs. 6.66 and 6.67 by $\widetilde{n}_g$ as x-axis. It can be seen that the alternative factor $\widetilde{n}_g$ has a critical effect

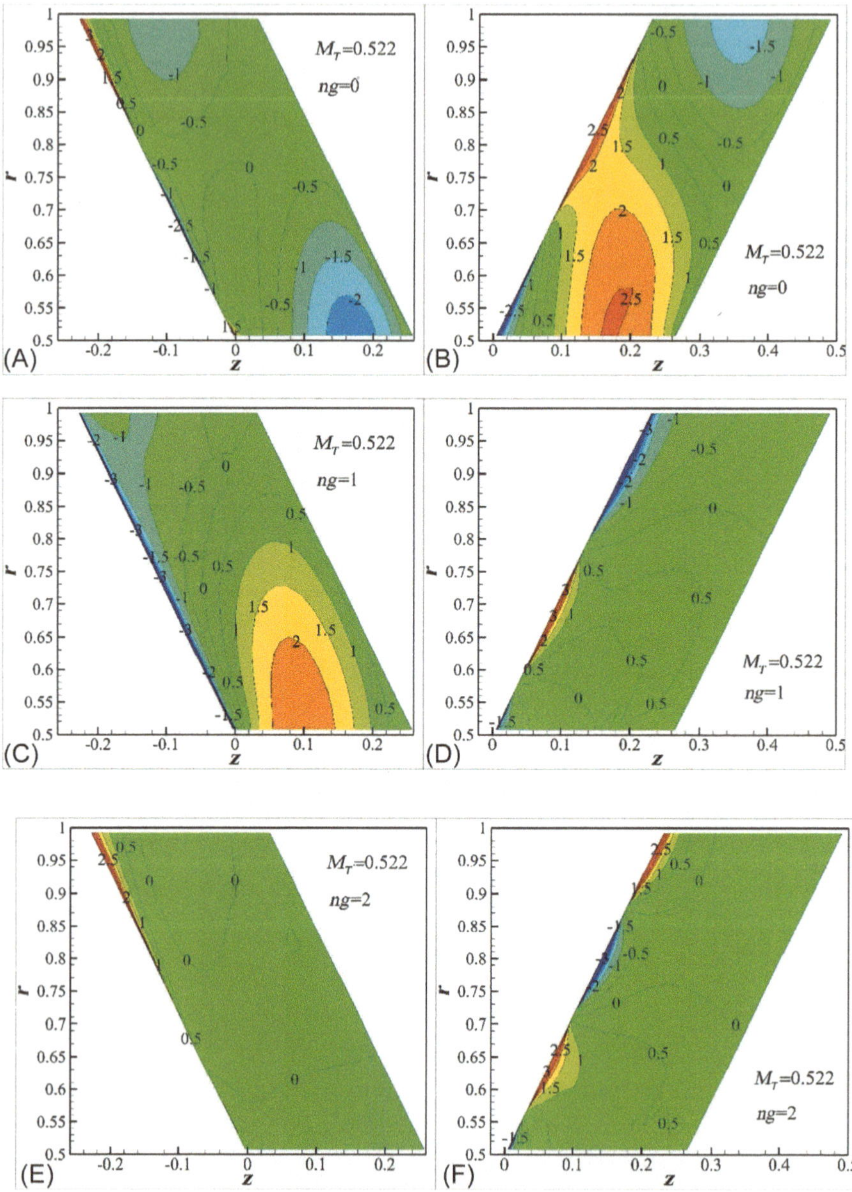

Fig. 6.63 Real part of the unsteady pressure jump across the swept vane for $M_T = 0.522$: (A) $\tilde{n}_g = -1.24$, (B) $\tilde{n}_g = 1.24$, (C) $\tilde{n}_g = -0.24$, (D) $\tilde{n}_g = 2.24$, (E) $\tilde{n}_g = 0.76$, (F) $\tilde{n}_g = 3.24$.

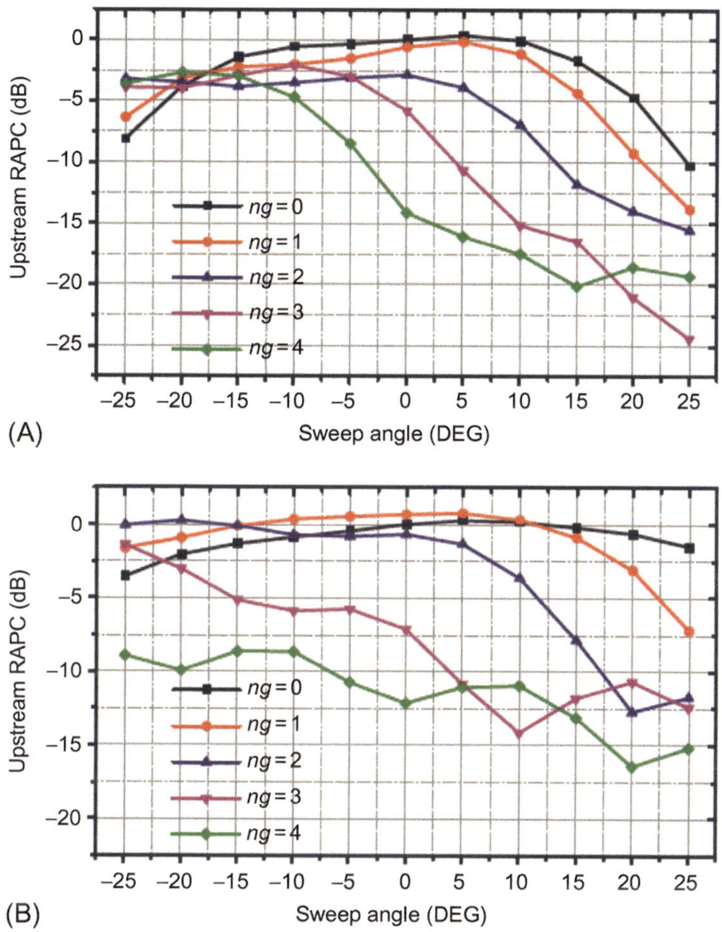

Fig. 6.64 The relative acoustic power changes with different combinations of sweep angle and radial wavenumber n_g, $M_T = 0.783$: (A) upstream, (B) downstream.

on the relative sound power changes. But the lines with different n_g do not overlap perfectly. This implies the interactions of the unsteady surface loadings in different radial slices. Fig. 6.68 gives the upstream and downstream acoustic pressure coefficients of different radial acoustic modes with $M_T = 0.783$. For a radial stator, the amplitudes of the acoustic modes are larger when their radial mode orders are close to the radial wavenumber n_g. If the vanes are swept, the similar phenomenon can be roughly identified by using $\widetilde{n}_g$.

For the application of the 2-D strip theory to study the passive control effects of vanes sweep and lean, the key point is actually the wake skewing.

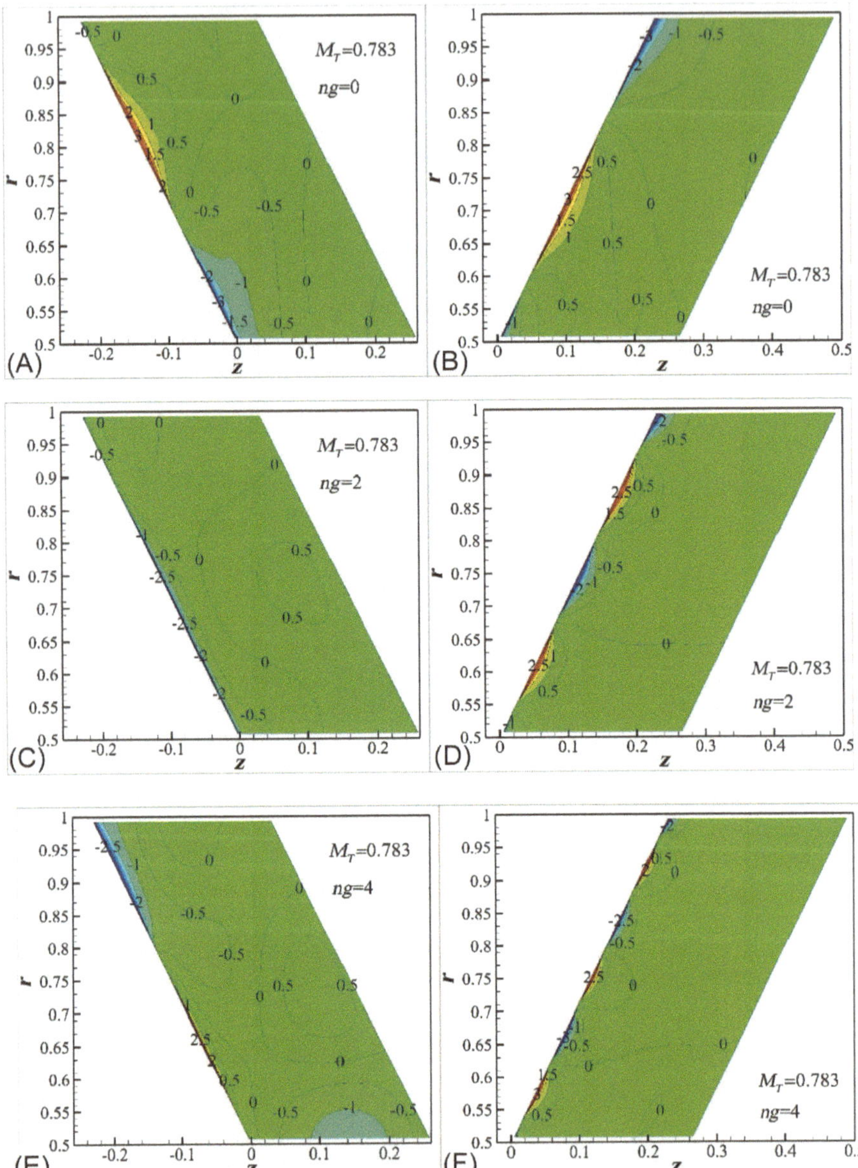

Fig. 6.65 Real part of the unsteady pressure jump across the swept vane for $M_T = 0.783$: (A) $\widetilde{n}_g = -1.86$, (B) $\widetilde{n}_g = 1.86$, (C) $\widetilde{n}_g = 0.14$, (D) $\widetilde{n}_g = 3.86$, (E) $\widetilde{n}_g = 2.14$, (F) $\widetilde{n}_g = 5.86$.

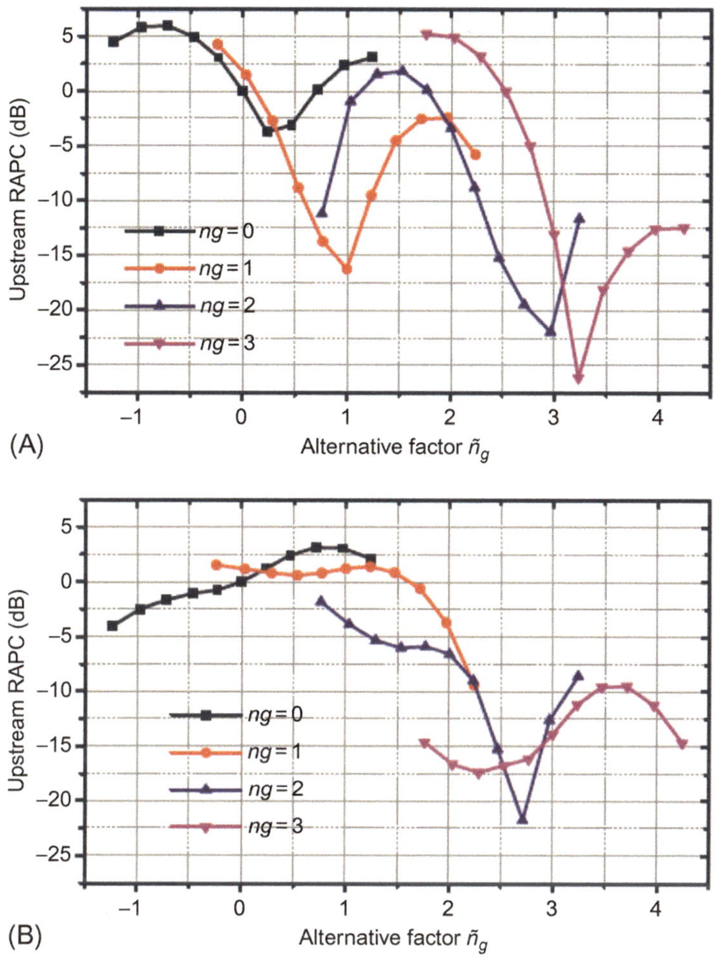

Fig. 6.66 The relative acoustic power changes for $M_T = 0.522$ with the alternative factor $\widetilde{n}_g = n_g + \frac{BM_T(1-h)}{\pi M} \tan\theta_S$: (A) upstream, (B) downstream.

Envia and Nallasamy [65] considered that the proper selection of swept-and-leaned vanes is effective to reduce the fan tonal noise, which enhanced the number of wake intersections per vane. This is quite a simple and practical conclusion, with little consideration of the coupling between the unsteady surface loadings across the whole vanes and acoustic field in the annular duct. In this chapter, the parameter $\widetilde{n}_g$ is used to characterize the radial phase variation as the rotor wakes reach the leading edge of the stator and intersect the vanes. After the previous discussion on the three-dimensional effect, it is clear that $\widetilde{n}_g$ should be selected more carefully. Also, the leaned vanes can

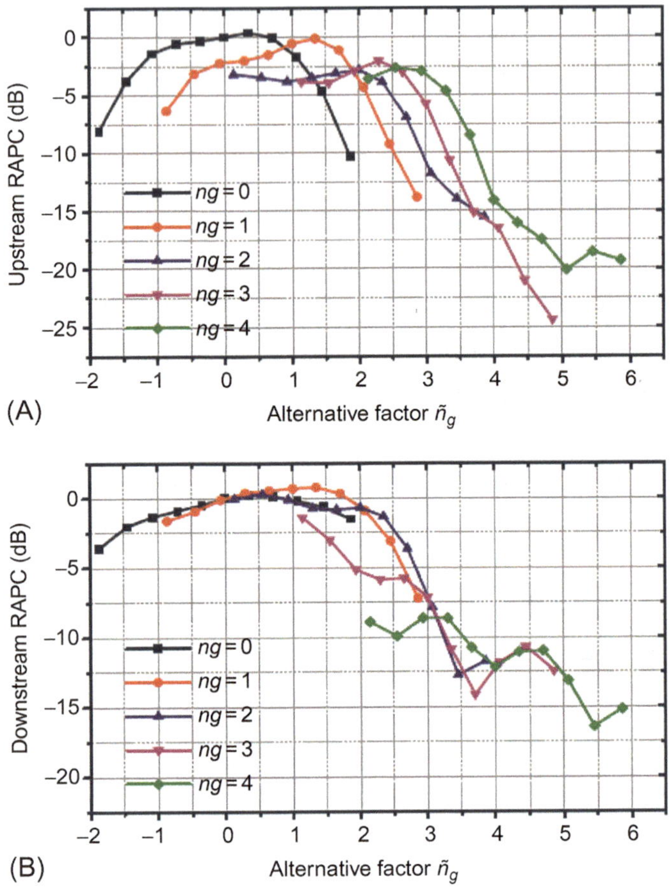

Fig. 6.67 The relative acoustic power changes for $M_T = 0.783$ with the alternative factor $\tilde{n}_g = n_g + \frac{BM_T(1-h)}{\pi M}\tan\theta_S$: (A) upstream, (B) downstream.

also change the spanwise phase variation of the incident wake. And more importantly, this effect is also in a very close connection with the acoustic propagation in an annular duct, as the radial disturbance velocity u_r' plays a part in the integral equation for the unsteady loadings. The combination of the sweep and lean should not be so simple and needs further discussion.

6.4.2.2 Vanes' lean effect on the tonal interaction noise

Vane lean effect would also change the radial phase variation of the upwash at the leading edge. Its effects on tonal interaction noise involve more complex mechanisms.

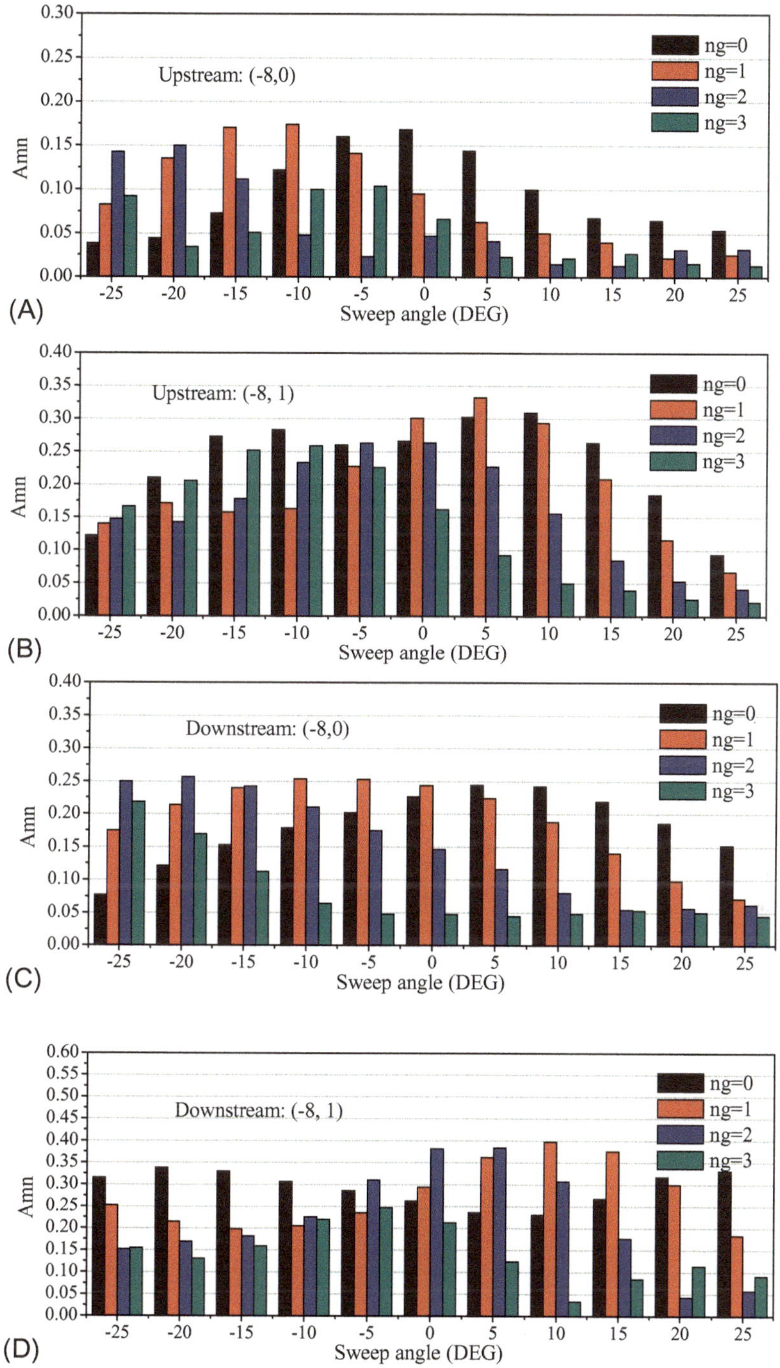

Fig. 6.68 The sweep effect on the acoustic modes coefficient with different radial wavenumbers n_g: (A) upstream, (−8.0), (B) upstream, (−8.1), (C) downstream, (−8.0), and (D) downstream, (−8.1).

Table 6.6 Input parameters of the case proposed by Schulten [66] for vane lean discussion.

Parameter description	Value
Number of rotor blades	19
Number of stator vanes	20
Hub/tip ratio	0.4
Vane chord length	0.1
Axial Mach number	0.3
Rotor circumferential tip Mach number	0.7
Rotor-stator spacing	0.2

Schulten [66] theoretically found that the vanes' lean effect would increase the fan interaction noise under some conditions. Table 6.6 gives the parameters. Fig. 6.69 gives the verification results. In this case, the generated circumferential mode is -1, and there are three radial modes that can propagate in the pipeline. Fig. 6.70 shows the acoustic mode coefficients with different lean angles. It can be seen that, with the given configurations, the leaned vanes have good control effect on the low-order radial modes and scatter the acoustic energy into the higher-order radial modes.

Elhadidi and Atassi [63] also gave a computing case to study the vanes' lean effect by solving the Euler equation. Table 6.7 gives the input parameters. The definition of RAPC is slightly different from the previous one:

$$\mathrm{RAPC} = 100 \times \frac{\wp - \wp_0}{\wp_0}.$$

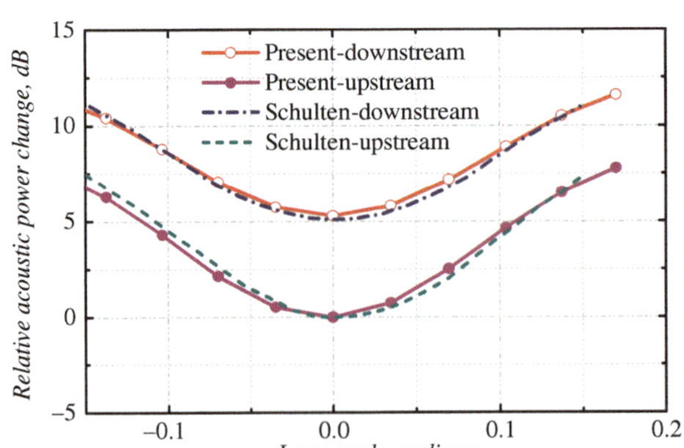

Fig. 6.69 Relative acoustic power changes (RAPC, dB) with different lean angles in the case proposed by Schulten (see [66] for Schulten's result).

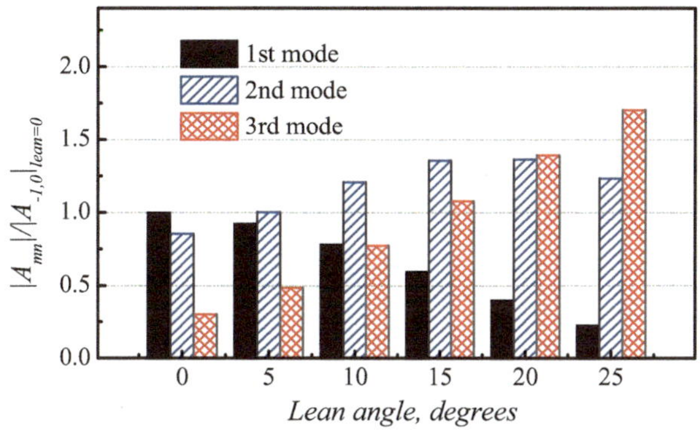

Fig. 6.70 Changes of radial acoustic mode coefficients with different lean angles.

Table 6.7 Input parameters of the case proposed by Elhadidi and Atassi for vane lean discussion.

Parameter description	Value
Hub/tip ratio	0.5
Vane chord length	$\frac{2\pi r_d}{V}$
Axial Mach number	0.5
Rotor tip Mach number	0.79
Reduced frequency	3π

Fig. 6.71A gives the upstream and downstream RAPC with different lean angles, and Fig. 6.71B gives the calculation results when $B = 23$ and $V = 24$. The prediction based on the model agrees well with the trend of numerical calculation results. When the number of rotor blades is 23, the generated circumferential mode is -1, and the negative influence of noise reduction is more obvious. Compared with the analysis of Elhadidi and Atassi [63], the present analytical method is more theoretically clear of the influence of the radial acoustic disturbance velocity when discussing the vanes' lean effect that is the second term in the function of Eq. (6.254). It contains the derivative term of the eigenfunction, which is affected by the acoustic mode and the geometric parameters of the three-dimensional annular duct. When the circumferential mode is small, the relative weight of the second term increases, whereas the second term of the radial higher-order mode also increases with the increase of the radial order, and this term will have the opposite effect through theoretical calculation.

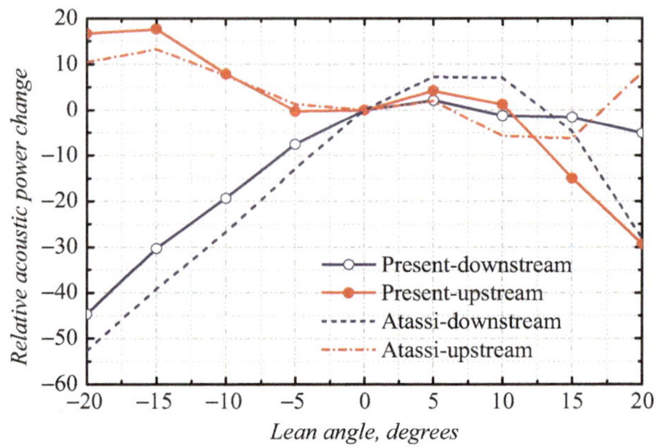

$B = 16, \quad V = 24$

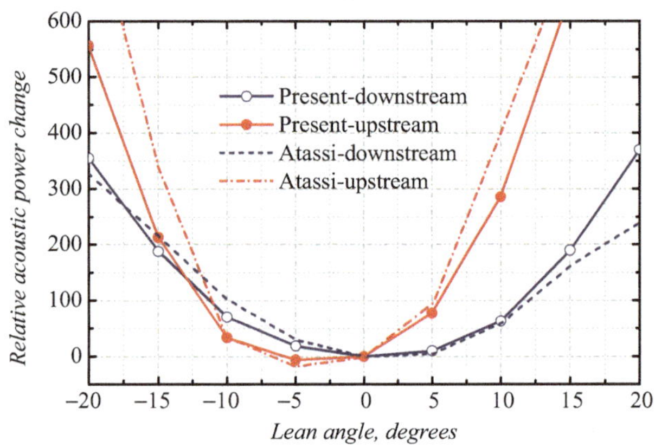

$B = 23, \quad V = 24$

Fig. 6.71 Relative acoustic power changes (RAPC, dB) with different lean angles (see [63] for Atassi's result).

Therefore, in general, when the low-order circumferential modes and the high-order radial modes are generated, it is difficult to achieve a good control effect by the lean design of the vanes only.

It can be seen from this analysis that the control mechanism of the vanes lean is more complicated, and the three-dimensional effect is more obvious. The geometric parameters of the annular duct and the acoustic propagation modes in the duct all have an important influence on the control effect of the leaned vanes. When discussing the control effect of swept-and-leaned vanes, the three-dimensional effect should be considered more seriously.

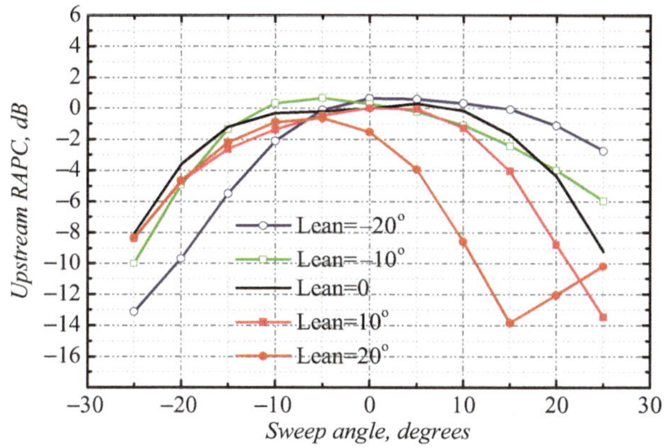

Fig. 6.72 Upstream RAPC with different selections of swept-and-leaned vanes.

6.4.2.3 Control effect of swept-and-leaned vanes

For the convenience of discussion, the example of the swept-and-leaned design of vanes with the same structural parameters as shown by Elhadidi and Atassi [63] is considered. The number of rotor blades is also selected to be $B=16$, and the number of stator vanes is $V=24$. Figs. 6.72 and 6.73, respectively, show the noise reduction effects of the upstream and downstream acoustic energy with different sweep and lean angles combinations.

The solid black line in the figure represents the control effect of swept-only vanes. It can be seen that a suitable selection of vanes sweep and lean

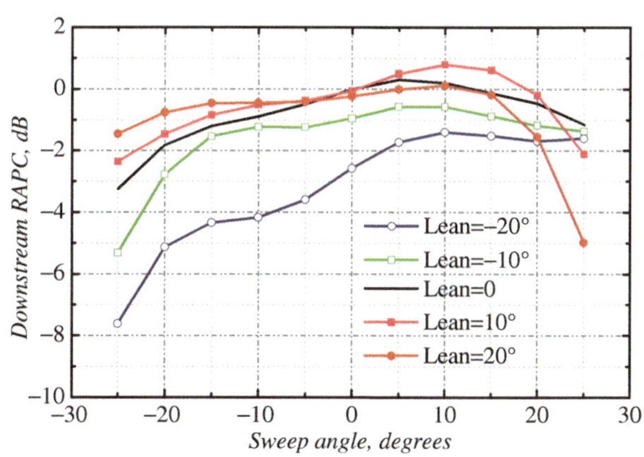

Fig. 6.73 Downstream RAPC with different selections of swept-and-leaned vanes.

angles can greatly improve the suppression of fan noise compared with swept-only vanes. It should be pointed out that the calculation of RAPC in the upstream and downstream did not consider the scattering effect of the upstream rotor blade row. The control effect of swept-and-leaned vanes is mainly due to the changes of the unsteady surface loadings' response on the vanes' surface. The sound propagation mode and the wavenumber are the same both for baseline vanes and swept-and-leaned vanes. So, the scattering effects of the rotor should have a relatively small effect on this variation trend.

6.4.3 Sound source prediction model for a finite region

As shown in Fig. 6.74, there are two ideal interfaces, "a" and "d", outside the rotor cascade. These interfaces do not have any influence on the acoustic and vorticial waves. The acoustic pressure and disturbance velocity are continuous at both sides of the interfaces. According to the TEM, three equations can be established on the interface,

$$p^{+}=p^{-},\ v^{+}=v^{-},\ w^{+}=w^{-}. \tag{6.258}$$

The part between the two interfaces containing the rotor cascade is called a transfer element, which requires all the unknowns on the interfaces. As will be seen in the later derivation, the size of the element is not limited, but the two interfaces cannot be selected to overlap the rotor cascade. Considering the origin of the axial coordinate is on the interface $a-a$, the acoustic wave propagating both upstream and downstream and the vorticial disturbance wave can only be convected downstream in the transfer element. They can be expressed in the stationary coordinate system as follows,

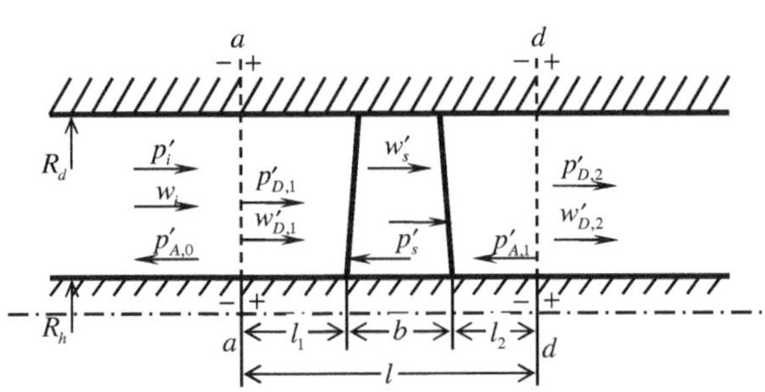

Fig. 6.74 Transfer element for rotor cascade.

$$\widetilde{p'}_{A,1} = e^{i\omega t} \sum_{m=-\infty}^{\infty} \sum_{\mu=1}^{\infty} D_{m\mu}^{1p} \cdot \phi_m\left(k_{m\mu} r\right) \cdot e^{i\alpha_1 z} \cdot e^{im\overline{\varphi}},$$

$$\widetilde{p'}_{A,1} = e^{i\omega t} \sum_{m=-\infty}^{\infty} \sum_{\mu=1}^{\infty} A_{m\mu}^{1p} \cdot \phi_m\left(k_{m\mu} r\right) \cdot e^{i\alpha_2 (z-l)} \cdot e^{im\overline{\varphi}}, \qquad (6.259a,b,c)$$

$$\widetilde{w'}_{\overline{\varphi},D,1} = e^{i\omega t} \sum_{m=-\infty}^{\infty} \sum_{\mu=1}^{\infty} D_{m\mu}^{1w} \cdot \frac{\phi_m\left(k_{m\mu} r\right)}{r} \cdot e^{i\alpha_3 z} \cdot e^{im\overline{\varphi}}.$$

When there is no acoustic component inside the element, the acoustic and vorticial field on the interfaces can be directly given by these three expressions by combining with the disturbance field outside the element.

$$p'_{A,0} = \sum_{q=-\infty}^{\infty} \sum_{\mu=1}^{\infty} A_{m\mu}^{0p} \cdot \phi_m\left(k_{m\mu} r\right) \cdot e^{i\alpha_2 z} \cdot e^{im\varphi} \quad ,$$

$$p'_{D,2} = \sum_{q=-\infty}^{\infty} \sum_{\mu=1}^{\infty} D_{m\mu}^{2p} \cdot \phi_m\left(k_{m\mu} r\right) \cdot e^{i\alpha_1 (z-l)} \cdot e^{im\varphi} \quad , \qquad (6.260a,b,c)$$

$$w'_{D,2} = \sum_{m=-\infty}^{\infty} \sum_{\mu=1}^{\infty} D_{m\mu}^{2w} \cdot \frac{\phi_m\left(k_{m\mu} r\right)}{r} \cdot e^{i\alpha_3 (z-l)} \cdot e^{im\varphi} \quad .$$

p'_i and w'_i are the foregone incident disturbance. Based on the previous deviation, we can establish the system of equations for the element as follows,

$$\begin{aligned}
p'_{A,0} - p'_{D,1} - p'_{A,1} &= -p'_i \,, \quad (interface\ a-a) \\
u'_{A,0} - u'_{D,1} - u'_{A,1} &= -u'_i \,, \quad (interface\ a-a) \\
w'_{D,1} &= -w'_i \,, \quad (interface\ a-a) \\
p'_{D,1} + p'_{A,1} - p'_{D,2} &= 0 \,, \quad (interface\ d-d) \\
u'_{D,1} + u'_{A,1} - u'_{D,2} &= 0 \,, \quad (interface\ d-d) \\
w'_{D,1} - w'_{D,2} &= 0 \,, \quad (interface\ d-d)
\end{aligned} \qquad (6.261)$$

In this system of equations, the acoustic particle velocity can be calculated according to the momentum equation in the z direction. Therefore, the number of unknowns in the whole equation group is $6 \times M \times N$. M and N are the number of circumferential and radial modes retained in the matching process, respectively. The equations are closed and can be solved directly.

However, when the element contains rotor or stator cascade, there will be scattering acoustic and vorticial waves by the blades. The scattering acoustic waves p_S' can propagate to the upstream interface $a - a$ and the downstream interface $d - d$. Also the vorticial waves w_S' only propagates to the downstream. The system of the matching equations should be written as follows,

$$
\begin{aligned}
p'_{A,0} - p'_{D,1} - p'_{A,1} - p'^{+}_{s} &= -p'_i\,, \quad (\textit{interface } a - a)\\
u'_{A,0} - u'_{D,1} - u'_{A,1} - u'^{+}_{S} &= -u'_i\,, \quad (\textit{interface } a - a)\\
w'_{D,1} &= -w'_i\,, \quad (\textit{interface } a - a)\\
p'_{D,1} + p'_{A,1} + p'^{-}_{S} - p'_{D,2} &= 0\,, \quad (\textit{interface } d - d)\\
u'_{D,1} + u'_{A,1} + u'^{-}_{S} - u'_{D,2} &= 0\,, \quad (\textit{interface } d - d)\\
w'_{D,1} + w'_S - w'_{D,2} &= 0\,, \quad (\textit{interface } d - d)
\end{aligned}
\tag{6.262}
$$

The equations can be closed by writing p_S' and w_S' as explicit expressions of incident disturbance $p'_{D,\ 1}$, $p'_{A,\ 1}$, and $w'_{D,\ 1}$, which is the focal point of the TEM. The following part is mainly about this deviation.

Eq. (6.263) gives the modal wave form at the interfaces, m represents the circumferential mode number, and n represents the radial mode order. In fact, for the rotor/stator interaction problem, the circumferential mode m is given by $m = sB - qV$ as mentioned before. It can be seen that the circumferential mode depends not only on the number of rotor and stator blades (B and V) but also on the orders of harmonics (s and q). Therefore, the incident waves at the interfaces are no longer defined by the general form with (m, n) but by the definition with (s, q, μ), which can more intuitively reflect the correspondence between the scattering waves and the incident waves. For example,

$$
\widetilde{p'}_{D,1} = e^{i\omega t} \sum_{s=-\infty}^{\infty} \sum_{q=-\infty}^{\infty} \sum_{\mu=1}^{\infty} D^{1p}_{m\mu} \cdot \phi_m\left(k_{m\mu} r\right) \cdot e^{i\alpha_1 z} \cdot e^{im\overline{\varphi}}. \tag{6.263}
$$

According to the linear superposition principle, the scattering field can be written as the sum of the three parts, which are the upstream and downstream incident acoustic waves and the upstream incident vorticial waves, respectively:

$$
\begin{aligned}
p'_S &= p'^{p,D}_S + p'^{p,A}_S + p'^{w,D}_S\,,\\
w'_S &= w'^{p,D}_S + w'^{p,A}_S + w'^{w,D}_S\,.
\end{aligned}
\tag{6.264}
$$

Based on the lifting surface theory introduced in the previous section, the relationship between the scattering waves and the incident waves can be expressed as $p'^{p,\ D}_S = F(p'_{D,\ 1})$. First, the integral equation should be established according to the boundary condition on the vanes to solve the unsteady loadings on the vanes surface:

$$\widetilde{v'}_{\varphi'',S}(r,z)+\widetilde{v'}_{\varphi'',D,1}(r,z)+\widetilde{v'}_{\varphi'',A,1}(r,z)+\widetilde{w'}_{\varphi'',D,1}(r,z)=0, \tag{6.265}$$

in which $\widetilde{v'}_{\varphi'',S}(r,z)$ represents the particle velocity normal to the vanes surface induced by the scattering waves. $\widetilde{v'}_{\varphi'',D,1}(r,z)$, $\widetilde{v'}_{\varphi'',A,1}(r,z)$ and $\widetilde{w'}_{\varphi'',D,1}(r,z)$ are the ones induced by the incident waves $p'_{D,\ 1}$, $p'_{A,\ 1}$ and $w'_{D,\ 1}$, respectively.

According to the momentum equation:

$$\widetilde{v}_{\varphi'',D,1}=-\frac{1}{\rho W}\sum_{q=-\infty}^{\infty}\sum_{\mu=1}^{\infty}D_{m\mu}^{1p}\cdot\phi_m(k_{m\mu}r)\cdot\frac{\frac{m}{r}-\alpha_1\frac{\Omega r}{U}}{\alpha_1+m\frac{\Omega}{U}+\frac{\lambda_1}{U}}e^{i\left(\alpha_1+m\frac{\Omega}{U}\right)z}e^{i\lambda_1 t},$$

$$\widetilde{v}_{\varphi'',A,1}=-\frac{1}{\rho W}\sum_{q=-\infty}^{\infty}\sum_{\mu=1}^{\infty}A_{m\mu}^{1p}\cdot\phi_m(k_{m\mu}r)\cdot\frac{\frac{m}{r}-\alpha_2\frac{\Omega r}{U}}{\alpha_2+m\frac{\Omega}{U}+\frac{\lambda_1}{U}}e^{i\alpha_2(z-l)}e^{im\frac{\Omega}{U}z}e^{i\lambda_1 t},$$

$$\widetilde{w'}_{\varphi'',D,1}=\sum_{m=-\infty}^{\infty}\sum_{\mu=1}^{\infty}D_{m\mu}^{1w}\cdot\frac{\phi_m(k_{m\mu}r)}{r}\cdot\left(\frac{U}{W}+\frac{\omega\cdot r}{U\cdot m}\cdot\frac{\Omega r}{W}\right)\cdot e^{i\alpha_3 z}\cdot e^{im\frac{\Omega}{U}z}\cdot e^{i\lambda_1 t}, \tag{6.266}$$

and, based on the linear superposition principle, Eq. (6.265) can be divided into three equations:

$$\widetilde{v'}^{p_D}_{\varphi'',S}(r,z)=-\widetilde{v'}_{\varphi'',D,1}(r,z),\ \widetilde{v'}^{p_A}_{\varphi'',S}(r,z)=-\widetilde{v'}_{\varphi'',A,1}(r,z),\ \widetilde{v'}^{w_D}_{\varphi'',S}(r,z)=-\widetilde{w'}_{\varphi'',D,1}(r,z). \tag{6.267}$$

Here, $\widetilde{v'}^{p_D}_{\varphi'',S}$, $\widetilde{v'}^{p_A}_{\varphi'',S}$ and $\widetilde{v'}^{w_D}_{\varphi'',S}$ represents the scattered particle velocity by the interaction of $p'_{D,\ 1}$, $p'_{A,\ 1}$, and $w'_{D,\ 1}$ with the rotor blades, respectively, and the unsteady surface loadings response on the blades are $\Delta\widetilde{p}^{p_D}$, $\Delta\widetilde{p}^{p_A}$, and $\Delta\widetilde{p}^{w_D}$.

In the previous section, the integral equations can be transformed into the linear equations. And by solving these equations, we can obtain the unsteady surface loadings on the blade surface induced by different incident waves,

$$\Delta p^{\upsilon}=\sum_{j'=1}^{J}\phi_{0j'}(r')\left[A_{1j'}^{\upsilon}\cot\frac{\vartheta'}{2}+\sum_{i'=2}^{I}A_{i'j'}^{\upsilon}\sin(i'-1)\vartheta'\right], \tag{6.268}$$

in which

$$A^{\upsilon}_{i'j'} = \sum_{i=1}^{I}\sum_{j=1}^{J} v^{\upsilon}_{ij} \cdot \mathfrak{I}^{-1}_{\upsilon(ij,\, i'j')}. \tag{6.269}$$

Substituting the previous expression for unsteady loadings into the sound field expression, the sound field generated by the corresponding incident waves in the stationary coordinate system can be obtained, and the upstream-propagating acoustic waves are:

$$p'_s(\vec{x}, t) = \frac{-B}{4\pi}\sum_{s'=-\infty}^{\infty}\sum_{n=1}^{\infty}\frac{\phi_{m'}(k_{m'n}r)e^{im'\varphi}}{\kappa_{n,m'}}$$
$$\int_{s(\tau)}\Delta p^{\upsilon}\phi_{m'n}(k_{m'n}r')e^{-im'\varphi'}\left(\frac{m'}{r'} - \alpha_2\frac{\Omega r'}{U}\right)e^{i\alpha_2(z-z')}ds(\vec{y}). \tag{6.270}$$

The downstream-propagating acoustic waves are:

$$p'_s(\vec{x}, t) = \frac{-B}{4\pi}\sum_{s'=-\infty}^{\infty}\sum_{n=1}^{\infty}\frac{\phi_{m'}(k_{m'n}r)e^{im'\varphi}}{\kappa_{n,m'}}$$
$$\int_{s(\tau)}\Delta p^{\upsilon}\phi_{m'n}(k_{m'n}r')e^{-im'\varphi'}\left(\frac{m'}{r'} - \alpha_1\frac{\Omega r'}{U}\right)e^{i\alpha_1(z-z')}ds(\vec{y}). \tag{6.271}$$

The downstream-propagating vorticial waves are:

$$w'_{\overline{\varphi},s}(\vec{x}, t) = \frac{-B}{2\pi\rho_0 U}\sum_{s'=-\infty}^{\infty}\sum_{n=1}^{\infty}\phi_{m'}(k_{m'n}r)e^{im'\varphi}$$
$$\int_{s(\tau)}\Delta p^{\upsilon}\phi_{m'}(k_{m'n}r')e^{-im'\varphi'}\left(\frac{m'}{r'} - \alpha_3\frac{\Omega r'}{U}\right)\frac{m'}{r}\frac{M^2e^{i\alpha_3(z-z')}}{k_0^2 + M^2k^2_{m'n}}ds(\vec{y}). \tag{6.272}$$

The generated upstream-propagating acoustic waves at the interface $a - a$ are:

$$p'^{D_p}_{S,a-a}(r) = \sum_{s'=-\infty}^{\infty}\sum_{n=1}^{\infty}\phi_{m'}(k_{m'n}r)\sum_i\sum_j\left[\sum_{s=-\infty}^{\infty}\sum_{q=-\infty}^{\infty}\sum_{\mu=1}^{\infty}D^{1p}_{m\mu}\cdot\aleph^1_{ij,m\mu}\right]\cdot\mathfrak{R}^{m'n,2}_{ij},$$

$$p'^{A_p}_{S,a-a}(r) = \sum_{s'=-\infty}^{\infty}\sum_{n=1}^{\infty}\phi_{m'}(k_{m'n}r)\sum_i\sum_j\left[\sum_{s=-\infty}^{\infty}\sum_{q=-\infty}^{\infty}\sum_{\mu=1}^{\infty}A^{1p}_{m\mu}\cdot\aleph^2_{ij,m\mu}\right]\cdot\mathfrak{R}^{m'n,2}_{ij},$$

$$p'^{D_w}_{S,a-a}(r) = \sum_{s'=-\infty}^{\infty}\sum_{n=1}^{\infty}\phi_{m'}(k_{m'n}r)\sum_i\sum_j\left[\sum_{s=-\infty}^{\infty}\sum_{q=-\infty}^{\infty}\sum_{\mu=1}^{\infty}D^{1w}_{m\mu}\cdot\aleph^3_{ij,m\mu}\right]\cdot\mathfrak{R}^{m'n,2}_{ij}.$$

(6.273a,b,c)

The generated downstream-propagating acoustic waves at the interface $d-d$ are:

$$p'^{D_p}_{s,d-d}(r)=\sum_{s'=-\infty}^{\infty}\sum_{n=1}^{\infty}\phi_{m'}(k_{m'n}r)\sum_{i}\sum_{j}\left[\sum_{s=-\infty}^{\infty}\sum_{q=-\infty}^{\infty}\sum_{\mu=1}^{\infty}D^{1p}_{m\mu}\cdot\aleph^{1}_{ij,m\mu}\right]\cdot\mathfrak{R}^{m'n,1}_{ij},$$

$$p'^{A_p}_{s,d-d}(r)=\sum_{s'=-\infty}^{\infty}\sum_{n=1}^{\infty}\phi_{m'}(k_{m'n}r)\sum_{i}\sum_{j}\left[\sum_{s=-\infty}^{\infty}\sum_{q=-\infty}^{\infty}\sum_{\mu=1}^{\infty}A^{1p}_{m\mu}\cdot\aleph^{2}_{ij,q\mu}\right]\cdot\mathfrak{R}^{m'n,1}_{ij},$$

$$p'^{D_w}_{s,a-a}(r)=\sum_{s'=-\infty}^{\infty}\sum_{n=1}^{\infty}\phi_{m'}(k_{m'n}r)\sum_{i}\sum_{j}\left[\sum_{s=-\infty}^{\infty}\sum_{q=-\infty}^{\infty}\sum_{\mu=1}^{\infty}D^{1w}_{m\mu}\cdot\aleph^{3}_{ij,m\mu}\right]\cdot\mathfrak{R}^{m'n,1}_{ij}.$$

(6.274a,b,c)

The generated downstream vortex waves at the interface $d-d$ are:

$$w'^{D_p}_{s,d-d}(r)=\sum_{s'=-\infty}^{\infty}\sum_{n=1}^{\infty}\frac{\phi_{m'}(k_{m'n}r)}{r}\sum_{i}\sum_{j}\left[\sum_{s=-\infty}^{\infty}\sum_{q=-\infty}^{\infty}\sum_{\mu=1}^{\infty}D^{1p}_{m\mu}\cdot\aleph^{1}_{ij,m\mu}\right]\cdot\mathfrak{R}^{m',n,3}_{ij},$$

$$w'^{A_p}_{s,d-d}(r)=\sum_{s'=-\infty}^{\infty}\sum_{n=1}^{\infty}\frac{\phi_{m'}(k_{m'n}r)}{r}\sum_{i}\sum_{j}\left[\sum_{s=-\infty}^{\infty}\sum_{q=-\infty}^{\infty}\sum_{\mu=1}^{\infty}A^{1p}_{m\mu}\cdot\aleph^{2}_{ij,m\mu}\right]\cdot\mathfrak{R}^{m',n,3}_{ij},$$

$$w'^{D_w}_{s,d-d}(r)=\sum_{s'=-\infty}^{\infty}\sum_{n=1}^{\infty}\frac{\phi_{m'}(k_{m'n}r)}{r}\sum_{i}\sum_{j}\left[\sum_{s=-\infty}^{\infty}\sum_{q=-\infty}^{\infty}\sum_{\mu=1}^{\infty}D^{1w}_{m\mu}\cdot\aleph^{3}_{ij,m\mu}\right]\cdot\mathfrak{R}^{m',n,3}_{ij}.$$

(6.275a,b,c)

where

$$\aleph'^{1}_{i'j',m\mu}=-\frac{1}{\rho W}\sum_{i=1}^{I}\sum_{j=1}^{J}\cdot\phi_{m}(k_{m\mu}r_j)\cdot\frac{\dfrac{m}{r_j}-\alpha_1\dfrac{\Omega r_j}{U}}{\alpha_1+\dfrac{\omega}{U}}e^{i\left(\alpha_1+m\frac{\Omega}{U}\right)z_i}\cdot\mathfrak{I}^{-1}_{\nu(ij,\,i'j')},$$

$$\aleph'^{2}_{i'j',m\mu}=-\frac{1}{\rho W}\sum_{i=1}^{I}\sum_{j=1}^{J}\phi_{m}(k_{m\mu}r_j)\cdot\frac{\dfrac{m}{r_j}-\alpha_2\dfrac{\Omega r_j}{U}}{\alpha_2+\dfrac{\omega}{U}}e^{i\alpha_2(z_i-l)}e^{im\frac{\Omega}{U}z_i}\cdot\mathfrak{I}^{-1}_{\nu(ij,\,i'j')},$$

$$\aleph'^{3}_{i'j',m\mu}=\sum_{i=1}^{I}\sum_{j=1}^{J}\frac{\phi_{m}(k_{m\mu}r_j)}{r_j}\cdot\left(\frac{U}{W}+\frac{\omega\cdot r_j}{U\cdot m}\cdot\frac{\Omega r_j}{W}\right)\cdot e^{i\left(\alpha_3+m\frac{\Omega}{U}\right)z_i}\cdot\mathfrak{I}^{-1}_{\nu(ij,\,i'j')}.$$

(6.276a,b,c)

$$\mathfrak{R}_{ij}^{m'n,1} = \frac{-B}{4\pi}\frac{\phi_{m'n}\left(k_{m'n}r'_j\right)}{\kappa_{n,m'}}\left(\frac{m'}{r'_j} - \alpha_1\frac{\Omega r'_j}{U}\right)e^{i\alpha_1\left(l-z'_i\right)}e^{-i\frac{m'\Omega}{U}z'_i}\frac{\pi}{I}\frac{R_d - R_h}{J},$$

$$\mathfrak{R}_{ij}^{m'n,2} = \frac{-B}{4\pi}\frac{\phi_{mn}\left(k_{mn}r'_j\right)}{\kappa_{n,m'}}\left(\frac{m'}{r'_j} - \alpha_2\frac{\Omega r'_j}{U}\right)e^{-i\left(\alpha_2 + \frac{m'\Omega}{U}\right)z'_i}\frac{\pi}{I}\frac{R_d - R_h}{J},$$

$$\mathfrak{R}_{i'j'}^{n,3} = \frac{-B}{2\pi\rho_0 U}\phi_0\left(k_{0l'}r'_j\right)\frac{m'}{r'_j}\left(\frac{m'}{r'} - \alpha_3\frac{\Omega r'}{U}\right)$$
$$\frac{M^2}{k_0^2 + M^2k_{m'n}^2}e^{i\alpha_3 l}e^{-i\left(\alpha_3 + m\frac{\Omega}{U}\right)z'_i}\frac{\pi}{I}\frac{R_d - R_h}{J}.$$

(6.277a,b,c)

In this way, the scattering wave at the interfaces in the stationary coordinate system is expressed as a function of the unknown coefficients of the incident modal waves. So, the sound field in the duct can be obtained by solving the equations established by the TEM.

According to the momentum equation, the particle disturbance velocity u' on the interface can be written as a function of sound pressure p'. Then, Eq. (6.262) can be written as a system of algebraic equations:

$$\begin{pmatrix} SS^{p_{a-a}}_{A^{0p}_{m\mu}} & SS^{p_{a-a}}_{D^{1p}_{m\mu}} & SS^{p_{a-a}}_{A^{1p}_{m\mu}} & SS^{p_{a-a}}_{D^{1w}_{m\mu}} & 0 & 0 \\ SS^{u_{a-a}}_{A^{0p}_{m\mu}} & SS^{u_{a-a}}_{D^{1p}_{m\mu}} & SS^{u_{a-a}}_{A^{1p}_{m\mu}} & SS^{u_{a-a}}_{D^{1w}_{m\mu}} & 0 & 0 \\ 0 & 0 & 0 & SS^{w_{a-a}}_{D^{1w}_{m\mu}} & 0 & 0 \\ 0 & SS^{p_{d-d}}_{D^{1p}_{m\mu}} & SS^{p_{d-d}}_{A^{1p}_{m\mu}} & SS^{p_{d-d}}_{D^{1w}_{m\mu}} & SS^{p_{d-d}}_{D^{2p}_{m\mu}} & 0 \\ 0 & SS^{u_{d-d}}_{D^{1p}_{m\mu}} & SS^{u_{d-d}}_{A^{1p}_{m\mu}} & SS^{u_{d-d}}_{D^{1w}_{m\mu}} & SS^{u_{d-d}}_{D^{2p}_{m\mu}} & 0 \\ 0 & SS^{w_{d-d}}_{D^{1p}_{m\mu}} & SS^{w_{d-d}}_{A^{1p}_{m\mu}} & SS^{w_{d-d}}_{D^{1w}_{m\mu}} & 0 & SS^{w_{d-d}}_{D^{2w}_{m\mu}} \end{pmatrix}\begin{pmatrix} A^{0p}_{m\mu} \\ D^{1p}_{m\mu} \\ A^{1p}_{m\mu} \\ D^{1w}_{m\mu} \\ D^{2p}_{m\mu} \\ D^{2w}_{m\mu} \end{pmatrix} = \begin{pmatrix} I^{p}_{m\mu} \\ I^{u}_{m\mu} \\ I^{w}_{0\mu} \\ 0 \\ 0 \\ 0 \end{pmatrix}. \tag{6.278}$$

The middle part in the matrix is called the transfer element,

$$\begin{bmatrix} SS^{p_{a-a}}_{D^{1p}_{m\mu}} & SS^{p_{a-a}}_{A^{1p}_{m\mu}} & SS^{p_{a-a}}_{D^{1w}_{m\mu}} \\ SS^{u_{a-a}}_{D^{1p}_{m\mu}} & SS^{u_{a-a}}_{A^{1p}_{m\mu}} & SS^{u_{a-a}}_{D^{1w}_{m\mu}} \\ 0 & 0 & SS^{w_{a-a}}_{D^{1w}_{m\mu}} \\ SS^{p_{d-d}}_{D^{1p}_{m\mu}} & SS^{p_{d-d}}_{A^{1p}_{m\mu}} & SS^{p_{d-d}}_{D^{1w}_{m\mu}} \\ SS^{u_{d-d}}_{D^{1p}_{m\mu}} & SS^{u_{d-d}}_{A^{1p}_{m\mu}} & SS^{u_{d-d}}_{D^{1w}_{m\mu}} \\ SS^{w_{d-d}}_{D^{1p}_{m\mu}} & SS^{w_{d-d}}_{A^{1p}_{m\mu}} & SS^{w_{d-d}}_{D^{1w}_{m\mu}} \end{bmatrix}_{[6\cdot(2M+1)\cdot N]\times[3\cdot(2M+1)\cdot N]}. \tag{6.279}$$

This part focuses on how to establish a sound source prediction model in a limited region and gives a basic solution for the sound field. This model has no advantage for solving the wake/stator interaction noise. But when we consider the rotor shield effect on the rotor/stator interaction noise and the mutual effect between the sound sources and the acoustic treatments, this model will be the only one suitable for these problems. In the next section, these problems will be discussed further.

6.5 Interaction effect

As shown in Fig. 6.75, the sound generation, propagation, and control in an aeroengine nacelle are really complex. For a turbofan engine, the rotor-stator interaction is a dominated source, which is generated by unsteady forces on the stator vanes when periodic fan rotor wakes impinge on the stator cascade. The upstream and downstream acoustic waves propagate through the liner and the rotor cascade, and finally radiate from the nacelle inlet and outlet. Physically, these liners, rotor blades, and the duct ends cause the reflection and scattering of the acoustic waves. In this situation, the sound prediction and liner design based on the assumption of an infinite-length duct do not conform to a practical application. Therefore, it is necessary to develop a new model that can handle such interaction. In this section, emphasis will be put on the interaction effect between various acoustic elements.

6.5.1 The interaction between rotor and Stator cascades

In Chapter 3 we have shown how to use the lifting surface theory to calculate the sound generation by wake-stator interaction. However, the existence of rotor cascade can reflect and scatter the acoustic waves, which will further result in the interaction effect between rotor cascade and stator

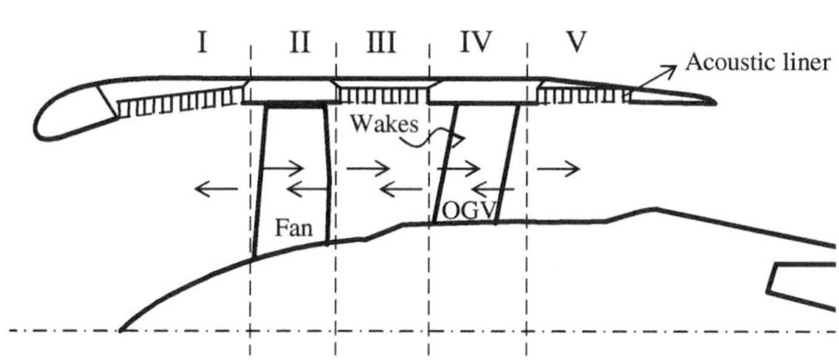

Fig. 6.75 A schematic of aeroengine nacelle.

cascade. In this section, the rotor-stator interaction instead of wake-stator interaction will be investigated.

We have discussed the frequency and acoustic mode characteristic of rotor-stator interaction in Chapter 3. Based on this analysis, the sound generated by rotor-stator interaction can be predicted by TEM.

First, the wake of upstream rotor blades can be expressed as:

$$w_i' = \sum_{s=-\infty}^{\infty} W_s \cdot e^{isB[\Omega t + \varphi - \Omega z/U]}, \tag{6.280}$$

where s is the order of the wake, and W_s denotes the amplitude of the wake. According to the linear superposition principle, we can consider the wake components of different orders separately. Here we consider the first-order wake component, i.e. $s = 1$. This wake component impinges on the stator cascade and generates unsteady force on the stator vanes. This unsteady loading can be expressed as:

$$\Delta\widetilde{p}^S = \Delta p_s e^{i\left[\lambda_S \tau + \sigma_S \frac{2\pi}{V}(k-1)\right]}, \ \lambda_S = sB\Omega\ , \quad \sigma_S = sB, \tag{6.281}$$

where Δp_s is the amplitude of the unsteady force, and σ_S is the interblade phase angle. Based on the discussion in Chapter 3, the sound generated by the unsteady loading with specific frequency and interblade phase angle has a form of:

$$\widetilde{p_1'}(\vec{x}, t) = \sum P_1 e^{i\omega_1 t} e^{im_1\varphi}, \ (\omega_1 = sB\Omega\ , \quad m_1 = qV + \sigma_S = sB + qV). \tag{6.282}$$

This acoustic wave propagates upstream through the rotor cascade. Considering a potential interaction, the unsteady loading on the rotor blades induced by the upstream acoustic wave can be described by:

$$\Delta\widetilde{p}^R = \Delta p_r e^{i\lambda_R \tau} e^{\sigma_R \frac{2\pi}{B}(k-1)}, \tag{6.283}$$

where $\lambda_R = \omega_1 - m_1\Omega = -qV$ is the frequency of unsteady loading on the rotor blades, and $\sigma_R = m_1 = sB + qV$ stands for the interblade phase angle for rotor cascade. Then, the acoustic field generated by this unsteady force can be expressed as:

$$\widetilde{p_2'}(\vec{x}, t) = \sum P_2 e^{i\omega_2 t} e^{im_2\varphi}, \tag{6.284}$$

where $\omega_2 = \lambda_R + m_2\Omega = (s' + s)B\Omega$ is the frequency of the acoustic field generated by rotor cascade, whereas $m_2 = sB + \sigma_R = (s' + s)B + qV$ represents the corresponding acoustic mode.

According to this description, it is clearly shown that the acoustic field generated by rotor-stator interaction instead of wake-stator interaction is comprised of specific frequency and mode components, i.e., $\omega = sB\Omega$ and $m = sB + qV$. Based on the characteristics of the acoustic mode and frequency described earlier, we can derive a uniform equation to solve the rotor-stator interaction problem.

It is also clear that the acoustic frequency only depends on s, but the acoustic mode depends on s and q. When the acoustic wave propagates through the rotor cascade, both the acoustic frequency and acoustic mode may change, whereas when the acoustic wave propagates through the stator cascade, only the acoustic mode can may change.

Considering one stage case that includes one rotor cascade and one stator cascade, as shown in Fig. 6.76, the corresponding equations can be expressed as:

$$\begin{pmatrix}
ss_A^{p_{a-a}} & & & & & & & \\
ss_A^{v_{a-a}} & & & & & & & \\
 & B - C & & & & & & \\
 & & ss_D^{p_{d-d}} & ss_E^{p_{d-d}} & 0 & & & \\
 & & ss_D^{v_{d-d}} & ss_E^{v_{d-d}} & 0 & & & \\
 & & 0 & 0 & ss_{w_D}^{w_{d-d}} & & & \\
 & & ss_D^{p_{e-e}} & ss_E^{p_{e-e}} & 0 & & & \\
 & & ss_D^{v_{e-e}} & ss_E^{v_{e-e}} & 0 & & & \\
 & & 0 & 0 & ss_{w_D}^{w_{e-e}} & & & \\
 & & & & & F - G & ss_H^{p_{h-h}} & \\
 & & & & & & ss_H^{v_{h-h}} & \\
 & & & & & & & ss_{w_H}^{w_{h-h}}
\end{pmatrix}
\begin{pmatrix}
A_{mn}^p \\ B_{mn}^p \\ C_{mn}^p \\ B_{0n}^w \\ D_{mn}^p \\ E_{mn}^p \\ D_{0n}^w \\ F_{mn}^p \\ G_{mn}^p \\ F_{0n}^w \\ H_{mn}^p \\ H_{0n}^w
\end{pmatrix}
=
\begin{pmatrix}
0 \\ 0 \\ \vdots \\ I_{mn}^p \\ I_{mn}^v \\ I_{0n}^w \\ \vdots \\ 0 \\ 0 \\ 0 \\ 0 \\ 0
\end{pmatrix}
\tag{6.285}$$

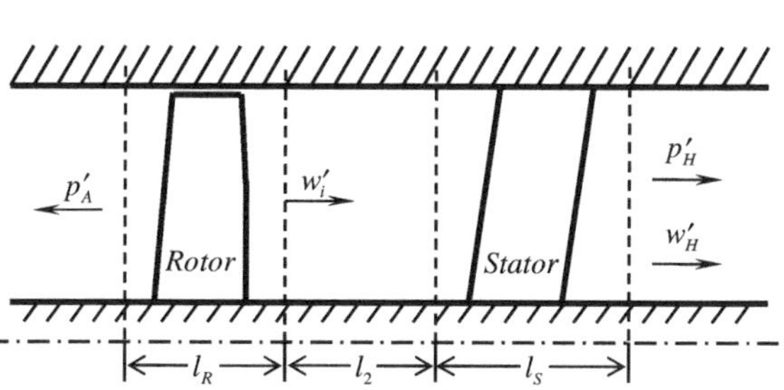

Fig. 6.76 The model of rotor-stator interaction.

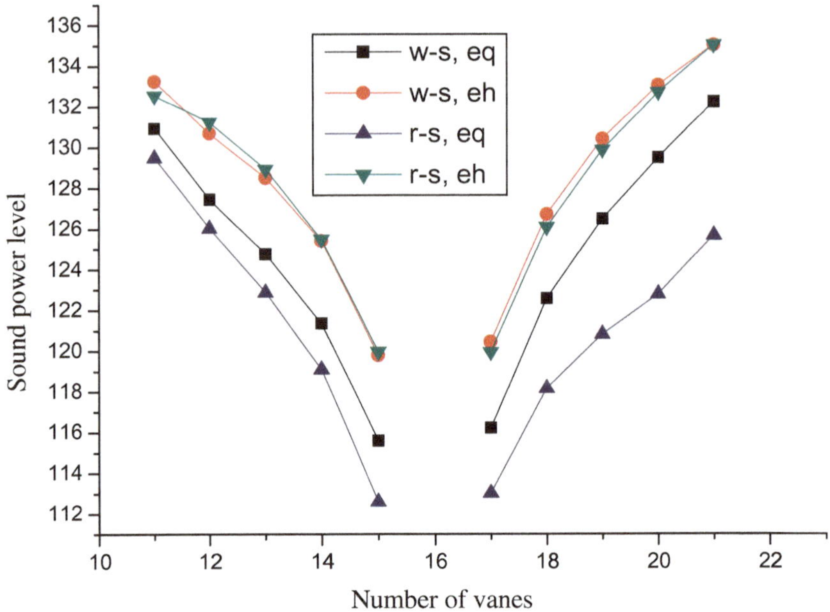

Fig. 6.77 Sound power level for rotor-stator interaction noise.

where $B - C$ is the transfer element for the rotor cascade, and $F - G$ is the transfer element for the stator cascade.

Fig. 6.77 gives the comparison results between rotor-stator interaction and wake-stator interaction for upstream and downstream acoustic power. "w-s" and "r-s" represent wake-stator interaction and rotor-stator interaction, respectively, whereas "eq" and "eh" are the sound power level of upstream and downstream waves. The parameters in this figure are rotor blade number $B = 16$, chord of blade $b = 0.15$, rotor tip speed $M_t = 0.406$, uniform Mach number $M_a = 0.245$, and hub-tip ratio $R_h/R_d = 0.6$.

As shown in Fig. 6.77, it can be noted that the lower upstream noise is predicted when the rotor cascade is considered. Meanwhile, the downstream noise does not have a big change when the rotor-stator interaction is compared with the wake-stator interaction. To explain this phenomenon, we will identify and discuss some possible mechanisms for the rotor-stator interaction problem.

The first mechanism for the reduction of the upstream sound radiation can be attributed to the rotor-shielding effect. Fig. 6.78 shows the transmission loss of sound through a rotor. The circumferential periodicity of the cut-on modes is $m \in [-5, 5]$, which could be realized by the variation of

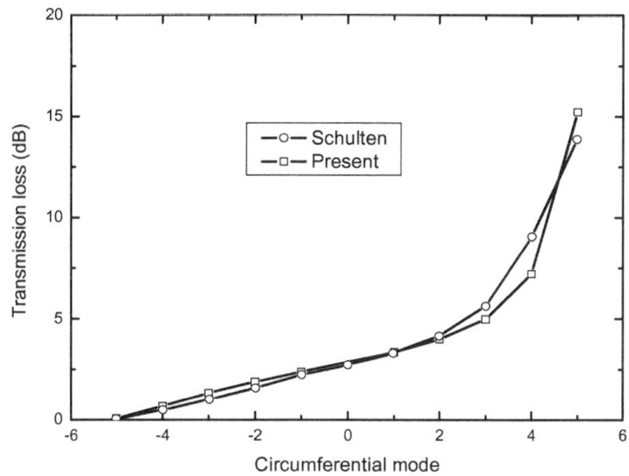

Fig. 6.78 Transmission loss of acoustic power through a rotor cascade (see [67] for Schulten's result).

the stator vane number $V \in [11, 21]$ in previous example. The first radial acoustic mode ($n = 1$) is the only cut-on radial mode for the previous circumferential acoustic mode at this frequency. Comparison of the results with Schulten's calculation [67] shows good agreement for the rotor-shielding effect for the different acoustic modes of the same acoustic frequency. According to Schulten's analysis [67], the reflection effect of rotor cascade on the acoustic wave mainly depends on two factors. The first is angle α between the incident modal wave fronts and the blade normal, as shown in Fig. 6.79. For a complete 3-D blade row, however, angle α is not a constant along the blade span because the blade chord direction α_r (as shown in Fig. 6.79) varies at different radial positions. It is obvious that the acoustic energy of the first radial mode mainly locates at the maximal radius. Thus, in the present case, it may be used the angle at tip of rotor as an approximate angle. From Fig. 6.79, it can be seen that the acoustic wave energy reflected by rotor blades increases with the decreasing angle. The second dominating factor is the chord/gap ratio. The rotor cascade with many wide-chord blades can reflect much of the incident acoustic waves.

The second mechanism is associated with frequency scattering. According to the previous derivation, there are frequency and mode scattering when the acoustic wave propagates through the rotor cascade. Fig. 6.80 shows the scattering effect for the first BPF incident wave, which is indicative of the acoustic energy transfer from the first BPF to the second BPF

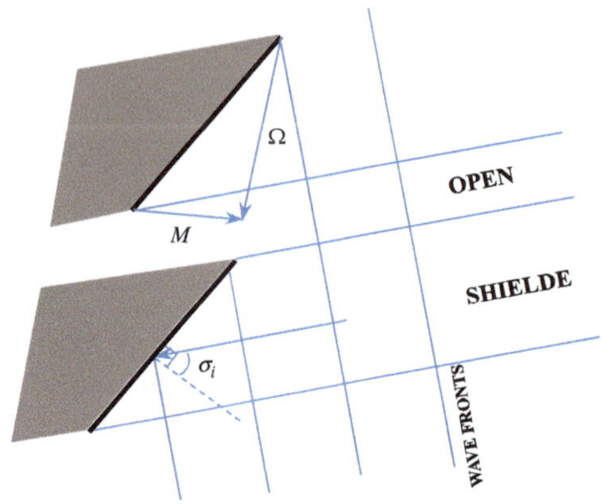

Fig. 6.79 Shielded effect of rotor cascade [67].

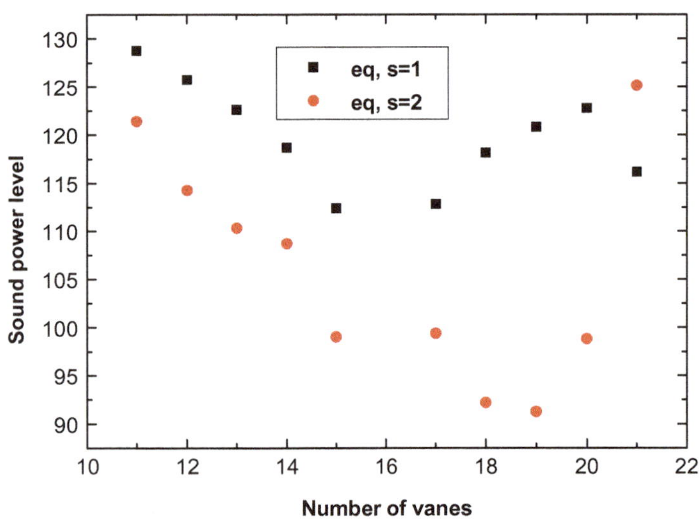

Fig. 6.80 Acoustic frequency scattering by rotor cascade.

through the rotor cascade. This phenomenon may result in the inefficiency of a designed liner that is optimal for the first BPF.

The third mechanism is the coupling effect between unsteady vortex and acoustic waves. The unsteady vorticity wave is generated by rotor blades or stator vanes when the acoustic wave propagates through a cascade, which means the acoustic energy is dissipated by the coupling effect.

From Fig. 6.77, it can be concluded that the coupling effect is strong when the acoustic wave propagates through multistage cascades.

It is obvious that capturing rotor-stator interaction, instead of considering only wake-stator interaction, is very important for the prediction of fan inlet noise. Meanwhile, it is also imperative for the acoustic liner design of an aeroengine nacelle when the shielding, scattering, and dissipation effect are considered.

6.5.2 The interaction between source and liner

Acoustic liner design in an aeroengine nacelle is a key problem for aircraft noise control. In general, for the purpose of suppressing the rotor-stator interaction noise by acoustic treatments, the design of the acoustic treatments consists of two steps. The first step is to determine the characteristics of the sound source, such as dominating propagating modes and frequencies, in terms of the number of rotor and stator, flow Mach number, and duct geometry. The second step is to design the geometrical parameters of an acoustic liner by combining the wall impedance model and the related acoustic features. It is known that the efficiency of an acoustic liner completely depends on the variation of an acoustic field. Therefore, a good liner design needs to tolerate various possible variations of an acoustic field. Otherwise, the design aimed at a single factor will not consider effective suppression of noise propagation. Especially, if we consider the interaction of acoustic treatments and the sound source, the design of acoustic treatments will face more challenges compared with that for a strength-fixed sound source. Here, we will make use of the present model to discuss how to achieve more sound attenuation by adjusting the spacing of the fan stator and acoustic treatment shown in following figure.

In this simple case, the effect of stator vanes on the acoustic liner design is considered first, as shown in Fig. 6.81. The transfer element for fan stator and

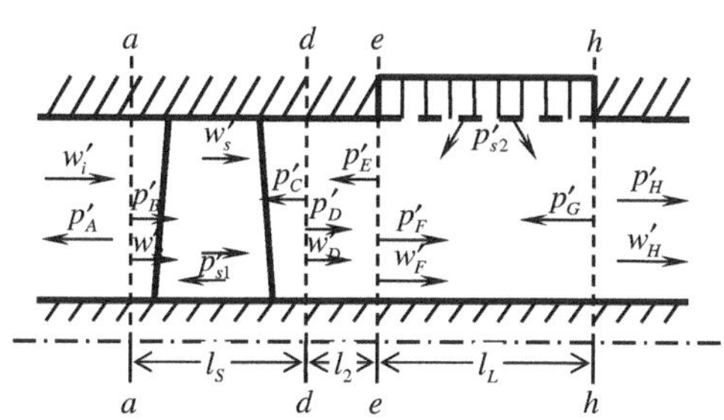

Fig. 6.81 The interaction effect between stator cascade and acoustic treatment.

acoustic treatment that have been established in the previous section allow us to investigate the interaction effect. As shown in Fig. 6.81, the related matrix equation can be derived as:

$$\begin{pmatrix}
SS_A^{p_{a-a}} & & & & & & & & & & & \\
SS_A^{v_{a-a}} & & & & & & & & & & & \\
 & B & - & C & & & & & & & & \\
 & & & & SS_D^{p_{d-d}} & SS_E^{p_{d-d}} & 0 & & & & & \\
 & & & & SS_D^{v_{d-d}} & SS_E^{v_{d-d}} & 0 & & & & & \\
 & & & & 0 & 0 & SS_{W_D}^{w_{d-d}} & & & & & \\
 & & & & SS_D^{p_{e-e}} & SS_E^{p_{e-e}} & 0 & & & & & \\
 & & & & SS_D^{v_{e-e}} & SS_E^{v_{e-e}} & 0 & & & & & \\
 & & & & 0 & 0 & SS_{W_D}^{w_{e-e}} & & & & & \\
 & & & & & & & F & - & G & SS_H^{p_{h-h}} & \\
 & & & & & & & & & & SS_H^{v_{h-h}} & \\
 & & & & & & & & & & & SS_{W_H}^{w_{h-h}}
\end{pmatrix}
\begin{pmatrix}
A_{mn}^p \\ B_{mn}^p \\ C_{mn}^p \\ B_{0n}^w \\ D_{mn}^p \\ E_{mn}^p \\ D_{0n}^w \\ F_{mn}^p \\ G_{mn}^p \\ F_{0n}^w \\ H_{mn}^p \\ H_{0n}^w
\end{pmatrix}
=
\begin{pmatrix}
I_{mn}^p \\ I_{mn}^v \\ I_{0n}^w \\ 0 \\ 0 \\ 0 \\ 0 \\ 0 \\ 0 \\ 0 \\ 0 \\ 0
\end{pmatrix}
\tag{6.286}$$

where $B-C$ represents the transfer element of the fan stator , whereas $F-G$ is the transfer element of the acoustic treatment. The middle part between the stator cascade and the liner describe the element of hard-walled duct segment with nothing inside.

Consider the wake-stator interaction first, and rotor and stator blade numbers are $B=16$ and $V=24$, respectively, and the blade chord $b=0.2\,\text{m}$, Gap/chord $=1.0$ at the tip, flow Mach number $M=0.5$, and rotor tip Mach number $M_T=0.783$. The radius can be determined by Gap/chord $=1.0$ at the tip, i.e., $2\pi R_d/V=b$. The hub-tip ratio is $R_h/R_d=0.98$. In addition, the upwash velocity is given by:

$$w'_\varphi(r,z,t)=e^{iB\left[\Omega t-\frac{\Omega}{U}z\right]}, \tag{6.287}$$

This simplified description of the rotor wake does not include the variation in the radial direction. Besides, the sound power level is defined by:

$$SWL=10\log\frac{E_H}{E_0}, \tag{6.288}$$

where $E_0=10^{-12}J$, the downstream sound energy, E_H is the exit sound energy calculated by Goldstein [68]:

$$E=\int_s I_s ds=\int_s\left[\left(1+M^2\right)p'v^*+\frac{M}{\rho_0 c_0}p'p'^*+M\rho_0 c_0 vv^*\right]ds, \tag{6.289}$$

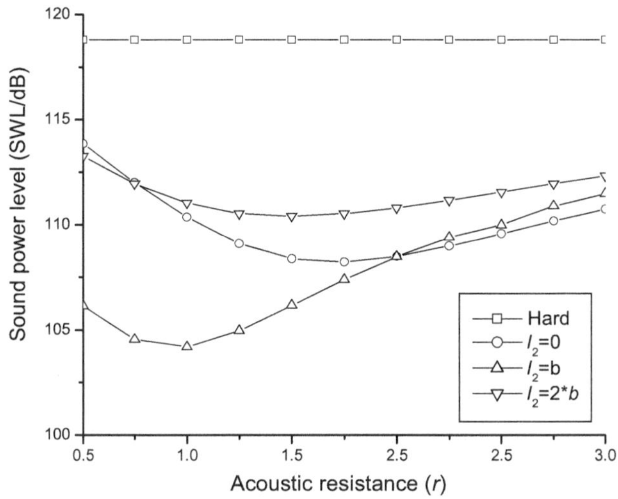

Fig. 6.82 Effect of fan stator on the optimization of acoustic liner, $Z_{zk} = r_{resistance} + ix$ ($x = -0.5$).

where v represents the velocity for particle displacement match, $s$stands for the integral surface, i.e., cross-section, and superscript "*" represents "conjugate".

As shown in Fig. 6.82, we assume that $l_s = b$, $l_L = 3b$, and $l_2 = 0, b, 2b$. Besides, when we fix the reactance as -0.5, emphasis will be placed on the effect of the variation of the resistance on the sound propagation. Fig. 6.82 shows the variation of SWL with the spacing and also the results for hard wall case for comparison. Under this circumstance, there are two propagating modes, i.e., $(-8, 1)$ and $(-8, 2)$. It is found that the optimum sound attenuation will be quite different for different spaces. When the spacing equals zero, the acoustic resistance corresponding to the optimum sound attenuation is 1.75, and when the spacing equals the blade chord, the associated optimal acoustic resistance r is 1.0, and when the spacing increases to 2b, the optimum result for r is 1.5. It is concluded that, when the interaction between the fan stator and acoustic treatments is considered, the determination of the optimum acoustic resistance is closely associated with the design spacing between the fan stator and acoustic treatment.

In addition, from the computational results, it can also be noted that the different spacing design will result in different sound attenuation. Specifically, when the spacing between the vane stator and acoustic treatment equals to the blade chord, the corresponding noise reduction is two times higher than that for the case when $l_2 = 2b$. Therefore, it can be further concluded that the spacing between the fan stator and acoustic treatment is one of the main design parameters in enhancing sound attenuation.

On the basis of what we found here, we extend our model to a more complicated case as shown in Fig. 6.75, i.e., two acoustic treatments are placed on the front and aft of the fan stator, respectively. This undoubtedly introduces more factors to the acoustic treatment design. To simplify the problem, we assume the length and resistance of the two acoustic treatments are same, the spacing between the liner and fan stator is 0.5b, and the other parameters are as described in Fig. 6.81. Fig. 6.83A–C is a schematic diagram related to various possible interactions. Fig. 6.84A and B show the comparison of upstream and downstream sound energy with the variation of acoustic resistance for three different element combinations corresponding to Fig. 6.83. From the comparison between case (b) and the rigid wall case,

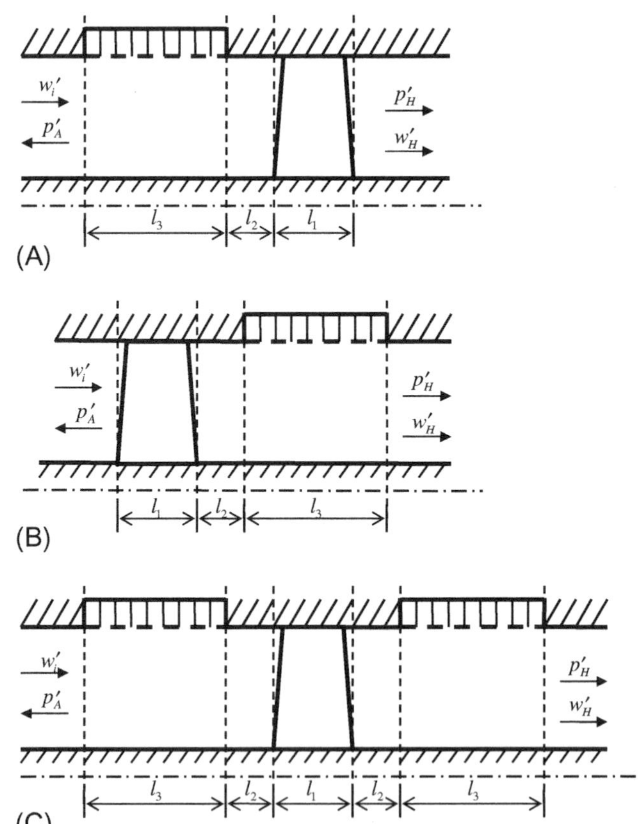

Fig. 6.83 A schematic of various element combinations. (A) stator cascade with upstream liner (B) stator cascade with downstream liner (C) stator cascade with upstream and downstream liner.

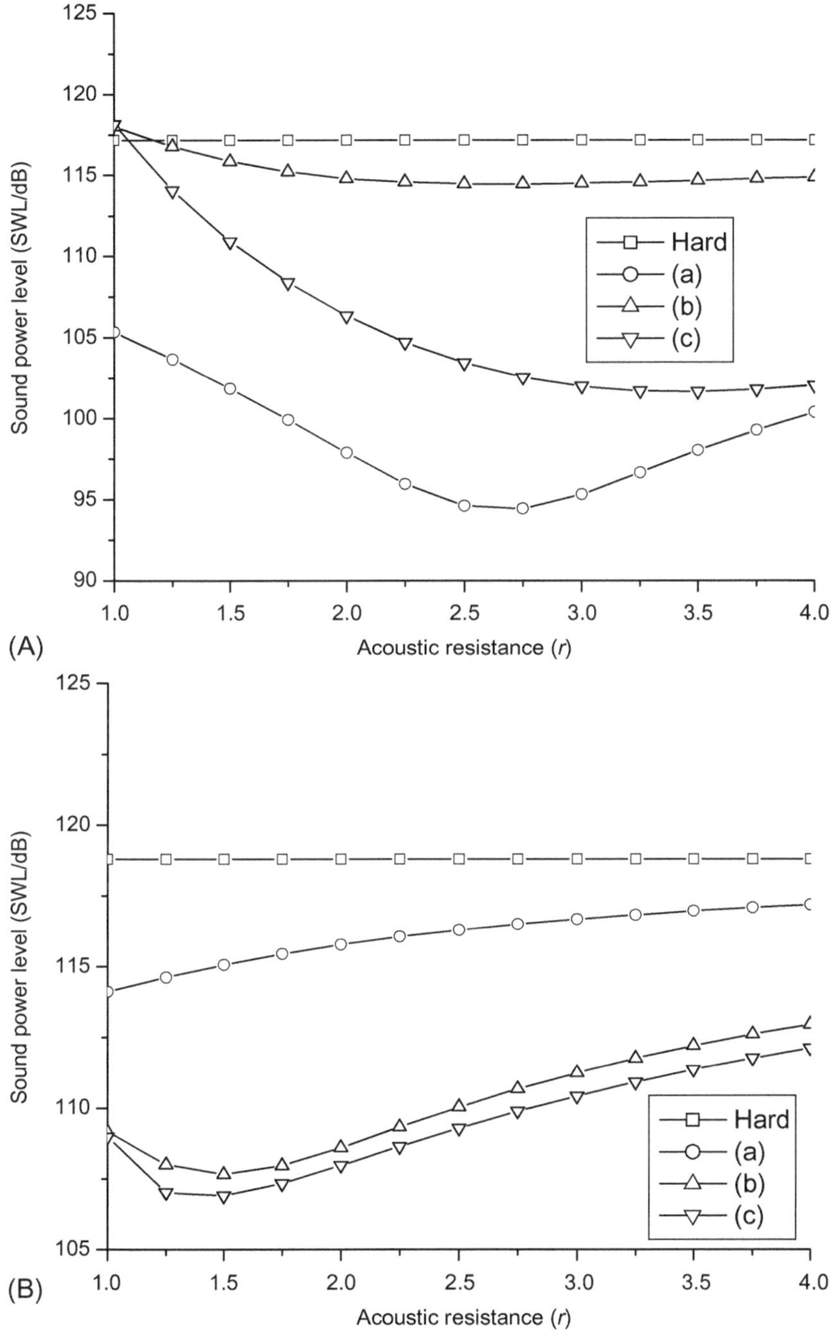

Fig. 6.84 Comparison of three cases. (A) Upstream, (B) downstream.

and the comparison between case (a) and case (c) in Fig. 6.84A, it is seen that the aft acoustic treatment of the vane stator has great effect on the upcoming propagating waves. Obviously, the noise reduction for case (c) is 7 dB, which is less than that for case (a) due to the negative effect caused by the aft acoustic treatment. On the other hand, it can be seen that the noise reduction for case (b) and case (c) in Fig. 6.84B show no obvious differences. Actually, this means that the front acoustic treatment is less effective for attenuating the downstream waves. So, the designs of the front and aft acoustic treatments of the vane stator are not independent, and the design of optimal noise reduction should consider the interaction between these elements.

In these examples, we have not deliberately aimed at the optimization of acoustic treatment design with the change of a single factor like the acoustic resistance; in contrast, we just want to show that, with different acoustic treatment combinations, the sound energy transmitting upstream and downstream will depend on many factors, such as the spacing between the fan stator and acoustic treatment, liner length and wall impedance, and so on. Therefore, an effective design strategy of acoustic liners should take into account the effect of various interaction mechanisms on the sound generation and propagation to optimize the sound attenuation in the wider range of design variables.

6.5.3 Far-field sound radiation of an Aeroengine nacelle

As mentioned before, we want to calculate the noise generation in, sound propagation through, and sound radiation from an aeroengine duct, as shown in Fig. 6.85. The interaction between various acoustic elements is expected to be included in preliminary liner design. In the previous section, based on the transfer elements of liner, rotor cascade, and stator cascade, we have discussed the interaction between sound source and various liners. In this section, to emphasize the sound radiation from the duct, i.e., the effect of the inlet and outlet of an aeroengine nacelle on the liner design, the problem about sound radiation from a duct, which is shown in Fig. 6.85, will be investigated. Regarding the sound source as a rotating point source, and only considering the locally reacting liner, the sound radiation model will be established in this section. It is noticed that all transfer elements given before can be combined with the model in this section to solve the problem about sound radiation from the duct depicted in Fig. 6.85.

To predict sound radiation from a finite-length duct of uniform cross-section, a BIEM is applied instead of using the mode reflection coefficients

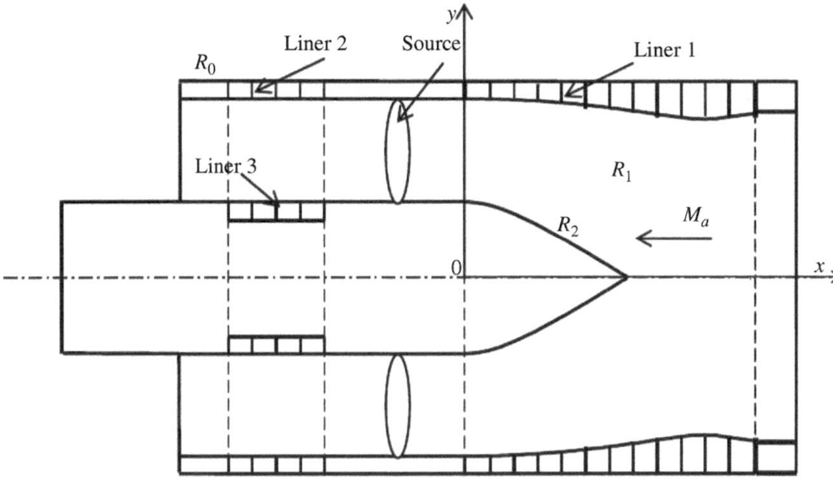

Fig. 6.85 Schematic of sound radiation of an aeroengine.

defined by Dunn et al. [39] and Yang and Wang [40]. Based on the potential theory, when the boundary conditions $p(Q), \frac{\partial p(Q)}{\partial n}$ on the exterior surface are known, the sound radiation from the duct can be obtained by:

$$C(P)p(P) = \int_S \left[p(Q)\frac{\partial G(P, Q)}{\partial n} - \frac{\partial p(Q)}{\partial n} G(P, Q) \right] dS(Q), \qquad (6.290)$$

where P, Q are points on the surface S, n is a outwards-drawn unit normal at a point on the surface, and p is the surface pressure. The function G is the free-space Green's function $G(P, Q) = \exp(-ik_0R)/4\pi R$, where $R = |r_p - r_q|$. As a function of the position of P, the value of $C(P)$ is given by:

$$C(P) = 1 + \int_S \frac{\partial}{\partial n_q}\left[\frac{1}{4\pi R}\right] dS(Q). \qquad (6.291)$$

Obviously we have:

$$\begin{aligned} &C(p) = 0, \text{for } P \quad \text{in duct,} \\ &C(p) = 1, \text{for } P \quad \text{out of duct,} \\ &C(p) = 0.5, \text{for } P \quad \text{on a smooth surface of duct.} \end{aligned} \qquad (6.292)$$

Considering the properties of the axisymmetric duct, and using the expansion of the boundary conditions and the surface distribution functions in Fourier series with respect to the angle of revolution, the surface integral

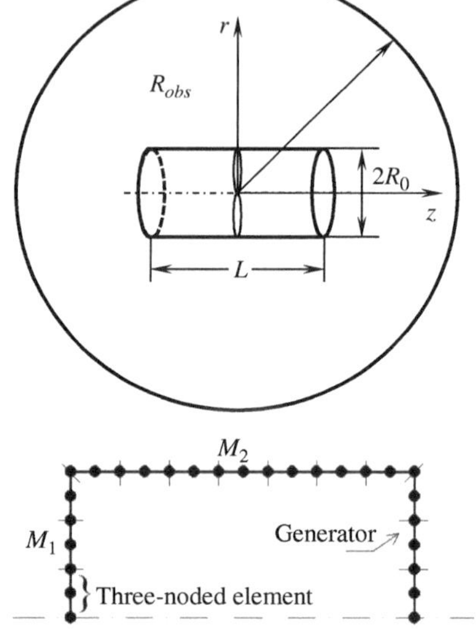

Fig. 6.86 Geometry of duct and observation point.

of Eq. (6.290) is reduced to a line integral along the generator of the duct [69, 70]. The evaluation of Eq. (6.290) can be performed numerically by discretizing the generator of the duct into a series of curvilinear isoparametric elements [69, 70]. As Fig. 6.86 shows, there are $2M_1 + M_2$ nodes along the generator. Each element has three nodes, i.e., $(2M_1 + M_2 - 3)/2$ elements here. Discretizing the boundary integral of Eq. (6.290) leads to $2M_1 + M_2$ algebraic equations:

$$c(P_j)p_{mj} - \sum_{n=1}^{N_e}\sum_{a=1}^{3} p_{mna}A^a_{mnj} + \sum_{n=1}^{N_e}\sum_{a=1}^{3} p^n_{mna}C^a_{mnj} = 0 \quad j = 1,2,3,\ldots 2M_1 + M_2, N_e = 2M_1 + M_2 - 3. \tag{6.293}$$

Here, the method mentioned by Wang et al. and Sun et al. [69, 70] is used to obtain the correlative coefficients matrixes A^a_{mnj}, C^a_{mnj}. m denotes the circumferential mode of the sound source, and p_{mna} and p^n_{mna} are the values of $p, \frac{\partial p}{\partial n}$ at the ath node of nth element. Obviously, there are $4M_1 + 2M_2$ unknowns in Eq. (6.293). On the solid exterior surface of the duct, the boundary condition is given by $\frac{\partial p}{\partial n} = 0$. Then $4M_1 + M_2$ unknowns remain in Eq. (6.293). To determine the values of these unknowns, $2M_1$ equations are also needed.

It should be noted that Eq. (6.290) is based on the moving-medium wave equation with the source term $\alpha\left(\vec{r'}, \tau\right) = 0$. When there is only z-axis uniform flow, we can rewrite wave equation as:

$$\frac{\partial^2 p}{\partial x^2} + \frac{\partial^2 p}{\partial y^2} + \left(1 - M^2\right)\frac{\partial^2 p}{\partial z^2} - 2\frac{i\omega M}{c_0}\frac{\partial p}{\partial z} + k_0^2 p = 0, \tag{6.294}$$

With $p(x, y, z) = \Psi(x, y, \zeta)e^{ik_0' M\zeta}$, $k_0' = k_0/\beta$ and $z = \beta\zeta$, Eq. (6.294) will be rearranged as:

$$\frac{\partial^2 \Psi}{\partial x^2} + \frac{\partial^2 \Psi}{\partial y^2} + \frac{\partial^2 \Psi}{\partial \zeta^2} + k_0'^2 \Psi = 0, \tag{6.295}$$

Eq. (6.295) is a Helmholtz equation. It can be solved by Eq. (6.293) to get the values of Ψ, $\frac{\partial \Psi}{\partial z}$. The values of Ψ and $\frac{\partial \Psi}{\partial z}$ can be determined by solving Eq. (6.293). Then the values of p and $\frac{\partial p}{\partial n}$ can be obtained as follows:

$$\frac{\partial p}{\partial n} = \frac{\partial p}{\partial z} = \frac{e^{ik_0' M\zeta}}{\beta}\left(\Psi_n + ik_0' M\Psi\right), \text{ for Q on } R_{diet}\text{section.} \tag{6.296a}$$

$$\frac{\partial p}{\partial n} = -\frac{\partial p}{\partial z} = \frac{e^{ik_0' M\zeta}}{\beta}\left(-\Psi_n + ik' M\Psi\right), \text{ for Q on } L_{diet}\text{section.} \tag{6.296b}$$

It is well known that if the duct has zero-thickness walls, there is a singularity at the leading edge of the duct. To avoid this problem, we assume the duct has a finite thickness h_d, and $h_d \ll r_1$. In spite of this, as an approximation treatment, we still regard the flow both inside and outside the duct as uniform flow. In addition, the Kutta condition is imposed on the trailing edge of the duct during the process of discretizing the equation.

In the present model, the incident sound is produced by a collection of M_s points or line monopoles and/or dipoles. The duct is divided into n_d sections, i.e., has $n_d + 1$ cross-sectional interfaces, as shown Fig. 6.87, and the interior surface of each section has different boundary conditions. The sound sources are assumed to be located in the jth section. The sound pressure field can be described by:

$$p_i^j = \sum_{n=1}^{N} p_{in}^j \phi_m(k_{m,n} r) e^{i\gamma_{m,n}^- z}, \tag{6.297}$$

$$p_B^j = \sum_{n=1}^{N} B_n^j \phi_m(k_{m,n} r) e^{i\gamma_{m,n}^- z}, \tag{6.298}$$

$$p_C^j = \sum_{n=1}^{N} C_n^j \phi_m(k_{m,n} r) e^{i\gamma_{m,n}^+ (z - l_j)}, \tag{6.299}$$

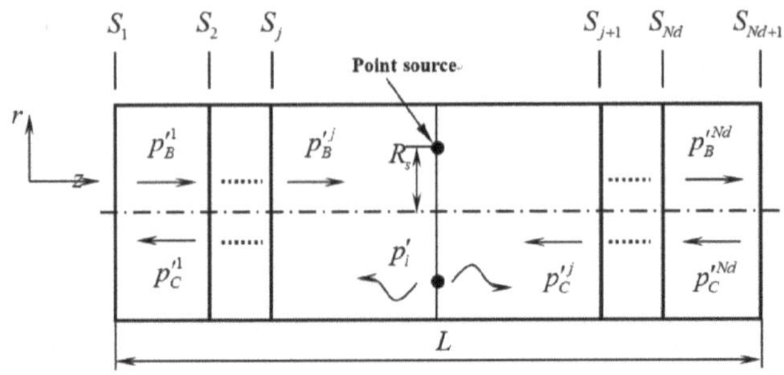

Fig. 6.87 Geometry of a finite duct with *nd* transfer elements.

$$p_d^j = \frac{\rho_0}{2}\sum_{n=1}^{N}\frac{\phi_m(k_{m,n}r)e^{i\gamma_{m,n}^{\pm}z}}{\Gamma_{m,n}}\left\{\sum_{\mu=1}^{N}\left[B_\mu^j Q_{n\pm}^{B_\mu} + C_\mu^j Q_{n\pm}^{C_\mu}\right]\right\}. \tag{6.300}$$

where superscript $j$denotes the number of the section. N denotes how many radial modes are considered. $\phi_m(k_{m,\,n}r)$ is the eigenfunction. For a circular duct, $\phi_m(k_{m,\,n}r)$ is the first kind Bessel function $J_m(k_{m,\,n}r)$, where $k_{m,\,n}$ is the eigenvalue given by $J_m{}'(k_{m,\,n}r) = 0$. If we let $N = M_1$, from Eqs. (6.293) and (6.297) through (6.300), we know that $(4 + 2nd)M_1 + M_2$ unknowns have to be determined to describe the whole sound field. Based on TEM, one can obtain the algebraic equations in the form of:

$$\left\{\begin{array}{cccccccc}
{}_{ss}B_p^{2-} & {}_{ss}C_p^{2-} & {}_{ss}B_p^{2+} & {}_{ss}C_p^{2+} & & & & \\
{}_{ss}B_v^{2-} & {}_{ss}C_v^{2-} & {}_{ss}B_v^{2+} & {}_{ss}C_v^{2+} & & & & \\
 & & {}_{ss}B_p^{3-} & {}_{ss}C_p^{3-} & \cdots & & & \\
 & & {}_{ss}B_v^{3-} & {}_{ss}C_v^{3-} & \cdots & & & \\
 & & & & \cdots\ {}_{ss}B_p^{(n_d-1)+} & {}_{ss}C_p^{(n_d-1)+} & & \\
 & & & & \cdots\ {}_{ss}B_v^{(n_d-1)+} & {}_{ss}C_v^{(n_d-1)+} & & \\
 & & & & {}_{ss}B_p^{n_d-} & {}_{ss}C_p^{n_d-} & {}_{ss}B_p^{n_d+} & {}_{ss}C_p^{n_d+} \\
 & & & & {}_{ss}B_v^{n_d-} & {}_{ss}C_v^{n_d-} & {}_{ss}B_v^{n_d+} & {}_{ss}C_v^{n_d+}
\end{array}\right\}_{2\cdot(n_d-1)\cdot M_1 \times 2\cdot n_d\cdot M_1}
\begin{pmatrix} B^1 \\ C^1 \\ \vdots \\ \vdots \\ \vdots \\ \vdots \\ B^{n_d} \\ C^{n_d} \end{pmatrix}
= \begin{pmatrix} 0 \\ \vdots \\ pp_i^{n+} \\ vv_i^{n+} \\ pp_i^{(n+1)-} \\ vv_i^{(n+1)-} \\ \vdots \\ 0 \end{pmatrix}_{2\cdot(n_d-1)\cdot M_1}. \tag{6.301}$$

The known quantities on the right side are related to incident waves. Superscript of $ss_{B_p}^{2-}$ denotes on the left side of S_2. As mentioned before, $2M_1 + M_2$ and $2(n_d - 1)M_1$ algebraic equations have been obtained by using BIEM and TEM, respectively. The continuity conditions on the inlet and outlet terminations lead to $4M_1$ equations in terms of p and $\frac{\partial p}{\partial n}$,

$$\begin{Bmatrix} BB_p^{1-} & AA_{Bp}^{1+} & AA_{Cp}^{1+} & \\ BB_{p''}^{1-} & AA_{Bp''}^{1+} & AA_{Cp''}^{1+} & \\ & AA_{Bp}^{(nd+1)-} & AA_{Cp}^{(n_d+1)-} & CC_p^{(n_d+1)+} \\ & AA_{Bp''}^{(nd+1)-} & AA_{Cp''}^{(n_d+1)-} & CC_{p''}^{(n_d+1)+} \end{Bmatrix}_{4M_1\times 6M_1} = \begin{Bmatrix} pp_{ip}^{1+} \\ pp_{ip''}^{1+} \\ pp_{ip}^{(n_d+1)-} \\ pp_{ip''}^{(n_d+1)-} \end{Bmatrix}_{4M_1\times 1}, \tag{6.302}$$

where the subscripts p and p'' denote that the matrix with respect to p and $\frac{\partial p}{\partial n}$, respectively. Matrixes BB and CC are associated with the values of p and $\frac{\partial p}{\partial n}$ of each nodes on the radius, respectively. They are defined by:

$$\int_0^{r_0} p\phi_m(k_{m,n}r)rdr, \tag{6.303}$$

$$\int_0^{r_0} \frac{\partial p}{\partial n}\phi_m(k_{m,n}r)rdr. \tag{6.304}$$

By solving the simultaneous linear equations (6.293), (6.301), and (6.302), the $(4 + 2nd)M_1 + M_2$ unknowns can be determined, i.e., the sound boundary condition of the duct can be obtained. It is therefore concluded that the sound radiation from a finite duct can be determined by combining TEM with BIEM, which allows the effect of both locally and nonlocally reacting liners to be included.

It is easy to see that the sound generation in, propagation through, and generation from a duct can be solved by using TEM and BEM. Here, two examples will be given to investigate the acoustic attenuation problem.

As shown in Fig. 6.88, first, the sound radiation from a finite-length duct with locally reacting liner and nonlocally reacting liner has been investigated based on the example given by Dunn [39] but with the locally reacting liner partly replaced by the nonlocally reacting liner.

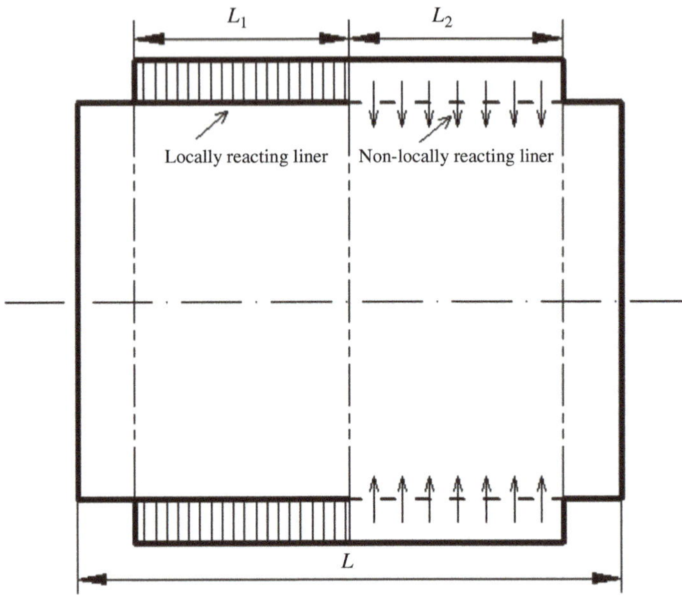

Fig. 6.88 Sketch of segmented liners.

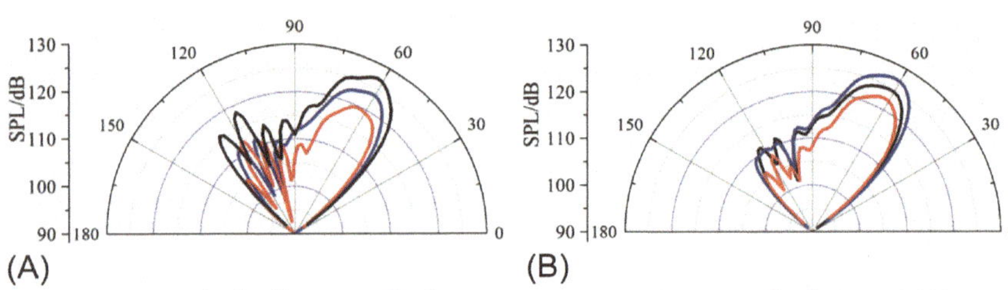

(A) —— hard wall; —— local: $Z_1 = Z_2 = (1.0, 1.5)$; —— non-local: $M_h = 0.023$.

(B) —— local: $Z_2 = (0.5, 0.5)$; —— non-local: $M_h = 0.005$; —— non-local: $M_h = 0.05$.

Fig. 6.89 Sound pressure level at surface of a hemisphere whose radius is equal to 10 m.

As shown in Fig. 6.89A, there is an obvious difference when the non-locally reacting liner with bias flow is considered. The present results show that the TEM has obvious advantages over the BIEM and FEM for the far-field prediction when the nonlocally reacting liner is used. This kind of liner is particularly efficient for sound attenuation in different states of an aeroengine, as we mentioned earlier.

Finally, we will give an example of the acoustic radiation from a finite-length duct with acoustic treatment, which is closer to the actual case of sound radiation from an aeroengine than the previous example. As shown in Fig. 6.85, in this simplified model the sound source was considered as a spinning point dipole source. The axial length of the duct with inlet liner 1 ($Z_1 = (2, -1)$, $x \in (0, 1.86393)$) and exhaust liner 2 and 3 ($Z_2 = Z_3 = (2, -1)$, $x \in (-1.5, -0.5)$) is 2 m. The frequency of spinning dipole source at position $x_i = -0.25$ is 865.8 Hz. The source is described by the following formula:

$$p_i = -\frac{BF_T}{2} \sum_{m=-\infty}^{\infty} \sum_{n=1}^{\infty} \frac{\phi_{mn}(k_{mn}r) e^{-im\varphi} e^{i\gamma_{mn}^{\pm} z}}{\Gamma_{mn}} \frac{\gamma_{mn}^{\pm}}{\kappa_{mn}} \phi_{mn}(k_{mn} r_1). \tag{6.305}$$

Here, $r_1 = 0.9$ and $F_T = 50$. The outer radius R_1 and inner radius R_2 inside the duct are represented as:

$$R_1 = \begin{cases} 1 & x \in (-2, 0) \\ 1 - 0.18453 x'^2 + 0.10158 \dfrac{e^{-11(1-x')} - e^{-11}}{1 - e^{-11}} & x \in (0, 1.86393) \\ 1 - 0.18453 - 0.10158 \dfrac{e^{-11}}{1 - e^{-11}} & x \in (1.86393, 2) \end{cases}, \tag{6.306}$$

$$R_2 = \begin{cases} 0.64212 - 0.04777^{1/2} & x \in (-2.6, 0) \\ \max\left[0\ ,\ 0.64212 - (0.04777 + 0.98234 x')^{1/2}\right] & x \in (0, 1.86393), \\ 0 & x \in (1.86393, 2) \end{cases} \tag{6.307}$$

Here, $x' = x/1.86393$. The outer radius of duct is $R_0 = 1.02$m, and the radiation radius is $r = 10$m.

In Fig. 6.90, "Hard wall" denotes hard inlet and hard exhaust; "imp1" represents that there is only liner 1 in an aeroengine duct; "imp2" denotes the case where only liner1 and liner2 are included; and "imp3" denotes that all three liners exist in the duct depicted by Fig. 6.85. The result plotted in Fig. 6.90 shows the effect of acoustic treatment on the sound radiation. The inlet noise has been reduced substantially due to liner 1, but the exhaust noise increased slightly due to the reflection effect of liner 1 by considering the interaction between the inlet and outlet duct openings. When the exhaust liner exists, the sound radiation is reduced in both inlet and outlet directions.

However, comparing case "imp2" with case "imp3", it is found that the difference is very slight due to the absence of liner optimization and swirling

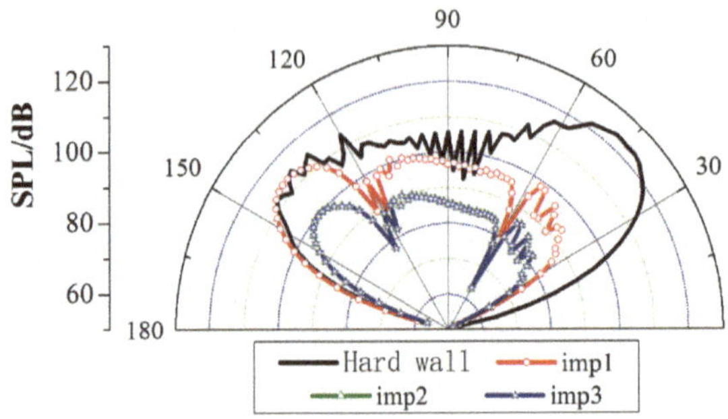

Fig. 6.90 Effect of liner on the sound radiation.

flow. On the other hand, the effect of liner 3 on the sound attenuation is small when the hub-tip ratio is less than 0.5, so it is possible that the radiation results of both situations are almost similar.

Once again, some discovery and analysis by using TEM have been introduced in this part, but how to realize an optimal acoustic design of an aeroengine nacelle still requires considerable work along with the consideration of various interaction mechanisms.

References

[1] E. Envia, Emerging community noise reduction approaches, in: 3rd AIAA Atmospheric Space Environments Conference, paper no. 2011-3532, June 2011, 2011.
[2] E. Envia, Fan noise reduction: an overview, Int. J. Aeroacoust. 1 (2002) 43–64.
[3] D.L. Huff, Noise reduction technologies for turbofan engines, (2006)NASA/TM 2007-214495, September 2007.
[4] G.W. Bielak, J.W. Premo, A.S. Hersh, Advanced Turbofan Duct Liner Concepts, (1999)NASA/CR 1999-209002, February 1999.
[5] J. Yu, E. Nesbitt, H.W. Kwan, S. Uellenberg, E. Chien, J. Premo, M. Ruiz, M. Czech, Quiet technology demonstrator 2 intake liner design and validation, in: 27th AIAA Aeroacoustics Conference, paper no. 2006-2458, May 2006, 2006.
[6] J. Mardjono, R. Georges, B. Jean-Michel, F. Boubila, EPSL static tests demonstration of liners noise reduction concepts, in: 19th AIAA/CEAS Aeroacoustics Conference, paper no. 2013-2174, May 2013, 2013.
[7] M.J.T. Smith, Aircraft noise, Cambridge University Press, New York, 1989.
[8] J.M. Tyler, T.G. Sofrin, Axial flow compressor noise studies, Soc. Automot. Eng. Trans. 70 (1962) 309–332.
[9] M.K. Myers, Boundary integral formulations for ducted fan radiation calculations, CEAS/AIAA Paper 1995-076, 1995, 1995.
[10] M.K. Myers, Radiation of sound form a point source in a short duct, in: NASA CP-3352, 1997, 1997.
[11] M.H. Dunn, J. Tweed, F. Farassat, The application of a boundary integral equation method to the prediction of ducted fan engine noise, J. Sound Vib. 277 (5) (1999) 1019–1048.

[12] E. Quaranta, D. Drikakis, Noise radiation from a ducted rotor in a swirling-translating flow, J. Fluid Mech. 641 (2009) 463–473.
[13] H. Choi, L. Joo, Development of the numerical method for calculating sound radiation form a rotating dipole source in an opened thin duct, J. Sound Vib. 295 (2006) 739–752.
[14] Y. Ozyoruk, Numerical prediction of aft radiation of turbofan tones through exhaust jets, J. Sound Vib. 325 (2009) 122–144.
[15] A. Iob, R. Arina, Frequency-domain linearized Euler model for turbomachinery noise radiation through engine exhaust, AIAA J. 48 (2010) 848–858.
[16] R.J. Astley, W. Eversman, Finite element formulations for acoustical radiation, J. Sound Vib. 88 (1983) 47–64.
[17] R.I. Danda, W. Eversman, Improved finite element modeling of the turbofan engine inlet radiation problem, J. Sound Vib. 117 (1995) 109–115.
[18] W. Rienstra, W. Eversman, A numerical comparison between the multiple-scales and finite-element solution for sound propagation in lined flow ducts, J. Fluid Mech. 437 (2001) 367–384.
[19] C.K.W. Tam, Computational aeroacoustics—issues and methods, AIAA J. 33 (1995) 1788–1796.
[20] C.K.W. Tam, Computational aeroacoustics: an overview of computational challenges and applications, Int. J. Comput. Fluid Dyn. 18 (2004) 547–567.
[21] Y. Ozyoruk, V. Ahujia, N. Long, Time domain simulation of radiation from ducted fans with liners, in: AIAA-2001-2171, 2001.
[22] X. Li, C. Scemel, U. Michel, On the azimuthal mode propagation in axisymmetric duct flows, in: 8th AIAA/CEAS Aeroacoustcs Conference, Paper no. 2002-2521, June 2002, 2002.
[23] U. Ingard, H. Ising, Acoustic nonlineartry of an orifice, J. Acoust. Soc. Am. 41 (1967) 1582–1583.
[24] X. Jing, X. Sun, Experimental investigations of perforated liners with bias flow, J. Acoust. Soc. Am. 106 (1999) 2436–2441.
[25] X. Jing, X. Sun, Effect of plate thickness on impedance of perforated plates with Bias flow, AIAA J. 38 (2000) 1573–1578.
[26] X. Jing, X. Sun, Sound-excited flow and acoustic nonlinearity at an orifice, Phys. Fluids 14 (2002) 268–276.
[27] X. Sun, X. Jing, H. Zhang, Y.W. Shi, Effect of grazing-bias flow interaction on acoustic impedance of perforated plates, J. Sound Vib. 254 (2002) 557–573.
[28] S.H. Ko, Sound attenuation in acoustically lined circular ducts in the presentce of uniform flow and shear flow, J. Sound Vib. 22 (1972) 193–210.
[29] B.J. Tester, The propagation and attenuation of sound in lined ducts containing uniform or plug flow, J. Sound Vib. 28 (1973) 151–203.
[30] B. Noble, Method based on the Winer-Hopf technique for the solution of partial differential equations, Phys. Today 12 (9) (1959) 50.
[31] W.E. Zorumski, Acoustic theory of axisymmetric multisection ducts, in: NASA TR R-419, 1974.
[32] W. Eversman, Erratum: computation of axial and transverse wave numbers for uniform two-dimensional ducts with flow using a numerical integration scheme, J. Sound Vib. 41 (1975) 252–255.
[33] W. Eversman, Initial values for the integration scheme to compute the eigenvalues for propagation in ducts, J. Sound Vib. 50 (1977) 159–162.
[34] X. Sun, L. Du, V. Yang, A homotopy method for determining the eigenvalues of locally or nonlocally reacting acoustic liners in flow ducts, J. Sound Vib. 303 (2007) 277–286.
[35] A.H. Nayfeh, J.E. Kaiser, D.P. Telionis, Acoustics of aircraft engine-duct systems, AIAA J. 13 (1975) 130–153.

[36] W.E. Zorumski, Generalized radiation impedance and reflection coefficient of circular and annular ducts, J. Acoust. Soc. Am. 54 (1973) 1667–1673.
[37] G.W. Johnston, K. Ogimoto, Sound radiation from a finite length unflanged circular duct with uniform axial flow. II. Computed radiation characteristics, J. Acoust. Soc. Am. 68 (1980) 1871–1883.
[38] G.W. Johnston, K. Ogimoto, Sound radiation from a finite length unflanged circular duct with uniform axial flow. I. Theoretical analysis, J. Acoust. Soc. Am. 68 (1980) 1858–1870.
[39] M. Dunn, J. Tweed, F. Farassat, The prediction of ducted fan engine noise via a boundary integral equation method, in: 2nd AIAA/CEAS Aeroacoustics Conference, Paper no. 1996-1770, May 1996, 1996.
[40] B. Yang, T. Wang, An approach to predict ducted fan noise by boundary integral equation method, in: 11th AIAA/CEAS Aeroacoustics Conference, Paper no. 2005-3068, May 2005, 2005.
[41] R.J. Astley, W. Eversman, A finite element method for transmission in non-uniform duct without flow: comparison with the method of weighted residuals, J. Sound Vib. 57 (1978) 367–388.
[42] W. Eversman, Turbofan noise propagation and radiation at high frequencies, in: NASA CR-2003-212323, 2003.
[43] C.K.W. Tam, Recent advances in computational aeroacoustics, Fluid Dyn. Res. 38 (2006) 591–615.
[44] C. Richter, F. Thiele, X. Li, M. Zhuang, Comparison of timedomain impedance boundary conditions for lined duct flows, AIAA J. 45 (2007) 1333–1345.
[45] M.K. Myers, On the acoustic boundary condition in the presence of flow, J. Sound Vib. 71 (1980) 429–434.
[46] W. Eversman, A.V. Parrett, J.S. Preisser, R.J. Silcox, Contributions to the finite element solution of the fan noise radiation problem, J. Vib. Acoust., Trans. ASME 107 (1985) 216–223.
[47] M. Namba, K. Fukushige, Application of the equivalent surface source method to the acoustics of duct systems with non-uniform wall impedance, J. Sound Vib. 73 (1980) 125–146.
[48] J.D. Eldredge, A.P. Dowling, The absorption of axial acoustic waves by a perforated liner with bias flow, J. Fluid Mech. 485 (2003) 307–335.
[49] M. Utsumi, An efficient method for sound transmission in non-uniform circular ducts, J. Sound Vib. 277 (4) (1999) 735–748.
[50] M. Utsumi, Sound transmission in circular ducts of continuously varying cross-section area, J. Sound Vib. 242 (2) (2001) 369–376.
[51] A. Selamet, M.B. Xu, I.J. Lee, N.T. Huff, Analytical approach for sound attenuation in perforated dissipative silencers, J. Acoust. Soc. Am. 115 (2004) 2091–2099.
[52] D. Zhou, X. Wang, X. Sun, Investigation on the broadband noise absorption performance of multiple cavity resonance liner, in: Proceedings of the 22nd International Congress on Sound and Vibration, Florence, Italy, 2015.
[53] J.D. Eldredge, On the interaction of higher duct modes with a perforated liner system with Bias flow, J. Fluid Mech. 510 (2004) 303–331.
[54] M.S. Howe, On the theory of unsteady high Reynolds number flow through a circular aperture, Proc. R. Soc. Lond. A366 (1979) 205–233.
[55] L.M. Shaw, J.R. Balombin, Rotor wave characteristics relevant to rotor-stator interaction noise generation, in: 7th AIAA Areoacoustics Conference, Paper no. 1981-2031, October 1981, 1981.
[56] H.W. Liepmann, J. Laufer, K. Liepmann, On the spectrum of isotropic turbulence, Tech. Rep. Arch. Image Libr. 18 (1951) 263–275.

[57] T. Von Karman, Progress in the statistical theory of turbulence, Proc. Natl. Acad. Sci. U. S. A. 34 (1948) 530–539.
[58] E.J. Kerschen, P.R. Gliebet, Noise caused by the interaction of a rotor with anisotropic turbulence, AIAA J. 19 (1981) 717–723.
[59] D.B. Hanson, K.P. Horan, Turbulence/cascade interaction: spectra of inflow, cascade response, and noise, in: 4th AIAA/CEAS Aeroacoustics Conference, AIAA-1998-2319, 1998.
[60] M. Nallasamy, E. Envia, Computation of rotor wake turbulence noise, J. Sound Vib. 282 (2005) 649–678.
[61] H. Posson, S. Moreau, M. Roger, Broadband noise prediction of fan outlet guide vane using a cascade response function, J. Sound Vib. 330 (2011) 6153–6183.
[62] J. Schulten, Vane sweep effects on rotor/stator interaction noise, AIAA J. 35 (1997) 945–951.
[63] B. Elhadidi, H.M. Atassi, Passive control for turbofan tonal noise, AIAA J. 43 (2005) 2279–2292.
[64] H.M. Atassi, A.A. Ali, O.V. Atassi, I.V. Vinogradov, Scattering of incident disturbances by an annular cascade in a swirling flow, J. Fluid Mech. 499 (2004) 111–138.
[65] E. Envia, M. Nallasamy, Design selection and analysis of a swept and leaned stator concept, J. Sound Vib. 228 (1999) 793–836.
[66] J.B.H.M. Schulten, Sound generated by rotor wakes interacting with a leaned vane Stator, AIAA J. 20 (2012) 1352–1358.
[67] J.B.H.M. Schulten, Transmission of sound through a rotor, in: 14th DGLR/AIAA Aeroacoustics Conference, DGLR/AIAA Paper 92-02-085, 1992.
[68] M.E. Goldstein, Aeroacoustics, McGraw-Hill International Book Company, New York, 1976.
[69] W. Wang, N. Atalla, J. Nicolas, A boundary integral approach for acoustic radiation of axisymmetric bodies with arbitrary boundary conditions valid for all wave numbers, J. Acoust. Soc. Am. 101 (1997) 1468–1478.
[70] X. Sun, X. Wang, L. Du, et al., A new model for the prediction of turbofan noise with the effect of locally and nonlocally reacting liners, J. Sound Vib. 316 (2008) 50–68.

CHAPTER 7
Thermoacoustic instability

7.1 Basic concepts of thermoacoustics

Thermoacoustic instability, also known as combustion instability, has been encountered in the development and operation of practical devices including rocket engines, gas turbines, boilers, and aeroengines, manifesting itself with large-amplitude pressure oscillations, which may cause detrimental consequences. The instabilities are excited by the feedback loop between the unsteady combustion and natural acoustic mode of combustors, as shown in Fig. 7.1. The heat transfer in the combustors is enhanced spontaneously when the instabilities occur. Therefore, the components in the combustors are more prone to be melted, and the systems then lose efficiency or become disabled.

The deadly catastrophic effect and imperative control of combustion instability have drawn more and more research interest. Many investigations have been carried out since it was noticed. However, research exploration began long before the birth of propulsion and power devices. In 1777, Higgins [1] found that the flame in a tube emitted a sound that was called "singing flame," and the Sondhauss tube and Rijke tube phenomena were discovered later.

Rayleigh criterion is widely known and frequently quoted for his physical elucidation about the necessary condition for the onset of self-sustained oscillations in a Rijke tube: "If heat is periodically communicated to, and abstracted from, a mass of air vibrating in a cylinder bounded by a piston, the effect produced will depend upon the phase of the vibration at which the transfer of heat takes place. If heat is given to the air at the moment of greatest condensation, or taken from it at the moment of greatest rarefaction, or abstracted at the moment of greatest condensation, the vibration is discouraged." [2] However, this does not necessarily imply the combustion is unstable, because the portion of the acoustic energy dissipated at the boundaries plays an important role as well. Only if the acoustic energy enhanced from the heat exceeds that dissipated at the

Fundamentals of Aeroacoustics with Applications to Aeropropulsion Systems
https://doi.org/10.1016/B978-0-12-408069-0.00007-5

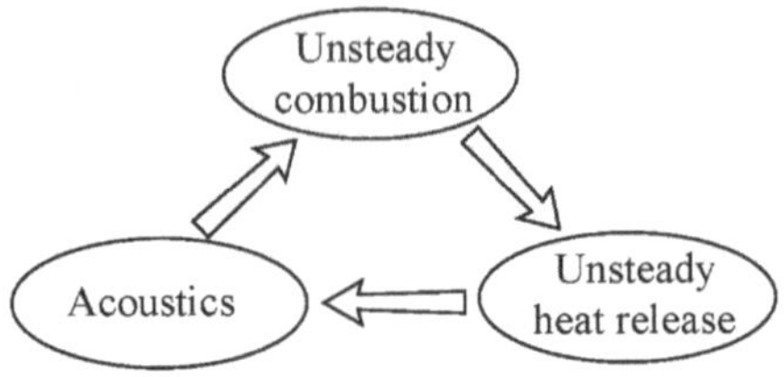

Fig. 7.1 Feedback loop between unsteady combustion and acoustics.

boundaries does the instability occur. This can be expressed by the following expression [3]:

$$\int_V \int_T p'(x,t) q'(x,t)\, dt dV \geq \int_V \int_T \sum_i L_i(x,t)\, dt dV, \tag{7.1}$$

where $p'(x, t)$, $q'(x, t)$, V, T, and L_i are the combustor pressure oscillations, heat addition oscillations, combustor volume, period of the oscillations, and the ith acoustic energy loss process (e.g., viscous dissipation, radiation of acoustic energy out of the combustor through its boundaries), respectively.

The research of combustion instability moved beyond purely academic interest with the arrival of large-amplitude oscillations in industrial equipment. Extensive efforts have been made on the characteristics of combustion instabilities in operational or laboratory systems and the methods for predicting and suppression in both theoretical and experimental approaches [4–8]. The mechanism is grouped into linear and nonlinear categories. A linearly unstable system is one that is unstable for infinitesimally small disturbances, thus the properties of the flow field can be decomposed into a time-mean and fluctuation quantity. The stability analysis is based on the decomposition. Linear stability analysis makes it convenient to solve the stability problem mathematically, although it is only an ideal assumption. In real unstable systems, there exists a process of evolution of the disturbance amplitude from small to the threshold valueA_T; this differs from the linear unstable system and is a nonlinearly unstable system. Zinn and Lieuwen [3] studied the characteristics of nonlinearities of thermoacoustic systems. However, the real system is always accompanied by many factors, and it is still very hard to analyze a thermoacoustic system using nonlinear methods, although it is mature and feasible to study the mechanisms and control methods of combustion instability through linear approaches. In this chapter, the second part will briefly introduce the one-dimensional thermoacoustic calculation methods, and the three-dimensional thermoacoustic derivation will be given in the third part. Some calculation results will be shown at the same time.

In the fourth part, some control methods of combustion instability will be presented, and conclusions will be summarized in the last section.

7.2 One-dimensional calculation method

The one-dimensional thermoacoustic calculation method is often easy to carry out and physically clear. Most investigations incorporate the Rijke tube. In this section, the method is briefly introduced based on the study of Dowling [9, 10] and Heckl [11].

Consider one-dimensional thermoacoustic oscillations in the geometry in Fig. 7.2; the heat source is located at $x = l$ within a length of Δ, which is small compared with the length of the tube. The density, pressure, and velocity can be decomposed into a time-mean property and a fluctuation property in the following forms:

$$\begin{aligned} p(x,t) &= \bar{p}(x) + p'(x,t), \\ u(x,t) &= \bar{u}(x) + u'(x,t), \\ \rho(x,t) &= \bar{\rho}(x) + \rho'(x,t). \end{aligned} \tag{7.2}$$

In a region added by heat, the density of the fluid is a function of two variables, i.e., the pressure and the specific entropy. Therefore,

$$\frac{D\rho}{Dt} = \frac{1}{c^2}\frac{Dp}{Dt} + \left.\frac{\partial\rho}{\partial s}\right|_p \frac{Ds}{Dt}, \tag{7.3}$$

where c is the speed of sound. The fluid is assumed to be inviscid, and the heat conduction effect is ignored, hence,

$$ds = \frac{\delta q_{rev}}{T}, \tag{7.4}$$

$$\rho ds = \frac{Q}{T}, \tag{7.5}$$

$$\rho T Ds/Dt = q(\mathbf{x}, t), \tag{7.6}$$

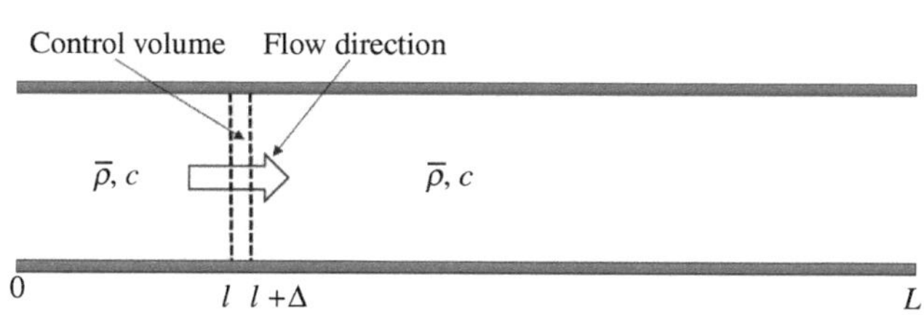

Fig. 7.2 Geometry of the Rijke tube.

where q is the heat addition per unit volume, and T is the absolute temperature. For perfect gas, $p = \rho RT$, $\left(\frac{\partial T}{\partial s}\right)_p = \frac{T}{c_p}$, $c_p = \frac{\gamma}{\gamma-1}R$, c_p is the specific heat at constant pressure and γ is the ratio of specific heats. Hence,

$$\left.\frac{\partial \rho}{\partial s}\right|_p = \frac{-\rho T(\gamma-1)}{c^2}. \tag{7.7}$$

Substituting Eqs. (7.4) and (7.5) into Eq. (7.3) yields:

$$\frac{D\rho}{Dt} = \frac{1}{c^2}\left(\frac{Dp}{Dt} - (\gamma-1)q\right). \tag{7.8}$$

For a linear perturbation region where there is no mean velocity and mean heat release, Eq. (7.8) can be linearized,

$$\frac{D\rho'}{Dt} = \frac{1}{\bar{c}^2}\left(\frac{\partial p'}{\partial t} - (\gamma-1)q'\right), \tag{7.9}$$

where the overbar denotes the mean property, and the prime denotes the fluctuation property.

The linearized mass equation and momentum equation are:

$$\frac{D\rho'}{Dt} + \bar{\rho}\frac{\partial u'}{\partial x} = 0, \tag{7.10}$$

$$\bar{\rho}\frac{\partial u'}{\partial t} + \frac{\partial p'}{\partial x} = 0, \tag{7.11}$$

respectively.

By differentiating Eq. (7.10) with respect to t, and Eq. (7.11) with respect to x, and subtracting, we obtain:

$$\frac{D^2\rho'}{Dt^2} - \frac{\partial^2 p'}{\partial x^2} = 0. \tag{7.12}$$

Considering Eqs. (7.9) and (7.12), an inhomogeneous wave equation is obtained:

$$\frac{1}{\bar{c}^2}\frac{\partial^2 p'}{\partial t^2} - \frac{\partial^2 p'}{\partial x^2} = \frac{\gamma-1}{\bar{c}^2}\frac{\partial q'}{\partial t}. \tag{7.13}$$

According to the wave equation, the acoustic field is determined by definition:

$$p'(\mathbf{x}, t) = \begin{cases} \left(Ae^{-i\omega(x-l)/\bar{c}_1} + Be^{i\omega(x-l)/\bar{c}_1}\right)e^{i\omega t}, & x < l \\ \left(Ce^{-i\omega(x-l)/\bar{c}_2} + De^{i\omega(x-l)/\bar{c}_2}\right)e^{i\omega t}, & x \geq l' \end{cases} \tag{7.14}$$

$$u'(x,t)=\begin{cases}\dfrac{1}{\bar{\rho}c}\left(Ae^{-i\omega(x-l)/\bar{c}_1}-Be^{i\omega(x-l)/\bar{c}_1}\right)e^{i\omega t}, & x<l\\ \dfrac{1}{\bar{\rho}c}\left(Ce^{-i\omega(x-l)/\bar{c}_2}-De^{i\omega(x-l)/\bar{c}_2}\right)e^{i\omega t}, & x\geq l\end{cases}, \tag{7.15}$$

where A, B and C, D are the complex amplitudes of the upstream and downstream traveling pressure waves in the tube. They are related by the conservation laws at the heat addition interface and the end boundary conditions.

Integration of Eq. (7.13) across the region $x=l$, we obtain:

$$p'_{x=l^+}=p'_{x=l^-}, \tag{7.16}$$

$$\left.\frac{1}{\bar{\rho}}\frac{\partial p'}{\partial x}\right|_{x=l^-}^{x=l^+}=-\frac{(\gamma-1)}{\gamma\bar{p}}\frac{Q}{S}\quad i.e.\quad u'|_{x=l^-}^{x=l^+}=-\frac{(\gamma-1)}{\gamma\bar{p}}\frac{Q}{S}. \tag{7.17}$$

Eq. (7.16) is the continuous pressure condition, and Eq. (7.17) is velocity jump condition at the interface, where Q is the heat addition and S is the cross area. For simplicity, only the fundamental mode is considered. Reflection of the fundamental mode occurs at the two ends of the tube by the coefficients R_0, R_L, respectively. Therefore,

$$Ae^{-i\omega(x-l)/\bar{c}_1}=R_0Be^{i\omega(x-l)/\bar{c}_1}, \tag{7.18}$$

$$De^{i\omega(x-l)/\bar{c}_2}=R_lCe^{-i\omega(x-l)/\bar{c}_2}. \tag{7.19}$$

The heat release function is quoted from Heckl [11],

$$Q=\beta e^{-i\omega\tau}(A-B)/\bar{\rho}c. \tag{7.20}$$

By substituting Q from Eq. (7.20), p' from Eq. (7.14), and u' from Eq. (7.15), then Eqs. (7.16)–(7.19) can be written as a homogeneous set of four equations for the unknowns A, B, C, D,

$$X(\omega)\begin{pmatrix}A\\B\\C\\D\end{pmatrix}=\mathbf{0}. \tag{7.21}$$

The equations have non-trivial solutions if and only if the determinant of the matrix $X(\omega)$ equals zero, that is to say, the disturbances can propagate only if the determinant of $X(\omega)$ vanishes. As a consequence, an eigenvalue problem is established, and the resonance frequency and the growth rate of the system can be calculated. The specific of the matrix can be easily obtained as:

$$X(\omega)=\begin{pmatrix} e^{i\omega l/c} & -R_0 e^{-i\omega l/c} & 0 & 0 \\ 0 & 0 & R_L e^{-i\omega(L-l)/c} & -e^{-i\omega(L-l)/c} \\ 1 & 1 & -1 & -1 \\ -1-\dfrac{(\gamma-1)\beta e^{-i\omega\tau}}{\bar{\rho}c^2 S} & 1+\dfrac{(\gamma-1)\beta e^{-i\omega\tau}}{\bar{\rho}c^2 S} & 1 & -1 \end{pmatrix}. \tag{7.22}$$

The eigenvalues are the solutions of $\det X(\omega) = 0$. With the specific time dependence $e^{i\omega t}$ of the fluctuation component, the stability of the Rijke tube can be determined by examining the imaginary part of the complex eigenfrequency.

Dowling [9] gave a systematic investigation of the various factors that have an influence on the one-dimensional thermoacoustic frequency oscillations. It is found that the form of coupling between the heat input and unsteady flow has a crucial effect on the frequency of oscillation. Actually, by a mean temperature ratio of six, there is about a 60% difference between the lowest frequency predicted for Case II (when the instantaneous rate of heat input is proportional to the instantaneous mass flow rate) and for Case I (when there is no unsteady rate of heat input). The mean flow effects can become significant even at modest inlet Mach numbers, and Dowling indicated a distributed heat input treated as a concentrated one requires not only $\omega d/\bar{c} \ll 1$ but also $\omega d/\bar{u} \ll 1$. It means the entropy wave will be affected by modest spatial distributed heat input. Then, the frequency of the thermoacoustic resonance is significantly altered when the entropy wave is coupled with the acoustic wave.

The one-dimensional calculation method is a prevalent choice to investigate the "reheat buzz," which is an instability phenomenon of the reheat system of a jet engine. The low-frequency instability is caused by the coupling of longitudinal acoustic waves and unsteady combustion [12]. In addition, the model combustor or real aeroengine is often modeled by some individual straight ducts that can be described by one-dimensional disturbances [13–15], as shown in Fig. 7.3. By relating the disturbances in two adjacent ducts through appropriate conservation laws, a closed set of equations is established. Thus, the thermoacoustic dynamics of the system can be analyzed in the preliminary design stage.

The one-dimensional linear calculation is able to analyze the mechanism and characteristic of combustion instability with the advantage of

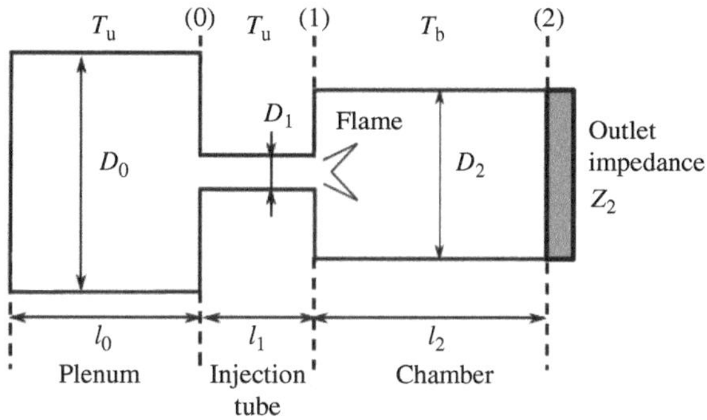

Fig. 7.3 Schematic view of a combustor modeled by three coupled cavities with an arbitrary outlet impedance [13].

convenience and fastness. However, it is deficient for treating more complex situations, for example, it is unable to solve the problem with non-uniform mean velocity or heat distribution, and it cannot solve the eigenvalue problems that involve complex geometries and soft wall boundary conditions. Also, the vorticity plays an important role in the issue of combustion instability, whereas the one-dimensional approach fails to do that. Therefore, researchers have devoted a lot of time and resources toward developing three-dimensional methods of combustion instability.

7.3 Three-dimensional linear combustion instability analysis method

7.3.1 Analytical approach

Although the numerical approach can solve the problem caused by oversimplification, it creates other problems. Actually, whether it is chosen or not depends on the tended object and available resources. For industrial applications, however, fast and reliable calculations for prediction are widely used, which have the following benefits: (1) less cost of both cost and time; (2) more direct physical understanding; and (3) promoting the development of numerical methods. Thus, aiming at the issue of combustion instability, it is necessary to develop a three-dimensional analytical method (AM) for prediction and prevention. This section is based on the work of You et al. [16, 17]. The formulation is created on the premise of uniform mean flow and uniform heat release.

7.3.1.1 Formulation

In the previous section, the basic equations that encompass the circumferential non-uniformities were established. For simplicity, the mean flow is assumed to be one-dimensional homogeneous. The three-dimensional linear equations in cylindrical coordinates of x, r, θ are:

$$\frac{\partial \rho'}{\partial t} + U\frac{\partial \rho'}{\partial x} + \rho_0\left(\frac{\partial u'}{\partial x} + \frac{\partial v'}{r\partial \theta} + \frac{\partial (rw')}{r\partial r}\right) = 0, \tag{7.23}$$

$$\frac{\partial u'}{\partial t} + U\frac{\partial u'}{\partial x} = -\frac{1}{\rho_0}\frac{\partial p'}{\partial x}, \tag{7.24}$$

$$\frac{\partial v'}{\partial t} + U\frac{\partial v'}{\partial x} = -\frac{1}{\rho_0}\frac{\partial p'}{r\partial \theta}, \tag{7.25}$$

$$\frac{\partial w'}{\partial t} + U\frac{\partial w'}{\partial x} = -\frac{1}{\rho_0}\frac{\partial p'}{\partial r}, \tag{7.26}$$

$$\frac{1}{P_0}\left(\frac{\partial p'}{\partial t} + U\frac{\partial p'}{\partial x}\right) - \frac{\gamma}{\rho_0}\left(\frac{\partial \rho'}{\partial t} + U\frac{\partial \rho'}{\partial x}\right) = 0, \tag{7.27}$$

where U is the mean flow velocity, and P_0, ρ_0 are the mean pressure and density, respectively. ρ', u', v', w', p' are the fluctuation quantity of the density, axial velocity, circumferential velocity, radial velocity, and pressure, respectively.

According to Eqs. (7.23) and (7.27), we have:

$$\frac{1}{P_0}\left(\frac{\partial p'}{\partial t} + U\frac{\partial p'}{\partial x}\right) + \gamma\left(\frac{\partial u'}{\partial x} + \frac{\partial v'}{r\partial \theta} + \frac{\partial (rw')}{r\partial r}\right) = 0. \tag{7.28}$$

By differentiating Eq. (7.28) with respect to t and x, we obtain:

$$\frac{1}{P_0}\left(\frac{\partial^2 p'}{\partial t^2} + U\frac{\partial^2 p'}{\partial x\partial t}\right) + \gamma\left(\frac{\partial^2 u'}{\partial x\partial t} + \frac{\partial^2 v'}{r\partial \theta\partial t} + \frac{\partial^2 (rw')}{r\partial r\partial t}\right) = 0, \tag{7.29}$$

$$\frac{1}{P_0}\left(\frac{\partial^2 p'}{\partial t\partial x} + U\frac{\partial^2 p'}{\partial x^2}\right) + \gamma\left(\frac{\partial^2 u'}{\partial x^2} + \frac{\partial^2 v'}{r\partial \theta\partial x} + \frac{\partial^2 (rw')}{r\partial r\partial x}\right) = 0. \tag{7.30}$$

Differentiate Eqs. (7.24), (7.25), and (7.26) with respect to x, r, θ, and add them to Eq. (7.26), then the following partial differential equation is achieved:

$$\begin{aligned}&\frac{U}{P_0\gamma}\left(\frac{\partial^2 p'}{\partial x\partial t} + U\frac{\partial^2 p'}{\partial x^2}\right) - \frac{1}{\rho_0}\left(\frac{\partial^2 p'}{\partial x^2} + \frac{\partial^2 p'}{r^2\partial \theta^2} + \frac{\partial^2 p'}{\partial r^2} + \frac{\partial p'}{r\partial r}\right)\\ &-\left(\frac{\partial^2 u'}{\partial t\partial x} + \frac{\partial^2 v'}{r\partial \theta\partial t} + \frac{\partial (rw')}{r\partial t\partial r}\right) = 0.\end{aligned} \tag{7.31}$$

Then, considered together with Eqs. (7.29) and (7.30), we have the final wave equation:

$$\left(M_a^2-1\right)\frac{\partial^2 p'}{\partial x^2}-\frac{\partial^2 p'}{r^2\partial\theta^2}-\frac{\partial^2 p'}{\partial r^2}-\frac{\partial p'}{r\partial r}+\frac{1}{C_0^2}\frac{\partial^2 p'}{\partial t^2}+\frac{2M_a}{C_0}\frac{\partial^2 p'}{\partial x\partial t}=0, \tag{7.32}$$

where M_a is the Mach number of the mean flow, and C_0 is the mean sound speed of the flow field.

According to the theory of mathematical physics, the general solutions of the wave equation can be written as:

$$p'(x,r,\theta,t)=P(x)\psi(r)H(\theta)e^{i\omega t}. \tag{7.33}$$

By substituting Eq. (7.33) into Eq. (7.32), and separating the variables, we have:

$$r^2\left(M_a^2-1\right)\frac{1}{P}\frac{\partial^2 P}{\partial x^2}-r^2\frac{1}{\psi}\frac{\partial^2\psi}{\partial r^2}-\frac{r}{\psi}\frac{\partial\psi}{\partial r}+\frac{2i\omega M_a r^2}{C_0}\frac{1}{P}\frac{\partial P}{\partial x}-\frac{r^2\omega^2}{C_0^2}=\frac{1}{H}\frac{\partial^2 H}{\partial\theta^2}, \tag{7.34}$$

which satisfies:

$$\frac{1}{H}\frac{\partial^2 H}{\partial\theta^2}=\zeta, \tag{7.35}$$

$$r^2\left(M_a^2-1\right)\frac{1}{P}\frac{\partial^2 P}{\partial x^2}-r^2\frac{1}{\psi}\frac{\partial^2\psi}{\partial r^2}-\frac{r}{\psi}\frac{\partial\psi}{\partial r}+\frac{2i\omega M_a r^2}{C_0}\frac{1}{P}\frac{\partial P}{\partial x}-\frac{r^2\omega^2}{C_0^2}=\zeta, \tag{7.36}$$

where ζ denotes a constant. The general solution of Eq. (7.35) is:

$$H=e^{im\theta}, m=\ldots,-2,-1,0,1,2,3,\ldots \tag{7.37}$$

The periodic boundary condition requires that m should be an integer. Therefore,

$$\zeta=-m^2. \tag{7.38}$$

Eq. (7.34) is then equivalent to:

$$\left(M_a^2-1\right)\frac{1}{P}\frac{\partial^2 P}{\partial x^2}-\frac{1}{\psi}\frac{\partial^2\psi}{\partial r^2}-\frac{1}{r\psi}\frac{\partial\psi}{\partial r}+\frac{2i\omega M_a}{C_0}\frac{1}{P}\frac{\partial P}{\partial x}-\frac{\omega^2}{C_0^2}+\frac{m^2}{r^2}=0. \tag{7.39}$$

Separating the variables again gives:

$$\left(M_a^2-1\right)\frac{1}{P}\frac{\partial^2 P}{\partial x^2}+\frac{2i\omega M_a}{a_0}\frac{1}{P}\frac{\partial P}{\partial x}-\frac{\omega^2}{C_0^2}=\frac{1}{\psi}\frac{\partial^2\psi}{\partial r^2}+\frac{1}{r\psi}\frac{\partial\psi}{\partial r}-\frac{m^2}{r^2}. \tag{7.40}$$

It is reasonable only if the following relation is satisfied:

$$\frac{1}{\psi}\frac{\partial^2\psi}{\partial r^2}+\frac{1}{r\psi}\frac{\partial\psi}{\partial r}-\frac{m^2}{r^2}=-\mu. \tag{7.41}$$

If $\mu < 0$, the general solutions of Eq. (7.41) are the virtual Bessel equations. However, these kinds of solutions are not significant because the homogenous boundary conditions are not satisfied in this situation. Hence, the situation of $\mu \geq 0$ is what we are going to consider.

If $\mu = 0$, to satisfy the natural boundary condition at the center of the cylinder, the general solutions of Eq. (7.41) are:

$$\psi(r)=\begin{cases} E, (m=0) \\ Er^m, (m=1,2,..) \end{cases}. \tag{7.42}$$

The solutions Er^m, $(m = 1,2,..)$ should be abandoned because of the homogeneous boundary condition. Therefore, $\psi(r) = E$, $(m = 0)$, as this is the solution of a plane wave. Then, it comes to the situation of $\mu > 0$. In this situation, the general solutions of Eq. (7.41) are:

$$\psi(r)=AJ_m(k_{mn}r)+BY_m(k_{mn}r), \tag{7.43}$$

where $\mu = k_{mn}^2, J_m(k_{mn}r)$, $Y_m(k_{mn}r)$ are m order Bessel function and $-m$ order Bessel function, respectively, and m, n represent the circumferential wave number and the radial wave number, respectively. For the natural boundary condition consideration, the general solutions are reduced to:

$$\psi(r)=AJ_m(k_{mn}r). \tag{7.44}$$

If the wall boundary condition is $\psi'(R) = 0$, then:

$$\frac{d}{dr}J_m(k_{mn}R)=0. \tag{7.45}$$

The previous equation determines the eigenvaluek_{mn}. When $m = 0$, zero order Bessel function has only one zero eigenvalue, which corresponds to the plane wave. In addition, the primary wave equation has only one unknown function $P(x)$ as follows:

$$\left(M_a^2-1\right)\frac{1}{P}\frac{\partial^2 P}{\partial x^2}+\frac{2i\omega M_a}{C_0}\frac{1}{P}\frac{\partial P}{\partial x}-\frac{\omega^2}{C_0^2}+k_{mn}^2=0. \tag{7.46}$$

As stated before, the disturbance is assumed to be harmonic for simplicity consideration, thus, $P(x)$ can be written as:

$$P(x)=P_{mn}e^{i\alpha x}, \tag{7.47}$$

where P_{mn} is the complex amplitude. Substituting Eq. (7.47) into Eq. (7.46) leads to:

$$-\left(M_a^2-1\right)\alpha^2-\frac{2\omega M_a}{C_0}\alpha-\left(\frac{\omega^2}{C_0^2}-k_{mn}^2\right)=0. \tag{7.48}$$

Therefore,

$$\alpha_{mn}^{+,-}=\frac{M_a^2\left(\frac{\omega}{C_0}\right)\pm\sqrt{\left(\frac{\omega}{C_0}\right)^2-\left(1-M_a^2\right)k_{mn}^2}}{1-M_a^2}, \tag{7.49}$$

where "+" denotes the downstream direction, and "−" denotes the upstream direction. Hence, the general solution is:

$$P(x)=P_{mn}^{+}e^{i\alpha_{mn}^{+}x}+P_{mn}^{-}e^{i\alpha_{mn}^{-}x}. \tag{7.50}$$

If the waves propagate at the position of x^j, hence,

$$P(x)=P_{mn}^{+}e^{i\alpha_{mn}^{+}(x-x^j)}+P_{mn}^{-}e^{i\alpha_{mn}^{-}(x-x^j)}. \tag{7.51}$$

Consequently, we get the general solution of Eq. (7.32),

$$p^j(x,\theta,r,t)=\sum_{m=-\infty}^{+\infty}\sum_{n=1}^{\infty}\left(P_{mn}^{+j}\psi_{mn}^{+j}(r)e^{i\alpha_{mn}^{+j}(x-x^j)}+P_{mn}^{-j}\psi_{mn}^{-j}(r)e^{i\alpha_{mn}^{-j}(x-x^j)}\right)e^{i(m\theta+\omega t)}. \tag{7.52}$$

The fluctuating velocity solution can be found by solving Eqs. (7.24), (7.25), and (7.26). Also, the velocity can be decomposed into two parts, the velocity induced by pressure fluctuation and the velocity induced by vorticity. Thus,

$$\begin{cases} u'=u_p+u_v \\ v'=v_p+v_v \\ w'=w_p+w_v \end{cases}. \tag{7.53}$$

As the fluctuating pressure is already obtained, according to Eq. (7.24), the first part of the axial velocity fluctuation component is:

$$u_p^j(x,\theta,r,t)=-\sum_{m=-\infty}^{+\infty}\sum_{n=1}^{\infty}\frac{1}{\rho_0^j}\left(\frac{\alpha_{mn}^{+j}P_{mn}^{+j}\psi_{mn}^{+j}(r)e^{i\alpha_{mn}^{+j}(x-x^j)}}{\omega+\alpha_{mn}^{+j}U^j}+\frac{\alpha_{mn}^{-j}P_{mn}^{-j}\psi_{mn}^{-j}(r)e^{i\alpha_{mn}^{-j}(x-x^j)}}{\omega+\alpha_{mn}^{-j}U^j}\right)e^{i(m\theta+\omega t)}, \tag{7.54}$$

which is the same as the other two components,

$$v_p^j(x,\theta,r,t) = -\sum_{m=-\infty}^{+\infty}\sum_{n=1}^{\infty}\frac{1}{\rho_0^j}\left(\frac{mP_{mn}^{+j}\psi_{mn}^{+j}(r)e^{i\alpha_{mn}^{+j}(x-x^j)}}{r\left(\omega+\alpha_{mn}^{+j}U^j\right)}+\frac{mP_{mn}^{-j}\psi_{mn}^{-j}(r)e^{i\alpha_{mn}^{-j}(x-x^j)}}{r\left(\omega+\alpha_{mn}^{-j}U^j\right)}\right)e^{i(m\theta+\omega t)}, \quad (7.55)$$

$$w_p^j(x,\theta,r,t) = \sum_{m=-\infty}^{+\infty}\sum_{n=1}^{\infty}\frac{1}{\rho_0^j}\left(\frac{ik_{mn}P_{mn}^{+j}\phi_{mn}^{+j}(z)e^{i\alpha_{mn}^{+j}(x-x^j)}}{\omega+\alpha_{mn}^{+j}U^j}+\frac{ik_{mn}P_{mn}^{-j}\phi_{mn}^{-j}(z)e^{i\alpha_{mn}^{-j}(x-x^j)}}{\omega+\alpha_{mn}^{-j}U^j}\right)e^{i(m\theta+\omega t)}, \quad (7.56)$$

where $z = k_{mn}r$, $\phi_{mn}^{+j}(z) = \frac{d\psi_{mn}^{+j}(r)}{dz}$, $\phi_{mn}^{-j}(z) = \frac{d\psi_{mn}^{-j}(r)}{dz}$.

On the other hand, we know $Tds = C_p dT + \frac{1}{\rho}dp$, which is equivalent to $d\rho = \frac{1}{c^2}dp - \frac{\rho}{C_p}ds$. From this relation, it can be concluded that the density fluctuation is induced by two factors, the pressure oscillation and the entropy oscillation. Obviously, the density fluctuation induced by pressure oscillation is $\rho_p = \frac{1}{C_0^2}p$, thereby,

$$\rho_p^j(x,\theta,r,t) = \frac{1}{C_0^2}\sum_{m=-\infty}^{+\infty}\sum_{n=1}^{\infty}\left(P_{mn}^{+j}\psi_{mn}^{+j}(r)e^{i\alpha_{mn}^{+j}(x-x^j)}+P_{mn}^{-j}\psi_{mn}^{-j}(r)e^{i\alpha_{mn}^{-j}(x-x^j)}\right)e^{i(m\theta+\omega t)}. \quad (7.57)$$

If the pressure oscillation is zero, Eqs. (7.24), (7.25), and (7.26) have different forms,

$$\frac{\partial u'}{\partial t} + U\frac{\partial u'}{\partial x} = 0, \quad (7.58)$$

$$\frac{\partial v'}{\partial t} + U\frac{\partial v'}{\partial x} = 0, \quad (7.59)$$

$$\frac{\partial w'}{\partial t} + U\frac{\partial w'}{\partial x} = 0. \quad (7.60)$$

The solution of these equations is the vorticity wave that is independent of the pressure wave. The general solution can be written as:

$$u_v^j = \sum_{m=-\infty}^{+\infty}\sum_{n=1}^{+\infty}\overline{u}_{vmn}^j\psi_{vmn}^j(z)e^{-i\frac{\omega}{U^j}(x-x^j)}e^{i(m\theta+\omega t)}, \quad (7.61)$$

$$v_v^j = \sum_{m=-\infty}^{+\infty}\sum_{n=1}^{+\infty}\overline{v}_{vmn}^j\psi_{vmn}^j(z)e^{-i\frac{\omega}{U^j}(x-x^j)}e^{i(m\theta+\omega t)}, \quad (7.62)$$

$$w_v^j = \sum_{m=-\infty}^{+\infty} \sum_{n=1}^{+\infty} \overline{w}_{vmn}^j \phi_{vmn}^j(z) e^{-i\frac{\omega}{U^j}(x-x^j)} e^{i(m\theta + \omega t)}, \tag{7.63}$$

where $\psi_{vmn}^j(z)$ and $\phi_{vmn}^j(z)$ are the corresponding eigenfunctions, respectively.

To obtain the entropy wave, we look back to Eq. (7.27), thus,

$$\frac{\partial s^j}{\partial t} + U\frac{\partial s^j}{\partial x} = 0. \tag{7.64}$$

Because of the adiabatic condition of the wall boundary, we have:

$$\left.\frac{\partial T}{\partial r}\right|_{r=R} = 0. \tag{7.65}$$

Therefore, the general solution of Eq. (7.64) can be written as:

$$s^j = \sum_{m=-\infty}^{+\infty} \sum_{n=1}^{\infty} S_{vmn}^j \psi_{vmn}^j(z) e^{-i\frac{\omega}{U^j}(x-x^j)} e^{i(m\theta + \omega t)}. \tag{7.66}$$

Thereby, we obtain the fluctuating density induced by the entropy oscillation $\rho_s = -\frac{\rho}{C_p} ds$,

$$\rho_s^j = -\frac{\rho_0}{C_p} \sum_{m=-\infty}^{+\infty} \sum_{n=1}^{\infty} S_{vmn}^j \psi_{vmn}^j(z) e^{-i\frac{\omega}{U^j}(x-x^j)} e^{i(m\theta + \omega t)}. \tag{7.67}$$

Finally, we have the fluctuating density in the form of,

$$\rho^j(x, \theta, r, t) = \frac{1}{C_0^2} \sum_{m=-\infty}^{+\infty} \sum_{n=1}^{\infty} \begin{pmatrix} P_{mn}^{+j} \psi_{mn}^{+j}(r) e^{i\alpha_{mn}^{+j}(x-x^j)} \\ +P_{mn}^{-j} \psi_{mn}^{-j}(r) e^{i\alpha_{mn}^{-j}(x-x^j)} \end{pmatrix} e^{i(m\theta + \omega t)}$$

$$-\frac{\rho_0}{C_p} \sum_{m=-\infty}^{+\infty} \sum_{n=1}^{\infty} S_{vmn}^j \psi_{vmn}^j(z) e^{-i\frac{\omega}{U^j}(x-x^j)} e^{i(m\theta + \omega t)}. \tag{7.68}$$

As the pressure, velocity, and density fluctuation quantities are obtained, the discontinuities of temperature or geometry in the flow field can be treated by matching relations. That is to say, the geometry can be divided into as many cells as needed, and the mean flow field can be regarded as approximately even in each cell, which is a good simplification. The different properties in adjacent cells are related by matching conditions, the conservation of mass, momentum, and energy.

As shown in Fig. 7.4, two adjacent cells are separated by a control volume where the discontinuity is located. The matching equations are established at the control volume.

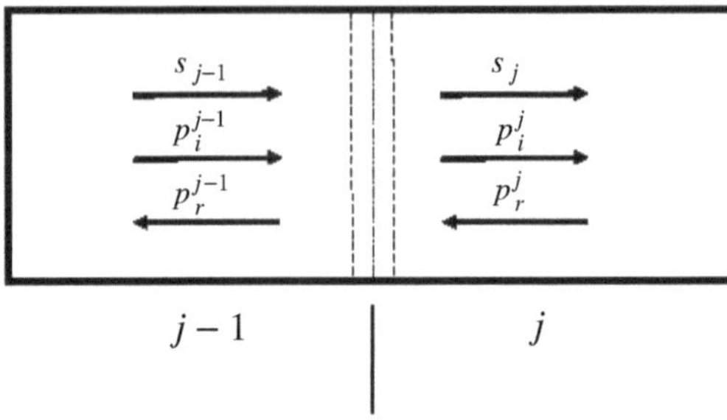

Fig. 7.4 Two adjacent cells.

According to the conservation of mass, for the control volume, it should satisfy,

$$\oint_s \rho \boldsymbol{v} \cdot d\boldsymbol{S} = -\frac{\partial}{\partial t}\int_V \rho d\tau. \tag{7.69}$$

The left-hand side of the equation is the surface integral of the control volume with the infinitesimal $d\boldsymbol{S}$, and the right-hand side of the equation is the volume integral, with the infinitesimal $d\tau$. By linearizing the integral equation, and if the sectional area is assumed to be constant, we obtain:

$$\left(\rho_0^j u^j + \rho^j U^j\right) - \left(\rho_0^{j+1} u^{j+1} + \rho^{j+1} U^{j+1}\right) = 0. \tag{7.70}$$

Substituting Eqs. (7.54) and (7.68) into Eq. (7.70) gives:

$$\begin{aligned}
\rho_0^j u^j + \rho^j U^j = & \sum_{m=-\infty}^{+\infty}\sum_{n=1}^{\infty} -\left(\frac{\alpha_{mn}^{+\,j} P_{mn}^{+\,j} \psi_{mn}^{+\,j}(r) e^{i\alpha_{mn}^{+\,j}(x-x^j)}}{\omega + \alpha_{mn}^{+\,j} U^j} + \frac{\alpha_{mn}^{-\,j} P_{mn}^{-\,j} \psi_{mn}^{-\,j}(r) e^{i\alpha_{mn}^{-\,j}(x-x^j)}}{\omega + \alpha_{mn}^{-\,j} U^j}\right) \\
& + \frac{1}{C_0^2}\sum_{m=-\infty}^{+\infty}\sum_{n=1}^{\infty}\Bigg[\Big(P_{mn}^{+\,j}\psi_{mn}^{+\,j}(r)e^{i\alpha_{mn}^{+\,j}(x-x^j)} \\
& + P_{mn}^{-\,j}\psi_{mn}^{-\,j}(r)e^{i\alpha_{mn}^{-\,j}(x-x^j)}\Big) - \frac{\rho_0}{C_p} S_{vmn}^j \psi_{vmn}^j(z) e^{-i\frac{\omega}{U^j}(x-x^j)}\Bigg] U^j \\
= & \sum_{m=-\infty}^{+\infty}\sum_{n=1}^{\infty} \left(\frac{U_j}{C_0^2} - \frac{\alpha_{mn}^{+\,j}}{\omega + \alpha_{mn}^{+\,j} U^j}\right) P_{mn}^{+\,j}\psi_{mn}^{+\,j}(r) e^{i\alpha_{mn}^{+\,j}(x-x^j)} \\
& + \left(\frac{U_j}{C_0^2} - \frac{\alpha_{mn}^{-\,j}}{\omega + \alpha_{mn}^{-\,j} U^j}\right) P_{mn}^{-\,j}\psi_{mn}^{-\,j}(r) e^{i\alpha_{mn}^{-\,j}(x-x^j)} \\
& + U^j \rho_{vmn}^j \psi_{vmn}^j(z) e^{-i\frac{\omega}{U^j}(x-x^j)},
\end{aligned} \tag{7.71}$$

$$\rho_0^j u^j + \rho^j U^j = \sum_{m=-\infty}^{+\infty}\sum_{n=1}^{\infty} \left(-\frac{\alpha_{mn}^{+j} P_{mn}^{+j} \psi_{mn}^{+j}(r) e^{i\alpha_{mn}^{+j}(x-x^j)}}{\omega + \alpha_{mn}^{+j} U^j} + \frac{\alpha_{mn}^{-j} P_{mn}^{-j} \psi_{mn}^{-j}(r) e^{i\alpha_{mn}^{-j}(x-x^j)}}{\omega + \alpha_{mn}^{-j} U^j} \right)$$

$$+ \sum_{m=-\infty}^{+\infty}\sum_{n=1}^{\infty} \left[\frac{1}{C_0^2} \left(P_{mn}^{+j} \psi_{mn}^{+j}(r) e^{i\alpha_{mn}^{+j}(x-x^j)} + P_{mn}^{-j} \psi_{mn}^{-j}(r) e^{i\alpha_{mn}^{-j}(x-x^j)} \right) - \frac{\rho_0}{C_p} S_{vmn}^j \psi_{vmn}^j(z) e^{-i\frac{\omega}{U^j}(x-x^j)} \right] U^j$$

$$= \sum_{m=-\infty}^{+\infty}\sum_{n=1}^{\infty} \left(\begin{array}{l} \left(\frac{U_j}{C_0^2} - \frac{\alpha_{mn}^{+j}}{\omega + \alpha_{mn}^{+j} U^j} \right) P_{mn}^{+j} \psi_{mn}^{+j}(r) e^{i\alpha_{mn}^{+j}(x-x^j)} \\ + \left(\frac{U_j}{C_0^2} - \frac{\alpha_{mn}^{-j}}{\omega + \alpha_{mn}^{-j} U^j} \right) P_{mn}^{-j} \psi_{mn}^{-j}(r) e^{i\alpha_{mn}^{-j}(x-x^j)} \end{array} \right) + U^j \rho_{vmn}^j \psi_{vmn}^j(z) e^{-i\frac{\omega}{U^j}(x-x^j)},$$

where $\rho_{vmn}{}^j = -\frac{\rho_0}{C_p} S_{vmn}^j$, which is the fluctuating density induced by the entropy fluctuation. Let $L^j = x - x^j$. Because every term contains the dependent $e^{i(m\theta+\omega t)}$, Eq. (7.71) is then reduced to:

$$\rho_0^j u^j + \rho^j U^j = \sum_{m=-\infty}^{+\infty}\sum_{n=1}^{\infty} \left(\frac{U^j}{C_0^2} - \frac{\alpha_{mn}^{+j}}{\omega + \alpha_{mn}^{+j} U^j} \right) P_{mn}^{+j} \psi_{mn}^{+j}(r) e^{i\alpha_{mn}^{+j} L^j} + \left(\frac{U_j}{C_0^2} - \frac{\alpha_{mn}^{-j}}{\omega + \alpha_{mn}^{-j} U^j} \right) P_{mn}^{-j} \psi_{mn}^{-j}(r) e^{i\alpha_{mn}^{-j} L^j} + U^j \rho_{vmn}^j \psi_{vmn}^j(z) e^{-i\frac{\omega}{U^j} L^j}. \tag{7.72}$$

Similarly,

$$\begin{aligned} &\rho_0^{j+1} u^{j+1} + \rho^{j+1} U^{j+1} \\ &= \sum_{m=-\infty}^{+\infty}\sum_{n=1}^{\infty} \left(\frac{U^{j+1}}{C_0^2} - \frac{\alpha_{mn}^{+j+1}}{\omega + \alpha_{mn}^{+j+1} U^{j+1}} \right) P_{mn}^{+j+1} \psi_{mn}^{+j+1}(r) \\ &+ \left(\frac{U^{j+1}}{C_0^2} - \frac{\alpha_{mn}^{-j+1}}{\omega + \alpha_{mn}^{-j+1} U^{j+1}} \right) P_{mn}^{-j+1} \psi_{mn}^{-j+1}(r) + U^{j+1} \rho_{vmn}^{j+1} \psi_{vmn}^{j+1}(z). \end{aligned} \tag{7.73}$$

In fact, for every circumferential mode, according to Eq. (7.70), we obtain,

$$\begin{aligned} &\sum_{n=1}^{\infty} [A_{mn}^{+j} \psi_{mn}^{+j}(r) P_{mn}^{+j} + A_{mn}^{-j} \psi_{mn}^{-j}(r) P_{mn}^{-j} + A_{vmn}^{+j} \psi_{vmn}^{+j}(r) \rho_{vmn}^{+j} \\ &+ A_{mn}^{+j+1} \psi_{mn}^{+j+1}(r) P_{mn}^{+j+1} + A_{mn}^{-j+1} \psi_{mn}^{-j+1}(r) P_{mn}^{-j+1} \\ &+ A_{vmn}^{+j+1} \psi_{vmn}^{+j+1}(r) \rho_{vmn}^{+j+1}] = 0. \end{aligned} \tag{7.74}$$

The coefficients are listed in Appendix A.
According to the momentum conservation of the control volume,

$$\int_{v}\frac{\partial\rho\boldsymbol{v}}{\partial t}d\tau+\oint_{s}\rho u_{n}\boldsymbol{v}d\boldsymbol{s}+\oint_{s}n_{u}\cdot p d\boldsymbol{s}=0. \tag{7.75}$$

Similarly, by taking the linearization and expansion, we obtain:

$$\left(2U^{j}\rho_{0}^{j}u^{j}+U^{j^{2}}\rho^{j}\right)-\left(2U^{j+1}\rho_{0}^{j+1}u^{j+1}+U^{(j+1)^{2}}\rho^{j+1}\right)=p^{j+1}-p^{j}. \tag{7.76}$$

Substituting Eqs. (7.54) and (7.68) into Eq. (7.76) leads to:

$$\begin{aligned}&\sum_{n=1}^{\infty}[B_{mn}^{+j}\psi_{mn}^{+j}(r)P_{mn}^{+j}+B_{mn}^{-j}\psi_{mn}^{-j}(r)P_{mn}^{-j}+B_{vmn}^{+j}\psi_{vmn}^{+j}(r)\rho_{vmn}^{+j}\\&+B_{mn}^{+j+1}\psi_{mn}^{+j+1}(r)P_{mn}^{+j+1}+B_{mn}^{-j+1}\psi_{mn}^{-j+1}(r)P_{mn}^{-j+1}\\&+B_{vmn}^{+j+1}\psi_{vmn}^{+j+1}(r)\rho_{vmn}^{+j+1}]=0.\end{aligned} \tag{7.77}$$

The coefficients are presented in Appendix A.

For the conservation of energy, the following relation is to be satisfied,

$$\int_{v}\frac{\partial}{\partial t}\rho\left(e+\frac{u^{2}}{2}\right)d\tau+\oint_{s}\left(e+\frac{p}{\rho}+\frac{u^{2}}{2}\right)\rho u ds=0. \tag{7.78}$$

Similarly, we can have the following form equation,

$$\begin{aligned}&U^{j}\left(\frac{\gamma}{\gamma-1}+\frac{\left(M_{a}^{j}\right)^{2}}{2}\right)p^{j}+\left(\frac{\gamma P_{0}^{j}}{\gamma-1}+\frac{3}{2}\rho_{0}^{j}\left(U^{j}\right)^{2}\right)u^{j}\\&-\frac{\rho_{0}^{j}\left(U^{j}\right)^{3}}{2C_{p}}s^{j}=U^{j+1}\left(\frac{\gamma}{\gamma-1}+\frac{\left(M_{a}^{j+1}\right)^{2}}{2}\right)p^{j+1}\\&+\left(\frac{\gamma P_{0}^{j+1}}{\gamma-1}+\frac{3}{2}\rho_{0}^{j+1}\left(U^{j+1}\right)^{2}\right)u^{j+1}-\frac{\rho_{0}^{j+1}\left(U^{j+1}\right)^{3}}{2C_{p}}s^{j+1}-Q_{h}^{j},\end{aligned} \tag{7.79}$$

where Q_{h}^{j} is the heat release at the interface, and $s^{j}=\frac{C_{p}}{\rho_{0}}\frac{1}{c_{0}^{2}}p^{j}-\frac{C_{p}}{\rho_{0}}\rho^{j}$. Substituting the fluctuation quantities into this equation results in:

$$\begin{aligned}&\sum_{n=1}^{\infty}[C_{mn}^{+j}\psi_{mn}^{+j}(r)P_{mn}^{+j}+C_{mn}^{-j}\psi_{mn}^{-j}(r)P_{mn}^{-j}+C_{vmn}^{+j}\psi_{vmn}^{+j}(r)\rho_{vmn}^{+j}\\&+C_{mn}^{+j+1}\psi_{mn}^{+j+1}(r)P_{mn}^{+j+1}+C_{mn}^{-j+1}\psi_{mn}^{-j+1}(r)P_{mn}^{-j+1}\\&+C_{vmn}^{+j+1}\psi_{vmn}^{+j+1}(r)\rho_{vmn}^{+j+1}]=Q_{h}^{j}.\end{aligned} \tag{7.80}$$

The coefficients are also listed in Appendix A. To close the equations, the boundary conditions and heat release functions should be given. Here, the inlet and the outlet boundary conditions are given as impedance forms:

$$\frac{p^j}{u^j} = z_0, \frac{p^{j+1}}{u^{j+1}} = z_{end}, \tag{7.81}$$

where z_0 and z_{end} are the impedance of the inlet and outlet of the combustor, respectively. The two boundary conditions can also be expanded as:

$$\begin{aligned}\sum_{n=1}^{\infty} [&D_{mn}^{+j}\psi_{mn}^{+j}(r)P_{mn}^{+j} + D_{mn}^{-j}\psi_{mn}^{-j}(r)P_{mn}^{-j} + D_{vmn}^{+j}\psi_{vmn}^{+j}(r)\rho_{vmn}^{+j} \\ &+D_{mn}^{+j+1}\psi_{mn}^{+j+1}(r)P_{mn}^{+j+1} + D_{mn}^{-j+1}\psi_{mn}^{-j+1}(r)P_{mn}^{-j+1} \\ &+D_{vmn}^{+j+1}\psi_{vmn}^{+j+1}(r)\rho_{vmn}^{+j+1}] = 0,\end{aligned} \tag{7.82}$$

$$\begin{aligned}\sum_{n=1}^{\infty} [&E_{mn}^{+j}\psi_{mn}^{+j}(r)P_{mn}^{+j} + E_{mn}^{-j}\psi_{mn}^{-j}(r)P_{mn}^{-j} + E_{vmn}^{+j}\psi_{vmn}^{+j}(r)\rho_{vmn}^{+j} \\ &+E_{mn}^{+j+1}\psi_{mn}^{+j+1}(r)P_{mn}^{+j+1} + E_{mn}^{-j+1}\psi_{mn}^{-j+1}(r)P_{mn}^{-j+1} \\ &+E_{vmn}^{+j+1}\psi_{vmn}^{+j+1}(r)\rho_{vmn}^{+j+1}] = 0.\end{aligned} \tag{7.83}$$

The unsteady heat release is written as a function of pressure oscillation and velocity oscillation,

$$Q_h^j = R_p p^j + R_v u^j, \tag{7.84}$$

which is similarly expanded as:

$$\begin{aligned}\sum_{n=1}^{\infty} [&F_{mn}^{+j}\psi_{mn}^{+j}(r)P_{mn}^{+j} + F_{mn}^{-j}\psi_{mn}^{-j}(r)P_{mn}^{-j} + F_{vmn}^{+j}\psi_{vmn}^{+j}(r)\rho_{vmn}^{+j} \\ &+F_{mn}^{+j+1}\psi_{mn}^{+j+1}(r)P_{mn}^{+j+1} + F_{mn}^{-j+1}\psi_{mn}^{-j+1}(r)P_{mn}^{-j+1} \\ &+F_{vmn}^{+j+1}\psi_{vmn}^{+j+1}(r)\rho_{vmn}^{+j+1}] = Q_h^j.\end{aligned} \tag{7.85}$$

The coefficients D_{mn}, E_{mn}, F_{mn} are listed in Appendix A.

In summary, three conservation equations are established on every interface without heat release when the cooling air and dissipation of the wall are ignored. The two boundary conditions need to be satisfied at the inlet and outlet of the combustor. Therefore, when one more interface is added for consideration, three new unknowns are produced, and three more conservation equations of the new unknowns of P^+, P^-, ρ^+ are established. For the interface added with heat release, apart from the three conservation equations, the heat release function equation is formed. Hence, four more equations with four more unknowns can be established. This also implies that, no matter how many interfaces are concerned, the equations are closed.

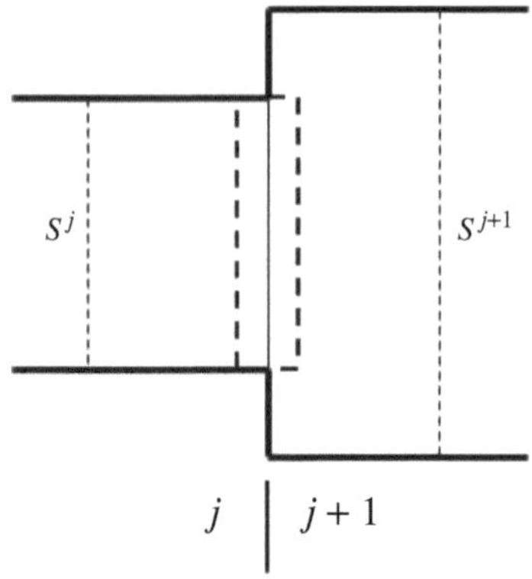

Fig. 7.5 Break section.

7.3.1.2 Solution to variable cross-section cases

In the previous model, the variation of cross-section of the tube is ignored, thus, it is only valid for the straight tube. However, in practical use, few combustors are designed without cross-section variation. To update the model, we use the method developed by Alfredson [18], which is initially applied to study the characteristic of acoustics propagation in a variable cross-section.

As shown in Fig. 7.5, the cross-section of the right side is larger than that of the left side. By applying the conservation of mass and linearizing it, we obtain:

$$\left(\rho_0^j u^j + \rho^j U^j\right) S^j - \left(\rho_0^{j+1} u^{j+1} + \rho^{j+1} U^{j+1}\right) S^{j+1} = 0, \tag{7.86}$$

where S^j is the area of cross-section of the left straight tube, and S^{j+1} is the area of that of the right straight tube. Because of the different areas, the radial eigenfunction $\psi^j_{mn}(r)$ in Eq. (7.72) and $\psi^{j+1}_{mn}(r)$ in Eq. (7.73) are not the same. Substitute the fluctuation quantities into Eq. (7.86), we get,

$$\begin{aligned}
&\sum_{n=1}^{\infty}\left[A^{+j}_{mn}\psi^{+j}_{mn}(r)P^{+j}_{mn} + A^{-j}_{mn}\psi^{-j}_{mn}(r)P^{-j}_{mn} + A^{+j}_{vmn}\psi^{+j}_{vmn}(r)\rho^{+j}_{vmn}\right. \\
&+A^{+j+1}_{mn}\psi^{+j+1}_{mn}(r)P^{+j+1}_{mn} + A^{-j+1}_{mn}\psi^{-j+1}_{mn}(r)P^{-j+1}_{mn} \\
&\left.+A^{+j+1}_{vmn}\psi^{+j+1}_{vmn}(r)\rho^{+j+1}_{vmn}\right] = 0.
\end{aligned} \tag{7.87}$$

Although Eq. (7.87) is similar to Eq. (7.74), the difference between $\psi^j_{mn}(r)$ and $\psi^{j+1}_{mn}(r)$ makes them different. As a result, we cannot separate the radial mode from each other as before. They are tangled with each other. Without consideration of the circumferential modes, Alfredson presented an approach to figure it out. For example, if $m = 0$, multiply Eq. (7.87) by r, and integrate it from 0 to r_j that is the radius of the left tube, then we have:

$$\begin{aligned}
&A_{01}^{+j} P_{01}^{+j} \int_0^{r_j} r dr + A_{01}^{-j} P_{01}^{-j} \int_0^{r_j} r dr + A_{v01}^{+j} \rho_{v01}^{+j} \int_0^{r_j} r dr \\
&+ \sum_{n=2}^{N} \left[A_{mn}^{+j} P_{mn}^{+j} \int_0^{r_j} r\psi_{mn}^{j}(r) dr + A_{mn}^{-j} P_{mn}^{-j} \int_0^{r_j} r\psi_{mn}^{j}(r) dr + A_{vmn}^{+j} \rho_{vmn}^{+j} \int_0^{r_j} r\psi_{mn}^{j}(r) dr \right] \\
&+ A_{01}^{+j+1} P_{01}^{+j+1} \int_0^{r_j} r dr + A_{01}^{-j+1} P_{01}^{-j+1} \int_0^{r_j} r dr + A_{v01}^{+j+1} \rho_{v01}^{+j+1} \int_0^{r_j} r dr \\
&+ \sum_{n=2}^{N} [A_{mn}^{+j+1} P_{mn}^{+j+1} \int_0^{r_j} r\psi_{mn}^{j+1}(r) dr + A_{mn}^{-j+1} P_{mn}^{-j+1} \int_0^{r_j} r\psi_{mn}^{j+1}(r) dr \\
&+ A_{vmn}^{+j+1} \rho_{vmn}^{+j+1} \int_0^{r_j} r\psi_{mn}^{j+1}(r) dr] = 0.
\end{aligned} \tag{7.88}$$

Because in the range of (r_j, r_{j+1}) in the control volume, $u = 0$. Therefore, $\rho_0^{j+1} u^{j+1} = 0$. Multiplying it by r and then integrating it from r_j to r_{j+1} gives:

$$\begin{aligned}
&\frac{\alpha_{01}^{+j+1} P_{01}^{+j+1}}{\omega + \alpha_{mn}^{+j+1} U^{j+1}} \int_{r_j}^{r_{j+1}} r dr + \frac{\alpha_{01}^{-j+1} P_{01}^{-j+1}}{\omega + \alpha_{mn}^{-j+1} U^{j+1}} \int_{r_j}^{r_{j+1}} r dr \\
&+ \sum_{n=2}^{\infty} \left(\frac{\alpha_{mn}^{+j+1} P_{mn}^{+j+1} \int_{r_j}^{r_{j+1}} r\psi_{mn}^{j+1}(r) dr}{\omega + \alpha_{mn}^{+j+1} U^{j+1}} + \frac{\alpha_{mn}^{-j+1} P_{mn}^{-j+1} \int_{r_j}^{r_{j+1}} r\psi_{mn}^{j+1}(r) dr}{\omega + \alpha_{mn}^{-j+1} U^{j+1}} \right) = 0,
\end{aligned} \tag{7.89}$$

Add Eqs. (7.88) and (7.89) together, and we get:

$$\begin{aligned}
&Ab_{101}^{+j} P_{01}^{+j} + Ab_{101}^{-j} P_{01}^{-j} + Ab_{v101}^{+j} \rho_{v01}^{+j} \\
&+ \sum_{n=2}^{N} [Ab_{1mn}^{+j} P_{0n}^{+j} + Ab_{1mn}^{-j} P_{0n}^{-j} + Ab_{v1mn}^{+j} \rho_{v0n}^{+j}] \\
&+ Ab_{101}^{+j+1} P_{01}^{+j+1} + Ab_{101}^{-j+1} P_{01}^{-j+1} + Ab_{v101}^{+j+1} \rho_{v01}^{+j+1} \\
&+ \sum_{n=2}^{N} \left[Ab_{1mn}^{+j+1} P_{0n}^{+j+1} + Ab_{1mn}^{-j+1} P_{0n}^{-j+1} + Ab_{v1mn}^{+j+1} \rho_{vmn}^{+j+1} \right] = 0.
\end{aligned} \tag{7.90}$$

where Ab_{10n} is an expression including A and integration of ψ. "1" in the subscript represents the first kind of integration, distinguishing it from the other kind to be presented next. "n" means the radial mode number.

The circumferential mode number is zero. The specific coefficients are shown in Appendix B.

Similarly, for the momentum integral,

$$\begin{aligned}&\left(2U^j\rho_0^j u^j + U^{j^2}\rho^j\right)S^j - \left(2U^{j+1}\rho_0^{j+1}u^{j+1} + U^{(j+1)^2}\rho^{j+1}\right)S^j \\ &= p^{j+1}S^j - p^j S^j.\end{aligned} \tag{7.91}$$

It can be written as the same form with Eq. (7.90) as:

$$\begin{aligned}&\sum_{n=1}^{\infty}\left[B_{mn}^{+\,j}\psi_{mn}^{+\,j}(r)P_{mn}^{+\,j} + B_{mn}^{-\,j}\psi_{mn}^{-\,j}(r)P_{mn}^{-\,j} + B_{vmn}^{+\,j}\psi_{vmn}^{+\,j}(r)\rho_{vmn}^{+\,j}\right. \\ &+B_{mn}^{+\,j+1}\psi_{mn}^{+\,j+1}(r)P_{mn}^{+\,j+1} + B_{mn}^{-\,j+1}\psi_{mn}^{-\,j+1}(r)P_{mn}^{-\,j+1} \\ &\left.+B_{vmn}^{+\,j+1}\psi_{vmn}^{+\,j+1}(r)\rho_{vmn}^{+\,j+1}\right] = 0.\end{aligned} \tag{7.92}$$

By the same token, the circumferential mode number $m = 0$, when $n = 1$, $K_{mn} = 0$, then $\psi_{01}^{+j}(r) = 1$. Thus, if Eq. (7.92) is multiplied by r, and integrated from 0 to r_j, we obtain,

$$\begin{aligned}&B_{01}^{+\,j}P_{01}^{+\,j}\int_0^{r_j} r\,dr + B_{01}^{-\,j}P_{01}^{-\,j}\int_0^{r_j} r\,dr + B_{v01}^{+\,j}\rho_{v01}^{+\,j}\int_0^{r_j} r\,dr \\ &+\sum_{n=2}^{N}\left[\begin{matrix}B_{mn}^{+\,j}P_{mn}^{+\,j}\displaystyle\int_0^{r_j} r\psi_{mn}^{j}(r)\,dr + B_{mn}^{-\,j}P_{mn}^{-\,j}\displaystyle\int_0^{r_j} r\psi_{mn}^{j}(r)\,dr \\ +B_{vmn}^{+\,j}\rho_{vmn}^{+\,j}\displaystyle\int_0^{r_j} r\psi_{mn}^{j}(r)\,dr\end{matrix}\right] \\ &+B_{01}^{+\,j+1}P_{01}^{+\,j+1}\int_0^{r_j} r\,dr + B_{01}^{-\,j+1}P_{01}^{-\,j+1}\int_0^{r_j} r\,dr + B_{v01}^{+\,j+1}\rho_{v01}^{+\,j+1}\int_0^{r_j} r\,dr \\ &+\sum_{n=2}^{N}\left[\begin{matrix}B_{mn}^{+\,j+1}P_{mn}^{+\,j+1}\displaystyle\int_0^{r_j} r\psi_{mn}^{j+1}(r)\,dr + B_{mn}^{-\,j+1}P_{mn}^{-\,j+1}\displaystyle\int_0^{r_j} r\psi_{mn}^{j+1}(r)\,dr \\ +B_{vmn}^{+\,j+1}\rho_{vmn}^{+\,j+1}\displaystyle\int_0^{r_j} r\psi_{mn}^{j+1}(r)\,dr\end{matrix}\right] \\ &= 0.\end{aligned} \tag{7.93}$$

Because the axial velocity in the range of (r_j, r_{j+1}) is zero, therefore, $2U^{j+1}\rho_0^{j+1}u^{j+1} = 0$. Multiplying it by r and integrating it from r_j to r_{j+1} results in:

$$2U^{j+1}\sum_{n=1}^{\infty}\left[\begin{matrix}\dfrac{\alpha_{mn}^{+\,j+1}P_{mn}^{+\,j+1}\displaystyle\int_{r^j}^{r^{j+1}} r\psi_{mn}^{j+1}(r)\,dr}{\omega + \alpha_{mn}^{+\,j+1}U^{j+1}} \\ +\dfrac{\alpha_{mn}^{-\,j+1}P_{mn}^{-\,j+1}\displaystyle\int_{r^j}^{r^{j+1}} r\psi_{mn}^{j+1}(r)\,dr}{\omega + \alpha_{mn}^{-\,j+1}U^{j+1}}\end{matrix}\right] = 0. \tag{7.94}$$

Adding Eq. (7.93) to Eq. (7.94), we obtain,

$$\begin{aligned}
&Bb_{101}^{+j}P_{01}^{+j} + Bb_{101}^{-j}P_{01}^{-j} + Bb_{\nu101}^{+j}\rho_{\nu01}^{+j} \\
&+\sum_{n=2}^{N}\left[Bb_{1mn}^{+j}P_{mn}^{+j} + Bb_{1mn}^{-j}P_{mn}^{-j} + Bb_{\nu1mn}^{+j}\rho_{\nu mn}^{+j}\right] \\
&+Bb_{101}^{+j+1}P_{01}^{+j+1} + Bb_{101}^{-j+1}P_{01}^{-j+1} + Bb_{\nu101}^{+j+1}\rho_{\nu01}^{+j+1} \\
&+\sum_{n=2}^{N}\left[Bb_{1mn}^{+j+1}P_{mn}^{+j+1} + Bb_{1mn}^{-j+1}P_{mn}^{-j+1} + Bb_{\nu1mn}^{+j+1}\rho_{\nu mn}^{+j+1}\right] = 0,
\end{aligned} \tag{7.95}$$

The explanation about the subscripts in Bb_{101}^{j} is the same with that in Ab_{10n}. When Eq. (7.92) is multiplied by $r\psi_{0\nu}^{+j}(r)_{(}\nu \neq 1_{)}$ and integrated from 0 to r_j, we have:

$$\begin{aligned}
&B_{01}^{+j}P_{01}^{+j}\int_0^{r_j} r\psi_{m\nu}^{j}(r)dr + B_{01}^{-j}P_{01}^{-j}\int_0^{r_j} r\psi_{m\nu}^{j}(r)dr \\
&+B_{\nu01}^{+j}\rho_{\nu01}^{+j}\int_0^{r_j} r\psi_{m\nu}^{j}(r)dr \\
&+\sum_{n=2}^{N}\left[\begin{array}{l}
B_{mn}^{+j}P_{mn}^{+j}\int_0^{r_j} r\psi_{mn}^{j}(r)\psi_{m\nu}^{j}(r)dr \\
+B_{mn}^{-j}P_{mn}^{-j}\int_0^{r_j} r\psi_{mn}^{j}(r)\psi_{m\nu}^{j}(r)dr \\
+B_{\nu mn}^{+j}\rho_{\nu mn}^{+j}\int_0^{r_j} r\psi_{mn}^{j}\psi_{m\nu}^{j}(r)(r)dr
\end{array}\right] \\
&+B_{01}^{+j+1}P_{01}^{+j+1}\int_0^{r_j} r\psi_{m\nu}^{j}(r)dr + B_{01}^{-j+1}P_{01}^{-j+1}\int_0^{r_j} r\psi_{m\nu}^{j}(r)dr \\
&+B_{\nu01}^{+j+1}\rho_{\nu01}^{+j+1}\int_0^{r_j} r\psi_{m\nu}^{j}(r)dr \\
&+\sum_{n=2}^{N}\left[\begin{array}{l}
B_{mn}^{+j+1}P_{mn}^{+j+1}\int_0^{r_j} r\psi_{mn}^{j+1}(r)\psi_{m\nu}^{j}(r)dr \\
+B_{mn}^{-j+1}P_{mn}^{-j+1}\int_0^{r_j} r\psi_{mn}^{j+1}(r)\psi_{m\nu}^{j}(r)dr \\
+B_{\nu mn}^{+j+1}\rho_{\nu mn}^{+j+1}\int_0^{r_j} r\psi_{mn}^{j+1}(r)\psi_{m\nu}^{j}(r)dr
\end{array}\right] = 0,
\end{aligned} \tag{7.96}$$

In the range of (r_j, r_{j+1}) in the control volume, $u = 0$. Therefore, $2U^{j+1}\rho_0^{j+1}u^{j+1} = 0$. Multiplying it by $r\psi_{0\nu}^{+j}(r)$ and then integrating it from r_j to r_{j+1}, and then combined with Eq. (7.93), we obtain:

$$\begin{aligned}
&Bb_{201}^{+\,j}P_{01}^{+\,j} + Bb_{201}^{-\,j}P_{01}^{-\,j} + Bb_{\nu 201}^{+\,j}\rho_{\nu 01}^{+\,j} \\
&+\sum_{n=2}^{N}\left[Bb_{2mn}^{+\,j}P_{mn}^{+\,j} + Bb_{2mn}^{-\,j}P_{mn}^{-\,j} + Bb_{\nu 2mn}^{+\,j}\rho_{\nu mn}^{+\,j}\right] \\
&+Bb_{201}^{+\,j+1}P_{01}^{+\,j+1} + Bb_{201}^{-\,j+1}P_{01}^{-\,j+1} + Bb_{\nu 201}^{+\,j+1}\rho_{\nu 01}^{+\,j+1} \\
&+\sum_{n=2}^{N}\left[Bb_{2mn}^{+\,j+1}P_{mn}^{+\,j+1} + Bb_{2mn}^{-\,j+1}P_{mn}^{-\,j+1} + Bb_{\nu 2mn}^{+\,j+1}\rho_{\nu mn}^{+\,j+1}\right] = 0
\end{aligned} \tag{7.97}$$

The coefficients are presented in Appendix B.

For energy conservation, we have,

$$\begin{aligned}
&U^j\left(\frac{\gamma}{\gamma-1} + \frac{\left(M_a^j\right)^2}{2}\right)p^j + \left(\frac{\gamma P_0^j}{\gamma-1} + \frac{3}{2}\rho_0^j\left(U^j\right)^2\right)u^j - \frac{\rho_0^j\left(U^j\right)^3}{2C_p}s^j \\
&= U^{j+1}\left(\frac{\gamma}{\gamma-1} + \frac{\left(M_a^{j+1}\right)^2}{2}\right)p^{j+1} + \left(\frac{\gamma P_0^{j+1}}{\gamma-1} + \frac{3}{2}\rho_0^{j+1}\left(U^{j+1}\right)^2\right)u^{j+1} \\
&\quad -\frac{\rho_0^{j+1}\left(U^{j+1}\right)^3}{2C_p}s^{j+1}.
\end{aligned} \tag{7.98}$$

Substituting the fluctuation quantities into Eq. (7.98) leads to:

$$\sum_{n=1}^{\infty}\begin{bmatrix}
C_{mn}^{+\,j}\psi_{mn}^{+\,j}(r)P_{mn}^{+\,j} + C_{mn}^{-\,j}\psi_{mn}^{-\,j}(r)P_{mn}^{-\,j} \\
+C_{\nu mn}^{+\,j}\psi_{\nu mn}^{+\,j}(r)\rho_{\nu mn}^{+\,j} \\
+C_{mn}^{+\,j+1}\psi_{mn}^{+\,j+1}(r)P_{mn}^{+\,j+1} + C_{mn}^{-\,j+1}\psi_{mn}^{-\,j+1}(r)P_{mn}^{-\,j+1} \\
+C_{\nu mn}^{+\,j+1}\psi_{\nu mn}^{+\,j+1}(r)\rho_{\nu mn}^{+\,j+1}
\end{bmatrix} = Q_h^j. \tag{7.99}$$

By the same token, the circumferential mode number $m = 0$, when $n = 1$, $K_{mn} = 0$, then $\psi_{01}^{+j}(r) = 1$. Thus, if Eq. (7.99) is multiplied by r, and integrated from 0 to r_j, we obtain:

$$
\begin{aligned}
&C_{01}^{+\,j}P_{01}^{+\,j}\int_0^{r_j} rdr + C_{01}^{-\,j}P_{01}^{-\,j}\int_0^{r_j} rdr + C_{\nu 01}^{+\,j}\rho_{\nu 01}^{+\,j}\int_0^{r_j} rdr\\
&+\sum_{n=2}^{N}\left[C_{mn}^{+\,j}P_{mn}^{+\,j}\int_0^{r_j} r\psi_{mn}^{j}(r)dr + C_{mn}^{-\,j}P_{mn}^{-\,j}\int_0^{r_j} r\psi_{mn}^{j}(r)dr + C_{\nu mn}^{+\,j}\rho_{\nu mn}^{+\,j}\int_0^{r_j} r\psi_{mn}^{j}(r)dr\right]\\
&+C_{01}^{+\,j+1}P_{01}^{+\,j+1}\int_0^{r_j} rdr + C_{01}^{-\,j+1}P_{01}^{-\,j+1}\int_0^{r_j} rdr\\
&+C_{\nu 01}^{+\,j+1}\rho_{\nu 01}^{+\,j+1}\int_0^{r_j} rdr\\
&+\sum_{n=2}^{N}[C_{mn}^{+\,j+1}P_{mn}^{+\,j+1}\int_0^{r_j} r\psi_{mn}^{j+1}(r)dr + C_{mn}^{-\,j+1}P_{mn}^{-\,j+1}\int_0^{r_j} r\psi_{mn}^{j+1}(r)dr\\
&+C_{\nu mn}^{+\,j+1}\rho_{\nu mn}^{+\,j+1}\int_0^{r_j} r\psi_{mn}^{j+1}(r)dr] = 0.
\end{aligned}
\tag{7.100}
$$

Because the axial velocity in the range of (r_j, r_{j+1}) is zero, therefore, $\left(\frac{\gamma P_0^{j+1}}{\gamma-1} + \frac{3}{2}\rho_0^{j+1}U^{j+1^2}\right)\rho_0^{j+1}u^{j+1} = 0$, multiplying it by r, and integrating it from r_j to r_{j+1}gives:

$$
\left(\frac{\gamma P_0^{j+1}}{\gamma-1} + \frac{3}{2}\rho_0^{j+1}U^{j+1^2}\right)\sum_{n=1}^{\infty}\left[\frac{\alpha_{mn}^{+\,j+1}P_{mn}^{+\,j+1}\int_{r^j}^{r^{j+1}} r\psi_{mn}^{j+1}(r)dr}{\omega+\alpha_{mn}^{+\,j+1}U^{j+1}} + \frac{\alpha_{mn}^{-\,j+1}P_{mn}^{-\,j+1}\int_{r^j}^{r^{j+1}} r\psi_{mn}^{j+1}(r)dr}{\omega+\alpha_{mn}^{-\,j+1}U^{j+1}}\right] = 0.
\tag{7.101}
$$

Adding Eqs. (7.100) and (7.101) up, we have:

$$
\begin{aligned}
&Cb_{101}^{+\,j}P_{01}^{+\,j} + Cb_{101}^{-\,j}P_{01}^{-\,j} + Cb_{\nu 101}^{+\,j}\rho_{\nu 01}^{+\,j}\\
&+\sum_{n=2}^{N}\left[Cb_{1mn}^{+\,j}P_{mn}^{+\,j} + Cb_{1mn}^{-\,j}P_{mn}^{-\,j} + Cb_{\nu 1mn}^{+\,j}\rho_{\nu mn}^{+\,j}\right]\\
&+Cb_{101}^{+\,j+1}P_{01}^{+\,j+1} + Cb_{101}^{-\,j+1}P_{01}^{-\,j+1} + Cb_{\nu 101}^{+\,j+1}\rho_{\nu 01}^{+\,j+1}\\
&+\sum_{n=2}^{N}\left[Cb_{1mn}^{+\,j+1}P_{mn}^{+\,j+1} + Cb_{1mn}^{-\,j+1}P_{mn}^{-\,j+1} + Cb_{\nu 1mn}^{+\,j+1}\rho_{\nu mn}^{+\,j+1}\right] = 0,
\end{aligned}
\tag{7.102}
$$

Therefore, we have obtained three conservation equations of mass, momentum, and energy. If n radial modes are considered, then the number of conservation equations is $3n$. Thus, one interface corresponds to $3n$ conservation equations and $3n$ unknowns. For example, if $n = 2$, then, for an individual interface,

$$\begin{pmatrix} Ab^j & Ab^{j+1} \\ Bb^j & Bb^{j+1} \\ Cb^j & Cb^{j+1} \end{pmatrix} \begin{pmatrix} P^j \\ P^{j+1} \end{pmatrix}, \tag{7.103}$$

where Ab, Bb, and Cb are the coefficients of mass equation, momentum equation, and energy equation, respectively, in the form of:

$$Ab^j = \begin{pmatrix} Ab_{101}^{+j} & Ab_{101}^{-j} & Ab_{\nu 101}^{+j} & Ab_{102}^{+j} & Ab_{102}^{-j} & Ab_{\nu 102}^{+j} \\ Ab_{201}^{+j} & Ab_{201}^{-j} & Ab_{\nu 201}^{+j} & Ab_{202}^{+j} & Ab_{202}^{-j} & Ab_{\nu 202}^{+j} \end{pmatrix} \tag{7.104}$$

$$Bb^j = \begin{pmatrix} Bb_{101}^{+j} & Bb_{101}^{-j} & Bb_{\nu 101}^{+j} & Bb_{102}^{+j} & Bb_{102}^{-j} & Bb_{\nu 102}^{+j} \\ Bb_{201}^{+j} & Bb_{201}^{-j} & Bb_{\nu 201}^{+j} & Bb_{202}^{+j} & Bb_{202}^{-j} & Bb_{\nu 202}^{+j} \end{pmatrix} \tag{7.105}$$

$$Cb^j = \begin{pmatrix} Cb_{101}^{+j} & Cb_{101}^{-j} & Cb_{\nu 101}^{+j} & Cb_{102}^{+j} & Cb_{102}^{-j} & Cb_{\nu 102}^{+j} \\ Cb_{201}^{+j} & Cb_{201}^{-j} & Cb_{\nu 201}^{+j} & Cb_{202}^{+j} & Cb_{202}^{-j} & Cb_{\nu 202}^{+j} \end{pmatrix} \tag{7.106}$$

$$P^j = \begin{pmatrix} P_{01}^{+j} & P_{01}^{-j} & \rho_{\nu 01}^{+j} & P_{02}^{+j} & P_{02}^{-j} & \rho_{\nu 02}^{+j} \end{pmatrix} \tag{7.107}$$

The coefficients Ab^{j+1}, Bb^{j+1}, Bb^{j+1}, P^{j+1} have the same form. Actually, the previous discussion is only valid for the situation of Fig. 7.5, which describes the cross-section becoming larger. For the other situation, in which, the cross-section becomes smaller, the equations can be derived similarly.

It should be pointed out that the previous formulation only considers that the circumferential mode number equals zero, that is to say, the circumferential instability is ignored. However, if there exists the circumferential instability, the method for a variable cross-section problem can't be directly applied. In contrast, all matching relations must be re-deduced with a different circumferential mode number to set up a new eigenvalue equation.

7.3.2 Numerical calculation method

7.3.2.1 Formulation

When the viscosity and the sources apart from the heat source are ignored, the equations of mass, momentum, and energy are written as:

$$\frac{\partial \rho}{\partial t} + \nabla \cdot (\rho \boldsymbol{v}) = q_m(\mathbf{x}, t), \tag{7.108}$$

$$\rho\frac{d\boldsymbol{v}}{dt}+\nabla p=\mathbf{0}, \tag{7.109}$$

$$\rho T\frac{ds}{dt}=q_h(\mathbf{x},t), \tag{7.110}$$

where ρ, $\boldsymbol{v}$, s, T are the density, velocity, entropy, and absolute temperature, respectively. The bold fonts mean they are vectors. q_m denotes the mass input per unit volume, whereas q_h is the heat input per unit volume. As the density varies through two variables, the pressure and the entropy,

$$\rho=\rho(p,s). \tag{7.111}$$

Hence, the complete differential of the density is,

$$d\rho=\left.\frac{\partial\rho}{\partial p}\right|_s dp+\left.\frac{\partial\rho}{\partial s}\right|_p ds. \tag{7.112}$$

For perfect gas,

$$p=R_g\rho T, \tag{7.113}$$

where R_g is the constant of ideal gas. According to Eq. (7.113), we obtain:

$$dp=R_gTd\rho+R_g\rho dT, \tag{7.114}$$

thus,

$$dT=\frac{dp}{R_g\rho}-\frac{Td\rho}{\rho}, \tag{7.115}$$

according to the definition of entropy s,

$$dQ=Tds, \tag{7.116}$$

and the first law of thermodynamics,

$$dQ=C_pdT-\frac{1}{\rho}dp, \tag{7.117}$$

where Q is the quantity of heat, and C_p is the specific heat capacity at constant pressure. The differential of s, T, and p are related by Eqs. (7.116) and (7.117),

$$Tds=C_pdT-\frac{1}{\rho}dp, \tag{7.118}$$

Substituting Eq. (7.114) into Eq. (7.118), we have,

$$Tds=\frac{C_v}{R_g\rho}dp-\frac{C_pT}{\rho}d\rho, \tag{7.119}$$

where C_v is specific heat capacity at constant volume. Therefore,

$$\left.\frac{\partial\widetilde{\rho}}{\partial\widetilde{s}}\right|_p = -\frac{\widetilde{\rho}}{C_p}, \tag{7.120}$$

The definition of the sound speed is,

$$\left.\frac{\partial\rho}{\partial p}\right|_s = \frac{1}{c^2}, \tag{7.121}$$

where c is the sound speed. Substitute Eqs. (7.120) and (7.121) into Eq. (7.112), we obtain,

$$d\rho = \frac{1}{c^2}dp - \frac{\rho}{C_p}ds \tag{7.122}$$

The previous equation can also be written as:

$$ds = \frac{C_p}{\rho c^2}dp - \frac{C_p}{\rho}d\rho \tag{7.123}$$

Substituting Eq. (7.123) into Eq. (7.110), we have,

$$\frac{C_p T}{c^2}\frac{dp}{dt} - C_p T\frac{d\rho}{dt} = q_h(\mathbf{x}, t) \tag{7.124}$$

because, for perfect gas, $C_p\widetilde{T} = \frac{\gamma R_g \widetilde{T}}{\gamma - 1} = \frac{\widetilde{c}^2}{\gamma - 1}$, where γ is the specific heat ratio. Thus, Eq. (7.124) is equivalent to,

$$\frac{dp}{dt} - c^2\frac{d\rho}{dt} = (\gamma - 1)q_h(\mathbf{x}, t) \tag{7.125}$$

Therefore, the basic equations involving the unknowns of the density, velocity, pressure, and sound speed are obtained.

As shown by the one-dimensional part, the unknowns can be decomposed into two quantities, the mean quantity and the fluctuation quantity.

$$\begin{aligned} \rho &= \overline{\rho} + \rho' \\ \boldsymbol{v} &= \overline{\boldsymbol{v}} + \boldsymbol{v}' \\ p &= \overline{p} + p' \end{aligned} \tag{7.126}$$

It is the same for the source terms of mass and heat,

$$\begin{aligned} q_m &= \overline{q}_m + q'_m \\ q_h &= \overline{q}_h + q'_h \end{aligned} \tag{7.127}$$

The overbar means the mean quantity, whereas the prime means the fluctuation quantity. Substitute Eqs. (7.126) and (7.127) into Eq. (7.108), we obtain:

$$\frac{\partial\overline{\rho}}{\partial t}+\frac{\partial\rho'}{\partial t}+\nabla\cdot(\overline{\rho}\overline{\boldsymbol{v}})+\nabla\cdot(\overline{\rho}\boldsymbol{v}')+\nabla\cdot(\rho'\overline{\boldsymbol{v}})+\nabla\cdot(\rho'\boldsymbol{v}')=\overline{q}_m+q_m'. \quad (7.128)$$

The mean values satisfy Eq. (7.108),

$$\frac{\partial\overline{\rho}}{\partial t}+\nabla\cdot(\overline{\rho}\overline{\boldsymbol{v}})=\overline{q}_m(\mathbf{x},t). \quad (7.129)$$

Subtracting Eqs. (7.128) and (7.129), and ignoring the high-order small quantities yield:

$$\frac{\partial\rho'}{\partial t}+\nabla\cdot(\overline{\rho}\boldsymbol{v}')+\nabla\cdot(\rho'\overline{\boldsymbol{v}})=q_m'. \quad (7.130)$$

When the same treatment is applied to Eqs. (7.109) and (7.125), the following two equations are obtained:

$$\overline{\rho}\frac{\partial\boldsymbol{v}'}{\partial t}+\rho'\frac{\partial\overline{\boldsymbol{v}}}{\partial t}+\overline{\rho}\overline{\boldsymbol{v}}\cdot\nabla\boldsymbol{v}'+\overline{\rho}\boldsymbol{v}'\cdot\nabla\overline{\boldsymbol{v}}+\rho'\overline{\boldsymbol{v}}\cdot\nabla\overline{\boldsymbol{v}}+\nabla p'=\mathbf{0}, \quad (7.131)$$

$$\begin{aligned}&\frac{\partial p'}{\partial t}+\overline{\boldsymbol{v}}\cdot\nabla p'+\boldsymbol{v}'\cdot\nabla\overline{p}-\overline{c}^2\frac{\partial\rho'}{\partial t}-c'^2\frac{\partial\overline{\rho}}{\partial t}\\&-\overline{c}^2\overline{\boldsymbol{v}}\cdot\nabla\rho'-\overline{c}^2\boldsymbol{v}'\cdot\nabla\overline{\rho}-c'^2\overline{\boldsymbol{v}}\cdot\nabla\overline{\rho}=(\gamma-1)q_h'.\end{aligned} \quad (7.132)$$

Because the perfect gas equation requires:

$$\overline{\rho}c'^2+\rho'\overline{c}^2=\gamma p'. \quad (7.133)$$

Substituting Eq. (7.133) into Eq. (7.132) leads to:

$$\begin{aligned}&\frac{\partial p'}{\partial t}+\overline{\boldsymbol{v}}\cdot\nabla p'+\boldsymbol{v}'\cdot\nabla\overline{p}-\overline{c}^2\frac{\partial\rho'}{\partial t}-\frac{\gamma}{\overline{\rho}}p'\frac{\partial\overline{\rho}}{\partial t}+\frac{\overline{c}^2}{\overline{\rho}}\rho'\frac{\partial\overline{\rho}}{\partial t}-\overline{c}^2\overline{\boldsymbol{v}}\cdot\nabla\rho'\\&-\overline{c}^2\boldsymbol{v}'\cdot\nabla\overline{\rho}-\frac{\gamma}{\overline{\rho}}p'\overline{\boldsymbol{v}}\cdot\nabla\overline{\rho}+\frac{\overline{c}^2}{\overline{\rho}}\rho'\overline{\boldsymbol{v}}\cdot\nabla\overline{\rho}=(\gamma-1)q_h'.\end{aligned} \quad (7.134)$$

Now we obtain the linearized equations of mass, momentum, and energy as:

$$\begin{aligned}&\frac{\partial\rho'}{\partial t}+\nabla\cdot(\overline{\rho}\boldsymbol{v}')+\nabla\cdot(\rho'\overline{\boldsymbol{v}})=q_m',\\&\overline{\rho}\frac{\partial\boldsymbol{v}'}{\partial t}+\rho'\frac{\partial\overline{\boldsymbol{v}}}{\partial t}+\overline{\rho}\overline{\boldsymbol{v}}\cdot\nabla\boldsymbol{v}'+\overline{\rho}\boldsymbol{v}'\cdot\nabla\overline{\boldsymbol{v}}+\rho'\overline{\boldsymbol{v}}\cdot\nabla\overline{\boldsymbol{v}}+\nabla p'=\mathbf{0},\\&\frac{\partial p'}{\partial t}+\overline{\boldsymbol{v}}\cdot\nabla p'+\boldsymbol{v}'\cdot\nabla\overline{p}-\overline{c}^2\frac{\partial\rho'}{\partial t}-\frac{\gamma}{\overline{\rho}}p'\frac{\partial\overline{\rho}}{\partial t}+\frac{\overline{c}^2}{\overline{\rho}}\rho'\frac{\partial\overline{\rho}}{\partial t}-\overline{c}^2\overline{\boldsymbol{v}}\cdot\nabla\rho'\\&-\overline{c}^2\boldsymbol{v}'\cdot\nabla\overline{\rho}-\frac{\gamma}{\overline{\rho}}p'\overline{\boldsymbol{v}}\cdot\nabla\overline{\rho}+\frac{\overline{c}^2}{\overline{\rho}}\rho'\overline{\boldsymbol{v}}\cdot\nabla\overline{\rho}=(\gamma-1)q_h'.\end{aligned} \quad (7.135)$$

The linearized equations have the unknowns of ρ', $\boldsymbol{v}'$, and p'. If the mean quantities, mass input, and heat input are given, it is viable to investigate the issue of combustion instability, which is induced by the thermoacoustic coupling.

For simplicity, the fluctuation quantity is expanded in the harmonic forms:

$$\begin{aligned} \rho' &= \widetilde{\rho}e^{i\omega t} \\ p' &= \widetilde{p}e^{i\omega t} \\ \boldsymbol{v}' &= \widetilde{\boldsymbol{v}}e^{i\omega t} \end{aligned} \tag{7.136}$$

Then, Eq. (7.135) can be written as:

$$\begin{aligned} & i\omega\widetilde{\rho} + \nabla\cdot(\widetilde{\rho}\overline{\boldsymbol{v}}) + \nabla\cdot(\overline{\rho}\widetilde{\boldsymbol{v}}) = \widetilde{q}_m, \\ & \widetilde{\rho}\left(\frac{\partial\overline{\boldsymbol{v}}}{\partial t} + \overline{\boldsymbol{v}}\cdot\nabla\overline{\boldsymbol{v}}\right) + i\omega\overline{\rho}\widetilde{\boldsymbol{v}} + \overline{\rho}\overline{\boldsymbol{v}}\cdot\nabla\widetilde{\boldsymbol{v}} + \overline{\rho}\widetilde{\boldsymbol{v}}\cdot\nabla\overline{\boldsymbol{v}} + \nabla\widetilde{p} = \boldsymbol{0}, \\ & \left(-i\omega\overline{c}^2 + \frac{\overline{c}^2}{\overline{\rho}}\frac{\partial\overline{\rho}}{\partial t} + \frac{\overline{c}^2}{\overline{\rho}}\overline{\boldsymbol{v}}\cdot\nabla\overline{\rho}\right)\widetilde{\rho} - \overline{c}^2\overline{\boldsymbol{v}}\cdot\nabla\widetilde{\rho} + \widetilde{\boldsymbol{v}}\cdot\nabla\overline{p} - \overline{c}^2\widetilde{\boldsymbol{v}}\cdot\nabla\overline{\rho} \\ & + \left(i\omega - \frac{\gamma}{\overline{\rho}}\frac{\partial\overline{\rho}}{\partial t} - \frac{\gamma}{\overline{\rho}}\overline{\boldsymbol{v}}\cdot\nabla\overline{\rho}\right)\widetilde{p} + \overline{\boldsymbol{v}}\cdot\nabla\widetilde{p} = (\gamma-1)\widetilde{q}_h. \end{aligned} \tag{7.137}$$

If the unsteady effects of the mean flow are ignored, then we obtain,

$$\begin{aligned} & i\omega\widetilde{\rho} + \nabla\cdot(\widetilde{\rho}\overline{\boldsymbol{v}}) + \nabla\cdot(\overline{\rho}\widetilde{\boldsymbol{v}}) = \widetilde{q}_m, \\ & \widetilde{\rho}\overline{\boldsymbol{v}}\cdot\nabla\overline{\boldsymbol{v}} + i\omega\overline{\rho}\widetilde{\boldsymbol{v}} + \overline{\rho}\overline{\boldsymbol{v}}\cdot\nabla\widetilde{\boldsymbol{v}} + \overline{\rho}\widetilde{\boldsymbol{v}}\cdot\nabla\overline{\boldsymbol{v}} + \nabla\widetilde{p} = \boldsymbol{0}, \\ & \left(-i\omega\overline{c}^2 + \frac{\overline{c}^2}{\overline{\rho}}\overline{\boldsymbol{v}}\cdot\nabla\overline{\rho}\right)\widetilde{\rho} - \overline{c}^2\overline{\boldsymbol{v}}\cdot\nabla\widetilde{\rho} + \widetilde{\boldsymbol{v}}\cdot\nabla\overline{p} - \overline{c}^2\widetilde{\boldsymbol{v}}\cdot\nabla\overline{\rho} \\ & + \left(i\omega - \frac{\gamma}{\overline{\rho}}\overline{\boldsymbol{v}}\cdot\nabla\overline{\rho}\right)\widetilde{p} + \overline{\boldsymbol{v}}\cdot\nabla\widetilde{p} = (\gamma-1)\widetilde{q}_h. \end{aligned} \tag{7.138}$$

Because most combustion instabilities take place in cylindrical or annular combustors, therefore, we expand Eq. (7.138) in cylindrical coordinates of x, r, θ, thus:

$$\begin{aligned} & \left(i\omega + \frac{\partial\overline{v}_r}{\partial r} + \frac{\overline{v}_r}{r} + \frac{1}{r}\frac{\partial\overline{v}_\theta}{\partial\theta} + \frac{\partial\overline{v}_x}{\partial x}\right)\widetilde{\rho} \\ & + \overline{v}_r\frac{\partial\widetilde{\rho}}{\partial r} + \frac{\overline{v}_\theta}{r}\frac{\partial\widetilde{\rho}}{\partial\theta} + \overline{v}_x\frac{\partial\widetilde{\rho}}{\partial x} \\ & + \overline{\rho}\frac{\partial\widetilde{v}_r}{\partial r} + \frac{\overline{\rho}}{r}\frac{\partial\widetilde{v}_\theta}{\partial\theta} + \overline{\rho}\frac{\partial\widetilde{v}_x}{\partial x} + \left(\frac{\overline{\rho}}{r} + \frac{\partial\overline{\rho}}{\partial r}\right)\widetilde{v}_r + \frac{1}{r}\frac{\partial\overline{\rho}}{\partial\theta}\widetilde{v}_\theta + \frac{\partial\overline{\rho}}{\partial x}\widetilde{v}_x = \widetilde{q}_m, \end{aligned} \tag{7.139}$$

$$\left(\bar{v}_r\frac{\partial\bar{v}_r}{\partial r}+\frac{\bar{v}_\theta}{r}\frac{\partial\bar{v}_r}{\partial\theta}+\bar{v}_x\frac{\partial\bar{v}_r}{\partial x}\right)\widetilde{\rho}+i\omega\bar{\rho}\widetilde{v}_r+\bar{\rho}\bar{v}_r\frac{\partial\widetilde{v}_r}{\partial r}+\frac{\bar{\rho}\bar{v}_\theta}{r}\frac{\partial\widetilde{v}_r}{\partial\theta}$$
$$+\bar{\rho}\bar{v}_x\frac{\partial\widetilde{v}_r}{\partial x}+\bar{\rho}\frac{\partial\bar{v}_r}{\partial r}\widetilde{v}_r+\frac{\bar{\rho}}{r}\frac{\partial\bar{v}_r}{\partial\theta}\widetilde{v}_\theta+\bar{\rho}\frac{\partial\bar{v}_r}{\partial x}\widetilde{v}_x+\frac{\partial\widetilde{p}}{\partial r}=0, \tag{7.140}$$

$$\left(\bar{v}_r\frac{\partial\bar{v}_\theta}{\partial r}+\frac{\bar{v}_\theta}{r}\frac{\partial\bar{v}_\theta}{\partial\theta}+\bar{v}_x\frac{\partial\bar{v}_\theta}{\partial x}\right)\widetilde{\rho}+i\omega\bar{\rho}\widetilde{v}_\theta+\bar{\rho}\bar{v}_r\frac{\partial\widetilde{v}_\theta}{\partial r}+\frac{\bar{\rho}\bar{v}_\theta}{r}\frac{\partial\widetilde{v}_\theta}{\partial\theta}$$
$$+\bar{\rho}\bar{v}_x\frac{\partial\widetilde{v}_\theta}{\partial x}+\bar{\rho}\frac{\partial\bar{v}_\theta}{\partial r}\widetilde{v}_r+\frac{\bar{\rho}}{r}\frac{\partial\bar{v}_\theta}{\partial\theta}\widetilde{v}_\theta+\bar{\rho}\frac{\partial\bar{v}_\theta}{\partial x}\widetilde{v}_x+\frac{1}{r}\frac{\partial\widetilde{p}}{\partial\theta}=0, \tag{7.141}$$

$$\left(\bar{v}_r\frac{\partial\bar{v}_x}{\partial r}+\frac{\bar{v}_\theta}{r}\frac{\partial\bar{v}_x}{\partial\theta}+\bar{v}_x\frac{\partial\bar{v}_x}{\partial x}\right)\widetilde{\rho}+i\omega\bar{\rho}\widetilde{v}_x+\bar{\rho}\bar{v}_r\frac{\partial\widetilde{v}_x}{\partial r}+\frac{\bar{\rho}\bar{v}_\theta}{r}\frac{\partial\widetilde{v}_x}{\partial\theta}$$
$$+\bar{\rho}\bar{v}_x\frac{\partial\widetilde{v}_x}{\partial x}+\bar{\rho}\frac{\partial\bar{v}_x}{\partial r}\widetilde{v}_r+\frac{\bar{\rho}}{r}\frac{\partial\bar{v}_x}{\partial\theta}\widetilde{v}_\theta+\bar{\rho}\frac{\partial\bar{v}_x}{\partial x}\widetilde{v}_x+\frac{\partial\widetilde{p}}{\partial x}=0, \tag{7.142}$$

$$\left[-i\omega\bar{c}^2+\frac{\bar{c}^2}{\bar{\rho}}\left(\bar{v}_r\frac{\partial\bar{\rho}}{\partial r}+\frac{\bar{v}_\theta}{r}\frac{\partial\bar{\rho}}{\partial\theta}+\bar{v}_x\frac{\partial\bar{\rho}}{\partial x}\right)\right]\widetilde{\rho}-\bar{c}^2\bar{v}_r\frac{\partial\widetilde{\rho}}{\partial r}-\bar{c}^2\frac{\bar{v}_\theta}{r}\frac{\partial\widetilde{\rho}}{\partial\theta}$$
$$-\bar{c}^2\bar{v}_x\frac{\partial\widetilde{\rho}}{\partial x}+\left(\frac{\partial\bar{p}}{\partial r}-\bar{c}^2\frac{\partial\bar{\rho}}{\partial r}\right)\widetilde{v}_r+\frac{1}{r}\left(\frac{\partial\bar{p}}{\partial\theta}-\bar{c}^2\frac{\partial\bar{\rho}}{\partial\theta}\right)\widetilde{v}_\theta+\left(\frac{\partial\bar{p}}{\partial x}-\bar{c}^2\frac{\partial\bar{\rho}}{\partial x}\right)\widetilde{v}_x$$
$$+\left[i\omega-\frac{\gamma}{\bar{\rho}}\left(\bar{v}_r\frac{\partial\bar{\rho}}{\partial r}+\frac{\bar{v}_\theta}{r}\frac{\partial\bar{\rho}}{\partial\theta}+\bar{v}_x\frac{\partial\bar{\rho}}{\partial x}\right)\right]\widetilde{p}$$
$$+\bar{v}_r\frac{\partial p}{\partial r}+\frac{\bar{v}_\theta}{r}\frac{\partial\widetilde{p}}{\partial\theta}+\bar{v}_x\frac{\partial\widetilde{p}}{\partial x}=(\gamma-1)\widetilde{q}_h, \tag{7.143}$$

where v_x, v_r, and v_θ are the axial, radial, and circumferential velocity components, respectively. Because the primary issue we care about is the effect of the circumferential non-uniformities, thus, the radial and the tangential components are ignored, which is reasonable. Therefore, Eqs. (7.139)–(7.143) are simplified as the following equations:

$$r\left(i\omega+\frac{\partial\widetilde{v}_x}{\partial x}\right)\widetilde{\rho}+r\bar{v}_x\frac{\partial\widetilde{\rho}}{\partial x}+r\bar{\rho}\frac{\partial\widetilde{v}_r}{\partial r}+\bar{\rho}\frac{\partial\widetilde{v}_\theta}{\partial\theta}+r\bar{\rho}\frac{\partial\widetilde{v}_x}{\partial x}$$
$$+\left(\bar{\rho}+r\frac{\partial\bar{\rho}}{\partial r}\right)\widetilde{v}_r+\frac{\partial\bar{\rho}}{\partial\theta}\widetilde{v}_\theta+r\frac{\partial\bar{\rho}}{\partial x}\widetilde{v}_x=r\widetilde{q}_m, \tag{7.144}$$

$$i\omega\bar{\rho}\widetilde{v}_r+\bar{\rho}\bar{v}_x\frac{\partial\widetilde{v}_r}{\partial x}+\frac{\partial\widetilde{p}}{\partial r}=0, \tag{7.145}$$

$$i\omega r\bar{\rho}\widetilde{v}_\theta+r\bar{\rho}\bar{v}_x\frac{\partial\widetilde{v}_\theta}{\partial x}+\frac{\partial\widetilde{p}}{\partial\theta}=0, \tag{7.146}$$

$$
\begin{aligned}
&r\overline{v}_x\frac{\partial\widetilde{v}_x}{\partial x}\widetilde{\rho}+r\left(i\omega\overline{\rho}+\overline{\rho}\frac{\partial\overline{v}_x}{\partial x}\right)\widetilde{v}_x+r\overline{\rho}\overline{v}_x\frac{\partial\widetilde{v}_x}{\partial x}\\
&+r\overline{\rho}\frac{\partial\overline{v}_x}{\partial r}\widetilde{v}_r+\overline{\rho}\frac{\partial\overline{v}_x}{\partial\theta}\widetilde{v}_\theta+r\frac{\partial\widetilde{p}}{\partial x}=0,
\end{aligned}
\tag{7.147}
$$

$$
\begin{aligned}
&\left(-ir\omega+\frac{r\overline{v}_x}{\overline{\rho}}\frac{\partial\overline{\rho}}{\partial x}\right)\widetilde{\rho}-r\overline{v}_x\frac{\partial\widetilde{\rho}}{\partial x}+r\left(\frac{1}{\overline{c}^2}\frac{\partial\overline{p}}{\partial r}-\frac{\partial\overline{\rho}}{\partial r}\right)\widetilde{v}_r\\
&+\left(\frac{1}{\overline{c}^2}\frac{\partial\overline{p}}{\partial\theta}-\frac{\partial\overline{\rho}}{\partial\theta}\right)\widetilde{v}_\theta+r\left(\frac{1}{\overline{c}^2}\frac{\partial\overline{p}}{\partial x}-\frac{\partial\overline{\rho}}{\partial x}\right)\widetilde{v}_x\\
&+\frac{r}{\overline{c}^2}\left(i\omega-\frac{\gamma\overline{v}_x}{\overline{\rho}}\frac{\partial\overline{\rho}}{\partial x}\right)\widetilde{p}+r\overline{v}_x\frac{1}{\overline{c}^2}\frac{\partial\widetilde{p}}{\partial x}=\frac{r}{\overline{c}^2}(\gamma-1)\widetilde{q}_h.
\end{aligned}
\tag{7.148}
$$

The constants that are only related to the mean values can be written in reduced form. Then the renewed equations are:

$$
\begin{aligned}
&A_1\widetilde{\rho}+A_2\frac{\partial\widetilde{\rho}}{\partial x}+A_3\frac{\partial\widetilde{v}_r}{\partial r}+A_4\frac{\partial\widetilde{v}_\theta}{\partial\theta}+A_5\frac{\partial\widetilde{v}_x}{\partial x}+A_6\widetilde{v}_r+A_7\widetilde{v}_\theta+A_8\widetilde{v}_x=\widetilde{q}_m,\\
&B_1\widetilde{v}_r+B_2\frac{\partial\widetilde{v}_r}{\partial x}+B_3\frac{\partial\widetilde{p}}{\partial r}=0,\\
&C_1\widetilde{v}_\theta+C_2\frac{\partial\widetilde{v}_\theta}{\partial x}+C_3\frac{\partial\widetilde{p}}{\partial\theta}=0,\\
&D_1\widetilde{\rho}+D_2\widetilde{v}_x+D_3\frac{\partial\widetilde{v}_x}{\partial x}+D_4\widetilde{v}_r+D_5\widetilde{v}_\theta+D_6\frac{\partial\widetilde{p}}{\partial x}=0,\\
&E_1\widetilde{\rho}+E_2\frac{\partial\widetilde{\rho}}{\partial x}+E_3\widetilde{v}_r+E_4\widetilde{v}_\theta+E_5\widetilde{v}_x+E_6\widetilde{p}+E_7\frac{\partial\widetilde{p}}{\partial x}=(\gamma-1)\widetilde{q}_h.
\end{aligned}
\tag{7.149}
$$

And the renewed constants are presented in Appendix C. Thus far, the first-order differential equations have been presented as the basic equations for investigating the combustion instability, which is able to account for the circumferential non-uniform effects. In the formulation, the radial and circumferential mean flow are ignored for simplicity. Actually, if we can also account for the effects of these two velocity components, it only makes the equations more complex without enlarging the computation cost.

7.3.2.2 Discretization method

The spectral method is adopted here for solving the eigenvalue problem, which is superior to the finite difference method for higher accuracy and faster convergence. This method has been widely used in solving the eigenvalue problems of flow instability and other mechanics. The method is briefly introduced here, and more detail can be found in Peyret [19].

In Eq. (7.149), the unknowns are discretized by Fourier methods in the circumferential direction and Chebyshev methods in the axial direction,

$$\chi(x, r, \theta) = \sum_{i=0}^{N_x} \sum_{j=0}^{N_r} \sum_{k=1}^{N_\theta} F_{ijk} T_i(x) T_j(r) e^{ik\theta}, \tag{7.150}$$

where χ represents unknowns; F_{ijk} denotes the unknown coefficients; N_x, N_r, and N_θ are numbers of collocation points in axial, radial, and circumferential directions, respectively; and $T_i(x)$ is the Chebyshev polynomial of degree i defined for x in the range of $[-1, 1]$. Thus, the derivatives of the variables at the collocation points can be obtained,

$$\left.\frac{\partial \chi}{\partial x}\right|_{i,j,k} = \chi_x^1(x_i, r_j, \theta_k) = \sum_{m=0}^{N_x} d_{i,m}^1 \chi(x_m, r_j, \theta_k), \tag{7.151}$$

$$\left.\frac{\partial \chi}{\partial r}\right|_{i,j,k} = \chi_r^1(x_i, r_j, \theta_k) = \sum_{n=0}^{N_r} d_{j,n}^1 \chi(x_i, r_n, \theta_k), \tag{7.152}$$

$$\left.\frac{\partial \chi}{\partial \theta}\right|_{i,j,k} = \chi_\theta^1(x_i, r_j, \theta_k) = \sum_{p=1}^{N_\theta} e_{k,p}^1 \chi(x_i, r_j, \theta_p), \tag{7.153}$$

where $i = 0, \ldots, N_x, j = 0, \ldots, N_r$ and $k = 0, \ldots, N_\theta$. $e_{k,\,p}^1$, $d_{i,\,m}^1$ and $d_{j,\,n}^1$ are the first-order Fourier and Chebyshev derivative matrix given by Peyret [19].

7.3.2.3 Treatment of heat source

In our model, the mass source is not included, i.e., $q_m = 0$. For the heat source, the combustion takes place in a region,

$$q_h = q_h(x, r, \theta). \tag{7.154}$$

The heat source will be treated in two ways. First and usually, the flame is assumed to be compacted in a plate, then,

$$q_h = \delta(x - x_0)\left(\lambda_p p + \lambda_v v_x\right). \tag{7.155}$$

It is convenient to make such assumptions, especially for an analytical approach. Eq. (7.155) means that the unsteady heat release is related to the axial velocity fluctuation and the pressure fluctuation. δ is the delta function. This kind of heat release function means in the two regions upstream, the heat release plate, and downstream of the plate, the heat source q_h is zero. Most heat release function can be written in this form. Thus, the energy equation in Eq. (7.149) becomes:

$$E_1\widetilde{\rho}+E_2\frac{\partial\widetilde{\rho}}{\partial x}+E_3\widetilde{v}_r+E_4\widetilde{v}_\theta+E_5\widetilde{v}_x+E_6\widetilde{p}+E_7\frac{\partial\widetilde{p}}{\partial x}=0, \tag{7.156}$$

where E_1 E_2 E_3 E_4 E_7 remain unchanged, whereas E_5 and E_6 become:

$$E_5=\frac{1}{\overline{c}^2}\frac{\partial\overline{p}}{\partial x}-\frac{\partial\overline{\rho}}{\partial x}-\frac{1}{\overline{c}^2}\delta(x-x_0)\lambda_p, \tag{7.157}$$

and

$$E_6=i\omega\frac{1}{\overline{c}^2}-\frac{\gamma\overline{v}_x}{\overline{c}^2\overline{\rho}}\frac{\partial\overline{\rho}}{\partial x}-\overline{c}^2\delta(x-x_0)\lambda_p, \tag{7.158}$$

respectively.

This reveals a problem. Because of the heat input, Eq. (7.156) is different across the heat release plate, and the flow parameters change after the heat release plate. The discretization method is unable to conquer it directly. Thus, we need to discretize the equations for different zones separately as shown in Fig. 7.6, and relate the flow parameters with the conservation laws.

The subscripts 1 and 2 in Fig. 7.6 denote the variables of up- and downstream regions of the heat release region. For the control volume, applying conservation of mass, momentum, and energy, the following relations are obtained,

Mass equation is:

$$\oint_s \widetilde{\rho\boldsymbol{v}}\cdot d\boldsymbol{s}=0. \tag{7.159}$$

By linearizing the equation, we obtain,

$$\overline{\rho}_1 v_{xA}+\overline{v}_{x1}\rho_A=\overline{\rho}_2 v_{xB}+\overline{v}_{x2}\rho_B, \tag{7.160}$$

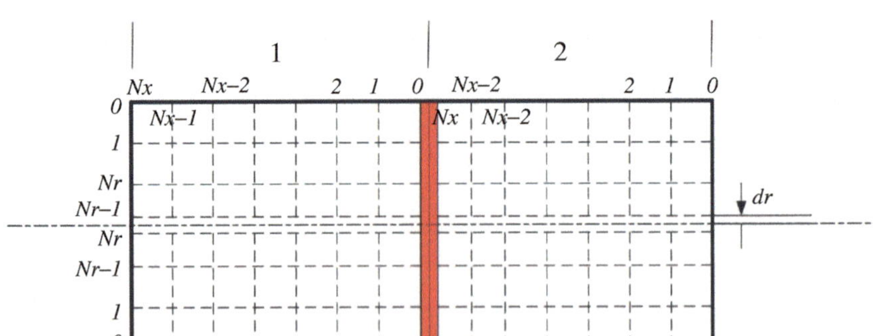

Fig. 7.6 Schematic of spectral discretization method.

where the subscripts 1 and 2 represent the region, and A and B denote the right boundary of region 1 and the left boundary of region 2, respectively.

Momentum equation is:

$$\oint_s \widetilde{\rho v}_n \widetilde{\boldsymbol{v}} d\boldsymbol{s} + \oint_s n_v \cdot \widetilde{p} d\boldsymbol{s} = 0. \tag{7.161}$$

The linearized equations are:

$$p_A + \rho_A \overline{v}_{x1}^2 + 2\overline{\rho}_1 \overline{v}_{x1} v_{xA} = p_B + \rho_B \overline{v}_{x2}^2 + 2\overline{\rho}_2 \overline{v}_{x2} v_{xB}, \tag{7.162}$$

$$\overline{\rho}_1 \overline{v}_{x1} v_{rA} = \overline{\rho}_2 \overline{v}_{x2} v_{rB}, \tag{7.163}$$

$$\overline{\rho}_1 \overline{v}_{x1} v_{\theta A} = \overline{\rho}_2 \overline{v}_{x2} v_{\theta B}. \tag{7.164}$$

Energy is:

$$\delta \widetilde{Q} + \oint_s \left(e + \frac{\widetilde{p}}{\widetilde{\rho}} + \frac{\widetilde{u}^2}{2} \right) \rho \widetilde{v}_n ds = 0. \tag{7.165}$$

By linearizing the equation, we have:

$$\begin{aligned} & Q' + \left(c_p \overline{T}_{01} \overline{\rho}_1 + \overline{\rho}_1 \overline{v}_{x1}^2\right) v_{xA} + \left(c_p \overline{T}_{01} \overline{v}_{x1} - c_p T_1 \overline{v}_{x1}\right) \rho_A + \frac{c_p \overline{v}_{x1}}{Rg} p_A \\ & = \left(c_p \overline{T}_{02} \overline{\rho}_2 + \overline{\rho}_2 \overline{v}_{x2}^2\right) v_{xB} + \left(c_p \overline{T}_{02} \overline{v}_{x2} - c_p T_2 v_{x2}\right) \rho_B + \frac{c_p v_{x2}}{Rg} p_B, \end{aligned} \tag{7.166}$$

where Q' is the unsteady heat release per unit area, $\overline{T}_{01}$ and $\overline{T}_{02}$ are the total temperature of region 1 and region 2, respectively, $\overline{T}_0 = \overline{T} + \frac{\overline{u}^2}{2C_p}$.

If the unsteady heat release can be expressed as:

$$Q' = \lambda_p p + \lambda_v v_x. \tag{7.167}$$

Then, it is obtained:

$$\begin{aligned} & \left(\lambda_v + c_p \overline{T}_{01} \overline{\rho}_1 + \overline{\rho}_1 \overline{v}_{x1}^2\right) v_{xA} + \left(c_p \overline{T}_{01} \overline{v}_{x1} - c_p T_1 \overline{v}_{x1}\right) \rho_A + \left(\lambda_p + \frac{c_p \overline{v}_{x1}}{Rg}\right) p_A \\ & = \left(c_p \overline{T}_{02} \overline{\rho}_2 + \overline{\rho}_2 \overline{v}_{x2}^2\right) v_{xB} + \left(c_p \overline{T}_{02} \overline{v}_{x2} - c_p T_2 v_{x2}\right) \rho_B + \frac{c_p v_{x2}}{Rg} p_B. \end{aligned} \tag{7.168}$$

The five linearized conservation equations can be written as:

$$\begin{aligned} & K_{a1} \rho_{x1} + K_{a2} v_{x1} + K_{a3} \rho_{x2} + K_{a4} v_{x2} = 0, \\ & K_{b1} \rho_1 + K_{b2} v_{x1} + K_{b3} p_1 + K_{b4} \rho_2 + K_{b5} v_{x2} + K_{b6} p_2 = 0, \\ & K_{c1} v_{r1} + K_{c2} v_{r2} = 0, \\ & K_{d1} v_{\theta 1} + K_{d1} v_{\theta 2} = 0, \\ & K_{e1} \rho_1 + K_{e2} v_{x1} + K_{e3} p_1 + K_{e4} \rho_2 + K_{e5} v_{x2} + K_{e6} p_2 = 0, \end{aligned} \tag{7.169}$$

where the coefficients K can be found in Appendix D.

For a specific problem, it should satisfy some kind of boundary conditions, for example, for a hard-wall condition, the following equation should be met,

$$\left.\frac{\partial p}{\partial r}\right|_{r=R_o} = 0, \tag{7.170}$$

where R_o is the radius of the tube. If the wall has a damping effect, we need to know the impedance of the wall, with the impedance boundary condition as:

$$\frac{p}{v_r} = Z_n, \tag{7.171}$$

where Z_n is the impedance of the wall. With all of the conditions taken into account, Eq. (7.149) can be discretized as:

$$\begin{pmatrix} \zeta_1 & \zeta_3 \\ \zeta_4 & \zeta_2 \end{pmatrix} \begin{pmatrix} \xi_1 \\ \xi_2 \end{pmatrix} = 0, \tag{7.172}$$

Where ζ_1, ζ_2, ζ_3, and ζ_4 are submatrices of size $[5(N+1)]^2$ with only the complex frequency ω as the unknown, where $N = (N_x + 1)(N_r + 1)N_t - 1$. The column vectors ξ_1 and ξ_2 of length $5(N+1)$ contain the discrete unknowns in regions 1 and 2, respectively. If Eq. (7.149) has non-zero solutions, the determinant of the matrix must be zero, and the complex frequency ω can be obtained by solving the equation:

$$\begin{vmatrix} \zeta_1 & \zeta_3 \\ \zeta_4 & \zeta_2 \end{vmatrix} = 0. \tag{7.173}$$

7.3.3 Effect of vorticity waves on azimuthal instabilities in annular chambers

In this section, the effect of vorticity disturbance on azimuthal instabilities in an annular geometry is investigated analytically. The investigated problem is depicted in Fig. 7.7. In an annular chamber, a compacted flame is located at $x = L_f$. The flame separates uniform regions of cold reactants (region 1) and hot products (region 2). For simplicity, the following assumptions are made about the problem: (1) the flame is thin enough that the thickness is considered to be zero; (2) the effect of the flame is embodied in unsteady heat release, but the mechanisms are not discussed here, e.g., the flame speed and flame area disturbances; (3) because the geometry is axisymmetric, no circumferential modal coupling occurs, so the azimuthal modes of two directions are considered separately; (4) only axial mean velocity is

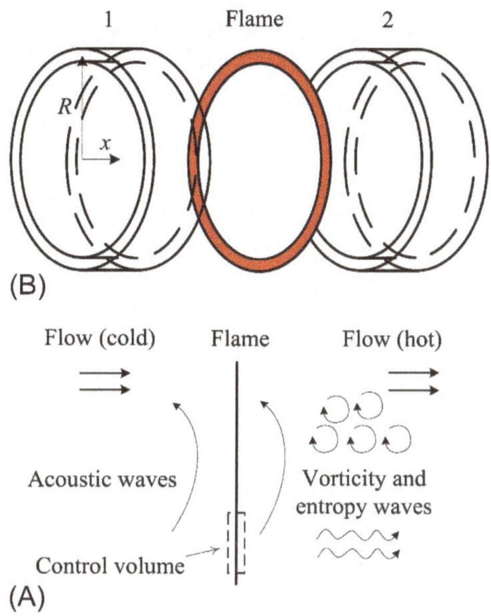

Fig. 7.7 Schematic of the annular chamber [20]. (A) Fluctuating waves in the chamber; (B) Geometry of the chamber.

considered, which is given to be uniform; (5) the radial gap is much smaller than axial and circumferential length so the radial dependence is ignored; (6) we focus on the azimuthal modes, hence the acoustically closed and choked boundary conditions are taken; and (7) the following analysis is limited in the linear range so that each variable can be written as a sum of its mean (denoted by superscript −) and fluctuating component (denoted by ′), and the higher-order terms are ignored (e.g., $p=\overline{p}+p'$).

To validate this analytical model with the effect of the vorticity wave, the linear Euler equation method (LEEM) introduced in Section 7.3.2 is adopted. For the annular geometry, the radial dependence is ignored for the reason that the radial gap is much smaller than the axial and circumferential length. Therefore, the analytical description for the annular geometry can be recovered from the three-dimensional AMs introduced in Section 7.3.1. Analytical results without the vorticity wave are also given on the analogy of the network model and finite element method. A comparison of the results with and without vorticity waves shows that vorticity disturbance has a significant effect on the azimuthal instabilities in a chamber with mean flow. And the results also show that calculations of azimuthal instabilities using the network model and finite element method may lead to significant prediction errors when the mean flow is considered. The work can be seen in the work of Li and Sun [20] in detail.

7.3.3.1 Description of disturbances

The flow parameters are given as axial velocity u, pressure p, density ρ, speed of sound c, and Mach number M. The acoustic disturbances can be written in the following forms using standard techniques:

$$p' = \left(P_m^+ e^{i\alpha_m^+ x} + P_m^- e^{i\alpha_m^- x}\right) e^{i(\omega t + m\theta)}, \tag{7.174}$$

$$u_a' = -\frac{1}{\bar{\rho}}\left(\frac{\alpha_m^+ P_m^+ e^{i\alpha_m^+ x}}{\omega + \alpha_m^+ \bar{u}} + \frac{\alpha_m^- P_m^- e^{i\alpha_m^- x}}{\omega + \alpha_m^- \bar{u}}\right) e^{i(\omega t + m\theta)}, \tag{7.175}$$

$$v_a' = -\frac{1}{\bar{\rho}}\left(\frac{m P_m^+ e^{i\alpha_m^+ x}}{R(\omega + \alpha_m^+ \bar{u})} + \frac{m P_m^- e^{i\alpha_m^- x}}{R(\omega + \alpha_m^- \bar{u})}\right) e^{i(\omega t + m\theta)}, \tag{7.176}$$

$$\rho_a' = \frac{1}{\bar{c}^2}\left(P_m^+ e^{i\alpha_m^+ x} + P_m^- e^{i\alpha_m^- x}\right) e^{i(\omega t + m\theta)}, \tag{7.177}$$

where p', u', v', ρ', and R denote pressure, axial velocity, circumferential velocity disturbances, and radius of the chamber, respectively. The subscript a denotes the acoustic quantities. P is the unknown parameter that denotes the magnitude of the acoustic disturbances. m is an integral, and it denotes the circumferential mode number. + and - denote the waves that propagate upstream and downstream, respectively. R is the mean radius of the annular chamber. x and θ denote axial and circumferential coordinates, respectively. α is the axial wave number that has the form:

$$\alpha_m^{+,-} = \frac{\overline{M}\left(\frac{\omega}{\bar{c}}\right) \pm \sqrt{\left(\frac{\omega}{\bar{c}}\right)^2 - (1 - \overline{M}^2)\left(\frac{m}{R}\right)^2}}{1 - \overline{M}^2}, \tag{7.178}$$

where $\omega = \omega_r + i\omega_i$ denotes complex frequency, the real part ω_r represents the radial frequency of oscillation, the imaginary part ω_i is the damping coefficient, and i is the imaginary unit. The value of ω_i determines the decay rate of an acoustic mode. The instability characteristics can be investigated through ω.

Density fluctuation can also be induced by an entropy wave, which has the form:

$$\rho_e' = -\frac{\bar{\rho}}{C_p} S_{em} e^{-i\frac{\omega}{\bar{u}}x} e^{i(m\theta + \omega t)}, \tag{7.179}$$

where C_p is specific heat at constant pressure, and S_{em} is the unknown parameter. Similarly, the velocity fluctuations induced by vorticity waves can be written as:

$$u'_v = V_{xm} e^{-i\frac{\omega}{\overline{u}}x} e^{i(m\theta + \omega t)}, \tag{7.180}$$

$$v'_v = V_{\theta m} e^{-i\frac{\omega}{\overline{u}}x} e^{i(m\theta + \omega t)}. \tag{7.181}$$

It should be pointed out that the divergence of the vorticity is zero, which leads to,

$$V_{\theta m} = \frac{\omega R}{\overline{u} m} V_{xm}. \tag{7.182}$$

Thus, the total density disturbance can be written as:

$$\rho' = \rho'_a + \rho'_e = \frac{1}{\overline{c}^2}\left(P_m^+ e^{i\alpha_m^+ x} + P_m^- e^{i\alpha_{mn}^- x} - \frac{\overline{\rho}\overline{c}^2}{C_p} S_{em} e^{-i\frac{\omega}{\overline{u}}x} \right) e^{i(\omega t + m\theta)}. \tag{7.183}$$

The total velocity disturbances are:

$$u' = u'_a + u'_v = -\frac{1}{\overline{\rho}}\left(\frac{\alpha_m^+ P_m^+ e^{i\alpha_m^+ x}}{\omega + \alpha_m^+ \overline{u}} + \frac{\alpha_m^- P_m^- e^{i\alpha_m^- x}}{\omega + \alpha_m^- \overline{u}} - \overline{\rho} V_{xm} e^{-i\frac{\omega}{\overline{u}}x} \right) e^{i(\omega t + m\theta)}, \tag{7.184}$$

$$v' = v'_a + v'_v = -\frac{1}{\overline{\rho}} \left[\begin{array}{l} \dfrac{m P_m^+ e^{i\alpha_m^+ x}}{R(\omega + \alpha_m^+ \overline{u})} + \dfrac{m P_m^- e^{i\alpha_m^- x}}{R(\omega + \alpha_m^- \overline{u})} \\ -\dfrac{\omega R \rho_0}{\overline{u} m} V_{xm} e^{-i\frac{\omega}{\overline{u}}x} \end{array} \right] e^{i(m\theta + \omega t)}. \tag{7.185}$$

It can be seen that, in the mean flow, the acoustic, vorticity, and entropy waves propagate independently. The pressure perturbation is dependent on acoustic disturbance only. But the density perturbation is the superposition of acoustic and entropy disturbances, and the velocity perturbation is induced by both acoustic and vorticity disturbances.

Eqs. (7.174) and (7.183)–(7.185) describe the unsteady flow fields of the up- and downstream regions of the flame. Actually, these two regions are coupled through several matching conditions. Applying conservation law of mass, momentum, and energy on the interface of two regions (as shown in Fig. 7.7) and employing a linearization procedure yields the following equations that describe the correlations of the disturbance quantities across the flame, where subscripts 1 and 2 denote the variables of up- and downstream regions of the flame.

Mass equation:

$$\overline{\rho}_1 u_1' + \overline{u}_1 \rho_1' = \overline{\rho}_2 u_2' + \overline{u}_2 \rho_2'. \tag{7.186}$$

Axial momentum equation:

$$\left(2\overline{u}_1\overline{\rho}_1 u_1' + \overline{u}_1^2 \rho_1'\right) - \left(2\overline{u}_2\overline{\rho}_2 u_2' + \overline{u}_2^2 \rho_2'\right) = p_2' - p_1'. \tag{7.187}$$

Circumferential momentum equation:

$$v_1' = v_2'. \tag{7.188}$$

Energy equation:

$$\begin{aligned} &\frac{\gamma \overline{u}_1}{\gamma - 1} p_1' + \left(\frac{\gamma \overline{p}_1}{\gamma - 1} + \frac{3}{2}\overline{\rho}_1 \overline{u}_1^2\right) u_1' + \frac{\overline{u}_1^3}{2}\rho_1' \\ &= \frac{\gamma \overline{u}_2}{\gamma - 1} p_2' + \left(\frac{\gamma \overline{p}_2}{\gamma - 1} + \frac{3}{2}\overline{\rho}_2 \overline{u}_2^2\right) u_2' + \frac{\overline{u}_2^3}{2}\rho_2' - Q', \end{aligned} \tag{7.189}$$

where γ is the ratio of specific heats. Q' denotes unsteady heat release per unit area.

Because we focus on the azimuthal modes, the inlet boundary condition is taken to be an acoustic closed end, which can be expressed as:

$$u_1'|_{x=0} = 0. \tag{7.190}$$

In the traditional method for analysis of combustion instabilities, impedance boundary conditions usually are given for the inlet and outlet. However, when it has a mean flow, there will be three types of waves, recognized as acoustic, entropy, and vorticity waves, propagating in the system. It has been discussed that the acoustic disturbance propagates up- and downstream, whereas the entropy and vorticity disturbances only travel downstream with the flow. Thus, in all three, inlet boundary conditions and one outlet boundary condition are needed to determine all the quantities of the four waves. Stow et al. have discussed the boundary conditions corresponding to choked outlet and choked inlet cases [15]. To observe the generation of the vorticity disturbance by the interaction of acoustic wave and heat source, the choked inlet boundary condition is not adopted here. We assume that the inlet entropy and vorticity disturbances are zero. Therefore, no entropy and vorticity waves propagate in region 1. These two complementary boundary conditions can be given as:

$$\rho_{e1}' = 0, \tag{7.191}$$

$$u_{v1}' = 0. \tag{7.192}$$

For the outlet, two forms of boundary conditions are considered, respectively, for the acoustically closed end:

$$u'_2|_{x=L} = 0. \tag{7.193}$$

and for the choked end [21]:

$$-\frac{p'_2}{\overline{p}_2} + \frac{\rho'_2}{\overline{\rho}_2} + 2\frac{u'_2}{\overline{u}_2}\bigg|_{x=L} = 0. \tag{7.194}$$

The azimuthal modes can be formed with either the closed or the choked outlet. For the closed boundary condition in Eq. (7.193), the sum of acoustic and vorticity axial velocity disturbances is constrained to be zero, which denotes that the acoustic and vorticity waves are coupled at the outlet, but the entropy wave will freely propagate out of the duct. So, with this boundary condition, we can investigate the coupling of vorticity and acoustic waves at the outlet without the effect of an entropy wave. The choked boundary is also investigated because it is coherent with the outlet of real annular combustors connected to a nozzle. For such a choked outlet boundary condition in Eq. (7.194), the three types of waves are all coupled at the outlet. When the flow velocity tends to zero, Eq. (7.193) can be recovered from Eq. (7.194).

With these assumptions, it can be found that, for the present case, there are no entropy and vorticity waves in region 1, and in region 2 they are generated by the interaction between acoustic waves and the heat source. The generation process of such disturbances is described by Eqs. (7.186)–(7.189).

For general cases, the unsteady heat release is given as:

$$Q' = K_u u'_1|_{x=L_1} + K_p p'_1|_{x=L_1}, \tag{7.195}$$

where L_1 is the length of the region 1. This equation denotes that the unsteady heat release is induced by axial velocity or pressure disturbances in region 1 at the place of the flame. In this chapter, the effect of unsteady heat release depending on velocity, pressure disturbance, and no unsteady heat release will be investigated, respectively.

Substituting Eqs. (7.174), (7.183)–(7.185) to (7.186)–(7.189), (7.190)–(7.194) and considering Eq. (7.195) yields:

$$\mathbf{X}\begin{pmatrix} P_{m1}^{+} \\ P_{m1}^{-} \\ P_{m2}^{+} \\ P_{m2}^{-} \\ s_{em2} \\ V_{xm2} \\ Q' \end{pmatrix} = 0, \tag{7.196}$$

where **X** is a 7 × 7 matrix, and it has the following form:

$$\begin{pmatrix} \left(\frac{\bar{u}_1}{\bar{c}_1^2}-A_1\right)e_1 & \left(\frac{\bar{u}_1}{\bar{c}_1^2}-A_2\right)e_2 & -\left(\frac{\bar{u}_2}{\bar{c}_2^2}-A_3\right) & -\left(\frac{\bar{u}_2}{\bar{c}_2^2}-A_4\right) & -\bar{u}_2 & -\bar{\rho}_2 & 0 \\ -\left(1+\overline{M}_1^2-2\bar{u}_1A_1\right)e_1 & -\left(1+\overline{M}_1^2-2\bar{u}_1A_2\right)e_2 & \left(1+\overline{M}_2^2-2\bar{u}_2A_3\right) & \left(1+\overline{M}_2^2-2\bar{u}_2A_4\right) & \bar{u}_2^2 & 2\bar{u}_2\bar{\rho}_2 & 0 \\ \left(B_1\bar{u}_1-\frac{B_3A_1}{\bar{\rho}_1}\right)e_1 & \left(B_1\bar{u}_1-\frac{B_3A_2}{\bar{\rho}_1}\right)e_2 & -\left(B_2\bar{u}_2-\frac{B_4A_3}{\bar{\rho}_2}\right) & -\left(B_2\bar{u}_2-\frac{B_4A_4}{\bar{\rho}_2}\right) & -\frac{\bar{u}_2^3}{2} & -B_4 & 1 \\ A_5e_1 & A_6e_2 & -A_7 & -A_8 & 0 & \frac{\omega R\bar{\rho}_2}{m} & 0 \\ A_1 & A_2 & 0 & 0 & 0 & 0 & 0 \\ 0 & 0 & A_3e_3 & A_4e_4 & 0 & -\bar{\rho}_2e_5 & 0 \\ -\left(K_p-\frac{K_v}{\bar{\rho}_1}A_1\right)e_1 & -\left(K_p-\frac{K_v}{\bar{\rho}_1}A_2\right)e_2 & 0 & 0 & 0 & 0 & 1 \end{pmatrix}, \quad (7.197)$$

with

$$A_1=\frac{\alpha_{m1}^{+}}{\omega+\alpha_{m1}^{+}\bar{u}_1},\quad A_2=\frac{\alpha_{m1}^{-}}{\omega+\alpha_{m1}^{-}\bar{u}_1},\quad A_3=\frac{\alpha_{m2}^{+}}{\omega+\alpha_{m2}^{+}\bar{u}_2},\quad A_4=\frac{\alpha_{m2}^{-}}{\omega+\alpha_{m2}^{-}\bar{u}_2}$$

$$A_5=\frac{m\bar{u}_1}{R(\omega+\alpha_{m1}^{+}\bar{u}_1)},\ A_6=\frac{m\bar{u}_1}{R(\omega+\alpha_{m1}^{-}\bar{u}_1)},\ A_7=\frac{m\bar{u}_2}{R(\omega+\alpha_{m2}^{+}\bar{u}_2)},\ A_8=\frac{m\bar{u}_2}{R(\omega+\alpha_{m2}^{-}\bar{u}_2)}$$

$$B_1=\frac{\gamma}{\gamma-1}+\frac{\overline{M}_1^2}{2},\quad B_2=\frac{\gamma}{\gamma-1}+\frac{\overline{M}_2^2}{2},\quad B_3=\frac{\gamma\bar{p}_1}{\gamma-1}+\frac{3}{2}\bar{\rho}_1\bar{u}_1^2,\quad B_4=\frac{\gamma\bar{p}_2}{\gamma-1}+\frac{3}{2}\bar{\rho}_2\bar{u}_2^2$$

$$e_1=e^{\mathrm{i}\alpha_{m1}^{+}L_1},\quad e_2=e^{\mathrm{i}\alpha_{m1}^{-}L_1},\quad e_3=e^{\mathrm{i}\alpha_{m2}^{+}L_2},\quad e_4=e^{\mathrm{i}\alpha_{m2}^{-}L_2},\quad e_5=e^{-\mathrm{i}\frac{\omega}{\bar{u}_2}L_2},$$

where L_2 is the length of the region 2. The matrix is for the boundary condition in Eq. (7.193). For choked boundary condition, it should substitute the sixth row to the corresponding terms. Eq. (7.196) has a zero solution unless the determinant of $\mathbf{X}$ vanishes. Therefore, only disturbances with which the determinant of $\mathbf{X}$ equal zero can exist. Because the mean flow parameters are known, it forms an equation for ω, which can be solved numerically. And the relative values of the unknown parameters of the flow disturbances in Eqs. (7.174), (7.183)–(7.185) can be determined through Eq. (7.196).

To investigate the effect of the vorticity wave, the results of AM without vorticity disturbance should also be given for comparison. If the vorticity wave is not considered, the flow disturbances are described by Eqs. (7.174)–(7.176) and (7.183). It should be noted that the entropy waves are included in the model all the same. The jump conditions are given by Eqs. (7.186), (7.187), and (7.189), which have been commonly used in the analysis of thermoacoustic instabilities. Because the vorticity disturbance is not included, the circumferential jump condition is not needed. Then the eigenfrequency can be obtained through the same process introduced earlier.

It has been discussed that vorticity waves exist in azimuthal instabilities, and an AM in which three types of disturbances are included has been given. With this analytical model, it is convenient to present the differences with or without the vorticity waves. However, whether the results from the analytical model are correct also needs to be validated. For this reason, LEEM as introduced in Section 7.3.2 will be used. In LEEM, the disturbances in regions 1 and 2 (as shown in Fig. 7.7) are described by a linear Euler equation, therefore all forms of flow disturbances are included. The heat source will be treated in two manners, which are compacted in a plate and distributed in a thin region, respectively. In the first case, the same jump conditions using the AM are also needed. In the second case, such conditions are no longer needed. The AM including vorticity disturbance can be validated by comparison of the results, and the jump conditions can also be validated by comparison of these methods.

Using the assumptions made in previous sections, the linearized equations for perturbations can be written in the following forms:

$$\frac{\partial \rho'}{\partial t} + \nabla \cdot (\overline{\rho}\boldsymbol{u}') + \nabla \cdot (\rho'\overline{\boldsymbol{u}}) = 0, \tag{7.198}$$

$$\left(\overline{\rho}\frac{\partial}{\partial t}+\overline{\rho}\overline{\boldsymbol{u}}\cdot\nabla\right)\boldsymbol{u}'+\overline{\rho}\boldsymbol{u}'\cdot\nabla\overline{\boldsymbol{u}}+\left(\frac{\partial\overline{\boldsymbol{u}}}{\partial t}+\overline{\boldsymbol{u}}\cdot\nabla\overline{\boldsymbol{u}}\right)\rho'+\nabla p'=0, \tag{7.199}$$

$$\begin{aligned}&\left(\frac{\partial}{\partial t}+\overline{\boldsymbol{u}}\cdot\nabla\right)p'-\left(\frac{\gamma}{\overline{\rho}}\frac{\partial\overline{\rho}}{\partial t}+\frac{\gamma}{\overline{\rho}}\overline{\boldsymbol{u}}\cdot\nabla\overline{\rho}\right)p'-\left(\overline{c}^2\frac{\partial}{\partial t}+\overline{c}^2\overline{\boldsymbol{u}}\cdot\nabla\right)\rho'\\&+\left(\frac{\overline{c}^2}{\overline{\rho}}\frac{\partial\overline{\rho}}{\partial t}+\frac{\overline{c}^2}{\overline{\rho}}\overline{\boldsymbol{u}}\cdot\nabla\overline{\rho}\right)\rho'-\overline{c}^2\boldsymbol{u}'\cdot\nabla\overline{\rho}+\boldsymbol{u}'\cdot\nabla\overline{p}=(\gamma-1)Q',\end{aligned} \tag{7.200}$$

where $u=(u,\ v)$ is the velocity vector. The perturbation variables can be written in forms of time harmonics:

$$\begin{aligned}\rho'&=\hat{\rho}e^{\mathrm{i}\omega t},\\ p'&=\hat{p}e^{\mathrm{i}\omega t},\\ \boldsymbol{u}'&=\hat{\boldsymbol{u}}e^{\mathrm{i}\omega t},\end{aligned} \tag{7.201}$$

Here, the overhat ○ denotes a complex function of spatial coordinates. The equations are next expanded in cylindrical coordinates. Note that the mean flow of the circumferential direction is ignored, and then equations of such forms can be obtained that:

$$C_1\hat{\rho}+C_2\frac{\partial\hat{\rho}}{\partial x}+C_3\frac{\partial\hat{v}}{\partial\theta}+C_4\frac{\partial\hat{u}}{\partial x}+C_5\hat{v}+C_6\hat{u}=0, \tag{7.202}$$

$$D_1\hat{v}+D_2\frac{\partial\hat{v}}{\partial x}+D_3\frac{\partial\hat{p}}{\partial\theta}=0, \tag{7.203}$$

$$E_1\hat{\rho}+E_2\hat{u}+E_3\frac{\partial\hat{u}}{\partial x}+E_4\hat{v}+E_5\frac{\partial\hat{p}}{\partial x}=0, \tag{7.204}$$

$$F_1\hat{\rho}+F_2\frac{\partial\hat{\rho}}{\partial x}+F_3\hat{v}+F_4\hat{u}+F_5\hat{p}+F_6\frac{\partial\hat{p}}{\partial x}=(\gamma-1)\hat{Q}, \tag{7.205}$$

where the coefficients C, D, E, and F are functions of the mean flow parameters, which are given in Appendix C.

Spectral method is adopted here for solving the eigenvalue problem. In Eqs. (7.202) through (7.205), the unknowns are discretized by Fourier methods in the circumferential direction and Chebyshev methods in the axial direction:

$$\chi(x,\theta)=\sum_{i=0}^{N_x}\sum_{k=1}^{N_\theta}F_{ik}T_i(x)\mathrm{e}^{\mathrm{i}k\theta}, \tag{7.206}$$

where χ represents unknowns; F_{ik} denotes the unknown coefficients; N_x and N_θ are numbers of collocation points in axial and circumferential directions, respectively; and $T_i(x)$ is the Chebyshev polynomial of degree i defined for x in the range $[-1.1]$. Thus, the derivatives of the variables at the collocation points can be obtained:

$$\left.\frac{\partial \chi}{\partial x}\right|_{i,k} = \chi_x^1(x_i, \theta_k) = \sum_{m=0}^{N_x} d_{i,m}^1 \chi(x_m, \theta_k), \tag{7.207}$$

$$\left.\frac{\partial \chi}{\partial \theta}\right|_{i,k} = \chi_\theta^1(x_i, \theta_k) = \sum_{p=1}^{N_\theta} e_{k,p}^1 \chi(x_i, \theta_p), \tag{7.208}$$

where $i = 0, \ldots, N_x$, and $k = 0, \ldots, N_\theta$. $e_{k,\ p}^1$, $d_{i,\ m}^1$ are the first Fourier and Chebyshev derivative matrix given by Peyret [19]. Eqs. (7.202) through (7.205) can be discretized using the technique introduced earlier on both regions 1 and 2 shown in Fig. 7.7. Next, the treatment of the heat source will be introduced.

The first kind of heat source treatment that the heat input is assumed to be concentrated on a plate was introduced in Section 7.3.2.3. Here, another kind of heat input treatment will be introduced, that is, when the flame is distributed in a thin region.

Such method for treating longitudinal modes with distributed heat source is detailed in Li et al. [22]. In this case, the whole duct is divided into three regions, as shown in Fig. 7.8. In zone 1 and zone 2, there is no heat source, and the flow is uniform. Therefore, Q' in Eq. (7.200) should also be set to zero for these two regions. In zone f, the relevant equations are also Eqs. (7.202) through (7.205). However, the flow is axially non-uniform for the existence of the heat release. It is assumed that the mean heat is uniformly distributed over an axial length d. Assume that the center of the heat source is

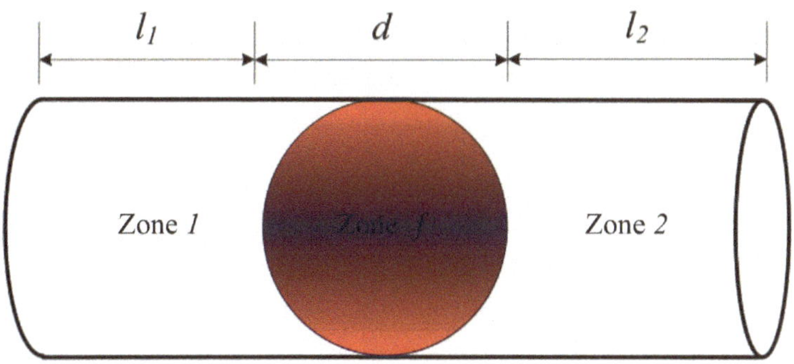

Fig. 7.8 The distributed heat source [20].

at $x = b$, where x is the axial coordinate. The mean heat input is given in the following form:

$$\overline{q}(x) = \begin{cases} c_p\overline{\rho}_1\overline{u}_1\left(\overline{T}_{02} - \overline{T}_{01}\right)/d & \text{for } |x-b| < \frac{1}{2}d \\ 0 & \text{for } |x-b| > \frac{1}{2}d \end{cases}. \tag{7.209}$$

Then it is easier to obtain the flow through the duct by integrating the one-dimensional conservation equations of mass, momentum, and energy. The distributed unsteady heat release can be written as:

$$Q'(x) = K_u(x)u_1'|_{x=L_1} + K_p(x)p_1'|_{x=L_1}. \tag{7.210}$$

Now the equations for these zones are discretized separately. In fact, the flow parameters are continuous across the interface of the two zones. Thus, we have:

$$\begin{aligned} p_1' &= p_f', \\ \rho_1' &= \rho_f', \\ u_1' &= u_f', \\ v_1' &= v_f', \end{aligned} \tag{7.211}$$

where the subscript f denotes zone f. The same relations are given for zone f and zone 2. It can be seen that, when zone f is thin enough, the case of distributed heat source is coherent with compacted heat source.

The basic equations and discretization method are particularly introduced in Section 7.3.2. The same boundary conditions as Eqs. (7.190)–(7.194) are needed for the LEEM. Actually, Eqs. (7.190), (7.193), and (7.194) can be used directly. However, no analytical form for entropy and vorticity disturbances can be found in linear Euler equations. Note that the difference between pressure and density disturbances is brought by entropy waves. Hence, a zero inlet entropy disturbance condition can be obtained by equation:

$$\left.\rho_1' - \frac{p_1'}{\overline{c}_1^2}\right|_{x=0} = 0. \tag{7.212}$$

Similarly, a zero inlet vorticity disturbance condition can also be obtained by equation:

$$\left.\frac{1}{R}\frac{\partial u_1'}{\partial \theta} - \frac{\partial v_1'}{\partial x}\right|_{x=0} = 0. \tag{7.213}$$

In fact, substituting analytical expressions of the disturbances in Eqs. (7.174), (7.183)–(7.185) to Eqs. (7.212) and (7.213), we can obtain Eqs. (7.191) and (7.192). Hence, they are coherent.

For the compacted heat source, imposing the inlet and outlet boundary conditions, and jump conditions across the flame plane, the discrete forms of Eq. (7.202) through (7.205) can be written as:

$$\begin{pmatrix} \boldsymbol{\zeta}_1 & \boldsymbol{\zeta}_3 \\ \boldsymbol{\zeta}_4 & \boldsymbol{\zeta}_2 \end{pmatrix} \begin{pmatrix} \boldsymbol{\xi}_1 \\ \boldsymbol{\xi}_2 \end{pmatrix} = 0. \tag{7.214}$$

Here, $\boldsymbol{\zeta}_1$, $\boldsymbol{\zeta}_2$, $\boldsymbol{\zeta}_3$, and $\boldsymbol{\zeta}_4$ are submatrices of size $[4(N+1)]^2$ with only the complex frequency ω as the unknown, where $N=(N_x+1)N_t$–1. The column vectors $\boldsymbol{\xi}_1$ and $\boldsymbol{\xi}_2$ of length $4(N+1)$ contain the discrete unknowns in region 1 and 2, respectively. If Eq. (7.214) has non-zero solutions, the determinant of the matrix must be zero, and the complex frequency ω can be obtained by solving the equation:

$$\begin{vmatrix} \boldsymbol{\zeta}_1 & \boldsymbol{\zeta}_3 \\ \boldsymbol{\zeta}_4 & \boldsymbol{\zeta}_2 \end{vmatrix} = 0, \tag{7.215}$$

and with the frequency known, the spatial distributions in the flow can be acquired from Eq. (7.214).

For the distributed heat source, the system equations have the following form:

$$\begin{pmatrix} \boldsymbol{\zeta}_1 & \boldsymbol{\zeta}_3 & 0 \\ \boldsymbol{\zeta}_4 & \boldsymbol{\zeta}_f & \boldsymbol{\zeta}_5 \\ 0 & \boldsymbol{\zeta}_6 & \boldsymbol{\zeta}_2 \end{pmatrix} \begin{pmatrix} \boldsymbol{\xi}_1 \\ \boldsymbol{\xi}_f \\ \boldsymbol{\xi}_2 \end{pmatrix} = 0, \tag{7.216}$$

where $\boldsymbol{\zeta}$ and $\boldsymbol{\xi}$ are defined the same as Eq. (7.214). The eigenfrequency can also be solved through the same process.

As shown in Fig. 7.7, the geometry parameters are given as the entire length of the chamber $L=0.2$ m, length of region 1 and region 2 $L1=0.1$ m, $L2=0.1$ m, respectively, and the mean radius of the annular chamber $R=0.3$ m. For the mean flow of the region 1, the pressure is 4 Mpa, and the stagnation temperature is 700 K. Therefore, with the given inlet flow Mach number and the temperature ratio across the flame, the mean flow parameters of region 2 can be calculated through conservation relations. Some of these parameters are in reference to Stow and Dowling [23]. The mean heat input per unit area from the flame can be given as:

$$\overline{Q}=C_p\left(K_t\overline{T}_{01}-\overline{T}_{01}\right)\overline{\rho}_1\overline{u}_1, \tag{7.217}$$

with

$$\overline{T}_{02} = K_t \overline{T}_{01},$$

where the subscript 0 denotes the stagnation temperature, and K_t denotes the stagnation temperature ratio across the flame. Three form of unsteady heat release are given as:

$$Q' = 0, \tag{7.218}$$

$$\frac{Q'}{\overline{Q}} = k_u \frac{u_1'}{\overline{u}_1} e^{-i\omega\tau} |_{x=L_1}, \tag{7.219}$$

$$\frac{Q'}{\overline{Q}} = k_p \frac{p_1'}{\overline{p}_1} e^{-i\omega\tau} |_{x=L_1}, \tag{7.220}$$

where τ denotes the time lag between the unsteady heat release and reference disturbance, and k_u and k_{p} are coefficients. Eq. (7.218 denotes no unsteady heat release. Eqs. (7.219) and (7.220) denote that the unsteady heat release is induced by velocity and pressure disturbances at the place of flame in region 1, respectively. For distributed heat source used in LEEM, the unsteady heat release can be written as:

$$\begin{aligned} &Q'(x) = 0, \\ &\frac{Q'(x)}{\overline{q}(x)} = k_u \frac{u_1'(L_1)}{\overline{u}_1} e^{-i\omega\tau}, \\ &\frac{Q'(x)}{\overline{q}(x)} = k_p \frac{p_1'(L_1)}{\overline{p}_1} e^{-i\omega\tau}, \end{aligned} \tag{7.221}$$

where $q(x)$ is given by Eq. (7.209).

Lieuwen [24] discussed the main causes of unsteady combustion. It shows that in a real lean premixed combustor, the unsteady heat release is greatly influenced by the velocity disturbance near the fuel bars rather than by the unsteady pressure. However, all three forms of the unsteady heat release are used here. On one hand, we want to investigate how the unsteady heat release affects the generation of vorticity waves. On the other hand, we also want to know whether it can reach a general conclusion by using these different forms of unsteady heat release.

7.3.3.2 Results and discussion

Results for the acoustically closed outlet boundary condition are given in the following section. Fig. 7.9 depicts the real and imaginary part of the frequency of azimuthal mode $m = 1$ as a function of temperature ratio across

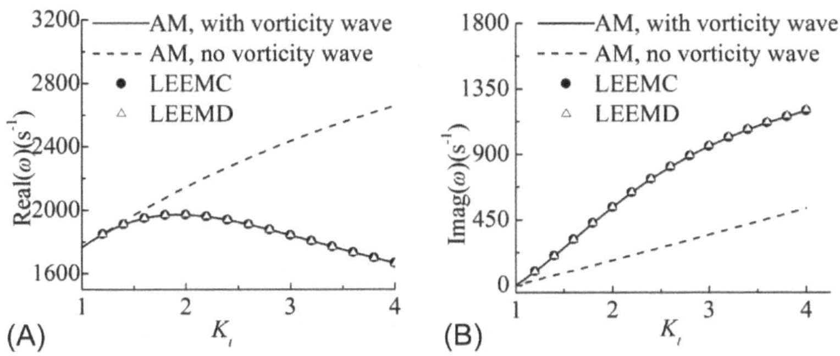

Fig. 7.9 The real and imaginary part of the frequency as a function of K_t, for no unsteady heat release, $m = 1$, no axial modes, and inlet flow Mach number 0.1 [20]. (A) Real part of the frequency; (B) Imaginary part of the frequency.

the flame K_t for no unsteady heat release (Eq. (7.213)) and inlet flow Mach number 0.1. Four types of results are given. They are the results from the AM with and without the effect of vorticity waves, LEEM with a compact heat source (denoted by LEEMC), and distributed heat source (LEEMD). For the distributed heat source, the length of the three regions is given as region 1 0.099 m, region f 0.02 m, and region 2 0.099 m. Good agreement is observed between results of AM with the vorticity disturbance and LEEM, which shows that the solutions of the unsteady flow field (Eqs. 7.174, 7.183–7.185) are correct. Moreover, the results of LEEMC and LEEMD are also the same. It indicates the jump conditions given by Eqs. (7.186)–(7.189) are reliable. Furthermore, for the real part of the frequency, when the temperature ratio across the flame K_t is small (<1.5), the results of the AM without vorticity waves are almost the same as the results with vorticity wave. However, with the increase of the value of K_t, a significant deviation appears between results with and without the vorticity wave. For the imaginary part of the frequency, the differences between the results with and without vorticity wave are large. Note that, because the unsteady heat release is zero, the imaginary part of the frequency is positive, which denotes that no instabilities occur. When the temperature and the mean velocity in regions 1 and 2 are not equal, the reflections of inlet and outlet are not symmetrical, which would bring damping to the system.

Fig. 7.10 depicts the real and imaginary part of the frequency of azimuthal mode $m = 1$ as a function of temperature ratio across the flame K_t for the unsteady heat release related to axial velocity disturbance (Eq. (7.48), $k_u = 1$ and $\tau = 0$) and inlet flow Mach number 0.1. Fig. 7.11 shows the results for the unsteady heat release related to pressure disturbance

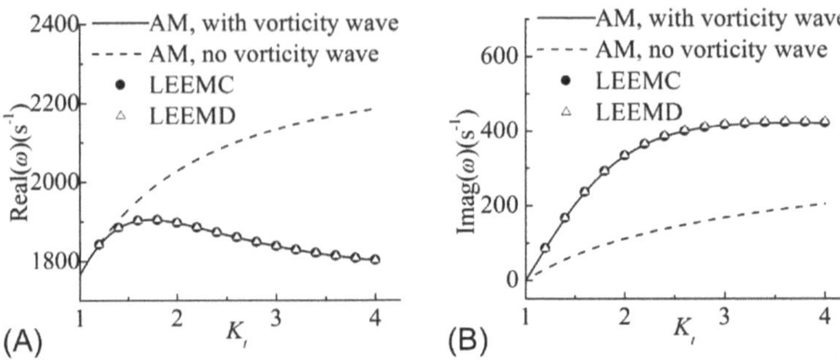

Fig. 7.10 The real and imaginary part of the frequency as a function of K_t, for unsteady heat release related to axial velocity disturbance $k_u = 1$, $\tau = 0$, $m = 1$, no axial modes, and inlet flow Mach number 0.1 [20]. (A) Real part of the frequency; (B) Imaginary part of the frequency.

(Eq. (7.49), $k_p = 2$, $\tau = 0.0003$ s) and flow Mach number 0.05. If the angular frequency is given to be 2000 s^{-1}, the value of the time delay is about one-tenth of the period. Good agreements are also found between the results of AM with vorticity wave and LEEM in Figs. 7.10 and 7.11. When the value of K_t is small, the real part of the frequencies obtained by AM with vorticity wave almost equals the results without the vorticity wave, which is the same as Fig. 7.9. When K_t becomes large, the difference is significant. For the imaginary part of the frequency, the results with and without vorticity waves are not the same at all. In Fig. 7.11, it is found that the imaginary part of the frequency obtained by AM without the vorticity wave is below zero, which denotes that this mode is unstable. But the results with the effect of the vorticity wave are positive, which shows the mode is stable.

The outlet boundary condition is given to be acoustically closed end, where the coupling between axial velocity and vorticity disturbances occurs. The coupling of the acoustic and vorticity disturbances significantly changes the reflection of the outlet boundary. Therefore, in Figs. 7.9–7.11, we can observe that, for the real part of the frequency, the results with and without the vorticity waves are different. It can also be observed from Figs. 7.9–7.11 that the imaginary parts of the frequencies obtained by AM with the vorticity wave are significantly greater than the results without the vorticity waves. In Fig. 7.11, the calculated results without the vorticity wave show the system is unstable, but the results with the vorticity wave denote that the system is very stable. In fact, the generation of a vorticity disturbance from the interaction of an acoustic disturbance and heat release transfers energy from the acoustic mode to vertical mode, which causes damping to the acoustic perturbation. Such vorticity waves propagate out of the boundary with the

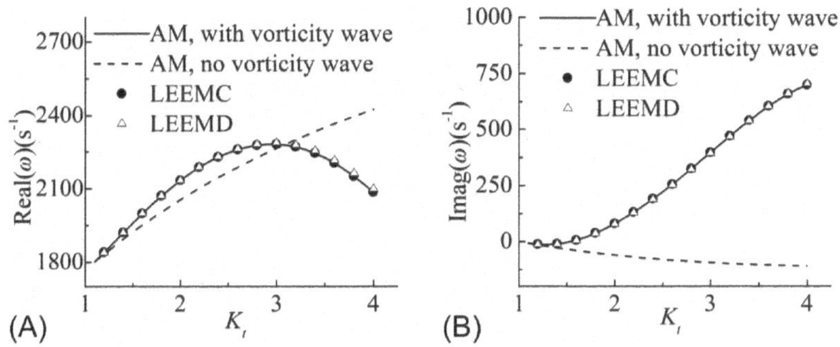

Fig. 7.11 The real and imaginary part of the frequency as a function of K_t, for unsteady heat release related to pressure disturbance, $k_p = 2$, $\tau = 0.0003$ s, $m = 1$, no axial modes, and inlet flow Mach number 0.05[20]. (A) Real part of the frequency; (B) Imaginary part of the frequency.

flow. Then, the interaction between vorticity disturbances and the outlet boundary generates acoustic wave propagating upstream. All the results show that the system is more stable when the vorticity waves are considered. It means that the vorticity disturbances cause damping to the system. From Figs. 7.9–7.11, it is also found that vorticity waves are generated with all forms of unsteady heat release, even when the unsteady heat release is zero.

Fig. 7.12 depicts the frequencies as a function of inlet flow Mach number for $m = 1$, no axial modes, the unsteady heat release is related to the axial velocity disturbances ($k_u = 1$, $\tau = 0$), and inlet temperature ratio across flame $K_t = 3$. Good agreements are observed between the results of AM with the vorticity wave and LEEM. The results of LEEMC and LEEMD are the same. The differences between results with and without the vorticity

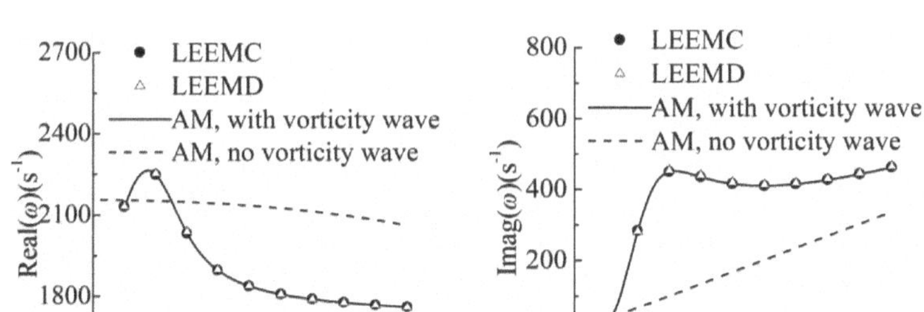

Fig. 7.12 The real and imaginary part of the frequency as a function of inlet flow Mach number, for unsteady heat release related to axial velocity disturbance, $k_u = 1$, $\tau = 0$, $m = 1$, no axial modes, and inlet temperature ratio across flame $K_t = 3$ [20]. (A) Real part of the frequency; (B) Imaginary part of the frequency.

wave are significant. For the real part of the frequency, the results without the vorticity wave decrease with an increase of flow Mach number. This phenomenon accords with the results that the frequency of axial mode decrease as the flow Mach number is increased [9]. With the effect of the vorticity disturbance, when the flow Mach number increases, the frequencies also increase if the Mach number is not large. However, when the flow Mach number is greater than some value (e.g., 0.05), the frequencies decrease. If the values are given as $K_t = 3$, it can be estimated that $\omega = 2200$ Hz, and the sound speed of the region 2 is about 900 m/s. The wavelength of vorticity wave in region 2 is determined by $2\pi u_2/\omega$. If the wavelength is 0.1 m, which is the length of the region 2, the value of the mean flow velocity is about 35 m/s, and the flow Mach number is about 0.04. The Mach number of the flow in region 1 is the same order with this. When the flow Mach number is greater than this value, there will not be an entire wavelength of vorticity wave in region 2. But if the flow Mach number is less than the value, there will be one more wavelength in region 2. This is why the reflection points exist in the curves of frequency-flow Mach number. When the wavelength is equal to 0.1 m, it can be deduced that the vorticity disturbance at the outlet boundary is just in phase with the acoustic wave. At this condition, the frequency will not be affected by the vorticity disturbances.

Next, the eigenfrequency with the choked outlet boundary condition will be discussed. For the choked outlet boundary, the acoustic, vorticity, and entropy waves are coupled. This is different from the closed outlet boundary condition, for which only acoustic and vorticity disturbance are constrained, but the entropy wave is free. Fig. 7.13 depicts the real and imaginary part of the frequency of azimuthal mode $m = 1$ as a function of

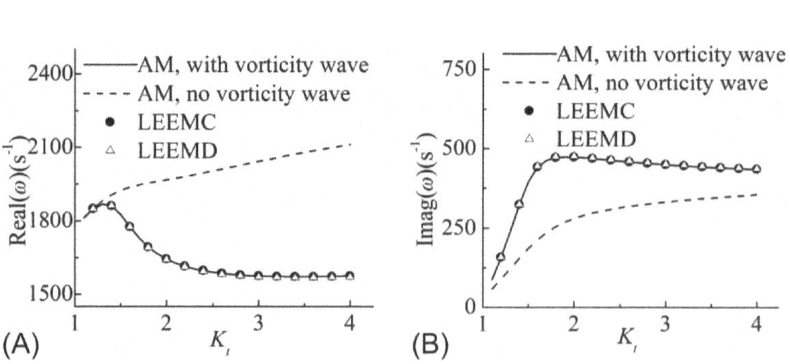

Fig. 7.13 The real and imaginary part of the frequency as a function of K_t, for unsteady heat release related to axial velocity disturbance, $k_u = 0.2$, $\tau = 0$, $m = 1$, no axial modes, and inlet flow Mach number 0.1 [20]. (A) Real part of the frequency; (B) Imaginary part of the frequency.

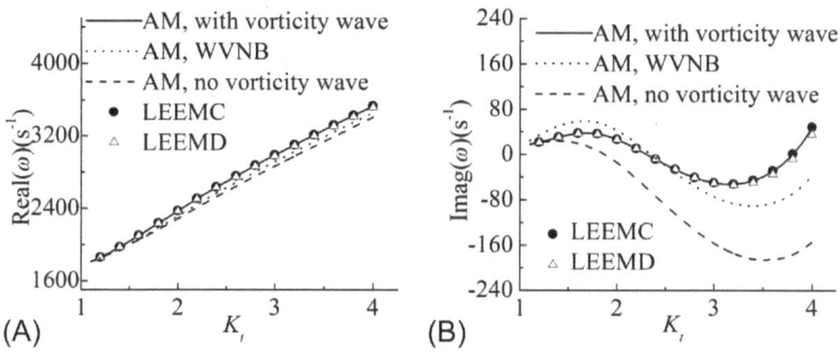

Fig. 7.14 The real and imaginary part of the frequency as a function of K_t, for unsteady heat release related to axial velocity disturbance, $k_u = 0.2$, $\tau = 0.0015$ s, $m = 1$, no axial modes, and inlet flow Mach number 0.05; WVNB denotes the results with vorticity wave but no boundary coupling with vorticity wave [20]. (A) Real part of the frequency; (B) Imaginary part of the frequency.

temperature ratio across the flame K_t for unsteady heat release related with axial velocity disturbances ($k_u = 0.2$, $\tau = 0$) and inlet flow Mach number 0.1. It also gives results of four methods. The results with and without vorticity waves have large discrepancies. The agreement between results of AM with vorticity wave and LEEM is good. And the results of LEEMC and LEEMD are the same, as well. Fig. 7.14 shows the eigenfrequency of azimuthal mode $m = 1$ for $k_u = 0.1$, $\tau = 0.0015$ s, and inlet flow Mach number 0.05. For angular frequency 3000 s^{-1}, the period is about 0.002 s. The time lag is 3/4 of the period. From the results without the vorticity wave, it can be seen that the imaginary part of the frequency is negative, which denotes the system is unstable. But with the vorticity wave, the imaginary part of frequency is positive for some range of K_t. And the value with the vorticity wave is always greater than the value without the vorticity wave. Moreover, the real part of a frequency obtained with and without vorticity disturbance are almost the same. This is because, under this condition, the wavelength of such a disturbance is just equal to length of the region 2 (see Fig. 7.7).

Fig. 7.14 gives the results with the vorticity wave, but no boundary coupling of the vorticity wave is given. Under this condition, the vorticity disturbances propagate out of the system without reflection. It means that all the energy transferred from the acoustics to the vertical mode is damped. The imaginary part of the frequency is greater than the results without the vorticity wave. When K_t is small, the imaginary part is also greater than the results with the coupling of the vorticity wave at the boundary. However, when K_t gets larger, the value is less than that with the coupling effect at the boundary. It denotes that, under this condition, the coupling between acoustic, vorticity, and entropy waves reflects less perturbation energy than that without coupling effect of the vorticity waves. The mode structures of the pressure, density, and velocity disturbances and the specific explanation of the generation of vorticity wave can be seen in a study by Lieuwen [24].

In this section, the problem of the azimuthal instabilities in an annular chamber is investigated. With the assumptions of mean flow and compacted heat release, the effects of vorticity waves are studied in an analytical model. The LEEM is applied to validate the analytical model with the vorticity wave. In the Euler equation method, the heat source is assumed to be compacted in a plate or distributed in a thin region. The results of AM without the vorticity wave are also given on the analogy of a network model and finite element method. After the calculations, good agreements are observed between results of the AM with the effect of the vorticity wave and the linear Euler equations method. And results of the LEEM with compacted and distributed heat sources are also the same. However, for pure azimuthal modes, significant differences are found between the results with and without vorticity waves in AM. For the mixed mode of azimuthal and axial modes, the results with and without vorticity waves are almost the same, which shows that the effects of vorticity waves are weak for mixed modes. Moreover, for longitudinal modes, the results without vorticity disturbances and LEEM are the same. From these investigations, the following conclusions can be obtained: (a) The comparison of AM and LEEM shows that the description of the flow disturbances in AM is correct; (b) The comparison of the heat source treatments shows that the jump conditions used for a compacted heat source are reliable; (c) For azimuthal and axial mixed modes, the vorticity disturbance indeed exists; however, its effect on the instabilities is very weak and can be even ignored; (d) For longitudinal modes with mean flow and uniform heat source, no vorticity waves exist.; and (e) For pure azimuthal mode, the vorticity waves are generated by the interaction between acoustic waves and the heat source. Also, the circumferential velocity disturbance continuity across the flame sheet is the mechanism of vorticity disturbance generation in the present case. In conclusion, two kinds of effects of vorticity waves are observed: (i) the coupling of axial acoustic and vorticity disturbances at the acoustically closed end affects the frequencies of the system and (ii) the generation of vorticity disturbance causes damping to the system. For the pure azimuthal modes, the effects of the vorticity disturbances are significant. When the flow Mach number is small, such effects become weak. However, in the combustor of a gas turbine, the effect of mean flow is inevitable; therefore. it can be concluded that the calculations of azimuthal instabilities using network model and finite element method may lead to a severe prediction error if the mean flow is considered. This also indicates the essence and necessity of the three-dimensional AM.

7.4 Control of thermoacoustic instability in a Rijke tube

The control methods of combustion instability can be grouped into passive control methods and active control methods. Because of the intrinsic disadvantages of an active control method, in practical applications, a passive control method is widely adopted, for example, the baffles in rocket engines, the Helmholtz resonators in gas turbines, and the perforated liners in afterburners of aeroengines, and so on. Although the mathematical description of Helmholtz resonators is easy by analytical approach, they are not widely used in rocket engines and aeroengines for the shortcoming of bulk and narrow frequency bandwidth. The cooling system in the afterburners of aeroengines can be models as perforated liners with bias flow. It is severely wrong to model the distributed damping system as a lumped parameter system. Thus, only a three-dimensional model is able to do that. The traditional way to consider the effect of the wall boundary is to provide the boundary condition of the wall, then the impedance of the wall is obtained. For example, the boundary condition for a hard wall is:

$$\left.\frac{\partial p}{\partial r}\right|_{r=R} = 0, \tag{7.222}$$

The eigenvalue of the radial eigenfunction can be obtained by solving the following equation:

$$J_m'(k_{mn}R) = 0, \tag{7.223}$$

where $J_m(k_{mn}r)$ is the m order Bessel function. Obviously, the eigenvalue k_{mn} is only related to the radius R. However, for the soft wall boundary condition, the impedance is always related to the frequency, $Z = Z(\omega)$. It is very difficult to find the impedance corresponding to some specific frequency. Moreover, the eigenvalue k_{mn} is related to the impedance of the wall. As a consequence, the eigenvalue k_{mn} is indirectly related to the frequency. If we solve the problem by the traditional way, it is needed to carry out two complex processes of iteration because Fig. 7.15 shows which is unreliable and unstable in numerical calculation, and the solutions are not accurate. Therefore, the traditional way of solving the thermoacoustic problems associated with soft wall boundary is not feasible. Sun and Wang [26–28]

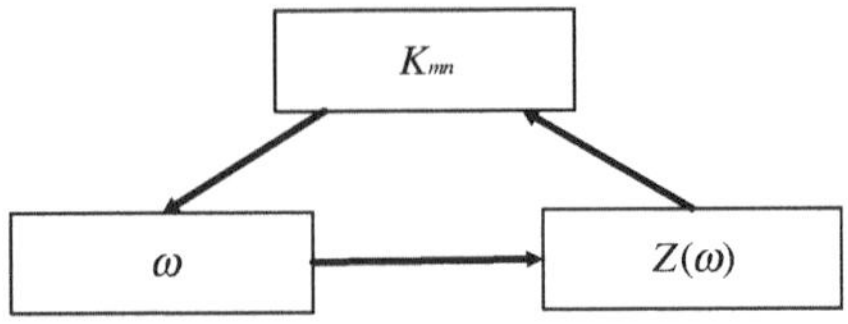

Fig. 7.15 The double iteration [25].

posed a new method named the "Transfer Element Method (TEM)" that is used to consider the interaction between the acoustic source and the soft acoustic treatment, which can be adopted here to avoid obstacles.

7.4.1 Perforated liner with bias flow

In this part, the control method using perforated liner with bias flow is presented. First, the main idea of the TEM will be briefly introduced.

As sketched in Fig. 7.16, a section of a circular tube is installed with a combination of perforated liner and a back cavity. In the view of acoustics, it is called nonlocally reacting liner. The main flow parameters are respectively denoted with the pressurep_0, the density ρ_0, the speed of sound C_0, and the Mach number M_a. For simplicity, we assume $M_a < 1$. As shown in Fig. 7.16, the pressure disturbance in the tube can be regarded as the sum of the incident acoustic pressure and the scattered acoustic pressure of the liner,

$$p = p_i + p_s, \tag{7.224}$$

where p_i is the acoustic pressure without the liner. At the hole on the liner, the displacement of the fluid particles is not zero. Thus, the motion of the particle can be seen as a mass source, which will induce the scattered acoustic

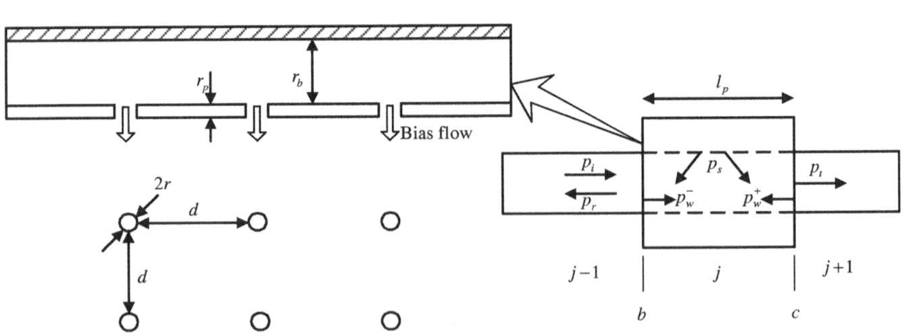

Fig. 7.16 Perforated liner with back cavity [25].

pressure p_s. If the acoustic pressure in the cavity and in the tube are respectively denoted as p^+ and p^-, then in the annular cavity with bias flow, according to the Rayleigh conductivity, we have:

$$c = \frac{i\omega_0 \rho_0 Q}{p^+ - p^-}, \tag{7.225}$$

where $p^+ = p_s^+$, $p^- = p_s^- + p_i^-$, and Q is the volume flux in the hole. Therefore, we obtain the oscillation equation:

$$p_s^- - p_s^+ + \frac{i\omega_0 \rho_0 Q}{c} = -p_i^-. \tag{7.226}$$

With the help of sine transform, Eq. (7.226) can be written as:

$$\begin{aligned} &\frac{2}{l_p}\int_0^{l_p} \left(p_s^- - p_s^+\right) \sin\frac{j'\pi x}{l_p} dx + \frac{2}{l_p}\int_0^{l_p} \frac{i\omega_0 \rho_0 Q}{c} \sin\frac{j'\pi x}{l_p} dx \\ &= -\frac{2}{l_p}\int_0^{l_p} p_i \sin\frac{j'\pi x}{l_p} dx, \end{aligned} \tag{7.227}$$

where $j' = 1, 2, 3, 4, \cdots$, $\frac{Q}{c} = \frac{V_n}{\eta}$, $V_n = \sum_{j'} V_{j'} \sin\frac{j'\pi x'}{l_p}$ represents the averaged velocity fluctuation of the mass source in the hole, and l_p is the length of the liner. Thus, Eq. (7.227) can be written as:

$$\sum_k \left(z_{j'k}^- - z_{j'k}^+ + \delta_{j'k} \frac{i\omega_0 \rho_0}{\eta} \right) V_{j'} = -I_{j'}, \tag{7.228}$$

where $k = 1, 2, 3, 4, \ldots$, and η is given by Eldredge & Dowling [29],

$$\frac{1}{\eta} = \frac{\pi r_o^2}{\sigma} \frac{1}{K_a} + \frac{r_p}{\sigma}, \tag{7.229}$$

where r_o is the radius of the hole, r_p is the thickness of the perforated liner, $\sigma = \pi r_o^2 / d^2$, $K_a = 2r_o(\gamma + i\delta)$, and γ and δ are both given by Howe [30].

To solve this problem, the relationship between the scattered acoustic pressure and the velocity fluctuation of the mass source in the hole need to be found. By applying the Green function method [26], the acoustic field induced by the mass source is obtained:

$$\tilde{p}(\vec{x}, t) = -\int_{-T}^{T} \int_{s(\tau)} \rho_0 \tilde{V}_n' \frac{D_0 G}{D\tau} ds(\vec{y}) d\tau, \tag{7.230}$$

where $\frac{D_0}{D\tau}=\frac{\partial}{\partial\tau}+U\frac{\partial}{\partial x}$ is the total derivative, and n denotes the outer normal direction. What needs to be pointed out is the outer normal direction in the tube, whereas the cavity is opposite. G is the Green's function, which has a different form in the tube and in the cavity respectively as:

$$G=\frac{-i}{4\pi}\sum_{m=-\infty}^{+\infty}\sum_{n=1}^{+\infty}\left[\begin{array}{c}\dfrac{\psi_m(k_{mn}r)e^{im\theta}\psi_m(k_{mn}r')e^{-im\theta'}}{\Gamma_{nm}}\times\\ \displaystyle\int_{-\infty}^{\infty}\frac{1}{\kappa_{mn}}e^{i\omega(t-\tau)+i\alpha^{\pm}_{m,n}(x-x')}d\omega\end{array}\right], \tag{7.231}$$

$$G=-\frac{1}{2\pi}\sum_{m=-\infty}^{+\infty}\sum_{n=1}^{+\infty}\sum_{q=0}^{+\infty}\left[\begin{array}{c}\dfrac{\phi_m(k_{mn}r)e^{im\theta}\cos\dfrac{q\pi x}{l_p}\phi_m(k_{mn}r')e^{-im\theta'}\cos\dfrac{q\pi x'}{l_p}}{\Gamma_{mnq}}\times\\ \displaystyle\int_{-\infty}^{\infty}\frac{e^{i\omega(t-\tau)}}{k_0^2-k_{mnq}^2}d\omega\end{array}\right], \tag{7.232}$$

where $\psi_m(k_{mn}r)$ is the radial eigenfunction of the tube, i.e., the first kind of Bessel function. $\phi_m(k_{mn}r)$ is the radial eigenfunction of the annular tube, which is the linear combination of the first kind of Bessel function and the second kind of Bessel function. $k_0=\frac{\omega}{C_0}$, $\alpha^{\pm}_{mn}$ is the axial wave numbers as stated before,

$$\alpha^{\pm}_{mn}=\frac{M_a^2k_0\pm\sqrt{k_0^2-(1-M_a^2)k_{mn}^2}}{1-M_a^2}. \tag{7.233}$$

Γ_{mn}, Γ_{mnq} and κ_{mn} are given as:

$$\Gamma_{mn}=\int_A\left|\psi_m(k_{mn}r)e^{im\theta}\right|^2drd\theta, \tag{7.234}$$

$$\Gamma_{mnq}=\begin{cases}\Gamma_{mn}\cdot l & q=0\\ \Gamma_{mn}\cdot\dfrac{l}{2} & q\neq 0\end{cases}, \tag{7.235}$$

$$\kappa_{mn}=\begin{cases}\sqrt{k_0^2-(1-M_a^2)k_{mn}^2} & k_0^2>(1-M_a^2)k_{mn}^2\\ -i\sqrt{(1-M_a^2)k_{mn}^2-k_0^2} & k_0^2<(1-M_a^2)k_{mn}^2\end{cases}. \tag{7.236}$$

Substitute the Green function and $\lim_{T\to\infty}\int_{-T}^{T}e^{i(\omega_0-\omega)\tau}d\tau=2\pi\delta(\omega_0-\omega)$ into Eq. (7.230), we get:

$$\widetilde{p_s^-} = \frac{\rho_0 e^{i\omega_0 t}}{2} \sum_{m=-\infty}^{\infty} \sum_{n=1}^{\infty} \frac{\psi_m(k_{mn}r)e^{im\theta}}{\Gamma_{mn}} \int_{s(\tau)} V_n' \psi_m(k_{mn}r')e^{-im\theta'} \frac{\left[\omega + U\alpha_{mn}^{\pm}\right]}{\kappa_{mn}} e^{i\alpha_{mn}^{\pm}(x-x')} ds(\vec{\gamma}), \tag{7.237}$$

$$\widetilde{p_s^+} = i\rho_0\omega_0 e^{i\omega_0 t} \sum_{m=-\infty}^{+\infty} \sum_{n=1}^{+\infty} \sum_{q=0}^{+\infty} \frac{\phi_m(k_{mn}r)e^{im\theta} \cos\frac{q\pi x}{l_p}}{\Gamma_{mnq}\left(k_0^2 - k_{mnq}^2\right)} \int_{s(\tau)} V_n \phi_m(k_{mn}r')e^{-im\theta'} \cos\frac{q\pi x'}{l_p} dS(\vec{\gamma}). \tag{7.238}$$

According to the continuity of the displacement of the particle, the velocity fluctuation at the hole can be expanded as sine forms along the axial direction,

$$V_n' = \frac{\partial\eta}{\partial\tau} + U\frac{\partial\eta}{\partial x'} = V_n + \frac{U}{i\omega}\frac{\partial V_n}{\partial x'} = \left(1 + \frac{U}{i\omega}\frac{\partial}{\partial x'}\right) \sum_{k=1}^{\infty} V_k^{\theta\prime} \sin\frac{k\pi x'}{l_p}. \tag{7.239}$$

Substitute $\int^{2\pi} V_k^{\theta\prime} e^{-im\theta'} d\theta' = 2\pi V_k^m$ into the acoustic pressure equation, then take the^0 sine transformation, and then the radiation impedance equation can be expressed as:

$$z_{j'k}^- = \frac{\rho_0 e^{i\omega_0 t}}{l} \sum_{m=-\infty}^{\infty} \sum_{n=1}^{\infty} \frac{\psi_m(k_{mn}r_h)\psi_m(k_{mn}r_h)}{\Gamma_{mn}\kappa_{nm}} \int_0^l \int_0^l \left[\omega + U\alpha_{mn}^{\pm}\right] e^{i\alpha_{mn}^{\pm}(x-x')} \left(1 + \frac{U}{i\omega}\frac{\partial}{\partial x'}\right) \sin\frac{k\pi x'}{l} dx' \sin\frac{j'\pi x}{l} dx, \tag{7.240}$$

$$z_{j'k}^+ = \frac{4i\pi\rho_0\omega_0 r_h e^{i\omega_0 t}}{l} \sum_m \sum_n \sum_q \frac{\phi_m(k_{mn}r_h)\phi_m(k_{mn}r_h)}{\Gamma_{mnq}\left(k_0^2 - k_{mnq}^2\right)} \int_0^l \cos\frac{q\pi x}{l} \sin\frac{k\pi x}{l} dx \int_0^l \cos\frac{q\pi x'}{l} \sin\frac{j'\pi x'}{l} dx'. \tag{7.241}$$

Substitute Eqs. (7.240) and (7.241) into Eq. (7.228), then the velocity fluctuation quantity at the hole is solved. The scattered acoustic pressure can now be expressed with the coefficients of the incident acoustic pressure. As shown in Fig. 7.16, the section with perforated liner is denoted by j, the left side and the right side of the liner section are respectively denoted by b and c, p_i and p_r are respectively the incident and reflected acoustic pressure at the b interface, and P_w^+,P_w^- and p_s are respectively the propagating acoustic pressure and the scattered acoustic pressure. P_w^+ and P_w^- can be written as:

$$P_w^m = \sum_n P_{mn}^{m\ j} \psi_m(k_{mn} r) e^{i\alpha_{mn}^m x + im\theta}. \tag{7.242}$$

Eq. (7.228) can be written as:

$$V_{j'} = -I_{j'} Z_{j'k,\mu,P}^{-1}, \tag{7.243}$$

where $Z_{j'k,\mu,P} = \sum_k \left(z_{j'k}^- - z_{j'k}^+ + \delta_{j'k} \frac{i\omega_0 \rho_0}{\eta} \right)$. And the scattered acoustic pressure is expressed as:

$$p_s = \pi \rho_0 r_0 \sum_{n=1}^{\infty} \frac{\psi_m(k_{mn} r) \psi_m(k_{mn} r_0) e^{im\theta + i\alpha_{m,n}^{\pm} x}}{\Gamma_{mn}} \frac{\left[\omega + U\alpha_{m,n}^{\pm}\right]}{\kappa_{mn}}$$
$$\left\{ \sum_{\mu=1}^{\infty} \psi_m(k_{m\mu} r_0) \left[P_{m\mu}^{-\ j} Q_{n-}^{P_{m\mu}^{-\ j}} + P_{m\mu}^{+\ j} Q_{n+}^{P_{m\mu}^{+\ j}} \right] \right\}, \tag{7.244}$$

where P_{mn}^{-j} and P_{mn}^{+j} are the unknown amplitudes, moreover,

$$Q_{n\mp}^{P_{m\mu}^{\mp\ j}} = \sum_{k=1}^{\infty} z_{j'k,\mu,P}^{-1} I_{j'}^{P_{m\mu}^{\mp\ j}} S_k^{n\mp}, \tag{7.245}$$

$$S_k^{n\pm} = \int_0^1 e^{-i\alpha_{mn}^{\pm} x'} \left(1 + \frac{U}{i\omega} \frac{\partial}{\partial x'} \right) \sin \frac{k\pi x'}{l} dx'. \tag{7.246}$$

Finally, the total acoustic pressure in the tube is obtained,

$$\begin{aligned} p = P_w^+ + P_w^- + p_s = &\sum_n P_{mn}^{-\ j} \psi_m(k_{mn} r) e^{i\alpha_{mn}^- x + im\theta} \\ &+ \sum_n P_{mn}^{+\ j} \psi_m(k_{mn} r) e^{i\alpha_{mn}^+ x + im\theta} \\ &+ \pi \rho_0 r_0 \sum_{n=1}^{\infty} \frac{\psi_m(k_{mn} r) \psi_m(k_{mn} r_0) e^{im\theta + i\alpha_{mn}^{\pm} x}}{\Gamma_{m,n}} \frac{\left[\omega + U\alpha_{m,n}^{\pm}\right]}{\kappa_{n,m}} \\ &\left\{ \sum_{\mu=1}^{\infty} \psi_m(k_{m\mu} r_0) \left[P_{m\mu}^{-\ j} Q_{n-}^{P_{m\mu}^{-\ j}} + P_{m\mu}^{+\ j} Q_{n+}^{P_{m\mu}^{+\ j}} \right] \right\} \\ = &\sum_{n=1}^{\infty} \psi_{mn}(r) \sum_{\mu=1}^{\infty} \left(P_{m\mu}^{+\ j} p_{\mu n}^+ + P_{m\mu}^{-\ j} p_{\mu n}^- \right) e^{im\theta}, \end{aligned} \tag{7.247}$$

where $\psi_{mn}(r)$ is written as the form of $\psi_m(k_{m,n}r)$, $p_{\mu n}^{+}$ and $p_{\mu n}^{-}$ can be written as the following forms,

$$p_{\mu n}^{\pm} = \pi \rho_0 r_0 \frac{\psi_m(k_{m,n} r_0) e^{i\alpha_{m,n}^{\pm} x}}{\Gamma_{m,n}} \times \frac{\left[\omega + U\alpha_{m,n}^{\pm}\right]}{\kappa_{n,m}} \psi_m(k_{m\mu} r_0) Q_{n\pm}^{P_{m\mu}^{\pm j}} + \delta_{n\mu} e^{i\alpha_{mn}^{\pm} x}. \tag{7.248}$$

Both the velocity fluctuation and density fluctuation in the tube can be expressed similarly as the pressure fluctuation. Therefore, the derivation aforementioned is aimed at achieving the specific forms of the TEM used on the three-dimensional analytical model.

As the fluctuation quantities are obtained, as shown in Fig. 7.16, the equations of mass, momentum, and energy are formed. For example, at the interface b, we have:

$$\sum_{n=1}^{\infty} [A_{mn}^{+\,j-1} \psi_{mn}^{+\,j-1}(r) P_{mn}^{+\,j-1} + A_{mn}^{-\,j-1} \psi_{mn}^{-\,j-1}(r) P_{mn}^{-\,j-1} + A_{vmn}^{+\,j-1} \psi_{vmn}^{+\,j-1}(r) \rho_{vmn}^{+\,j-1}$$
$$+\psi_{mn}^{+\,j}(r) \sum_{\mu=1}^{\infty} A_{mn\mu}^{+\,j} P_{m\mu}^{+\,j} + \psi_{mn}^{-\,j}(r) \sum_{\mu=1}^{\infty} A_{mn\mu}^{-\,j} P_{m\mu}^{-\,j} + A_{vmn}^{+\,j} \psi_{vmn}^{+\,j}(r) \rho_{vmn}^{+\,j}] = 0, \tag{7.249}$$

$$\sum_{n=1}^{\infty} [B_{mn}^{+\,j-1} \psi_{mn}^{+\,j-1}(r) P_{mn}^{+\,j-1} + B_{mn}^{-\,j-1} \psi_{mn}^{-\,j-1}(r) P_{mn}^{-\,j-1} + B_{vmn}^{+\,j-1} \psi_{vmn}^{+\,j-1}(r) \rho_{vmn}^{+\,j-1}$$
$$+\psi_{mn}^{+\,j}(r) \sum_{\mu=1}^{\infty} B_{mn\mu}^{+\,j} P_{m\mu}^{+\,j} + \psi_{mn}^{-\,j}(r) \sum_{\mu=1}^{\infty} B_{mn\mu}^{-\,j} P_{m\mu}^{-\,j} + B_{vmn}^{+\,j} \psi_{vmn}^{+\,j}(r) \rho_{vmn}^{+\,j}] = 0, \tag{7.250}$$

$$\sum_{n=1}^{\infty} [C_{mn}^{+\,j-1} \psi_{mn}^{+\,j-1}(r) P_{mn}^{+\,j-1} + C_{mn}^{-\,j-1} \psi_{mn}^{-\,j-1}(r) P_{mn}^{-\,j-1} + C_{vmn}^{+\,j-1} \psi_{vmn}^{+\,j-1}(r) \rho_{vmn}^{+\,j-1}$$
$$+\psi_{mn}^{+\,j}(r) \sum_{\mu=1}^{\infty} C_{mn\mu}^{+\,j} P_{m\mu}^{+\,j} + \psi_{mn}^{-\,j}(r) \sum_{\mu=1}^{\infty} C_{mn\mu}^{-\,j} P_{m\mu}^{-\,j} + C_{vmn}^{+\,j} \psi_{vmn}^{+\,j}(r) \rho_{vmn}^{+\,j}] = 0, \tag{7.251}$$

where A_{mn}, B_{mn}, C_{mn} are the coefficients related to the equation of mass, momentum, and energy, respectively. Obviously, they are different from those from the hard-wall situation. For the hard-wall boundary condition, each radial mode at one side of the interface is matched with the same radial mode at the other side of the interface. However, when the soft treatment such as a liner is included in the tube, each incident radial mode is matched with infinite radial modes at the other side.

At the interface c with the same pattern, similar expressions are obtained. At last, a series of equations are obtained including the end boundary conditions and heat release functions. Although the radial eigenfunction $\psi_{mn}(r)$ is still orthogonal, after the orthogonal transformation the modes still stay tangled. Thus, the equations need to be solved with all the modes considered, which is different from that of a hard-wall condition. If the calculation is only aimed at low-order modes, we set $\mu = N_\mu$ approximately; in fact, for the fundamental incident acoustic wave, it is precise enough to assume $N_\mu = 3$. For the situation shown in Fig. 7.16, for the three parts of $j-1, j$, and $j+1$, assuming that $N_\mu = 2$, when the two interfaces of b, c are considered, we have:

$$\begin{pmatrix} D_{11} & D_{12} & 0 \\ 0 & D_{22} & D_{23} \end{pmatrix} \begin{pmatrix} P_1 \\ P_2 \\ P_3 \end{pmatrix} = 0, \tag{7.252}$$

where the submatrix D and P have the forms as:

$$D_{11} = \begin{pmatrix} A_{m1}^{+\,j-1} & A_{m1}^{-\,j-1} & A_{vm1}^{+\,j-1} & 0 & 0 & 0 \\ B_{m1}^{+\,j-1} & B_{m1}^{-\,j-1} & B_{vm1}^{+\,j-1} & 0 & 0 & 0 \\ C_{m1}^{+\,j-1} & C_{m1}^{-\,j-1} & C_{vm1}^{+\,j-1} & 0 & 0 & 0 \\ 0 & 0 & 0 & A_{m2}^{+\,j-1} & A_{m2}^{-\,j-1} & A_{vm2}^{+\,j-1} \\ 0 & 0 & 0 & B_{m2}^{+\,j-1} & B_{m2}^{-\,j-1} & B_{vm2}^{+\,j-1} \\ 0 & 0 & 0 & C_{m2}^{+\,j-1} & C_{m2}^{-\,j-1} & C_{vm2}^{+\,j-1} \end{pmatrix}, \tag{7.253}$$

$$D_{12} = \begin{pmatrix} A_{Lm11}^{+\,j} & A_{Lm11}^{-\,j} & A_{Lvm12}^{+\,j} & A_{Lm12}^{+\,j} & A_{Lm12}^{-\,j} & 0 \\ B_{Lm11}^{+\,j} & B_{Lm11}^{-\,j} & B_{Lvm1}^{+\,j} & B_{Lm12}^{+\,j} & B_{Lm12}^{-\,j} & 0 \\ C_{Lm11}^{+\,j} & C_{Lm11}^{-\,j} & C_{Lvm1}^{+\,j} & C_{Lm12}^{+\,j} & C_{Lm12}^{-\,j} & 0 \\ A_{Lm21}^{+\,j} & A_{Lm21}^{-\,j} & 0 & A_{Lm22}^{+\,j} & A_{Lm22}^{-\,j} & A_{Lvm22}^{+\,j} \\ B_{Lm21}^{+\,j} & B_{Lm21}^{-\,j} & 0 & B_{Lm22}^{+\,j} & B_{Lm22}^{-\,j} & B_{Lvm22}^{+\,j} \\ C_{Lm21}^{+\,j} & C_{Lm21}^{-\,j} & 0 & C_{Lm22}^{+\,j} & C_{Lm22}^{-\,j} & C_{Lvm22}^{+\,j} \end{pmatrix}, \tag{7.254}$$

$$D_{22} = \begin{pmatrix} A_{Rm11}^{+\,j} & A_{Rm11}^{-\,j} & A_{Rvm12}^{+\,j} & A_{Rm12}^{+\,j} & A_{Rm12}^{-\,j} & 0 \\ B_{Rm11}^{+\,j} & B_{Rm11}^{-\,j} & B_{Rvm1}^{+\,j} & B_{Rm12}^{+\,j} & B_{Rm12}^{-\,j} & 0 \\ C_{Rm11}^{+\,j} & C_{Rm11}^{-\,j} & C_{Rvm1}^{+\,j} & C_{Rm12}^{+\,j} & C_{Rm12}^{-\,j} & 0 \\ A_{Rm21}^{+\,j} & A_{Rm21}^{-\,j} & 0 & A_{Rm22}^{+\,j} & A_{Rm22}^{-\,j} & A_{Rvm22}^{+\,j} \\ B_{Rm21}^{+\,j} & B_{Rm21}^{-\,j} & 0 & B_{Rm22}^{+\,j} & B_{Rm22}^{-\,j} & B_{Rvm22}^{+\,j} \\ C_{Rm21}^{+\,j} & C_{Rm21}^{-\,j} & 0 & C_{Rm22}^{+\,j} & C_{Rm22}^{-\,j} & C_{Rvm22}^{+\,j} \end{pmatrix}, \tag{7.255}$$

$$D_{23}=\begin{pmatrix} A_{m1}^{+\,j+1} & A_{m1}^{-\,j+1} & A_{vm1}^{+\,j+1} & 0 & 0 & 0 \\ B_{m1}^{+\,j+1} & B_{m1}^{-\,j+1} & B_{vm1}^{+\,j+1} & 0 & 0 & 0 \\ C_{m1}^{+\,j+1} & C_{m1}^{-\,j+1} & C_{vm1}^{+\,j+1} & 0 & 0 & 0 \\ 0 & 0 & 0 & A_{m2}^{+\,j+1} & A_{m2}^{-\,j+1} & A_{vm2}^{+\,j+1} \\ 0 & 0 & 0 & B_{m2}^{+\,j+1} & B_{m2}^{-\,j+1} & B_{vm2}^{+\,j+1} \\ 0 & 0 & 0 & C_{m2}^{+\,j+1} & C_{m2}^{-\,j+1} & C_{vm2}^{+\,j+1} \end{pmatrix}, \tag{7.256}$$

$$P_1=\begin{pmatrix} P_{m1}^{+\,j-1} \\ P_{m1}^{-\,j-1} \\ \rho_{vm1}^{+\,j-1} \\ P_{m2}^{+\,j-1} \\ P_{m2}^{-\,j-1} \\ \rho_{vm2}^{+\,j-1} \end{pmatrix}, P_2=\begin{pmatrix} P_{m1}^{+\,j} \\ P_{m1}^{-\,j} \\ \rho_{vm1}^{+\,j} \\ P_{m2}^{+\,j} \\ P_{m2}^{-\,j} \\ \rho_{vm2}^{+\,j} \end{pmatrix}, P_3=\begin{pmatrix} P_{m1}^{+\,j+1} \\ P_{m1}^{-\,j+1} \\ \rho_{vm1}^{+\,j+1} \\ P_{m2}^{+\,j+1} \\ P_{m2}^{-\,j+1} \\ \rho_{vm2}^{+\,j+1} \end{pmatrix}, \tag{7.257}$$

where D_{11} corresponds to the interface b of the $j-1$ section, D_{12} corresponds to the interface b of the j section, D_{22} corresponds to the interface c of the j section, and D_{23} corresponds to the interface c of the section $j+1$. Therefore, the liner section is integrally included in the three-dimensional analytical model.

To validate the effectiveness of this method, once again, the model presented by Heckl [11] is used here; what is different is we will add a section of liner and back cavity to the original model to investigate the effect of the liner with bias flow. The parameters are not changed. The length of the tube is $l_j = 1\,\text{m}$, the radius of the tube is $r_j = 0.0223\,\text{m}$, the temperature of the heated wire is $T_w = 700\,\text{K}$, and the heated gauze is placed in the tube 0.3 m from the left end. The reflection coefficients of the end are $R_o = R_L = -0.986 + 0.12i$, and the heat release function is:

$$Q'(t) = \beta u(t-\tau), \tag{7.258}$$

where β is the amplitude with value of 187 kg m s^{-2}, and τ is the delay time. Under these conditions, for the hard-wall situation, the complex frequency varies through only the delay time τ; the resonance frequency calculated with the analytical model is shown in Fig. 7.17.

The imaginary of the complex frequency represents the stability of the system, as shown in Fig. 7.18. There is a phase delay between the real part of the resonance frequency and the imaginary. When the delay time lays in the range of $0 < \tau < T/2$, the instability occurs.

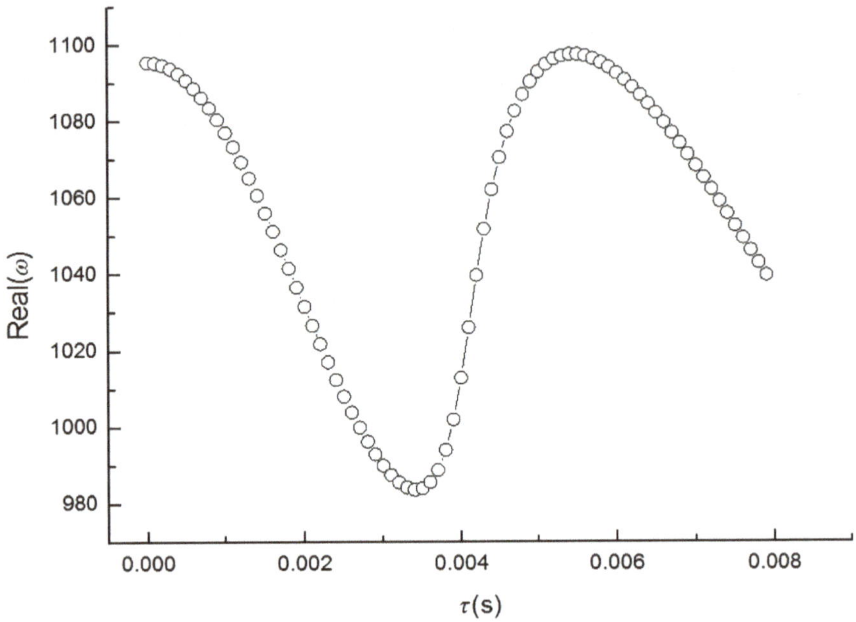

Fig. 7.17 The variation of the fundamental frequency through the delay time.

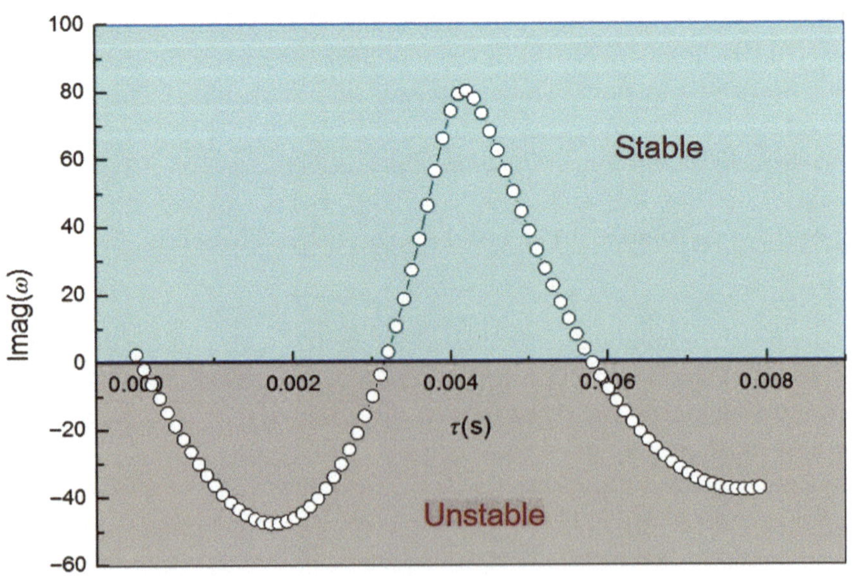

Fig. 7.18 The imaginary part of the complex resonance frequency.

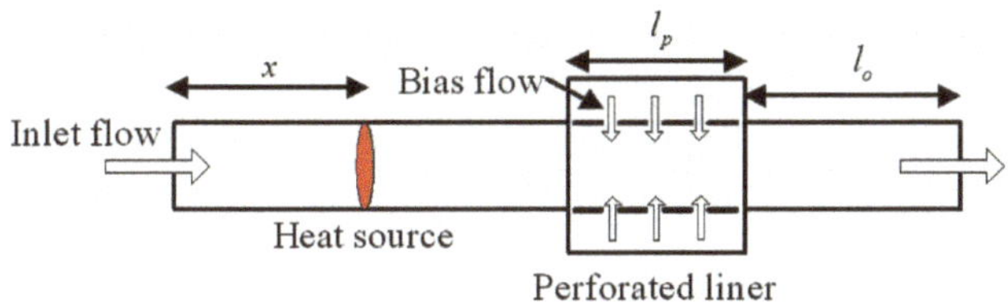

Fig. 7.19 Control of thermoacoustic instability occurred in a Rijke tube with perforated liner.

The length of each part of the tube is shown in the sketch diagram in Fig. 7.19. Specifically, the length of the liner or the back cavity is l_p, the radius of the perforated hole is r_o, the distance between neighboring holes is d, the cavity depth is r_b, and the position of the liner can be determined by the length of l_0. The perforation ratio depends on two parameters, r_0 and d, $\sigma = \pi r_o^2/d^2$.

In the following section, the parameters of the liner, the back cavity, and the bias flow are investigated for the control effect on the instability with the three-dimensional analytical model.

Fig. 7.20 shows the variation of the real part of the resonance frequency with the nondimensional length of the liner. Three lines represent different Mach numbers of 0.000, 0.001, and 0.005. The radius of the hole is

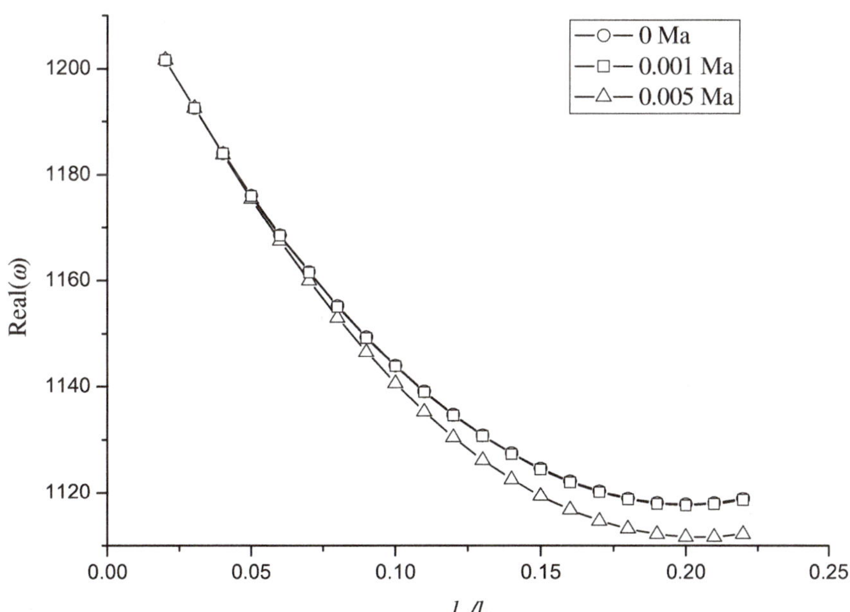

Fig. 7.20 Variation of the real part of the complex frequency with the nondimensional length of the liner.

$r_o = 0.001$m, the perforation ratio is $\sigma = 0.01$, the depth of the back cavity is $r_b = 0.01$m, and $l_p + l_0 = 0.4$m. It is found that the length of the liner has little effect on the resonance frequency with the maximum change less than 10%. However, it greatly affects the stability of the system from Fig. 7.21, which reveals that, at the highest Mach number circumstance, when the nondimensional length reaches 15% of the tube, the system becomes stable, whereas the on the other two conditions, it is still unstable. It is concluded that the bias flow enhances the dissipation of acoustic energy, and it is really effective to increase the length of the liner.

The results of the investigation of the perforation ratio is displayed in Figs. 7.22 and 7.23; at high bias Mach number, when $\sigma < 0.012$, the imaginary part of the resonance frequency is positive, which means the system is stable.

The influence of the depth of the back cavity is also investigated with the results shown in Figs. 7.24 and 7.25. The resonance frequency is comparatively sensitive because the volume of the back cavity plays an important role in changing the reactance of the impedance of the wall.

From the calculation results mentioned before, the bias flow in the hole indeed has a positive effect on suppressing the instability. The variation of

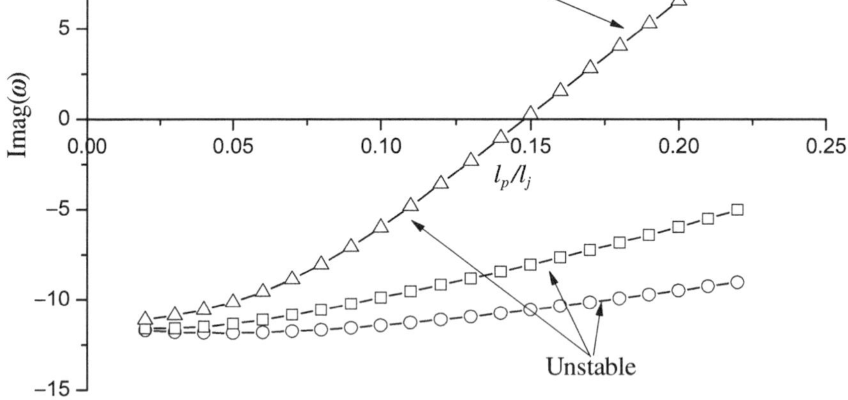

Fig. 7.21 Variation of the imaginary part of the complex frequency with the nondimensional length of the liner.

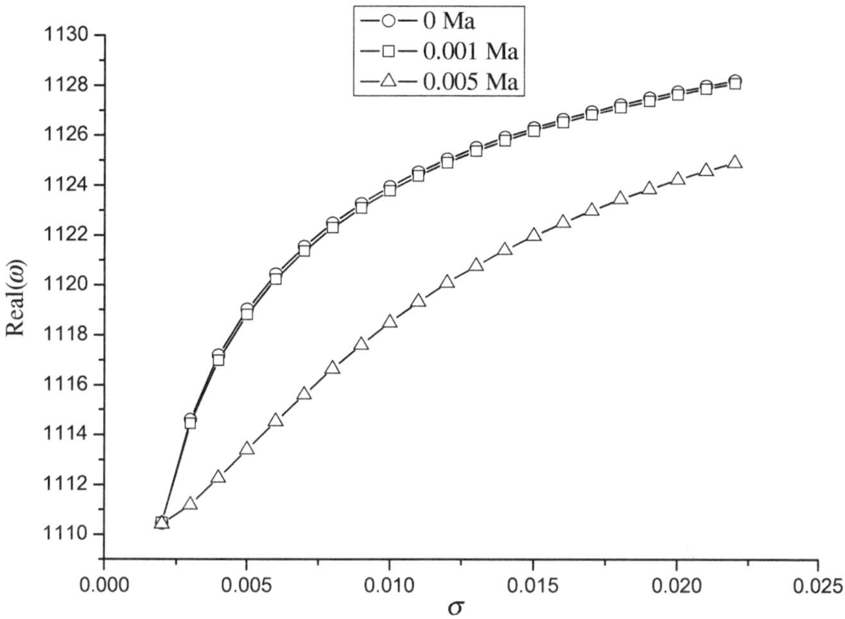

Fig. 7.22 Variation of the real part of the complex frequency with the perforation ratio.

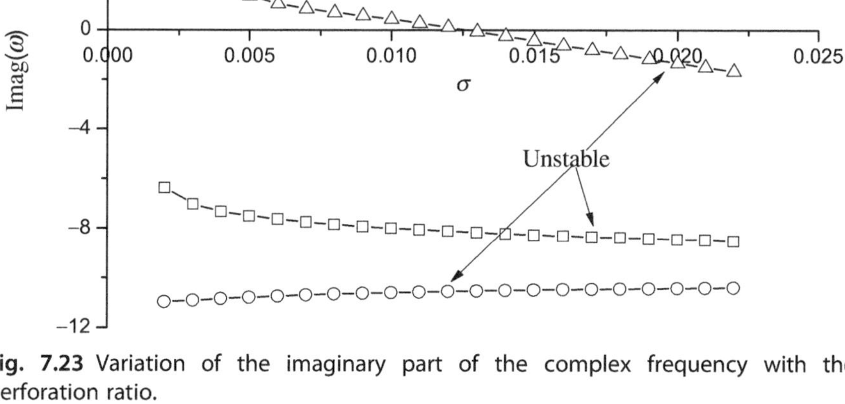

Fig. 7.23 Variation of the imaginary part of the complex frequency with the perforation ratio.

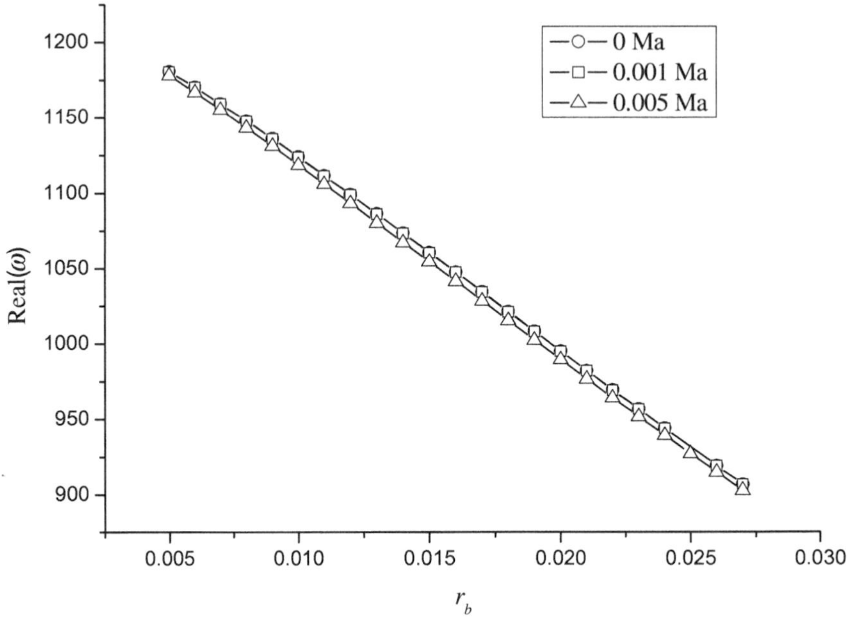

Fig. 7.24 Variation of the real part of the complex frequency with the depth of the back cavity.

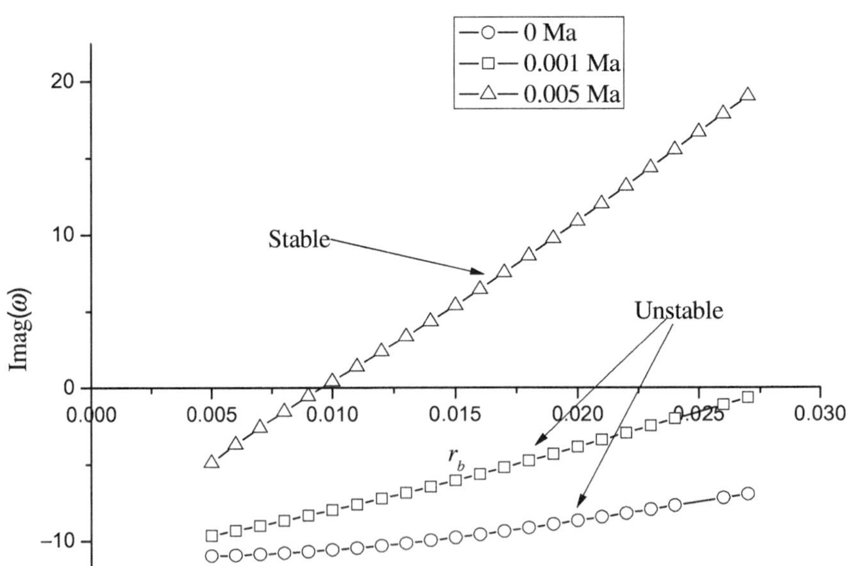

Fig. 7.25 Variation of the imaginary part of the complex frequency with the depth of the back cavity.

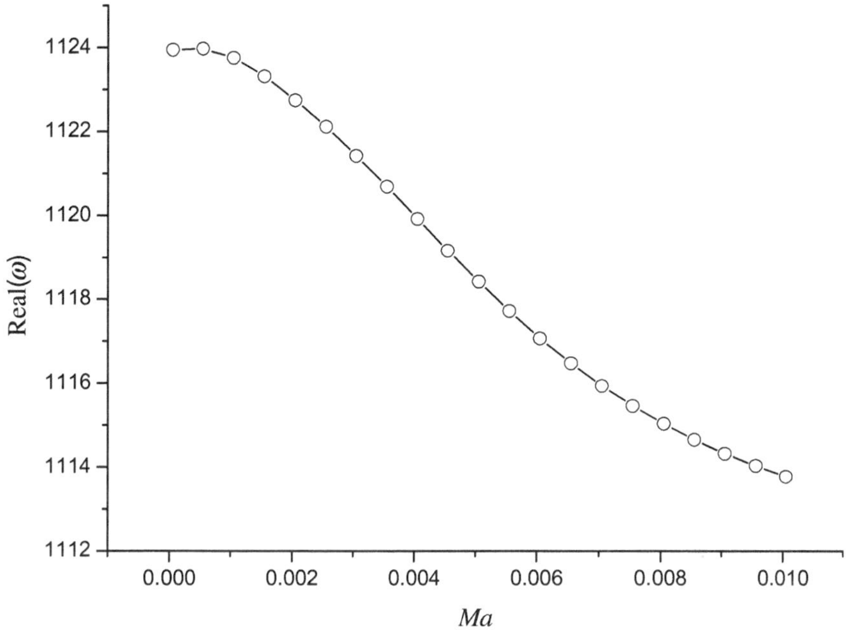

Fig. 7.26 Variation of the real part of the complex frequency with the Mach number of the bias flow.

the complex resonance frequency with the Mach number of the bias flow is displayed in Figs. 7.26 and 7.27, with $\sigma = 0.01$, $l_p/l_j = 0.15$, and $r_b = 0.01\text{m}$. The real part of the frequency changes little, yet the system falls into stable region from unstable region when the Mach number reaches 0.0045. Thus, the bias flow really helps to suppress instability and enhance the energy dissipation.

According to the calculated results, it is valid to suppress thermoacoustic instability by the liner with bias flow. This kind of liner actually works as an acoustic damping structure in the system, which dissipated the acoustic energy, as a result, making the dissipated acoustic energy larger than that enhanced, thus, the instability will never occur. In the meantime, the bias flow will produce vortexes at the hole, interacting with the acoustic pressure, and making the acoustic energy converted to the energy of vortex that is to be dissipated. In a word, the existence of the bias flow indeed enhanced the control effectiveness. In the operation of propulsion systems, the working condition is changing continuously; at some specific state, it works on the boundary of the stable region. The passive control methods are unable to change according to the working condition as soon as the control devices

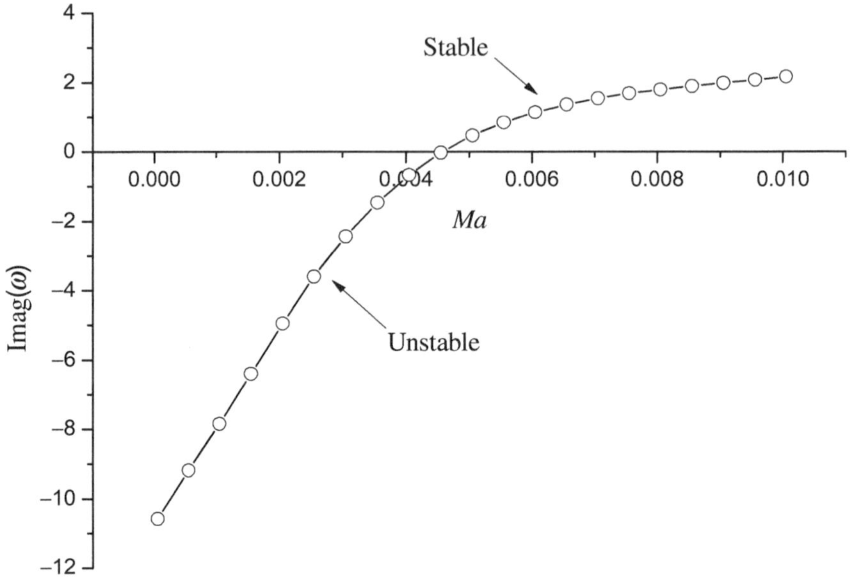

Fig. 7.27 Variation of the imaginary part of the complex frequency with the Mach number of the bias flow.

are established. However, from the model presented before, the bias flow can change the impedance of the wall, thus it can enhance the damping effect by increasing the bias flow rate, consequently enlarging the stable working region. Therefore, in a practical application, we can change the bias flow rate to avert the instability, which is called the hybrid control method. The result is shown in Fig. 7.28, from which the contour of zero, which is the dividing line of the instability region and stability region, is obtained as Fig. 7.29 shows. According to this control line, the stable working region is obtained and, according to which, we know how to change the bias flow to control the instability. Li et al. [25] conducted experiments to validate this idea of hybrid control method.

7.4.2 Drum-like silencer

A perforated liner with bias flow functions by adding a damping source into the thermoacoustic system. However, based on the fact that the unsteady heat release contributes to the instability through the interaction with acoustic pressure, the acoustic energy converted from the heat release depends on the phase difference between pressure and heat release. Thus, we proposed a method that is helpful in controlling thermoacoustic instability by a

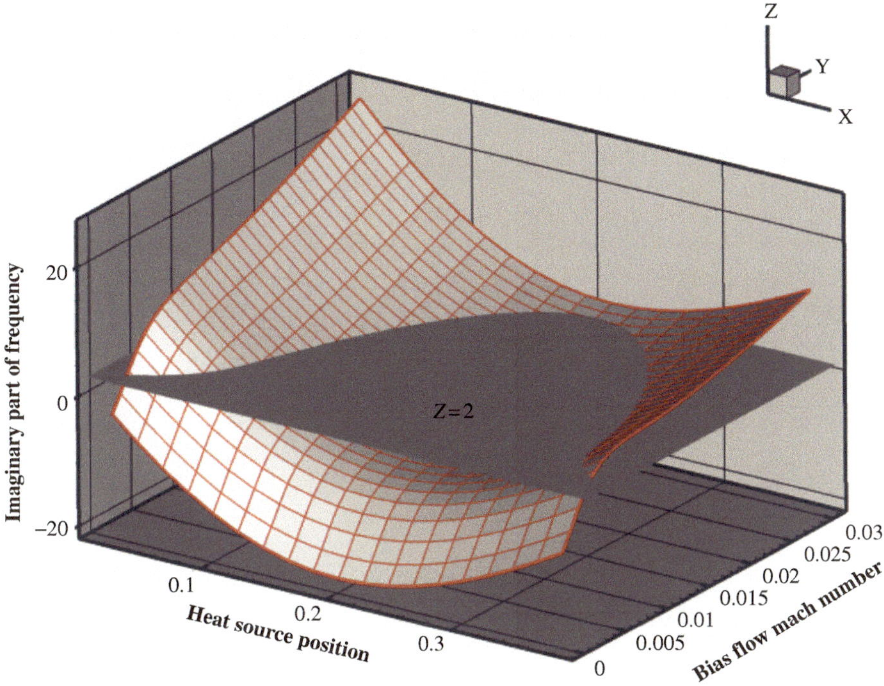

Fig. 7.28 Variation of the imaginary part of the complex frequency through the heat source position and the bias flow Mach number.

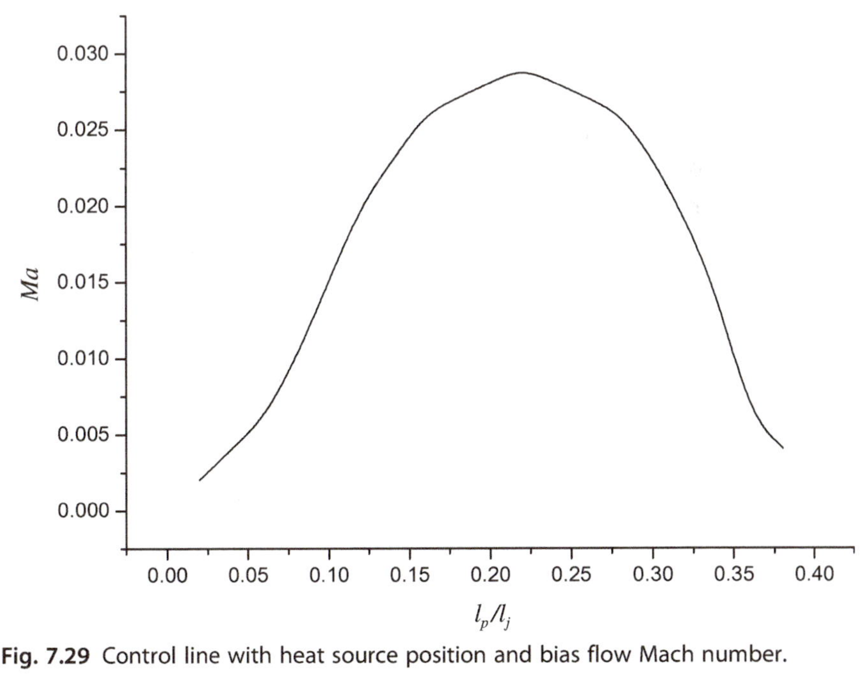

Fig. 7.29 Control line with heat source position and bias flow Mach number.

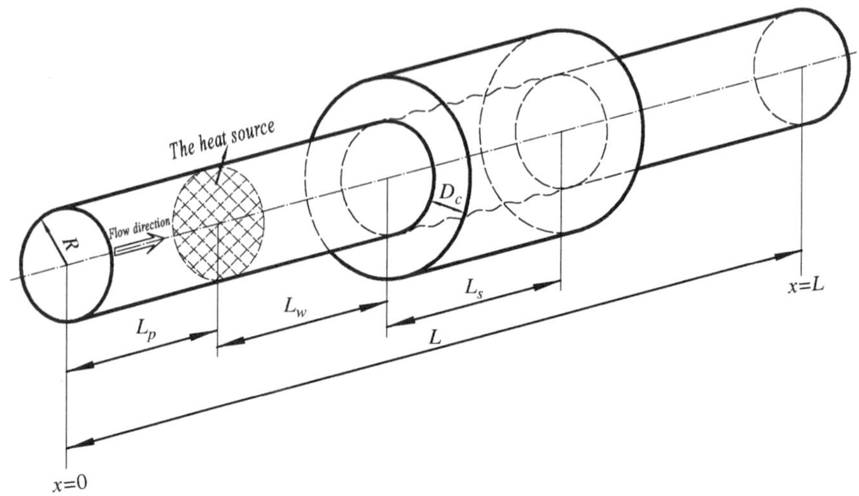

Fig. 7.30 Sketch of the control configuration [31].

drum-like silencer. Similar to the liner, TEM is used to consider the interaction between the soft boundary and the acoustics in the tube. The control configuration is illustrated by Fig. 7.30.

The modeling process is similar to that of the perforated liner in Section 7.4.1. The only difference is the impedance equation of the perforated liner is replaced by the oscillation equation of the membrane.

The symbols representing the relative positions of the components are specified in Fig. 7.30. The Rijke tube, of length $L = 1$ m and radius $R = 0.0223$ m, is lined in part by an annular membrane of length L_s, and the position of the silencer is defined by L_w. The membrane is enclosed by an annular rigid-wall cavity of depth D_c and length L_s. The heated gauze is displaced inside the pipe, with a distance of L_p from the upper end. The tension of the membrane per unit length is denoted by T_x, and the mass of the membrane per unit area is denoted by m_x.

Detailed modeling and validation can be referred to Zhang et al. [31]. Actually, for this kind of Rijke tube with radius of 0.0223 m, the higher modes reflected by the drum-like silencer cannot propagate in a resonance condition. Therefore, the following results are limited to the planar mode.

It is shown from Figs. 7.31 and 7.32 that incorporation of a drum-like silencer decreases the natural frequency of the system. And, as a result, the fundamental mode becomes stable. Note, however, the time delay is assumed to be constant, which is physically reasonable for the heated wire

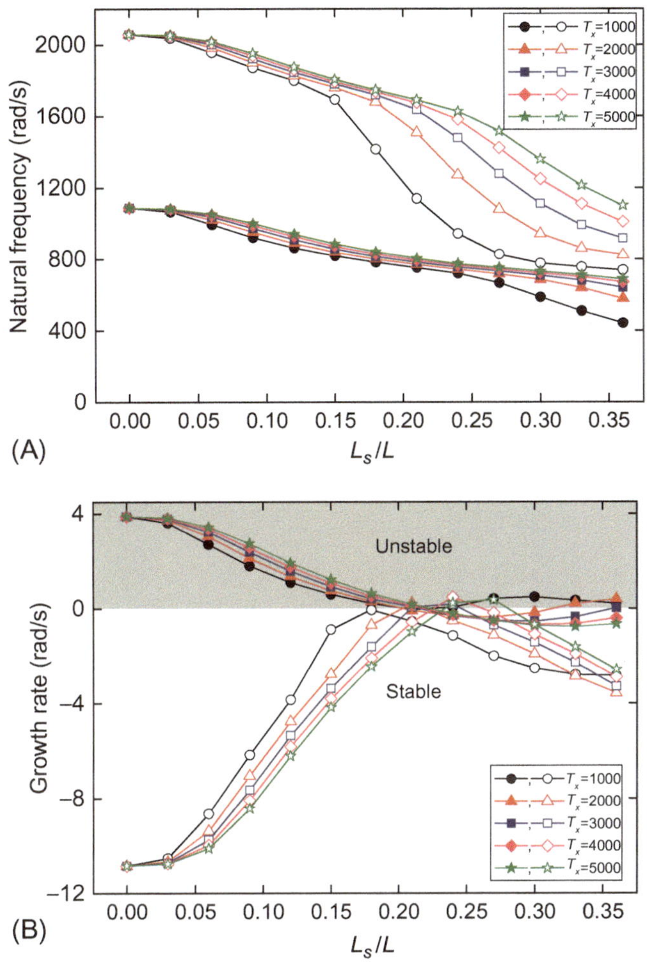

Fig. 7.31 Complex frequency against length of drum-like silencer with $D_c/R=1$.: (A) Natural frequency (rad/s); (B) growth rate (rad/s) [31].

case as depicted by Heckl [11]. When the frequency is decreased, the phase between the unsteady heat release and the velocity is decreased. In the tube, because of the approximately open-end condition, the velocity leads the pressure by 90 degrees. Thus, the phase difference between the pressure and the heat release is increased as the frequency decreases. Consequently, the acoustic energy converted from the heat release is reduced according to Rayleigh criterion [2]. Thus, the acoustic energy quantity transferred from the heat can be dissipated at the ends of the tube. Consequently, the thermoacoustic instability is controlled.

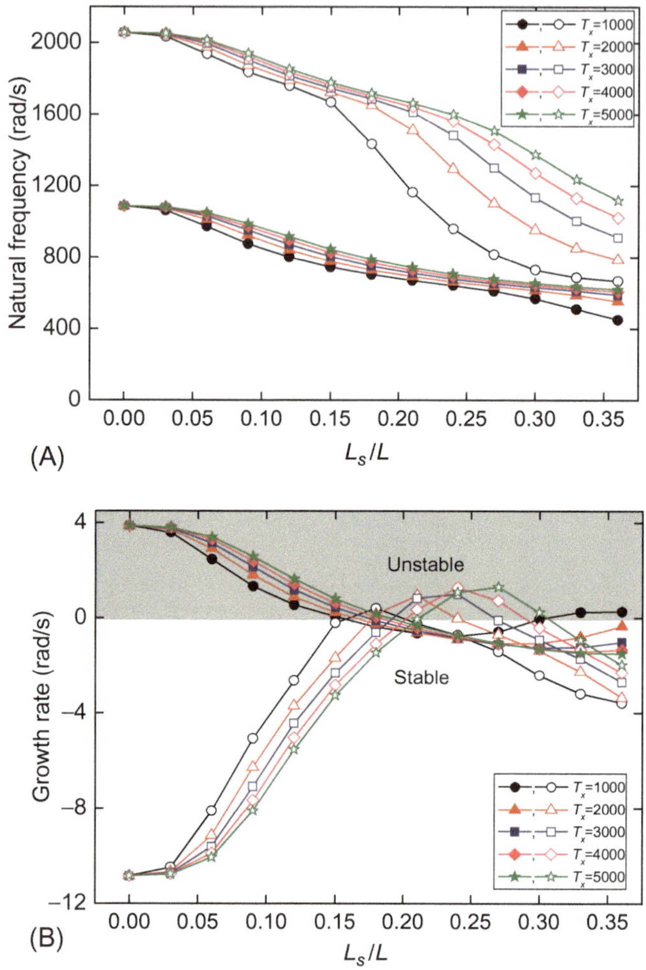

Fig. 7.32 Complex frequency against length of drum-like silencer with $D_c/R = 1.3$: (A) Natural frequency (rad/s); (B) growth rate (rad/s) [31]

Appendix A Coefficients of the matching conditions

$$A_{mn}^{+\,j} = \left(\frac{U^j}{C_0^2} - \frac{\alpha_{mn}^{+\,j}}{\omega + \alpha_{mn}^{+\,j} U^j}\right) e^{i\alpha_{mn}^{+\,j} L^j}$$

$$A_{mn}^{-\,j} = \left(\frac{U^j}{C_0^2} - \frac{\alpha_{mn}^{-\,j}}{\omega + \alpha_{mn}^{-\,j} U^j}\right) e^{i\alpha_{mn}^{-\,j} L^j}$$

$$A_{vmn}^{+\,j} = U^j e^{-i\frac{\omega}{U^j} L^j}$$

$$A_{mn}^{+\,j+1} = -\left(\frac{U^{j+1}}{C_0^2} - \frac{\alpha_{mn}^{+\,j+1}}{\omega + \alpha_{mn}^{+\,j+1} U^{j+1}}\right)$$

$$A_{mn}^{-\,j+1} = -\left(\frac{U^{j+1}}{C_0^2} - \frac{\alpha_{mn}^{-\,j+1}}{\omega + \alpha_{mn}^{-\,j+1} U^{j+1}}\right)$$

$$A_{vmn}^{+\,j} = -U^{j+1}$$

$$B_{mn}^{+\,j} = \left(1 + \left(M_a^j\right)^2 - \frac{2U^j \alpha_{mn}^{+\,j}}{\omega + \alpha_{mn}^{+\,j} U^j}\right) e^{i\alpha_{mn}^{+\,j} L^j}$$

$$B_{mn}^{+\,j} = \left(1 + \left(M_a^j\right)^2 - \frac{2U^j \alpha_{mn}^{-\,j}}{\omega + \alpha_{mn}^{-\,j} U^j}\right) e^{i\alpha_{mn}^{-\,j} L^j}$$

$$B_{vmn}^{+\,j} = \left(U^j\right)^2 e^{-i\frac{\omega}{U^j} L^j}$$

$$B_{mn}^{+\,j+1} = -1 - \left(\left(M_a^{j+1}\right)^2 - \frac{2U^{j+1} \alpha_{mn}^{+\,j+1}}{\omega + \alpha_{mn}^{+\,j+1} U^{j+1}}\right)$$

$$B_{mn}^{-\,j+1} = -1 - \left(\left(M_a^{j+1}\right)^2 - \frac{2U^{j+1} \alpha_{mn}^{-\,j+1}}{\omega + \alpha_{mn}^{-\,j+1} U^{j+1}}\right)$$

$$B_{vmn}^{+\,j} = -\left(U^{j+1}\right)^2$$

$$C_{mn}^{+\,j} = \left[U^j \left(\frac{\gamma}{\gamma - 1} + \frac{\left(M_a^j\right)^2}{2}\right) - \frac{1}{\rho_0^j}\left(\frac{\gamma P_0^j}{\gamma - 1} + \frac{3}{2}\rho_0^j \left(U^j\right)^2\right) \frac{\alpha_{mn}^{+\,j}}{\omega + \alpha_{mn}^{+\,j} U^j}\right] e^{i\alpha_{mn}^{+\,j} L^j}$$

$$C_{mn}^{-\,j} = \left[U^j \left(\frac{\gamma}{\gamma - 1} + \frac{\left(M_a^j\right)^2}{2}\right) - \frac{1}{\rho_0^j}\left(\frac{\gamma P_0^j}{\gamma - 1} + \frac{3}{2}\rho_0^j \left(U^j\right)^2\right) \frac{\alpha_{mn}^{-\,j}}{\omega + \alpha_{mn}^{-\,j} U^j}\right] e^{i\alpha_{mn}^{-\,j} L^j}$$

$$C_{vmn}^{+\,j} = \frac{\left(U^j\right)^3}{2} e^{-i\frac{\omega}{U^j} L^j}$$

$$C_{mn}^{+\,j+1} = -U^{j+1}\left(\frac{\gamma}{\gamma - 1} + \frac{\left(M_a^{j+1}\right)^2}{2}\right) + \frac{1}{\rho_0^{j+1}}\left(\frac{\gamma P_0^{j+1}}{\gamma - 1} + \frac{3}{2}\rho_0^{j+1}\left(U^{j+1}\right)^2\right) \frac{\alpha_{mn}^{+\,j+1}}{\omega + \alpha_{mn}^{+\,j+1} U^{j+1}}$$

$$C_{mn}^{-\,j+1} = -U^{j+1}\left(\frac{\gamma}{\gamma - 1} + + \frac{\left(M_a^{j+1}\right)^2}{2}\right) + \frac{S^{j+1}}{\rho_0^{j+1}}\left(\frac{\gamma P_0^{j+1}}{\gamma - 1} + \frac{3}{2}\rho_0^{j+1}\left(U^{j+1}\right)^2\right) \frac{\alpha_{mn}^{-\,j+1}}{\omega + \alpha_{mn}^{-\,j+1} U^{j+1}}$$

$$C_{vmn}^{+\,j+1} = -\frac{\left(U^{j+1}\right)^3}{2}$$

$$D_{mn}^{+\,j} = \frac{\rho_0^j}{z_0} + \frac{\alpha_{mn}^{+\,j}}{\omega + \alpha_{mn}^{+\,j} U^j}$$

$$D_{mn}^{-j} = \frac{\rho_0^j}{z_0} + \frac{\alpha_{mn}^{-j}}{\omega + \alpha_{mn}^{-j} U^j}$$

$$D_{vmn}^{+j} = 0$$

$$D_{mn}^{+j+1} = 0$$

$$D_{mn}^{-j+1} = 0$$

$$D_{vmn}^{+j+1} = 0$$

$$E_{mn}^{+j} = 0$$

$$E_{mn}^{-j} = 0$$

$$E_{vmn}^{+j} = 0$$

$$E_{mn}^{+j+1} = \left(\frac{\rho_0^{j+1}}{z_{end}} + \frac{\alpha_{mn}^{+j+1}}{\omega + \alpha_{mn}^{+j+1} U^{j+1}} \right) e^{i\alpha_{mn}^{+j+1} L^{j+1}}$$

$$E_{mn}^{-j+1} = \left(\frac{\rho_0^{j+1}}{z_0} + \frac{\alpha_{mn}^{-j+1}}{\omega + \alpha_{mn}^{-j+1} U^{j+1}} \right) e^{i\alpha_{mn}^{-j+1} L^{j+1}}$$

$$E_{vmn}^{+j+1} = 0$$

$$F_{mn}^{+j} = \left(R_p - \frac{R_v}{\rho_0^j} \frac{\alpha_{mn}^{+j}}{\omega + \alpha_{mn}^{+j} U^j} \right) e^{i\alpha_{mn}^{+j} L^j}$$

$$F_{mn}^{-j} = \left(R_p - \frac{R_v}{\rho_0^j} \frac{\alpha_{mn}^{-j}}{\omega + \alpha_{mn}^{-j} U^j} \right) e^{i\alpha_{mn}^{-j} L^j}$$

$$F_{vmn}^{+j} = 0$$

$$F_{mn}^{+j+1} = 0$$

$$F_{mn}^{-j+1} = 0$$

$$F_{vmn}^{+j+1} = 0$$

Appendix B Coefficients of the matching conditions for variable cross-sections cases

$$Ab_{101}^{+j} = A_{01}^{+j} \int_0^{r_j} r dr$$

$$Ab_{101}^{-\,j} = A_{01}^{-\,j}\int_0^{r_j} rdr$$

$$Ab_{\nu 101}^{+\,j} = A_{\nu 01}^{+\,j}\int_0^{r_j} rdr$$

$$Ab_{1mn}^{+\,j} = A_{mn}^{+\,j}\int_0^{r_j} r\psi_{mn}^{j}(r)dr$$

$$Ab_{1mn}^{-\,j} = A_{mn}^{-\,j}\int_0^{r_j} r\psi_{mn}^{j}(r)dr$$

$$Ab_{\nu 1mn}^{+\,j} = A_{\nu mn}^{+\,j}\int_0^{r_j} r\psi_{mn}^{j}(r)dr$$

$$Ab_{101}^{+\,j+1} = -\frac{U^{j+1}}{C_0^2}\int_0^{r_j} rdr + \frac{\alpha_{mn}^{+\,j+1}}{\omega + \alpha_{mn}^{+\,j+1}U^{j+1}}\int_0^{r_{j+1}} rdr$$

$$Ab_{101}^{-\,j+1} = -\frac{U^{j+1}}{C_0^2}\int_0^{r_j} rdr + \frac{\alpha_{mn}^{-\,j+1}}{\omega + \alpha_{mn}^{-\,j+1}U^{j+1}}\int_0^{r_{j+1}} rdr$$

$$Ab_{\nu 101}^{+\,j+1} = A_{\nu 01}^{+\,j+1}\int_0^{r_j} rdr$$

$$Ab_{1mn}^{+\,j+1} = -\frac{U^{j+1}}{C_0^2}\int_0^{r_j} r\psi_{mn}^{j+1}(r)dr + \frac{\alpha_{mn}^{+\,j+1}}{\omega + \alpha_{mn}^{+\,j+1}U^{j+1}}\int_0^{r_{j+1}} r\psi_{mn}^{j+1}(r)dr$$

$$Ab_{mn}^{-\,j+1} = -\frac{U^{j+1}}{C_0^2}\int_0^{r_j} r\psi_{mn}^{j+1}(r)dr + \frac{\alpha_{mn}^{-\,j+1}}{\omega + \alpha_{mn}^{-\,j+1}U^{j+1}}\int_0^{r_{j+1}} r\psi_{mn}^{j+1}(r)dr$$

$$Ab_{\nu 1mn}^{+\,j+1} = A_{\nu mn}^{+\,j+1}\int_0^{r_j} r\psi_{mn}^{j+1}(r)dr$$

$$\int rJ_m(K_{mn}r)J_m(K_{m\nu}r)dr = \frac{r(K_{mn}J_{m+1}(K_{mn}r)J_m(K_{m\nu}r) - K_{m\nu}J_m(K_{mn}r)J_{m+1}(K_{m\nu}r))}{K_{mn}^2 - K_{m\nu}^2},\quad n \neq \nu$$

$$\int rJ_m^2(K_{mn}r)dr = \frac{1}{2}r^2\left\{\left(1 - \frac{m^2}{K_{mn}^2 r^2}\right)J_m^2(K_{mn}r) + J'^2_m(K_{mn}r)\right\}$$

For $m = 0$:

$$\int_0^{r_j} rdr = \frac{1}{2}r_j^2$$

$$\int_0^{r_j} r\psi_{mn}^{j}(r)dr = 0; \; n > 1$$

$$\int_0^{r_j} r\psi_{mn}^{j+1}(r)dr = \frac{r_j J_1\left(K_{mn}^{j+1} r_j\right)}{K_{mn}^{j+1}}$$

$$Ab_{201}^{+j} = A_{01}^{+j}\int_0^{r_j} r\psi_{mv}^{j}(r)dr$$

$$Ab_{201}^{-j} = A_{01}^{-j}\int_0^{r_j} r\psi_{mv}^{j}(r)dr$$

$$Ab_{v201}^{+j} = A_{v01}^{+j}\int_0^{r_j} r\psi_{mv}^{j}(r)dr$$

$$Ab_{2mn}^{+j} = A_{mn}^{+j}\int_0^{r_j} r\psi_{mn}^{j}(r)\psi_{mv}^{j}(r)dr$$

$$Ab_{2mn}^{-j} = A_{mn}^{-j}\int_0^{r_j} r\psi_{mn}^{j}(r)\psi_{mv}^{j}(r)dr$$

$$Ab_{v2mn}^{+j} = A_{vmn}^{+j}\int_0^{r_j} r\psi_{mn}^{j}(r)\psi_{mv}^{j}(r)dr$$

$$Ab_{201}^{+j+1} = -\frac{U^{j+1}}{a_0^2}\int_0^{r_j} r\psi_{mv}^{j}(r)dr + \frac{\alpha_{mn}^{+j+1}}{\omega + \alpha_{mn}^{+j+1}U^{j+1}}\int_0^{r_{j+1}} r\psi_{mv}^{j}(r)dr$$

$$Ab_{201}^{-j+1} = -\frac{U^{j+1}}{a_0^2}\int_0^{r_j} r\psi_{mv}^{j}(r)dr + \frac{\alpha_{mn}^{-j+1}}{\omega + \alpha_{mn}^{-j+1}U^{j+1}}\int_0^{r_{j+1}} r\psi_{mv}^{j}(r)dr$$

$$Ab_{v201}^{+j+1} = A_{v01}^{+j+1}\int_0^{r_j} r\psi_{mv}^{j}(r)dr$$

$$\begin{aligned} Ab_{2mn}^{+j+1} &= -\frac{U^{j+1}}{C_0^2}\int_0^{r_j} r\psi_{mn}^{j+1}(r)\psi_{mv}^{j}(r)dr \\ &+ \frac{\alpha_{mn}^{+j+1}}{\omega + \alpha_{mn}^{+j+1}U^{j+1}}\int_0^{r_{j+1}} r\psi_{mn}^{j+1}(r)\psi_{mv}^{j}(r)dr \end{aligned}$$

$$\begin{aligned} Ab_{2mn}^{-j+1} &= -\frac{U^{j+1}}{a_0^2}\int_0^{r_j} r\psi_{mn}^{j+1}(r)\psi_{mv}^{j}(r)dr \\ &+ \frac{\alpha_{mn}^{-j+1}}{\omega + \alpha_{mn}^{-j+1}U^{j+1}}\int_0^{r_{j+1}} r\psi_{mn}^{j+1}(r)\psi_{mv}^{j}(r)dr \end{aligned}$$

$$Ab_{v2mn}^{+j+1} = A_{vmn}^{+j+1}\int_0^{r_j} r\psi_{mn}^{j+1}(r)\psi_{mv}^{j}(r)dr$$

where, $m=0$:

$$\int_0^{r_j} r\psi_{m\nu}^j(r)dr = 0$$

$$\int_0^{r_j} r\psi_{mn}^j(r)\psi_{m\nu}^j(r)dr = \begin{cases} 0; \; n \neq \nu \\ \frac{1}{2} r_j^2 J_0^2(K_{mn} r_j); \; n = \nu \end{cases}$$

$$\int_0^{r_j} r\psi_{mn}^{j+1}(r)\psi_{m\nu}^j(r)dr = \frac{r_j K_{mn}^{j+1} J_1(K_{mn}^{j+1} r_j) J_0(K_{m\nu}^j r)}{\left(K_{mn}^{j+1}\right)^2 - \left(K_{m\nu}^j\right)^2}$$

$$Bb_{101}^{+j} = B_{01}^{+j} \int_0^{r_j} rdr$$

$$Bb_{101}^{-j} = B_{01}^{-j} \int_0^{r_j} rdr$$

$$Bb_{\nu 101}^{+j} = B_{\nu 01}^{+j} \int_0^{r_j} rdr$$

$$Bb_{1mn}^{+j} = B_{mn}^{+j} \int_0^{r_j} r\psi_{mn}^j(r)dr$$

$$Bb_{1mn}^{-j} = B_{mn}^{-j} \int_0^{r_j} r\psi_{mn}^j(r)dr$$

$$Bb_{\nu 1mn}^{+j} = B_{\nu mn}^{+j} \int_0^{r_j} r\psi_{mn}^j(r)dr$$

$$Bb_{101}^{+j+1} = \left(-1 - \left(M_a^{j+1}\right)^2\right) \int_0^{r_j} rdr + \frac{2U^{j+1}\alpha_{mn}^{+j+1}}{\omega + \alpha_{mn}^{+j+1} U^{j+1}} \int_0^{r_{j+1}} rdr$$

$$Bb_{101}^{-j+1} = \left(-1 - \left(M_a^{j+1}\right)^2\right) \int_0^{r_j} rdr + \frac{2U^{j+1}\alpha_{mn}^{-j+1}}{\omega + \alpha_{mn}^{-j+1} U^{j+1}} \int_0^{r_{j+1}} rdr$$

$$Bb_{\nu 101}^{+j} = -U^{j+1^2} \int_0^{r_j} rdr$$

$$Bb_{1mn}^{+j+1} = \left(-1 - \left(M_a^{j+1}\right)^2\right) \int_0^{r_j} r\psi_{mn}^{j+1}(r)dr + \frac{2U^{j+1}\alpha_{mn}^{+j+1}}{\omega + \alpha_{mn}^{+j+1} U^{j+1}} \int_0^{r_{j+1}} r\psi_{mn}^{j+1}(r)dr$$

$$Bb_{1mn}^{-j+1} = \left(-1 - \left(M_a^{j+1}\right)^2\right) \int_0^{r_j} r\psi_{mn}^{j+1}(r)dr + \frac{2U^{j+1}\alpha_{mn}^{-j+1}}{\omega + \alpha_{mn}^{-j+1} U^{j+1}} \int_0^{r_{j+1}} r\psi_{mn}^{j+1}(r)dr$$

$$Bb^{+\,j}_{v1mn} = -U^{j+1^2}\int_0^{r_j} r\psi^{j+1}_{mn}(r)\,dr$$

$$Bb^{+\,j}_{201} = B^{+\,j}_{01}\int_0^{r_j} r\psi^{j}_{mv}(r)\,dr$$

$$Bb^{-\,j}_{201} = B^{-\,j}_{01}\int_0^{r_j} r\psi^{j}_{mv}(r)\,dr$$

$$Bb^{+\,j}_{v201} = B^{+\,j}_{v01}\int_0^{r_j} r\psi^{j}_{mv}(r)\,dr$$

$$Bb^{+\,j}_{2mn} = B^{+\,j}_{mn}\int_0^{r_j} r\psi^{j}_{mn}(r)\psi^{j}_{mv}(r)\,dr$$

$$Bb^{-\,j}_{2mn} = B^{-\,j}_{mn}\int_0^{r_j} r\psi^{j}_{mn}(r)\psi^{j}_{mv}(r)\,dr$$

$$Bb^{+\,j}_{v2mn} = B^{+\,j}_{vmn}\int_0^{r_j} r\psi^{j}_{mn}(r)\psi^{j}_{mv}(r)\,dr$$

$$\begin{aligned}Bb^{+\,j+1}_{201} &= \left(-1-\left(M_a^{j+1}\right)^2\right)\int_0^{r_j} r\psi^{j}_{mv}(r)\,dr\\ &\quad+\frac{2U^{j+1}\alpha^{+\,j+1}_{mn}}{\omega+\alpha^{+\,j+1}_{mn}U^{j+1}}\int_0^{r_{j+1}} r\psi^{j}_{mv}(r)\,dr\end{aligned}$$

$$\begin{aligned}Bb^{-\,j+1}_{201} &= \left(-1-\left(M_a^{j+1}\right)^2\right)\int_0^{r_j} r\psi^{j}_{mv}(r)\,dr\\ &\quad+\frac{2U^{j+1}\alpha^{-\,j+1}_{mn}}{\omega+\alpha^{-\,j+1}_{mn}U^{j+1}}\int_0^{r_{j+1}} r\psi^{j}_{mv}(r)\,dr\end{aligned}$$

$$Bb^{+\,j+1}_{v201} = -U^{j+1^2}\int_0^{r_j} r\psi^{j}_{mv}(r)\,dr$$

$$\begin{aligned}Bb^{+\,j+1}_{2mn} &= \left(-1-M^{j+1^2}\right)\int_0^{r_j} r\psi^{j+1}_{mn}(r)\psi^{j}_{mv}(r)\,dr\\ &\quad+\frac{2U^{j+1}\alpha^{+\,j+1}_{mn}}{\omega+\alpha^{+\,j+1}_{mn}U^{j+1}}\int_0^{r_{j+1}} r\psi^{j+1}_{mn}(r)\psi^{j}_{mv}(r)\,dr\end{aligned}$$

$$\begin{aligned}Bb^{-\,j+1}_{2mn} &= \left(-1-\left(M_a^{j+1}\right)^2\right)\int_0^{r_j} r\psi^{j+1}_{mn}(r)\psi^{j}_{mv}(r)\,dr\\ &\quad+\frac{2U^{j+1}\alpha^{-\,j+1}_{mn}}{\omega+\alpha^{-\,j+1}_{mn}U^{j+1}}\int_0^{r_{j+1}} r\psi^{j+1}_{mn}(r)\psi^{j}_{mv}(r)\,dr\end{aligned}$$

$$Bb_{v2mn}^{+\,j+1} = -U^{j+1^2}\int_0^{r_j} r\psi_{mn}^{j}(r)\psi_{mv}^{j}(r)dr$$

$$Cb_{101}^{+\,j} = C_{01}^{+\,j}\int_0^{r_j} rdr$$

$$Cb_{101}^{-\,j} = C_{01}^{-\,j}\int_0^{r_j} rdr$$

$$Cb_{v101}^{+\,j} = C_{v01}^{+\,j}\int_0^{r_j} rdr$$

$$Cb_{1mn}^{+\,j} = C_{mn}^{+\,j}\int_0^{r_j} r\psi_{mn}^{j}(r)dr$$

$$Cb_{1mn}^{-\,j} = C_{mn}^{-\,j}\int_0^{r_j} r\psi_{mn}^{j}(r)dr$$

$$Cb_{v1mn}^{+\,j} = C_{vmn}^{+\,j}\int_0^{r_j} r\psi_{mn}^{j}(r)dr$$

$$Cb_{101}^{+\,j+1} = -U^{j+1}\left(\frac{\gamma}{\gamma-1}+\frac{\left(M_a^{j+1}\right)^2}{2}\right)\int_0^{r_j} rdr + \frac{1}{\rho_0^{j+1}}\left(\frac{\gamma P_0^{j+1}}{\gamma-1}+\frac{3}{2}\rho_0^{j+1}U^{j+1^2}\right)\frac{\alpha_{mn}^{+\,j+1}}{\omega+\alpha_{mn}^{+\,j+1}U^{j+1}}\int_0^{r_{j+1}} rdr$$

$$Cb_{101}^{-\,j+1} = -U^{j+1}\left(\frac{\gamma}{\gamma-1}+\frac{\left(M_a^{j+1}\right)^2}{2}\right)\int_0^{r_j} rdr + \frac{1}{\rho_0^{j+1}}\left(\frac{\gamma P_0^{j+1}}{\gamma-1}+\frac{3}{2}\rho_0^{j+1}U^{j+1^2}\right)\frac{\alpha_{mn}^{-\,j+1}}{\omega+\alpha_{mn}^{-\,j+1}U^{j+1}}\int_0^{r_{j+1}} rdr$$

$$Cb_{v01}^{+\,j} = -\frac{U^{j+1^3}}{2}\int_0^{r_j} rdr$$

$$Cb_{1mn}^{+\,j+1} = -U^{j+1}\left(\frac{\gamma}{\gamma-1}+\frac{\left(M_a^{j+1}\right)^2}{2}\right)\int_0^{r_j} r\psi_{mn}^{j+1}(r)dr + \frac{1}{\rho_0^{j+1}}\left(\frac{\gamma P_0^{j+1}}{\gamma-1}+\frac{3}{2}\rho_0^{j+1}U^{j+1^2}\right)\frac{\alpha_{mn}^{+\,j+1}}{\omega+\alpha_{mn}^{+\,j+1}U^{j+1}}\int_0^{r_{j+1}} r\psi_{mn}^{j+1}(r)dr$$

$$Cb_{1mn}^{-\,j+1} = -U^{j+1}\left(\frac{\gamma}{\gamma-1} + \frac{\left(M_a^{j+1}\right)^2}{2}\right)\int_0^{r_j} r\psi_{mn}^{j+1}(r)dr$$
$$+\frac{1}{\rho_0^{j+1}}\left(\frac{\gamma P_0^{j+1}}{\gamma-1} + \frac{3}{2}\rho_0^{j+1}U^{j+1^2}\right)\frac{\alpha_{mn}^{-\,j+1}}{\omega + \alpha_{mn}^{-\,j+1}U^{j+1}}\int_0^{r_{j+1}} r\psi_{mn}^{j+1}(r)dr$$

$$Cb_{201}^{+\,j} = C_{01}^{+\,j}\int_0^{r_j} r\psi_{mv}^{j}(r)dr$$

$$Cb_{201}^{-\,j} = C_{01}^{-\,j}\int_0^{r_j} r\psi_{mv}^{j}(r)dr$$

$$Cb_{v201}^{+\,j} = C_{v01}^{+\,j}\int_0^{r_j} r\psi_{mv}^{j}(r)dr$$

$$Cb_{2mn}^{+\,j} = C_{mn}^{+\,j}\int_0^{r_j} r\psi_{mn}^{j}(r)\psi_{mv}^{j}(r)dr$$

$$Cb_{2mn}^{-\,j} = C_{mn}^{-\,j}\int_0^{r_j} r\psi_{mn}^{j}(r)\psi_{mv}^{j}(r)dr$$

$$Cb_{v2mn}^{+\,j} = -\frac{U^{j+1^3}}{2}\int_0^{r_j} r\psi_{mn}^{j}(r)dr$$

$$Cb_{v2mn}^{+\,j} = C_{vmn}^{+\,j}\int_0^{r_j} r\psi_{mn}^{j}(r)\psi_{mv}^{j}(r)dr$$

$$Cb_{201}^{+\,j+1} = -U^{j+1}\left(\frac{\gamma}{\gamma-1} + \frac{\left(M_a^{j+1}\right)^2}{2}\right)\int_0^{r_j} r\psi_{mv}^{j}(r)dr$$
$$+\frac{1}{\rho_0^{j+1}}\left(\frac{\gamma P_0^{j+1}}{\gamma-1} + \frac{3}{2}\rho_0^{j+1}U^{j+1^2}\right)\frac{\alpha_{mn}^{+\,j+1}}{\omega + \alpha_{mn}^{+\,j+1}U^{j+1}}\int_0^{r_{j+1}} r\psi_{mv}^{j}(r)dr$$

$$Cb_{201}^{-\,j+1} = -U^{j+1}\left(\frac{\gamma}{\gamma-1} + \frac{\left(M_a^{j+1}\right)^2}{2}\right)\int_0^{r_j} r\psi_{mv}^{j}(r)dr$$
$$+\frac{1}{\rho_0^{j+1}}\left(\frac{\gamma P_0^{j+1}}{\gamma-1} + \frac{3}{2}\rho_0^{j+1}U^{j+1^2}\right)\frac{\alpha_{mn}^{-\,j+1}}{\omega + \alpha_{mn}^{-\,j+1}U^{j+1}}\int_0^{r_{j+1}} r\psi_{mv}^{j}(r)dr$$

$$Cb_{v201}^{+\,j} = -\frac{U^{j+1^3}}{2}\int_0^{r_j} r\psi_{mv}^{j}(r)dr$$

$$Cb_{2mn}^{+\ j+1} = -U^{j+1}\left(\frac{\gamma}{\gamma-1} + \frac{\left(M_a^{j+1}\right)^2}{2}\right)\int_0^{r_j} r\psi_{mv}^{j}(r)\psi_{mn}^{j}(r)dr$$

$$+\frac{1}{\rho_0^{j+1}}\left(\frac{\gamma P_0^{j+1}}{\gamma-1} + \frac{3}{2}\rho_0^{j+1}U^{j+1^2}\right)\frac{\alpha_{mn}^{+\ j+1}}{\omega + \alpha_{mn}^{+\ j+1}U^{j+1}}\int_0^{r_{j+1}} r\psi_{mn}^{j}(r)\psi_{mv}^{j}(r)dr$$

$$Cb_{2mn}^{-\ j+1} = -U^{j+1}\left(\frac{\gamma}{\gamma-1} + \frac{\left(M_a^{j+1}\right)^2}{2}\right)\int_0^{r_j} r\psi_{mn}^{j+1}(r)\psi_{mv}^{j}(r)dr$$

$$+\frac{1}{\rho_0^{j+1}}\left(\frac{\gamma P_0^{j+1}}{\gamma-1} + \frac{3}{2}\rho_0^{j+1}U^{j+1^2}\right)\frac{\alpha_{mn}^{-\ j+1}}{\omega + \alpha_{mn}^{-\ j+1}U^{j+1}}\int_0^{r_{j+1}} r\psi_{mn}^{j+1}(r)\psi_{mv}^{j}(r)dr$$

$$Cb_{v2mn}^{+\ j} = -\frac{U^{j+1^3}}{2}\int_0^{r_j} r\psi_{mn}^{j+1}(r)\psi_{mv}^{j}(r)dr$$

Appendix C Coefficients in Eq. (7.149)

$$A_1 = \frac{(r+1)}{C_w}\left(i\omega + \frac{\partial \overline{v}_x}{\partial x}\right)$$

$$A_2 = \frac{(r+1)\overline{v}_x}{C_w}$$

$$A_3 = \frac{\overline{\rho}(r+1)}{C_w}$$

$$A_4 = \frac{\overline{\rho}}{C_w}$$

$$A_5 = \frac{\overline{\rho}(r+1)}{C_w}$$

$$A_6 = \frac{\overline{\rho}}{C_w} + \frac{\partial \overline{\rho}}{\partial r}\frac{(r+1)}{C_w}$$

$$A_7 = \frac{1}{C_w}\frac{\partial \overline{\rho}}{\partial \theta}$$

$$A_8 = \frac{1}{C_w}\frac{\partial \overline{\rho}}{\partial x}(r+1)$$

$$B_1 = \frac{i\omega\overline{\rho}}{C_w}$$

$$B_2 = \frac{\overline{\rho}\overline{v}_x}{C_w}$$

$$B_3 = \frac{1}{C_w}$$

$$C_1 = \frac{i\omega\overline{\rho}(r+1)}{C_w}$$

$$C_2 = \frac{\overline{\rho}\overline{v}_x(r+1)}{C_w}$$

$$C_3 = \frac{1}{C_w}$$

$$D_1 = \frac{\overline{v}_x}{C_w}\frac{\partial \overline{v}_x}{\partial x}(r+1)$$

$$D_2 = \left(i\omega\overline{\rho} + \overline{\rho}\frac{\partial \overline{v}_x}{\partial x}\right)\frac{(r+1)}{C_w}$$

$$D_3 = \frac{\overline{\rho}\overline{v}_x(r+1)}{C_w}$$

$$D_4 = \frac{\overline{\rho}}{C_w}\frac{\partial \overline{v}_x}{\partial r}(r+1)$$

$$D_5 = \frac{\overline{\rho}}{C_w}\frac{\partial \overline{v}_x}{\partial \theta}$$

$$D_6 = \frac{r+1}{C_w}$$

$$E_1 = \left(-i\omega + \frac{\overline{v}_x}{\overline{\rho}}\frac{\partial \overline{\rho}}{\partial x}\right)\frac{(r+1)}{C_w}$$

$$E_2 = -\frac{\overline{v}_x}{C_w}(r+1)$$

$$E_3 = \left(\frac{1}{\overline{c}^2}\frac{\partial \overline{p}}{\partial r} - \frac{\partial \overline{\rho}}{\partial r}\right)\frac{(r+1)}{C_w}$$

$$E_4 = \left(\frac{1}{\overline{c}^2}\frac{\partial \overline{p}}{\partial \theta} - \frac{\partial \overline{\rho}}{\partial \theta}\right)\frac{1}{C_w}$$

$$E_5 = \left(\frac{1}{\overline{c}^2}\frac{\partial \overline{p}}{\partial x} - \frac{\partial \overline{\rho}}{\partial x}\right)\frac{(r+1)}{C_w}$$

$$E_6 = \left(i\omega\frac{1}{\overline{c}^2} - \frac{\gamma \overline{v}_x}{\overline{c}^2 \overline{\rho}}\frac{\partial \overline{p}}{\partial x}\right)\frac{(r+1)}{C_w}$$

$$E_7 = \frac{(r+1)}{\overline{c}^2}\frac{\overline{v}_x}{C_w}$$

C_w is a dimensionless number included to adjust the determinant of the matrix for calculation need.

Appendix D Coefficients in Eq. (7.169)

$$K_{a1} = \overline{v}_{x1}$$
$$K_{a2} = \overline{\rho}_1$$
$$K_{a3} = -\overline{v}_{x2}$$
$$K_{a4} = -\overline{\rho}_2$$

$$K_{b1} = \overline{v}_{x1}^2$$
$$K_{b2} = 2\overline{\rho}_1\overline{v}_{x1}$$
$$K_{b3} = 1$$
$$K_{b4} = -\overline{v}_{x2}^2$$
$$K_{b5} = -2\overline{\rho}_2\overline{v}_{x2}$$
$$K_{b6} = -1$$

$$K_{c1} = \overline{\rho}_1 \overline{v}_{x1}$$

$$K_{c2} = -\overline{\rho}_2 \overline{v}_{x2}$$

$$K_{d1} = \overline{\rho}_1 \overline{v}_{x1}$$

$$K_{d2} = -\overline{\rho}_2 \overline{v}_{x2}$$

$$K_{e1} = \left(c_p \overline{T}_{01} \overline{v}_{x1} - c_p T_1 \overline{v}_{x1}\right)$$

$$K_{e2} = \left(\lambda_v + c_p \overline{T}_{01} \overline{\rho}_1 + \overline{\rho}_1 \overline{v}_{x1}^2\right)$$

$$K_{e3} = \lambda_p + \frac{c_p \overline{v}_{x1}}{Rg}$$

$$K_{e4} = -\left(c_p \overline{T}_{02} \overline{v}_{x2} - c_p T_2 v_{x2}\right)$$

$$K_{e5} = -\left(c_p \overline{T}_{02} \overline{\rho}_2 + \overline{\rho}_2 \overline{v}_{x2}^2\right)$$

$$K_{e6} = -\frac{c_p v_{x2}}{Rg}$$

References

[1] J. Tyndall, Sound, D. Appleton, 1897.
[2] L. Rayleigh, Theory of Sound (Two Volumes), Dover Publications, New York, 1945 1877, re-issued 1945.
[3] B.T. Zinn, T.C. Lieuwen, Combustion instabilities: basic concepts, in: Combustion Instabilities in Gas Turbine Engines: Operational Experience, Fundamental Mechanisms, and Modeling, 210 2005, pp. 3–26.
[4] T.C. Lieuwen, V. Yang, Combustion Instabilities in Gas Turbine Engines: Operational Experience, Fundamental Mechanisms, and Modeling, American Institute of Aeronautics and Astronautics, 2005.
[5] T. Poinsot, Prediction and control of combustion instabilities in real engines, Proc. Combust. Inst. 36 (1) (2017) 1–28.
[6] K.R. McManus, T. Poinsot, S.M. Candel, A review of active control of combustion instabilities, Prog. Energy Combust. Sci. 19 (1) (1993) 1–29.
[7] B. Zinn, Y. Neumeier, B. Zinn, Y. Neumeier, An overview of active control of combustion instabilities, in: 35th Aerospace Sciences Meeting and Exhibit, 1997, p. 461.
[8] A.P. Dowling, A.S. Morgans, Feedback control of combustion oscillations, Annu. Rev. Fluid Mech. 37 (2005) 151–182.
[9] A.P. Dowling, The calculation of thermoacoustic oscillations, J. Sound Vib. 180 (4) (1995) 557–581.
[10] A.P. Dowling, J.E. Ffowcs Williams, Sound and Sources of Sound, Horwood, 1983.
[11] M.A. Heckl, Active control of the noise from a Rijke tube, J. Sound Vib. 124 (1) (1988) 117–133.
[12] G.J. Bloxsidge, A.P. Dowling, P.J. Langhorne, Reheat buzz: an acoustically coupled combustion instability. Part 2. Theory, J. Fluid Mech. 193 (1988) 445–473.
[13] T. Schuller, D. Durox, P. Palies, S. Candel, Acoustic decoupling of longitudinal modes in generic combustion systems, Combust. Flame 159 (5) (2012) 1921–1931.

[14] A.P. Dowling, S.R. Stow, Acoustic analysis of gas turbine combustors, J. Propuls. Power 19 (5) (2003) 751–764.
[15] S.R. Stow, A.P. Dowling, T.P. Hynes, Reflection of circumferential modes in a choked nozzle, J. Fluid Mech. 467 (2002) 215–239.
[16] D. You, X. Sun, V. Yang, A three-dimensional linear acoustic analysis of gas turbine combustion instability, in: 41st Aerospace Sciences Meeting and Exhibit, 2003, p. 118.
[17] D. You, X. Sun, V. Yang, Three-dimensional linear stability analysis of gas turbine combustion dynamics, in: Combustion Instabilities in Gas Turbine Engines: Operational Experience, Fundamental Mechanisms, and Modeling, Progress in Astronautics and Aeronautics, 2005.
[18] R.J. Alfredson, The propagation of sound in a circular duct of continuously varying cross-sectional area, J. Sound Vib. 23 (4) (1972) 433–442.
[19] R. Peyret, Spectral Methods for Incompressible Viscous Flow, Vol. 148, Springer Science & Business Media, 2013.
[20] L. Li, X. Sun, Effect of vorticity waves on azimuthal instabilities in annular chambers, Combust. Flame 162 (3) (2015) 628–641.
[21] F.E. Marble, S.M. Candel, Acoustic disturbance from gas non-uniformities convected through a nozzle, J. Sound Vib. 55 (2) (1977) 225–243.
[22] L. Li, L. Yang, X. Sun, Effect of distributed heat source on low frequency thermoacoustic instabilities, J. Sound Vib. 332 (12) (2013) 3098–3111.
[23] S.R. Stow, A.P. Dowling, Modelling of circumferential modal coupling due to Helmholtz resonators, in: ASME Turbo Expo 2003, Collocated with the 2003 International Joint Power Generation Conference, American Society of Mechanical Engineers, 2003, January, pp. 129–137.
[24] T. Lieuwen, Modeling premixed combustion-acoustic wave interactions: a review, J. Propuls. Power 19 (5) (2003) 765–781.
[25] L. Li, Z. Guo, C. Zhang, X. Sun, A passive method to control combustion instabilities with perforated liner, Chin. J. Aeronaut. 23 (6) (2010) 623–630.
[26] X. Sun, X. Wang, L. Du, X. Jing, A new model for the prediction of turbofan noise with the effect of locally and non-locally reacting liners, J. Sound Vib. 316 (1–5) (2008) 50–68.
[27] X. Wang, X. Sun, A new segmentation approach for sound propagation in non-uniform lined ducts with mean flow, J. Sound Vib. 330 (10) (2011) 2369–2387.
[28] X. Wang, X. Sun, Transfer element method with application to acoustic design of aero-engine nacelle, Chin. J. Aeronaut. 28 (2) (2015) 327–345.
[29] J.D. Eldredge, A.P. Dowling, The absorption of axial acoustic waves by a perforated liner with bias flow, J. Fluid Mech. 485 (2003) 307–335.
[30] M.S. Howe, On the theory of unsteady high Reynolds number flow through a circular aperture, Proc. R. Soc. Lond. A 366 (1725) (1979) 205–223 M.E. Goldstein, Aeroacoustics, McGraw-Hill International Book Co., New York, 1976.
[31] G. Zhang, X. Wang, L. Li, X. Jing, X. Sun, Control of thermoacoustic instability with a drum-like silencer, J. Sound Vib. 406 (2017) 253–276.

CHAPTER 8

Impedance eduction for acoustic liners

8.1 Introduction

It is known from the previous three chapters that the applications of an acoustic liner deal with two kinds of problems: one is how to establish its physical and mathematical models, as discussed in Chapter 5, and the other is how to estimate or enhance its effects on reducing noise propagation and suppressing various instabilities, as described in Chapters 6 and 7. To realize these objectives, the remaining key problem we have to solve is how to experimentally measure the impedance distribution corresponding to the acoustic liner. In fact, only by such experimental validation can the acoustic liner be applied effectively to various engineering practices. This chapter is thus dedicated to briefly introducing developing history and presenting some new investigations in this regard.

As a kind of typical engineering application, the locally reacting acoustic liner flush-mounted in the engine nacelles, as shown in Fig. 8.1, is currently the major acoustic treatment to significantly reduce the fan noise, a dominant source of modern commercial aircraft propulsion noise at both take-off and approach operations [1]. Two frequently used types of acoustic liners are shown in Fig. 8.2. Further improvement in noise reduction performance of such an acoustic liner is to optimize its acoustic impedance [2], i.e., the ratio of the sound pressure to the normal component of sound particle velocity at the liner surface. Therefore, an accurate knowledge of this quantity is critical in the development of more efficient liners and quieter aircrafts. One of the key techniques in acoustic liner design is a suitable impedance eduction method, which can accurately determine the liner impedance.

A classical approach to measure the acoustic impedance is to use an impedance tube with a sample of an acoustic liner exposed to a normal incident sound wave. However, this method has difficulty in introducing the grazing flow over the liner surface [3]. The in-situ method (ISM) was then developed by Dean in 1974 for impedance measurement in the presence of grazing flow [3]. Two-microphone method is the most frequently used

Fundamentals of Aeroacoustics with Applications to Aeropropulsion Systems
https://doi.org/10.1016/B978-0-12-408069-0.00008-7

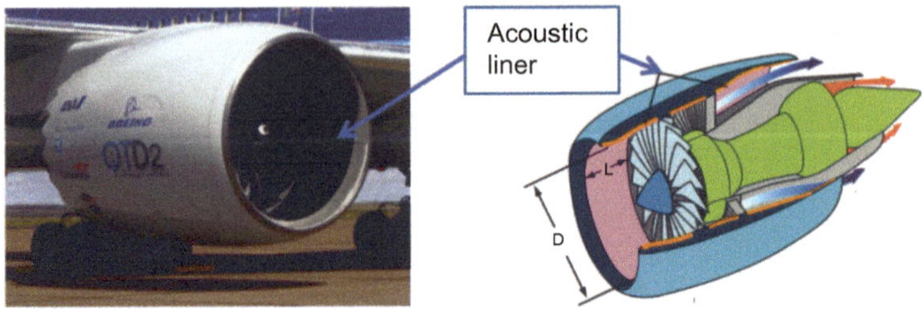

Fig. 8.1 Acoustic liners used in the turbofan (from Internet).

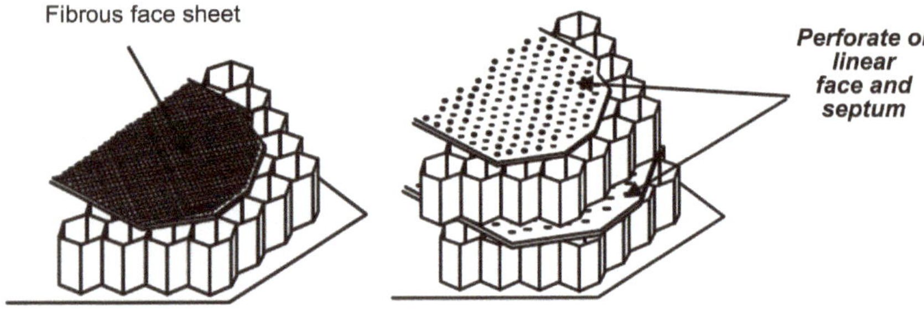

Fig. 8.2 Single and double degree of freedom acoustic liners [2].

ISM, which requires measuring complex sound pressures at two measuring points, one at the liner surface and the other at the rear wall of the cavity. The transfer function between these two pressure measurements is then used to evaluate the acoustic impedance of a locally reacting acoustic liner according to a simple relationship between cavity sound pressure and particle velocity. The distance between the first measuring point and the liner surface must be carefully designed, otherwise, the estimated impedance is incorrect [3]. In addition, the ISM is intrusive and may damage the ordered structure of the liner sample. At almost the same time, another line of thought was proposed by Armstrong in 1971 [4] that the liner impedance can be deduced by using the sound field information in a flow duct with the help of a waveguide model. His method is known as the single mode method (SMM) because only a single unidirectional mode is assumed to propagate over the axial extent of the test liner. SMM utilizes the wall impedance boundary condition together with the dispersion relation between the transversal and axial wavenumbers to educe the impedance. The axial wavenumber is related to the measured sound and phase decay rates over the liner.

Consequently, an axial array of microphones is flush-mounted on the rigid duct wall opposite or adjacent to the test liner to measure the wall complex sound pressure profile in the direction of axis. SMM was later extended to account for shear flow [5–7]. Evidently, SMM breaks down when the single mode assumption is not satisfied, i.e., there is no region where the slopes of the sound pressure level (SPL) and phase decay rates are constant. This restricts the waveguide method from being a general method for decades until a so-called straightforward method (SFM) was proposed by Jing et al. [8]. SFM was later developed to consider a shear flow [9] by introducing the Pridmore-Brown equation, and the original 2-D SFM has recently been extended to handle a 3-D sound field in the flow duct [10]. The highlight of SFM is representing the axial sound pressure profile as a sum of complex exponentials, corresponding to different acoustic modes, and employing Prony's method for modal extraction. The axial wavenumber of the dominant mode is used to deduce the liner impedance in the same manner as SMM. The outstanding progress of SFM is to make the waveguide method capable of handling multimode acoustic field such that the end reflection in a flow-on case or mode scattering from abrupt impedance transition effects at the leading and trailing edges of liner are no longer obstacles. Thanks to the revolution of computer science and the improvement of numerical methods, educing the liner impedance in an iterative way becomes accessible. Such kind of iterative method was first put forward by Watson et al. [11, 12] and was later extended to consider a more realistic mean flow profile [13] and to use a different sound propagation model [14–17]. In an iterative method, also known as the objective function method (OFM), the unknown impedance of the liner is searched by minimizing an objective function defined as the discrepancy between the measured and predicted acoustic parameters. The acoustic parameters chosen to determine the impedance can be either the axial sound pressure profile or the poles of scattering matrix of the test liner. The OFM has a convergence problem when the objective function becomes flat [18], and the accuracy of the educed results is sensitive to the duct exit boundary condition [19]. Besides, it is reported that OFM is typically one to two orders of magnitude more expensive to implement than SMM and SFM [18].

From the previous review on impedance eduction techniques, it can be noticed that SFM is an important method with many advantages over other approaches. Therefore, this chapter will describe the basic theory and development of SFM in detail.

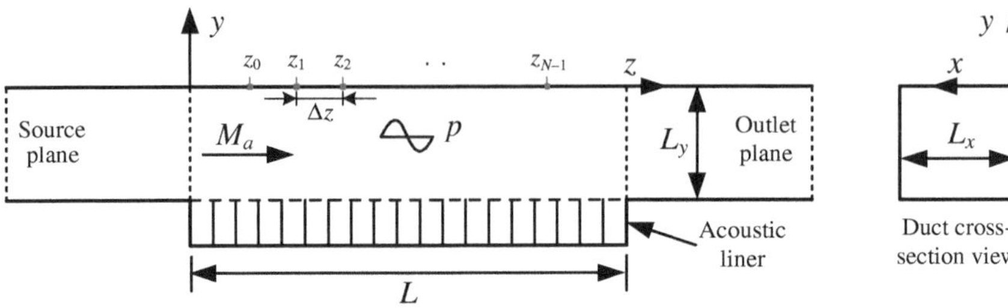

Fig. 8.3 Illustration of the straightforward impedance eduction in a flow duct.

8.2 Straightforward method of acoustic impedance eduction

8.2.1 Model description

As depicted in Fig. 8.3, a grazing flow environment is formed in a rectangular duct whose width and height are L_x and L_y, respectively. An acoustic liner of length L is placed in the lower central portion of the otherwise rigid duct wall. The test liner is treated as locally reacting with the unknown uniform impedance ζ (normalized by the characteristic impedance $Z_0 = \rho_0 c_0$ in the air, and ρ_0 and c_0 being the air density and sound speed, respectively) to be determined. It is reasonable to assume that the subsonic flow is uniform, with the Mach number being M_a. In the flow duct, when the time-harmonic plane sound wave generated in the hard wall section is incident upon the acoustic liner, multimode reflected and transmitted sound waves are excited.

To facilitate the theoretical development, a coordinate system is introduced, with z being along the duct axis and originating from the liner leading edge, and putting $y = 0$ on the upper duct wall. The axial sound pressure distribution is measured at a sequence of equally spaced locations on the upper wall region directly over the acoustic liner, as shown in Fig. 8.3. The axial coordinates of measuring points are z_s, $s = 0, 1, \ldots, N_z - 1$ where N_z is the number of measuring points.

8.2.2 Sound field in the flow duct

The harmonic sound propagation in the flow duct is governed by the convective Helmholtz equation for the sound pressure p,

$$\nabla^2 p - M_a^2 \frac{\partial^2 p}{\partial z^2} - 2ikM_a \frac{\partial p}{\partial z} + k^2 p = 0, \tag{8.1}$$

where $i=\sqrt{-1}$, k is the free-space wavenumber, $k = 2\pi f/c_0$, f is the frequency, and c is the sound speed. The boundary conditions at the rigid walls are that the normal particle velocity vanishes or, equivalently,

$$\frac{\partial p}{\partial y}=0. \tag{8.2}$$

On the lined duct wall, the following impedance boundary condition is imposed,

$$\frac{\partial p}{\partial y}=\frac{1}{\zeta}\left(ikp+2M_a\frac{\partial p}{\partial z}+\frac{M_a^2}{ik}\frac{\partial^2 p}{\partial z^2}\right). \tag{8.3}$$

It can be intuitively deduced and mathematically proven that the uniformity of both the wall impedance and the incident sound pressure in x direction renders an essentially two-dimensional sound field in the flow duct. Further analysis based on the classical mode-decomposition theory indicates that the sound pressure p inside the lined segment can be expressed in the following form:

$$p(y,z)=\sum_{n=0}^{\infty}\left[A_n^+\cos\left(k_{yn}^+y\right)e^{ik_{zn}^+z}+A_n^-\cos\left(k_{yn}^-y\right)e^{ik_{zn}^-z}\right], \tag{8.4}$$

where A is the complex mode amplitude, k_y is the complex eigenvalue or the wavenumber in the y direction, and k_z is the complex axial wavenumber. n is the mode number, and the superscript + or − denotes the mode wave propagating along the positive or negative z axis, respectively. Here, the time dependent $e^{i\omega t}$ where $\omega = 2\pi f$ is adopted.

The axial wavenumber $k_{zn}^{\pm}$ is related to the eigenvalue $k_{yn}^{\pm}$ through the dispersion equation:

$$k_{zn}^{\pm}=\frac{M_a k\mp\sqrt{k^2-\left(1-M_a^2\right)\left(k_{yn}^{\pm}\right)^2}}{1-M_a^2}. \tag{8.5}$$

Here, the square root with a non-positive imaginary part is chosen so that the forward propagating mode waves decay as they propagate along the positive z axis, whereas the left-moving mode waves decay as they propagate along the negative z axis.

After applying the Ingard-Myers boundary condition [20] and making some manipulations, we obtain the equation relating $k_{zn}^{\pm}$ and $k_{yn}^{\pm}$ with the liner impedance ζ as:

$$\frac{k_{yn}^{\pm}}{k}\tan\left(k_{yn}^{\pm}L_y\right)-\frac{i}{\zeta}\left(1+M_a\frac{k_{zn}^{\pm}}{k}\right)^2=0. \tag{8.6}$$

So, once one axial wavenumber $k_{zn}^{\pm}$ of a particular mode is calculated from the measured sound pressure data, the corresponding eigenvalue $k_{yn}^{\pm}$ can be computed through Eq. (8.5), so that the liner impedance ζ can be educed by using Eq. (8.6). Note that the Ingard-Myers boundary condition, together with uniform flow assumption, only describes the refraction effect of the vanishingly thin boundary layer [21].

To simplify the illustration and for practical reasons, the infinite series in Eq. (8.4) is truncated and rewritten as:

$$p(y,z)=\sum_{n=0}^{N-1}\left[A_n\cos\left(k_{yn}y\right)e^{ik_{zn}z}\right]. \tag{8.7}$$

Here, the right- and left-moving modes are distinguished by their vertical mode numbers ns rather than the superscripts $\pm$ as in Eq. (8.4). For instance, the first two vertical mode wavenumbers $k_{y0}^{\pm}$ in Eq. (8.4) correspond to k_{y0} and k_{y1} in Eq. (8.7). The infinite summations are truncated at the maximum mode number of $N-1$ ($n=0, 1, \ldots, N-1$). Note that N should be large enough to ensure the convergence of the series.

8.2.3 Mode decomposition by using Prony's method

Substituting the sound pressure data measured at the measuring points into Eq. (8.7) yields:

$$p(0,z_s)=\sum_{n=0}^{N-1}\left[A_ne^{ik_{zn}z_s}\right],\quad s=0,1,\ldots,N_z-1 \tag{8.8}$$

Denote the measured complex sound pressure at the measuring point with $z=z_s$ as $p_s=p(0,z_s)$ and substituting $z_s=z_0+s\Delta z$, $s=0, 1, \ldots, N_z-1$ into Eq. (8.8) obtains:

$$p_s=\sum_{n=0}^{N-1}\left[A_ne^{ik_{zn}(z_0+s\Delta z)}\right],\quad s=0,1,\ldots,N_z-1 \tag{8.9}$$

where Δz is the measuring point spacing. This equation set can be directly solved by Prony's method [8] or Prony-like methods such as the KT method

[22]. Here, Prony's method is adopted, and the detailed procedure is illustrated in the following.

By introducing an exponential function:

$$X_n = e^{ik_{zn}\Delta z}, \quad n = 0, 1, \ldots, N-1, \tag{8.10}$$

Eq. (8.9) can be rewritten as:

$$p_s = \sum_{n=0}^{N-1} A'_n X_n^s, \quad s = 0, 1, \ldots, N_z - 1, \tag{8.11}$$

where

$$A'_n = A_n e^{ik_{zn} z_0}. \tag{8.12}$$

Eq. (8.11) can be written in the following form by introducing another index r:

$$\begin{cases} p_r = A'_0 X_0^r + A'_1 X_1^r + \cdots A'_{N-1} X_{N-1}^r \\ p_{r+1} = A'_0 X_0^{r+1} + A'_1 X_1^{r+1} + \cdots A'_{N-1} X_{N-1}^{r+1} \\ \vdots \\ p_{r+N} = A'_0 X_0^{r+N} + A'_1 X_1^{r+N} + \cdots A'_{N-1} X_{N-1}^{r+N} \end{cases}, \quad r = 0, 1, \ldots, N_z - N - 1. \tag{8.13}$$

Assume $X_0 \sim X_{N-1}$ are the zeros of a polynomial $F(X)$ of degree N:

$$F(X) = 1 + E_1 X + E_2 X^2 + \cdots + E_N X^N, \tag{8.14}$$

where $E_1 \sim E_N$ are the polynomial coefficients.

Manipulating Eq. (8.13) leads to:

$$\begin{cases} p_r = A'_0 X_0^r + A'_1 X_1^r + \cdots A'_{N-1} X_{N-1}^r \\ E_1 p_{r+1} = A'_0 X_0^r \cdot E_1 X_0 + A'_1 X_1^r \cdot E_1 X_1 + \cdots A'_{N-1} X_{N-1}^r \cdot E_1 X_{N-1} \\ \vdots \\ E_N p_{r+N} = A'_0 X_0^r \cdot E_N X_0^N + A'_1 X_1^r \cdot E_N X_1^N + \cdots A'_{N-1} X_{N-1}^r \cdot E_N X_{N-1}^N, \end{cases}$$
$$r = 0, 1, \ldots, N_z - N - 1 \tag{8.15}$$

Adding up the equations of the equation sets in Eq. (8.15) yields:

$$\begin{aligned} & p_r + E_1 p_{r+1} + \cdots + E_N p_{r+N} \\ & = A'_0 X_0^r \left(1 + E_1 X_0 + \cdots + E_N X_0^N\right) + A'_1 X_1^r \left(1 + E_1 X_1 + \cdots + E_N X_1^N\right) \\ & \quad + A'_{N-1} X_{N-1}^r \left(1 + E_1 X_{N-1} + \cdots + E_N X_{N-1}^N\right), \\ & \qquad r = 0, 1, \ldots, N_z - N - 1 \end{aligned} \tag{8.16}$$

As $X_0 \sim X_{N-1}$ are the zeros of $F(X)$, Eq. (8.16) becomes:

$$E_1 p_{r+1} + \cdots + E_N p_{r+N} = -p_r, \quad r = 0, 1, \ldots, N_z - N - 1. \tag{8.17}$$

Rewrite Eq. (8.11) and Eq. (8.17) in matrix form,

$$[C]_{N_z \times N} \{A'\}_{N \times 1} = \{P\}_{N_z \times 1}, \tag{8.18}$$

and

$$\left[P^1\right]_{(N_z - N) \times N} \{E\}_{N \times 1} = -\left\{P^2\right\}_{(N_z - N) \times 1}, \tag{8.19}$$

where

$$[C]_{N_z \times N} = \begin{bmatrix} (X_0)^0 & (X_1)^0 & \cdots & (X_{N-1})^0 \\ (X_0)^1 & (X_1)^1 & \cdots & (X_{N-1})^1 \\ \vdots & \vdots & \ddots & \vdots \\ (X_0)^{N_z-1} & (X_1)^{N_z-1} & \cdots & (X_{N-1})^{N_z-1} \end{bmatrix}, \tag{8.20}$$

$$\{A'\}_{N \times 1} = \left[A_0 e^{ik_{z0} z_0} \;\; A_0 e^{ik_{z1} z_0} \;\; \cdots \;\; A_{N-1} e^{ik_{z.N-1} z_0}\right]^T, \tag{8.21}$$

$$\{P\}_{N_z \times 1} = \left[p_0 \;\; p_1 \;\; \cdots \;\; p_{N_z-1}\right]^T, \tag{8.22}$$

$$\left[P^1\right]_{(N_z - N) \times N} = \begin{bmatrix} p_1 & p_2 & \cdots & p_N \\ p_2 & p_3 & \cdots & p_{N+1} \\ \vdots & \vdots & \ddots & \vdots \\ p_{N_z-N} & p_{N_z-N+1} & \cdots & p_{N_z-1} \end{bmatrix}, \tag{8.23}$$

$$\{E\}_{N \times 1} = \left[E_1 \;\; E_2 \;\; \cdots \;\; E_N\right]^T, \tag{8.24}$$

$$\left\{P^2\right\}_{(N_z - N) \times 1} = \left[p_0 \;\; p_1 \;\; \cdots \;\; p_{N_z-N-1}\right]^T. \tag{8.25}$$

Here, $[\cdot]$ and $\{\cdot\}$ denote the 2-D matrix and vector, respectively. The superscript T represents matrix transpose.

The unknown vector $\{E\}_{N \times 1}$ of polynomial coefficients of $F(X)$ can be obtained by solving the linear equation set of Eq. (8.19), so that N zeros of $F(X)$, i.e., $X_0 \sim X_{N-1}$, can be calculated. The vector $\{A'\}_{N \times 1}$ can be obtained by solving Eq. (8.18). From Eq. (8.19), we can see that it should be required that $N_z \geq 2N$.

According to Eqs. (8.10) and (8.12), the axial wavenumber k_{zn} and modal amplitude A_n can be finally computed as:

$$k_{zn} = \frac{\log_e(X_n)}{i\Delta z}, \quad n = 0, 1, \ldots, N - 1, \tag{8.26}$$

and

$$A_n = \frac{A'_n}{e^{ik_{zn}z_0}}, \quad n = 0, 1, \cdots, N-1. \tag{8.27}$$

Note that a multisolution problem will arise in Eq. (8.26), and the detailed discussion will be made in Section 8.4.6.

8.2.4 Impedance eduction

In principle, we can use any of the identified axial wavenumbers to determine the unknown impedance. However, limited by the numerical accuracy of Prony's method, it is very important to choose the most appropriate wavenumber in practice. For this consideration, a simple and convenient selection criterion is proposed that the most reliable wavenumber must be associated with the dominant mode with the largest amplitude. Denote the mode number of the dominant mode as n_D.

Because the sound waves come from upstream of the lined section, it can be confirmed that the chosen mode must be the right-moving mode. Therefore, the specific impedance ζ can be calculated as:

$$\zeta = \frac{ik}{k_{yn_D} \tan\left(k_{yn_D} L_y\right)} \left[\frac{1 - M_a \sqrt{1 - \left(1 - M_a^2\right) \frac{k_{yn_D}^2}{k^2}}}{1 - M_a^2} \right]^2, \tag{8.28}$$

where

$$k_{yn_D} = \pm \sqrt{\frac{k^2 - \left[M_a k - \left(1 - M_a^2\right) k_{zn_D}\right]^2}{1 - M_a^2}}, \tag{8.29}$$

according to Eqs. (8.5) and (8.6).

8.2.5 Model validation

Here, the present method is applied to the situation of a real experiment [19], which provided the widely accepted benchmark data for verification of various duct aeroacoustic simulations and measurements. The experiment was conducted in the NASA Langley Flow Impedance Test Facility, which could also be sketched as Fig. 8.3 by changing the coordinate system. The cross-section of the flow duct is a $5.1 \times 5.1\ \text{cm}^2$, such that $L_y = 5.1$ cm. A kind of ceramic tubular liner, labeled CT57, was used as the calibration liner because it is known to be linear, i.e., its acoustic impedance is independent of mean flow and SPL. The liner length is $L = 40.6$ cm, and the experimental frequency is below the cut-off frequency of the first higher-order mode. Thirty-one microphones mounted on the upper wall (opposite the liner) were used to acquire sound pressure data giving inputs into a 2-D inverse impedance eduction method by Jones et al. [19], one of the OFMs. Some other conditions are that the sound speed is 344.3 m/s and the SPL at the source plane is kept at 130 dB.

Although the sound pressure data given in the reference are distributed at almost evenly spaced measurement locations, this data format does not completely meet the requirement of using Prony's method. Hence, to prepare the input for the present method, data preprocessing is performed with an interpolation scheme to create the equal-distance data sequence over the z coordinate range from 0.03 to 38.13 cm. The data point number used in this validation is $N_z = 16$ or, correspondingly, the spacing $\Delta z = 2.54$ cm is chosen so that most of the sample points coincide with the original measurement locations, thus minimizing the interpolation error.

A thorough comparison between the present straightforward method and the 2-D inverse method of Jones et al. [19] is made in Table 8.1 for a variety of flow Mach numbers from 0 to 0.4 at two test frequencies, 1.5 and 2.5 kHz, respectively. As seen in Table 8.1, the impedances educed by both methods agree well with each other.

For $M_a = 0.4$ and $f = 2.5$ kHz, the measured SPL and the sampled input data are depicted as points in Fig. 8.4. The FEM (Finite Element Method) simulations based on the impedance educed by the present method and the 2-D inverse method are also presented. It can be observed that the two

Table 8.1 Comparison between the present method and the previous 2-D inverse method [19] for the educed impedance at different Mach numbers when $f = 1.5$ and 2.5 kHz, respectively.

	1.5 kHz		2.5 kHz	
M_a	Present	2-D inverse	Present	2-D inverse
0.000	$0.98 + i1.28$	$1.02 + i1.30$	$1.43 - i1.65$	$1.54 - i1.60$
0.079	$1.00 + i1.22$	$1.01 + i1.25$	$1.37 - i1.57$	$1.42 - i1.39$
0.172	$1.07 + i1.22$	$1.25 + i1.18$	$1.17 - i1.47$	$1.19 - i1.55$
0.255	$1.12 + i1.32$	$1.26 + i1.26$	$1.04 - i1.43$	$1.02 - i1.46$
0.335	$1.05 + i1.23$	$1.18 + i1.27$	$0.98 - i1.36$	$0.93 - i1.43$
0.400	$1.07 + i1.22$	$1.21 + i1.08$	$0.89 - i1.30$	$0.75 - i1.29$

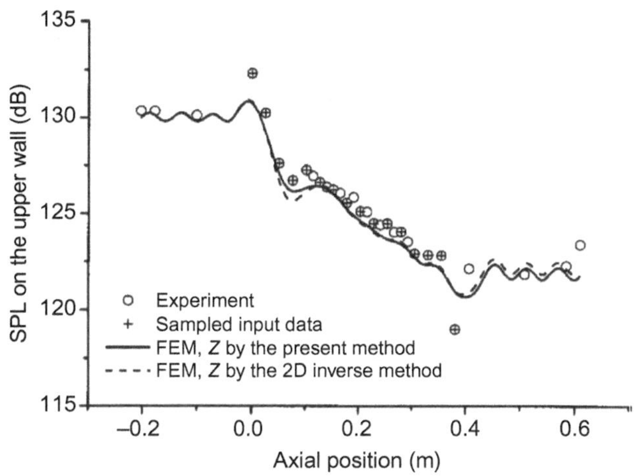

Fig. 8.4 The upper-wall sound pressure distribution presented by the experiment, the sampled input data, and the FEM simulations based on the impedance educed by the present method and the 2-D inverse method [19]; $M_a = 0.4$, $f = 2.5$ kHz, and the specific impedance on the exit plane is $0.92 - i0.14$.

educed impedances afford almost equally good FEM predictions of duct sound attenuation when compared with the measured data.

Thus, the present straightforward method is satisfactorily validated for impedance measurement under practical flow condition.

8.3 Shear flow effect on the impedance eduction

In the nacelle and inner passages of an aircraft engine or other similar flow ducts such as the silencers of internal engines, ventilations, etc., grazing flow past the liner surface can possibly have significant effects on the liner impedance. The original SFM assumes uniform grazing flow over the liner surface, and an infinitely thin boundary layer is considered by using an Ingard-Myers boundary condition. However, the real grazing flow in the duct is actually a shear flow with a finite boundary layer.

In this section, SFM is extended to be applied in the presence of shear flow, and the effect of different boundary layer profiles is discussed on the impedance eduction procedure. Moreover, the educed impedances by shear-flow SFM are compared with those by original SFM to confirm the reasonability of the uniform flow assumption.

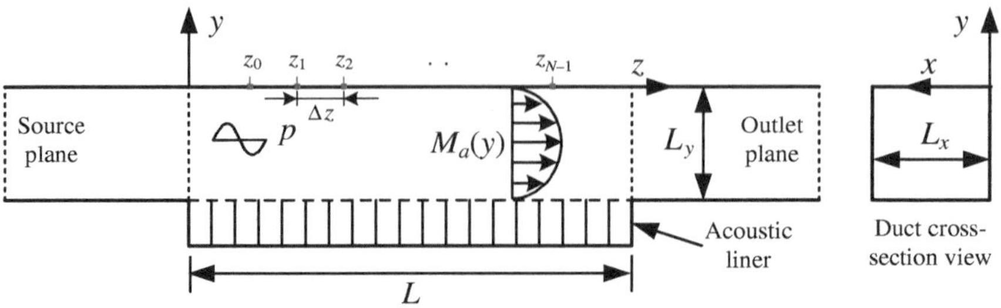

Fig. 8.5 Illustration of the flow duct for impedance eduction in the presence of shear flow.

8.3.1 Model description

As shown in Fig. 8.5, the duct geometry and measuring points arrangement are the same, but there is a parallel shear flow of 2-D Mach number profile, $M_a(y)$, inside the duct. The problem is how to obtain the liner impedance from the measured sound pressure sequence, p_s, $s = 0, 1, \ldots, N_z - 1$, when the shear flow effect is considered.

The 2-D mean flow profile, $M_a(y)$, assumes that flow is uniform in the spanwise direction. However, in the real situation of a rectangular duct, there is considerable flow non-uniformity in the spanwise direction. The flow profile away from the midspan not only has increasingly lower velocity as the boundary layer effect increases but also gradually changes its shape due to the corner flow effect. To adequately reflect the flow non-uniformity in the rectangular duct, a spanwise averaging factor is defined to multiply the entire midspan flow profile, which is,

$$\eta_q = \frac{M_{ave} L_y}{\int_{-L_y}^{0} M_{a,m}(y)\, dy}, \tag{8.30}$$

where M_{ave} is the average Mach number over the cross-section of the rectangular duct and $M_{a,\,m}(y)$ is the midspan ($x = L_x/2$) Mach number profile.

So, when $M_{a,\,m}(y)$ is known, the cross-sectional average Mach number profile used in the present model can be calculated as:

$$M_a(y) = \eta_q M_{a,m}(y). \tag{8.31}$$

8.3.2 Sound field in the flow duct

The propagation of harmonic sound waves in a parallel shear flow is governed by the linearized Euler equations (LEEs) of the following form:

$$ik\vec{q} + A\frac{\partial \vec{q}}{\partial z} + B\frac{\partial \vec{q}}{\partial y} + C\vec{q} = \vec{0}, \tag{8.32}$$

where

$$\vec{q} = \begin{bmatrix} u \\ v \\ p \end{bmatrix}, \quad A = \begin{bmatrix} M_a & 0 & 1 \\ 0 & M_a & 0 \\ 1 & 0 & M_a \end{bmatrix}, \quad B = \begin{bmatrix} 0 & 0 & 0 \\ 0 & 0 & 1 \\ 0 & 1 & 0 \end{bmatrix}, \quad C = \begin{bmatrix} 0 & dM_a/dy & 0 \\ 0 & 0 & 0 \\ 0 & 0 & 0 \end{bmatrix}. \tag{8.33}$$

Here, u and v are the acoustic particle velocity components, and p is the sound pressure. Eqs. (8.32) and (8.33) are written in nondimensional form, where c_0, L_y, and air density ρ_0 are used as the reference scales of velocity, length, and density, respectively. Both acoustic and flow quantities are assumed to be two-dimensional (2-D) in the $y-z$ plane. Three-dimensional (3-D) non-uniformity of mean flow in the rectangular duct is only taken into account through a spanwise averaging factor applied to the 2-D Mach number profile, $M_a(y)$, as discussed before.

The solution of the previous governing equation can be sought in the following modal forms:

$$p = \sum_{n=0}^{\infty} P_n(y)e^{ik_{zn}z}, \quad u = \sum_{n=0}^{\infty} U_n(y)e^{ik_{zn}z}, \quad v = \sum_{n=0}^{\infty} V_n(y)e^{ik_{zn}z}, \tag{8.34}$$

where k_{zn} s are the axial wavenumbers, and P_n, U_n, and V_n are the modal profile functions. The substitution of Eq. (8.34) into Eq. (8.32) leads to the well-known Pridmore-Brown equation [23]:

$$\frac{d^2 P_n}{dy^2} - \frac{2k_{zn}}{k + k_{zn}M_a}\frac{dM_a}{dy}\frac{dP_n}{dy} + \left[(k + k_{zn}M_a)^2 - k_{zn}^2\right]P_n = 0. \tag{8.35}$$

The hard-wall boundary condition is zero normal particle velocity, or equivalently,

$$\frac{dP_n}{dy} = 0, \quad \text{at the hard wall.} \tag{8.36}$$

At the liner surface, there is a finitely thin boundary layer beneath which the no-slip boundary condition is satisfied. So, when the boundary layer profile is directly incorporated into the sound propagation model, the classical impedance boundary condition can be imposed as:

$$\frac{dP_n}{dy} = ik\frac{P_n}{\zeta}, \quad y = -L_y. \tag{8.37}$$

On the other hand, it is also a common practice to approximately replace the boundary layer with a zero-thickness impermeable vortex sheet and only consider the non-uniformity of mean flow outside the vortex sheet, i.e., a slip condition is applied. When such an approximation is assumed, the Ingard–Myers' boundary condition is well accepted as appropriate for the lined wall, and for a single mode, which is,

$$\frac{dP_n}{dy} = ik\left(1 + M_s\frac{k_{zn}}{k}\right)^2 \frac{P_n}{\zeta}, \quad y = -L_y, \tag{8.38}$$

where M_s is the slip Mach number at the wall. Both the slip and no-slip boundary conditions for the lined wall will be investigated.

8.3.3 Mode decomposition

When the infinite summation in Eq. (8.34) is truncated up to N leading modes, the upper-wall sound pressure can be expressed as,

$$p(0, z) = \sum_{n=0}^{N-1} A_n e^{ik_{zn}z}, \tag{8.39}$$

where $A_n = P_n(0)$. After substituting the measured sound pressure sequence $p_s = p(0, z_s)$ and the axial coordinates of the measuring locations $z_s = z_0 + s\Delta z$, $s = 0, 1, \cdots, N_z - 1$ ($N_z \geq 2N$), into Eq. (8.39), a set of nonlinear equations for k_{zn} and A_n, is obtained:

$$p_s = \sum_{n=0}^{N-1} \left[A_n e^{ik_{zn}(z_0 + s\Delta z)}\right], \quad s = 0, 1, \ldots, N_z - 1, \tag{8.40}$$

which is exactly the same as Eq. (8.9). So, by using the Prony algorithm described in Section 8.2.3, the complex wavenumbers, k_{zn}, and their amplitudes, A_n can be extracted from the measured sound pressure data.

8.3.4 Impedance eduction

Just like the original SFM, the dominant mode with the largest amplitude is chosen for an impedance eduction purpose. The corresponding mode number is denoted as n_D.

With uniform mean flow assumption, the liner impedance ζ can be directly educed from the wavenumber k_{zn_D} through applying the Ingard-Mores boundary condition, as demonstrated in Section 8.2.4. However, for shear flow situation, the impedance ζ is dependent on the sound pressure modal profile function P_{n_D}, which obeys the Pridmore-Brown equation in

Eq. (8.35). Thus, P_{n_D} must be solved before impedance eduction in the presence of shear flow.

Eq. (8.35) can be rewritten as the following first-order ordinary differential equation:

$$\begin{cases} \dfrac{dF}{dy} = G \\ \dfrac{dG}{dy} = \dfrac{2k_{zn_D}}{k + k_{zn_D} M_a} \dfrac{dM_a}{dy} G - \left[(k + k_{zn_D} M_a)^2 - k_{zn_D}^2 \right] F \end{cases}, \tag{8.41}$$

where $F = P_{n_D}$ is the eigenfunction of the dominant mode, and $G = dP_{n_D}/dy$ is its derivative.

The boundary conditions are: $F|_{y=0} = C$ and $G|_{y=0} = 0$ on the upper hard wall ($y = 0$), where C is an arbitrarily selected initial value, for instance, taking $C = 1$. Therefore, Eq. (8.41) can be numerically integrated by the use of the fourth-order Runge-Kutta scheme to obtain the values of the eigenfunction and its derivative at the liner surface: $P_{n_D}|_{y=-1}$ and $dP_{n_D}/dy|_{y=-1}$, from which, the liner impedance ζ can be determined as,

$$\zeta = ik \left. \frac{P_{n_D}}{dP_{n_D}/dy} \right|_{y=-L_y}, \tag{8.42}$$

for no-slip boundary condition, or,

$$\zeta = ik \left(1 + M_s \frac{k_{zn_D}}{k} \right)^2 \left. \frac{P_{n_D}}{dP_{n_D}/dy} \right|_{y=-L_y}, \tag{8.43}$$

for slip boundary condition.

8.3.5 Impedance eduction example in the presence of shear flow

The measured acoustic data for CT57 acoustic liner in NASA's benchmark experiment [19] is also used to give input of the shear-flow SFM. The measured mean flow profiles are depicted as dots in Fig. 8.6 where the Mach number data is normalized by the duct centerline Mach number M_{max}. Three different midspan mean flow profiles for the propagation model are given as follows:

(a) Parabola with a zero-thickness boundary layer:

$$\frac{M_{a,m}(y)}{M_{max}} = S \left[-\frac{4}{L_y^2} \left(1 - \frac{M_s}{M_{max}} \right) y (y + L_y) + \frac{M_s}{M_{max}} \right], \tag{8.44}$$

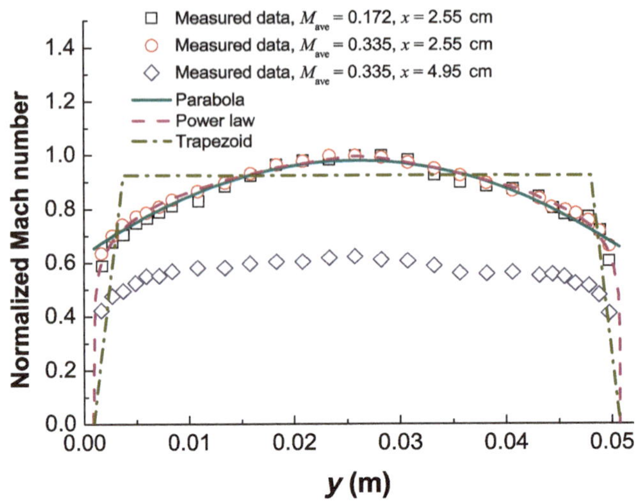

Fig. 8.6 Mean flow profiles in the square tube, *x* is the coordinate perpendicular to the (*y*, *z*) plane; the *dots* stand for the measured mean flow data [19].

(b) One-seventh power law with a linear sublayer:

$$\frac{M_{a,m}(y)}{M_{\max}} = \begin{cases} S\left[\dfrac{2(y+L_y)}{L_y}\right]^{1/7} & (y \geq \delta_l - L_y) \\ S\left(\dfrac{2\delta_l}{L_y}\right)^{1/7}\dfrac{y+L_y}{\delta_l} & (y < \delta_l - L_y) \end{cases}, \tag{8.45}$$

(c) Uniform core flow with a linear boundary layer ("trapezoid" for brevity):

$$\frac{M_{a,m}(y)}{M_{\max}} = \begin{cases} M_u = \dfrac{S}{1-(e/L_y)} & (y \geq e - L_y) \\ M_u\dfrac{y+L_y}{e} & (y < e - L_y) \end{cases}, \tag{8.46}$$

where $M_{a,\,m}$ is the 2-D mean flow profile at the midspan ($x = L_x/2$). Here, y ranges from $-L_y$ to $-L_y/2$, on the assumption that the flow profiles are symmetric about $y = -L_y/2$. S is the scaling factor to scale the flow profile such that the area under each curve is the same as for the measured midspan flow profile. In this way, the boundary layers of the three flow profiles have equal displacement thickness at midspan. δ_l is the thickness of the linear laminar sublayer and e is the thickness of the linear boundary layer for the trapezoid profile.

In Fig. 8.6, Eqs. (8.44)–(8.46) are compared with the flow profiles measured in NASA's Grazing Incidence Tube (GIT) facility. It can be noticed that the normalized Mach number profile shapes are almost the same for different M_{ave}s. The fitting parameter of the parabola, i.e., the slip velocity M_s is set to be $0.67M_{\max}$. For the other two flow profiles, δ_l and e are taken to be $0.06L_y$ and 2.8 mm, respectively. The scaling factors Ss are determined as 0.98, 1 and 0.875 for parabola, power law, and trapezoid profiles, respectively. The midspan displacement thickness $\delta^* = 0.063L_y$ which is close to the measured value.

As shown in Fig. 8.6, the flow profile away from the midspan ($x = 4.95$ cm) has relatively lower velocities. To reflect the spanwise flow non-uniformity, a spanwise averaging factor η_q defined in Section 8.3.1 is applied and $\eta_q = 0.93$ according to the measured data. The cross-sectional average Mach number $M_a(y) = \eta_q M_{a,\,m}(y)$ is used to represent the 2-D flow profile in shear-flow SFM.

Fig. 8.7 shows the impedance spectra educed by the shear-flow SFM for the three flow profiles given in Eqs. (8.44)–(8.46) when $M_{ave} = 0.173$ and SPL = 130 dB. As can be seen, despite their significantly different boundary layers, these flow profiles give almost coincident impedance spectra over the entire test frequency range of 0.5–3.0 kHz. And the same observation is obtained for a higher average Mach number of 0.335, as shown in Fig. 8.8.

To gain more insights into the shear flow effect, a problem of great concern is whether or not the impedance results educed with the shear-flow model are consistent with those from the uniform flow assumption. Only the power law flow profile is considered in this discussion as the other boundary layers essentially give the same results. Fig. 8.9 shows that the no-slip shear-flow method is quite consistent with two uniform-flow methods except for some obvious deviations occurring around the antiresonance frequency where it is always difficult to accurately educe the liner impedance. Similar comparisons are also performed at other Mach numbers tested in the NASA experiment. The conclusions are basically the same as before, and the detailed results are not plotted here.

8.4 3-D straightforward method of acoustic impedance eduction

Like most existing impedance measurement methods, the original SFM can only be applied below 3000 Hz in a small-sized (5.1×5.1 cm^2) duct like the GIT designed by Jones et al. [24]. This is because there is an implicit

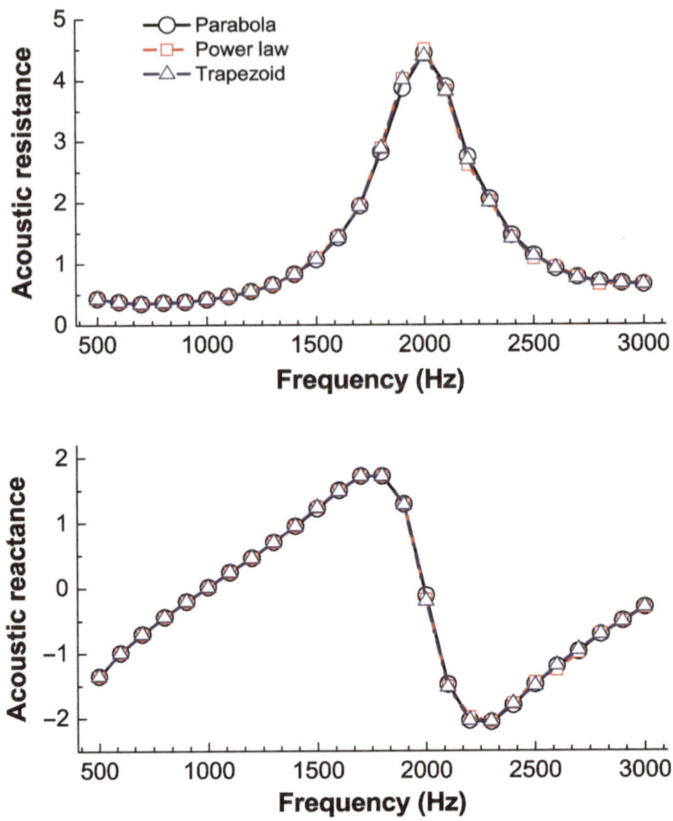

Fig. 8.7 Effect of different boundary layers on the educed impedance spectra; $M_{ave} = 0.172$, SPL = 130 dB.

requirement in these methods that the sound pressure field should be independent of the spanwise ($x-$) coordinate, which is satisfied by controlling the frequency lower than the cut-on frequency of the first higher-order mode to achieve plane wave incidence. In other words, these methods are indeed two-dimensional (2-D) methods. However, broadband liner concepts are recently attractive to engineers because the fan broadband noise becomes more and more important as harmonic noise is well attenuated. As a result, there is a highlighted need for measuring the liner impedance over a broader frequency range (up to as high as around 8000 Hz [25]). In addition, researchers are more willing to study the acoustic behavior of the liner in a large duct whose cross-section is close to the size of an aft bypass duct of a

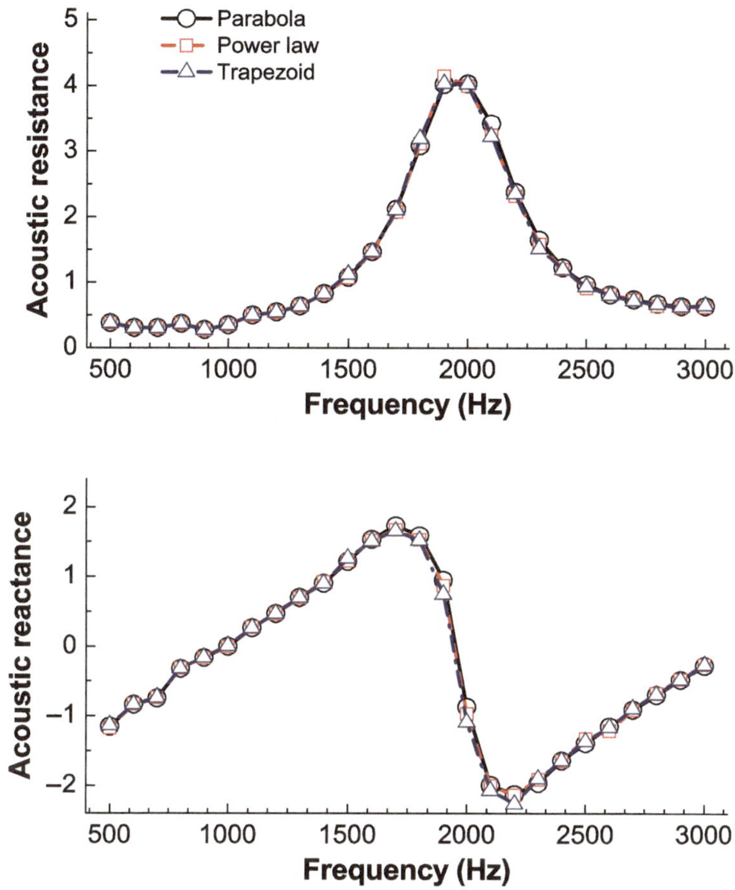

Fig. 8.8 Effect of different boundary layers on the educed impedance spectra; $M_{ave} = 0.335$, SPL = 130 dB.

full-scale engine, such as the Curved Duct Test Rig (CDTR) at NASA Langley Research Center (LaRC) [26–28]. So, it is necessary to extend the original SFM into a 3-D method.

There are two kinds of 3-D straightforward methods (3-D SFMs) proposed by Qiu et al. [29] and Chen et al. [10], respectively. Because the latter has some obvious advantages over the former, such as requiring fewer measuring points and without introducing an optimization algorithm, the 3-D SFM by Chen et al. is presented in this section to solve the high frequency impedance eduction issue. A solution selection process is also proposed to completely solve the multisolution problem existing in all SFMs.

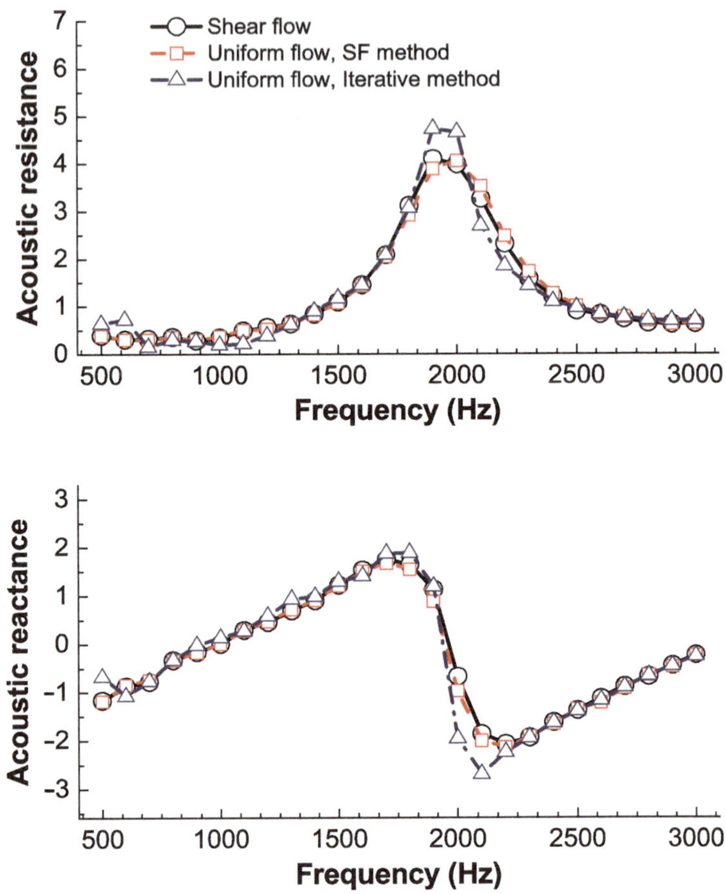

Fig. 8.9 Effect of the shear flow on the educed impedance spectra, in comparison with the uniform-flow results of the original SFM [8] and the iterative method [19], respectively; $M_{ave} = 0.335$, SPL = 130 dB.

8.4.1 Model description

Fig. 8.10 shows the schematic diagram of a rectangular flow duct of width L_x and height L_y for liner impedance eduction purposes. An L-long locally reacting test liner with uniform normalized impedance ζ is flush-mounted over a portion of the bottom sidewall, and all the other sidewalls are acoustically rigid. There is a uniform mean flow of Mach number M_a in the duct, and the lined section contains a three-dimensional multimodal sound field generated by a sound source plane upstream from the leading edge of the test

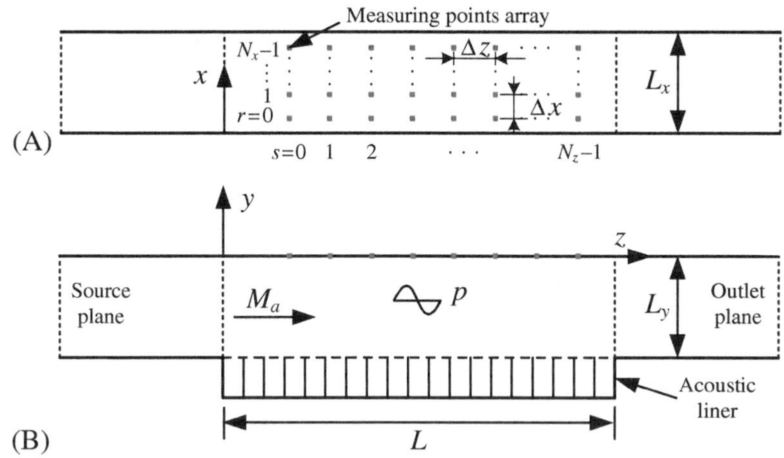

Fig. 8.10 (A) Top and (B) cross-sectional views of the rectangular flow duct for 3-D impedance eduction.

liner. A right-handed Cartesian coordinate system (x, y, z) is introduced with x, y being the spanwise (parallel to liner surface and perpendicular to duct axis) and vertical (perpendicular to liner surface) coordinates and z being along the axial direction. The origin of the coordinate system is located at the top-right corner (left view) of the flow duct in the transverse plane separating the inlet and lined sections, as shown in Fig. 8.10. A rectangular array of measuring microphones is deployed in the wall directly opposite the test liner. As illustrated in Fig. 8.10A, there are N_x and N_z evenly spaced measuring points in the spanwise and axial directions, respectively. The spanwise and axial spacing distances between adjacent measuring points are Δx and Δz, respectively.

8.4.2 Sound field in the flow duct

The governing equation is also the convective Helmholtz equation in Eq. (8.1) where p is a function of (x, y, z) in this case. By solving the equation, the sound pressure field in the lined section can be expanded as a series of duct modes:

$$p(x, y, z) = \sum_{m=0}^{\infty}\sum_{n=0}^{\infty}\left[A_{mn}^{+}\cos\left(k_{ymn}^{+}y\right)e^{ik_{zmn}^{+}z} + A_{mn}^{-}\cos\left(k_{ymn}^{-}y\right)e^{ik_{zmn}^{-}z}\right]\cos\left(\frac{m\pi}{L_x}x\right). \tag{8.47}$$

The dispersion relation and the Ingard-Myers boundary condition equation are:

$$k_{zmn}^{\pm} = \frac{M_a k \mp \sqrt{k^2 - \left(1 - M_a^2\right)\left[\left(k_{ymn}^{\pm}\right)^2 + \left(\frac{m\pi}{L_x}\right)^2\right]}}{1 - M_a^2}, \tag{8.48}$$

and

$$\frac{k_{ymn}^{\pm}}{k} \tan\left(k_{ymn}^{\pm} L_y\right) - \frac{i}{\zeta}\left(1 + M_a \frac{k_{zmn}^{\pm}}{k}\right)^2 = 0, \tag{8.49}$$

respectively.

Similar with the original SFM, the infinite series in Eq. (8.47) can be truncated and rewritten as:

$$p(x, y, z) = \sum_{m=0}^{M-1} \sum_{n=0}^{N-1} \left[A_{mn} \cos\left(k_{ymn} y\right) e^{ik_{zmn} z}\right] \cos\left(\frac{m\pi}{L_x} x\right), \tag{8.50}$$

through distinguishing the right- and left-moving modes by their vertical mode numbers *n*s rather than the superscript $\pm$ as in Eq. (8.47). For instance, the first four vertical mode wavenumbers $k_{ymn}^{\pm}$, $n = 0, 1$ in Eq. (8.47) correspond to k_{ymn}, $n = 0, 1, 2, 3$ in Eq. (8.50). The infinite summations are truncated at the maximum spanwise mode number of $M - 1$ ($m = 0, 1, \ldots, M - 1$) and maximum vertical mode number of $N - 1$ ($n = 0, 1, \ldots, N - 1$). Note that M and N should be large enough to ensure the convergence of the series.

8.4.3 Spanwise mode decomposition

Substituting the measured sound pressure data into Eq. (8.50) yields:

$$p(x_r, 0, z_s) = \sum_{m=0}^{M-1} B_{ms} \cos\left(\frac{m\pi}{L_x} x_r\right) = p_{rs}, \quad r = 0, 1, \ldots, N_x - 1,$$
$$s = 0, 1, \ldots, N_z - 1, \tag{8.51}$$

where

$$B_{ms} = \sum_{n=0}^{N-1} A_{mn} e^{ik_{zmn} z_s}, \quad s = 0, 1, \ldots, N_z - 1. \tag{8.52}$$

Here, p_{rs} represents the measured complex sound pressure at the (r, s) measuring point whose spanwise and axial coordinates are x_r and z_s, respectively. As the measuring points are equally spaced, $x_r = x_0 + r\Delta x$ and $z_s = z_0 + s\Delta z$.

It is more convenient to describe Eq. (8.51) in the matrix form as:

$$[T]_{N_x \times M}[B]_{M \times N_z} = [p]_{N_x \times N_z}, \tag{8.53}$$

where

$$[T]_{N_x \times M} = \begin{bmatrix} 1 & \cos\left(\dfrac{\pi}{L_x}x_0\right) & \cdots & \cos\left[\dfrac{(M-1)\pi}{L_x}x_0\right] \\ 1 & \cos\left(\dfrac{\pi}{L_x}x_1\right) & \cdots & \cos\left[\dfrac{(M-1)\pi}{L_x}x_1\right] \\ \vdots & \vdots & \ddots & \vdots \\ 1 & \cos\left(\dfrac{\pi}{L_x}x_{N_z-1}\right) & \cdots & \cos\left[\dfrac{(M-1)\pi}{L_x}x_{N_z-1}\right] \end{bmatrix}, \tag{8.54}$$

$$[B]_{M \times N_z} = \begin{bmatrix} B_{00} & B_{01} & \cdots & B_{0,N_z-1} \\ B_{10} & B_{11} & \cdots & B_{1,N_z-1} \\ \vdots & \vdots & \ddots & \vdots \\ B_{M-1,0} & B_{M-1,1} & \cdots & B_{M-1,N_z-1} \end{bmatrix}, \tag{8.55}$$

and

$$[p]_{N_x \times N_z} = \begin{bmatrix} p_{00} & p_{01} & \cdots & p_{0,N_z-1} \\ p_{10} & p_{11} & \cdots & p_{1,N_z-1} \\ \vdots & \vdots & \ddots & \vdots \\ p_{N_x-1,0} & p_{N_x-1,1} & \cdots & p_{N_x-1,N_z-1} \end{bmatrix}. \tag{8.56}$$

The matrix $[p]_{N_x \times Nz}$ consists of the measured sound pressure data of the microphone array, and matrix $[T]_{N_x \times M}$ is only associated with the spanwise coordinates x_rs of the measuring points. By using the least square method, we obtain unknown matrix as:

$$[B]_{M \times N_z} = \left[\left([T]_{N_x \times M}\right)^H [T]_{N_x \times M}\right]^{-1} \left([T]_{N_x \times M}\right)^H [p]_{N_x \times N_z}, \tag{8.57}$$

where $(\cdot)^{-1}$ and $(\cdot)^H$ denote matrix inversion and conjugate transpose, respectively. Actually, Eq. (8.57) gives the results of spanwise mode decomposition.

Due to measurement and computation errors, it is necessary to choose the dominant spanwise mode for impedance eduction calculation. The dominant spanwise mode number m_D is determined by comparing the 1-norms of all row vectors of matrix $[B]_{M\times N_z}$ that correspond to different spanwise modes: the row vector $\{B^D\}_{1\times N_z}$ of matrix with $m = m_D$ having the largest 1-norm. Substituting the row vector:

$$\{B^D\}_{1\times N_z} = \left[B_{m_D,0} \quad B_{m_D,1} \quad \cdots \quad B_{m_D,N_z-1}\right] \tag{8.58}$$

into Eq. (8.52) gives:

$$B_{m_D,s} = \sum_{n=0}^{N-1} A_{m_D,n} e^{ik_{z,m_D,n}(z_0 + s\Delta z)}, \quad s = 0, 1, \ldots, N_z - 1, \tag{8.59}$$

which can be directly solved by Prony's method.

8.4.4 Vertical mode decomposition

Essentially, solving Eq. (8.59) is to carry on vertical mode decomposition, as Eq. (8.59) is in fact the superposition of multiple vertical modes with unknown amplitudes and wavenumbers (unknown axial wavenumbers $k_{z,\ m_D,\ n}$s are dependent on the wavenumber $k_{y,\ m_D,\ n}$ in the y direction through the dispersion relation). Unlike the mode decomposition in the spanwise direction where the wavenumbers k_{xmn} in the x direction are known as $m\pi/L_x$, the vertical mode decomposition cannot be conducted in the same way, but it can be realized with the help of Prony's method; the detailed procedure is illustrated in Section 8.2.3. The axial wavenumber $k_{z,\ m_D,\ n}$ can be finally computed as:

$$k_{z,m_D,n} = \frac{\log_e(X_n)}{i\Delta z}, \quad n = 0, 1, \ldots, N-1, \tag{8.60}$$

where X_ns are the zeros of a polynomial $F(X)$ defined in Eq. (8.14).

To maximize the accuracy of this method, a dominant vertical mode is chosen to determine the liner impedance, which is also the common strategy in 2-D impedance eduction methods. The dominant vertical mode with $n = n_D$ should be the one having the largest $|A_{m_D,\ n}|$.

8.4.5 Impedance eduction

Because the dominant (m_D, n_D) mode must be the right-moving mode, the liner impedance ζ can be calculated as:

$$\zeta = \frac{ik}{k_{y,m_D,n_D}\tan\left(k_{y,m_D,n_D}L_y\right)}\left[\frac{1-M_a\sqrt{1-\frac{\left(1-M_a^2\right)}{k^2}\left[k_{y,m_D,n_D}^2+\left(\frac{m_D\pi}{L_x}\right)^2\right]}}{1-M_a^2}\right]^2, \tag{8.61}$$

where

$$k_{y,m_D,n_D} = \pm\sqrt{\frac{k^2-\left[M_a k-\left(1-M_a^2\right)k_{z,m_D,n_D}\right]^2}{1-M_a^2}-\left(\frac{m_D\pi}{L_x}\right)^2}, \tag{8.62}$$

according to Eqs. (8.48) and (8.49).

From Eqs. (8.53) and (8.19), we can see that it should be required that $N_x \geq M$ and $N_z \geq 2N$. So, when $M \times N$ modes are necessary to ensure Eq. (8.50) well represents the sound pressure field in the lined section, at least $2MN$ measuring microphones should be mounted to give enough information for impedance eduction ($2MN$ independent unknowns in total: A_{mn} and k_{zmn}, $m = 0, 1, \ldots, M-1$, $n = 0, 1, \ldots, N-1$).

Based on the experience of the author, it is recommended to choose $N_x = M$ and $N_z = 2N + 4$ for reducing the number of measuring microphones as much as possible and ensuring the robustness of the method at the same time. The number M of considered spanwise modes is suggested empirically as the number of cut-on spanwise modes, although it is difficult to give a general principle on the minimum number N of included vertical modes that varies with frequency, duct size and impedance ζ. However, it is definite that N should be greater than the number of all cut-on vertical modes (in a rigid duct with same geometry), and N becomes larger at a higher frequency or in a larger flow duct. For convenience, M refers to the number of cut-on spanwise modes in the following narration.

8.4.6 Multisolution problem

The calculation formula of k_{z,m_D,n_D}, Eq. (8.60), is deduced from Eq. (8.10). A multisolution problem can be found by determining k_{z,m_D,n_D}; if k_{z,m_D,n_D} is the solution, then $k_{z,m_D,n_D} \pm s\frac{2\pi}{\Delta z}$ ($s = 1, 2, \ldots$) also satisfies Eq. (8.10). But only one solution is correct to educe the true liner impedance.

In Chen et al. [10], a solution selection process has been proposed to completely solve the multisolution problem, i.e.,

$$\mathrm{Re}\left(k_{z,m_D,n_D}\right) \cdot \Delta z \in (-2\pi + \delta, 0). \tag{8.63}$$

where

$$\delta = \left(k_{z,m_D,0}^{rigid} - k_{z,0,0}^{rigid}\right) \cdot \Delta z, \tag{8.64}$$

is used to account for the situation when the spanwise mode number m_D of the dominant mode is not 0. Here,

$$k_{z,m_D,0}^{rigid} = \frac{M_a k - \sqrt{k^2 - \left(1 - M_a^2\right)\left(\frac{m_D \pi}{L_x}\right)^2}}{1 - M_a^2} \tag{8.65}$$

and

$$k_{z,0,0}^{rigid} = -\frac{k}{1 + M_a} \tag{8.66}$$

are the axial wavenumbers of $(m_D, 0)$ and $(0, 0)$ modes in the rigid flow duct with the same size.

Conservatively, apart from conducting the solution selection process, the requirement in [8]:

$$\Delta z < \frac{2\pi(1 + M_a)}{k}, \tag{8.67}$$

on axial spacing distance Δz is still applied.

8.4.7 Impedance eduction example beyond the cut-off frequency

To verify the effectiveness of the 3-D SFM, the numerical simulation experiments are conducted to synthesize the sound pressure field in the flow duct, the sketch of which is depicted in Fig. 8.11. The duct width L_x and height L_y are both set as 5.1 cm and the liner length $L = 60$ cm.

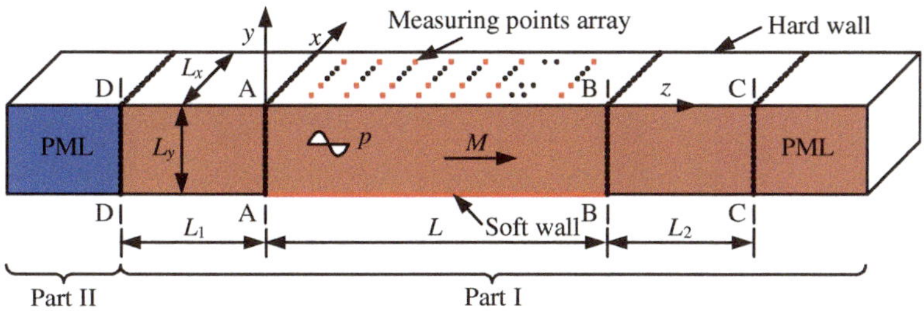

Fig. 8.11 Diagrammatic sketch of the flow duct used in numerical simulations.

The frequency range is from 500 to 9000 Hz in 100 Hz increments. The corresponding range of Helmholtz number kL_x is 0.47–8.41. The maximum frequency is extremely higher than the normal frequency limit (3000 Hz) by using 2-D impedance eduction methods. The axial positions of measuring microphones are $z_s = (3.5 + 3.5s)$ cm, $s = 0, 1, \ldots, 15$ ($\Delta z = 3.5$ cm), and the number of considered vertical modes is chosen as $N = 7$ to make sure of the accuracy of the educed impedance up to 9000 Hz. The number of spanwise measuring points is determined as $N_x = 3$ because there are three cut-on spanwise modes ($M = 3$: $m = 0, 1, 2$) at the maximum frequency. The spanwise locations are set as $x_r = (1 + 1.5s)$ cm, $r = 0, 1, 2$ ($\Delta x = 1.5$ cm). Thus, there are totally $3 \times 16 = 48$ measuring points.

The frequently used validation liner, CT57 [19], is still used as the test liner. However, the measured impedance data of CT57 are limited below 3000 Hz. So, an impedance model is established in [10] to characterize the impedance spectrum of CT57. The impedance model of CT57 is proved to predict impedance spectra which agree with the measurement data, as shown in Fig. 8.12.

In the simulations, the test acoustic liner is modeled as a soft wall with imposed impedance according to the impedance model of CT57, and the Ingard-Myers boundary condition is applied. From the results of numerical simulations, sound pressure data can be extracted at the measuring point array opposite to the test liner to give the input into the 3-D method. The uncertainty analysis regarding the measurement perturbation has been studied by using the Monte Carlo method (MCM), and the range of random errors of pressure amplitude and phase are ± 0.5 dB and $\pm 1^{\circ}$, respectively.

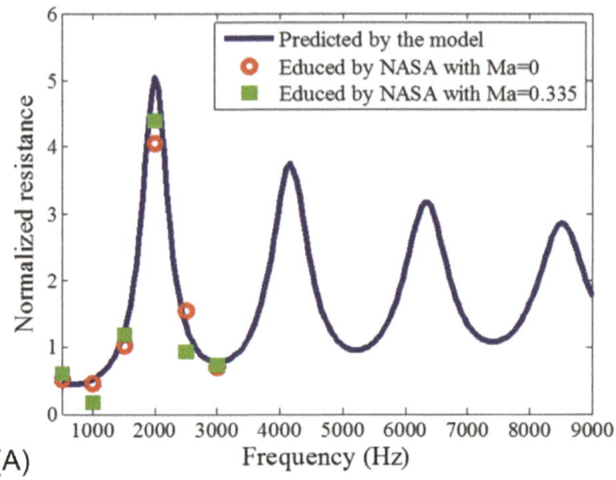

(A)

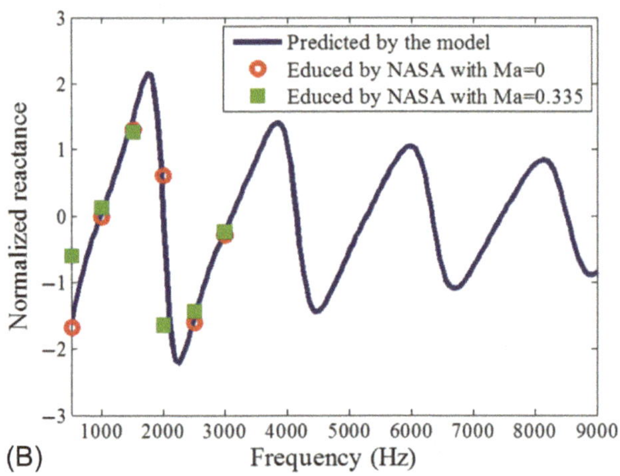

(B)

Fig. 8.12 The (A) normalized resistance and (B) normalized reactance spectra of liner CT57 predicted by the impedance model [10] and by measurement [19].

The comparison between the imposed and educed impedances for $M_a = 0.3$ is shown in Fig. 8.13 where the frequency range is divided into several parts according to the number M of cut-on modes. For clean data without perturbation, the educed impedance tracks the imposed impedance spectrum extremely well in the whole frequency range. For perturbed data with typical random error, the uncertainty range is apparently acceptable, implying the present method can well reproduce the correct impedance in practice. The width of the uncertainty range seems to have negative correlation with the attenuation level, as is demonstrated by comparing

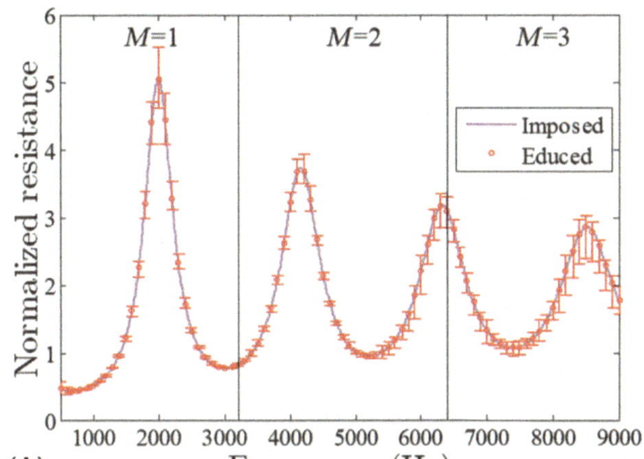

(A)

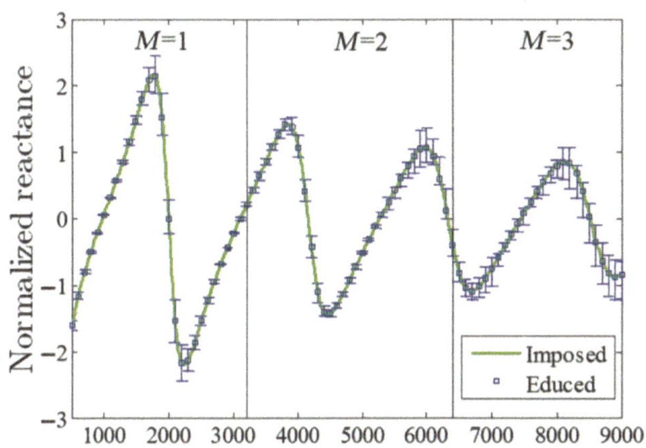

(B)

Fig. 8.13 Comparison between the imposed *(solid line)* and educed impedances *(dots)* in a lined small-sized duct for $M_a = 0.3$: (A) the normalized resistance and (B) the normalized reactance; the *dots* stand for educed impedances without perturbation, whereas the error bars present the 95.45% confidence intervals.

Fig. 8.13 with Fig. 8.14 where the transmission losses of the test liner are drawn. The transmission loss near the antiresonance (i.e., around 2000 Hz, 4000 Hz, 6000 Hz, or 8000 Hz) reaches local minimum so that the uncertainty range gets slightly wider, indicating the impedance eduction may be more likely contaminated by the measurement error near antiresonant points. Even so, the uncertainty range of the educed results near antiresonances is still acceptable.

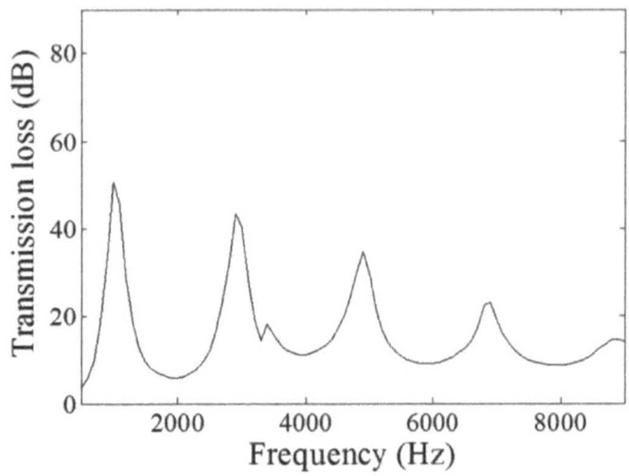

Fig. 8.14 The transmission losses of the test liner.

References

[1] E. Envia, Fan noise reduction: an overview, Int. J. Aeroacoust. 1 (2002) 43–64.
[2] G.W. Bielak, J.W. Premo, A.S. Hersh, Advanced turbofan duct liner concepts, (1999) NASA/CR-1999-209002.
[3] P. Dean, An in situ method of wall acoustic impedance measurement in flow ducts, J. Sound Vib. 34 (1974) 97-IN96.
[4] D. Armstrong, Acoustic grazing flow impedance using waveguide principles, (1971) NASA-CR-120848.
[5] W.R. Watson, A new method for determining acoustic-liner admittance in a rectangular duct with grazing flow from experimental data, (1984), NASA-TP-2310.
[6] W.R. Watson, A method for determining acoustic-liner admittance in ducts with sheared flow in two-cross-sectional directions, (1985), NASA-TP-2518.
[7] M. Jones, W. Watson, M. Tracy, T. Parrott, Comparison of two waveguide methods for educing liner impedance in grazing flow, AIAA J. 42 (2004) 232–240.
[8] X. Jing, S. Peng, X. Sun, A straightforward method for wall impedance eduction in a flow duct, J. Acoust. Soc. Am. 124 (2008) 227–234.
[9] X. Jing, S. Peng, L. Wang, X. Sun, Investigation of straightforward impedance eduction in the presence of shear flow, J. Sound Vib. 335 (2015) 89–104.
[10] L. Chen, L. Du, X. Wang, X. Jing, X. Sun, A three-dimensional straightforward method for liner impedance eduction in uniform grazing flow, J. Sound Vib. 468 (2020) 115119.
[11] W.R. Watson, M.G. Jones, S.E. Tanner, T.L. Parrott, A finite element propagation model for extracting normal incidence impedance in nonprogressive acoustic wave fields, J. Comput. Phys. 125 (1996) 177–186.
[12] W.R. Watson, M.G. Jones, T.L. Parrott, Validation of an impedance eduction method in flow, AIAA J. 37 (1999) 818–824.
[13] W. Watson, M. Tracy, M. Jones, T. Parrott, Impedance eduction in the presence of shear flow, in: 7th AIAA/CEAS Aeroacoustics Conference and Exhibit, Maastricht, Netherlands, AIAA Paper 2001-2263, 2001.
[14] Y. Aurégan, M. Leroux, V. Pagneux, Measurement of liner impedance with flow by an inverse method, in: 10th AIAA/CEAS Aeroacoustics Conference, Manchester, Great Britain, AIAA Paper 2004-2838, 2004.

[15] W. Watson, M. Jones, Comparison of a convected Helmholtz and Euler model for impedance eduction in flow, in: 12th AIAA/CEAS Aeroacoustics Conference, 27th AIAA Aeroacoustics Conference, Cambridge, MA, 2006.
[16] T. Elnady, H. Bodén, B. Elhadidi, Validation of an inverse semi-analytical technique to educe liner impedance, AIAA J. 47 (2009) 2836–2844.
[17] W.R. Watson, M.G. Jones, Impedance eduction in a duct using linearized Euler equations, in: 2018 AIAA/CEAS Aeroacoustics Conference, Atlanta, GA, USA, AIAA Paper 2018-3773, 2018.
[18] W.R. Watson, M.H. Carpenter, M.G. Jones, Performance of Kumaresan and Tufts algorithm in liner impedance eduction with flow, AIAA J. 53 (2015) 1091–1102.
[19] M. Jones, W. Watson, T. Parrott, Benchmark data for evaluation of aeroacoustic propagation codes with grazing flow, in: 11th AIAA/CEAS Aeroacoustics Conference, Monterey, CA, USA, AIAA Paper 2005-2853, 2005.
[20] M. Myers, On the acoustic boundary condition in the presence of flow, J. Sound Vib. 71 (1980) 429–434.
[21] S.W. Rienstra, Fundamentals of Duct Acoustics, Von Karman Institute Lecture Notes, 2015.
[22] R. Kumaresan, D. Tufts, Estimating the parameters of exponentially damped sinusoids and pole-zero modeling in noise, IEEE Trans. Acoust. Speech Signal Process. 30 (1982) 833–840.
[23] D.C. Pridmore-Brown, Sound propagation in a fluid flowing through an attenuating duct, J. Fluid Mech. 4 (1958) 393–406.
[24] M. Jones, W. Watson, T. Parrott, C. Smith, Design and evaluation of modifications to the NASA Langley flow impedance tube, in: 10th AIAA/CEAS Aeroacoustics Conference, Manchester, Great Britain, AIAA Paper 2004-2837, 2004.
[25] D.M. Nark, M.G. Jones, Further development and assessment of a broadband liner optimization process, in: 22nd AIAA/CEAS Aeroacoustics Conference, Lyon, France, AIAA Paper 2016-2784, 2016.
[26] W. Watson, M. Jones, New numerical procedure for impedance eduction in ducts containing mean flow, AIAA J. 49 (2011) 2109–2122.
[27] W.R. Watson, M.G. Jones, Impedance eduction in large ducts containing high-order modes and grazing flow, in: 23rd AIAA/CEAS Aeroacoustics Conference, Denver, CO, AIAA Paper 2017-3183, 2017.
[28] W.R. Watson, M.G. Jones, J.C. June, Single mode theory for impedance eduction in large-scale ducts with grazing flow, in: 20th AIAA/CEAS Aeroacoustics Conference, Atlanta, GA, AIAA Paper 2014-3351, 2014.
[29] X. Qiu, B. Xin, X. Jing, Straightforward impedance eduction method for non-grazing incidence wave with multiple modes, J. Sound Vib. 432 (2018) 1–16.

Index

Note: Page numbers followed by *f* indicate figures and *t* indicate tables.

D

E

F

M

R

Kolibri 360 GmbH
Ferdinand-Jühlke-Straße 7
99095 Erfurt, Deutschland
produktsicherheit@kolibri360.de